T0137153

Stochastic Numerics for Mathematical Physics

Scientific Computation

More information about this series at https://link.springer.com/bookseries/718

Grigori N. Milstein · Michael V. Tretyakov

Stochastic Numerics for Mathematical Physics

Second Edition

 Springer

Grigori N. Milstein
Chappaqua, NY, USA

Michael V. Tretyakov
School of Mathematical Sciences
University of Nottingham
Nottingham, UK

ISSN 1434-8322   ISSN 2198-2589   (electronic)
Scientific Computation
ISBN 978-3-030-82042-8   ISBN 978-3-030-82040-4   (eBook)
https://doi.org/10.1007/978-3-030-82040-4

This Springer imprint is published by the registered company Springer Nature Switzerland AG
The registered company address is: Gewerbestrasse 11, 6330 Cham, Switzerland

Preface to the Second Edition

This book is a substantially revised and expanded edition reflecting major developments in stochastic numerics since the 1st edition [314] was published in 2004. The new topics include, in particular, mean-square and weak approximations in the case of nonglobally Lipschitz coefficients of stochastic differential equations (SDEs); conditional probabilistic representations and their application to practical variance reduction using regression methods; the multi-level Monte Carlo method; computing ergodic limits and additional classes of geometric integrators used in molecular dynamics; numerical methods for forward-backward stochastic differential equations (FBSDEs); approximation of parabolic stochastic partial differential equations (SPDEs) based on the method of characteristics.

In Chap. 1 we added Sect. 1.1.6 on almost sure convergence of SDEs' approximations. We extended the section (Sect. 1.4) on mean-square approximations in the case of nonglobally Lipschitz coefficients of SDEs. SDEs with nonglobally Lipschitz coefficients possessing unique solutions make up a very important class in applications. Section 1.4 includes a version of the fundamental mean-square convergence theorem for SDEs with nonglobally Lipschitz conditions and examples of explicit and implicit schemes convergent in the nonglobally Lipschitz case.

Chapter 2 "Weak approximation of stochastic differential equations" of the 1st edition is now split into two chapters: Chap. 2 "Weak approximation for stochastic differential equations: foundations" and Chap. 3 "Weak approximation for stochastic differential equations: special cases".

Chapter 2 "... foundations" includes the procedure how to construct weak approximations, the main weak-sense convergence theorem, and examples of weak schemes for general SDEs. It also contains three new sections. The new Sect. 2.2.5 addresses the problem of weak-sense numerical integration of SDEs in the nonglobally Lipschitz case via the concept of rejecting exploding trajectories. Due to the concept of rejecting "bad" trajectories, we can choose a suitable method for solving a system of SDEs with nonglobally Lipschitz coefficients, taking into account all the known weak methods presented in this book. The new Sect. 2.4 introduces conditional probabilistic representations and, based on them, practical variance reduction. The new Sect. 2.5 is a brief survey of quasi-Monte Carlo and multi-level Monte Carlo methods.

Chapter 3 "... special cases" includes weak schemes for SDEs with additive noise and colored noise and for computing Wiener integrals. Section 3.4 on conditional Wiener integral contains some new material. New Sect. 3.6 illustrates how stochastic numerics is used in financial mathematics.

Chapter 4 is mostly unchanged Chapter 3 of the 1st edition.

Chapter 5 "Geometric integrators and computing ergodic limits" is a revision of Chapter 4 from the 1st edition. The new material includes more examples of geometric integrators, in particular for such models as stochastic rigid body dynamics in Sect. 5.9 and the stochastic Landau-Lifshitz equation in Sect. 5.10. Geometric integrators considered in Chap. 5 demonstrate superiority over long time intervals in comparison with standard schemes for SDEs, which make them particularly suitable for computing ergodic limits. The new Sect. 5.11 is dedicated to computing the mean of a given function with respect to the invariant law of the diffusion, i.e., the ergodic limits. This topic is important for many applications in Statistics and Molecular Dynamics.

Chapter 6 is mostly unchanged Chapter 5 of the 1st edition.

In Chap. 7 (Chapter 6 in the 1st edition), Sect. 7.6 on weak approximation methods for solving the Neumann/Robin boundary value problem for linear PDEs is restructured and extended.

Chapters 8 and 9 are revised Chapters 7 and 8 of the 1st edition.

Chapters 10 and 11 are new. In Chap. 10 FBSDEs are considered. They have numerous applications in stochastic control theory and mathematical finance. Solutions of FBSDEs are associated with the semilinear or quasilinear PDEs. In turn, the corresponding solutions of the PDEs have probabilistic representations by means of the FBSDEs, which are a generalization of the Feynman-Kac formula. The chapter presents numerical algorithms for solving FBSDEs in the mean-square sense. In both cases of semilinear and quasilinear PDEs, the four-step scheme for solution of FBSDEs via the solution of the corresponding PDE [243] is used. For numerical solution of the semilinear PDE, layer methods and the corresponding algorithms from Chap. 8 are exploited. For the quasilinear case, layer methods are given in Chap. 10. Along with FBSDEs on a fixed time interval, FBSDEs with random terminal time are also considered in this chapter. They are associated with the Dirichlet boundary value problem for semilinear PDE. Algorithms for FBSDEs with random terminal time are constructed using mean-square approximations of diffusions in bounded domains from Chap. 6 and layer methods for the Dirichlet problem from Chap. 9.

SPDEs are used for modeling in physics, financial engineering, biology and chemistry, and they are also closely related to the nonlinear filtering problem. The aim of Chap. 11 is to exploit the method of characteristics (the generalized Feynman-Kac formula) and numerical integration of (ordinary) SDEs together with the Monte Carlo technique to propose numerical methods for linear SPDEs of parabolic type. Layer methods for linear and semilinear SPDEs are also presented and the corresponding convergence theorems are proved. The last section of this chapter contains a brief tutorial on nonlinear filtering and its links to SPDEs and PDEs with random coefficients.

A number of additional books and papers related to the subject of this book are added to the bibliography. Some misprints are corrected.

This edition uses a different hierarchical numbering system in comparison with the first edition [314], namely the k-th equation (or the k-th theorem, figure, table, etc) in Section j of Chapter i is labeled and cited as $(i.j.k)$ (or Theorem $i.j.k$, etc) throughout the whole book. The equation (theorem, figure, etc) counter is reset at the beginning of each section.

We would like to thank the Ural Federal University (Ekaterinburg, Russia) and the University of Nottingham (Nottingham, UK) for their support.

New York and Nottingham   Grigori N. Milstein
2019–2021   Michael V. Tretyakov

Preface to the First Edition

Using stochastic differential equations (SDEs), we can successfully model systems that function in the presence of random perturbations. Such systems are among the basic objects of modern control theory, signal processing and filtering, physics, biology and chemistry, economics and finance, to mention just a few. However the very importance acquired by SDEs lies, to a large extent, in the strong connection they have with the equations of mathematical physics. It is well known that problems of mathematical physics involve "curse of dimensionality", often leading to severe difficulties in solving boundary value problems. A way out is provided by the employment of probabilistic representations together with Monte Carlo methods. As a result, a complex multi-dimensional problem for partial differential equations (PDEs) reduces to the Cauchy problem for a system of SDEs. The last system can naturally be regarded as one-dimensional, since it contains only one independent variable; it arises as a characteristic system of the considered problem for PDEs. The importance of this approach, while enabling the reduction of a multi-dimensional boundary value problem to the one-dimensional Cauchy problem, cannot be underestimated for computational mathematics.

Two books:

Milstein G.N., Numerical Integration of Stochastic Differential Equations. Kluwer, 1995 (English translation from Russian 1988),
Kloeden P.E. and Platen E., Numerical Solution of Stochastic Differential Equations. Springer, 1992,

present a systematic treatment of mean-square and weak numerical schemes for SDEs. These approximations represent two fundamental aspects in the contemporary theory of numerical integration of SDEs. Mean-square methods are useful for direct simulation of stochastic trajectories which, for instance, can give information on general behavior of a stochastic model. They are the basis for construction of weak methods which are important for many practical applications. Weak methods are sufficient for evaluation of mean values and solving problems of mathematical physics by the Monte Carlo technique, and they are simpler than mean-square ones.

In the present book, numerical integration of SDEs receives a large developmental effort in two directions: a lot of new special schemes are constructed for a number of stochastic systems which are important for applications and for the first time numerical methods for SDEs in bounded domains are proposed. The second part of the book is devoted to construction of stochastic numerical algorithms for solving complicated problems for PDEs, both linear and nonlinear.

The first two chapters contain essentially revised material of the previously mentioned book by the first author, with broad supplements. For instance, a number of effective numerical methods for systems with colored noise are included in these chapters. Another example of the new material is construction of fully implicit mean-square schemes for SDEs with multiplicative noise.

Many difficulties arise with realizing numerical methods for SDEs of a general form. At the same time methods adapted to specific systems can be more efficient than general methods. Very often fluctuations, which affect a physical system, are small. Fortunately, as shown in Chap. 3, in the case of stochastic systems with small noise, it is possible to construct special numerical methods. The errors of these methods are estimated in terms of products $h^i \varepsilon^j$, where h is the step-size of discretization and ε is a small noise parameter. Usually, the global errors in these methods have the form $O(h^j + \varepsilon^k h^l)$, where $j > l, k > 0$. Thanks to the fact that the accuracy order l of such methods is comparatively small, they are not too complicated, while due to the large j and the small factor ε^k at h^l, their errors are fairly low. This allows us to construct effective (high-exactness) methods with low time-step order which nevertheless have small errors.

In Chap. 4, specific methods for stochastic Hamiltonian systems and Langevin type equations are proposed. Stochastic Hamiltonian systems, like deterministic Hamiltonian systems, possess the property of preserving symplectic structure (symplecticness). For instance, Hamiltonian systems with additive noise are a rather wide and important class of equations having this property. It is well known from deterministic numerical analysis that an effective numerical solution of deterministic Hamiltonian systems on long time intervals requires symplectic methods. It turns out that symplectic methods for stochastic Hamiltonian systems, which are proposed in the first part of Chap. 4, have significant advantages over standard schemes for SDEs.

In the second part of Chap. 4 we construct special numerical methods (we call them quasi-symplectic) for Langevin type equations which have widespread occurrence in models from physics, chemistry, and biology. The proposed methods are such that they degenerate to symplectic methods when the system degenerates to a Hamiltonian one and their law of phase volume contractivity is close to the exact one. The presented numerical tests of both symplectic and quasi-symplectic methods clearly demonstrate superiority of the proposed methods over very long time intervals in comparison with standard ones.

Our probabilistic methods for solving boundary value problems ensure that the proposed approximations of solutions of the corresponding SDEs belong to a bounded domain. Such mean-square approximations are considered in Chap. 5.

A numerical method for simulation of an autonomous diffusion process in a space bounded domain is based on a space discretization using a random walk over small spheres. The algorithm gives points which are close to the points of the real phase trajectory for SDEs. To realize the algorithm, the exit point of the Wiener process from a d-dimensional ball has to be constructed at each step. Due to independence of the first exit time and the first exit point of the Wiener process from the ball, they can be simulated separately. It is known that the exit point is distributed uniformly on the sphere, but simulation of the exit time is a fairly laborious problem. Consequently, the algorithm gives only the phase component of the approximate trajectory without modeling the corresponding time component. The space-time point lies on the d-dimensional lateral surface of a semicylinder with spherical base in the $(d+1)$-dimensional semispace $[t_0, \infty) \times R^d$. The algorithm ensures smallness of the phase increments at each step, but the nonsimulated time increments can take arbitrary large values with some probability. "Ordinary" mean-square methods from Chap. 1, intended to solve SDEs on a finite time interval, are based on a time discretization. The space-time point, corresponding to an "ordinary" one-step approximation constructed at a time point t_k, lies on the d-dimensional plane $t = t_k$, which belongs to the $(d+1)$-dimensional semispace $[t_0, \infty) \times R^d$. The "ordinary" mean-square methods give both time and phase components of the approximate trajectory. They ensure smallness of time increments at each step, but space increments can take arbitrary large values with some probability. In Chap. 5 we also introduce mean-square approximations which control boundedness of both space increments and time increments. In addition they give approximate values for both phase and time components of the space-time diffusion in the space-time bounded domain. The space-time point lies on a bounded d-dimensional manifold. This problem is solved in a constructive manner by the implementation of a space-time discretization with a random walk over boundaries of small space-time parallelepipeds.

Chapter 6 is devoted to random walks related to linear Dirichlet and Neumann boundary value problems for PDEs of elliptic and parabolic type. These random walks are Markov chains. Using them together with the Monte Carlo technique, complex multi-dimensional problems for linear PDEs can be solved. The random walks are constructed on the basis of mean-square and weak approximations for the characteristic system of SDEs due to the corresponding probabilistic representations of the solution to the considered boundary value problem. As in Chap. 5, a certain boundedness of the simulated increments of the Markov chains is necessary here. The proposed algorithms are accompanied by convergence theorems and numerical tests.

Nonlinear PDEs are suggested as mathematical models of problems in many fields such as fluid dynamics, combustion, biochemistry, dynamics of populations, finance, etc. They are mostly investigated by numerical methods, which are traditionally based on deterministic approaches. A probabilistic approach to construction of new numerical methods for solving initial and boundary value problems for nonlinear parabolic PDEs is developed in Chaps. 7 and 8. The approach is based on making use of the well-known probabilistic representations of solutions of linear PDEs and the idea of SDE numerical integration in the weak sense. Despite their probabilistic nature these methods are nevertheless deterministic. The probabilistic approach takes into account a coefficient dependence on the space variables and a relationship between diffusion and advection in an intrinsic manner. In particular, the layer methods derived allow us to avoid difficulties stemming from essentially changing coefficients and strong advection. A lot of computer experiments were made using the numerical algorithms proposed in Chaps. 7 and 8. Among them are numerical tests on the Burgers equation with small viscosity and on the FKPP-equation. Their results are in a good agreement with the theory. We also present a comparison analysis of the layer methods and the well-known finite-difference schemes demonstrating some of the advantages of the proposed methods.

Chapter 9 is devoted to applications of stochastic numerics. Among lots of possibilities, we select applications of constructed stochastic simulation algorithms to such models of stochastic dynamics as systems with stochastic resonance and stochastic ratchets. We demonstrate here both mean-square methods for simulating trajectories of the considered models and weak methods for solving a number of boundary value problems.

An overwhelming majority of the methods proposed in this book are brought to numerical algorithms. Then it only remains to write a computer program, which is usually not complicated, and to use the method in practice. We give some illustrations in the Appendix how our methods can be implemented.

The field of stochastic numerics and its applications is too broad for one book. For example, such important topics as numerical integration of stochastic partial differential equations and of backward stochastic differential equations, being close to interests of the authors, are not covered in the book. The authors do not aim to provide an exhaustive review of literature. As a rule, the references are cited in the course of presentation. Some references are given without comments. On the whole, the content of the book is mainly based on the results obtained by the authors.

Throughout the book we use the hierarchical numbering system: the k-th equation (or the k-th theorem, figure, table, etc) in Sect. j of Chap. i is labeled at the place, where it occurs, and is cited within Chap. i as $(j.k)$ (or Theorem $j.k$, etc); it is cited as $(i.j.k)$ (or Theorem $i.j.k$, etc) outside Chap. i. The equation (theorem, figure, etc) counter is reset at the beginning of each section. The only exception is listings in the Appendix: the k-th listing is labeled and cited as Listing A.k.

We would like to thank the institutions which have made this work possible – the Ural State University and the Institute of Mathematics and Mechanics of the Russian Academy of Sciences (Ekaterinburg, Russia), the Weierstrass Institute for Applied Analysis and Stochastics (Berlin, Germany), and the University of Leicester (Leicester, UK).

Berlin and Leicester   Grigori N. Milstein
2002–2003   Michael V. Tretyakov

Contents

Frequently Used Notation

a.s.	Almost surely
FBSDEs	Forward-backward stochastic differential equations
FKPP	Fisher-Kolmogorov-Petrovskii-Piskunov
i.i.d.	Independent identically distributed
MC	Monte Carlo
PDE	Partial differential equation
PRK	Partitioned Runge-Kutta
RK	Runge-Kutta
RKN	Runge-Kutta-Nyström
RNG	Random number generator
SDE	Stochastic differential equation
SPDE	Stochastic partial differential equation
$:=$	Equal by definition
$\doteq$ or $\simeq$ or $\approx$	Approximately equal
$\cdots \circ dw(t)$	Stratonovich SDE
δ_{ij}	Kronecker delta symbol
$\chi_A(\omega)$ or $I_A(x)$	Indicator function of the set A
$\mathbf{R}^d$	d-dimensional Euclidean space
$x = (x^1, \ldots, x^d)^{\mathbf{T}}$	Column vector $x \in \mathbf{R}^d$ with ith-component x^i
$x^{\mathbf{T}}y$ or (x, y)	Scalar product of vectors $x, y \in \mathbf{R}^d$
$\|x\|$	Euclidean norm of vector $x \in \mathbf{R}^d$
$A = \{a^{ij}\}$	Matrix A with ijth-component a^{ij}
$\|\sigma\| = (tr\sigma\sigma^{\mathbf{T}})^{1/2}$	Euclidean norm of matrix σ
$O(\rho)$	Expression being divided by ρ remains bounded as $\rho \to 0$
$(\Omega, \mathcal{F}, P)$	Probability space
$\mathcal{F}_t$	Nondecreasing family of σ-subalgebras of $\mathcal{F}$
$E(X)$ or EX	Expectation of X
$Var(X)$ or $D(X)$	Variance of X

$\bar{D}_M$	Sample variance
$E(X\|\mathcal{F}_t)$	Conditional expectation of X under $\mathcal{F}_t$
$\mathcal{N}(\mu,\sigma^2)$	Gaussian distribution with mean μ and variance σ^2
$w_r(t)$ or $W_r(t)$	Standard Wiener process
$I_{i\ldots j}, J_{i\ldots j}, I_{i_1,\ldots,i_j}$	Ito integrals
$X_{t,x}(s), s \geq t$	Trajectory such that $X_{t,x}(t) = x$
h or h_k or Δt	Step of time discretization
$\Delta_k w_r(h) = w_r(t_k + h) - w_r(t_k)$	Increment of $w_r(t)$ on $[t_k, t_k + h]$
$\bar{X}(t_k)$ or $\bar{X}_k$ or X_k	Approximation of $X(t_k)$
$\bar{X}_{t,x}(t+h)$	One-step approximation
$\Delta = X_{t,x}(t+h) - x$	Increment of solution
$\bar{\Delta} = \bar{X}_{t,x}(t+h) - x$	Increment of approximation of solution
$\bar{X}^{(m)}, m = 1, \ldots, M$	Independent realizations
$\mathbf{F}$	Class of functions with polynomial growth
$C^k(\bar{D})$	Space of k times continuously differentiable functions on $\bar{D}$
$C^d_{0,0}$	Set of all d-dimensional continuous vector-functions $x(t)$ satisfying the condition $x(0) = 0$
$C^d_{0,a;T,b}$	Set of all d-dimensional continuous vector-functions $x(t)$ satisfying the conditions $x(0) = a$, $x(T) = b$
$\mu_{0,0}(x)$	Wiener measure corresponding to Brownian paths with fixed initial point $(0,0)$
$\mu^{T,b}_{0,a}(x)$	Conditional Wiener measure corresponding to Brownian paths $X^{T,b}_{0,a}(t)$ with fixed initial and final points
L, Λ_r	Differential operators
Δ	Laplace operator
$H(t,p,q), H_r(t,p,q)$	Hamiltonians
G	Bounded domain in $\mathbf{R}^d$
∂G	Boundary of domain G
Γ_δ	Interior of δ-neighborhood of ∂G belonging to G
$Q = [t_0, T] \times G$	Cylinder in $\mathbf{R}^{d+1}$
$\Gamma = \bar{Q} \backslash Q$	Part of boundary of Q consisting of upper base and lateral surface
$U_r \subset \mathbf{R}^d$	Open sphere of radius r with center at the origin of $\mathbf{R}^d$
$U^\sigma_r(x)$	Open ellipsoid obtained from U_r by linear transformation $\sigma(x)$ and shift x
$C_r \subset \mathbf{R}^d$	Cube with center at the origin of $\mathbf{R}^d$ and edges of the length $2r$ being parallel to the coordinate axes

$\Pi_{r,l} = [0, lr^2) \times C_r$ Space-time parallelepiped

$\bar{\nu}_x(D)$ First exit time of approximate trajectory from D

$\tau, \theta, \bar{\theta}$ Markov moments

$\nu, \bar{\nu}, \varkappa$ Random steps

Chapter 1
Mean-Square Approximation
for Stochastic Differential Equations

The simplest approximate method for solving the Ito system

$$dX = a(t, X)dt + \sum_{r=1}^{q} \sigma_r(t, X)dw_r(t) \qquad (1.0.1)$$

is *Euler's method*:

$$X_{k+1} = X_k + \sum_{r=1}^{q} \sigma_{rk} \Delta_k w_r(h) + a_k h , \qquad (1.0.2)$$

where $\Delta_k w_r(h) = w_r(t_{k+1}) - w_r(t_k)$, and the index k at σ_r and a indicates that these functions are evaluated at the point (t_k, X_k).

Marujama [257] showed the mean-square convergence of this method, while Gichman and Skorochod [123] proved that the order of accuracy of Euler's method is $1/2$, i.e.,

$$(E(X(t_k) - X_k)^2)^{1/2} \leq Ch^{1/2} , \qquad (1.0.3)$$

where C is a constant not depending on k and h.

If for some method we would have

$$(E(X(t_k) - X_k)^2)^{1/2} \leq Ch^p , \qquad (1.0.4)$$

then we say that *the mean-square order of accuracy* of the method is p.

© Springer Nature Switzerland AG 2021
G. N. Milstein and M. V. Tretyakov, *Stochastic Numerics for Mathematical Physics*,
Scientific Computation, https://doi.org/10.1007/978-3-030-82040-4_1

A method of first order of accuracy was first constructed in [269]. It is as follows:

$$X_{k+1} = X_k + \sum_{r=1}^{q} \sigma_{rk} \Delta_k w_r(h) + a_k h \tag{1.0.5}$$

$$+ \sum_{i=1}^{q} \sum_{r=1}^{q} (\Lambda_i \sigma_r)_k \int_{t_k}^{t_{k+1}} (w_i(\theta) - w_i(t_k)) dw_r(\theta),$$

where $\Lambda_i = (\sigma_i, \dfrac{\partial}{\partial x})$.

For a single noise $(q = 1)$, the iterated Ito integral in (1.0.5) can be expressed in terms of $\Delta_k w$ and the formula takes the following form, e.g. in the scalar case:

$$X_{k+1} = X_k + \sigma_k \Delta_k w + a_k h + \frac{1}{2} \left(\sigma \frac{\partial \sigma}{\partial x} \right)_k \Delta_k^2 w - \frac{1}{2} \left(\sigma \frac{\partial \sigma}{\partial x} \right)_k h. \tag{1.0.6}$$

The same paper [269] also contains methods of higher order of accuracy for the scalar case.

The authors of [339, 351, 373, 392, 426, 427] and many others obtained results that rest on those of [269]. More recent developments [65, 72, 248] include approximations of first order of accuracy which avoid exact simulation of iterated Ito integrals.

The technique by which a number of formulas were obtained in [269] is somewhat laborious. It uses the theory of Markov operator semigroups in combination with the method of undetermined coefficients. Wagner and Platen [459] gave a very simple proof (using only Ito's formula) of the expansion of the solution $X_{t,x}(t + h)$ of the system (1.0.1) in powers of h and in integrals depending on the increments $w_r(\theta) - w_r(t)$, $t \le \theta \le t + h$, $r = 1, \ldots, q$. This expansion generalizes (1.0.5). In the deterministic situation it comes down to Taylor's formula for $X_{t,x}(t + h)$ in powers of h in a neighborhood of the point (t, x). Theoretically, this expansion allows one to construct methods of arbitrary high order of accuracy (of course, with corresponding conditions on the coefficients of the system (1.0.1)).

We begin (Sect. 1.1) with the fundamental convergence theorem which establishes the mean-square order of convergence of a method resting on properties of its one-step approximation only [273, 274]. The conditions of this theorem use both properties of mean and mean-square deviation of one-step approximation. The theorem asserts that if p_1 and p_2 are orders of accuracy of the one-step approximation for the deviation of expectation and the mean-square deviation, respectively, and if $p_2 > 1/2$, $p_1 \ge p_2 + 1/2$, then the method converges and the order of the method is $p_2 - 1/2$. The proof of mean-square convergence of overwhelming majority of the methods considered in this book is based on this theorem. To derive a one-step approximation, which determines a method, many approaches can be used. In the deterministic theory, an expansion of the solution by Taylor's formula underlies all one-step methods, both implicit and explicit. The analogue of such an expansion for stochastic systems – Wagner–Platen expansion – is considered in Sect. 1.2. Here, on the basis of the same convergence theorem, we establish the order of a method depending on components in the expansion. Section 1.3 is devoted to implicit methods. Recall that in distinction to *explicit methods, implicit methods* have far better stability properties. In particular, they are necessary for integrating stiff systems of differential equations (see, e.g. [156, 157, 372]). Let us also note that in the case of general Hamiltonian systems symplectic Runge–Kutta methods are all implicit [154, 155, 291, 398]. We start Sect. 1.3 considering methods with general implicitness in drift terms (drift-implicit methods) and methods with implicitness of a special kind in stochastic terms. Drift-implicit methods were proposed in many papers. Stochastic implicit methods of the special kind (balanced implicit methods) were first considered in [285]. The balanced methods are linear with respect to the one-step approximation, and they cannot be used for constructing symplectic methods which we introduce for stochastic Hamiltonian systems (see Chap. 5). General stochastic implicit (fully implicit) methods [291] are also considered in Sect. 1.3. The increments of Wiener processes are substituted by some truncated random variables in these fully implicit schemes. The conditions of the fundamental convergence theorem of Sect. 1.1 are rather restrictive: the coefficients of (1.0.1) are assumed to be globally Lipschitz. At the same time, SDEs with nonglobally Lipschitz conditions, such that there exist unique extendable solutions for them, form a very broad and important class in applications. In Sect. 1.4 we consider a version of the fundamental convergence theorem under some relaxed assumptions and discuss numerical methods which are convergent under nonglobally Lipschitz conditions. In Sect. 1.5 we consider means for both exact and approximate modeling of Ito integrals depending on single or several noises. It turns out that the possibilities for exact modeling are very limited, especially in the case of several noises. Therefore, our main interest is in methods of approximate modelling. We develop such methods for those Ito integrals that are necessary for construction of mean-square approximations with first order of accuracy in the general case and with orders $3/2$ and 2 for systems with one noise. Stochastic differential equations with additive and colored noise are met very often in applications and that is why they are of great independent interest. In Sects. 1.6– 1.7, we construct efficient high-order methods because of special properties of such systems.

Note that the wish to construct a large amount of methods, both explicit and implicit, is related to the fact that distinct methods have various properties as regards their accuracy, stability, time and computational cost, etc.

1.1 Fundamental Theorem on the Mean-Square Order of Convergence

In Sect. 1.1.1 we state the fundamental theorem on mean-square convergence and prove it in Sect. 1.1.2. In Sect. 1.1.3 SDEs in the sense of Stratonovich are considered. The theorem is discussed further in Sect. 1.1.4. Section 1.1.5 deals with the explicit Euler method. Almost sure (a.s.) convergence is considered in Sect. 1.1.6.

1.1.1   Statement of the Theorem

Let $(\Omega, \mathcal{F}, P)$ be a probability space, $\mathcal{F}_t$, $t_0 \leq t \leq T$, be a nondecreasing family of σ-subalgebras of $\mathcal{F}$, and $(w_r(t), \mathcal{F}_t)$, $r = 1, \ldots, q$, be independent Wiener processes. Consider the system of SDEs in the sense of Ito

$$dX = a(t, X)dt + \sum_{r=1}^{q} \sigma_r(t, X)dw_r(t), \tag{1.1.1}$$

where X, a, σ_r are vectors of dimension d.

Assume that the functions $a(t, x)$ and $\sigma_r(t, x)$ are defined and continuous for $t \in [t_0, T]$, $x \in \mathbf{R}^d$, and satisfy the *globally Lipschitz condition*: for all $t \in [t_0, T]$, $x \in \mathbf{R}^d$, $y \in \mathbf{R}^d$ there is an inequality

$$|a(t, x) - a(t, y)| + \sum_{r=1}^{q} |\sigma_r(t, x) - \sigma_r(t, y)| \leq K|x - y|. \tag{1.1.2}$$

Here and below we denote by $|x|$ the Euclidean norm of the vector x and by $x^\mathsf{T} y$ or by (x, y) the scalar (inner) product of two vectors x and y.

Let $(X(t), \mathcal{F}_t)$, $t_0 \leq t \leq T$, be a solution of the system (1.1.1) with $E|X(t_0)|^2 < \infty$. The *one-step approximation* $\bar{X}_{t,x}(t + h)$, $t_0 \leq t < t + h \leq T$, depends on x, t, h, and $\{w_1(\theta) - w_1(t), \ldots, w_q(\theta) - w_q(t), t \leq \theta \leq t + h\}$ and is defined as follows:

$$\bar{X}_{t,x}(t + h) = x + A(t, x, h; w_i(\theta) - w_i(t), \ i = 1, \ldots, q, \ t \leq \theta \leq t + h). \tag{1.1.3}$$

Using the one-step approximation, we recurrently construct the approximations $(\bar{X}_k, \mathcal{F}_{t_k})$, $k = 0, \ldots, N$, $t_{k+1} - t_k = h_{k+1}$, $T_N = T$:

$$\bar{X}_0 = X_0 = X(t_0), \quad \bar{X}_{k+1} = \bar{X}_{t_k, \bar{X}_k}(t_{k+1})$$
$$= \bar{X}_k + A(t_k, \bar{X}_k, h_{k+1}; w_i(\theta) - w_i(t_k), \ i = 1, \ldots, q, \ t_k \leq \theta \leq t_{k+1}). \tag{1.1.4}$$

We will use the following notation. An approximation of $X(t_k)$ will be denoted by $\bar{X}(t_k)$, $\bar{X}_k$, or simply by X_k. Everywhere below we put $\bar{X}_0 = X(t_0)$. Further, let X be an $\mathcal{F}_{t_k}$-measurable random variable with $E|X|^2 < \infty$; as usual, $X_{t_k, X}(t)$ denotes the solution of the system (1.1.1) for $t_k \leq t \leq T$ satisfying the following initial condition at $t = t_k$: $X(t_k) = X$. By $\bar{X}_{t_k, X}(t_i)$, $t_i \geq t_k$, we denote an approximation of the solution at step i such that $\bar{X}_k = X$. Clearly,

$$\bar{X}_{k+1} = \bar{X}_{t_k, \bar{X}_k}(t_{k+1}) = \bar{X}_{t_0, \bar{X}_0}(t_{k+1}).$$

For simplicity, we assume that $t_{k+1} - t_k = h = (T - t_0)/N$.

Theorem 1.1.1 *Suppose the one-step approximation $\bar{X}_{t,x}(t+h)$ has order of accuracy p_1 for expectation of the deviation and order of accuracy p_2 for the mean-square deviation; more precisely, for arbitrary $t_0 \leq t \leq T - h$, $x \in \mathbf{R}^d$ the following inequalities hold:*

$$|E(X_{t,x}(t+h) - \bar{X}_{t,x}(t+h))| \leq K(1 + |x|^2)^{1/2} h^{p_1}, \qquad (1.1.5)$$

$$\left[E|X_{t,x}(t+h) - \bar{X}_{t,x}(t+h)|^2\right]^{1/2} \leq K(1 + |x|^2)^{1/2} h^{p_2}. \qquad (1.1.6)$$

Also let

$$p_2 \geq \frac{1}{2}, \ p_1 \geq p_2 + \frac{1}{2}. \qquad (1.1.7)$$

Then for any N and $k = 0, 1, \ldots, N$ the following inequality holds:

$$\left[E|X_{t_0,X_0}(t_k) - \bar{X}_{t_0,X_0}(t_k)|^2\right]^{1/2} \leq K(1 + E|X_0|^2)^{1/2} h^{p_2 - 1/2}, \qquad (1.1.8)$$

i.e., the order of accuracy of the method constructed using the one-step approximation $\bar{X}_{t,x}(t+h)$ is $p = p_2 - 1/2$.

We note that all the constants K mentioned above, as well as the ones that will appear in the sequel, depend in the final analysis on the system (1.1.1) and the approximation (1.1.3) only and do not depend on X_0 and N.

Remark 1.1.2 Often the notion of *strong order of accuracy* is used (see, e.g., [202]): if for some method

$$E\,|X(t_k) - X_k| \leq Kh^p,$$

where K is a positive constant independent of k and h, then we say that the strong order of accuracy of the method is equal to p. Clearly, if the mean-square order of a method is p, then the method has the same strong order. We prefer to use the notions of accuracy and convergence in the mean-square sense.

1.1.2 Proof of the Fundamental Theorem

Before proceeding to the proof of Theorem 1.1.1, we need three lemmas.

Lemma 1.1.3 *There is a representation*

$$X_{t,x}(t+h) - X_{t,y}(t+h) = x - y + Z \qquad (1.1.9)$$

for which

$$E|X_{t,x}(t+h) - X_{t,y}(t+h)|^2 \leq |x-y|^2(1+Kh),\qquad(1.1.10)$$

$$EZ^2 \leq K|x-y|^2h.\qquad(1.1.11)$$

Proof Ito's formula readily implies that for $0 \leq \theta \leq h$:

$$E|X_{t,x}(t+\theta) - X_{t,y}(t+\theta)|^2 = |x-y|^2$$
$$+2E\int_t^{t+\theta}(X_{t,x}(s) - X_{t,y}(s))^\mathsf{T}(a(s, X_{t,x}(s)) - a(s, X_{t,y}(s)))ds$$
$$+E\int_t^{t+\theta}\sum_{r=1}^q|\sigma_r(s, X_{t,x}(s)) - \sigma_r(s, X_{t,y}(s))|^2ds.$$

It follows from here and (1.1.2) that

$$E|X_{t,x}(t+\theta) - X_{t,y}(t+\theta)|^2 \leq |x-y|^2 + K\int_t^{t+\theta}E|X_{t,x}(s) - X_{t,y}(s)|^2ds.$$
$$(1.1.12)$$

In turn, this implies

$$E|X_{t,x}(t+\theta) - X_{t,y}(t+\theta)|^2 \leq |x-y|^2 \times e^{Kh},\ 0 \leq \theta \leq h,\qquad(1.1.13)$$

from which (1.1.10) follows. Further, noting that

$$Z = \int_t^{t+h}\sum_{r=1}^q(\sigma_r(s, X_{t,x}(s)) - \sigma_r(s, X_{t,y}(s)))dw_r(s)$$
$$+ \int_t^{t+h}(a(s, X_{t,x}(s)) - a(s, X_{t,y}(s)))ds,$$

and using (1.1.13), it is not difficult to obtain (1.1.11). $\square$

Remark 1.1.4 In the sequel we will use a *conditional version of the inequalities* (1.1.5), (1.1.6), (1.1.10), (1.1.11). In this conditional version the deterministic variables x, y are replaced by $\mathcal{F}_t$-measurable random variables X, Y. For example, the conditional version of (1.1.5) reads:

$$|E(X_{t,X}(t+h) - \bar{X}_{t,X}(t+h)|\mathcal{F}_t)| \leq K(1+|X|^2)^{1/2}h^{p_1}.\qquad(1.1.14)$$

We will also use simple consequences of these inequalities. For example, (1.1.14) implies

$$E|E(X_{t,X}(t+h) - \bar{X}_{t,X}(t+h)|\mathcal{F}_t)|^2 \leq K(1+E|X|^2)h^{2p_1}.\qquad(1.1.15)$$

The proof of these conditional versions rests on an assertion of the following kind: if ζ is $\tilde{\mathcal{F}}$-measurable, $\tilde{\mathcal{F}} \subset \mathcal{F}$, $f(x, \omega)$ does not depend on $\tilde{\mathcal{F}}$, $\omega \in \Omega$, and $Ef(x, \omega) = \phi(x)$, then $E(f(\zeta, \omega)|\tilde{\mathcal{F}}) = \phi(\zeta)$ (see [123, 213]). In the case under consideration *the increments of the Wiener processes* $w_1(\theta) - w_1(t), \ldots, w_q(\theta) - w_q(t)$, $t \le \theta \le t + h$, do not depend on $\mathcal{F}_t$ (see, e.g. [123]), and neither does $\bar{X}_{t,x}(t + h)$, which is formed so that it depends on x, t, h, and the mentioned increments only.

Lemma 1.1.5 *For all natural N and all $k = 0, \ldots, N$ the following inequality holds:*

$$E|\bar{X}_k|^2 \le K(1 + E|X_0|^2). \tag{1.1.16}$$

Proof Suppose that $E|\bar{X}_k|^2 < \infty$. Then using the conditional version of (1.1.6), we obtain

$$E|X_{t_k, \bar{X}_k}(t_{k+1}) - \bar{X}_{t_k, \bar{X}_k}(t_{k+1})|^2 \le K(1 + E|\bar{X}_k|^2)h^{2p_2}. \tag{1.1.17}$$

It is well known [123] that if an $\mathcal{F}_t$-measurable random variable X has bounded second moment, then the solution $X_{t,X}(t + \theta)$ also has bounded second moment. Therefore, $E|X_{t_k, \bar{X}_k}(t_{k+1})|^2 < \infty$. This and (1.1.17) readily imply that $E|\bar{X}_{k+1}|^2 < \infty$ (recall that $\bar{X}_{t_k, \bar{X}_k}(t_{k+1}) = \bar{X}_{k+1}$). Since $E|\bar{X}_0|^2 < \infty$, we have proved the existence of all $E|\bar{X}_k|^2 < \infty$, $k = 0, \ldots, N$.

Consider the equation

$$\begin{aligned} E|\bar{X}_{k+1}|^2 &= E|\bar{X}_k|^2 + E|X_{t_k, \bar{X}_k}(t_{k+1}) - \bar{X}_k|^2 \\ &\quad + E|X_{t_k, \bar{X}_k}(t_{k+1}) - \bar{X}_{t_k, \bar{X}_k}(t_{k+1})|^2 + 2E\bar{X}_k^{\mathsf{T}}(X_{t_k, \bar{X}_k}(t_{k+1}) - \bar{X}_k) \\ &\quad + 2E\bar{X}_k^{\mathsf{T}}(\bar{X}_{t_k, \bar{X}_k}(t_{k+1}) - X_{t_k, \bar{X}_k}(t_{k+1})) \\ &\quad + 2E(X_{t_k, \bar{X}_k}(t_{k+1}) - \bar{X}_k)^{\mathsf{T}}(\bar{X}_{t_k, \bar{X}_k}(t_{k+1}) - X_{t_k, \bar{X}_k}(t_{k+1})). \end{aligned} \tag{1.1.18}$$

We have (see [123])

$$E|X_{t_k, \bar{X}_k}(t_{k+1}) - \bar{X}_k|^2 \le K(1 + E|\bar{X}_k|^2)h. \tag{1.1.19}$$

Further, we obtain from (1.1.17) and (1.1.19):

$$\begin{aligned} 2|E(X_{t_k, \bar{X}_k}(t_{k+1}) - \bar{X}_k)^{\mathsf{T}}(\bar{X}_{t_k, \bar{X}_k}(t_{k+1}) - X_{t_k, \bar{X}_k}(t_{k+1}))| \\ \le 2(E|X_{t_k, \bar{X}_k}(t_{k+1}) - \bar{X}_k|^2)^{1/2}(E|X_{t_k, \bar{X}_k}(t_{k+1}) - \bar{X}_{t_k, \bar{X}_k}(t_{k+1})|^2)^{1/2} \\ \le K(1 + E|\bar{X}_k|^2)h^{p_2 + 1/2}. \end{aligned} \tag{1.1.20}$$

It is not difficult to prove the inequality

$$E|E(X_{t_k, \bar{X}_k}(t_{k+1}) - \bar{X}_k|\mathcal{F}_{t_k})|^2 \le K(1 + E|\bar{X}_k|^2)h^2. \tag{1.1.21}$$

Therefore

$$2|E\bar{X}_k^\mathsf{T}(X_{t_k,\bar{X}_k}(t_{k+1}) - \bar{X}_k)| = 2|E\bar{X}_k^\mathsf{T} E(X_{t_k,\bar{X}_k}(t_{k+1}) - \bar{X}_k|\mathcal{F}_{t_k})| \tag{1.1.22}$$
$$\leq 2(E|\bar{X}_k|^2)^{1/2}(E|E(X_{t_k,\bar{X}_k}(t_{k+1}) - \bar{X}_k|\mathcal{F}_{t_k})|^2)^{1/2} \leq K(1 + E|\bar{X}_k|^2)h \,.$$

Similarly, but referring to (1.1.15) instead of (1.1.21), we obtain

$$2|E\bar{X}_k^\mathsf{T}(\bar{X}_{t_k,\bar{X}_k}(t_{k+1}) - X_{t_k,\bar{X}_k}(t_{k+1}))| \leq K(1 + E|\bar{X}_k|^2)h^{p_1} \,. \tag{1.1.23}$$

Applying the inequalities (1.1.19), (1.1.17), (1.1.22), (1.1.23), and (1.1.20) to the equality (1.1.18) and recalling that $p_1 \geq 1$, $p_2 \geq 1/2$, we arrive at the inequality (taking, without loss of generality, $h \leq 1$) :

$$E|\bar{X}_{k+1}|^2 \leq E|\bar{X}_k|^2 + K(1 + E|\bar{X}_k|^2)h = (1 + Kh)E|\bar{X}_k|^2 + Kh \,. \tag{1.1.24}$$

Hence, using the well-known result which, for the sake of reference, is stated as Lemma 1.1.6 below, we obtain (1.1.16). ☐

Lemma 1.1.6 *Suppose that for arbitrary N and $k = 0, \ldots, N$ we have*

$$u_{k+1} \leq (1 + Ah)u_k + Bh^p \,, \tag{1.1.25}$$

where $h = T/N$, $A \geq 0$, $B \geq 0$, $p \geq 1$, $u_k \geq 0$, $k = 0, \ldots, N$. Then

$$u_k \leq e^{AT}u_0 + \frac{B}{A}(e^{AT} - 1)h^{p-1} \tag{1.1.26}$$

(where for $A = 0$ we put $(e^{AT} - 1)/A$ equal to T).

Let us proceed to the proof of Theorem 1.1.1 itself. We have

$$X_{t_0,X_0}(t_{k+1}) - \bar{X}_{t_0,X_0}(t_{k+1}) = X_{t_k,X(t_k)}(t_{k+1}) - \bar{X}_{t_k,\bar{X}_k}(t_{k+1}) \tag{1.1.27}$$
$$= (X_{t_k,X(t_k)}(t_{k+1}) - X_{t_k,\bar{X}_k}(t_{k+1})) + (X_{t_k,\bar{X}_k}(t_{k+1}) - \bar{X}_{t_k,\bar{X}_k}(t_{k+1})) \,.$$

The first difference in the right-hand side of (1.1.27) is the error of the solution arising due to the error in the initial data at time t_k, accumulated at the k-th step. The second difference is the one-step error at the $(k + 1)$-step. Taking the square of both sides of the equation, we obtain

$$E\left|X_{t_0,X_0}(t_{k+1}) - \bar{X}_{t_0,X_0}(t_{k+1})\right|^2$$
$$= EE(|X_{t_k,X(t_k)}(t_{k+1}) - X_{t_k,\bar{X}_k}(t_{k+1})|^2|\mathcal{F}_{t_k})$$
$$+ EE(|X_{t_k,\bar{X}_k}(t_{k+1}) - \bar{X}_{t_k,\bar{X}_k}(t_{k+1})|^2|\mathcal{F}_{t_k})$$
$$+ 2EE((X_{t_k,X(t_k)}(t_{k+1}) - X_{t_k,\bar{X}_k}(t_{k+1}))^\mathsf{T}(X_{t_k,\bar{X}_k}(t_{k+1}) - \bar{X}_{t_k,\bar{X}_k}(t_{k+1}))|\mathcal{F}_{t_k}) \,.$$
$$\tag{1.1.28}$$

By the conditional version of Lemma 1.1.3, we get

$$
\begin{aligned}
E E(|X_{t_k, X(t_k)}(t_{k+1}) - X_{t_k, \bar{X}_k}(t_{k+1})|^2 | \mathcal{F}_{t_k}) \\
\leq E|X(t_k) - \bar{X}_k|^2 \times (1 + Kh).
\end{aligned}
\tag{1.1.29}
$$

By the conditional version of (1.1.6) and Lemma 1.1.5, we have

$$
\begin{aligned}
E E(|X_{t_k, \bar{X}_k}(t_{k+1}) - \bar{X}_{t_k, \bar{X}_k}(t_{k+1})|^2 | \mathcal{F}_{t_k}) \\
\leq K(1 + E|\bar{X}_k|^2) h^{2p_2} \leq K(1 + E|X_0|^2) h^{2p_2}.
\end{aligned}
\tag{1.1.30}
$$

The difference $X_{t_k, X(t_k)}(t_{k+1}) - X_{t_k, \bar{X}_k}(t_{k+1})$ in the last summand in (1.1.28) can be treated using Lemma 1.1.3:

$$
X_{t_k, X(t_k)}(t_{k+1}) - X_{t_k, \bar{X}_k}(t_{k+1}) := X(t_k) - \bar{X}_k + Z.
$$

Then we obtain two terms each of which can be estimated individually. Using (1.1.15) and Lemma 1.1.5, we get

$$
\begin{aligned}
|E E((X(t_k) - \bar{X}_k)^{\mathsf{T}}(X_{t_k, \bar{X}_k}(t_{k+1}) - \bar{X}_{t_k, \bar{X}_k}(t_{k+1})) | \mathcal{F}_{t_k})| \\
= |E((X(t_k) - \bar{X}_k)^{\mathsf{T}} E(X_{t_k, \bar{X}_k}(t_{k+1}) - \bar{X}_{t_k, \bar{X}_k}(t_{k+1}) | \mathcal{F}_{t_k}))| \\
\leq (E|X(t_k) - \bar{X}_k|^2)^{1/2} K(1 + E|X_0|^2)^{1/2} h^{p_1}.
\end{aligned}
\tag{1.1.31}
$$

Finally, using Lemma 1.1.5 and the inequality (1.1.6), we obtain

$$
\begin{aligned}
|E(Z^{\mathsf{T}}(X_{t_k, \bar{X}_k}(t_{k+1}) - \bar{X}_{t_k, \bar{X}_k}(t_{k+1})))| \\
\leq (E E(|Z|^2 | \mathcal{F}_{t_k}))^{1/2} (E E(|X_{t_k, \bar{X}_k}(t_{k+1}) - \bar{X}_{t_k, \bar{X}_k}(t_{k+1})|^2 | \mathcal{F}_{t_k}))^{1/2} \\
\leq K(E|X(t_k) - \bar{X}_k|^2)^{1/2} (1 + E|X_0|^2)^{1/2} h^{p_2 + 1/2}.
\end{aligned}
\tag{1.1.32}
$$

Introduce the notation $\varepsilon_k^2 := E|X(t_k) - \bar{X}_k|^2$. The relations (1.1.28) – (1.1.32) and the condition $p_1 \geq p_2 + 1/2$ then lead to the inequality (we take $h < 1$):

$$
\varepsilon_{k+1}^2 \leq \varepsilon_k^2 (1 + Kh) + K(1 + E|X_0|^2)^{1/2} \varepsilon_k h^{p_2 + 1/2} + K(1 + E|X_0|^2) h^{2p_2}.
$$

Using the elementary relation

$$
(1 + E|X_0|^2)^{1/2} \varepsilon_k h^{p_2 + 1/2} \leq \frac{\varepsilon_k^2 h}{2} + \frac{1 + E|X_0|^2}{2} h^{2p_2},
$$

we get

$$
\varepsilon_{k+1}^2 \leq \varepsilon_k^2 (1 + Kh) + K(1 + E|X_0|^2) h^{2p_2}.
$$

The inequality (1.1.8) follows from this, taking into account Lemma 1.1.6 and the fact that $\varepsilon_0^2 = 0$. This proves Theorem 1.1.1. $\qquad \square$

In [274] a strengthening of the main convergence theorem (Theorem 1.1.1) is proved. Here we give this result without proof.

Theorem 1.1.7 *In addition to the conditions of Theorem 1.1.1, suppose that also*

$$\left[E|X_{t,x}(t+h) - \bar{X}_{t,x}(t+h)|^4\right]^{1/4} \le K(1+|x|^4)^{1/4}h^{p_2-1/4},$$

and let $p_2 \ge 3/4$. Then

$$\left[E \max_{0 \le k \le N} |X_{t_0,X_0}(t_k) - \bar{X}_{t_0,X_0}(t_k)|^2\right]^{1/2} \le K(1+E|X_0|^4)^{1/4}h^{p_2-1/2}.$$

For the Euler scheme (1.0.2), it was proved (see [187, 250] and also [139, 145]) that under global Lipschitz conditions on the coefficients of the SDEs (1.0.1) for any $p \ge 1$

$$\left[E \max_{0 \le k \le N} |X_{t_0,X_0}(t_k) - \bar{X}_{t_0,X_0}(t_k)|^{2p}\right]^{1/2p} \le K(1+E|X_0|^{2p})^{1/2p}h^{1/2}. \quad (1.1.33)$$

A proof of (1.1.33) is based on constructing the continuous-time approximation by piece-wise constant interpolation of the Euler scheme, which satisfies the SDEs system with piece-wise constant coefficients. Then the difference between the solution of (1.0.1) and the continuous-time approximation is analyzed using, in particular, Doob's martingale inequality. This approach to proving estimates of the form (1.1.33) works as long as a numerical scheme can be re-written, after an interpolation, as an SDEs solution, which is not always possible (e.g., it is not applicable to balanced schemes from Sect. 1.3.2). In contrast to this approach, Theorem 1.1.7 is valid for any one-step approximation as its proof does not require expressing the approximation as an SDEs solution [274].

1.1.3   The Fundamental Theorem for Equations in the Sense of Stratonovich

Consider the following *system in the sense of Stratonovich*:

$$dX = a(t, X)dt + \sum_{r=1}^{q} \sigma_r(t, X) \circ dw_r(t), \quad (1.1.34)$$

where we use the sign $\circ$ in distinction to (1.1.1). It is well known (see, e.g., [177]) that this system is equivalent to the following system in the sense of Ito:

$$dX = \left(a(t, X) + \frac{1}{2} \sum_{r=1}^{q} \frac{\partial \sigma_r}{\partial x}(t, X)\sigma_r(t, X) \right) dt + \sum_{r=1}^{q} \sigma_r(t, X)dw_r(t) . \quad (1.1.35)$$

In this system, $\partial \sigma_r / \partial x$ is the matrix with entry $\partial \sigma_r^i / \partial x^j$ at the intersection of the i-th row and j-th column. Here, of course, we assume the σ_r, $r = 1, \ldots, q$, to be not only differentiable but also require that the vectors $(\partial \sigma_r / \partial x)\sigma_r$ satisfy a global Lipschitz condition with respect to $x \in \mathbf{R}^d$, i.e., that system (1.1.35) satisfies the condition (1.1.2).

By the above it is not difficult to see that Theorem 1.1.1 remains true for solutions of equations understood in the sense of Stratonovich. Incidentally, we note that systems with additive noise (let us recall that one distinguishes between multiplicative and additive noise depending on whether the diffusion coefficients σ_r depend on state X or not) have evidently the same form for systems in both sense of Ito and Stratonovich. Another example of systems for which these two forms coincide is given by the second-order differential equation with noise

$$\ddot{X} = a(t, X, \dot{X})dt + \sum_{r=1}^{q} \gamma_r(t, X) \circ \dot{w}_r(t) , \quad (1.1.36)$$

which can be written as $2d$-dimensional system

$$dX = Ydt \quad (1.1.37)$$

$$dY = a(t, X, Y)dt + \sum_{r=1}^{q} \gamma_r(t, X) \circ dw_r(t) .$$

Clearly, for the $2d$-dimensional column-vector σ_r corresponding to this system, the first d components are equal to zero and the other d components are equal to those of γ_r. From here the $2d \times 2d$-matrix, which plays the role of $\partial \sigma_r / \partial x$, has nonzero entries at the intersection of last d rows and first d columns only, whence the correction term for (1.1.37) is equal to zero, i.e., the Ito form of (1.1.37) is the same.

1.1.4 Discussion

The rule:

"if, in a single step, the mean-square error has order h^{p_2} (i.e., inequality (1.1.6) holds), then it has order $h^{p_2-1/2}$ on the whole interval"

is not true without the additional condition $p_1 \geq p_2 + 1/2$. A simple example in which this can be seen is given by the method

$$\bar{X}_{k+1} = \bar{X}_k + \sigma(t_k, \bar{X}_k)\Delta_k w(h)$$

for the scalar version of system (1.1.1). It is easy to see that here $p_2 = 1$ while the method diverges for $a \neq 0$.

The rule:

"if, in a single step, the mean-square deviation has order h^{p_2}, then it has order h^{p_2-1} on the whole interval"

is true, but a bit rough. Following this rule, we cannot prove the convergence of the Euler method, in which $p_2 = 1$. Moreover, a more efficient rule for the mean-square deviation cannot be found if we are guided only by the mean-square characteristic of the one-step approximation. The rule following from Theorem 1.1.1 is based on properties of both the *mean* and the *mean-square deviation* of the one-step approximation. In particular, for Euler's method $p_1 = 2$, $p_2 = 1$ (see the next subsection), and so it follows from Theorem 1.1.1 that Euler's method has order of accuracy $1/2$.

Properties of the mean are used at a single place in the proof of the theorem (and at a "delicate" place indeed), more precisely, when deriving the inequality (1.1.31). If the left-hand side of this inequality is roughly estimated, without taking into account (1.1.5), using the Bunyakovsky–Schwarz inequality, we obtain

$$
\begin{aligned}
&|EE((X(t_k) - \bar{X}_k)^{\mathsf{T}}(X_{t_k,\bar{X}_k}(t_{k+1}) - \bar{X}_{t_k,\bar{X}_k}(t_{k+1}))|\mathcal{F}_{t_k})| \\
&\leq (E|X(t_k) - \bar{X}_k|^2)^{1/2}(E|X_{t_k,\bar{X}_k}(t_{k+1}) - \bar{X}_{t_k,\bar{X}_k}(t_{k+1})|^2)^{1/2} \\
&\leq (E|X(t_k) - \bar{X}_k|^2)^{1/2}K(1 + E|\bar{X}_k|^2)^{1/2}h^{p_2}.
\end{aligned}
$$

While at the right-hand side of (1.1.31) we had the factor h^{p_1}, here we only have h^{p_2}, which is not sufficient for concluding the theorem.

Let us consider two simple examples which warn against a noncritical use of "sufficiently natural" methods.

Example 1.1.8 Take a piecewise linear interpolation of the Wiener process, $w^h(t) = w(t_k) + \Delta_k w(h)(t - t_k)/h$, $t_k \leq t \leq t_{k+1}$, and consider, instead of the equation

$$
dX = a(t, X)dt + \sigma(t, X)dw(t),
$$

on each interval $t_k \leq t \leq t_{k+1}$ the equation

$$
dX^h = a(t, X^h)dt + \sigma(t, X^h)dw^h(t)
$$

and its solution $X_k = X^h(t_k)$ at the nodes, which is $\mathcal{F}_{t_k}$-measurable. For the equation

$$
dX = aXdt + \sigma Xdw, \quad 0 \leq t \leq T, \tag{1.1.38}
$$

we obtain the one-step approximation:

$$
\bar{X}_{t,x}(t + h) = x \exp(ah + \sigma(w(t + h) - w(t))). \tag{1.1.39}
$$

At the same time,

$$X_{t,x}(t + h) = x \exp\left(\left(a - \frac{\sigma^2}{2}\right)h + \sigma(w(t + h) - w(t))\right). \tag{1.1.40}$$

Since

$$E e^{\sigma(w(t+h)-w(t))} = e^{(\sigma^2/2)h},$$

we have

$$E(X_{t,x}(t + h) - \bar{X}_{t,x}(t + h)) = (e^{ah} - e^{(a+\sigma^2/2)h})x = O(h),$$

i.e., $p_1 = 1$ and Theorem 1.1.1 cannot be used to ensure convergence of the approximations. Actually, there is no convergence, since

$$EX_{k+1} = x \exp(ah + \sigma(w(t + h) - w(t))) \cdot EX_k = e^{(a+\sigma^2/2)h} EX_k,$$
$$E\bar{X}(T) = EX_N = e^{(a+\sigma^2/2)T} EX_0,$$

while

$$EX(T) = e^{aT} EX_0.$$

Note that if, instead of (1.1.38), we consider an equation in the sense of Stratonovich,

$$dX = aXdt + \sigma X \circ dw, \tag{1.1.41}$$

then the approximation (1.1.39) coincides with the solution of (1.1.41) (see the relation between (1.1.34) and (1.1.35)), i.e., the approximation (1.1.39) for (1.1.41) has infinite order of accuracy.

Example 1.1.9 We consider the implicit method (trapezoidal method) for Eq. (1.1.38):

$$X_{k+1} = X_k + a\frac{X_k + X_{k+1}}{2}h + \sigma\frac{X_k + X_{k+1}}{2}\Delta_k w(h). \tag{1.1.42}$$

So,

$$X_{k+1} = X_k + X_k\frac{ah + \sigma\Delta_k w(h)}{1 - ah/2 - \sigma\Delta_k w(h)/2}. \tag{1.1.43}$$

Since expectation of the right-hand side does not exist, it is clear that the approximation (1.1.42) cannot converge in mean-square to the solution of (1.1.38).

Consider the following one-step approximation:

$$\bar{X}_{t,x}(t+h) = x + axh + \sigma x \Delta w(h) + \frac{\sigma^2}{2} x \Delta^2 w(h), \qquad (1.1.44)$$

$$\Delta w(h) = w(t+h) - w(t),$$

which can be thought of as corresponding to (1.1.43). We have

$$E\bar{X}_{t,x}(t+h) = \left(1 + ah + \frac{\sigma^2}{2}h\right)x,$$

which implies that $p_1 = 1$. It is not difficult to prove that also in this case there is no convergence of the approximation (1.1.44) to the solution of Eq. (1.1.38). At the same time, an explicit computation gives that for the approximation (1.1.44) to the solution of (1.1.41) we have $p_1 = 2$, $p_2 = 3/2$, i.e., the method based on (1.1.44) has first order of accuracy for Eq. (1.1.41). The computation can be performed by expanding the solution of (1.1.41) $X_{t,x}(t+h) = x \exp(ah + \sigma \Delta w(h))$ in powers of h and $\Delta w(h)$ and retaining the powers up to h^2 and $(\Delta w(h))^4$ inclusive in all the relations.

1.1.5 The Explicit Euler Method

For the system (1.1.1) we consider the one-step approximation (1.1.3) of the form

$$\bar{X}_{t,x}(t+h) = x + a(t,x)h + \sum_{r=1}^{q} \sigma_r(t,x)\Delta_t w_r(h), \qquad (1.1.45)$$

where $\Delta_t w_r(h) = w_r(t+h) - w_r(t)$.

By (1.1.4), this approximation generates the *Euler method*:

$$X_0 = X(t_0), \quad X_{k+1} = X_k + a_k h + \sum_{r=1}^{q} \sigma_{rk}\Delta_k w_r(h), \qquad (1.1.46)$$

where a_k, σ_{rk} are the values of the coefficients a and σ_r at the point (t_k, X_k), and $\Delta_k w_r(h) = w_r(t_k + h) - w_r(t_k)$.

For (1.1.45) we can find p_1 and p_2 satisfying the estimates (1.1.5) and (1.1.6). To this end we term-wise subtract (1.1.46) from the identity

$$X_{t,x}(t+h) = x + \int_t^{t+h} a(s, X_{t,x}(s))ds + \sum_{r=1}^{q} \sigma_r(s, X_{t,x}(s))dw_r(s). \qquad (1.1.47)$$

We obtain

$$X_{t,x}(t+h) - \bar{X}_{t,x}(t+h) = \int_t^{t+h} (a(s, X_{t,x}(s)) - a(t, x))ds \qquad (1.1.48)$$

$$+ \sum_{r=1}^{q} \int_t^{t+h} (\sigma_r(s, X_{t,x}(s)) - \sigma_r(t, x))dw_r(s).$$

Calculation of the expectation leads to

$$E(X_{t,x}(t+h) - \bar{X}_{t,x}(t+h)) = E \int_t^{t+h} (a(s, X_{t,x}(s)) - a(t, x))ds. \qquad (1.1.49)$$

Taking the square of both sides of (1.1.48) and then evaluating expectation, we obtain

$$E|X_{t,x}(t+h) - \bar{X}_{t,x}(t+h)|^2 \qquad (1.1.50)$$

$$\leq 2E \left| \int_t^{t+h} (a(s, X_{t,x}(s)) - a(t, x))ds \right|^2$$

$$+ 2 \sum_{r=1}^{q} \int_t^{t+h} E|\sigma_r(s, X_{t,x}(s)) - \sigma_r(t, x)|^2 ds.$$

We assume that, in addition to (1.1.2), the functions $a(t, x)$ and $\sigma_r(t, x)$ have partial derivatives with respect to t that grow at most as linear functions of x as $|x| \to \infty$. Then a and σ_r satisfy an inequality of the form

$$|a(s, X_{t,x}(s)) - a(t, x)| \leq |a(s, X_{t,x}(s)) - a(s, x)| \qquad (1.1.51)$$
$$+ |a(s, x) - a(t, x)|$$
$$\leq K|X_{t,x}(s) - x| + K(1 + |x|^2)^{1/2}(s - t).$$

By the Bunyakovsky–Schwarz inequality we have

$$\left| \int_t^{t+h} (a(s, X_{t,x}(s)) - a(t, x))ds \right|^2 \leq h \int_t^{t+h} |a(s, X_{t,x}(s)) - a(t, x)|^2 ds,$$

and due to (1.1.51) and (1.1.19) the first term of the right-hand side of (1.1.50) is bounded by $K(1 + |x|^2)h^3$. Using the inequality (1.1.51) for σ_r and (1.1.19), we see that the second term of the right-hand side of (1.1.50) is bounded by $K(1 + |x|^2)h^2$. Thus $p_2 = 1$. Note that this value of p_2 is determined by the second term at the right-hand side of (1.1.50).

To find p_1, we turn to (1.1.49) and, using (1.1.51) and (1.1.19), we get (1.1.5) with $p_1 = 3/2 \geq p_2 + 1/2$. Then, by Theorem 1.1.1, under the given assumptions regarding the coefficients a and σ_r, *Euler's method is a method of order* $p = p_2 - 1/2 = 1/2$.

Now we consider Euler's method for the system with additive noise:

$$dX = a(t, X)dt + \sum_{r=1}^{q} \sigma_r(t)dw_r(t). \tag{1.1.52}$$

It can be readily seen that in this case the second term at the right-hand side of (1.1.50) can be bounded by $K(1 + |x|^2)h^3$. As a result, we have $p_2 = 3/2$. Now let us find p_1. Assume, in addition to (1.1.2) and to the functions $a(t, x)$ and $\sigma_r(t, x)$ having partial derivatives with respect to t which grow at most as linear functions of x as $|x| \to \infty$, that the derivatives $\partial a^i/\partial x^j$, $i, j = 1, \ldots, d$, are uniformly bounded and that $\partial^2 a^i/\partial x^j \partial x^k$, $i, j, k = 1, \ldots, d$, grow at most as linear functions of x as $|x| \to \infty$. To find p_1, we again turn to (1.1.49). Using Ito's formula, we obtain

$$a(s, X_{t,x}(s)) - a(t, x) \tag{1.1.53}$$

$$= \int_t^s \left[\frac{\partial}{\partial t} a(s', X_{t,x}(s')) + a^\top(s', X_{t,x}(s')) \frac{\partial}{\partial x} a(s', X_{t,x}(s')) \right.$$

$$\left. + \frac{1}{2} \sum_{r=1}^{q} \sum_{i,j=1}^{d} \sigma_r^i(s')\sigma_r^j(s') \frac{\partial^2}{\partial x^i \partial x^j} a(s', X_{t,x}(s')) \right] ds'$$

$$+ \sum_{r=1}^{q} \sum_{i=1}^{d} \sigma_r^i(s') \frac{\partial}{\partial x^i} a(s', X_{t,x}(s'))dw_r(s'),$$

where $\partial a/\partial x$ is the $d \times d$-matrix with entries $\partial a^i/\partial x^j$ at the intersection of the i-th row and j-th column. It follows from (1.1.49), (1.1.53) and the assumptions made that $p_1 = 2 \geq p_2 + 1/2$. Hence, Theorem 1.1.1 implies that *the order of Euler's method for systems with additive noise is* $p = p_2 - 1/2 = 1$.

Remark 1.1.10 The formula (1.1.46) approximates the solution $X(t)$ of (1.1.1) at the nodes t_k. Consider a piecewise linear Euler approximation on the entire interval $t_0 \leq t \leq T$:

$$\bar{X}(t) = X_k + a_k(t - t_k) + \sum_{r=1}^{q} \sigma_{rk} \Delta_k w_r(h) \frac{t - t_k}{h}, \quad t_k \leq t \leq t_{k+1}. \tag{1.1.54}$$

Note that now $\bar{X}(t)$ is not $\mathcal{F}_t$-measurable since $w_r(t_{k+1})$, $r = 1, \ldots, q$, participate in (1.1.54). Equation (1.1.54) immediately implies

$$E|\bar{X}(t) - X_k|^2 = O(h), \ t_k \le t \le t_{k+1}.$$

Therefore, we have on the interval $t_0 \le t \le T$:

$$(E|\bar{X}(t) - X(t)|^2)^{1/2} = O(h^{1/2}),$$

i.e., a piecewise linear approximation has the same order of accuracy on the entire interval as at the nodes. It is necessary to emphasize that this fact only holds for methods whose order of accuracy does not exceed $1/2$. In fact, consider the scalar equation

$$dX = dw, \ X(0) = 0.$$

Its Euler approximation

$$X_0 = 0, \ X_{k+1} = X_k + \Delta_k w(h)$$

is exact at nodes, i.e., it has infinite order of accuracy at nodes. At the same time, a piecewise linear approximation, e.g. for $t = (t_k + t_{k+1})/2$, has overall error of order $1/2$:

$$E\left(X\left(\frac{t_k + t_{k+1}}{2}\right) - \bar{X}\left(\frac{t_k + t_{k+1}}{2}\right)\right)^2$$

$$= E\left(w\left(\frac{t_k + t_{k+1}}{2}\right) - \frac{1}{2}(w(t_{k+1}) + w(t_k))\right)^2 = \frac{1}{4}h.$$

Remark 1.1.11 Consider the system of linear differential equations with additive noise

$$dX = A(t)X dt + \sum_{r=1}^{q} \sigma_r(t) dw_r(t). \tag{1.1.55}$$

In [286] a method is constructed that uses simulation of discrete processes $\Delta_k w_r(h)$ in an optimal manner. Already for the scalar equation

$$dX = aX dt + \sigma dw(t)$$

with constant coefficients $a \neq 0$, $\sigma \neq 0$, this method is precisely of first order of accuracy. Thus, there is no numerical integration method for (1.1.55) that uses only information about $w_r(t)$, $r = 1, \dots, q$, at discrete time moments t_k and would have order of accuracy exceeding $O(h)$ (see also [61]). We note that though the optimal method has the same order as the Euler method, it gives more accurate results in a number of cases (see [286] and also Sect. 5.4.3).

1.1.6 Almost Sure Convergence

As one can see from the definition (1.0.4), mean-square methods guarantee closeness of the exact trajectory and its approximation in the mean-square sense (L_2-norm). Mean-square methods are useful for direct simulation of stochastic trajectories, when one needs to look at qualitative behavior of the model. We can also consider a stronger type of convergence and require that for almost every trajectory $w(\cdot)$ the method converges *almost surely* with order γ, i.e.

$$|X(t_k) - X_k| \leq C(\omega)h^\gamma \quad a.s., \tag{1.1.56}$$

where C is an a.s. bounded random variable independent of k and h.

One can prove under some conditions that the following L_p-estimate holds for the Euler scheme (1.1.46) (see e.g. [145, 321, 448] and also Theorem 1.4.3 in Sect. 1.4): for some $p \geq 1$:

$$\left[E|X(t_k) - X_k|^{2p}\right]^{1/(2p)} \leq Kh^{1/2}. \tag{1.1.57}$$

Denote $R_k := |X(t_k) - X_k|$. The estimate (1.1.57) and Markov's inequality imply

$$P(R_k > h^\gamma) \leq \frac{E R_k^{2p}}{h^{2p\gamma}} \leq Kh^{p(1-2\gamma)}.$$

Then for any $\gamma = 1/2 - \varepsilon$ there is a sufficiently large $p \geq 1$ such that (recall that $h = T/N$)

$$\sum_{N=1}^{\infty} P\left(R_k > \frac{T^\gamma}{N^\gamma}\right) \leq KT^{p(1-2\gamma)} \sum_{N=1}^{\infty} \frac{1}{N^{p(1-2\gamma)}} < \infty.$$

Hence, due to the Borel-Cantelli lemma, the random variable

$$\varkappa := \sup_{h>0} h^{-\gamma}|R_k|$$

is a.s. finite which implies

$$|X(t_k) - X_k| \leq C(\omega)h^{1/2-\varepsilon} \quad a.s., \tag{1.1.58}$$

where C is an a.s. finite random variable independent of h. Note that $C(\omega)$ depends on ε.

There are other approaches to constructing and proving pathwise convergence. One can make use of the Doss-Sussmann formalism [383], which in some cases allows to express any trajectory $X(t)$ of SDEs by the solution of a system of ordinary differential equations that depends on the realization of the Wiener process $w(t)$. For instance, this approach is used in [294] for uniform simulation of the Cox-Ingersoll-

Ross process. Here the uniform approximation means that the pathwise error is uniformly bounded, i.e.

$$\sup_{t_0 \le t \le t_0 + T} |\bar{X}(t) - X(t)| \le r \quad a.s.,$$

where $r > 0$ is fixed in advance. Alternatively, in [295] the idea of simulating first-passage times [304, 314] (see also [29]) is used to construct uniform approximations of the Cox-Ingersoll-Ross process.

1.2 Methods Based on Taylor-Type Expansion

We start this section (Sect. 1.2.1) with recalling the use of Taylor's expansion for constructing numerical methods for deterministic ordinary differential equations. In Sect. 1.2.2 the Wagner–Platen expansion and its properties are considered. Based on the results of Sect. 1.2.2, explicit methods for SDEs are constructed in Sect. 1.2.3.

1.2.1 Taylor Expansion of Solutions of Ordinary Differential Equations

As is well known, the Taylor expansion lies at the basis of all one-step methods, both explicit and implicit (e.g., Runge–Kutta type methods). Here we will give it in a form convenient for our subsequent construction of its stochastic counterpart.

Consider the system of ordinary differential equations

$$\frac{dX}{dt} = a(t, X). \tag{1.2.1}$$

The right-hand side of (1.2.1) is assumed to be such that all the subsequent constructions can be performed. It suffices for this that the function $a(t, x)$, $t_0 \le t \le T$, $x \in \mathbf{R}^d$, is sufficiently smooth and that $a(t, x)$ grows at most as a linear function of $|x|$ as $|x| \to \infty$.

Let $f(t, x)$ be a scalar or vector function (of course, sufficiently smooth). We have along a solution $X(t)$ of (1.2.1):

$$\frac{d}{dt} f(t, X(t)) = \frac{\partial f}{\partial t}(t, X(t)) + \frac{\partial f}{\partial x}(t, X(t))a(t, X(t)) \tag{1.2.2}$$

$$= \frac{\partial f}{\partial t}(t, X(t)) + \sum_{i=1}^{d} \frac{\partial f}{\partial x^i}(t, X(t))a^i(t, X(t)).$$

Define by L the operator:

$$L = \frac{\partial}{\partial t} + a^\mathsf{T} \frac{\partial}{\partial x} = \frac{\partial}{\partial t} + \left(a, \frac{\partial}{\partial x} \right) = \frac{\partial}{\partial t} + \sum_{i=1}^{d} a^i \frac{\partial}{\partial x^i} \, .$$

Assuming $X(t) = x$, (1.2.2) implies

$$f(s, X(s)) = f(t, x) + \int_t^s Lf(\theta, X(\theta)) d\theta \, . \tag{1.2.3}$$

Let $f(t, x) = x$. Then $Lf(t, x) = a(t, x)$, $L^2 f(t, x) = La(t, x)$, etc. Therefore, (1.2.3) for $s = t + h$ implies

$$X(t + h) = x + \int_t^{t+h} a(s, X(s)) ds \tag{1.2.4}$$

(we have given these arguments for deriving the obvious identity (1.2.4) for consistency with the subsequent computations). Further, using (1.2.3) for $a(s, X(s))$, we obtain

$$X(t + h) = x + \int_t^{t+h} \left(a(t, x) + \int_t^s La(\theta, X(\theta)) d\theta \right) ds \tag{1.2.5}$$

$$= x + a(t, x)h + \int_t^{t+h} (t + h - s) La(s, X(s)) ds \, .$$

Again we use (1.2.3), but now for $La(s, X(s))$. We find

$$X(t + h) = x + a(t, x)h + La(t, x) \frac{h^2}{2} \tag{1.2.6}$$

$$+ \int_t^{t+h} \frac{(t + h - s)^2}{2} L^2 a(s, X(s)) ds \, .$$

Continuing in this way, we obtain the well-known *Taylor expansion* in powers of h in a neighborhood of t for the solution of (1.2.1). This expansion lies at the basis of creating explicit methods of various orders of accuracy and reads:

$$X(t+h) = x + a(t,x)h + La(t,x)\frac{h^2}{2} + \cdots \tag{1.2.7}$$

$$+ L^{m-1}a(t,x)\frac{h^m}{m!} + \int\limits_t^{t+h} \frac{(t+h-s)^m}{m!} L^m a(s, X(s))ds .$$

By (1.2.7), the *one-step approximation*

$$\bar{X}_{t,x}(t+h) = x + a(t,x)h + La(t,x)\frac{h^2}{2} + \cdots + L^{m-1}a(t,x)\frac{h^m}{m!} \tag{1.2.8}$$

has error of order $m+1$ at a single step, and the method based on (1.2.8) has the m-th global order of accuracy.

1.2.2 Wagner–Platen Expansion of Solutions of Stochastic Differential Equations

Let $X_{t,x}(s) = X(s)$ be the solution of the system (1.1.1), and let $f(t,x)$ be a sufficiently smooth (scalar or vector) function. By Ito's formula, we have for $t_0 \le t \le \theta \le T$:

$$f(\theta, X(\theta)) = f(t,x) \tag{1.2.9}$$

$$+ \sum_{r=1}^q \int\limits_t^\theta \Lambda_r f(\theta_1, X(\theta_1))dw_r(\theta_1) + \int\limits_t^\theta Lf(\theta_1, X(\theta_1))d\theta_1 ,$$

where the operators Λ_r, $r = 1, \ldots, q$, and L are given by:

$$\Lambda_r = \sigma_r^\mathsf{T}\frac{\partial}{\partial x} = \left(\sigma_r, \frac{\partial}{\partial x}\right) = \sum_{i=1}^d \sigma_r^i \frac{\partial}{\partial x^i} ,$$

$$L = \frac{\partial}{\partial t} + a^\mathsf{T}\frac{\partial}{\partial x} + \frac{1}{2}\sum_{r=1}^q\sum_{i=1}^d\sum_{j=1}^d \sigma_r^i\sigma_r^j \frac{\partial^2}{\partial x^i \partial x^j} .$$

The formula (1.2.9) is an analogue of formula (1.2.3).

Apply (1.2.9) to the functions $\Lambda_r f$ and Lf and then insert the expressions obtained for $\Lambda_r f(\theta_1, X(\theta_1))$ and $Lf(\theta_1, X(\theta_1))$ into (1.2.9). We find

$$f(s, X(s)) = f + \sum_{r=1}^{q} \Lambda_r f \int_t^s dw_r(\theta) + Lf \int_t^s d\theta \tag{1.2.10}$$

$$+ \sum_{r=1}^{q} \int_t^s \left(\sum_{i=1}^{q} \int_t^\theta \Lambda_i \Lambda_r f(\theta_1, X(\theta_1)) \, dw_i(\theta_1) \right) dw_r(\theta)$$

$$+ \sum_{r=1}^{q} \int_t^s \left(\int_t^\theta L\Lambda_r f(\theta_1, X(\theta_1)) d\theta_1 \right) dw_r(\theta)$$

$$+ \sum_{r=1}^{q} \int_t^s \left(\int_t^\theta \Lambda_r Lf(\theta_1, X(\theta_1)) dw_r(\theta_1) \right) d\theta$$

$$+ \int_t^s \left(\int_t^\theta L^2 f(\theta_1, X(\theta_1)) d\theta_1 \right) d\theta \, ,$$

where f, $\Lambda_r f$, and Lf are computed at (t, x).

Continuing in this way, we obtain an expansion for $f(t + h, X(t + h))$. As proved in the previous subsection, in the deterministic situation this expansion is the Taylor expansion in powers of h with remainder of integral type. In the stochastic situation the role of powers is played by random variables of the form (they are independent of $\mathcal{F}_t$):

$$I_{i_1, \dots, i_j}(h) = \int_t^{t+h} dw_{i_j}(\theta) \int_t^\theta dw_{i_{j-1}}(\theta_1) \int_t^{\theta_1} \cdots \int_t^{\theta_{j-2}} dw_{i_1}(\theta_{j-1}) \, , \tag{1.2.11}$$

where $i_1, \dots, i_j$ take values in the set $\{0, 1, \dots, q\}$, and $dw_0(\theta_i)$ is understood to mean $d\theta_i$.

It is obvious that $E I_{i_1, \dots, i_j}(h) = 0$ if at least one $i_k \neq 0$, $k = 1, \dots, j$, while $E I_{i_1, \dots, i_j}(h) = O(h^j)$ if all $i_k = 0$, $k = 1, \dots, j$.

Let us evaluate $E(I_{i_1, \dots, i_j})^2$.

Lemma 1.2.1 *We have*

$$(E(I_{i_1, \dots, i_j})^2)^{1/2} = O(h^{\sum_{k=1}^{j}(2 - i_k')/2}) \, , \tag{1.2.12}$$

where

$$i_k' = \begin{cases} 0, & i_k = 0, \\ 1, & i_k \neq 0. \end{cases}$$

In other words, when computing the *order of smallness of the integral* (1.2.11) we should be guided by the following rule: $d\theta$ contributes one to the order of smallness and $dw_r(\theta)$, $r = 1, \dots, q$, contributes one half.

Proof Suppose $i_j \neq 0$. Then (we can put $t = 0$ in (1.2.11) without loss of generality)

$$E(I_{i_1,\dots,i_j})^2 = \int_0^h E(I_{i_1,\dots,i_{j-1}}(\theta))^2 d\theta . \tag{1.2.13}$$

If $i_j = 0$, i.e. $dw_{i_j}(\theta) = d\theta$, then

$$E(I_{i_1,\dots,i_j})^2 = E\left(\int_0^h I_{i_1,\dots,i_{j-1}}(\theta)d\theta\right)^2 \leq h \int_0^h E(I_{i_1,\dots,i_{j-1}}(\theta))^2 d\theta . \tag{1.2.14}$$

Denote by $p(i_1,\dots,i_j)$ the order of smallness of $E(I_{i_1,\dots,i_j})^2$. Then the formulas (1.2.13) and (1.2.14) give the recurrence relation

$$p(i_1,\dots,i_j) = p(i_1,\dots,i_{j-1}) + (2 - i'_j) ,$$

which proves (1.2.12). $\qquad\qquad\square$

To clarify the general rule for establishing expansions of the form (1.2.10) using the integrals (1.2.11), we give the following formula, which can be obtained by a series of direct substitutions (we take s equal to $t + h$):

$$f(t+h, X(t+h)) = f + \sum_{r=1}^q \Lambda_r f \int_t^{t+h} dw_r(\theta) + Lf \int_t^{t+h} d\theta \tag{1.2.15}$$

$$+ \sum_{r=1}^q \sum_{i=1}^q \Lambda_i \Lambda_r f \int_t^{t+h} dw_r(\theta) \int_t^\theta dw_i(\theta_1)$$

$$+ \sum_{r=1}^q \sum_{i=1}^q \sum_{s=1}^q \Lambda_s \Lambda_i \Lambda_r f$$

$$\times \int_t^{t+h} dw_r(\theta) \int_t^\theta dw_i(\theta_1) \int_t^{\theta_1} dw_s(\theta_2)$$

$$+ \sum_{r=1}^q \Lambda_r Lf \int_t^{t+h} d\theta \int_t^\theta dw_r(\theta_1)$$

$$+ \sum_{r=1}^q L\Lambda_r f \int_t^{t+h} dw_r(\theta) \int_t^\theta d\theta_1$$

$$+ L^2 f \int_t^{t+h} d\theta \int_t^\theta d\theta_1 + \rho ,$$

where

$$\rho = \sum_{r=1}^{q}\sum_{i=1}^{q}\sum_{s=1}^{q}\sum_{j=1}^{q}\int_{t}^{t+h}\left(\int_{t}^{\theta}\left(\int_{t}^{\theta_1}\left(\int_{t}^{\theta_2}\Lambda_j\Lambda_s\Lambda_i\Lambda_r f(\theta_3, X(\theta_3))\right.\right.\right. \tag{1.2.16}$$

$$\left.\left.\left.\times dw_j(\theta_3)\right)dw_s(\theta_2)\right)dw_i(\theta_1)\right)dw_r(\theta)$$

$$+\sum_{r=1}^{q}\sum_{i=1}^{q}\int_{t}^{t+h}\left(\int_{t}^{\theta}\left(\int_{t}^{\theta_1}L\Lambda_i\Lambda_r f(\theta_2, X(\theta_2))d\theta_2\right)dw_i(\theta_1)\right)dw_r(\theta)$$

$$+\sum_{r=1}^{q}\sum_{i=1}^{q}\int_{t}^{t+h}\left(\int_{t}^{\theta}\left(\int_{t}^{\theta_1}\Lambda_iL\Lambda_r f(\theta_2, X(\theta_2))dw_i(\theta_2)\right)d\theta_1\right)dw_r(\theta)$$

$$+\sum_{r=1}^{q}\sum_{i=1}^{q}\int_{t}^{t+h}\left(\int_{t}^{\theta}\left(\int_{t}^{\theta_1}\Lambda_i\Lambda_rLf(\theta_2, X(\theta_2))dw_i(\theta_2)\right)dw_r(\theta_1)\right)d\theta$$

$$+\sum_{r=1}^{q}\sum_{i=1}^{q}\sum_{s=1}^{q}\int_{t}^{t+h}\left(\int_{t}^{\theta}\left(\int_{t}^{\theta_1}\left(\int_{t}^{\theta_2}L\Lambda_s\Lambda_i\Lambda_r f(\theta_3, X(\theta_3))d\theta_3\right.\right.\right.$$

$$\left.\left.\left.\times dw_s(\theta_2)\right)dw_i(\theta_1)\right)dw_r(\theta)$$

$$+\sum_{r=1}^{q}\int_{t}^{t+h}\left(\int_{t}^{\theta}\left(\int_{t}^{\theta_1}L^2\Lambda_r f(\theta_2, X(\theta_2))d\theta_2\right)d\theta_1\right)dw_r(\theta)$$

$$+\sum_{r=1}^{q}\int_{t}^{t+h}\left(\int_{t}^{\theta}\left(\int_{t}^{\theta_1}L\Lambda_rLf(\theta_2, X(\theta_2))d\theta_2\right)dw_r(\theta_1)\right)d\theta$$

$$+\sum_{r=1}^{q}\int_{t}^{t+h}\left(\int_{t}^{\theta}\left(\int_{t}^{\theta_1}\Lambda_rL^2f(\theta_2, X(\theta_2))dw_r(\theta_2)\right)d\theta_1\right)d\theta$$

$$+\int_{t}^{t+h}\left(\int_{t}^{\theta}\left(\int_{t}^{\theta_1}L^3f(\theta_2, X(\theta_2))d\theta_2\right)d\theta_1\right)d\theta.$$

The right-hand side of (1.2.15) consists of:

(a) a term of zero order of smallness, the term f;
(b) terms of order of smallness $1/2$, they make up the sum of all possible integrals of the form (1.2.11) of order $1/2$ with corresponding coefficients, each of these terms is $\Lambda_r f \int_{t}^{t+h} dw_r(\theta)$;
(c) terms of order of smallness 1, they make up the sum of all possible integrals of the form (1.2.11) of order 1 with corresponding coefficients, here the terms are of two kinds: $Lf \int_{t}^{t+h} d\theta$ and $\Lambda_i \Lambda_r f \int_{t}^{t+h} dw_r(\theta) \int_{t}^{\theta} dw_i(\theta_1)$;

(d) all possible terms of order of smallness $3/2$;

(e) one term of order of smallness 2, the term $L^2 f \int_t^{t+h} d\theta \int_t^\theta d\theta_1$;

(f) the remainder ρ.

It is easy to see that the coefficient at the integral $I_{i_1,\ldots,i_j}$, whose order of smallness is $\sum_{k=1}^j (2 - i_k')/2$ by Lemma 1.2.1, is equal to $\Lambda_{i_j} \cdots \Lambda_{i_1} f$, where Λ_0 means L.

In (1.2.15) (discarding ρ) all terms of order of smallness up to $3/2$ have been included, as well as one term of order of smallness 2. This term is characteristic in that the integral in it does not involve Wiener processes, and so its expectation is not equal to zero (of course, if $L^2 f \neq 0$). The term $L^2 f \int_t^{t+h} d\theta \int_t^\theta d\theta_1 = L^2 f (h^2/2)$ has been included in the main part of (1.2.15) for convenience of reference; the true reason for its inclusion will be revealed later.

Lemma 1.2.2 *Suppose*

$$|\Lambda_{i_j} \cdots \Lambda_{i_1} f(t, x)| \leq K(1 + |x|^2)^{1/2}. \tag{1.2.17}$$

Then the quantity

$$I_{i_1,\ldots,i_j}(f, h) = \int\limits_t^{t+h} dw_{i_j}(\theta) \int\limits_t^\theta dw_{i_{j-1}}(\theta_1) \int\limits_t^{\theta_1} \cdots \tag{1.2.18}$$

$$\cdots \int\limits_t^{\theta_{j-2}} \Lambda_{i_j} \cdots \Lambda_{i_1} f(\theta_{j-1}, X(\theta_{j-1})) dw_{i_1}(\theta_{j-1})$$

satisfies the inequality

$$E|I_{i_1,\ldots,i_j}(f, h)|^2 \leq K(1 + E|X(t)|^2) h^{\sum_{k=1}^j (2 - i_k')}, \tag{1.2.19}$$

i.e., in particular, its order of smallness is the same as that of $I_{i_1,\ldots,i_j}(h)$. Furthermore, if at least one index i_k, $k = 1, \ldots, q$, is not equal to zero, then

$$E I_{i_1,\ldots,i_j}(f, h) = 0, \quad \sum_{k=1}^j i_k \neq 0. \tag{1.2.20}$$

Proof The proof of this lemma does, in essence, not differ from that of Lemma 1.2.1. We only have to estimate $E|\Lambda_{i_j} \cdots \Lambda_{i_1} f(\theta_{j-1}, X(\theta_{j-1}))|^2$ after the last step. Using (1.2.17) we obtain

$$E|\Lambda_{i_j} \cdots \Lambda_{i_1} f(\theta_{j-1}, X(\theta_{j-1}))|^2 \leq K(1 + E|X(\theta_{j-1})|^2),$$

whence follows (1.2.19) in view of the inequality $E|X(\theta_{j-1})|^2 \leq K(1 + E|X(t)|^2)$ for $t < \theta_{j-1}$. $\qquad\square$

Corollary 1.2.3 *Lemma 1.2.2 implies that each term in the remainder ρ (see (1.2.15) and (1.2.16)) has order of smallness at most two. Moreover, the expectation of all the terms of order of smallness 2 and 5/2 from ρ vanishes by (1.2.20). So,*

$$|E\rho| = O(h^3).$$

This is true if all the integrands in ρ satisfy, e.g., (1.2.17).

1.2.3   Construction of Explicit Methods

Now we substitute x for $f(t, x)$ in (1.2.15) and (1.2.16). Note that in this case

$$\Lambda_r f = \sigma_r, \quad Lf = a.$$

Therefore

$$X_{t,x}(t+h) = x + \sum_{r=1}^{q} \sigma_r \int_t^{t+h} dw_r(\theta) + ah \tag{1.2.21}$$

$$+ \sum_{r=1}^{q} \sum_{i=1}^{q} \Lambda_i \sigma_r \int_t^{t+h} (w_i(\theta) - w_i(t))dw_r(\theta)$$

$$+ \sum_{r=1}^{q} L\sigma_r \int_t^{t+h} (\theta - t)dw_r(\theta) + \sum_{r=1}^{q} \Lambda_r a \int_t^{t+h} (w_r(\theta) - w_r(t))d\theta$$

$$+ \sum_{r=1}^{q} \sum_{i=1}^{q} \sum_{s=1}^{q} \Lambda_s \Lambda_i \sigma_r \int_t^{t+h} \left(\int_t^{\theta} (w_s(\theta_1) - w_s(t))dw_i(\theta_1) \right) dw_r(\theta)$$

$$+ La\frac{h^2}{2} + \rho.$$

In this formula all the coefficients σ_r, a, $\Lambda_i \sigma_r$, $L\sigma_r$, $\Lambda_r a$, $\Lambda_s \Lambda_i \sigma_r$, and La are computed at the point (t, x), while the remainder ρ is equal to

$$\rho = \sum_{r=1}^{q}\sum_{i=1}^{q}\sum_{s=1}^{q}\sum_{j=1}^{q}\int_t^{t+h}\left(\int_t^{\theta}\left(\int_t^{\theta_1}\left(\int_t^{\theta_2}\Lambda_j\Lambda_s\Lambda_i\sigma_r(\theta_3, X(\theta_3))\right.\right.\right. \tag{1.2.22}$$

$$\left.\left.\left.\times dw_j(\theta_3)\right)dw_s(\theta_2)\right)dw_i(\theta_1)\right)dw_r(\theta)$$

$$+\sum_{r=1}^{q}\sum_{i=1}^{q}\int_t^{t+h}\left(\int_t^{\theta}\left(\int_t^{\theta_1}L\Lambda_i\sigma_r(\theta_2, X(\theta_2))d\theta_2\right)dw_i(\theta_1)\right)dw_r(\theta)$$

$$+\sum_{r=1}^{q}\sum_{i=1}^{q}\int_t^{t+h}\left(\int_t^{\theta}\left(\int_t^{\theta_1}\Lambda_i L\sigma_r(\theta_2, X(\theta_2))dw_i(\theta_2)\right)d\theta_1\right)dw_r(\theta)$$

$$+\sum_{r=1}^{q}\sum_{i=1}^{q}\int_t^{t+h}\left(\int_t^{\theta}\left(\int_t^{\theta_1}\Lambda_i\Lambda_r a(\theta_2, X(\theta_2))dw_i(\theta_2)\right)dw_r(\theta_1)\right)d\theta$$

$$+\sum_{r=1}^{q}\sum_{i=1}^{q}\sum_{s=1}^{q}\int_t^{t+h}\left(\int_t^{\theta}\left(\int_t^{\theta_1}\left(\int_t^{\theta_2}L\Lambda_s\Lambda_i\sigma_r(\theta_3, X(\theta_3))d\theta_3\right.\right.\right.$$

$$\left.\left.\left.\times dw_s(\theta_2)\right)dw_i(\theta_1)\right)dw_r(\theta)\right.$$

$$+\sum_{r=1}^{q}\int_t^{t+h}\left(\int_t^{\theta}\left(\int_t^{\theta_1}L^2\sigma_r(\theta_2, X(\theta_2))d\theta_2\right)d\theta_1\right)dw_r(\theta)$$

$$+\sum_{r=1}^{q}\int_t^{t+h}\left(\int_t^{\theta}\left(\int_t^{\theta_1}L\Lambda_r a(\theta_2, X(\theta_2))d\theta_2\right)dw_r(\theta_1)\right)d\theta$$

$$+\sum_{r=1}^{q}\int_t^{t+h}\left(\int_t^{\theta}\left(\int_t^{\theta_1}\Lambda_r La(\theta_2, X(\theta_2))dw_r(\theta_2)\right)d\theta_1\right)d\theta$$

$$+\int_t^{t+h}\left(\int_t^{\theta}\left(\int_t^{\theta_1}L^2 a(\theta_2, X(\theta_2))d\theta_2\right)d\theta_1\right)d\theta\,.$$

In connection with the formulas (1.2.21), (1.2.22), we consider the following *one-step approximations*:

$$\bar{X}_{t,x}^{(1)}(t+h) = x + \sum_{r=1}^{q}\sigma_r(w_r(t+h) - w_r(t))\,, \tag{1.2.23}$$

$$\bar{X}_{t,x}^{(2)}(t+h) = \bar{X}_{t,x}^{(1)}(t+h) + ah\,, \tag{1.2.24}$$

$$\bar{X}_{t,x}^{(3)}(t+h) = \bar{X}_{t,x}^{(2)}(t+h) + \sum_{r=1}^{q}\sum_{i=1}^{q}\Lambda_i\sigma_r \int_t^{t+h}(w_i(\theta) - w_i(t))dw_r(\theta), \quad (1.2.25)$$

$$\bar{X}_{t,x}^{(4)}(t+h) = \bar{X}_{t,x}^{(3)}(t+h) + \sum_{r=1}^{q}L\sigma_r \int_t^{t+h}(\theta - t)dw_r(\theta) \qquad (1.2.26)$$

$$+ \sum_{r=1}^{q}\Lambda_r a \int_t^{t+h}(w_r(\theta) - w_r(t))d\theta$$

$$+ \sum_{r=1}^{q}\sum_{i=1}^{q}\sum_{s=1}^{q}\Lambda_s \Lambda_i \sigma_r$$

$$\times \int_t^{t+h}\left(\int_t^{\theta}(w_s(\theta_1) - w_s(t))dw_i(\theta_1)\right)dw_r(\theta),$$

$$\bar{X}_{t,x}^{(5)}(t+h) = \bar{X}_{t,x}^{(4)}(t+h) + La\frac{h^2}{2}. \qquad (1.2.27)$$

To each of these approximations, we associate the error

$$\rho^{(i)} = X_{t,x}^{(i)}(t+h) - \bar{X}_{t,x}^{(i)}(t+h), \quad i = 1, \ldots, 5.$$

Using Lemmas 1.2.1 and 1.2.2, it can be readily shown that (under the condition (1.2.17) on the corresponding functions)

$$|E\rho^{(1)}| = O(h), \quad E|\rho^{(1)}|^2 = O(h^2),$$

i.e., $p_1 = 1$, $p_2 = 1$. Therefore, in order to satisfy the conditions of Theorem 1.1.1 (more precisely, the condition $p_1 \geq p_2 + 1/2$), we should take $p_2 = 1/2$. As a result, Theorem 1.1.1 does not guarantee convergence and, as it is already noted earlier, the method (1.2.23) clearly does not converge.

We have for $\rho^{(2)}$:

$$|E\rho^{(2)}| = O(h^2), \quad E|\rho^{(2)}|^2 = O(h^2),$$

i.e., $p_1 = 2$, $p_2 = 1$. Therefore, the second method (Euler's method) has order of convergence equal to $1/2$.

We have for $\rho^{(3)}$:

$$|E\rho^{(3)}| = O(h^2), \quad E|\rho^{(3)}|^2 = O(h^3),$$

i.e., $p_1 = 2$, $p_2 = 3/2$. Therefore, the third method has order of convergence equal to 1.

We have for $\rho^{(4)}$:

$$|E\rho^{(4)}| = O(h^2), \quad E|\rho^{(4)}|^2 = O(h^4),$$

i.e., $p_1 = 2$, $p_2 = 2$. But to satisfy the conditions of Theorem 1.1.1, we have to put $p_2 = 3/2$. As a result, the fourth method is also of order one.

Finally, we have for $\rho^{(5)}$:

$$|E\rho^{(5)}| = O(h^3), \quad E|\rho^{(5)}|^2 = O(h^4),$$

i.e., $p_1 = 3$, $p_2 = 2$. Therefore, the fifth method has order of convergence equal to $3/2$.

The results concerning (1.2.24), (1.2.25), and (1.2.27) are stated as the following theorem.

Theorem 1.2.4 *Suppose the conditions (1.2.17) on the corresponding functions hold. Then the mean-square order of accuracy of the methods based on the approximations (1.2.24),(1.2.25), and (1.2.27) are equal to 1/2, 1, and 3/2, respectively.*

These examples of methods (which are of independent significance) readily give support to and enable the formulation of a general result.

Suppose that an expansion of the type under consideration includes all the terms of order m. Then the remainder ρ includes terms of half-integer order $m + 1/2$ and of integer order $m + 1$. Since expectation of any term of half-integer order vanishes, we have $|E\rho| = O(h^{m+1})$, i.e., $p_1 = m + 1$. At the same time, $E|\rho|^2 = O(h^{2m+1})$, i.e., $p_2 = m + 1/2$. By Theorem 1.1.1, the order of accuracy of such a method is m.

Let an expansion contain only all terms of half-integer order $m + 1/2$. Then among terms of order $m + 1$ in the remainder there is, in general, one term having nonzero expectation; to be precise, it is $\int_t^{t+h} d\theta \int_t^{\theta} d\theta_1 \cdots \int_t^{\theta_{m-1}} L^m a(\theta_m, X(\theta_m)) d\theta_m$. So, $|E\rho| = O(h^{m+1})$, $(E|\rho|^2)^{1/2} = O(h^{m+1})$. Hence, Theorem 1.1.1 can be applied with $p_2 = m + 1/2$ only and the order of accuracy of such a method is again m. So, if we add all terms of order $m + 1/2$ to all terms up to order m inclusive, then the order of accuracy of the method does not increase. However, if we add the single term of order $m + 1$ mentioned above to all terms up to order $m + 1/2$ inclusive, then the order of accuracy of the method increases by $1/2$. In fact, expectation of all the remaining terms of order $m + 1$ is zero, and so $|E\rho| = O(h^{m+2})$, i.e., $p_1 = m + 2$, $p_2 = m + 1$, and $p = m + 1/2$. Thus, the following theorem holds (see also [364, 366, 459]).

Theorem 1.2.5 *Suppose that $\bar{X}_{t,x}(t + h)$ includes all terms of the form $\Lambda_{i_1} \cdots \Lambda_{i_j} f I_{i_1,\ldots,i_j}$, where $f \equiv x$, up to order m inclusive. Let all functions $\Lambda_{i_1} \cdots \Lambda_{i_j} f(t, x)$, where $f \equiv x$, $\sum_{k=1}^{j}(2 - i'_k)/2 \leq m + 1$, satisfy the inequality (1.2.17). Then the mean-square order of accuracy of the method based on this approximation is equal to m.*

Suppose that $\bar{X}_{t,x}(t+h)$ includes all terms of the form $\Lambda_{i_1} \cdots \Lambda_{i_j} f I_{i_1,\dots,i_j}$, where $f \equiv x$, up to order $m+1/2$ inclusive, as well as the term

$$L^m a \int\limits_t^{t+h} d\theta \int\limits_t^\theta d\theta_1 \cdots \int\limits_t^{\theta_{m-1}} d\theta_m = L^m a \frac{h^{m+1}}{(m+1)!}.$$

Suppose that all functions $\Lambda_{i_1} \cdots \Lambda_{i_j} f(t,x)$, where $f \equiv x$, $\sum_{k=1}^j (2 - i'_k)/2 \le m + 2$, satisfy the inequality (1.2.17). Then the mean-square order of accuracy of the method based on this approximation is equal to $m + 1/2$.

Remark 1.2.6 The sufficient conditions on the drift and diffusion coefficients in Theorems 1.2.4 and 1.2.5 consist in existence and boundedness of their derivatives up to a certain order. They are rather restrictive. Nevertheless these theorems are very useful since they allow us to derive various methods of numerical integration. Most likely the methods derived can be applied more widely. Of course, the issues of convergence, study of different properties of methods (for example, their relation to stiffness, ergodicity and so on) require some additional investigations and developing of corresponding recommendations concerning their practical applications (see Sect. 1.4).

Example 1.2.7 Consider the linear system of stochastic differential equations

$$dX = A(t)X dt + \sum_{r=1}^q B_r(t) X dw_r(t), \quad t_0 \le t \le T. \tag{1.2.28}$$

Here, $A(t)$ and $B_r(t)$ are $d \times d$-matrices with entries that are smooth on $[t_0, T]$, and $a(t,x) = A(t)x$, $\sigma_r(t,x) = B_r(t)x$. Therefore, the conditions of Theorem 1.2.5 hold if $A(t)$ and $B_r(t)$ are sufficiently smooth.

Let us consider the method of first order of accuracy given by the approximation (1.2.25). Since

$$\Lambda_i \sigma_r(t,x) = \left(B_i(t)x, \frac{\partial}{\partial x} \right) B_r(t)x = B_r(t) B_i(t)x,$$

this method has the form

$$X_{k+1} = X_k + \sum_{r=1}^q B_r(t_k) X_k \Delta_k w_r(h) + A(t_k) X_k h \tag{1.2.29}$$

$$+ \sum_{r=1}^q \sum_{i=1}^q B_r(t_k) B_i(t_k) X_k \int\limits_{t_k}^{t_k+h} (w_i(\theta) - w_i(t_k)) dw_r(\theta).$$

It is easy to verify the formula

$$\int\limits_{t_k}^{t_k+h} (w_i(\theta) - w_i(t_k)) dw_r(\theta) = \Delta_k w_i(h) \Delta_k w_r(h)$$

$$-\int\limits_{t_k}^{t_k+h} (w_r(\theta) - w_r(t_k)) dw_i(\theta).$$

If all $B_r(t)$ commute, then we have for $i \neq r$:

$$B_r B_i X \int\limits_{t_k}^{t_k+h} (w_i(\theta) - w_i(t_k)) dw_r(\theta)$$

$$= B_i B_r X \Delta_k w_i(h) \Delta_k w_r(h) - B_i B_r X \int\limits_{t_k}^{t_k+h} (w_r(\theta) - w_r(t_k)) dw_i(\theta).$$

Then the formula (1.2.29) takes the form

$$X_{k+1} = X_k + \sum_{r=1}^{q} B_r(t_k) X_k \Delta_k w_r(h) + A(t_k) X_k h \qquad (1.2.30)$$

$$+ \frac{1}{2} \sum_{r=1}^{q} B_i^2(t_k) X_k (\Delta_k^2 w_i(h) - h)$$

$$+ \sum_{r=2}^{q} \sum_{i=1}^{r-1} B_i(t_k) B_r(t_k) X_k \Delta_k w_i(h) \Delta_k w_r(h).$$

Thus, in the *commutative situation* we can construct a method of first order of accuracy by modeling the increments of the Wiener processes only.

We also note that in the nonlinear case the approximation (1.2.25) can be simplified in a similar manner if

$$\Lambda_i \sigma_r(t, x) = \Lambda_r \sigma_i(t, x). \qquad (1.2.31)$$

Example 1.2.8 Consider the following system which is a generalization of (1.1.37):

$$dX = a(t, X, Y) dt \qquad (1.2.32)$$

$$dY = b(t, X, Y) dt + \sum_{r=1}^{q} \gamma_r(t, X) dw_r(t).$$

For this system, the first d components of σ_r are zeros, the last d components of σ_r are γ_r which do not depend on y, and, consequently, $\Lambda_i \sigma_r = 0$. The approximation (1.2.25) gives

$$\bar{X}_{t,x,y}(t+h) = x + a(t, x, y)h \tag{1.2.33}$$

$$\bar{Y}_{t,x,y}(t+h) = y + b(t, x, y)h + \sum_{r=1}^{q} \gamma_r(t, x) \ (w_r(t+h) - w_r(t)) \ ,$$

i.e., the approximation (1.2.25) coincides with the Euler method which is of order one in the case of the system (1.2.32).

1.3   Implicit Mean-Square Methods

We first (Sect. 1.3.1) construct methods with implicitness present in the drift term only (drift-implicit methods). In Sect. 1.3.2 we introduce balanced methods [285] (see an explanation why such methods are called balanced in the discussion of the method (1.3.17)). Fully implicit (i.e., implicit both in the drift and diffusion) general methods are considered in Sects. 1.3.3–1.3.5.

1.3.1   Construction of Drift-Implicit Methods

To clarify the matter, we start with construction of implicit methods for ordinary differential equations. Rearranging (1.2.3), we can write the formula

$$f(s, X(s)) = f(t+h, X(t+h)) - \int_{s}^{t+h} Lf(\theta, X(\theta))d\theta . \tag{1.3.1}$$

Using (1.3.1), we replace $a(s, X(s))$ in (1.2.4) and obtain

$$X(t+h) = x + a(t+h, X(t+h))h - \int_{t}^{t+h} (s-t)La(s, X(s))ds . \tag{1.3.2}$$

Continuing in this way, we find

$$X(t+h) = x + a(t+h, X(t+h))h - La(t+h, X(t+h))\frac{h^2}{2} + \cdots \tag{1.3.3}$$

$$+ (-1)^{m-1} L^{m-1} a(t+h, X(t+h)) \frac{h^m}{m!}$$

$$+ \int_{t}^{t+h} (-1)^m \frac{(s-t)^m}{m!} L^m a(s, X(s))ds .$$

If we discard the integral in this formula, then we obtain the implicit one-step approximation on which we can base a method of m-th order of accuracy.

Using a simple trick, we can obtain a whole class of implicit methods. We illustrate this trick by deriving a class of implicit methods of second order of accuracy.

Introducing a parameter α, we write (1.2.4) in the following form:

$$X(t + h) = x + \alpha \int_t^{t+h} a(s, X(s))ds + (1 - \alpha) \int_t^{t+h} a(s, X(s))ds . \qquad (1.3.4)$$

Now we replace $a(s, X(s))$ in the integral at α by (1.2.3) and in the integral at $(1 - \alpha)$ by (1.3.1). We obtain

$$X(t + h) = x + \alpha a(t, x)h + (1 - \alpha)a(t + h, X(t + h))h \qquad (1.3.5)$$

$$+ \int_t^{t+h} (t + \alpha h - s)La(s, X(s))ds .$$

Further, we introduce a parameter β and rewrite the integral in (1.3.5) as the sum of two integrals:

$$\int_t^{t+h} (t + \alpha h - s)La(s, X(s))ds = \beta \int_t^{t+h} (t + \alpha h - s)La(s, X(s))ds$$

$$+ (1 - \beta) \int_t^{t+h} (t + \alpha h - s)La(s, X(s))ds .$$

Substituting $La(s, X(s)) = La(t, x) + O(h)$ in the first integral and $La(s, X(s)) = La(t + h, X(t + h)) + O(h)$ in the second integral, we obtain the following *implicit one-step approximation*

$$\bar{X}_{t,x}(t + h) = x + \alpha a(t, x)h + (1 - \alpha)a(t + h, \bar{X}_{t,x}(t + h))h \qquad (1.3.6)$$

$$+ \beta(2\alpha - 1) La(t, x)\frac{h^2}{2} + (1 - \beta)(2\alpha - 1)La(t + h, \bar{X}_{t,x}(t + h))\frac{h^2}{2} .$$

Using this approximation, we construct a two-parameter family of implicit methods of second order of accuracy:

$$X_{k+1} = X_k + \alpha a_k h + (1 - \alpha)a_{k+1}h \qquad (1.3.7)$$

$$+ \beta(2\alpha - 1) (La)_k\frac{h^2}{2} + (1 - \beta)(2\alpha - 1)(La)_{k+1}\frac{h^2}{2} ,$$

where the functions with index k are computed at (t_k, X_k), while those with index $k + 1$ are computed at (t_{k+1}, X_{k+1}).

Now we turn to construction of drift-implicit methods for SDEs. One of the simplest and popular *drift-implicit methods* has the form

$$X_{k+1} = X_k + \alpha a_k h + (1 - \alpha) a_{k+1} h + \sum_{r=1}^{q} \sigma_{rk} \Delta_k w_r(h). \qquad (1.3.8)$$

This method (in fact, the one-parameter family of methods) is of mean-square order $1/2$ for general systems and of order 1 for systems with additive noise. For $\alpha = 1$ it coincides with the explicit Euler method.

To construct more accurate methods, let us consider, next to (1.2.9), the formula $(t \le \theta_1 \le \theta)$

$$Lf(\theta_1, X(\theta_1)) = Lf(\theta, X(\theta)) - \sum_{r=1}^{q} \int_{\theta_1}^{\theta} \Lambda_r Lf(\theta_2, X(\theta_2)) dw_r(\theta_2) \qquad (1.3.9)$$

$$- \int_{\theta_1}^{\theta} L^2 f(\theta_2, X(\theta_2)) d\theta_2 .$$

As in the deterministic case, we substitute (1.3.9) in (1.2.9). Then putting $\theta = t + h$, we obtain

$$f(t + h, X(t + h)) = f(t, x) + \sum_{r=1}^{q} \int_{t}^{t+h} \Lambda_r f(\theta_1, X(\theta_1)) dw_r(\theta_1) \qquad (1.3.10)$$

$$+ Lf(t + h, X(t + h))h - \sum_{r=1}^{q} \int_{t}^{t+h} \left(\int_{\theta_1}^{t+h} \Lambda_r Lf(\theta_2, X(\theta_2)) dw_r(\theta_2) \right) d\theta_1$$

$$- \int_{t}^{t+h} \left(\int_{\theta_1}^{t+h} L^2 f(\theta_2, X(\theta_2)) d\theta_2 \right) d\theta_1 .$$

It would be rather unwise to use a formula of the form (1.3.9) in order to represent $\Lambda_r f(\theta_1, X(\theta_1))$ with its subsequent substitution in (1.3.10). Indeed, although the function $\Lambda_r f(\theta_1, X(\theta_1))$ does not depend itself on the future, all the terms in its representation of the form (1.3.9) do depend on the future and, e.g., the integral $\int_t^{t+h} (\int_{\theta_1}^{t+h} L \Lambda_r f(\theta_2, X(\theta_2)) d\theta_2) dw_r(\theta_1)$ does not make sense without some additional clarification. This complication can be overcome using the following trick. Consider, e.g., the formula (1.2.15) and represent the coefficient $\Lambda_r f$ in it as follows:

$$\Lambda_r f(t, x) = \Lambda_r f(t + h, X(t + h)) \tag{1.3.11}$$

$$- \sum_{i=1}^{q} \int_t^{t+h} \Lambda_i \Lambda_r f(\theta, X(\theta)) dw_i(\theta) - \int_t^{t+h} L \Lambda_r f(\theta, X(\theta)) d\theta.$$

After substitution of (1.3.11) in (1.2.15), the right-hand side contains a dependency on $X(t + h)$ which is a factor not at h, as in (1.3.10), but at $\Delta w_r(h)$. The appearance of this implicitness may lead to a method which is not acceptable a priori. We clarify this by considering, e.g. the following equation:

$$dX = aX dt + \sigma X dw(t).$$

We have

$$X_{t,x}(t + h) = x + \sigma x \Delta w(h) + axh + \rho,$$

where $E\rho = O(h^2)$, $E\rho^2 = O(h^2)$.

Further, as in (1.3.11), σx can be written as

$$\sigma x = \sigma X(t + h) - \int_t^{t+h} \sigma X(\theta) dw(\theta) - \int_t^{t+h} aX(\theta) d\theta.$$

As a result, we find

$$X(t + h) = x + \sigma X(t + h) \Delta w(h) + axh + \rho_1. \tag{1.3.12}$$

Omitting ρ_1 (at this moment we are not concerned with justifying this), we obtain the method

$$X_{k+1} = X_k + \sigma X_{k+1} \Delta w(h) + aX_k h, \tag{1.3.13}$$

which gives X_{k+1} with infinite second moment like the method (1.1.43) in Example 1.1.9. Generally, if the right-hand side of an equation contains terms with factors $\Delta w(h)$ (i.e., quantities taking arbitrary large values) and coefficients depending on X_{k+1}, then the solvability with respect to X_{k+1} of this relation is questionable. At the same time, if the expressions containing X_{k+1} involve a factor h with positive power, then for sufficiently small h the solvability is guaranteed under natural assumptions. Therefore, a certain caution is necessary when an implicitness is introduced by expressions occurring in stochastic integrals. However, we may hope that if an implicitness is introduced owing to expressions occurring in nonstochastic integrals then it is possible to get a suitable stability of the methods which is actually the reason for constructing implicit methods. Let us return to (1.2.15) again and write the coefficient Lf as

$$Lf = Lf(t+h, X(t+h)) \tag{1.3.14}$$

$$-\sum_{r=1}^{q} \int_{t}^{t+h} \Lambda_r Lf(\theta, X(\theta))dw_r(\theta) - \int_{t}^{t+h} L^2 f(\theta, X(\theta))d\theta$$

$$= Lf(t+h, X(t+h)) - \sum_{r=1}^{q} \Lambda_r Lf \int_{t}^{t+h} dw_r(\theta) - L^2 f \int_{t}^{t+h} d\theta + \rho_1 ,$$

where, as can be readily shown,

$$|E\rho_1 h| \le K(1+|x|^2)^{1/2}h^3, \quad E\rho_1^2 h^2 \le K(1+|x|^2)^{1/2}h^4 .$$

The other coefficients of (1.2.15) can be treated in the same manner (e.g., the above reasoning is immediately applicable to the coefficient $L^2 f$). Moreover, similar transformations can be performed repeatedly over individual terms containing, say, Lf, since a formula of the type (1.3.14) can be written for any smooth function. As a result, we can obtain a large amount of various representations for $f(t+h, X(t+h))$ using the integrals $I_{i_1,...,i_j}$ or using products of such integrals with coefficients depending on the points (t, x) and $(t+h, X(t+h))$. As in the deterministic situation, the amount of such representations can be increased by considering splittings of, e.g., the term $Lf \int_{t}^{t+h} d\theta$ into the sum of two integrals with coefficients α and $(1-\alpha)$ such that the term $\alpha Lf \int_{t}^{t+h} d\theta$ remains unchanged while the term Lf in $(1-\alpha)Lf \int_{t}^{t+h} d\theta$ is replaced by (1.3.14). In Sect. 1.6 we give a number of concrete implicit methods obtained on the basis of such representations for systems with additive noise.

In this subsection we have introduced implicitness in deterministic terms only. It is natural to call such methods as *drift-implicit* (or deterministically implicit) methods. Drift-implicit methods are well adapted, for instance, for stiff systems with additive noise (see Sect. 1.6). But when the stochastic part plays an essential role, the application of fully implicit (stochastically implicit) methods, which also involve implicit stochastic terms, is unavoidable. A good illustration of a situation when fully implicit methods should be applied is given in the next subsection.

1.3.2   The Balanced Method

First we consider the following one-dimensional Ito equation with multiplicative noise:

$$dX = \sigma X dw, \quad X(0) = x. \tag{1.3.15}$$

The solution of (1.3.15) decreases rapidly to zero for $|\sigma| \gg 1$ because its Lyapunov exponent $\lambda = -\sigma^2/2$ is negative. The one-dimensional equation (1.3.15) cannot be simply called stiff, but it can be an equation for one component in a stiff multidimensional problem. For large values of $|\sigma|$ in (1.3.15), one observes that explicit mean-square methods are unreliable and have large errors for not too small time step sizes. They can even lead to computer overflow. On the other hand, using very small time step sizes may require too much computational time. In a stiff situation this is the crucial point at which one has to look for other more suitable methods. For example, these difficulties occur in estimation of the Lyapunov exponent, which require 'good' long-time behavior of a numerical solution.

Obviously, one cannot apply drift-implicit schemes to improve the stability of the numerical solution of the stochastic equation (1.3.15), which does not contain any drift component. Thus, we have to construct fully implicit methods which involve implicitness in the deterministic as well as in the stochastic terms.

The Euler method for (1.3.15) has the form

$$X_{k+1} = X_k + \sigma X_k \Delta_k w(h), \quad X_0 = x.$$

There is no simple stochastic counterpart of the deterministic implicit Euler method, i.e., the method
$$X_{k+1} = X_k + \sigma X_{k+1} \Delta_k w(h)$$

fails because we have $E|(1 - \sigma \Delta_k w)^{-1}| = \infty$. Nevertheless, a way to introduce implicitness in the numerical treatment for this special equation could be to look at a higher-order explicit mean-square method and try to introduce implicitness there. For this purpose, we start from the scheme

$$X_{k+1} = X_k + \sigma X_k \Delta_k w(h) + \frac{1}{2}\sigma^2 X_k (\Delta_k^2 w(h) - h), \quad X_0 = x,$$

which represents a numerical method of mean-square order 1. Again, introduction of implicitness in $\sigma X_k \Delta_k w(h)$ as above fails, but one can analyze the term $\sigma^2 X_k (\Delta_k^2 w(h) - h)/2$ and introduce a partial implicitness. This leads to the scheme

$$X_{k+1} = X_k + (\sigma \Delta_k w(h) + \frac{1}{2}\sigma^2 \Delta_k^2 w(h))X_k - \frac{1}{2}\sigma^2 X_{k+1} h. \tag{1.3.16}$$

The numerical approximation described by the scheme (1.3.16) converges to the exact solution with order 1. This statement can be verified by Theorem 1.1.1. We note that no random term in (1.3.16) is implicit.

We derive a method with random implicit terms using heuristic reasoning. Let us denote the result due to the Euler method by $X = x + \sigma x \Delta w$. Let $x > 0$ for definiteness. If $\Delta w < 0$, then X is very often much less than the value $X(h)$ of the exact solution. Therefore, X needs a correction which has to be positive. It is natural to take this correction proportionally to σ, $|\Delta w|$, and the difference $x - X(h)$, i.e., to take the balanced term of the form $\sigma(x - X(h))|\Delta w|$. We come to the same conclusion if $\Delta w > 0$. Taking X_k instead of x and X_{k+1} instead of $X(h)$, we propose the following scheme

$$X_{k+1} = X_k + \sigma X_k \Delta_k w(h) + \sigma(X_k - X_{k+1})|\Delta_k w(h)|. \qquad (1.3.17)$$

The method (1.3.17) belongs to the class of balanced methods. We will prove that the balanced method converges with the same order $1/2$ as the Euler method does.

Several numerical experiments for the linear equation (1.3.15), which has the explicit solution

$$X(t) = x \exp\left\{\sigma w(t) - \frac{\sigma^2}{2}t\right\}, \ t \geq 0,$$

are performed in [285]. They demonstrate superiority of the method (1.3.17) over very long time in comparison with the Euler method and the method (1.3.16). The method (1.3.16) works slightly better than the Euler method but in spite of the fact that its order of convergence is higher than the order of (1.3.17), it works worse on long time intervals in comparison with (1.3.17). This can be clarified on a heuristic level as follows. The stochastic implicit term in (1.3.16) is of order 1 and in (1.3.17) is of order $1/2$, i.e., the level of implicitness in (1.3.16) is insufficient for an adequate behavior on long time intervals. Besides, we note that the scheme (1.3.17) preserves the positiveness property of solutions of Eq. (1.3.15) in contrast to (1.3.16) and to the Euler method.

Now we consider the d-dimensional system of SDEs (see (1.1.1))

$$dX = a(t, X)dt + \sum_{r=1}^{q} \sigma_r(t, X)dw_r(t), \qquad (1.3.18)$$

under the conditions of Theorem 1.1.1. Let us introduce the family of balanced methods. A balanced method applied to (1.3.18) can be written in the general form:

$$X_{k+1} = X_k + a(t_k, X_k)h + \sum_{r=1}^{q} \sigma_r(t_k, X_k)\Delta_k w_r(h) + C_k(X_k - X_{k+1}), \quad (1.3.19)$$

where

$$C_k = c_0(t_k, X_k)h + \sum_{r=1}^{q} c_r(t_k, X_k)|\Delta_k w_r(h)|. \qquad (1.3.20)$$

Here $c_0, \dots, c_r$ represent $d \times d$-matrix-valued functions. We assume that for any sequence of real numbers (α_i) with $\alpha_0 \in [0, \bar{\alpha}]$, $\alpha_1 \geq 0, \dots, \alpha_r \geq 0$, where $\bar{\alpha} \geq h$ for all step sizes h considered and $(t, x) \in [0, \infty) \times \mathbf{R}^d$, the matrix

$$M(t, x) := I + \alpha_0 c_0(t, x) + \sum_{r=1}^{q} \alpha_r c_r(t, x)$$

has the inverse which satisfies the condition

$$|(M(t, x))^{-1}| \le K < \infty.\tag{1.3.21}$$

Here I is the unit matrix. Obviously, (1.3.21) can easily be fulfilled by keeping $c_0, \dots, c_r$ all positive semidefinite. Thus, under these conditions one directly obtains the one-step increment $X_{k+1} - X_k$ of the balanced method via the solution of a system of linear algebraic equations. Furthermore, we suppose that the components of the matrices $c_0, \dots, c_r$ are uniformly bounded. The latter condition will be necessary to prove the convergence of the balanced method.

We remark that in the purely deterministic case the method (1.3.19)–(1.3.20) covers, for instance, the implicit Euler method with one or more Newton iteration steps.

Now we are able to state the corresponding convergence theorem for the general balanced method.

Theorem 1.3.1 *Under the above assumptions the balanced method (1.3.19)–(1.3.20) converges with mean-square order* $1/2$.

Proof First we show that the estimate (1.1.5) holds for the balanced method with $p_1 = 3/2$. For this purpose, let us introduce the local Euler approximation step

$$X^E = x + a(t, x)h + \sum_{r=1}^{q} \sigma_r(t, x)\Delta w_r(h)$$

and the local approximation corresponding to the method (1.3.19)–(1.3.20)

$$\bar{X} = x + a(t, x)h + \sum_{r=1}^{q} \sigma_r(t, x)\Delta w_r(h) + C(t, x)(x - \bar{X}),$$

where

$$C(t, x) = c_0(t, X)h + \sum_{r=1}^{q} c_r(t, X)|\Delta w_r(h)|.$$

Since $p_1 = 2$ for the Euler method, we have

$$
\begin{aligned}
|E(X_{t,x}(t+h) - \bar{X}_{t,x}(t+h))| &= |E(X_{t,x}(t+h) - X^E) + E(X^E - \bar{X})| \\
&\le K(1 + |x|^2)^{1/2}h^2 + |E(X^E - \bar{X})|.
\end{aligned}
$$

Further,

$$|E(X^E - \bar{X})| = |E(I - (I + C)^{-1})(a(t, x)h + \sum_{r=1}^{q} \sigma_r(t, x)\Delta w_r(h))|$$

$$= |E(I + C)^{-1}C(a(t, x)h + \sum_{r=1}^{q} \sigma_r(t, x)\Delta w_r(h))|.$$

Exploiting above the symmetry property of $\Delta w_r(h)$, $r = 1, \ldots, q$, in those expressions involving these zero-mean Gaussian variables, we get

$$|E(X^E - \bar{X})| = |E(I + C)^{-1}Ca(t, x)h|.$$

Now using (1.3.21) and boundedness of the components of the matrices $c_0, \ldots, c_r$, we obtain

$$|E(X^E - \bar{X})| \leq KE|Ca(t, x)h| \leq K(1 + |x|^2)^{1/2}h^{3/2}.$$

Hence the assumption (1.1.5) of Theorem 1.1.1 is satisfied with $p_1 = 3/2$ for the balanced method.

Similarly, we check the assumption (1.1.6) by standard arguments:

$$\left[E|X_{t,x}(t + h) - \bar{X}_{t,x}(t + h)|^2\right]^{1/2} \leq (E|X_{t,x}(t + h) - X^E|^2)^{1/2}$$

$$+ (E|X^E - \bar{X})|^2)^{1/2} \leq K(1 + |x|^2)^{1/2}h.$$

Thus we can choose the exponents p_2 and p_1 in Theorem 1.1.1 equal to 1 and 3/2, respectively, and apply Theorem 1.1.1 to finally prove the mean-square order $p_2 - 1/2 = 1/2$ of the balanced method, as is claimed in Theorem 1.3.1.   □

Remark 1.3.2 Theorem 1.3.1 remains evidently true for a more general method. Namely, the matrix C_k in (1.3.20) can be taken in the form

$$C_k = c_0(t_k, X_k)h + \sum_{r=1}^{q} c_r(t_k, X_k)|\Delta_k w_r(h)| + c_{q+1}(t_k, X_k)h^{1/2}. \qquad (1.3.22)$$

Some numerical experiments for two-dimensional systems are presented in [285]. They show that numerical methods, which also involve implicit random terms, can be successfully implemented. For the balanced methods, the type and degree of implicitness can be chosen by appropriate weights. An appropriate choice of these weights depends on the underlying dynamics and requires further investigations (see also Sect. 1.4). This problem is closely connected with the problem of determining a suitable test equation for such methods.

1.3.3 Fully Implicit Mean-Square Methods: The Main Idea

Construction of implicit methods for stochastic systems with additive noise does not cause any difficulties in principle. However, as we saw in the previous subsections, all is much more intricate in the case of stochastic systems with multiplicative noise. The balanced methods from Sect. 1.3.2 are of a very special form. In particular, this form does not allow us to construct symplectic methods for stochastic Hamiltonian systems with multiplicative noise (see Chap. 5). In this section we construct a sufficiently large class of fully implicit methods of mean-square order $1/2$ for general stochastic systems.

Let us start with an example. Consider the Ito scalar equation

$$dX = \sigma X dw(t). \tag{1.3.23}$$

The one-step approximation of the Euler method $\hat{X}$ for (1.3.23) is

$$\hat{X} = x + \sigma x \Delta w(h). \tag{1.3.24}$$

We can represent this approximation in the form

$$\hat{X} = x + \sigma \hat{X} \Delta w + \sigma(x - \hat{X})\Delta w = x - \sigma^2 x (\Delta w)^2 + \sigma \hat{X} \Delta w.$$

As h is small, $(\Delta w)^2 \sim h$ and we obtain the following "natural" implicit method

$$\tilde{X} = x - \sigma^2 x h + \sigma \tilde{X} \Delta w(h). \tag{1.3.25}$$

However, this method cannot be realized since $1 - \sigma \Delta w(h)$ can vanish for any small h. Further, for the formal value of $\tilde{X}$ from (1.3.25):

$$\tilde{X} = \frac{x(1 - \sigma^2 h)}{1 - \sigma \Delta w(h)},$$

we have $E|\tilde{X}| = \infty$ (see also Example 1.1.9). Clearly, the method (1.3.25) is not suitable. The reason for this is the unboundedness of the random variable $\Delta w(h)$ for any arbitrarily small h.

Our basic idea consists in replacement of $\Delta w(h) = \xi\sqrt{h}$, where ξ is an $\mathcal{N}(0, 1)$-distributed random variable, by another random variable $\zeta\sqrt{h} = \zeta_h\sqrt{h}$ such that $\zeta\sqrt{h}$ is bounded and the Euler-type method

$$\check{X} = x + \sigma x \zeta\sqrt{h} \tag{1.3.26}$$

is of the mean-square order $1/2$ as well. To achieve this, it is sufficient to require:

$$E(\check{X} - \hat{X}) = O(h^{3/2}), \ \ E(\check{X} - \hat{X})^2 = O(h^2). \tag{1.3.27}$$

We take a symmetric ζ. Then $E(\check{X} - \hat{X}) = 0$. To satisfy the second Eq. in (1.3.27), the condition $E(\zeta_h - \xi)^2 = O(h)$ is sufficient.

We shall require a stronger inequality

$$E(\zeta_h - \xi)^2 \le h^k, \ \ k \ge 1. \tag{1.3.28}$$

Let for $A_h > 0$

$$\zeta_h = \begin{cases} \xi, \ |\xi| \le A_h, \\ A_h, \ \xi > A_h, \\ -A_h, \ \xi < -A_h. \end{cases} \tag{1.3.29}$$

Since

$$E(\zeta_h - \xi)^2 = \frac{2}{\sqrt{2\pi}} \int\limits_{A_h}^{\infty} (x - A_h)^2 e^{-x^2/2} dx$$

$$= \frac{2}{\sqrt{2\pi}} e^{-A_h^2/2} \int\limits_{0}^{\infty} y^2 e^{-y^2/2} e^{-A_h y} dy < e^{-A_h^2/2},$$

the inequality (1.3.28) is fulfilled if $e^{-A_h^2/2} \le h^k$, i.e. $A_h^2 \ge 2k|\ln h|$. Thus, if

$$A_h = \sqrt{2k|\ln h|}, \ \ k \ge 1,$$

then the method based on the one-step approximation (1.3.26) has the mean-square order $1/2$.

Lemma 1.3.3 *Let $A_h = \sqrt{2k|\ln h|}$, $k \ge 1$, and ζ_h be defined by (1.3.29). Then the following inequality holds:*

$$0 \le E(\xi^2 - \zeta_h^2) = 1 - E\zeta_h^2 \le (1 + 2\sqrt{2k|\ln h|})h^k. \tag{1.3.30}$$

Proof We have

$$1 - E\zeta_h^2 = \frac{2}{\sqrt{2\pi}} \int_{A_h}^{\infty} (x^2 - A_h^2) e^{-x^2/2} dx$$

$$= \frac{2}{\sqrt{2\pi}} \int_{A_h}^{\infty} \left[(x - A_h)^2 + 2A_h(x - A_h) \right] e^{-x^2/2} dx$$

$$\leq e^{-A_h^2/2} + \frac{4A_h}{\sqrt{2\pi}} \int_{A_h}^{\infty} x e^{-x^2/2} dx$$

$$= e^{-A_h^2/2} \left(1 + \frac{4A_h}{\sqrt{2\pi}} \right) \leq (1 + 2A_h) e^{-A_h^2/2},$$

whence (1.3.30) follows.   □

Now consider the following implicit method (for definiteness we put $k = 1$ and $A_h = \sqrt{2|\ln h|}$) :

$$\bar{X} = x - \sigma^2 x h + \sigma \bar{X} \zeta_h \sqrt{h}, \tag{1.3.31}$$

$$\bar{X} = \frac{x(1 - \sigma^2 h)}{1 - \sigma \zeta_h \sqrt{h}}.$$

Since $|\zeta_h| \leq \sqrt{2|\ln h|}$, this method is realizable for all h satisfying the inequality

$$2h|\ln h| < \frac{1}{\sigma^2}. \tag{1.3.32}$$

Proposition 1.3.4 *The method (1.3.31) is of the mean-square order $1/2$.*

Proof Let us compare the method (1.3.31) with the Euler method (1.3.24). We get

$$E\bar{X} = x(1 - \sigma^2 h) E \sum_{m=0}^{\infty} \sigma^m \zeta_h^m h^{m/2} = x(1 - \sigma^2 h) E \sum_{m=0}^{\infty} \sigma^{2m} \zeta_h^{2m} h^m.$$

It is obvious from here that the principal term in the expansion of $E(\bar{X} - \hat{X})$ is equal to $x\sigma^2 h(E\zeta_h^2 - 1)$. Due to Lemma 1.3.3, we obtain for all sufficiently small h :

$$|E(\bar{X} - \hat{X})| \leq C|x|\sigma^2(1 + 2\sqrt{2|\ln h|})h^2, \tag{1.3.33}$$

where C is a positive constant.

Further

$$E(\bar{X} - \hat{X})^2 = E(-\sigma^2 x h + \sigma\bar{X}\zeta_h\sqrt{h} - \sigma x \xi\sqrt{h})^2 \tag{1.3.34}$$
$$\leq 2\sigma^4 x^2 h^2 + 2E(\sigma\bar{X}\zeta_h\sqrt{h} - \sigma x \xi\sqrt{h})^2$$
$$= 2\sigma^4 x^2 h^2 + 2E(\sigma \cdot (x - \sigma^2 x h + \sigma\bar{X}\zeta_h\sqrt{h})\zeta_h\sqrt{h} - \sigma x \xi\sqrt{h})^2$$
$$\leq 2\sigma^4 x^2 h^2 + 2\sigma^2 x^2 h E(\zeta_h - \xi)^2 + C_1 x^2 h^2 \leq C_2 x^2 h^2$$

for all sufficiently small h and some positive constants C_1 and C_2. The inequalities (1.3.33) and (1.3.34) imply mean-square convergence of the implicit method (1.3.31) with order $1/2$. □

Introduction of implicitness in the stochastic term leads to appearance of the compensating term $-\sigma^2 x h$ in (1.3.31). This can be explained in the following way. Since $\bar{X}$ must be close to $x + \sigma x \zeta_h\sqrt{h}$, the expression $x + \sigma\bar{X}\zeta_h\sqrt{h}$ is close to $x + \sigma x \zeta_h\sqrt{h} + \sigma^2 x \zeta_h^2 h$. Consequently, making use of the compensating term results in $x + \sigma\bar{X}\zeta_h\sqrt{h} - \sigma^2 x h = x + \sigma x \zeta_h\sqrt{h} + \sigma^2 x(\zeta_h^2 - 1)h \approx x + \sigma x \zeta_h\sqrt{h}$, i.e., we get the correct result.

Now let us consider the expression $\sigma((1 - \beta)x + \beta\bar{X})\zeta_h\sqrt{h}$ which introduces implicitness in the stochastic term with the parameter $0 \leq \beta \leq 1$. Clearly, the compensating term in this case is equal to $-\sigma^2 \beta x h$. Thus, we derive the method:

$$\bar{X} = x - \sigma^2 \beta x h + \sigma((1 - \beta)x + \beta\bar{X})\zeta_h\sqrt{h}, \ 0 \leq \beta \leq 1. \tag{1.3.35}$$

The following proposition can be proved analogously to Proposition 1.3.4.

Proposition 1.3.5 *The method (1.3.35) as well as the methods*

$$\bar{X} = x - \sigma^2 \beta x \zeta_h^2 h + \sigma((1 - \beta)x + \beta\bar{X})\zeta_h\sqrt{h}, \ 0 \leq \beta \leq 1, \tag{1.3.36}$$

$$\bar{X} = x - \sigma^2 \beta((1 - \alpha)x + \alpha\bar{X})h + \sigma((1 - \beta)x + \beta\bar{X})\zeta_h\sqrt{h}, \ 0 \leq \alpha, \beta \leq 1, \tag{1.3.37}$$

are of mean-square order $1/2$.

1.3.4 Convergence Theorem for Fully Implicit Methods

Now we are in position to introduce fully implicit methods for general systems of SDEs. For simplicity in writing, we deal here with the scalar Ito SDE:

$$dX = a(t, X)dt + \sigma(t, X)dw(t). \tag{1.3.38}$$

We assume that $a(t, x)$, $\sigma(t, x)$, $\dfrac{\partial\sigma}{\partial x}(t, x)$ are continuous for $t_0 \leq t \leq T$, $x \in \mathbf{R}$, and there exists a positive constant L such that

$$|a(t, y) - a(t, x)| \leq L|y - x|, \quad \left|\frac{\partial \sigma}{\partial x}(t, x)\right| \leq L, \ t_0 \leq t \leq T, \ x, y \in \mathbf{R}. \quad (1.3.39)$$

Recall that the same letter L (or K, or C) is used for various constants.

Consider the implicit one-step approximation (cf. (1.3.31))

$$\bar{X} = x + a(t, \bar{X})h - \sigma(t, x)\frac{\partial \sigma}{\partial x}(t, x)h + \sigma(t, \bar{X})\zeta_h \sqrt{h}, \quad (1.3.40)$$

where ζ_h is defined by (1.3.29) with $A_h = \sqrt{2|\ln h|}$ for definiteness.

Lemma 1.3.6 *There exist constants $K > 0$ and $h_0 > 0$ such that for any $h \leq h_0$, $t_0 \leq t \leq T$, $x \in \mathbf{R}$ Eq. (1.3.40) has a unique solution $\bar{X}$ which satisfies the inequality*

$$|\bar{X} - x| \leq K(1 + |x|)(|\zeta_h|\sqrt{h} + h). \quad (1.3.41)$$

The solution $\bar{X}$ of Eq. (1.3.40) can be found by the method of simple iteration with x as the initial approximation.

Proof For any fixed t, x, and h, let us introduce the function

$$\varphi(z) = x + a(t, z)h - \sigma(t, x)\frac{\partial \sigma}{\partial x}(t, x)h + \sigma(t, z)\zeta_h \sqrt{h}.$$

Then (1.3.40) can be written as
$$\bar{X} = \varphi(\bar{X}).$$

There is a positive constant C such that for any $z \in \mathbf{R}$

$$|\varphi(z) - x| \leq |a(t, x)|h + |a(t, z) - a(t, x)|h + |\sigma(t, x)||\zeta_h|\sqrt{h}$$
$$+ |\sigma(t, z) - \sigma(t, x)||\zeta_h|\sqrt{h} + |\sigma(t, x)\frac{\partial \sigma}{\partial x}(t, x)|h$$
$$\leq C(1 + |x|)(|\zeta_h|\sqrt{h} + h) + L|z - x|(|\zeta_h|\sqrt{h} + h).$$

Further, for any z_1, $z_2 \in \mathbf{R}$

$$|\varphi(z_2) - \varphi(z_1)| \leq L|z_2 - z_1|(|\zeta_h|\sqrt{h} + h).$$

Clearly, there exist positive constants K and h_0 such that for any $h \leq h_0$, $x \in \mathbf{R}$

$$L(|\zeta_h|\sqrt{h} + h) < 1,$$

and

$$|\varphi(z) - x| \leq K(1 + |x|)(|\zeta_h|\sqrt{h} + h)$$

if

$$|z - x| \le K(1 + |x|)(|\zeta_h|\sqrt{h} + h).$$

Let us note that the constants K in the last two inequalities are the same. Now the lemma follows from the contraction mapping principle. □

In addition to (1.3.39) we assume that there exist continuous $\partial a/\partial t$, $\partial \sigma/\partial t$, and $\partial^2 \sigma/\partial x^2$ and the inequalities

$$\left|\frac{\partial a}{\partial t}(t, x)\right| \le L(1 + |x|), \quad \left|\frac{\partial \sigma}{\partial t}(t, x)\right| \le L(1 + |x|), \quad t_0 \le t \le T, \ x \in \mathbf{R},$$
$$(1.3.42)$$

hold.

Theorem 1.3.7 *Assume (1.3.39) and (1.3.42). Let there exist $\delta > 0$ such that if $|y - x| \le \delta(1 + |x|)$, the inequality*

$$|\sigma(t, x)\frac{\partial^2 \sigma}{\partial x^2}(t, y)| \le L, \quad t_0 \le t \le T, \tag{1.3.43}$$

holds.
Then the implicit method based on the one-step approximation (1.3.40) converges in mean-square with the order 1/2.

We omit the proof of this theorem, see it in [291].

Remark 1.3.8 The condition (1.3.43) is satisfied if, for instance,

$$|\sigma(t, x)| \le L, \quad \left|\frac{\partial^2 \sigma}{\partial x^2}(t, x)\right| \le L, \ t_0 \le t \le T, \ x \in \mathbf{R}, \tag{1.3.44}$$

or

$$\left|\frac{\partial^2 \sigma}{\partial x^2}(t, x)\right| \le \frac{L}{1 + |x|}, \quad t_0 \le t \le T, \ x \in \mathbf{R}, \tag{1.3.45}$$

holds.

Let us underline that the conditions of Theorem 1.3.7 are not necessary and the method is applicable more widely. Indeed, convergence of implicit methods can also be proved under nonglobally Lipschitz conditions on the coefficients (see Sect. 1.4 and also [448] and references therein).

Remark 1.3.9 Let the function $c(t, x) := \sigma(t, x)\frac{\partial \sigma}{\partial x}(t, x)$ satisfy the condition

$$|c(t, y) - c(t, x)| \le L|y - x|. \tag{1.3.46}$$

Consider the implicit one-step approximation

$$\bar{X} = x + a(t, \bar{X})h - \sigma(t, \bar{X})\frac{\partial \sigma}{\partial x}(t, \bar{X})h + \sigma(t, \bar{X})\zeta_h\sqrt{h}. \qquad (1.3.47)$$

It is not difficult to prove that Theorem 1.3.7 is valid for the implicit method based on (1.3.47) provided (1.3.46) is fulfilled.

1.3.5 General Construction of Fully Implicit Methods

Let

$$dX^i = a^i(t, X)dt + \sum_{r=1}^{m} \sigma_r^i(t, X)dw_r(t), \ i = 1, \ldots, d. \qquad (1.3.48)$$

Introduce the one-step approximation:

$$\bar{X}^i = x^i + \sum_{k=1}^{l} \lambda_k^i a^i(t + \nu_k^i h, (1 - \alpha_{k1}^i)x^1 + \alpha_{k1}^i \bar{X}^1, \ldots, \qquad (1.3.49)$$

$$(1 - \alpha_{kd}^i)x^d + \alpha_{kd}^i \bar{X}^d)h$$

$$+ \sum_{r=1}^{m} \sum_{k=1}^{l} \mu_{rk}^i \sigma_r^i(t + \nu_{rk}^i h, (1 - \beta_{rk1}^i)x^1 + \beta_{rk1}^i \bar{X}^1,$$

$$\ldots, (1 - \beta_{rkd}^i)x^d + \beta_{rkd}^i \bar{X}^d)\zeta_{rh}\sqrt{h} + A^i,$$

where $0 \le \nu, \alpha, \beta \le 1$, $\lambda, \mu \ge 0$, $\sum_{k=1}^{l} \lambda_k^i = 1$, $\sum_{k=1}^{l} \mu_{rk}^i = 1$, $i = 1, \ldots, d$, l is a positive integer, and A^i are some expressions to be found. Substituting the Euler-type approximation

$$\hat{X}^j = x^j + a^j(t, x)h + \sum_{s=1}^{m} \sigma_s^j(t, x)\zeta_{sh}\sqrt{h}$$

instead of $\bar{X}^j$, $j = 1, \ldots, d$, in σ_r^i, we obtain

$$\sigma_r^i(t + \nu_{rk}^i h, (1 - \beta_{rk1}^i)x^1 + \beta_{rk1}^i \bar{X}^1, \ldots, (1 - \beta_{rkd}^i)x^d + \beta_{rkd}^i \bar{X}^d)$$

$$\approx \sigma_r^i(t, x) + \sum_{j=1}^{d} \frac{\partial \sigma_r^i}{\partial x^j}(t, x)\beta_{rkj}^i \sum_{s=1}^{m} \sigma_s^j(t, x)\zeta_{sh}\sqrt{h}.$$

It is clear from here that either

$$A^i = -\sum_{r=1}^{m} \sum_{k=1}^{l} \mu_{rk}^i \sum_{j=1}^{d} \frac{\partial \sigma_r^i}{\partial x^j}(t, x)\beta_{rkj}^i \sum_{s=1}^{m} \sigma_s^j(t, x)\zeta_{sh}\sqrt{h}\zeta_{rh}\sqrt{h} \qquad (1.3.50)$$

or

$$A^i = -\sum_{r=1}^{m}\sum_{k=1}^{l}\mu_{rk}^i \sum_{j=1}^{d}\frac{\partial \sigma_r^i}{\partial x^j}(t,x)\beta_{rkj}^i \sigma_r^j(t,x)h \qquad (1.3.51)$$

can be put in (1.3.49).

Substituting one of these expressions in (1.3.49), we obtain a multi-parameter family of implicit methods. Here we will not precisely indicate assumptions on the coefficients a and σ_r assuming that appropriate conditions on the coefficients hold.

Theorem 1.3.10 *Under appropriate conditions of smoothness and boundedness on the coefficients of (1.3.48) the method based on the one-step approximation (1.3.49) with A^i as in (1.3.50) or (1.3.51) is of mean-square order $1/2$.*

We omit the proof here (see details in [291]).

Remark 1.3.11 It is also possible to introduce implicitness in A^i by changing t, x as it was done in the terms related to a^i. Moreover, the family can be extended if some a^i or σ_r^i are represented as sums of terms. In this case the coefficients λ, ν, α, μ, β can differ for different terms.

Let us give an example of fully implicit methods:

$$\bar{X} = x + a(t,\bar{X})h - \sum_{r=1}^{m}\sum_{j=1}^{d}\frac{\partial \sigma_r}{\partial x^j}(t,\bar{X})\sigma_r^j(t,\bar{X})h + \sum_{r=1}^{m}\sigma_r(t,\bar{X})\zeta_{rh}\sqrt{h}.$$

Further, in the case of SDEs in the sense of Stratonovich

$$dX = a(t,X)dt + \sum_{r=1}^{m}\sigma_r(t,X)\circ dw_r(t) \qquad (1.3.52)$$

we construct the derivative-free fully implicit method (*midpoint method*):

$$X_{k+1} = X_k + a\left(t_k + \frac{h}{2}, \frac{X_k+X_{k+1}}{2}\right)h + \sum_{r=1}^{m}\sigma_r\left(t_k, \frac{X_k+X_{k+1}}{2}\right)(\zeta_{rh})_k\sqrt{h}.$$

$$(1.3.53)$$

For $\sigma_r^i = 0$, this method coincides with the well-known deterministic midpoint scheme, which has the second order of convergence. In the general case the method (1.3.53) is of the mean-square order $1/2$. In the commutative case, i.e., when $\Lambda_i\sigma_r = \Lambda_r\sigma_i$ (here the operator $\Lambda_r := (\sigma_r, \partial/\partial x)$) or in the case of a system with one noise (i.e., $m = 1$) the midpoint method (1.3.53) has the first mean-square order of convergence which is stated in the next theorem [291].

Theorem 1.3.12 *Suppose that the commutative   conditions $\Lambda_i\sigma_r = \Lambda_r\sigma_i$, $i,r = 1,\ldots,m$, are fulfilled. Let ζ_{rh} be defined by (1.3.29) with $A_h = \sqrt{4|\ln h|}$. Then the method (1.3.53) for the system (1.3.52) has the first mean-square order of convergence.*

1.4 Nonglobally Lipschitz Conditions

The conditions of Theorem 1.1.1 on the coefficients are rather restrictive. Nevertheless this theorem is very useful since it allows us to derive various methods of numerical integration. Methods or some their modifications obtained on the basis of this theorem can be applied more widely. For practice, the globally Lipschitz conditions (1.1.2) are most restrictive since equations with nonglobally Lipschitz conditions such that there exist unique extendable solutions for them make up a very broad and important class in applications.

The early papers on mean-square convergence of numerical methods for SDEs under various forms of nonglobally Lipschitz conditions include e.g. [145, 148, 164, 173, 262]. There was a considerable interest in this topic in the 2010s, see [175, 176, 191, 251, 424, 448] and references therein.

To illustrate the problem, let us consider a very simple example:

$$dX = -X^3 dt + dw(t). \tag{1.4.1}$$

In [262, 431] it was shown that the explicit Euler method X_k for (1.4.1) diverges as the moments $E|X_k|^{2m}$, $m \geq 1$, are not uniformly bounded in the time step h. It can be heuristically explained in the following way. Let us start, for instance, at $X(0) = X_0 = 1/h$ (we can do so without loss of generality since we always reach a similar position with a positive probability). Then

$$X_1 = X_0 - X_0^3 h + \Delta_0 w(h) = \frac{1}{h} - \frac{1}{h^2} + \Delta_0 w(h)$$

with a large probability is a negative number which can be approximately considered as $\approx -1/h^2$. Then

$$X_2 = -\frac{1}{h^2} + \frac{1}{h^5} + \Delta_1 w(h) \approx \frac{1}{h^5}, \ X_3 \approx -\frac{1}{h^{14}},$$

and so on. In experiments one can observe that the explicit Euler method leads to computer overflows.

As it was suggested in [314], one of the natural modifications of the usual Euler method consists in using a variable step-size. For example, let us consider the explicit Euler method for (1.4.1), which is based on the following one-step approximation

$$\bar{X}_{t,x}(t + h_x) = x - x^3 h_x + \Delta_t w(h_x), \tag{1.4.2}$$

where $\Delta_t w(h_x) = w(t + h_x) - w(t)$ and

$$h_x = \begin{cases} h, & |x| \leq 1/\sqrt{h}, \\ 1/x^2, & |x| > 1/\sqrt{h}. \end{cases}$$

The numerical experiments show that a typical trajectory of the explicit Euler method for (1.4.1) blows up at times t between 2.5 and 800 for the time step $h = 0.36$, $X_0 = 0$. A decrease of the time step improves the situation, e.g., for $h = 0.25$ a typical trajectory blows up at times from 33 to 8×10^4. At the same time, the approximation (1.4.2) with $h = 0.36$ remains stable on an arbitrary time interval. See [98, 191] where adaptive time-stepping strategies are used to obtain methods convergent under nonglobally Lipschitz conditions.

Other possibilities of methods convergent under nonglobally Lipschitz conditions consist in (i) using implicit schemes as in Sect. 1.3 (see [164, 175, 251, 432, 448] and also the discussion below, where implicit methods' convergence is treated under relaxed conditions on SDEs' coefficients) and (ii) constructing special explicit schemes (see [175, 176, 448] and also the balanced method below).

Let us weaken the conditions of Theorem 1.1.1 on the coefficients of the SDEs (1.1.1) and consider the following assumption.

Assumption 1.4.1 (i) The initial condition is such that

$$E|X_0|^{2m} \leq K < \infty, \quad \text{for all } m \geq 1. \tag{1.4.3}$$

(ii) For a sufficiently large $m_0 \geq 1$ there is a constant $c_1 \geq 0$ such that for $t \in [t_0, T]$ and $x, y \in \mathbf{R}^d$

$$(x - y, a(t, x) - a(t, y)) + \frac{2m_0 - 1}{2} \sum_{r=1}^{q} |\sigma_r(t, x) - \sigma_r(t, y)|^2 \leq c_1|x - y|^2.$$
$$\tag{1.4.4}$$

(iii) There exist $c_2 \geq 0$ and $\varkappa \geq 1$ such that for $t \in [t_0, T]$,

$$|a(t, x) - a(t, y)|^2 \leq c_2(1 + |x|^{2\varkappa - 2} + |y|^{2\varkappa - 2})|x - y|^2, \quad x, y \in \mathbf{R}^d. \tag{1.4.5}$$

It is known (see e.g. [194, 250]) that Assumption 1.4.1 guarantees existence of a unique solution to the SDEs (1.1.1), which has finite moments, i.e., there is $K > 0$:

$$E|X_{t_0, X_0}(t)|^{2m} < K(1 + E|X_0|^{2m}), \quad 1 \leq m \leq m_0 - 1, \quad t \in [t_0, T]. \tag{1.4.6}$$

The condition (1.4.4) occurs very often in applications (see e.g. [230, 250] and references therein).

Example 1.4.2 Let us illustrate Assumption 1.4.1 (ii) on a one-dimensional SDE:

$$dX = -\alpha X|X|^{r_1 - 1}dt + \mu X^{r_2}dw(t), \quad X(0) = x > 0,$$

with $\alpha, \mu > 0$, $r_1 \geq 1$, and $r_2 \geq 1$. If $r_1 + 1 > 2r_2$ or $r_1 = r_2 = 1$, then (1.4.4) is valid for any $m_0 \geq 1$. If $r_1 + 1 = 2r_2$ and $r_1 > 1$ then (1.4.4) is valid for $1 \leq m_0 \leq \alpha/\mu^2 + 1/2$. Note that since the solution of this SDE is unique under the above choice of the parameters, $X(t) > 0$ for all $t \geq 0$.

The following theorem is a generalization of fundamental Theorem 1.1.1 from the global to nonglobally Lipschitz case. It also has similarities with a mean-square convergence theorem in [164] proved for the case of nonglobally Lipschitz drift, global Lipschitz diffusion and Euler-type schemes.

Theorem 1.4.3 *Suppose (i) Assumption 1.4.1 holds;*

(ii) The one-step approximation $\bar{X}_{t,x}(t+h)$ from (1.1.3) has the following orders of accuracy: for sufficiently large $m_1 \geq 1$ there are $\lambda \geq 1$, $h_0 > 0$, and $K > 0$ such that for arbitrary $t_0 \leq t \leq T - h$, $x \in \mathbf{R}^d$, and all $0 < h \leq h_0$:

$$|E[X_{t,x}(t+h) - \bar{X}_{t,x}(t+h)]| \leq K(1+|x|^{2\lambda})^{1/2}h^{p_1}, \qquad (1.4.7)$$

$$\left[E|X_{t,x}(t+h) - \bar{X}_{t,x}(t+h)|^{2m_1}\right]^{1/(2m_1)} \leq K(1+|x|^{2\lambda m_1})^{1/(2m_1)}h^{p_2} \qquad (1.4.8)$$

with

$$p_2 \geq \frac{1}{2}, \ p_1 \geq p_2 + \frac{1}{2}; \qquad (1.4.9)$$

(iii) The approximation X_k from (1.1.4) has finite moments, i.e., for sufficiently large $m_2 \geq 1$ there are $\beta \geq 1$, $h_0 > 0$, and $K > 0$ such that for all $0 < h \leq h_0$ and all $k = 0, \ldots, N$:

$$E|X_k|^{2m_2} < K(1+E|X_0|^{2\beta m_2}). \qquad (1.4.10)$$

Then for any N and $k = 0, 1, \ldots, N$ the following inequality holds for some $l \geq 1$

$$\left[E|X_{t_0,X_0}(t_k) - \bar{X}_{t_0,X_0}(t_k)|^{2l}\right]^{1/(2l)} \leq K(1+E|X_0|^{2\gamma l})^{1/(2l)}h^{p_2-1/2}, \qquad (1.4.11)$$

where constants $K > 0$ and $\gamma \geq 1$ do not depend on h and k, i.e., the order of accuracy of the method (1.1.4) is $p = p_2 - 1/2$.

The theorem is proved in [448] based on the idea of the proof of Theorem 1.1.1 from Sect. 1.1.2 and it uses the following lemma (proved in [448]), which is an analogue of Lemma 1.1.3.

Lemma 1.4.4 *Suppose Assumption 1.4.1 holds. For the representation*

$$X_{t,x}(t+\theta) - X_{t,y}(t+\theta) = x - y + Z_{t,x,y}(t+\theta), \qquad (1.4.12)$$

we have for $1 \leq m \leq (m_0 - 1)/\varkappa$:

$$E|X_{t,x}(t+h) - X_{t,y}(t+h)|^{2m} \leq |x-y|^{2m}(1+Kh), \qquad (1.4.13)$$

$$E\left|Z_{t,x,y}(t+h)\right|^{2m} \leq K(1+|x|^{2\varkappa-2}+|y|^{2\varkappa-2})^{m/2}|x-y|^{2m}h^m. \qquad (1.4.14)$$

Let us discuss Theorem 1.4.3.

1. The constant K in (1.4.11) depends on l, t_0, T, c_1, and c_2. The constant γ in (1.4.11) depends on λ, β and $\varkappa$. In (1.4.11) the maximum possible l depends on the maximum values of m_1 and m_2 and on λ and $\varkappa$. Due to the bound (1.4.6) on the moments of the solution $X(t)$, it would be natural to require that β in (1.4.10) should be equal to 1. Indeed (see e.g. [448] and references therein), (1.4.10) with $\beta = 1$ holds for the drift-implicit method (1.3.8) and for fully implicit methods of Sect. 1.3.5. However, this is not the case for tamed-type methods (see [176]) or the balanced method (1.4.15) considered below.

2. As a rule, it is not difficult to check the conditions (1.4.7)–(1.4.8) following the usual routine calculations as in the global Lipschitz case. We note that in order to achieve the optimal p_1 and p_2 in (1.4.7)–(1.4.8) additional assumptions on smoothness of $a(t, x)$ and $\sigma_r(t, x)$ are usually needed. In contrast to the conditions (1.4.7)–(1.4.8), checking the condition (1.4.10) on moments of a method X_k is often rather difficult. In the case of global Lipschitz coefficients, boundedness of moments of X_k is direct implication of the boundedness of moments of the SDEs solution and the one-step properties of the method as shown in Lemma 1.1.5. There is no result of this type in the case of nonglobally Lipschitz SDEs and each scheme requires a special consideration. For a number of mean-square schemes, boundedness of moments in nonglobally Lipschitz cases were proved e.g. in [164, 173, 175, 176, 432, 448]. We show boundedness of moments for a particular balanced method in Lemma 1.4.7 below.

Roughly speaking, Theorem 1.4.3 says that if moments of X_k are bounded and the scheme was proved to be convergent with order p in the global Lipschitz case, then the scheme has the same convergence order p in the considered nonglobally Lipschitz case.

Example 1.4.5 Consider the drift-implicit scheme (1.3.8) with $\alpha = 0$ constructed in Sect. 1.3.1. Assume that the coefficients $a(t, x)$ and $\sigma_r(t, x)$ have continuous first-order partial derivatives in t and the coefficient $a(t, x)$ also has continuous first-order partial derivatives in x^i and that all these derivatives and the coefficients themselves satisfy inequalities of the form (1.4.5). It is not difficult to show that the one-step approximation corresponding to (1.3.8) satisfies (1.4.7) and (1.4.8) with $p_1 = 2$ and $p_2 = 1$, respectively. Its boundedness of moments, in particular, under the condition (1.4.4) for time steps $h \leq 1/(2c_1)$, is proved in [175, 251]. Then, due to Theorem 1.4.3, (1.3.8) converges with mean-square order $p = 1/2$ (note that for $p = 1/2$, it is sufficient to have $p_1 = 3/2$ which can be obtained under lesser smoothness of a). Further, in the case of additive noise $p_1 = 2$ and $p_2 = 3/2$ and (1.3.8) converges with mean-square order 1 due to Theorem 1.4.3. We recall that convergence of (1.3.8) with order $1/2$ in the global Lipschitz case was shown in Sect. 1.3.1.

Theorem 1.4.3 implies a.s. convergence (see Sect. 1.1.6).

Corollary 1.4.6 *In the setting of Theorem 1.4.3 for $l \geq 1/(2p)$ in (1.4.11), there is $0 < \varepsilon < p$ and an a.s. finite random variable $C(\omega) > 0$ such that*

$$|X_{t_0, X_0}(t_k) - X_k| \leq C(\omega)h^{p-\varepsilon},$$

i.e., the method (1.1.4) for (1.1.1) converges with order $p - \varepsilon$ a.s.

Let us consider a particular balanced scheme from [448] which belongs to the class of balanced methods introduced in Sect. 1.3.2 (see also [285]):

$$X_{k+1} = X_k + \frac{a(t_k, X_k)h + \sum_{r=1}^{q} \sigma_r(t_k, X_k)\xi_{rk}\sqrt{h}}{1 + h|a(t_k, X_k)| + \sqrt{h}\sum_{r=1}^{q} |\sigma_r(t_k, X_k)\xi_{rk}|}, \qquad (1.4.15)$$

where ξ_{rk} are Gaussian $\mathcal{N}(0, 1)$ i.i.d. random variables. We get (1.4.15) from (1.3.19)–(1.3.20) by taking

$$c_0(t, x) = |a(t, x)|, \quad c_r(t, x) = |\sigma_r(t, x)|.$$

Due to this specific choice of the coefficients $c_i(t, x)$, we easily resolve the implicitness in (1.3.19) and arrive at the scheme (1.4.15) which is natural to view as an explicit scheme.

The proof of convergence of the scheme (1.4.15) with mean-square order $1/2$ rests on Theorem 1.4.3, which conditions are verified by two lemmas. The first lemma is on boundedness of moments, which proof uses a stopping time technique, it is rather cumbersome and not presented here, see [448].

Lemma 1.4.7 *Suppose Assumption 1.4.1. For all natural N and all $k = 0, \ldots, N$ the following inequality holds for moments of the scheme (1.4.15):*

$$E|X_k|^{2m} \leq K(1 + E|X_0|^{2m\beta}), \quad 1 \leq m \leq \frac{m_0 - 1}{4(3\varkappa - 2)} - \frac{1}{2}, \qquad (1.4.16)$$

with some constants $\beta \geq 1$ and $K > 0$ independent of h and k.

The next lemma gives estimates for the one-step error of the balanced scheme (1.4.15).

Lemma 1.4.8 *Assume that (1.4.6) holds. Assume that the coefficients $a(t, x)$ and $\sigma_r(t, x)$ have continuous first-order partial derivatives in t and that these derivatives and the coefficients satisfy inequalities of the form (1.4.5). Then the scheme (1.4.15) satisfies the inequalities (1.4.7) and (1.4.8) with $p_1 = 3/2$ and $p_2 = 1$, respectively.*

The proof of this lemma is routine one-step error analysis (it can be found in [448] and see also Theorem 1.3.1 here). Lemmas 1.4.7 and 1.4.8 and Theorem 1.4.3 imply the following result.

Proposition 1.4.9 *Under the assumptions of Lemmas 1.4.7 and 1.4.8 the balanced scheme (1.4.15) has mean-square order $1/2$, i.e., for it the inequality (1.4.11) holds with $p = p_2 - 1/2 = 1/2$.*

Remark 1.4.10 In the additive noise case the mean-square order of the balanced scheme (1.4.15) does not improve (p_1 and p_2 remain $3/2$ and 1, respectively).

Some related numerical experiments can be found in [448]. Also, see Sect. 2.2.5 where weak-sense convergence of numerical methods for SDEs with nonglobally Lipschitz coefficients is discussed.

1.5 Modeling of Ito Integrals

Such integrals arise in numerical integration formulas when the system of equations has the form

$$dX = a(t, X)dt + \sigma(t, X)dw, \qquad (1.5.1)$$

where $w(t)$ is a scalar process ($q = 1$). We restrict ourselves to the integrals which are needed for methods of orders $3/2$ and 2 (see more details in [274]). We consider here explicit methods. Clearly, the integrals can be used for implicit methods as well.

1.5.1 Ito Integrals Depending on a Single Noise and Methods of Order 3/2 and 2

Due to Theorem 1.2.4, the method (1.2.27) is of order $3/2$. For (1.5.1) it acquires the form

$$\bar{X}_{t,x}(t + h) = x + \sigma(w(t + h) - w(t)) + \Lambda\sigma \int_t^{t+h} (w(\theta) - w(t))dw(\theta) \quad (1.5.2)$$

$$+ ah + L\sigma \int_t^{t+h} (\theta - t)dw(\theta) + \Lambda a \int_t^{t+h} (w(\theta) - w(t))d\theta$$

$$+ \Lambda^2\sigma \int_t^{t+h} \left(\int_t^\theta (w(\theta_1) - w(t))dw(\theta_1) \right) dw(\theta) + La\frac{h^2}{2} ,$$

where all the coefficients σ, a, $\Lambda\sigma$, $L\sigma$, Λa, $\Lambda^2\sigma$, La are evaluated at (t, x).

Since the distribution of random variables in (1.5.2) does not depend on t, we have to be able to model together with $w(h)$, $\int_0^h w(\theta)d\theta$ the following variables

$$\int_0^h w(\theta)dw(\theta) = \frac{1}{2}(w^2(h) - h),$$

$$\int_0^h \theta dw(\theta) = hw(h) - \int_0^h w(\theta)d\theta,$$

$$\int_0^h \left(\int_0^\theta w(\theta_1)dw(\theta_1)\right) dw(\theta) = \frac{1}{6}w^3(h) - \frac{1}{2}hw(h),$$

i.e., to construct the method of order $3/2$ it suffices at each step to model $w(h)$ and $\int_0^h w(\theta)d\theta$. The problem of modeling these variables can be solved very simply. In fact, their joint distribution is Gaussian. Write the integral $\int_0^h w(\theta)d\theta$ as

$$\int_0^h w(\theta)d\theta = \alpha w(h) + \left(\int_0^h w(\theta)d\theta - \alpha w(h)\right)$$

and choose α such that $w(h)$ and $\int_0^h w(\theta)d\theta - \alpha w(h)$ are independent. Clearly, α can be found from the condition that expectation of their product should vanish. Since $E(w(h)\int_0^h w(\theta)d\theta) = h^2/2$, we have $\alpha = h/2$. As a result, the integral can be written as a sum of two independent normally distributed random variables:

$$\int_0^h w(\theta)d\theta = \frac{1}{2}hw(h) + \left(\int_0^h w(\theta)d\theta - \frac{1}{2}hw(h)\right). \qquad (1.5.3)$$

The first term at the right-hand side of (1.5.3) is $\mathcal{N}(0, h^3/4)$-distributed, and the second term is $\mathcal{N}(0, h^3/12)$-distributed. Thus, from the point of view of modeling the random variables involved, the method (1.5.2) for the system (1.5.1) with a single noise is rather simple.

It is not difficult to prove that to construct a method of order 2, it suffices to model at each step three random variables $w(h)$, $\int_0^h w(\theta)d\theta$, and $\int_0^h w^2(\theta)d\theta$. In [274], the characteristic function of these random variables is found. However, it is very complicated and is not useful in practice. Thus, the exact modeling has bad perspectives, and therefore we need to be able to model these variables approximately.

Lemma 1.5.1 *Suppose that the one-step approximation (see (1.1.3))*

$$\bar{X}_{t,x}(t+h) = x + A(t, x, h; w_i(\theta) - w_i(t)), \ i = 1, \ldots, q, \ t \le \theta \le t + h)$$

(1.5.4)

generates a method with order of accuracy m. Suppose A contains terms of the form $P(t, x) \cdot \xi(w_i(\theta) - w_i(t)), \ i = 1, \ldots, q, \ t \le \theta \le t + h)$, where $|P(t, x)| \le K(1 + |x|^2)^{1/2}$ and ξ is a random variable, depending on the Wiener processes on the interval $[t, t + h]$, as indicated between the brackets. Let $\xi = \eta + \varsigma$, where η and ς are random variables depending on the same Wiener processes on the same interval. Finally, suppose

$$|E\varsigma| \le Kh^{m+1}, \ (E\varsigma^2)^{1/2} \le Kh^{m+1/2} \, .$$

(1.5.5)

Then the method based on the one-step approximation (1.5.4) and with $P \cdot \xi$ replaced by $P \cdot \eta$ has order of accuracy equal to m.

The proof of this lemma easily follows from fundamental Theorem 1.1.1.

This lemma makes it possible to replace random variables, which participate in the method and are difficult in modeling, by random variables for which modeling is simpler. A sufficiently general approach to the approximate modeling of random variables is as follows. Make up the system of SDEs whose solution at time h is given by the set of random variables to be modeled. If this system is integrable over the interval $[0, h]$ with sufficiently high accuracy (at the expense of a small integration step h_1, this can be done by a "rough" method in which simpler random variables are modeled), then we can construct the required approximations for the variables to be modeled. We note one important particular point that the needed random variables themselves have different orders of smallness with respect to h. And so we need approximations for them of a different order of accuracy. We clarify this using the variables $w(h)$, $\int_0^h w(\theta)d\theta$, and $\int_0^h w^2(\theta)d\theta$ as an example.

Introduce the new process

$$v(s) = \frac{w(sh)}{\sqrt{h}}, \ 0 \le s \le 1 \, .$$

(1.5.6)

It is obvious that $v(s)$ is a standard Wiener process. We have

$$w(h) = h^{1/2}v(1), \ \int_0^h w(\theta)d\theta = h^{3/2} \int_0^1 v(s)ds, \ \int_0^h w^2(\theta)d\theta = h^2 \int_0^1 v^2(s)ds \, .$$

(1.5.7)

Thus, the problem of modeling the variables $w(h)$, $\int_0^h w(\theta)d\theta$, and $\int_0^h w^2(\theta)d\theta$ can be reduced to that of modeling the variables $v(1)$, $\int_0^1 v(s)ds$, and $\int_0^1 v^2(s)ds$. These variables are the solution of the system of equations

$$dx = dv(s), \ x(0) = 0, \tag{1.5.8}$$
$$dy = x\,ds, \ y(0) = 0,$$
$$dz = x^2 ds, \ z(0) = 0,$$

at time $s = 1$.

Let $\bar{x}(s_k)$, $\bar{y}(s_k)$, $\bar{z}(s_k)$, $0 = s_0 < s_1 < \cdots < s_N = 1$, $s_{k+1} - s_k = h_1 = 1/N$, be an approximate solution of (1.5.8). If we are guided by Lemma 1.5.1 when constructing a method of order 2, then the random variables $w(h) = h^{1/2}x(1)$, $\int_0^h w(\theta)d\theta = h^{3/2}y(1)$, and $\int_0^h w^2(\theta)d\theta = h^2 z(1)$ participating in it can be replaced by $h^{1/2}\bar{x}(1)$, $h^{3/2}\bar{y}(1)$, and $h^2\bar{z}(1)$, if only the conditions

$$|E(v(1) - \bar{x}(1))| = O(h^{5/2}), \ (E(v(1) - \bar{x}(1))^2)^{1/2} = O(h^2), \tag{1.5.9}$$

$$\left| E\left(\int_0^1 v(s)ds - \bar{y}(1) \right) \right| = O(h^{3/2}), \ \left(E\left(\int_0^1 v(s)ds - \bar{y}(1) \right)^2 \right)^{1/2} = O(h),$$
$$\tag{1.5.10}$$

$$\left| E\left(\int_0^1 v^2(s)ds - \bar{z}(1) \right) \right| = O(h), \ \left(E\left(\int_0^1 v^2(s)ds - \bar{z}(1) \right)^2 \right)^{1/2} = O(h^{1/2})$$
$$\tag{1.5.11}$$

are fulfilled.

First we integrate (1.5.8) by the Euler method with step h_1 :

$$x_{k+1} = x_k + \Delta_k v(h_1), \ x_0 = 0, \tag{1.5.12}$$
$$y_{k+1} = y_k + x_k h_1, \ y_0 = 0,$$
$$z_{k+1} = z_k + x_k^2 h_1, \ z_0 = 0.$$

Since

$$x_k = v(s_k), \ x_N = \bar{x}(1) = v(1),$$
$$y_k = h_1 \sum_{i=0}^{k-1} v(s_i), \ \bar{y}(1) = h_1 \sum_{i=0}^{N-1} v(s_i),$$
$$z_k = h_1 \sum_{i=0}^{k-1} v^2(s_i), \ \bar{z}(1) = h_1 \sum_{i=0}^{N-1} v^2(s_i),$$

we have

$$Ev(1) = E\bar{x}(1) = 0, \tag{1.5.13}$$

$$E\int_0^1 v(s)ds = E\bar{y}(1) = 0,$$

$$E\int_0^1 v^2(s)ds = \frac{1}{2}, \quad E\bar{z}(1) = \frac{1}{2} - \frac{h_1}{2}.$$

Further, since the system (1.5.8) is a system with additive noise, Euler's method has order $O(h_1)$. Therefore, taking $h_1 = h$ we find that all the relations (1.5.9)–(1.5.11) hold (note that in our case $E(v(1) - \bar{x}(1))^2 = 0$). Indeed, direct computations (of quite some length in the case of the third equation) give (for $h_1 = h$) :

$$E\left(\int_0^1 v(s)ds - \bar{y}(1)\right)^2 = \frac{h^2}{3}, \quad \left|E\left(\int_0^1 v^2(s)ds - \bar{z}(1)\right)\right| = \frac{h}{2}, \tag{1.5.14}$$

$$E\left(\int_0^1 v^2(s)ds - \bar{z}(1)\right)^2 = \frac{11}{12}h^2 - \frac{h^3}{3}.$$

Thus, if we approximately model the random variables, needed for a method of order 2 in the case of a system with single noise, by applying Euler's method to (1.5.8), then at each step we have to model $\approx 1/h$ normally distributed random variables (here h is the integration step in both the original system and the auxiliary system (1.5.8)).

We will now use a method of order $3/2$ (see (1.5.2)) to integrate (1.5.8). Since

$$a = \begin{bmatrix} 0 \\ x \\ x^2 \end{bmatrix}, \quad \sigma = \begin{bmatrix} 1 \\ 0 \\ 0 \end{bmatrix}, \quad \Lambda\sigma = \Lambda^2\sigma = L\sigma = 0, \quad \Lambda a = \begin{bmatrix} 0 \\ 1 \\ 2x \end{bmatrix}, \quad La = \begin{bmatrix} 0 \\ 0 \\ 1 \end{bmatrix},$$

we get

$$x_{k+1} = x_k + \Delta_k v(h_1), \tag{1.5.15}$$

$$y_{k+1} = y_k + x_k h_1 + \int_{s_k}^{s_{k+1}} (v(\theta) - v(s_k))d\theta,$$

$$z_{k+1} = z_k + x_k^2 h_1 + 2x_k \int_{s_k}^{s_{k+1}} (v(\theta) - v(s_k))d\theta + \frac{h_1^2}{2}.$$

The method (1.5.15) has the following properties. First x_k and y_k are equal to $v(s_k)$ and $\int_0^{s_k} v(\theta)d\theta$, respectively (this is obvious) and, secondly, $E\bar{z}(1) = Ez(1)$ (this will be proved below). As a result, only the second relation in (1.5.11) among all the relations (1.5.9)–(1.5.11) has to be satisfied. However, since the method (1.5.15) is of order $3/2$, this relation reduces to the requirement

$$(E(z(1) - \bar{z}(1))^2)^{1/2} = O(h_1^{3/2}) = O(h^{1/2}). \qquad (1.5.16)$$

Thus, if we choose h_1 such that $h_1 = h^{1/3}$, then the conditions of Lemma 1.5.1 hold.

Let us we compare computational costs of the two procedures proposed above to approximately model the needed random variables. Suppose that we integrate the original system by a method of order 2 with the step $h = 0.001$. Then, if we use the first procedure to model the needed random variables, we have $h_1 = h$ and we need $\approx 1/h$, i.e., ≈ 1000 normally distributed random variables. In the second procedure we have $h_1 = h^{1/3} = 0.1$; however at each step we have to model not one but two random variables, $\Delta_k v(h_1)$ and $\int_{s_k}^{s_{k+1}} (v(\theta) - v(s_k))d\theta$. As a result, we need only 20 variables here instead of 1000 random variables. Of course, the second procedure is much more economical. Now we give a theorem (see its proof in [274, p. 88]) on second order methods for the system (1.5.1) related to approximate modeling by (1.5.15).

Theorem 1.5.2 *Suppose we solve the system (1.5.1) by a method of second order of accuracy with step h (for example, in accordance with Theorem 1.2.5), in which at each step the integrals $\int_{t_j}^{t_{j+1}} dw(\theta)$, $\int_{t_j}^{t_{j+1}} (w(\theta) - w(t_j))d\theta$, $\int_{t_j}^{t_{j+1}} (w(\theta) - w(t_j))^2 d\theta$ participate. If these random variables are replaced, independently at each step, by the random variables $h^{1/2}x_N$, $h^{3/2}y_N$, $h^2 z_N$, where x_N, y_N, z_N, can be found recurrently from (1.5.15) with step $h_1 = O(h^{1/3})$, then the order of accuracy of the obtained method remains the same, i.e., it is equal to 2.*

Remark 1.5.3 Application of the usual numerical integration formulas for modeling the integrals does not lead to a success. We will convince ourselves of this by the example of modeling the integral $\int_0^1 v(s)ds$ using the trapezoidal formula. It is well known that the trapezoidal formula has error $O(h^2)$. This is true for integrands having bounded second derivative. Here, however, $v(s)$ is a Wiener process with nonsmooth trajectories. Of course, we are interested in accuracy in the mean sense. More precisely, we are interested in mean and mean-square deviation of the approximation of the integral by the trapezoidal formula.

Applying the trapezoidal formula to the integral $\int_0^1 v(s)ds$, we get

$$\int_0^1 v(s)ds \doteq \frac{h}{2}(v(0) + 2v(s_1) + \cdots + 2v(s_{N-1}) + v(s_N)), \quad h = \frac{1}{N}. \qquad (1.5.17)$$

Here the mean deviation is zero, since expectation of both sides of (1.5.17) are zero. It is easy to compute that

$$
E\left(\int_0^1 v(s)ds - \sum_{k=0}^{N-1} \frac{v(s_k) + v(s_{k+1})}{2}h\right)^2
$$

$$
= \sum_{k=0}^{N-1} E\left(\int_{s_k}^{s_{k+1}}\left(v(s) - v(s_k) - \frac{v(s_{k+1}) - v(s_k)}{2}\right)ds\right)^2 = \frac{h^2}{12}.
$$

Thus, the mean-square deviation in the trapezoidal method is $O(h)$, which is of lower order than expected.

1.5.2   Modeling Ito Integrals by the Rectangle and Trapezoidal Methods

A method of first order of accuracy for the system (1.1.1) has the form

$$
X_0 = X(t_0), \quad X_{n+1} = X_n + \sum_{j=1}^q (\sigma_j)_n \Delta_n w_j(h) + a_n h \tag{1.5.18}
$$

$$
+ \sum_{i=1}^q \sum_{j=1}^q (\Lambda_i \sigma_j)_n \int_{t_n}^{t_{n+1}} (w_i(s) - w_i(t_n))dw_j(s),
$$

where $t_0 < t_1 < \cdots < t_N = T$, $h = t_{n+1} - t_n = (T - t_0)/N$. To realize the method (1.5.18), we have to model the set of random variables $\Delta_n w_j$, $\int_{t_n}^{t_{n+1}}(w_i(s) - w_i(t_n))dw_j(s)$, $i, j = 1, \ldots, q$, at each step. Since at different steps these sets are independent, the problem reduces to modeling the variables $w_j(h)$, $I_{ij} = \int_0^h w_i(s)dw_j(s)$, $i, j = 1, \ldots, q$. Due to

$$
\int_0^h w_j(s)dw_i(s) = w_i(h)w_j(h) - \int_0^h w_i(s)dw_j(s), \quad i \neq j, \tag{1.5.19}
$$

$$
\int_0^h w_i(s)dw_i(s) = \frac{w_i^2(h)}{2} - \frac{h}{2}, \tag{1.5.20}
$$

it suffices to model the set of variables

$$w_j(h), \ j = 1, \ldots, q, \tag{1.5.21}$$

$$I_{ij} = \int_0^h w_i(s)dw_j(s), \ i = 1, \ldots, q, \ j = i+1, \ldots, q.$$

Consider the *rectangle method*. We write the integral $\int_0^h w_i(s)dw_j(s)$ as a sum of l integrals:

$$\int_0^h w_i(s)dw_j(s) = \sum_{k=1}^l \int_{s_{k-1}}^{s_k} w_i(s)dw_j(s), \tag{1.5.22}$$

where $s_0 = 0$, $s_k - s_{k-1} = h/l$, $k = 1, \ldots, l$. We replace each of these integrals using the left rectangle formula and obtain

$$\int_0^h w_i(s)dw_j(s) \doteq \sum_{k=1}^l w_i(s_{k-1})(w_j(s_k) - w_j(s_{k-1})). \tag{1.5.23}$$

For the error

$$\Delta_{ij} = \sum_{k=1}^l \int_{s_{k-1}}^{s_k} (w_i(s) - w_i(s_{k-1}))dw_j(s)$$

in the approximate identity (1.5.23), we have

$$E\Delta_{ij} = 0, \ E\Delta_{ij}^2 = \sum_{k=1}^l E\left(\int_{s_{k-1}}^{s_k} (w_i(s) - w_i(s_{k-1}))dw_j(s) \right)^2 \tag{1.5.24}$$

$$= \sum_{k=1}^l \int_{s_{k-1}}^{s_k} E(w_i(s) - w_i(s_{k-1}))^2 ds = \frac{h^2}{2l}.$$

Thus, to approximately represent the variables (1.5.21) by (1.5.23), we have to model ql independent $\mathcal{N}(0, 1)$-distributed random variables ξ_{ik}, $i = 1, \ldots, q$, $k = 1, \ldots, l$ $(\sqrt{h/l}\xi_{ik} = w_j(s_k) - w_j(s_{k-1}))$ and put for $i = 1, \ldots, q$, $j = i+1, \ldots, q$:

$$w_i(h) = \sqrt{\frac{h}{l}} \sum_{r=1}^l \xi_{ir}, \ \int_0^h w_i(s)dw_j(s) \doteq \frac{h}{l} \sum_{k=1}^l \sum_{r=1}^{k-1} \xi_{ir}\xi_{jk}. \tag{1.5.25}$$

Note that if the integral $\int_0^h w_i(s)dw_i(s)$ is modeled according to (1.5.25), then the error involved is $h^2/(2l)$, while this integral can be computed exactly by (1.5.20). Further, the integrals for $i > j$ can be approximately computed either by using (1.5.19) after having approximately modeled (1.5.21), or by modeling them according to (1.5.23). In both cases the error will be the same. It is easy to see that the use of (1.5.19) is equivalent to the use of the right rectangle formula.

We now consider the *trapezoidal method*. Applying to each integral in the sum (1.5.22) the trapezoidal formula, we find

$$
\int_0^h w_i(s)dw_j(s) \doteq \sum_{k=1}^l \frac{1}{2}(w_i(s_{k-1}) + w_i(s_k))(w_j(s_k) - w_j(s_{k-1})). \qquad (1.5.26)
$$

For the error

$$
\Delta_{ij} = \frac{1}{2}\sum_{k=1}^l \int_{s_{k-1}}^{s_k} ((w_i(s) - w_i(s_{k-1})) - (w_i(s_{k-1}) - w_i(s)))dw_j(s)
$$

in the approximate identity (1.5.26), we have

$$
E\Delta_{ij} = 0, \qquad\qquad\qquad\qquad\qquad\qquad\qquad\qquad\qquad\qquad\qquad (1.5.27)
$$

$$
E\Delta_{ij}^2 = \frac{1}{4}\sum_{k=1}^l \int_{s_{k-1}}^{s_k} E((w_i(s) - w_i(s_{k-1})) - (w_i(s_{k-1}) - w_i(s)))^2 ds
$$

$$
= \frac{1}{4}\sum_{k=1}^l ((s - s_{k-1}) + (s_k - s))ds = \frac{1}{4}\frac{h^2}{l}.
$$

Here, having modeled ql independent $\mathcal{N}(0, 1)$-distributed random variables ξ_{ik}, we can set

$$
w_i(h) = \sqrt{\frac{h}{l}}\sum_{r=1}^l \xi_{ir}, \quad \int_0^h w_i(s)dw_j(s) \doteq \frac{h}{l}\sum_{k=1}^l \left(\sum_{r=1}^{k-1}\xi_{ir} + \frac{1}{2}\xi_{ik}\right)\xi_{jk}, \; i < j.
$$

$$
(1.5.28)
$$

It can be readily seen that (1.5.19) gives the same result as (1.5.28) for $i > j$. If we take $l \approx 1/h$, then (1.5.24), (1.5.27) imply that the mean-square error of the approximations of the integrals is $O(h^{3/2})$ in both the rectangle and the trapezoidal method, while the mean error is zero. By Lemma 1.5.1 this implies the following result.

Theorem 1.5.4 *If we replace the random variables at each step of the method (1.5.18) using either the rectangle or the trapezoidal formula with $l \approx 1/h$, then the order of accuracy of the obtained method remains the same, i.e., it is equal to 1.*

1.5.3   Modeling Ito Integrals by the Fourier Method

We now turn to a method which can be naturally called the Fourier method.

Consider the Fourier coefficients of the process $w_i(t) - (t/h)w_i(h)$, $i = 1, \ldots, q$, on the interval $0 \le t \le h$ with respect to the trigonometric system of functions 1, $\cos 2k\pi t/h$, $\sin 2k\pi t/h$, $k = 1, 2, \ldots$ (see [180, 344] for the *Wiener construction* of Brownian motion). We have

$$a_{ik} = \frac{2}{h} \int_0^h (w_i(s) - \frac{s}{h}w_i(h)) \cos \frac{2\pi ks}{h} ds, \ k = 0, 1, 2, \ldots, \quad (1.5.29)$$

$$b_{ik} = \frac{2}{h} \int_0^h (w_i(s) - \frac{s}{h}w_i(h)) \sin \frac{2\pi ks}{h} ds, \ k = 1, 2, \ldots. $$

The distribution of these coefficients is clearly Gaussian.

Lemma 1.5.5 *The following equalities hold:*

$$Ew_i(h)a_{ik} = 0, \ k = 0, 1, 2, \ldots, \ \ Ew_i(h)b_{ik} = 0, \ k = 1, 2, \ldots, \ (1.5.30)$$
$$Ea_{i0}^2 = \frac{h}{3}, \ Ea_{ik}^2 = Eb_{ik}^2 = \frac{h}{2k^2\pi^2}, \ k = 1, 2, \ldots, \quad (1.5.31)$$
$$Ea_{i0}a_{ik} = -\frac{h}{k^2\pi^2}, \ k = 1, 2, \ldots, \quad (1.5.32)$$
$$Ea_{i0}b_{ik} = Ea_{ik}a_{im} = Ea_{ik}b_{im} \quad (1.5.33)$$
$$= Eb_{ik}b_{im} = 0, \ k, m = 1, 2, \ldots, \ k \neq m.$$

Proof All these formulas can be obtained by direct computation. We give, as an example, a detailed proof of one of the formulas in (1.5.31). We have for $k \neq 0$:

$$Ea_{ik}^2 = \frac{4}{h^2} \int_0^h \int_0^h E((w(t) - \frac{t}{h}w(h))(w(s) - \frac{s}{h}w(h))) \cos \frac{2\pi kt}{h} \cos \frac{2\pi ks}{h} dt ds .$$

We evaluate:

$$E((w(t) - \frac{t}{h}w(h))(w(s) - \frac{s}{h}w(h))) = \begin{cases} t - ts/h, \ t \le s, \\ s - ts/h, \ t > s. \end{cases}$$

Therefore

$$Ea_{ik}^2 = \frac{4}{h^2} \int_0^h \left(\int_0^s t \cos \frac{2\pi kt}{h} dt \right) \cos \frac{2\pi ks}{h} ds \tag{1.5.34}$$

$$+ \frac{4}{h^2} \int_0^h \left(\int_s^h \cos \frac{2\pi kt}{h} dt \right) s \cos \frac{2\pi ks}{h} ds$$

$$- \frac{4}{h^2} \int_0^h \int_0^h \frac{st}{h} \cos \frac{2\pi kt}{h} \cos \frac{2\pi ks}{h} dt ds \,.$$

Further,

$$\int_0^s t \cos \frac{2\pi kt}{h} dt = s \sin \frac{2\pi ks}{h} \cdot \frac{h}{2\pi k} + \left(\frac{h}{2\pi k} \right)^2 \left(\cos \frac{2\pi ks}{h} - 1 \right) \,.$$

Then, in particular,

$$\int_0^h t \cos \frac{2\pi kt}{h} dt = 0 \,.$$

Hence the last term in (1.5.34) vanishes. Using these equations, we obtain

$$Ea_{ik}^2 = \frac{4}{h^2} \int_0^h \left(s \sin \frac{2\pi ks}{h} \cdot \frac{h}{2\pi k} + \left(\frac{h}{2\pi k} \right)^2 \left(\cos \frac{2\pi ks}{h} - 1 \right) \right) \cos \frac{2\pi ks}{h} ds$$

$$+ \frac{4}{h^2} \int_0^h \left(-\frac{h}{2\pi k} \sin \frac{2\pi ks}{h} \right) s \cos \frac{2\pi ks}{h} ds$$

$$= \frac{4}{h^2} \int_0^h \left(\frac{h}{2\pi k} \right)^2 \left(\cos \frac{2\pi ks}{h} - 1 \right) \cos \frac{2\pi ks}{h} ds = \frac{h}{2k^2\pi^2} \,.$$

The other formulas can be proved similarly. □

We replace the integrand $w_i(s)$ in the integral $I_{ij} = \int_0^h w_i(s) dw_j(s)$ by an expression containing a part of its Fourier series:

$$I_{ij} = \int_0^h w_i(s)dw_j(s) \doteq \bar{I}_{ij} \tag{1.5.35}$$

$$:= \int_0^h \left(\frac{s}{h} w_i(h) + \frac{a_{i0}}{2} + \sum_{k=1}^m \left(a_{ik} \cos \frac{2\pi ks}{h} + b_{ik} \sin \frac{2\pi ks}{h} \right) \right) dw_j(s).$$

Lemma 1.5.6 *For the error Δ_{ij} in the approximate identity (1.5.35), the following relations hold:*

$$E\Delta_{ij} = 0, \ E\Delta_{ij}^2 = \frac{h^2}{12} - \frac{h^2}{2\pi^2} \sum_{k=1}^m \frac{1}{k^2}, \ i \neq j. \tag{1.5.36}$$

Proof The relation $E\Delta_{ij} = 0$ is evident. We have for $i \neq j$:

$$E\Delta_{ij}^2 = E \left(\int_0^h (w_i(t) - \frac{t}{h}w_i(h) - \frac{a_{i0}}{2} \right.$$

$$\left. - \sum_{k=1}^m \left(a_{ik} \cos \frac{2\pi kt}{h} + b_{ik} \sin \frac{2\pi kt}{h} \right) \right) dw_j(t))^2$$

$$= E \int_0^h \left(w_i(t) - \frac{t}{h}w_i(h) - \frac{a_{i0}}{2} - \sum_{k=1}^m \left(a_{ik} \cos \frac{2\pi kt}{h} + b_{ik} \sin \frac{2\pi kt}{h} \right) \right)^2 dt.$$

Since the sum

$$\frac{a_{i0}}{2} + \sum_{k=1}^m \left(a_{ik} \cos \frac{2\pi kt}{h} + b_{ik} \sin \frac{2\pi kt}{h} \right)$$

is a part of the Fourier series of the function $w_i(t) - \frac{t}{h}w_i(h)$, we have

$$\int_0^h \left(w_i(t) - \frac{t}{h}w_i(h) - \frac{a_{i0}}{2} - \sum_{k=1}^m \left(a_{ik} \cos \frac{2\pi kt}{h} + b_{ik} \sin \frac{2\pi kt}{h} \right) \right)^2 dt$$

$$= \int_0^h \left(w_i(t) - \frac{t}{h}w_i(h) \right)^2 dt - \frac{1}{4}ha_{i0}^2 - \sum_{k=1}^m \frac{h}{2}(a_{ik}^2 + b_{ik}^2).$$

Then using Lemma 1.5.5, we find

$$E\Delta_{ij}^2 = \int_0^h E\left(w_i(t) - \frac{t}{h}w_i(h)\right)^2 dt - \frac{h}{4}Ea_{i0}^2 - \sum_{k=1}^m \frac{h}{2}E(a_{ik}^2 + b_{ik}^2)$$

$$= \int_0^h \left(t - \frac{2t^2}{h} + \frac{t^2}{h}\right) dt - \frac{1}{12}h^2 - \sum_{k=1}^m \frac{h}{2}\cdot\frac{h}{k^2\pi^2}.$$

This implies the second relation of (1.5.36). $\square$

Lemma 1.5.7 *The following formula is valid:*

$$\bar{I}_{ij} = \int_0^h \left(\frac{t}{h}w_i(h) + \frac{a_{i0}}{2} + \sum_{k=1}^m \left(a_{ik}\cos\frac{2\pi kt}{h} + b_{ik}\sin\frac{2\pi kt}{h}\right)\right) dw_j(t) \quad (1.5.37)$$

$$= \frac{1}{2}w_i(h)w_j(h) + \frac{a_{i0}}{2}w_j(h) - \frac{a_{j0}}{2}w_i(h) + \pi\sum_{k=1}^m k(a_{ik}b_{jk} - a_{jk}b_{ik}).$$

Proof We have

$$\int_0^h t\,dw_j(t) = hw_j(h) - \int_0^h w_j(t)dt, \qquad\qquad (1.5.38)$$

$$\int_0^h \cos\frac{2\pi kt}{h}dw_j(t) = w_j(h) + \frac{2\pi k}{h}\int_0^h w_j(t)\sin\frac{2\pi kt}{h}dt,$$

$$\int_0^h \sin\frac{2\pi kt}{h}dw_j(t) = -\frac{2\pi k}{h}\int_0^h w_j(t)\cos\frac{2\pi kt}{h}dt.$$

Further, we get (see the formulas (1.5.29) defining the Fourier coefficients):

$$a_{j0} = \frac{2}{h}\int_0^h w_j(t)dt - w_j(h), \qquad\qquad (1.5.39)$$

$$a_{jk} = \frac{2}{h}\int_0^h w_j(t)\cos\frac{2\pi kt}{h}dt, \ k \neq 0,$$

$$b_{jk} = \frac{2}{h}\int_0^h w_j(t)\sin\frac{2\pi kt}{h}dt + \frac{w_j(h)}{k\pi}.$$

Transforming first the expression for $\bar{I}_{ij}$ by using (1.5.38) and then by using (1.5.39), we come to (1.5.37). This proves the lemma.   □

In the right-hand side of (1.5.37) the coefficient a_{i0} depends on a_{ik}, the coefficient a_{j0} depends on a_{jk}, and all the remaining coefficients are mutually independent (see Lemma 1.5.5). Introduce the new random variable $a_{i0}^{(m)}$ by

$$a_{i0}^{(m)} = -\frac{a_{i0}}{2} - \sum_{k=1}^{m} a_{ik} .$$

We show that $a_{i0}^{(m)}$ does not depend on a_{ik}, $k = 1, \ldots, m$. Indeed, by Lemma 1.5.5 we have

$$E a_{i0}^{(m)} a_{ik} = -\frac{1}{2} E a_{i0} a_{ik} - E a_{ik}^2 = 0, \ k = 1, \ldots, m .$$

Then the independence follows from the fact that all the variables under consideration are Gaussian.

We can directly compute that

$$E(a_{i0}^{(m)})^2 = \frac{h}{12} - \frac{h}{2\pi^2} \sum_{k=1}^{m} \frac{1}{k^2} . \tag{1.5.40}$$

Substituting $a_{i0} = -2a_{i0}^{(m)} - 2\sum_{k=1}^{m} a_{ik}$ in (1.5.37), we obtain

$$\bar{I}_{ij} = \frac{1}{2} w_i(h) w_j(h) + a_{j0}^{(m)} w_i(h) - a_{i0}^{(m)} w_j(h) \tag{1.5.41}$$

$$+ \sum_{k=1}^{m} (a_{jk} w_i(h) - a_{ik} w_j(h)) + \pi \sum_{k=1}^{m} k(a_{ik} b_{jk} - a_{jk} b_{ik}) .$$

In (1.5.41) all $w_i(h)$, $w_j(h)$, $a_{i0}^{(m)}$, $a_{j0}^{(m)}$, a_{ik}, a_{jk}, b_{ik}, b_{jk} are independent Gaussian random variables.

We gather the results concerning the approximate modeling of the variables $w_i(h)$ and $I_{ij} = \int_0^h w_i(s) dw_j(s)$ obtained above in the following theorem.

Theorem 1.5.8 *Making up $\bar{I}_{ij}$, $i, j = 1, \ldots, q$, reduces to modeling $2(m+1)q$ independent $\mathcal{N}(0, 1)$-distributed random variables ξ_i, ξ_{ik}, $k = 0, \ldots, m$, η_{ik}, $k = 1, \ldots, m$. Here*

$$\xi_i = h^{-1/2} w_i(h), \ \xi_{i0} = \left(\frac{h}{12} - \frac{h}{2\pi^2} \sum_{k=1}^{m} \frac{1}{k^2} \right)^{-1/2} a_{i0}^{(m)},$$

$$\xi_{ik} = \sqrt{2}\pi k h^{-1/2} a_{ik}, \ \eta_{ik} = \sqrt{2}\pi k h^{-1/2} b_{ik} .$$

The quantities $w_i(h)$ and $I_{ij}(h)$ can be expressed in terms of these variables as

$$w_i(h) = h^{1/2}\xi_i, \quad I_{ii} = \bar{I}_{ii} = \frac{h}{2}(\xi_i^2 - 1), \quad i = 1, \ldots, q, \tag{1.5.42}$$

$$I_{ij} \doteq \bar{I}_{ij} = \frac{h}{2}\xi_i\xi_j + h\left(\frac{1}{12} - \frac{1}{2\pi^2}\sum_{k=1}^{m}\frac{1}{k^2}\right)^{1/2}(\xi_{j0}\xi_i - \xi_{i0}\xi_j)$$

$$+ \frac{h}{\pi\sqrt{2}}\sum_{k=1}^{m}\frac{1}{k}(\xi_{jk}\xi_i - \xi_{ik}\xi_j) + \frac{h}{2\pi}\sum_{k=1}^{m}\frac{1}{k}(\xi_{ik}\eta_{jk} - \xi_{jk}\eta_{ik}), \quad i \neq j.$$

The error $I_{ij} - \bar{I}_{ij}$ of the approximate modeling is characterized by the relations

$$E\Delta_{ij} = 0, \quad E\Delta_{ij}^2 = \frac{h^2}{12} - \frac{h^2}{2\pi^2}\sum_{k=1}^{m}\frac{1}{k^2}, \quad i \neq j. \tag{1.5.43}$$

If $m \approx 1/h$, then the following method (which is constructive from the point of view of modeling random variables):

$$X_{n+1} = X_n + \sum_{j=1}^{q}(\sigma_j)_n\xi_j^{(n)}h^{1/2} + a_nh + \sum_{i=1}^{q}\sum_{j=1}^{q}(\Lambda_i\sigma_j)_n\bar{I}_{ij}^{(n)}, \tag{1.5.44}$$

where the index n indicates that the random variables are modeled according to (1.5.42) independently at each step, is a method of the first order of accuracy for integrating (1.1.1).

Proof We only need to prove the last assertion, since all the previous ones follow from Lemmas 1.5.5 – 1.5.7. However, its proof does not differ at all from the proof of Theorem 1.5.4 if we take into account that (since $\sum_{k=1}^{\infty} 1/k^2 = \pi^2/6$)

$$\frac{1}{12} - \frac{1}{2\pi^2}\sum_{k=1}^{m}\frac{1}{k^2} = \frac{1}{2\pi^2}\sum_{k=m+1}^{\infty}\frac{1}{k^2} \leq \frac{1}{2\pi^2}\int_{m}^{\infty}\frac{dx}{x^2} = \frac{1}{2\pi^2 m}.$$

The theorem is proved. □

We will now compare results of modeling by the rectangle method, the trapezoidal method, and the Fourier method. It is clear from (1.5.24) and (1.5.27) that to achieve the same accuracy, the rectangle method requires twice as many independent $\mathcal{N}(0, 1)$-distributed random variables as does the trapezoidal method. We compare the Fourier method and the trapezoidal method from this point of view. To this end we put $l = 2(m + 1)$ in (1.5.27). Then for identical costs as regards forming random variables (since in the trapezoidal method we have to model $ql = 2(m + 1)q$ independent $\mathcal{N}(0, 1)$-distributed random variables in order to approximately form

Table 1.5.1 Coefficients of the error in the trapezoidal and the Fourier method

m	1	2	3	4	5	10	20
$\dfrac{1}{8(m+1)}$	0.0625	0.0417	0.0312	0.0250	0.0208	0.0114	0.0060
$\dfrac{1}{12} - \dfrac{1}{2\pi^2}\displaystyle\sum_{k=1}^{m}\dfrac{1}{k^2}$	0.0327	0.0200	0.0144	0.0112	0.0092	0.0048	0.0025

all the necessary Ito integrals by (1.5.28)) the error of the trapezoidal method can be computed by (1.5.27) and that of the Fourier method by (1.5.43). Table 1.5.1 gives some values of the coefficients at h^2 in the error $E\Delta_{ij}^2$: the second row of the table corresponds to the coefficients in the trapezoidal method; the third row corresponds to those in the Fourier method. For large m the Fourier method is 2.5 times more economical than the trapezoidal method, since the error of the Fourier method is close to $1/(20m)$ while that of the trapezoidal method is close to $1/(8m)$.

Remark 1.5.9 A rather large literature is devoted to theoretical and practical points of modeling Ito integrals (along with [202, 274] see e.g., [116, 204, 221, 463] and references therein).

We recall [274] that the integrals $I_{i_1,\dots,i_j}(h)$ (see (1.2.11)) needed for a method can be simulated approximately by numerical integration of a system of SDEs. Indeed, consider all Ito integrals of orders $m - 1/2$ and m. If $I_{i_1,\dots,i_j}(\theta)$ is an Ito integral of order $m - 1/2$, then $I_{i_1,\dots,i_k,i_{k+1}}$, where $i_{k+1} = 0$, is an Ito integral of order $m + 1/2$, and it satisfies the equation

$$dI_{i_1,\dots,i_k,i_{k+1}} = I_{i_1,\dots,i_k}\,d\theta\,. \tag{1.5.45}$$

If $I_{i_1,\dots,i_l}$ is an Ito integral of order m, then $I_{i_1,\dots,i_l,i_{l+1}}$, $i_{l+1} = 1, \dots, q$, satisfies the equation

$$dI_{i_1,\dots,i_l,i_{l+1}} = I_{i_1,\dots,i_l}\,dw_{i_{l+1}}(\theta)\,. \tag{1.5.46}$$

As a result, we have accounted for all Ito integrals of order $m + 1/2$. Adjoining the equations of the type (1.5.45) and (1.5.46) to the system of equations for the Ito integrals up to order m, we obtain a system of equations for the Ito integrals up to order $m + 1/2$ inclusively. It can be readily seen that this system is a linear autonomous system of SDEs. The initial data for each variable is zero. It is clear that the dimension of this system can be substantially reduced because of relations between the integrals.

1.6 Explicit and Implicit Methods of Order 3/2 for Systems with Additive Noise

From the point of view of numerical integration, the distinguishing feature of systems with additive noise is the absence of random variables of the form I_{i_1,i_2} and I_{i_1,i_2,i_3}, $i_1, i_2, i_3 \neq 0$, in the Taylor-type expansions, and therefore we are able to construct various constructive (with respect to modeling of random variables) methods with order of accuracy reaching $3/2$.

1.6.1 Explicit Methods Based on Taylor-Type Expansion

Consider the system of stochastic differential equations with additive noise

$$dX = a(t, X)dt + \sum_{r=1}^{q} \sigma_r(t)dw_r(t).\tag{1.6.1}$$

We use the formulas (1.2.21) and (1.2.22). We have

$$\Lambda_i \sigma_r = 0, \ L\sigma_r = \frac{d\sigma_r}{dt}(t), \ \Lambda_r a = \left(\sigma_r, \frac{\partial}{\partial x}\right)a(t, x), \ \Lambda_s \Lambda_i \sigma_r = 0,$$

$$La = \frac{\partial a}{\partial t}(t, x) + \left(a, \frac{\partial}{\partial x}\right)a(t, x) + \frac{1}{2}\sum_{r=1}^{q}\sum_{i=1}^{d}\sum_{j=1}^{d}\sigma_r^i\sigma_r^j\frac{\partial^2 a}{\partial x^i \partial x^j}(t, x).$$

By (1.2.21), we can write down the following numerical integration formula

$$X_{k+1} = X_k + \sum_{r=1}^{q}\sigma_{r_k}\Delta_k w_r(h) + a_k h\tag{1.6.2}$$

$$+ \sum_{r=1}^{q}(\Lambda_r a)_k \int_{t_k}^{t_{k+1}} (w_r(\theta) - w_r(t_k))d\theta$$

$$+ \sum_{r=1}^{q}\sigma_{r_k}'\int_{t_k}^{t_{k+1}} (\theta - t_k)dw_r(\theta) + (La)_k\frac{h^2}{2}.$$

Suppose the functions $a(t, x)$ and $\sigma_r(t)$ satisfy for $t_0 \leq t \leq T$, $x \in \mathbf{R}^d$ the following conditions:

(a) the function a and all its first- and second-order partial derivatives as well as the partial derivatives $\partial^3 a/\partial t\partial x^i\partial x^j$, $\partial^3 a/\partial x^i\partial x^j\partial x^k$, and $\partial^4 a/\partial x^i\partial x^j\partial x^k\partial x^l$ are continuous;

(b) the functions $\sigma_r(t)$ are twice continuously differentiable;

(c) the first-order partial derivatives of a with respect to x are uniformly bounded (so that the global Lipschitz condition is satisfied), while its remaining partial derivatives listed above, regarded as functions of x, grow at most as a linear function of $|x|$ as $|x| \to \infty$.

Then Theorem 1.2.4 holds (see (1.2.22) and Remark 1.2.6). Thus, the method (1.6.2) has order of accuracy equal to $3/2$.

In (1.6.2) we have the following random variables:

$$\Delta_k w_r(h), \quad \int_{t_k}^{t_{k+1}} (w_r(\theta) - w_r(t_k))d\theta, \quad \int_{t_k}^{t_{k+1}} (\theta - t_k)dw_r(\theta) .$$

Since

$$\int_{t_k}^{t_{k+1}} (\theta - t_k)dw_r(\theta) = h\Delta_k w_r(h) - \int_{t_k}^{t_{k+1}} (w_r(\theta) - w_r(t_k))d\theta , \tag{1.6.3}$$

to use (1.6.2) at the $(k+1)$-st step it suffices to model the random variables $\Delta_k w_r(h)$ and $\int_{t_k}^{t_{k+1}} (w_r(\theta) - w_r(t_k))d\theta$, $r = 1, \ldots, q$. Each of them has a Gaussian distribution. We can directly compute that

$$E\left[\Delta_k w_r(h) \left(\int_{t_k}^{t_{k+1}} (w_r(\theta) - w_r(t_k)) \, d\theta - \frac{1}{2}h\Delta_k w_r(h)) \right) \right] = 0, \tag{1.6.4}$$

$$E\left[\int_{t_k}^{t_{k+1}} (w_r(\theta) - w_r(t_k))d\theta - \frac{1}{2}h\Delta_k w_r(h)) \right]^2 = \frac{1}{12}h^3 . \tag{1.6.5}$$

Introduce the following independent $\mathcal{N}(0, 1)$-distributed random variables ξ_{rk} and η_{rk} :

$$\xi_{rk} = h^{-1/2}\Delta_k w_r(h), \tag{1.6.6}$$

$$\eta_{rk} = \sqrt{12}h^{-3/2} \left(\int_{t_k}^{t_{k+1}} (w_r(\theta) - w_r(t_k))d\theta - \frac{1}{2}h\Delta_k w_r(h) \right) .$$

Using these random variables, we obtain

$$\Delta_k w_r(h) = h^{1/2}\xi_{rk},\tag{1.6.7}$$

$$\int_{t_k}^{t_{k+1}} (w_r(\theta) - w_r(t_k))d\theta = h^{3/2}\left(\frac{1}{2}\xi_{rk} + \frac{1}{\sqrt{12}}\eta_{rk}\right).$$

As a result, the formula (1.6.2) takes the following form:

$$X_{k+1} = X_k + \sum_{r=1}^{q} \sigma_r(t_k)\xi_{rk}h^{1/2} + a(t_k, X_k)h \tag{1.6.8}$$

$$+ \sum_{r=1}^{q} \Lambda_r a(t_k, X_k)\left(\frac{1}{2}\xi_{rk} + \frac{1}{\sqrt{12}}\eta_{rk}\right)h^{3/2}$$

$$+ \sum_{r=1}^{q} \frac{d\sigma_r}{dt}(t_k)\left(\frac{1}{2}\xi_{rk} - \frac{1}{\sqrt{12}}\eta_{rk}\right)h^{3/2} + La(t_k, X_k)\frac{h^2}{2}.$$

We state the result in the next theorem. Here and below, as a rule, we will not precisely indicate conditions on the coefficients a and σ_r (see, for instance, conditions (a)–(c) above), and we will be satisfied by saying that these coefficients satisfy appropriate smoothness and boundedness conditions.

Theorem 1.6.1 *Suppose the coefficients $a(t, x)$ and $\sigma_r(t)$ of (1.6.1) satisfy appropriate smoothness and boundedness conditions. Then the method (1.6.8) has mean-square order of accuracy $3/2$.*

Remark 1.6.2 In (1.6.2) and (1.6.8) the terms containing $(d\sigma_r/dt)(t_k)$ appear because of the approximate representation of the integral $\int_{t_k}^{t_{k+1}} \sigma_r(\theta)dw_r(\theta)$ in the form

$$\int_{t_k}^{t_{k+1}} \sigma_r(\theta)dw_r(\theta) = \sigma_r(t_k)\Delta_k w_r(h) + \frac{d\sigma_r}{dt}(t_k)\int_{t_k}^{t_{k+1}}(\theta - t_k)dw_r(\theta) + \rho, \tag{1.6.9}$$

where ρ satisfies the relations $E\rho = 0$, $E\rho^2 = O(h^5)$.

The random variables $\int_{t_k}^{t_{k+1}} \sigma_r(\theta)dw_r(\theta)$, $r = 1, \ldots, q$, have a Gaussian distribution. If we model the integrals $\int_{t_k}^{t_{k+1}} \sigma_r(\theta)dw_r(\theta)$ exactly, then we can avoid the computation of $(d\sigma_r/dt)(t_k)$ and drop the requirement on smoothness of $\sigma_r(t)$ (see details in [274, pp. 40–41]).

1.6.2 Implicit Methods Based on Taylor-Type Expansion

First, we note that in the case of system (1.6.1) the formulas (1.2.21) and (1.2.22) take the simpler form:

$$X(t+h) = x + \sum_{r=1}^{q} \int_{t}^{t+h} \sigma_r(\theta) dw_r(\theta) + ah \qquad (1.6.10)$$

$$+ \sum_{r=1}^{q} \Lambda_r a \int_{t}^{t+h} (w_r(\theta) - w_r(t)) d\theta + La\frac{h^2}{2} + \rho,$$

$$\rho = \sum_{r=1}^{q} \sum_{i=1}^{q} \int_{t}^{t+h} \left(\int_{t}^{\theta} \left(\int_{t}^{\theta_1} \Lambda_i \Lambda_r a(\theta_2, X(\theta_2)) dw_i(\theta_2) \right) dw_r(\theta_1) \right) d\theta \qquad (1.6.11)$$

$$+ \sum_{r=1}^{q} \int_{t}^{t+h} \left(\int_{t}^{\theta} \left(\int_{t}^{\theta_1} L\Lambda_r a(\theta_2, X(\theta_2)) d\theta_2 \right) dw_r(\theta_1) \right) d\theta$$

$$+ \sum_{r=1}^{q} \int_{t}^{t+h} \left(\int_{t}^{\theta} \left(\int_{t}^{\theta_1} \Lambda_r La(\theta_2, X(\theta_2)) dw_r(\theta_2) \right) d\theta_1 \right) d\theta$$

$$+ \int_{t}^{t+h} \left(\int_{t}^{\theta} \left(\int_{t}^{\theta_1} L^2 a(\theta_2, X(\theta_2)) d\theta_2 \right) d\theta_1 \right) d\theta.$$

Following the recipe of Sect. 1.3.1, we write the term a in (1.6.10) as the sum $\alpha a + (1 - \alpha)a$. In the second term of this sum we replace a by

$$a(t, x) = a(t+h, X(t+h)) \qquad (1.6.12)$$

$$- \sum_{r=1}^{q} \int_{t}^{t+h} \Lambda_r a(\theta, X(\theta)) dw_r(\theta) - \int_{t}^{t+h} La(\theta, X(\theta)) d\theta$$

$$= a(t+h, X(t+h)) - \sum_{r=1}^{q} \Lambda_r a \int_{t}^{t+h} dw_r(\theta) - La \cdot h + \rho_1,$$

where

$$\rho_1 = -\sum_{r=1}^{q}\sum_{i=1}^{q}\int_{t}^{t+h}\left(\int_{t}^{\theta}\Lambda_i\Lambda_r a(\theta_1, X(\theta_1))\,dw_i(\theta_1)\right)dw_r(\theta) \tag{1.6.13}$$

$$-\sum_{r=1}^{q}\int_{t}^{t+h}\left(\int_{t}^{\theta}L\Lambda_r a(\theta_1, X(\theta_1))d\theta_1\right)dw_r(\theta)$$

$$-\sum_{r=1}^{q}\int_{t}^{t+h}\left(\int_{t}^{\theta}\Lambda_r La(\theta_1, X(\theta_1))\,dw_r(\theta_1)\right)d\theta$$

$$-\int_{t}^{t+h}\left(\int_{t}^{\theta}L^2 a(\theta_1, X(\theta_1))d\theta_1\right)d\theta\,.$$

Substitute (1.6.12) in (1.6.10). The relation obtained will contain a term $(2\alpha - 1)La \cdot h^2/2$. Again we write La as $\beta La + (1-\beta)La$ and replace La in the second term by

$$La(t, x) = La(t + h, X(t + h)) + \rho_2\,, \tag{1.6.14}$$

where

$$\rho_2 = -\int_{t}^{t+h}L^2 a(\theta, X(\theta))d\theta - \sum_{r=1}^{q}\int_{t}^{t+h}\Lambda_r La(\theta, X(\theta))dw_r(\theta)\,. \tag{1.6.15}$$

Combining all these expressions, we obtain

$$X(t + h) = x + \sum_{r=1}^{q}\int_{t}^{t+h}\sigma_r(\theta)dw_r(\theta) \tag{1.6.16}$$

$$+ \alpha a h + (1 - \alpha)a(t + h, X(t + h))h$$

$$- (1 - \alpha)h\sum_{r=1}^{q}\Lambda_r a\int_{t}^{t+h}dw_r(\theta) + \sum_{r=1}^{q}\Lambda_r a\int_{t}^{t+h}(w_r(\theta) - w_r(t))d\theta$$

$$+ \beta(2\alpha - 1)La\frac{h^2}{2} + (1 - \beta)(2\alpha - 1)La(t + h, X(t + h))\frac{h^2}{2}$$

$$+ \rho + (1 - \alpha)\rho_1 h + (1 - \beta)(2\alpha - 1)\rho_2\frac{h^2}{2}\,,$$

where ρ is defined by (1.6.11), ρ_1 by (1.6.13), and ρ_2 by (1.6.15).

It can be readily seen that $R = \rho + (1 - \alpha)\rho_1 h + (1 - \beta)(2\alpha - 1)\rho_2 h^2/2$ satisfies the inequalities (of course, under appropriate conditions on a; the σ_r are assumed to be continuous):

$$|ER| \le K(1+|x|^2)^{1/2}h^3, \quad (ER^2)^{1/2} \le K(1+|x|^2)^{1/2}h^2. \tag{1.6.17}$$

If we omit the term R in (1.6.16), we obtain an implicit one-step approximation whose realization requires modeling of the integrals $\int_t^{t+h} \sigma_r(\theta)dw_r(\theta)$. If $\sigma_r''(t)$ exist and are bounded, we can use the representation (1.6.9). Substituting this in (1.6.16), we obtain a new remainder, R_1, which, as can readily be seen, satisfies the same inequality as R (see above). Thus we obtain the one-step approximation

$$
\begin{aligned}
\bar{X}(t+h) = {}& x + \sum_{r=1}^{q} \sigma_r(t)(w_r(t+h) - w_r(t)) \tag{1.6.18} \\
& + \alpha a(t,x)h + (1-\alpha)a(t+h, \bar{X}(t+h))h \\
& - (1-\alpha)h \sum_{r=1}^{q} \Lambda_r a(t,x)(w_r(t+h) - w_r(t)) \\
& + \sum_{r=1}^{q} \Lambda_r a(t,x) \int_t^{t+h} (w_r(\theta) - w_r(t))d\theta \\
& + \sum_{r=1}^{q} \sigma_r'(t)\left((w_r(t+h) - w_r(t))h - \int_t^{t+h} (w_r(\theta) - w_r(t))d\theta\right) \\
& + \beta(2\alpha-1)La(t,x)\frac{h^2}{2} + (1-\beta)(2\alpha-1)La(t+h, \bar{X}(t+h))\frac{h^2}{2}.
\end{aligned}
$$

The following two-parameter implicit method corresponds to the approximation (1.6.18):

$$
\begin{aligned}
X_{k+1} = {}& X_k + \sum_{r=1}^{q} \sigma_r(t_k)\xi_{rk}h^{1/2} \tag{1.6.19} \\
& + \alpha a(t_k, X_k)h + (1-\alpha)a(t_{k+1}, X_{k+1})h \\
& + \sum_{r=1}^{q} \Lambda_r a(t_k, X_k)\left(\frac{2\alpha-1}{2}\xi_{rk} + \frac{1}{\sqrt{12}}\eta_{rk}\right)h^{3/2} \\
& + \sum_{r=1}^{q} \frac{d\sigma_r}{dt}(t_k)\left(\frac{1}{2}\xi_{rk} - \frac{1}{\sqrt{12}}\eta_{rk}\right)h^{3/2} \\
& + \beta(2\alpha-1)La(t_k, X_k)\frac{h^2}{2} + (1-\beta)(2\alpha-1)La(t_{k+1}, X_{k+1})\frac{h^2}{2},
\end{aligned}
$$

where ξ_{rk} and η_{rk} are the same as in the method (1.6.8).

Theorem 1.6.3 *Suppose that the coefficients $a(t,x)$ and $\sigma_r(t)$ of (1.6.1) satisfy appropriate smoothness and boundedness conditions (in particular, a and La satisfy a global Lipschitz condition). Then the implicit one-step approximation (1.6.18)* *satisfies the conditions of Theorem 1.1.1 with $p_1 = 3$, $p_2 = 2$, and so the order of accuracy of the method (1.6.19) is equal to $3/2$. The order of accuracy of the method based on one-step approximation according to formula (1.6.16) by omitting R is also equal to $3/2$.*

Proof Denote the right-hand side of (1.6.18) by $F(\bar{X}(t + h))$ regarding all the other variables as parameters and rewrite (1.6.18) as

$$\bar{X}(t + h) = F(\bar{X}(t + h)).\qquad(1.6.20)$$

Then we can write the equation

$$X(t + h) = F(X(t + h)) + R_1,\qquad(1.6.21)$$

where R_1 satisfies inequalities of the same type as R in (1.6.17).

Evaluate the difference $X(t + h) - \bar{X}(t + h)$ using (1.6.18), (1.6.20) – (1.6.21):

$$X(t + h) - \bar{X}(t + h)\qquad(1.6.22)$$
$$= (1 - \alpha)(a(t + h, X(t + h)) - a(t + h, \bar{X}(t + h)))h$$
$$+ (1 - \beta)(2\alpha - 1)(La(t + h, X(t + h)) - La(t + h, \bar{X}(t + h)))\frac{h^2}{2} + R_1.$$

Since a and La satisfy a global Lipschitz condition, we have

$$|X(t + h) - \bar{X}(t + h)|$$
$$\leq |1 - \alpha| \cdot h \cdot K|X(t + h) - \bar{X}(t + h)|$$
$$+ |1 - \beta| \cdot |2\alpha - 1| \cdot K\frac{h^2}{2}|X(t + h) - \bar{X}(t + h)| + |R_1|.$$

Hence, we get for sufficiently small $h > 0$:

$$|X(t + h) - \bar{X}(t + h)| \leq 2|R_1|,$$

and so (since $X(t) = \bar{X}(t) = x$)

$$E|X(t + h) - \bar{X}(t + h)|^2 \leq K(1 + |x|^2)h^4.\qquad(1.6.23)$$

Further, (1.6.22) implies

$$|E(X(t + h) - \bar{X}(t + h))|$$
$$\leq |1 - \alpha| \cdot h \cdot KE|X(t + h) - \bar{X}(t + h)|$$
$$+ |1 - \beta| \cdot |2\alpha - 1| \cdot K\frac{h^2}{2}E|X(t + h) - \bar{X}(t + h)| + |ER_1|,$$

whence, by (1.6.23), we have

$$|E(X(t + h) - \bar{X}(t + h))| \le K(1 + |x|^2)^{1/2}h^3 . \tag{1.6.24}$$

The inequalities (1.6.23), (1.6.24) and Theorem 1.1.1 imply that the method (1.6.19) has order of accuracy equal to 3/2. The second part of the theorem can be proved in a similar manner. $\qquad\square$

Example 1.6.4 For $\alpha = 1/2$ the method (1.6.19) becomes

$$X_{k+1} = X_k + \sum_{r=1}^{q} \sigma_r(t_k)\xi_{rk}h^{1/2} + \frac{a(t_k, X_k) + a(t_{k+1}, X_{k+1})}{2}h \tag{1.6.25}$$

$$+ \sum_{r=1}^{q} \Lambda_r a(t_k, X_k)\frac{1}{\sqrt{12}}\eta_{rk}h^{3/2} + \sum_{r=1}^{q} \frac{d\sigma_r}{dt}(t_k)(\frac{1}{2}\xi_{rk} - \frac{1}{\sqrt{12}}\eta_{rk})h^{3/2} .$$

In certain respects this method is even simpler than the explicit method (1.6.8) (it does not contain La). Moreover, as can be seen from the proof of the theorem, in this case it is not necessary to require La to satisfy a Lipschitz condition.

1.6.3 Stiff Systems of Stochastic Differential Equations with Additive Noise. A-Stability

As is well known, one often meets deterministic systems for which the application of explicit methods (e.g., Runge–Kutta methods of various orders of accuracy, Adams methods) requires the use of a very small step h on the whole interval of integration. In such a situation, on a relatively small interval of rapid change of the solution the choice of h is dictated by interpolation conditions, while on a substantially larger part of the interval there is no objective necessity for choosing a very small step, and this small step arises only as a consequence of some instability properties of the method itself. We clarify this by an example. Consider the two-dimensional system of linear differential equations with constant coefficients

$$\frac{dX}{dt} = AX, \; X(0) = X_0 , \tag{1.6.26}$$

where $\lambda_1 \ll \lambda_2 < 0$ are the eigenvalues of the matrix A with eigenvectors a_1 and a_2 (of course, in simulations λ_1, λ_2, a_1, a_2 are not known).

We apply Euler's method to the problem (1.6.26):

$$X_{k+1} = X_k + AX_kh . \tag{1.6.27}$$

Let $X_0 = \alpha^1 a_1 + \alpha^2 a_2$. Then the solution of (1.6.26) has the form

$$X(t) = \alpha^1 e^{\lambda_1 t} a_1 + \alpha^2 e^{\lambda_2 t} a_2 \,. \tag{1.6.28}$$

Since the first component of the solution decreases rapidly because of the condition $\lambda_1 \ll \lambda_2 < 0$, on an initial interval of small length $\sim 1/|\lambda_1|$, we have to choose a very small step in (1.6.27). On the rest of the interval (whose length we can compare in a natural way with $1/|\lambda_2|$), $X(t)$ changes slowly, and here the interpolation conditions do not require us to choose a small step h. We consider in more detail the nature of the method (1.6.27). We have

$$\begin{aligned} X_1 &= (I + Ah)X_0 = (I + Ah)(\alpha^1 a_1 + \alpha^2 a_2) \\ &= (1 + \lambda_1 h)\alpha^1 a_1 + (1 + \lambda_2 h)\alpha^2 a_2 \,. \end{aligned}$$

It can be readily seen that

$$X_{k+1} = (1 + \lambda_1 h)^{k+1} \alpha^1 a_1 + (1 + \lambda_2 h)^{k+1} \alpha^2 a_2 \,. \tag{1.6.29}$$

It is suitable to use (1.6.29) if h is chosen to satisfy $|1 + \lambda_1 h| \leq 1$, i.e.,

$$h \leq \frac{2}{|\lambda_1|} \,. \tag{1.6.30}$$

Moreover, the condition (1.6.30) must be satisfied on the whole interval of length $\sim 1/|\lambda_2|$. Even when the first component in (1.6.28) has practically damped out, a step choice $h > 2/|\lambda_1|$ would, because of inevitable errors in computations, again catch it, and a sharp increase in the error would result. Thus, using the method (1.6.27), we have to choose the step very small (in accordance with (1.6.30)) on the whole interval of integration (of course, on the initial interval of length $\approx 1/|\lambda_1|$ the step must even be smaller, but this is because of natural causes and, in view of the smallness of $1/|\lambda_1|$, does not lead to any complications). The necessity of choosing the integration step small not only implies increase in computational costs, but also, more importantly, that the computational error increases. As a result, for appropriate λ_1, λ_2 this error may become so large that the method (1.6.27) becomes inapplicable for solving (1.6.26).

For $\lambda_1 \ll \lambda_2 < 0$, the system (1.6.26) belongs to the class of so-called stiff systems [156, 157, 372]. There is no unique generally accepted notion of stiffness, and different authors have proposed various definitions. Here it is better to talk about the *phenomenon of stiffness*, which is characterized, from the point of view of physics, by the presence of both fast and slow processes described by the system of differential equations. When we solve such systems by explicit numerical integration methods, there arises a mismatch between the necessity of choosing a very small integration step on the whole interval and the objective possibility of interpolating the solution on a large part of the interval with a large step (since the solution changes slowly).

Moreover, when we use explicit methods, a small increase of the integration step within definite bounds leads to an explosion of the computational error.

Consider the implicit Euler method applied to the system (1.6.26):

$$X_{k+1} = X_k + AX_{k+1}h . \tag{1.6.31}$$

We have

$$X_{k+1} = (I - Ah)^{-1}X_k = (I - Ah)^{-(k+1)}X_0 \tag{1.6.32}$$

$$= \frac{1}{(1 - \lambda_1 h)^{k+1}}\alpha^1 a_1 + \frac{1}{(1 - \lambda_2 h)^{k+1}}\alpha^2 a_2 .$$

It is clear from (1.6.32) that the method (1.6.31) does not have the property of instability, even for arbitrary large h, i.e., choosing h in (1.6.31) we need only to worry about the error of the method. It is clear that in the end the major differences in properties between the methods (1.6.27) and (1.6.31) are related to the various means of interpolating the exponents $e^{\lambda_1 t}$. Of course, implicit methods are far more laborious than explicit methods, since in general they require to solve at each step a system of nonlinear equations in X_{k+1}.

A system of *linear equations*

$$\frac{dX}{dt} = AX + b(t) , \tag{1.6.33}$$

with a constant matrix A, which eigenvalues λ_k, $k = 1, \ldots, d$, have negative real parts, is called *stiff* (see [156, 157, 372]) if the following condition holds:

$$\frac{\max_k |\operatorname{Re} \lambda_k|}{\min_k |\operatorname{Re} \lambda_k|} \gg 1 . \tag{1.6.34}$$

A system of *nonlinear equations* is said to belong to the class of *stiff* systems if in a neighborhood of each point (t, x) in the domain under consideration its system of first approximation is stiff.

The quality, with respect to some measure, of a method is conveniently judged by the action of the method on some test system having a small number of parameters and a simple form. To clarify stability properties of a method, one chooses as a test system the equation

$$\frac{dX}{dt} = \lambda X , \tag{1.6.35}$$

where λ is a complex parameter with $\operatorname{Re} \lambda < 0$. The choice of (1.6.35) is related to the fact that every homogeneous system with constant coefficients having distinct eigenvalues with negative real parts can be decomposed into equations of the form (1.6.35).

The result of applying some method to (1.6.35) is a difference equation. For example, application of the explicit Euler method leads to the difference equation

$$X_{k+1} = X_k + \lambda h X_k, \qquad (1.6.36)$$

while application of the implicit Euler method leads to

$$X_{k+1} = X_k + \lambda h X_{k+1}. \qquad (1.6.37)$$

In such difference equations λh is a parameter. It is natural to require for computational purposes that the trivial solution of such systems be stable. And we have to require stability for all λh belonging to the left halfplane of the complex λh-plane in order to ensure applicability of the method for equations with arbitrary λ (requiring only $\operatorname{Re} \lambda < 0$) as h does not tend to zero.

Definition 1.6.5 The region of stability of a method is the set of values of λh satisfying the condition of asymptotic stability of the trivial solution of the difference equation arising when the test Eq. (1.6.35) is integrated by this method. A method is called A-stable (absolutely stable) if the halfplane $\operatorname{Re} \lambda h < 0$ belongs to its region of stability.

There are other definitions of stability, but we will not be concerned with them. It is well known that no explicit Runge–Kutta or Adams method is A-stable. Hence, numerical integration of stiff systems leads to the necessity of constructing implicit methods and of investigating their stability.

It can be seen from (1.6.37) that the implicit Euler method is A-stable, while (1.6.36) implies that the region of stability of the explicit Euler method is the interior of the disk with radius one and center at the point $\lambda h = -1$ in the complex λh-plane.

We now turn to the stochastic system with additive noise (1.6.1). It is natural to say that it is stiff if its deterministic part is stiff; as a test equation we can naturally take

$$dX = \lambda X dt + \sigma dw, \qquad (1.6.38)$$

where λ is a complex parameter with $\operatorname{Re} \lambda < 0$ and σ is an arbitrary real parameter.

The method (1.6.19) applied to Eq. (1.6.38) takes the form

$$X_{k+1} = \left(1 + \alpha \lambda h + \beta (2\alpha - 1) \frac{(\lambda h)^2}{2} \right) X_k \qquad (1.6.39)$$

$$+ \left((1 - \alpha)\lambda h + (1 - \beta)(2\alpha - 1)\frac{(\lambda h)^2}{2} \right) X_{k+1} + \sigma \xi_k h^{1/2}$$

$$+ \lambda \sigma \left(\frac{2\alpha - 1}{2} \xi_k + \frac{1}{\sqrt{12}} \eta_k \right) h^{3/2}.$$

Equation (1.6.39) is a difference equation with additive noise. If for $\sigma = 0$ the trivial solution of (1.6.39) is asymptotically stable, then, in particular, for any σ any solution of (1.6.39) with $E|X_0|^2 < \infty$ has second order moments that are uniformly bounded in k. It is readily verified that in the opposite case the second order moments tend to infinity as $k \to \infty$. Therefore the properties of the method (1.6.19) can be judged from the stability properties of (1.6.39) for $\sigma = 0$. Thus, e.g., to clarify the region of stability of a stochastic method, we have to apply the method to Eq. (1.6.38) with $\sigma = 0$, i.e., to (1.6.35), and then clarify the stability properties of the difference equation obtained; the latter is clearly deterministic. In relation with this, Definition 1.6.5, concerning the region of stability and A-stability of a method, can be transferred without modifications to stochastic numerical integration methods.

Example 1.6.6 Consider the method (1.6.25). Applying it to Eq. (1.6.35), we get the difference equation

$$X_{k+1} = \left(1 + \frac{\lambda h}{2}\right) X_k + \frac{\lambda h}{2} X_{k+1},$$

i.e.,

$$X_{k+1} = \frac{1 + \lambda h/2}{1 - \lambda h/2} X_k.$$

One can see that if $\mathrm{Re}\,\lambda h < 0$ then

$$\left|\frac{1 + \lambda h/2}{1 - \lambda h/2}\right| < 1,$$

i.e., the region of stability includes the whole left halfplane of the complex λh-plane, and so the method (1.6.25) is A-stable.

Example 1.6.7 Consider the method (1.6.19) for $\alpha = \beta = 0$. It can be readily computed that the region of stability of this method is given by the inequality

$$\frac{1}{|1 - \lambda h + (\lambda h)^2/2|} < 1,$$

or, setting $\lambda h = \mu + i\nu$, by

$$\left|1 - (\mu + i\nu) + \frac{(\mu + i\nu)^2}{2}\right|^2 > 1.$$

The left-hand side of this inequality can be rewritten as follows:

$$\left(1 - \mu + \frac{\mu^2 - \nu^2}{2}\right)^2 + \nu^2(-1 + \mu)^2$$

$$= (1 - \mu)^2 + \left(\frac{\mu^2 - \nu^2}{2}\right)^2 + \mu^2(1 - \mu) + \nu^2((1 - \mu)^2 - (1 - \mu)),$$

which clearly exceeds 1 for $\mu < 0$. Hence the method under consideration is A-stable.

Remark 1.6.8 In this section the concept of A-stability was naturally transferred from deterministic ordinary differential equations to SDEs with additive noise. The question of stability of (implicit or explicit) methods in the case of systems with diffusion coefficients depending on x is far more complicated including the fact that there are various meanings of stability in the stochastic setting: almost sure, mean-square, exponential, etc. [194]. A linear autonomous stochastic system

$$dX = AX dt + \sum_{r=1}^{q} B_r X dw_r(t) \tag{1.6.40}$$

can be regarded as stiff if, first, its trivial solution is asymptotically stable, e.g. in mean-square, and, secondly, among the negative eigenvalues for the system of second-order moments for (1.6.40) there are eigenvalues with large as well as small modulus. Systems like (1.6.40) have, in a certain sense, both fast and slow processes and can be used as a test system. Stiff SDEs arise in applications similarly how stiff ODEs do, but also discretization of stochastic PDEs (SPDEs) by the method of lines typically leads to stiff SDEs.

Two directions can be considered in this context. One is developing new numerical methods with good stability properties in the case of multiplicative noise (see Sects. 1.3.2 and 1.3.5 of this chapter and [1–3, 285, 291] and also references therein). The other direction is related to study of stability properties of known numerical methods for SDEs: if a system of SDEs has a stable solution in some sense, then one would want that a numerical method applied to this system also has the same stability properties (possibly with a restriction on the time step) as the continuous system, see e.g. [48, 162, 163] and references therein.

1.6.4   Runge–Kutta Type Methods

The most complicated term in the method (1.6.8) is usually $La(t_k, X_k)$. Using the idea of recalculation, we will construct a method in which $La(t_k, X_k)$ does not occur.

Introduce $\bar{X}^{(1)}(t + h)$ by Euler's method:

$$\bar{X}^{(1)}(t + h) = x + \sum_{r=1}^{q} \sigma_r(t)\xi_r h^{1/2} + a(t, x)h. \tag{1.6.41}$$

Under the conditions of Theorem 1.6.1 we have for $\rho_1 = X_{t,x}(t+h) - \bar{X}^{(1)}(t+h)$ (recall that we are considering a system with additive noise for which Euler's method has order of accuracy one):

$$|E\rho_1| \le K(1+|x|^2)^{1/2}h^2, \ (E\rho_1^2)^{1/2} \le K(1+|x|^2)^{1/2}h^{3/2}. \tag{1.6.42}$$

Further,

$$a(t+h, X_{t,x}(t+h)) = a(t,x) + \sum_{r=1}^{q} \Lambda_r a(t,x)\xi_r h^{1/2} + La(t,x)h + \rho_2, \tag{1.6.43}$$

where

$$|E\rho_2| \le K(1+|x|^2)^{1/2}h^2, \ (E\rho_2^2)^{1/2} \le K(1+|x|^2)^{1/2}h. \tag{1.6.44}$$

Put $\rho_3 = a(t+h, X_{t,x}(t+h)) - a(t+h, \bar{X}^{(1)}(t+h))$. Since a satisfies a Lipschitz condition, by the second relation in (1.6.42) we successively have

$$(E\rho_3^2)^{1/2} \le K(1+|x|^2)^{1/2}h^{3/2}, \ |E\rho_3| \le E|\rho_3| \le (E\rho_3^2)^{1/2}. \tag{1.6.45}$$

We can write

$$La(t,x)h = a(t+h, \bar{X}^{(1)}(t+h)) - a(t,x) - \sum_{r=1}^{q} \Lambda_r a(t,x)\xi_r h^{1/2} + \rho_4, \tag{1.6.46}$$

where $\rho_4 = \rho_3 - \rho_2$.

By (1.6.44) and (1.6.45) we have

$$|E\rho_4| \le K(1+|x|^2)^{1/2}h^{3/2}, \ (E\rho_4^2)^{1/2} \le K(1+|x|^2)^{1/2}h. \tag{1.6.47}$$

Consider the one-step approximation that is obtained from (1.6.8) by replacing $La(t,x)h$ in (1.6.8) by the right-hand side of (1.6.46) without the term ρ_4 (of course, in our context we also have to replace (t_k, X_k) by (t,x), ξ_{rk} by ξ_r, and η_{rk} by η_r) :

$$\bar{X}(t+h) = x + \sum_{r=1}^{q} \sigma_r(t)\xi_r h^{1/2} + (a(t,x) + a(t+h, \bar{X}^{(1)}(t+h)))\frac{h}{2} \tag{1.6.48}$$

$$+ \sum_{r=1}^{q} \Lambda_r a(t,x)\frac{1}{\sqrt{12}}\eta_r h^{3/2} + \sum_{r=1}^{q} \sigma_r'(t)\left(\frac{1}{2}\xi_r - \frac{1}{\sqrt{12}}\eta_r\right)h^{3/2}.$$

The value $\bar{X}(t+h)$ computed by (1.6.48) differs from $\bar{X}(t+h)$ of the one-step approximation (1.6.8) by $\rho_4 h/2$. Therefore, we have for the $\bar{X}(t+h)$ in (1.6.48):

$$X(t+h) - \bar{X}(t+h) = \rho + \rho_4 \frac{h}{2},$$

where ρ satisfies the relations (see the proof of (1.6.8) and of its order of accuracy):

$$|E\rho| \leq K(1+|x|^2)^{1/2}h^3, \quad (E\rho^2)^{1/2} \leq K(1+|x|^2)^{1/2}h^2.$$

By (1.6.47) we have

$$|E(\rho + \rho_4 \frac{h}{2})| = O(h^{5/2}), \quad \left(E\left(\rho + \rho_4 \frac{h}{2} \right)^2 \right)^{1/2} = O(h^2).$$

Since $p_1 = 5/2$, $p_2 = 2$, Theorem 1.1.1 implies that the method corresponding to the one-step approximation (1.6.48) has order of accuracy $3/2$. We state the result in the next theorem.

Theorem 1.6.9 *Suppose that the coefficients $a(t,x)$ and $\sigma_r(t)$ of (1.6.1) satisfy appropriate smoothness and boundedness conditions. Then the method*

$$X_{k+1} = X_k + \sum_{r=1}^{q} \sigma_r(t_k)\xi_{rk}h^{1/2} \tag{1.6.49}$$

$$+ (a(t_k, X_k) + a(t_{k+1}, X_k + \sum_{r=1}^{q} \sigma_r(t_k)\xi_{rk}h^{1/2} + a(t_k, X_k)h))\frac{h}{2}$$

$$+ \sum_{r=1}^{q} \Lambda_r a(t_k, X_k)\frac{1}{\sqrt{12}}\eta_{rk}h^{3/2} + \sum_{r=1}^{q} \frac{d\sigma_r}{dt}(t_k)\left(\frac{1}{2}\xi_{rk} - \frac{1}{\sqrt{12}}\eta_{rk} \right)h^{3/2},$$

where ξ_{rk} and η_{rk} are the same as in the method (1.6.8), has order of accuracy $3/2$.

Remark 1.6.10 In (1.6.48), $\Lambda_r a(t,x)$ contains first-order derivatives of a with respect to x:

$$\Lambda_r a^i(t,x) = \sum_{j=1}^{d} \sigma_r^j(t)(\partial a^i/\partial x^j)(t,x).$$

Therefore, the method (1.6.49), which is based on (1.6.48), is not a fully Runge–Kutta method. In fact, getting rid of computing the derivatives is not difficult. For example,

$$\frac{\partial a^i}{\partial x^j} \approx \frac{a^i(x^1, \ldots, x^j + \Delta x^j, \ldots, x^d) - a^i(x^1, \ldots, x^j - \Delta x^j, \ldots, x^d)}{2\Delta x^j}.$$

This is an equality up to $O\left(\left(\Delta x^j\right)^2\right)$ by the assumptions made with respect to the function a^i. Then replacement of all $\partial a^i/\partial x^j$ in (1.6.49) by their difference relations preserves the order of accuracy. However, this approach requires a large amount of recalculations. In the deterministic theory Runge–Kutta methods use a minimal amount of recalculations. Here we can also compute $\Lambda_r a(t, x)$ using only a single recalculation of vector a. For this it suffices to use the identity

$$\Lambda_r a(t, x) = \frac{a(t, x + \sigma_r(t)h) - a(t, x)}{h} + O(h). \tag{1.6.50}$$

Note that $x + \sigma_r(t)h$ is the Euler approximation for the Cauchy problem

$$\frac{dY_r}{dt} = \sigma_r(t), \quad Y_r(t) = x. \tag{1.6.51}$$

As a result, we write the following Runge–Kutta method instead of (1.6.49):

$$X_{k+1}^{(1)} = X_k + \sum_{r=1}^{q} \sigma_r(t_k)\xi_{rk}h^{1/2} + a(t_k, X_k)h, \tag{1.6.52}$$

$$Y_{r,k+1}^{(1)} = X_k + \sigma_r(t_k)h,$$

$$X_{k+1} = X_k + \sum_{r=1}^{q} \sigma_r(t_k)\xi_{rk}h^{1/2} + (a(t_k, X_k) + a(t_k, X_{k+1}^{(1)}))\frac{h}{2}$$

$$+ \sum_{r=1}^{q} (a(t_k, Y_{r,k+1}^{(1)}) - a(t_k, X_k))\frac{1}{\sqrt{12}}\eta_{rk}h^{1/2}$$

$$+ \sum_{r=1}^{q} \frac{d\sigma_r}{dt}(t_k)\left(\frac{1}{2}\xi_{rk} - \frac{1}{\sqrt{12}}\eta_{rk}\right)h^{3/2}.$$

The idea of invoking other systems of differential equations along with the original system, in the spirit of (1.6.50)–(1.6.51), to economize the amount of recalculations may turn out to be also useful in substantially more general situations. However, here we restrict ourselves to the remark.

Remark 1.6.11 It is also possible to construct fully Runge–Kutta schemes by the method of undetermined coefficients demonstrated in Sect. 5.3.1.2. Using this approach, we construct the following explicit Runge–Kutta method for the system (1.6.1):

$$X_{k+1} = X_k + \sum_{r=1}^{q} \sigma_r(t_k)\xi_{rk}h^{1/2} \tag{1.6.53}$$

$$+ \frac{h}{2} a \left(t_k, X_k + \sum_{r=1}^{q} \sigma_r(t_k) \left[\left(\frac{1}{2} + \frac{1}{\sqrt{6}} \right) \xi_{rk} + \frac{1}{\sqrt{12}}\eta_{rk} \right] h^{1/2} \right)$$

$$+ \frac{h}{2} (t_{k+1}, X_k + ha(t_k, X_k)$$

$$+ \sum_{r=1}^{q} \sigma_r(t_k) \left[\left(\frac{1}{2} - \frac{1}{\sqrt{6}} \right) \xi_{rk} + \frac{1}{\sqrt{12}}\eta_{rk} \right] h^{1/2}$$

$$+ \sum_{r=1}^{q} \frac{d\sigma_r}{dt}(t_k) \left(\frac{1}{2}\xi_{rk} - \frac{1}{\sqrt{12}}\eta_{rk} \right) h^{3/2} .$$

One can prove that under appropriate assumptions on the coefficients of (1.6.1) the method (1.6.53) is of mean-square order 3/2.

The method (1.6.49) is an explicit Runge–Kutta method. We can construct implicit Runge–Kutta methods by writing the implicit versions of formulas (1.6.41) and (1.6.43) and preserve all the remaining derivations. We can also substitute the right-hand side of (1.6.46) without ρ_4 in (1.6.19) putting $\beta = 1$ in (1.6.19). Having done, for example, the latter, we obtain a one-parameter family of implicit Runge–Kutta methods:

$$X_{k+1} = X_k + \sum_{r=1}^{q} \sigma_r(t_k)\xi_{rk}h^{1/2} + a(t_k, X_k)\frac{h}{2} \tag{1.6.54}$$

$$+ (1 - \alpha)a(t_{k+1}, X_{k+1})h + \left(\alpha - \frac{1}{2} \right) a(t_{k+1}, X_{k+1}^{(1)})h$$

$$+ \sum_{r=1}^{q} \Lambda_r a(t_k, X_k) \frac{1}{\sqrt{12}}\eta_{rk}h^{3/2} + \sum_{r=1}^{q} \frac{d\sigma_r}{dt}(t_k) \left(\frac{1}{2}\xi_{rk} - \frac{1}{\sqrt{12}}\eta_{rk} \right) h^{3/2} ,$$

where $X_{k+1}^{(1)}$ is the same as in (1.6.52). Following the proofs of Theorems 1.6.3 and 1.6.9, it is not difficult to prove that the method (1.6.54) has order of accuracy 3/2 (here the additional assumption that $La(t, x)$ has to satisfy a uniform Lipschitz condition in x can be dropped).

Another implicit Runge–Kutta method of order 3/2 for (1.6.1) is constructed in Sect. 5.3.1.2 (see (5.3.11)–(5.3.14)). Other Runge–Kutta type methods can be found in [1, 388] and also references therein.

1.6.5 Two-Step Difference Methods

In (1.6.19) we put $\beta = 0$, $\alpha = \alpha_1 \neq 1/2$, and we express $La(t_{k+1}, X_{k+1})$ in terms of X_k, X_{k+1}, ξ_{rk}, and η_{rk}. Then we take $k + 1$ instead of k in (1.6.19), put $\beta = 1$, $\alpha = \alpha_2$, and replace $La(t_{k+1}, X_{k+1})$ by the expression just found. As a result, we obtain

$$
\begin{aligned}
X_{k+2} = &\ \frac{1 - 2\alpha_2}{2\alpha_1 - 1} X_k + \frac{2(\alpha_1 + \alpha_2 - 1)}{2\alpha_1 - 1} X_{k+1} \\
&+ \frac{1 - 2\alpha_2}{2\alpha_1 - 1} \sum_{r=1}^{q} \sigma_r(t_k)\xi_{rk} h^{1/2} \\
&+ \sum_{r=1}^{q} \sigma_r(t_{k+1})\xi_{r(k+1)} h^{1/2} + \frac{\alpha_1(1 - 2\alpha_2)}{2\alpha_1 - 1} a(t_k, X_k)h \\
&+ \frac{4\alpha_1\alpha_2 - \alpha_1 - 3\alpha_2 + 1}{2\alpha_1 - 1} a(t_{k+1}, X_{k+1})h + (1 - \alpha_2)a(t_{k+2}, X_{k+2})h \\
&+ \frac{1 - 2\alpha_2}{2\alpha_1 - 1} \sum_{r=1}^{q} \Lambda r a(t_k, X_k) \left(\frac{2\alpha_1 - 1}{2} \xi_{rk} + \frac{1}{\sqrt{12}} \eta_{rk} \right) h^{3/2} \\
&+ \frac{1 - 2\alpha_2}{2\alpha_1 - 1} \sum_{r=1}^{q} \frac{d\sigma_r}{dt}(t_k) \left(\frac{1}{2}\xi_{rk} - \frac{1}{\sqrt{12}} \eta_{rk} \right) h^{3/2} \\
&+ \sum_{r=1}^{q} \Lambda_r a(t_{k+1}, X_{k+1}) \left(\frac{2\alpha_2 - 1}{2} \xi_{r(k+1)} + \frac{1}{\sqrt{12}} \eta_{r(k+1)} \right) h^{3/2} \\
&+ \sum_{r=1}^{q} \frac{d\sigma_r}{dt}(t_{k+1}) \left(\frac{1}{2}\xi_{r(k+1)} - \frac{1}{\sqrt{12}} \eta_{r(k+1)} \right) h^{3/2} .
\end{aligned}
\tag{1.6.55}
$$

For $\alpha_2 = 1/2$ the method (1.6.55) coincides with the implicit one-step method (1.6.25) from Example 1.6.4 (with index k increased by one). For $\alpha_2 = 1$, $\alpha_1 \neq 1/2$ this is a one-parameter family of explicit two-step difference methods. For other α_2 and $\alpha_1 \neq 1/2$ this is a two-parameter family of implicit two-step difference methods. The order of accuracy of the method (1.6.55) is stated in Theorem 1.6.12, a proof of which is available in [274]. Note that we cannot use Theorem 1.1.1 here, since it is highly accommodated to one-step methods only.

Theorem 1.6.12 *Suppose that the coefficients $a(t, x)$ and $\sigma_r(t)$ of (1.6.1) satisfy appropriate smoothness and boundedness conditions. Suppose*

$$
0 \leq \frac{1 - 2\alpha_2}{2\alpha_1 - 1} \leq 1 .
\tag{1.6.56}
$$

Then the method (1.6.55) has order of accuracy 3/2 (of course, under the assumptions that $X_0 = X(t_0)$, $X_1 = X(t_1)$).

Remark 1.6.13 The method (1.6.55) has the same features as difference methods in the deterministic situation. We do not compute La in it, while in comparison with the Runge–Kutta method it does not require recalculations. At the same time, to use it one has to find a value X_1 that is sufficiently close to $X(t_1)$. To this end, as in the deterministic situation, X_1 has to be found beforehand by using a one-step method that integrates the system (1.6.1) on the interval $[t_0, t_0 + h]$ with a small auxiliary step.

Example 1.6.14 Consider the method (1.6.55) with $\alpha_1 = -1/2$, $\alpha_2 = 1$ (it is explicit). We investigate its A-stability. Applying it to the test Eq. (1.6.35), we get the difference equation

$$X_{k+2} = \left(\frac{1}{2} - \frac{1}{4}\lambda h \right) X_k + \left(\frac{1}{2} + \frac{7}{4}\lambda h \right) X_{k+1} .$$

It is easy to convince oneself that negative λh with sufficiently large absolute values do not belong to the region of stability. Therefore this method is not A-stable.

Consider now the method (1.6.55) with $\alpha_1 = 1$, $\alpha_2 = 0$. The corresponding difference equation has the form

$$X_{k+2} = (1 + \lambda h) X_k + \lambda h X_{k+2} .$$

Its trivial solution is asymptotically stable for all λh in the left halfplane. Therefore this method is A-stable.

See other mean-square multi-step methods in [50, 51].

1.7   Numerical Schemes for Equations with Colored Noise

The simplest approximation of real fluctuations that affect a physical system is Gaussian white noise. However, Gaussian white noise, or a Gaussian delta-correlated random process, is a stochastic process with zero correlation time and infinite variance, so it is an unreal process. Such a random process may be considered only as the first approximation of real fluctuation with a short correlation time. This shortcoming is overcome by colored noise (finite-bandwidth noise) [118, 171].

Herein we consider differential equations with exponentially correlated colored noise

$$dY = f(Y)dt + G(Y)Zdt \qquad (1.7.1)$$

$$dZ = AZdt + \sum_{r=1}^{q} b_r dw_r(t) ,$$

where Y and f are l-dimensional vectors, Z and b_r are m-dimensional vectors, A is an $m \times m$ matrix, G is an $l \times m$ matrix, and w_r are uncorrelated standard Wiener processes. In the one-dimensional case equations (1.7.1) take the form

$$dy = f(y)dt + g(y)zdt \qquad (1.7.2)$$
$$dz = -azdt + bdw,$$

where z is the well-known Ornstein-Uhlenbeck process (a is supposed to be a positive number), or exponentially correlated colored noise, with the properties

$$Ez(t) = 0, \quad Ez(t)z(s) = \frac{b^2}{2a} \exp(-a|t - s|).$$

The system (1.7.1) is simpler than the general system of SDEs by two reasons: (1) (1.7.1) is a system with additive Gaussian white noise, (2) equations (1.7.1) are linear with respect to Z. That is why comparatively simple high-order methods can be constructed for this system. It is also possible to consider nonautonomous systems with colored noise, however, here we restrict ourselves to the system (1.7.1). In the earlier works (see, e.g. [110] and references therein), efficient explicit algorithms up to the second order were obtained. For the first time various methods for the system (1.7.1) were easily derived and justified on the basis of the general theory in [298]. Moreover, efficient implicit and Runge–Kutta schemes were also presented. From the point of view of numerical integration, the special features of the system (1.7.1) consist in the absence of random variables of the form I_{i_1,i_2}, I_{i_1,i_2,i_3}, and $I_{i_1,i_2,0}$, $i_1, i_2, i_3 \neq 0$, in the Taylor-type expansions, and therefore we are able to construct various constructive (with respect to modeling of random variables) methods with order of accuracy reaching $5/2$. Some numerical tests of methods given in this section are presented in [298].

1.7.1 Explicit Schemes of Orders 2 and 5/2

For the system (1.7.1), the coefficients a, σ_r and the operators L and Λ_r take the form

$$x = \begin{bmatrix} y \\ z \end{bmatrix}, \quad a = \begin{bmatrix} f(y) + G(y)z \\ Az \end{bmatrix}, \quad \sigma_r = \begin{bmatrix} 0 \\ b_r \end{bmatrix}, \qquad (1.7.3)$$

$$L = \left(f(y) + G(y)z, \frac{\partial}{\partial y} \right) + \left(Az, \frac{\partial}{\partial z} \right) + \frac{1}{2} \sum_{r=1}^{q} \sum_{i,j=1}^{m} b_r^i b_r^j \frac{\partial^2}{\partial z^i \partial z^j},$$

$$\Lambda_r = \left(b_r, \frac{\partial}{\partial z} \right).$$

We have

$$\Lambda_r a = \begin{bmatrix} G(y)b_r \\ Ab_r \end{bmatrix}, \quad La = \begin{bmatrix} [f'_y + (Gz)'_y](f + Gz) + GAz \\ A^2 z \end{bmatrix}, \quad (1.7.4)$$

where f'_y is the Jacobian matrix, $(Gz)'_y = [G'_{y^1}z \; G'_{y^2}z \cdots G'_{y^l}z]$ is an $l \times l$ matrix the columns of which are $G'_{y^1}z$, $G'_{y^2}z$, ..., $G'_{y^l}z$.

The method (1.6.8) in the case of the considered system (1.7.1) acquires the form

$$Y_{k+1} = Y_k + (f + Gz)_k h + \frac{1}{2}G_k \sum_{r=1}^{q} b_r \left(\xi_{rk} + \frac{1}{\sqrt{3}}\eta_{rk} \right) h^{3/2} \quad (1.7.5)$$

$$+ \frac{h^2}{2}([f'_y + (Gz)'_y]_k(f + Gz)_k + G_k AZ_k),$$

$$Z_{k+1} = Z_k + \sum_{r=1}^{q} b_r \xi_{rk} h^{1/2} + AZ_k h$$

$$+ \frac{1}{2}A \sum_{r=1}^{q} b_r \left(\xi_{rk} + \frac{1}{\sqrt{3}}\eta_{rk} \right) h^{3/2} + \frac{h^2}{2}A^2 Z_k ,$$

where, for example, $(f + Gz)_k = f(Y_k) + G(Y_k)Z_k$.

Recall that due to Theorem 1.6.1 the method (1.6.8) for the general system with additive noise is of order 3/2. It turns out that for the system (1.7.1) this method is of order 2.

Theorem 1.7.1 *Suppose the coefficients $f(y)$ and $G(y)$ of (1.7.1) satisfy appropriate smoothness and boundedness conditions. Then the method (1.7.5) has mean-square order of accuracy equal to 2.*

Proof Since $\Lambda_r a$ depends on y only (see (1.7.3)), all the $\Lambda_i \Lambda_r a$ are equal to zero. Consequently, the remainder ρ (see formula (1.6.11)) for method (1.7.5) does not contain the integrals of order 2, and we obtain

$$|E\rho| = O(h^3), \quad (E|\rho|^2)^{1/2} = O(h^{5/2}),$$

i.e., $p_1 = 3$, $p_2 = 5/2$. Now application of fundamental Theorem 1.1.1 proves the result.   □

Further expansion of the integrals of ρ in the formula (1.6.11) for the system (1.7.1) gives the following scheme

$$Y_{k+1} = \tilde{Y}_{k+1} + \sum_{r=1}^{q}[(Gb_r)'_y]_k(f + Gz)_k(I_{0r0})_k \tag{1.7.6}$$

$$+ \sum_{r=1}^{q}[f'_yGb_r + GAb_r + A_r\{(Gz)'_y(f + Gz)\}]_k(I_{r00})_k$$

$$+ \frac{h^3}{6}[L^2(f + Gz)]_k,$$

$$Z_{k+1} = \tilde{Z}_{k+1} + \sum_{r=1}^{q}A^2b_r(I_{r00})_k + \frac{h^3}{6}A^3Z_k.$$

In (1.7.6) $\tilde{Y}_{k+1}$ and $\tilde{Z}_{k+1}$ are the right-hand sides of (1.7.5), $(Gb_r)'_y = [G'_{y_1}b_r \; G'_{y_2}b_r \cdots G'_{y_l}b_r]$ is an $l \times l$ matrix the columns of which are $G'_{y_1}b_r, \; G'_{y_2}b_r, \; \ldots, \; G'_{y_l}b_r,$ I_{0r0} and I_{r00} are the known integrals (see the notation in Sect. 1.2.2). We have

$$I_{0r0} = 2J_r - hI_{r0}, \quad I_{r00} = hI_{r0} - J_r, \quad J_r = \int_0^h \theta w_r(\theta)d\theta,$$

These integrals can be simulated according to relations

$$I_{r0} = \frac{1}{2}h^{3/2}\left(\xi_r + \frac{1}{\sqrt{3}}\eta_r\right), \quad J_r = h^{5/2}\left(\frac{1}{3}\xi_r + \frac{1}{4\sqrt{3}}\eta_r + \frac{1}{12\sqrt{5}}\zeta_r\right),$$

where ξ_r, η_r, and ζ_r are independent random variables with standard Gaussian distribution $\mathcal{N}(0, 1)$ which are independently simulated at each step.

It is not difficult to prove that for this method $p_1 = 4$, $p_2 = 3$. Therefore, the following theorem holds.

Theorem 1.7.2 *Suppose the coefficients $f(y)$ and $G(y)$ of (1.7.1) satisfy appropriate smoothness and boundedness conditions. Then the method (1.7.6) has mean-square order of accuracy equal to $5/2$.*

Remark 1.7.3 As mentioned above, for a general system mean-square methods of order $1/2$ only may be obtained with easily simulated random variables. The higher-order methods need numerical solution of a special system of SDEs at each step for the simulation of the Ito integrals or some approximation of iterated Ito integrals in the case of the first-order scheme. However, for the system with colored noise (1.7.1) efficient mean-square methods up to the $5/2$ order are derived thanks to the special properties of the system (1.7.1). By the way, third-order schemes for (1.7.1) require calculation of iterated Ito integrals, and in the case of nonlinear functions f and G it is impossible to obtain an efficient third-order mean-square method with easily simulated random variables.

1.7.2 Runge–Kutta Schemes

To reduce calculations of derivatives, we propose the explicit second-order Runge–Kutta scheme

$$Y_{k+1} = Y_k + \frac{1}{2}h(f(Y_k) + G(Y_k)Z_k) + \frac{1}{2}h(f(\tilde{Y}_k) + G(\tilde{Y}_k)\tilde{Z}_k) \tag{1.7.7}$$

$$+ \sum_{r=1}^{q} G(Y_k)b_r h^{3/2}\eta_{rk}/\sqrt{12},$$

$$Z_{k+1} = Z_k + \sum_{r=1}^{q} b_r\xi_{rk}h^{1/2} + \frac{1}{2}h(Z_k + \tilde{Z}_k) + \sum_{r=1}^{q} Ab_r h^{3/2}\eta_{rk}/\sqrt{12},$$

where

$$\tilde{Y}_k = Y_k + (f(Y_k) + G(Y_k)Z_k)h, \tag{1.7.8}$$

$$\tilde{Z}_k = Z_k + \sum_{r=1}^{q} b_r\xi_{rk}h^{1/2} + AZ_k h.$$

·This algorithm has been derived by the substitution of the expansions

$$\frac{1}{2}(f(\tilde{Y}_k) + G(\tilde{Y}_k)\tilde{Z}_k) = \frac{1}{2}(f(Y_k) + G(Y_k)Z_k) + \frac{1}{2}\sum_{r=1}^{q} G(Y_k)b_r h^{1/2}\xi_{rk} \tag{1.7.9}$$

$$+ \frac{h}{2}([f'_y + (Gz)'_y]_k(f + Gz)_k + G(Y_k)AZ_k) + \rho_1,$$

$$\frac{1}{2}A\tilde{Z}_k = \frac{1}{2}AZ_k + \frac{1}{2}\sum_{r=1}^{q} Ab_r h^{1/2}\xi_{rk} + \frac{1}{2}A^2 Z_k h,$$

$$|E\rho_1| = O(h^2), \quad (E|\rho_1|^2)^{1/2} = O(h^{3/2}),$$

in the second-order scheme (1.7.5).

The Runge–Kutta scheme (1.7.7)–(1.7.8) does not include any derivatives. Thanks to the special properties of the system (1.7.1), it is a fully Runge–Kutta algorithm. The 5/2-order explicit method (1.7.6) can be simplified by the idea of attracting a subsidiary system of deterministic equations (see Remark 1.6.10). A method obtained in this way is available in [298].

1.7.3 Implicit Schemes

A family of the first-order implicit methods (implicit Euler schemes) has the form (cf. (1.3.8)):

$$Y_{k+1} = Y_k + \alpha h(f + Gz)_k + (1 - \alpha)h(f + Gz)_{k+1} \tag{1.7.10}$$

$$Z_{k+1} = Z_k + \sum_{r=1}^{q} b_r \xi_{rk} h^{1/2} + \alpha h A Z_k + (1 - \alpha)h A Z_{k+1},$$

where ξ_{rk} are independent normally distributed $\mathcal{N}(0, 1)$ random variables, and $0 \leq \alpha \leq 1$.

We present the two-parameter family of second-order implicit schemes (cf. (1.6.19)):

$$Y_{k+1} = Y_k + \alpha h(f + Gz)_k + (1 - \alpha)h(f + Gz)_{k+1} \tag{1.7.11}$$

$$+ h^{3/2} \sum_{r=1}^{q} G(Y_k) b_r ((2\alpha - 1)\xi_{rk}/2 + \eta_{rk}/\sqrt{12})$$

$$+ \beta(2\alpha - 1)h^2 [L(f + Gz)]_k/2 + (1 - \beta)(2\alpha - 1)h^2 [L(f + Gz)]_{k+1}/2,$$

$$Z_{k+1} = Z_k + \sum_{r=1}^{q} b_r \xi_{rk} h^{1/2} + \alpha h A Z_k + (1 - \alpha)h A Z_{k+1}$$

$$+ h^{3/2} \sum_{r=1}^{q} A b_r ((2\alpha - 1)\xi_{rk}/2 + \eta_{rk}/\sqrt{12})$$

$$+ \beta(2\alpha - 1)h^2 A^2 Z_k/2 + (1 - \beta)(2\alpha - 1)h^2 A^2 Z_{k+1}/2,$$

where $0 \leq \alpha, \; \beta \leq 1$. The family (1.7.11) is derived by representing the terms $(f + Gz)_k$ and AZ_k of (1.7.5) in the form

$$(f + Gz)_k = \alpha(f + Gz)_k + (1 - \alpha)(f + Gz)_{k+1} \tag{1.7.12}$$

$$- (1 - \alpha) \left(\sum_{r=1}^{q} G(Y_k) b_r h^{1/2} \xi_{rk} + h[L(f + Gz)]_k + \rho_1 \right),$$

$$AZ_k = \alpha A Z_k + (1 - \alpha) \left(A Z_{k+1} - \sum_{r=1}^{q} A b_r h^{1/2} \xi_{rk} + h A^2 Z_k + \rho_2 \right),$$

$$|E\rho_i| = O(h^2), \; (E|\rho_i|^2)^{1/2} = O(h^{3/2}), \; i = 1, 2,$$

and with the expressions

$$[L(f + Gz)]_k = \beta[L(f + Gz)]_k + (1 - \beta)[L(f + Gz)]_{k+1} + \rho_3, \tag{1.7.13}$$

$$A^2 Z_k = \beta A^2 Z_k + (1 - \beta)A^2 Z_{k+1} + \rho_4,$$

$$|E\rho_i| = O(h), \; (E|\rho_i|^2)^{1/2} = O(h^{1/2}), \; i = 3, 4.$$

If we choose $\alpha = 1/2$ in (1.7.11), we obtain the simplest scheme of the family (1.7.11), which is called the trapezoidal method (cf. (1.6.25)):

$$Y_{k+1} = Y_k + \frac{1}{2}h(f + Gz)_k + \frac{1}{2}h(f + Gz)_{k+1} \tag{1.7.14}$$

$$+ h^{3/2} \sum_{r=1}^{q} G(Y_k)b_r \eta_{rk}/\sqrt{12},$$

$$Z_{k+1} = Z_k + \sum_{r=1}^{q} b_r \xi_{rk} h^{1/2} + \frac{1}{2}hA(Z_k + Z_{k+1})$$

$$+ h^{3/2} \sum_{r=1}^{q} Ab_r \eta_{rk}/\sqrt{12}.$$

Chapter 2
Weak Approximation for Stochastic Differential Equations: Foundations

Using probabilistic representations together with Monte Carlo methods, a complex multi-dimensional problem for partial differential equations can be reduced to the Cauchy problem for a system of SDEs. This system, which contains one independent variable only, arises as a characteristic system of the considered problems for PDEs.

In its simplest form, the method of characteristics is as follows. Consider a system of d ordinary differential equations

$$dX = a(X)dt . \tag{2.0.1}$$

Let $X_x(t)$ be the solution of this system satisfying the initial condition $X_x(0) = 0$. Then we have for an arbitrary continuously differentiable function $u(x)$:

$$u(X_x(t)) - u(x) = \int_0^t (a(X_x(s)), \frac{\partial u}{\partial x}(X_x(s)))ds . \tag{2.0.2}$$

Consider the Cauchy problem for the first-order linear partial differential equation

$$(a(x), \frac{\partial u}{\partial x}) = 0 , \tag{2.0.3}$$

$$u_{|\gamma} = f(x) , \tag{2.0.4}$$

where γ is a curve in the d-dimensional space of the variable x. Let u be a solution of equation (2.0.3). Then (2.0.2) implies

$$u(x) = u(X_x(t)) . \tag{2.0.5}$$

© Springer Nature Switzerland AG 2021
G. N. Milstein and M. V. Tretyakov, *Stochastic Numerics for Mathematical Physics*,
Scientific Computation, https://doi.org/10.1007/978-3-030-82040-4_2

The formula (2.0.5) indicates the following way for solving the problem (2.0.3)–(2.0.4): starting at x, draw the trajectory $X_x(t)$ of the system (2.0.1) up to the moment τ of its intersection with γ. By (2.0.4), u is known on γ. Therefore

$$u(x) = u(X_x(\tau)) = f(X_x(\tau)). \tag{2.0.6}$$

Now we consider the system of stochastic differential equations

$$dX = a(X)dt + \sum_{r=1}^{q} \sigma_r(X)dw_r(t). \tag{2.0.7}$$

Applying Ito's formula to a sufficiently smooth function $u(x)$, we get the following analogue of (2.0.2):

$$u(X_x(\tau)) - u(x) = \int_0^\tau Lu(X_x(t))dt + \sum_{r=1}^{q} \int_0^\tau \Lambda_r u(X_x(t))dw_r(t). \tag{2.0.8}$$

In this formula, τ is a Markov moment and

$$L = (a, \frac{\partial}{\partial x}) + \frac{1}{2}\sum_{r=1}^{q}(\sigma_r, \frac{\partial}{\partial x})^2 = \sum_{i=1}^{d} a^i \frac{\partial}{\partial x^i} + \frac{1}{2}\sum_{r=1}^{q}\sum_{i,j=1}^{d} \sigma_r^i \sigma_r^j \frac{\partial^2}{\partial x^i \partial x^j},$$

$$\Lambda_r = (\sigma_r, \frac{\partial}{\partial x}) = \sum_{i=1}^{d} \sigma_r^i \frac{\partial}{\partial x^i},$$

where a^i, σ_r^i are the components of the vectors a, σ_r.

For an elliptic-type equation

$$Lu = 0, \tag{2.0.9}$$

we consider the Dirichlet problem in a domain G with boundary condition

$$u_{|\partial G} = f(x). \tag{2.0.10}$$

Let u be a solution of (2.0.9). Then (2.0.8) implies

$$u(x) = u(X_x(\tau)) - \sum_{r=1}^{q} \int_0^\tau \Lambda_r u(X_x(t))dw_r(t). \tag{2.0.11}$$

Taking τ as the time at which the trajectory $X_x(t)$ hits the boundary ∂G and averaging (2.0.11), we arrive at a probabilistic representation of the solution of (2.0.9)–(2.0.10):

$$u(x) = Eu(X_x(\tau)) = Ef(X_x(\tau)). \tag{2.0.12}$$

Using the Monte Carlo approach, we obtain

$$u(x) \simeq \frac{1}{M} \sum_{m=1}^{M} f(X_x^{(m)}(\tau_x^{(m)})), \qquad (2.0.13)$$

where $X_x^{(m)}(t)$, $m = 1, \ldots, M$, are independent realizations of the process $X_x(t)$ defined by the system (2.0.7).

Thus, the multi-dimensional boundary value problem (2.0.9)–(2.0.10) reduces to the Cauchy problem for the system (2.0.7). This system can be naturally regarded as one-dimensional, since it contains one independent variable only. The system (2.0.7) comes about as characteristic system of differential equations for the problem (2.0.9)–(2.0.10). This approach, which enables reduction of a multi-dimensional boundary value problem to a one-dimensional Cauchy problem, is of considerable importance for computational mathematics.

Let us give the well-known probabilistic representation to the solution of the Cauchy problem for the heat equation

$$\frac{\partial u}{\partial t} + \sum_{i=1}^{d} a^i(t, x) \frac{\partial u}{\partial x^i} + \frac{1}{2} \sum_{r=1}^{q} \sum_{i,j=1}^{d} \sigma_r^i(t, x) \sigma_r^j(t, x) \frac{\partial^2 u}{\partial x^i \partial x^j} = 0, \qquad (2.0.14)$$

$$u(T, x) = f(x), \qquad (2.0.15)$$

where $t_0 \leq t \leq T$, $x \in \mathbf{R}^d$.

The value of the unknown function u at a point (s, x) can be expressed as an expectation:

$$u(s, x) = Ef(X_{s,x}(T)), \qquad (2.0.16)$$

where $X_{s,x}(t)$ is the solution of the following system of SDEs (which is not autonomous in distinction to (2.0.7)):

$$dX = a(t, X)dt + \sum_{r=1}^{q} \sigma_r(t, X)dw_r(t), \qquad (2.0.17)$$

$$X_{s,x}(s) = x, \ s \leq t \leq T.$$

We note that the representations (2.0.12) and (2.0.16) are the well-known *Feynman–Kac formula*.

Application of the Monte Carlo technique gives

$$u(s, x) = Ef(X_{s,x}(T)) \simeq \frac{1}{M} \sum_{m=1}^{M} f(X_{s,x}^{(m)}(T)), \qquad (2.0.18)$$

where $X_{s,x}^{(m)}(T)$, $m = 1, \ldots, M$, are independent realizations of the random variable $X_{s,x}(T)$.

To be able to use (2.0.18) (see also (2.0.13)), we have to model the random variable $X_{s,x}(T)$. The exact computation of $X_{s,x}(T)$ is impossible by and large even in the deterministic situation. Therefore, we have to replace $X_{s,x}(T)$ by a nearly random variable $\bar{X}_{s,x}(T)$ that can be modeled. Instead of (2.0.18) we obtain

$$u(s, x) = Ef(X_{s,x}(T)) \simeq Ef(\bar{X}_{s,x}(T)) \simeq \frac{1}{M} \sum_{m=1}^{M} f(\bar{X}_{s,x}^{(m)}(T)) . \qquad (2.0.19)$$

The first approximate equality in (2.0.19) involves an error due to replacing X by $\bar{X}$ (an error is related to the approximate integration of the system (2.0.17)); in the second approximate equality the error comes from the Monte Carlo technique.

While modeling the solution of a system of SDEs is a prerequisite for using the Monte Carlo technique, it is not necessary at all to solve the very complicated problem of constructing mean-square approximations. Let $X(t)$ be the exact and $\bar{X}(t)$ be an approximate solution. In many problems of mathematical physics it is only required that the expectation $Ef(\bar{X}(T))$ is close to $Ef(X(T))$ for a sufficiently large class of functions f, i.e., that $\bar{X}(t)$ is close to $X(t)$ in a weak sense. If an approximation $\bar{X}$ is such that

$$|Ef(\bar{X}(T)) - Ef(X(T))| \leq Ch^p \qquad (2.0.20)$$

for f from a sufficiently large class of functions, then we say that the *weak order of accuracy* of the approximation $\bar{X}$ (the method $\bar{X}$) is p. We can prove, for example, that the weak order of accuracy of Euler's method is 1. Note that numerical integration in the mean-square sense with some order of accuracy guarantees an approximation in the weak sense with the same order of accuracy, since if $(E|\bar{X}(t) - X(t)|^2)^{1/2} = O(h^p)$ then for every function f satisfying a Lipschitz condition we have $E(f(\bar{X}(T) - f(X(T))) = O(h^p)$. Moreover, an increase in the order of accuracy in the mean-square sense does not, in general, imply an increase of the weak order of accuracy. For example, the method (1.0.5) has first weak order of accuracy as Euler's method does. At the same time, a crude method like (we give the formula for a scalar equation):

$$X_{k+1} = X_k + a_k h + \sigma_k \alpha_k h^{1/2} , \qquad (2.0.21)$$

where α_k, $k = 0, \ldots, N - 1$, are independent random variables taking the values $+1$ and -1 with probabilities $1/2$, also has first order of accuracy in the sense of weak approximation. The main interest in weak approximations lies in the hope to obtain simpler methods and, in particular, methods not requiring modeling of complicated random variables. We recall that, e.g., the mean-square method (1.0.5), which is of the first order only, requires to solve the difficult problem of modeling complicated random variables of the type $\int_0^h w_i(\theta) dw_j(\theta)$. These problems of modeling complicated random variables can be avoided by integrating in the weak sense, which gives an impetus for the development of methods for constructing weak approximations. In addition we note that while in the deterministic theory the one-dimensional case differs but little from the multi-dimensional one, for the numerical integration of stochastic differential equations the multi-dimensional case, especially when several noises are involved, is substantially more complicated than the one-dimensional case. Weak approximations were introduced for the first time in [270] (see also [272, 351, 428]). The fact that weak approximations suffice for the equations of mathematical physics shows that precisely this type of approximations are of the most interest in applications and should thus be at the center of investigations in numerical integration of SDEs. At the same time, it must be stressed that the construction of weak approximations uses the general theory of mean-square approximations in an essential way.

The weak approximation of SDEs for solving PDEs has numerous applications. The Monte Carlo simulation of option prices and its derivatives is a typical instance of such an application. Many works in financial mathematics are devoted to this approach (see among them [41, 44, 46, 82, 108, 109, 129, 223, 288, 293, 315, 382, 444]). SDEs are also extensively used in molecular dynamics (see [69–71, 181, 230, 265, 404] and references therein) for which weak approximation is again playing the crucial role (see Chap. 5 as well).

In this chapter we construct various methods of second order of accuracy in the weak sense for general systems of stochastic differential equations. These methods use random variables that are simple to model. They are also simpler than mean-square methods from another point of view: for the same mean-square and weak orders of accuracy the weak methods require calculating substantially fewer operators of coefficients of the considered system.

In Sect. 2.1 we give a detailed construction of a one-step approximation of third order of accuracy. It is the basis for constructing methods of second order of accuracy for stochastic systems of the general type.

In Sect. 2.2 we prove a theorem stating that if a one-step approximation has ($p + 1$)-th order of accuracy then the corresponding approximation on a finite time interval has p-th order of accuracy. This theorem plays the same role in the theory of weak approximation as the main convergence theorem of Sect. 1.1 does in the theory of mean-square approximation. Some Runge–Kutta type and implicit methods are constructed using this theorem. The theorem relies on boundedness of moments of approximations which can be violated when the SDEs' coefficients are not globally Lipschitz. In Sect. 2.2.5 we consider the concept of rejecting exploding trajectories, which allows us to apply any method of weak approximation to a very broad class of SDEs with nonglobally Lipschitz coefficients. The Talay–Tubaro extrapolation method is considered in Sect. 2.2.3.

Sections 2.3 and 2.4 are devoted to the important question of error reduction of the Monte Carlo method. Both the method of important sampling and the method of control variates as well as a combining method are considered in Sect. 2.3. To make variance reduction practical, in Sect. 2.4 we use conditional probabilistic representations together with the regression method.

In Sect. 2.5 we briefly consider other approaches to reduce computational cost in computing expectations, including quasi-Monte Carlo (QMC) methods and multi-level Monte Carlo (MLMC) methods.

Implementation of numerical methods for integrating SDEs requires a source of random numbers. On a computer random numbers are usually generated via iterative deterministic algorithms known as random number generators. They are discussed in Sect. 2.6.

In this chapter the word "weak" will be omitted if this does not lead to misunderstanding.

2.1 One-Step Approximation

A one-step weak approximation $\bar{X}_{t,x}(t+h)$ of the solution $X_{t,x}(t+h)$ can be constructed by computing some moments of the vector $\bar{X}_{t,x}(t+h) - x$ and the corresponding moments of the vector $X_{t,x}(t+h) - x$. In this case the order of accuracy of the one-step approximation depends on both the order of the moments under consideration and on the order of closeness of those. To construct the one-step approximations of third order of accuracy considered in this section, we have to take into account all moments up to order six inclusively. In the last subsection we give an expansion formula for $Ef(t+h, X_{t,x}(t+h))$ in powers of h.

2.1.1 Properties of Remainders and Ito Integrals

As before, we consider the system

$$dX = a(t, X)dt + \sum_{r=1}^{q} \sigma_r(t, X)dw_r(t), \qquad (2.1.1)$$

where X, a, and σ_r are vectors of dimension d with components X^i, a^i, σ_r^i. We assume that the functions $a(t, x)$ and $\sigma_r(t, x)$ are sufficiently smooth with respect to the variables t, x and satisfy a global Lipschitz condition with respect to x: for all $t \in [t_0, T]$, $x \in \mathbf{R}^d$, $y \in \mathbf{R}^d$ the following inequality holds:

$$|a(t, x) - a(t, y)| + \sum_{r=1}^{q} |\sigma_r(t, x) - \sigma_r(t, y)| \le K|x - y|. \qquad (2.1.2)$$

Here and below $|x|$ denotes the Euclidean norm of the vector x, and we denote by $x^\mathsf{T} y$ or by (x, y) the scalar (inner) product of two vectors x and y. We introduce the operators

$$\Lambda_r f = (\sigma_r, \frac{\partial}{\partial x})f = \sum_{i=1}^{d} \sigma_r^i \frac{\partial f}{\partial x^i},$$

$$Lf = (\frac{\partial}{\partial t} + a^\mathsf{T}\frac{\partial}{\partial x} + \frac{1}{2}\sum_{r=1}^{q}\sum_{i=1}^{d}\sum_{j=1}^{d}\sigma_r^i\sigma_r^j\frac{\partial^2}{\partial x^i\partial x^j})f,$$

where f may be a scalar function or a vector-function.

In the course of exposition we will impose additional conditions on a and σ_r. Note that the conditions on a and σ_r given in Theorem 1.2.5 are sufficient for all results in this section to hold. We recall that these conditions are related to the growth of functions of the form $\Lambda_{i_j}\cdots\Lambda_{i_1}f(t,x)$ for $f \equiv x$ as $|x| \to \infty$ (see (1.2.17)); more precisely, these functions grow with respect to x at most as a linear function of $|x|$ as $|x| \to \infty$. The indices $i_1,\ldots,i_j$ take the values $0, 1,\ldots, q$, and $\Lambda_0 = L$. We rewrite (1.2.21):

$$X_{t,x}(t+h) = x + \sum_{r=1}^{q}\sigma_r\int_t^{t+h}dw_r(\theta) + ah \tag{2.1.3}$$

$$+\sum_{r=1}^{q}\sum_{i=1}^{q}\Lambda_i\sigma_r\int_t^{t+h}(w_i(\theta)-w_i(t))dw_r(\theta)$$

$$+\sum_{r=1}^{q}L\sigma_r\int_t^{t+h}(\theta-t)dw_r(\theta) + \sum_{r=1}^{q}\Lambda_r a\int_t^{t+h}(w_r(\theta)-w_r(t))d\theta$$

$$+\sum_{r=1}^{q}\sum_{i=1}^{q}\sum_{s=1}^{q}\Lambda_s\Lambda_i\sigma_r\int_t^{t+h}(\int_t^{\theta}(w_s(\theta_1)-w_s(t))dw_i(\theta_1))dw_r(\theta)$$

$$+La\frac{h^2}{2}+\rho,$$

where the coefficients σ_r, a, $\Lambda_i\sigma_r$, $L\sigma_r$, $\Lambda_r a$, $\Lambda_s\Lambda_i\sigma_r$, La are calculated at the point (t, x), while the remainder ρ is given in (1.2.22) (we do not write it here).

Definition 2.1.1 We say that a function $f(x)$ belongs to the class $\mathbf{F}$, written as $f \in \mathbf{F}$, if we can find constants $K > 0$, $\kappa > 0$ such that for all $x \in \mathbf{R}^d$ the following inequality holds:

$$|f(x)| \le K(1 + |x|^\kappa). \tag{2.1.4}$$

If a function $f(s, x)$ depends not only on $x \in \mathbf{R}^d$ but also on a parameter $s \in S$, then we say that $f(s, x)$ belongs to $\mathbf{F}$ (with respect to the variable x) if an inequality of the type (2.1.4) holds uniformly in $s \in S$.

In the sequel we need that $\sigma_r, a, \Lambda_i \sigma_r, L\sigma_r, \Lambda_r a, \Lambda_s \Lambda_i \sigma_r, La$, etc. belong to the class $\mathbf{F}$. For example, in the proof of Lemma 2.1.2 (see below) we use the fact that all integrands participating in the remainder ρ as well as all functions obtained by applying the operators $\Lambda_l, l = 1, \ldots, q$, and L to the functions $\Lambda_j \Lambda_s \Lambda_i \sigma_r, L\Lambda_i \sigma_r, \Lambda_i L\sigma_r, \Lambda_i \Lambda_r a$ belong to the class $\mathbf{F}$. Clearly, it is sufficient for this to require that all partial derivatives up to order five, inclusively, of the coefficients a, σ_r with respect to t and x belong to $\mathbf{F}$. In such cases we assert that the coefficients $a, \sigma_r, r = 1, \ldots, q$, together with their partial derivatives of sufficiently high order belong to $\mathbf{F}$.

Lemma 2.1.2 *Suppose that the Lipschitz condition (2.1.2) holds and the functions* $a, \sigma_r, r = 1, \ldots, q$, *together with their partial derivatives of a sufficiently high order belong to* $\mathbf{F}$. *Then the following inequalities hold:*

$$|E\rho| \leq K(x)h^3, \quad K(x) \in \mathbf{F}, \tag{2.1.5}$$

$$E|\rho|^2 \leq K(x)h^4, \quad K(x) \in \mathbf{F}, \tag{2.1.6}$$

$$\left|E\rho \int_t^{t+h} dw_r(\theta)\right| \leq K(x)h^3, \quad K(x) \in \mathbf{F}. \tag{2.1.7}$$

Proof The form of the remainder ρ (see (1.2.22)) and the fact that $L^2 a \in \mathbf{F}$ imply that we can find an even number $2m$ and a number $K > 0$ such that

$$|E\rho| = \left|E \int_t^{t+h} \left(\int_t^{\theta} \left(\int_t^{\theta_1} L^2 a(\theta_2, X(\theta_2))d\theta_2\right)d\theta_1\right)d\theta\right| \tag{2.1.8}$$

$$\leq \left|\int_t^{t+h} \left(\int_t^{\theta} \left(\int_t^{\theta_1} K(1 + E|X(\theta_2)|^{2m})d\theta_2\right)d\theta_1\right)d\theta\right|.$$

But according to [123], $E|X(\theta_2)|^{2m}$ is bounded by a quantity $K(1 + |x|^{2m})$. Hence (2.1.8) implies (2.1.5). In the proof of (2.1.6) we use the fact that each term (more precisely, the expectation of the norm of each term in (1.2.22)) is, in any case, of second order of smallness with respect to h. To prove (2.1.7), we use Ito's formula to analyze each integral in the first four sums in (1.2.22). For example,

$$\int_t^{t+h}(\int_t^\theta(\int_t^{\theta_1}(\int_t^{\theta_2}\Lambda_j\Lambda_s\Lambda_i\sigma_r(\theta_3,X(\theta_3))dw_j(\theta_3))dw_s(\theta_2))dw_i(\theta_1))dw_r(\theta)$$

$$=\Lambda_j\Lambda_s\Lambda_i\sigma_r(t,x)\int_t^{t+h}(\int_t^\theta(\int_t^{\theta_1}(\int_t^{\theta_2}dw_j(\theta_3))dw_s(\theta_2))dw_i(\theta_1))dw_r(\theta)$$

$$+\sum_{l=1}^q\int_t^{t+h}(\int_t^\theta(\int_t^{\theta_1}(\int_t^{\theta_2}(\int_t^{\theta_3}\Lambda_l\Lambda_j\Lambda_s\Lambda_i\sigma_r(\theta_4,X(\theta_4))dw_l(\theta_4))dw_j(\theta_3))$$

$$\times dw_s(\theta_2))dw_i(\theta_1))dw_r(\theta)$$

$$+\int_t^{t+h}(\int_t^\theta(\int_t^{\theta_1}(\int_t^{\theta_2}(\int_t^{\theta_3}L\Lambda_j\Lambda_s\Lambda_i\sigma_r(\theta_4,X(\theta_4))d\theta_4)dw_j(\theta_3))$$

$$\times dw_s(\theta_2))dw_i(\theta_1))dw_r(\theta)\,.$$

As a result, ρ can be written as a sum of terms of second, or higher, order of smallness with respect to h. Moreover, the terms of second order look like one of the integrals

$$I_{risj}=\int_t^{t+h}(\int_t^\theta(\int_t^{\theta_1}(\int_t^{\theta_2}dw_j(\theta_3))dw_s(\theta_2))dw_i(\theta_1))dw_r(\theta)\,,$$

$$I_{r0i}=\int_t^{t+h}(\int_t^\theta(\int_t^{\theta_1}dw_i(\theta_2))d\theta_1)dw_r(\theta)\,,$$

$$I_{ri0}=\int_t^{t+h}(\int_t^\theta(\int_t^{\theta_1}d\theta_2)dw_i(\theta_1))dw_r(\theta)\,,$$

$$I_{0ri}=\int_t^{t+h}(\int_t^\theta(\int_t^{\theta_1}dw_i(\theta_2))dw_r(\theta_1))d\theta\,,$$

with nonrandom coefficients $\Lambda_j\Lambda_s\Lambda_i\sigma_r$, $\Lambda_iL\sigma_r$, $L\Lambda_i\sigma_r$, $\Lambda_i\Lambda_r a$, respectively. It is easy to prove that the expectation of the product of $\int_t^{t+h}dw_l(\theta)$ with any term of second order of smallness is zero. For example, let us show that

$$E(I_{risj}\cdot I_l)\tag{2.1.9}$$

$$=E(\int_t^{t+h}(\int_t^\theta(\int_t^{\theta_1}(\int_t^{\theta_2}dw_j(\theta_3))dw_s(\theta_2))dw_i(\theta_1))dw_r(\theta)\cdot\int_t^{t+h}dw_l(\theta))\,.$$

Indeed, by changing variables

$$v_k = -w_k, \quad k = 1, \ldots, q,$$

the v_k are independent Wiener processes. Since the Wiener processes participating in (2.1.9) are odd, we have

$$E(I_{risj} \cdot I_l)$$

$$= -E(\int_t^{t+h} (\int_t^{\theta} (\int_t^{\theta_1} (\int_t^{\theta_2} dv_j(\theta_3))dv_s(\theta_2))dv_i(\theta_1))dv_r(\theta) \cdot \int_t^{t+h} dv_l(\theta))$$

$$= -E(I_{risj} \cdot I_l),$$

which implies (2.1.9).

The other terms in ρ have order of smallness at least $5/2$. Using the Bunyakovsky-Schwarz inequality, we can readily show that expectation of absolute value of the product of each of such terms with $\int_t^{t+h} dw_l(\theta)$ is smaller than or equal to $K(x)h^3$ with $K(x) \in \mathbf{F}$. This proves (2.1.7) and hence the lemma. $\square$

Using the identity

$$\int_t^{t+h} (\theta - t)dw_r(\theta) = h \int_t^{t+h} dw_r(\theta) - \int_t^{t+h} (w_r(\theta) - w_r(t))d\theta$$

and the notation

$$I_j = \int_t^{t+h} dw_j(\theta) = w_j(t+h) - w_j(t),$$

$$I_{jp} = \int_t^{t+h} (w_j(\theta) - w_j(t))dw_p(\theta),$$

$$I_{sir} = \int_t^{t+h} (\int_t^{\theta} (w_s(\theta_1) - w_s(t))dw_i(\theta_1))dw_r(\theta),$$

$$J_r = \int_t^{t+h} (w_r(\theta) - w_r(t))d\theta,$$

we introduce $\tilde{X}$ by the formula

$$\tilde{X} = x + \sum_{r=1}^{q} \sigma_r I_r + ah + \sum_{r=1}^{q}\sum_{i=1}^{q} \Lambda_i \sigma_r I_{ir} \tag{2.1.10}$$

$$+ \sum_{r=1}^{q} L\sigma_r \cdot I_r h + \sum_{r=1}^{q} (\Lambda_r a - L\sigma_r) \cdot J_r + La \cdot \frac{h^2}{2}$$

and rewrite (2.1.3) as follows:

$$X = \tilde{X} + \sum_{r=1}^{q}\sum_{i=1}^{q}\sum_{s=1}^{q} \Lambda_s \Lambda_i \sigma_r \cdot I_{sir} + \rho. \tag{2.1.11}$$

Lemma 2.1.3 *The following identities hold:*

$$E I_{sir} = 0, \;\; E I_{sir} I_j = 0, \;\; E I_{sir} I_j I_p = 0, \tag{2.1.12}$$
$$E I_{sir} I_{jp} = 0, \;\; i, j, p, r, s = 1, \dots, q.$$

Proof The first, third, and fourth identities in (2.1.12) are obvious because of oddness. We prove the second identity. Without loss of generality we may put $t = 0$. For $j \neq s$, $j \neq i$, $j \neq r$ this identity follows from the independence of I_j and I_{sir}. To consider the other cases, we introduce the system of equations

$$dx(\theta) = w_s(\theta)dw_i(\theta), \;\; x(0) = 0,$$
$$dy(\theta) = x(\theta)dw_r(\theta), \;\; y(0) = 0.$$

Then $I_{sir} = y(h)$ and $E I_{sir} I_j = E y(h) w_j(h)$. Let, e.g., $j = s$. Then by Ito's formula,

$$d(y(\theta)w_s(\theta)) = w_s(\theta)x(\theta)dw_r(\theta) + y(\theta)dw_s(\theta) + x(\theta)\delta_{sr}d\theta,$$

where δ_{sr} is the Kronecker symbol. Hence,

$$dE(y(\theta)w_s(\theta)) = Ex(\theta)\delta_{sr}d\theta.$$

Since $Ex(\theta) = 0$ and $E(y(0)w_s(0)) = 0$, we have $E I_{sir} I_s = E y(h) w_s(h) = 0$. The cases $j = i$ and $j = r$ can be treated in a similar way. □

2.1.2 One-Step Approximations of Third Order

We introduce the notation $X = X(t + h)$, $\Delta = X - x$, $\tilde{\Delta} = \tilde{X} - x$, $\bar{\Delta} = \bar{X} - x$, and denote by x^i the i-th coordinate of the vector x. Our nearest goal is to form a random vector $\bar{X}$ such that the difference of all moments up to order five, inclusively, of the coordinates of the vectors Δ and $\bar{\Delta}$ would have third order of smallness with respect to h. More precisely,

$$|E(\prod_{j=1}^{s} \Delta^{i_j} - \prod_{j=1}^{s} \bar{\Delta}^{i_j})| \le K(x)h^3, \ i_j = 1, \ldots, d, \tag{2.1.13}$$

$$s = 1, \ldots, 5, \ K(x) \in \mathbf{F},$$

and, moreover,

$$E \prod_{j=1}^{s} |\bar{\Delta}^{i_j}| \le K(x)h^3, \ i_j = 1, \ldots, d, \ s = 6, \ K(x) \in \mathbf{F}. \tag{2.1.14}$$

First of all, we state the following lemma.

Lemma 2.1.4 *Under the conditions of Lemma 2.1.2 the following inequalities hold:*

$$|E(\prod_{j=1}^{s} \Delta^{i_j} - \prod_{j=1}^{s} \tilde{\Delta}^{i_j})| \le K(x)h^3, \ s = 1, \ldots, 5, \ K(x) \in \mathbf{F}. \tag{2.1.15}$$

Proof The proof of this lemma is based on Lemmas 2.1.2 and 2.1.3. In fact, by (2.1.11) each component Δ^{i_j} of Δ differs from the corresponding component $\tilde{\Delta}^{i_j}$ of $\tilde{\Delta}$ by a sum made up from the corresponding components of the vectors ρ and $\Lambda_s \Lambda_i \sigma_r \cdot I_{sir}$. Therefore, the difference $\prod_{j=1}^{s} \Delta^{i_j} - \prod_{j=1}^{s} \tilde{\Delta}^{i_j}$ consists of terms each of which must have a component of ρ or a component of the integral I_{sir} as at least one of its factors. If $s = 1$, i.e., we are considering first moments, then these terms do not have other factors and (2.1.15) follows from (2.1.5) and the first identity in (2.1.12). If $s = 2$ then the terms containing ρ either have I_r as a factor or they have a factor whose order is at least one. In the first case we use the estimate (2.1.7), and in the second case we use the Bunyakovsky-Schwarz inequality and, subsequently, (2.1.6). In other words, in the second case the order of a term is at least three because one factor has order at least one and the other factor (which is a component of ρ) has, by (2.1.6), order two, i.e., by the Bunyakovsky-Schwarz inequality we may sum the orders of the factors. This already makes clear that for $s = 3, 4, 5$ the terms containing at least one component of ρ as a factor have order of smallness at least three with respect to h. For $s = 2$ we return to the terms containing I_{sir} as a factor; I_{sir} has order of smallness $3/2$ with respect to h. Such terms either contain an expression of the form $I_{sir}I_j$, $I_{sir}h$, $I_{sir}I_{jp}$ as a factor and the expectation of such terms is zero (by the first, second, and fourth identity in (2.1.12)), or they have at least third order of smallness with respect to h. For $s = 3$, the terms containing I_{sir} and having a mean-square order of smallness with respect to h which is less than three are easily seen to contain an expression of the form $I_{sir}I_jI_p$, and their expectation is zero (see the third identity in (2.1.12)). For $s = 4, 5$, all terms containing I_{sir} are at least of order three. These considerations imply that the inequality (2.1.15) holds. □

We will form a random vector $\bar{X}$ such that the inequalities

$$|E(\prod_{j=1}^{s} \tilde{\Delta}^{ij} - \prod_{j=1}^{s} \bar{\Delta}^{ij})| \leq K(x)h^3, \; s = 1, \ldots, 5, \; K(x) \in \mathbf{F}, \tag{2.1.16}$$

as well as (2.1.14) hold. Since (2.1.15) and (2.1.16) imply (2.1.13), we will reach our aim in constructing the vector $\bar{X}$ satisfying the inequalities (2.1.13) and (2.1.14). We construct $\bar{X}$ similar to $\tilde{X}$ as follows:

$$\bar{X} = x + \sum_{r=1}^{q} \sigma_r \xi_r h^{1/2} + ah + \sum_{r=1}^{q} \sum_{i=1}^{q} \Lambda_i \sigma_r \xi_{ir} h \tag{2.1.17}$$

$$+ \sum_{r=1}^{q} L\sigma_r \xi_r h^{3/2} + \sum_{r=1}^{q} (\Lambda_r a - L\sigma_r)\eta_r h^{3/2} + La\frac{h^2}{2}.$$

Lemma 2.1.5 *Suppose the conditions of Lemma 2.1.2 are satisfied. Then the inequalities (2.1.13) and (2.1.14) hold if the random variables ξ_r, ξ_{ir}, η_r in (2.1.17) have finite moments up to order six, inclusively, and the following relations hold:*

$$E\xi_r h^{1/2} = EI_r = 0, \; E\xi_{ir}h = EI_{ir} = 0, \; E\eta_r h^{3/2} = EJ_r = 0; \tag{2.1.18}$$

$$E\xi_i\xi_r h = EI_i I_r = \delta_{ir}h, \; E\xi_i\xi_{rj}h^{3/2} = EI_i I_{rj} = 0, \tag{2.1.19}$$

$$E\xi_r\eta_j h^2 = EI_r J_j = \delta_{rj}\frac{h^2}{2},$$

$$E\xi_{ir}\xi_{js}h^2 = EI_{ir}I_{js} = \begin{cases} h^2/2 & \text{if } i = j, \; r = s, \\ 0 & \text{otherwise,} \end{cases}$$

$$E\xi_{ir}\eta_j h^{5/2} = EI_{ir}J_j = 0;$$

$$E\xi_i\xi_r\xi_j h^{3/2} = EI_i I_r I_j = 0, \tag{2.1.20}$$

$$E\xi_i\xi_r\xi_{js}h^2 = EI_i I_r I_{js}$$

$$= \begin{cases} h^2/2 & \text{if } j \neq s \text{ and either } i = j, \; r = s \text{ or } i = s, \; r = j, \\ h^2 & \text{if } i = r = j = s, \\ 0 & \text{otherwise,} \end{cases}$$

$$E\xi_i\xi_r\eta_j h^{5/2} = EI_i I_r J_j = 0, \; E\xi_i\xi_{jr}\xi_{sl}h^{5/2} = EI_i I_{jr} I_{sl} = 0;$$

$$E\xi_i\xi_r\xi_j\xi_s h^2 = E I_i I_r I_j I_s \tag{2.1.21}$$

$$= \begin{cases} h^2 & \text{if } \{i, r, j, s\} \text{ consists of two pairs of equal numbers,} \\ 3h^2 & \text{if } i = r = j = s, \\ 0 & \text{otherwise,} \end{cases}$$

$$E\xi_i\xi_r\xi_j\xi_{sl} h^{5/2} = E I_i I_r I_j I_{sl} = 0;$$

$$E\xi_i\xi_r\xi_j\xi_s\xi_l h^{5/2} = E I_i I_r I_j I_s I_l = 0. \tag{2.1.22}$$

Proof The inequality (2.1.14) for sixth moments of the absolute values of the coordinates of the vector $\bar{\Delta} = \bar{X} - x$ evidently follows from (2.1.17), since each term in $\bar{\Delta}$ has at least order of smallness $1/2$ with respect to h. Further, all the identities (2.1.18)–(2.1.22) consist of two parts: the right part and the left part. We prove the right parts below. The left parts of (2.1.18) are clearly sufficient to get (2.1.16) and (2.1.13) for $s = 1$, i.e., to prove that the first moments of the coordinates of the vectors Δ and $\bar{\Delta}$ coincide up to $O(h^3)$. The left parts of (2.1.18)–(2.1.19) suffice for the second moments; (2.1.18)–(2.1.20) suffice for the third moments; (2.1.18)–(2.1.21) suffice for the fourth moments; and (2.1.18)–(2.1.22) suffice for the fifth moments.

Almost all the right parts of (2.1.18)–(2.1.22) can be easily derived taking into account oddness and independence; only evaluation of the expectations $E I_{ir} I_{js}$ and $E I_i I_r I_{js}$ causes some difficulties. Without loss of generality, we set $t = 0$. To evaluate

$$E I_{ir} I_{js} = E \int_0^h w_i(\theta) dw_r(\theta) \int_0^h w_j(\theta) dw_s(\theta),$$

we introduce the system of equations

$$dx(\theta) = w_i(\theta) dw_r(\theta), \quad x(0) = 0,$$
$$dy(\theta) = w_j(\theta) dw_s(\theta), \quad y(0) = 0.$$

It is evident that $E I_{ir} I_{js} = E x(h) y(h)$. By Ito's formula, we have

$$dxy = y w_i dw_r + w_j dw_s + w_i w_j \delta_{rs} d\theta.$$

Therefore

$$dE(x(\theta) y(\theta)) = \delta_{rs} E(w_i(\theta) w_j(\theta)) d\theta,$$

which immediately implies the last of the identities (2.1.19).

To evaluate

$$E I_i I_r I_{js} = E w_i(h) w_r(h) \int_0^h w_j(\theta) dw_s(\theta),$$

we introduce the equation

$$dy(\theta) = w_j(\theta)dw_s(\theta), \ \ y(0) = 0.$$

It is obvious that $E I_i I_r I_{js} = E w_i(h) w_r(h) y(h)$. By Ito's formula,

$$d(w_i w_r y) = w_r y dw_i + w_i y dw_r + w_i w_r w_j dw_s$$
$$+ y \delta_{ir} d\theta + w_r w_j \delta_{is} d\theta + w_i w_j \delta_{rs} d\theta.$$

In view of $E y(\theta) = 0$ we obtain

$$dE(w_i w_r y) = \delta_{is} E(w_r w_j) d\theta + \delta_{rs} E(w_i w_j) d\theta,$$

which immediately implies the second identity in (2.1.20). □

Theorem 2.1.6 *Suppose the conditions of Lemma 2.1.2 hold. Let a function $f(x)$ and all its partial derivatives up to order six inclusively belong to the class* **F**. *Let ξ_i, η_i, ξ_{ij} be chosen such that (2.1.18)–(2.1.22) hold. Then $\bar{X}$ from (2.1.17) satisfies the inequality (recall that $X = X(t + h)$) :*

$$|Ef(X) - Ef(\bar{X})| \le K(x)h^3, \ \ K(x) \in \mathbf{F}, \tag{2.1.23}$$

i.e., the method (2.1.17) has the third order of accuracy on a single step in the sense of weak approximations.

Proof Lemmas 2.1.2–2.1.5 imply the inequalities (2.1.13) and (2.1.14). Moreover, similarly to the proof of (2.1.14) in Lemma 2.1.5, we can prove the inequality

$$E \prod_{j=1}^{s} |\Delta^{i_j}| \le K(x)h^3, \ i_j = 1, \ldots, d, \ s = 6, \ K(x) \in \mathbf{F}. \tag{2.1.24}$$

Now we write the Taylor expansion of $f(X)$ with respect to powers of $\Delta^i = X^i - x^i$ in a neighborhood of x and with Lagrange remainder term containing terms of order six. We similarly expand $f(\bar{X})$ with respect to the $\bar{\Delta}^i = \bar{X}^i - x^i$. Using (2.1.13), (2.1.14), and (2.1.24), we arrive at (2.1.23). □

Let us proceed to modeling of random variables and obtaining of constructive one-step approximations of third order of accuracy. There are various methods that satisfy the relations (2.1.18)–(2.1.22). For example (see [428]), let ξ_i, $i = 1, \ldots, q$, ζ_{ij}, $i = 2, \ldots, q$, $j = 1, \ldots, i - 1$, be mutually independent random variables, where ξ_i are distributed by the law $P(\xi = 0) = 2/3$, $P(\xi = -\sqrt{3}) = P(\xi = \sqrt{3}) = 1/6$, and ζ_{ij} are distributed by the law $P(\zeta = -1) = P(\zeta = 1) = 1/2$. Then if $\zeta_{ii} := -1$, $\zeta_{ij} = -\zeta_{ji}$, $j > i$, and

$$\eta_i = \frac{1}{2}\xi_i, \ \xi_{ij} = \frac{1}{2}\xi_i\xi_j + \frac{1}{2}\zeta_{ij},$$

the relations (2.1.18)–(2.1.22) are satisfied. This is easy to verify. This method requires modeling $q(q + 1)/2$ random variables. We propose a method (see [272]) which requires $2q$ random variables only.

Consider mutually independent random variables ξ_i and ζ_j, $i, j = 1, \ldots, q$, and put

$$\eta_i = \frac{1}{2}\xi_i, \ \xi_{ij} = \frac{1}{2}\xi_i\xi_j - \frac{1}{2}\gamma_{ij}\zeta_i\zeta_j, \ \gamma_{ij} = \begin{cases} -1, \ i < j, \\ 1 \ \ \ i \geq j. \end{cases} \tag{2.1.25}$$

We will assume that ξ_i and ζ_j have all moments needed. Below we will verify that if, in addition to the above-said, we require that

$$E\xi_i = E\xi_i^3 = E\xi_i^5 = 0, \ E\xi_i^2 = 1, \ E\xi_i^4 = 3, \tag{2.1.26}$$
$$E\zeta_i = E\zeta_i^3 = 0, \ E\zeta_i^2 = E\zeta_i^4 = 1,$$

then all the relations (2.1.18)–(2.1.22) are satisfied.

For example, we can model ξ_i by the law $\mathcal{N}(0, 1)$ and the ζ_i by the law $P(\zeta = -1) = P(\zeta = 1) = 1/2$. But ξ_i can be modeled by a much simpler law as above: $P(\xi = 0) = 2/3$, $P(\xi = \pm\sqrt{3}) = 1/6$. In practice (see the end of Sect. 2.6), the following modeling of ξ_i is of interest too: $P(\xi = \pm\sqrt{1 - \sqrt{6}/3}) = 3/8$, $P(\xi = \pm\sqrt{1 + \sqrt{6}}) = 1/8$.

We turn to the direct verification of (2.1.18)–(2.1.22) for the random variables (2.1.25). Many of them can be easily verified. Therefore, we verify more complicated ones only.

Lemma 2.1.7 *Let $\xi_i, \zeta_j, i, j = 1, \ldots, q$, be mutually independent random variables such that (2.1.25)–(2.1.26) hold. Then*

$$E\xi_{ir}\xi_{js} = \begin{cases} 1/2 \ if \ i = j, \ r = s, \\ 0 \ \ \ \ \ otherwise. \end{cases} \tag{2.1.27}$$

Proof We have

$$E\xi_{ir}\xi_{js} = \frac{1}{4}(E\xi_i\xi_r\xi_j\xi_s - \gamma_{js}E\xi_i\xi_r\zeta_j\zeta_s - \gamma_{ir}E\xi_j\xi_s\zeta_i\zeta_r + \gamma_{ir}\gamma_{js}E\zeta_i\zeta_r\zeta_j\zeta_s). \tag{2.1.28}$$

Let $i \neq j$. It is obvious that if r is not equal to j, then all expectations in the right-hand side of (2.1.28) are zero. So, for $i \neq j$ the right-hand side can be nonzero only if $r = i$ or $r = j$. Consider the case $i \neq j$, $r = i$. If also $s \neq j$, then the right-hand side is zero. Now we evaluate (2.1.28) for $i \neq j$, $r = i$, $s = j$. We have $\gamma_{ir} = \gamma_{js} = 1$ and each of the four expectations in the right-hand side of (2.1.28) is equal to 1. As a result, for $i \neq j$, $r = i$ the right-hand side vanishes. Consider now the case $i \neq j$, $r = j$. In this case the right-hand side of (2.1.28) can be nonzero only if $s = i$. So, let $i \neq j$, $r = j$, $s = i$. We have $E\xi_i\xi_r\xi_j\xi_s = E\zeta_i\zeta_r\zeta_j\zeta_s = 1$ but $E\xi_i\xi_r\zeta_j\zeta_s = E\xi_j\xi_s\zeta_i\zeta_r = 0$. For $i \neq j$, $r = j$, $s = i$ the product $\gamma_{ir}\gamma_{js}$ is always -1. Indeed, if $i < j$ then $i < r$

and $j > s$, since $r = j$, $s = i$. But $\gamma_{ir} = -1$ for $i < r$ and $\gamma_{js} = 1$ for $j > s$. Hence $\gamma_{ir}\gamma_{js} = -1$. The case $i > j$ can be treated similarly. So, if $i \neq j$, the right-hand side of (2.1.28) is always zero.

Now let $i = j$. Then the right-hand side of (2.1.28) can be nonzero for $r = s$ only. We distinguish three cases. In the first case $i < r$. Then $j < s$, $\gamma_{ir} = \gamma_{js} = -1$, $E\xi_i\xi_r\xi_j\xi_s = E\zeta_i\zeta_r\zeta_j\zeta_s = 1$, $E\xi_i\xi_r\zeta_j\zeta_s = E\xi_j\xi_s\zeta_i\zeta_r = 0$, and hence $E\xi_{ir}\xi_{js} = 1/2$. The second case, $i > r$, can be treated similarly. It differs by the relations $\gamma_{ir} = \gamma_{js} = 1$. The third case, $i = r$, gives $i = j = r = s$, $E\xi_i\xi_r\xi_j\xi_s = 3$, $E\xi_i\xi_r\zeta_j\zeta_s = E\xi_j\xi_s\zeta_i\zeta_r = E\zeta_i\zeta_r\zeta_j\zeta_s = 1$, $\gamma_{ir} = \gamma_{js} = 1$, and hence $E\xi_{ir}\xi_{js} = 1/2$. $\qquad\square$

We need this lemma to substantiate the penultimate relation in (2.1.19). The other relations in (2.1.19) as well as (2.1.18) can be verified in an obvious manner. In (2.1.20) the second relation presents some difficulty. To verify it, we prove the following lemma.

Lemma 2.1.8 *Let $\xi_i, \zeta_j, i, j = 1, \ldots, q$, be mutually independent random variables such that (2.1.25)–(2.1.26) hold. Then*

$$
E\xi_i\xi_r\xi_{js} = \begin{cases} 1/2 \text{ if } j \neq s \text{ and either } i = j, \ r = s \text{ or } i = s, \ r = j, \\ 1 \qquad\qquad\quad \text{if } i = r = j = s, \\ 0 \qquad\qquad\quad otherwise. \end{cases} \tag{2.1.29}
$$

Proof We have

$$
E\xi_i\xi_r\xi_{js} = \frac{1}{2}(E\xi_i\xi_r\xi_j\xi_s - \gamma_{js}E\xi_i\xi_r\zeta_j\zeta_s). \tag{2.1.30}
$$

For $j \neq s$ the right-hand side of (2.1.30) can be nonzero only for $i = j$, $r = s$ or $i = s$, $r = j$. In both these cases $E\xi_i\xi_r\xi_j\xi_s = 1$, $E\xi_i\xi_r\zeta_j\zeta_s = 0$, which proves (2.1.29) for $j \neq s$. If $j = s$ but $i \neq j$, then the right-hand side of (2.1.30) can be nonzero for $i = r$ only. But in this case $\gamma_{js} = 1$, $E\xi_i\xi_r\xi_j\xi_s = E\xi_i\xi_r\zeta_j\zeta_s = 1$ and, hence, the right-hand side of (2.1.30) is zero. Let $j = s$, $i = j$. Then (2.1.30) can be nonzero only if $i = r$, i.e. $i = r = j = s$. If $i = r = j = s$, then $E\xi_i\xi_r\xi_j\xi_s = 3$, $E\xi_i\xi_r\zeta_j\zeta_s = 1$, $\gamma_{js} = 1$, i.e. $E\xi_i\xi_r\xi_{js} = 1$. $\qquad\square$

The other relations (2.1.18)–(2.1.22) can be verified in a simple way. As a result, we can write the one-step approximation (2.1.17) as

$$
\bar{X} = x + \sum_{r=1}^{q} \sigma_r\xi_r h^{1/2} + ah + \sum_{r=1}^{q}\sum_{i=1}^{q} \Lambda_i\sigma_r\xi_{ir}h \tag{2.1.31}
$$

$$
+ \frac{1}{2}\sum_{r=1}^{q}(\Lambda_r a + L\sigma_r)\xi_r h^{3/2} + La\frac{h^2}{2},
$$

where ξ_{ir} satisfy (2.1.25), and ξ_i, ζ_j are independent random variables satisfying (2.1.26). We recall that ξ_i can be modeled by, e.g., the law $P(\xi = 0) = 2/3$, $P(\xi =$

$\pm\sqrt{3}) = 1/6$ or $P(\xi = \pm\sqrt{1 - \sqrt{6}/3}) = 3/8$, $P(\xi = \pm\sqrt{1 + \sqrt{6}}) = 1/8$ and ζ_j can be modeled by $P(\zeta = \pm 1) = 1/2$. The one-step approximation (2.1.31) has the third order of accuracy in the sense of weak approximation.

2.1.3 The Taylor Expansion of Expectations

Let us derive some expansions of $Ef(t + h, X(t + h))$ with respect to powers of h. By Ito's formula, we have for $u \geq t$:

$$f(u, X(u)) = f(t, X(t)) + \int_t^u Lf(\theta, X(\theta))d\theta \qquad (2.1.32)$$

$$+ \sum_{r=1}^q \int_t^u \Lambda_r f(\theta, X(\theta))dw_r(\theta) .$$

Applying (2.1.32) to $Lf(\theta, X(\theta))$, substituting the obtained expression in $\int_t^u Lf(\theta, X(\theta))d\theta$, and making a few simple transformations, we get

$$f(u, X(u)) = f(t, X(t)) + Lf(t, X(t))(u - t) + \int_t^u (u - \theta)L^2 f(\theta, X(\theta))d\theta$$

$$+ \sum_{r=1}^q \int_t^u (\Lambda_r f(\theta, X(\theta)) + (u - \theta)\Lambda_r Lf(\theta, X(\theta)))dw_r(\theta) .$$

Proceeding further in this way, we find

$$f(u, X(u)) = f(t, X(t)) + Lf(t, X(t))(u - t) + \cdots \qquad (2.1.33)$$

$$+ \frac{1}{m!}L^m f(t, X(t))(u - t)^m + \int_t^u \frac{(u - \theta)^m}{m!}L^{m+1} f(\theta, X(\theta))d\theta$$

$$+ \int_t^u \sum_{r=1}^q (\Lambda_r f(\theta, X(\theta))$$

$$+ \cdots + \frac{(u - \theta)^m}{m!}\Lambda_r L^m f(\theta, X(\theta)))dw_r(\theta) .$$

Lemma 2.1.9 *Suppose that the following expectations exist and are continuous with respect to θ :*

$$EL^k f(\theta, X(\theta)), \ k = 0, 1, \ldots, m + 1,$$
$$E(\Lambda_r L^k f(\theta, X(\theta)))^2, \ k = 0, 1, \ldots, m, \ r = 1, \ldots, q.$$

Then the following formulas hold for $t \le s \le t + h$:

$$E(f(t + h, X_{t,x}(t + h))|\mathcal{F}_s) = f(s, X_{t,x}(s)) \tag{2.1.34}$$
$$+(t + h - s)Lf(s, X_{t,x}(s))$$
$$+\cdots+ \frac{(t + h - s)^m}{m!} L^m f(s, X_{t,x}(s))$$
$$+ \int_s^{t+h} \frac{(t + h - \theta)^m}{m!}$$
$$\cdot E(L^{m+1} f(\theta, X_{t,x}(\theta))|\mathcal{F}_s)d\theta,$$

$$Ef(t + h, X_{t,x}(t + h)) = f(t, x) + hLf(t, x) + \cdots + \frac{h^m}{m!} L^m f(t, x) \tag{2.1.35}$$
$$+ \int_t^{t+h} \frac{(t + h - \theta)^m}{m!} EL^{m+1} f(\theta, X_{t,x}(\theta))d\theta.$$

The proof clearly follows from (2.1.33).

The formula (2.1.35) is related to the Taylor expansion of semigroups [165]. It is more convenient than the Taylor expansion of semigroups because, in particular, it is also applicable to unbounded functions f. Clearly, in (2.1.34) and (2.1.35) the remainders of integral type are $O(h^{m+1})$.

2.2 Global Errors of Weak Approximations

In this section we first prove a theorem which is the fundamental convergence theorem for weak approximations. It establishes the weak order of convergence of a method resting on properties of its one-step approximation only. Using this theorem, we prove, in particular, that the method based on the one-step approximation of third order constructed in the previous section is of weak order 2. Acting analogously, it is not difficult, in principle, to construct weak methods of any orders (see e.g. Chap. 3 where methods of weak order 3 are constructed for systems with additive noise). Later in this section we consider the Talay–Tubaro error expansion and we introduce the concept of rejecting exploding trajectories used in the case of SDEs with nonglobally Lipschitz coefficients.

2.2.1   The General Convergence Theorem

Along with the system (2.1.1), we consider the approximation

$$\bar{X}_{t,x}(t+h) = x + A(t, x, h; \xi),\qquad (2.2.1)$$

where ξ is a random variable (in general, a vector) having moments of a sufficiently high order, and A is a vector function of dimension d. Partition the interval $[t_0, T]$ into N equal parts with step $h = (T - t_0)/N : t_0 < t_1 < \cdots < t_N = T, \ t_{k+1} - t_k = h$. According to (2.2.1), we construct the sequence

$$\bar{X}_0 = X_0 = X(t_0), \ \bar{X}_{k+1} = \bar{X}_k + A(t, \bar{X}_k, h; \xi_k), \ k = 0, \dots, N-1, \quad (2.2.2)$$

where ξ_0 is independent of $\bar{X}_0$, while ξ_k for $k > 0$ is independent of $\bar{X}_0, \dots, \bar{X}_k$, $\xi_0, \dots, \xi_{k-1}$. As before, we write $\varDelta = X - x = X_{t,x}(t+h) - x, \ \bar{\varDelta} = \bar{X} - x = \bar{X}_{t,x}(t+h) - x$. Let $X(t) = X_{t_0, X_0}(t)$ be a solution of (2.1.1) and $\bar{X}_{t_0, X_0}(t_k) = \bar{X}_k$.

Theorem 2.2.1 *Suppose that*

(a) the coefficients of equation (2.1.1) are continuous, satisfy a Lipschitz condition (2.1.2) and together with their partial derivatives with respect to x of order up to $2p + 2$, inclusively, belong to **F**;

(b) the method (2.2.1) is such that

$$|E(\prod_{j=1}^{s} \varDelta^{i_j} - \prod_{j=1}^{s} \bar{\varDelta}^{i_j})| \le K(x)h^{p+1}, \ s = 1, \dots, 2p+1, \ K(x) \in \mathbf{F}, \quad (2.2.3)$$

$$E \prod_{j=1}^{2p+2} |\bar{\varDelta}^{i_j}| \le K(x)h^{p+1}, \ K(x) \in \mathbf{F}; \quad (2.2.4)$$

(c) the function $f(x)$ together with its partial derivatives of order up to $2p + 2$, inclusively, belong to **F**;

(d) for a sufficiently large m (specified below) the expectations $E|\bar{X}_k|^{2m}$ exist and are uniformly bounded with respect to N and $k = 0, 1, \dots, N$.

Then, for all N and all $k = 0, 1, \dots, N$ the following inequality holds:

$$|Ef(X_{t_0, X_0}(t_k)) - Ef(\bar{X}_{t_0, X_0}(t_k))| \le Kh^p, \quad (2.2.5)$$

i.e., the method (2.2.2) has order of accuracy p in the sense of weak approximation.

Proof First of all we note that the Lipschitz condition (2.1.2) implies that for any $m > 0$ the expectations $E|X(\theta)|^{2m}$ exist and are uniformly bounded with respect to $\theta \in [t_0, T]$ if only $E|X(t_0)|^{2m} < \infty$ (see [123]). Moreover, the same (2.1.2) implies

$$E \prod_{j=1}^{2p+2} |\Delta^{i_j}| \le K(x)h^{p+1}, \quad K(x) \in \mathbf{F}. \tag{2.2.6}$$

Further, suppose that $u(x)$ is a function that together with its partial derivatives of order up to $2p + 2$, inclusively, belongs to $\mathbf{F}$. Then

$$|Eu(X_{t,x}(t+h)) - Eu(\bar{X}_{t,x}(t+h))| \le K(x)h^{p+1}, \quad K(x) \in \mathbf{F}. \tag{2.2.7}$$

Thanks to (2.2.3), (2.2.4), (2.2.6), the proof of (2.2.7) is analogous to the proof of Theorem 2.1.6.

We introduce the function

$$u(s, x) = Ef(X_{s,x}(t_{k+1})).$$

By the requirements (a) and (c), u has partial derivatives with respect to x of order up to $2p + 2$, inclusively; moreover, these derivatives belong to $\mathbf{F}$ (see [123]). Therefore, the function $u(s, x)$ satisfies an estimate of the form (2.2.7) uniformly with respect to $s \in [t_0, t_{k+1}]$.

Further, since $\bar{X}_0 = X_0$, $X_{t_0, X_0}(t_1) = X(t_1)$, $X_{t_1, X_{t_0, \bar{X}_0}(t_1)}(t_{k+1}) = X(t_{k+1})$, we have

$$Ef(X(t_{k+1})) = Ef(X_{t_1, X_{t_0, \bar{X}_0}(t_1)}(t_{k+1})) - Ef(X_{t_1, \bar{X}_1}(t_{k+1})) \tag{2.2.8}$$
$$+ Ef(X_{t_1, \bar{X}_1}(t_{k+1})).$$

Similarly, since $X_{t_1, \bar{X}_1}(t_{k+1}) = X_{t_2, X_{t_1, \bar{X}_1}(t_2)}(t_{k+1})$, we have

$$Ef(X_{t_1, \bar{X}_1}(t_{k+1})) = Ef(X_{t_2, X_{t_1, \bar{X}_1}(t_2)}(t_{k+1})) - Ef(X_{t_2, \bar{X}_2}(t_{k+1})) \tag{2.2.9}$$
$$+ Ef(X_{t_2, \bar{X}_2}(t_{k+1})).$$

Now (2.2.8) and (2.2.9) imply

$$Ef(X(t_{k+1})) = Ef(X_{t_1, X_{t_0, \bar{X}_0}(t_1)}(t_{k+1})) - Ef(X_{t_1, \bar{X}_1}(t_{k+1}))$$
$$+ Ef(X_{t_2, X_{t_1, \bar{X}_1}(t_2)}(t_{k+1})) - Ef(X_{t_2, \bar{X}_2}(t_{k+1}))$$
$$+ Ef(X_{t_2, \bar{X}_2}(t_{k+1})).$$

Proceeding further, we obtain

$$Ef(X(t_{k+1})) = \sum_{i=0}^{k-1} Ef(X_{t_{i+1}, X_{t_i, \bar{X}_i}(t_{i+1})}(t_{k+1})) \tag{2.2.10}$$
$$- \sum_{i=0}^{k-1} Ef(X_{t_{i+1}, \bar{X}_{i+1}}(t_{k+1})) + Ef(X_{t_k, \bar{X}_k}(t_{k+1})).$$

This immediately implies the identity (recall that $\bar{X}_{i+1} = \bar{X}_{t_i, \bar{X}_i}(t_{i+1})$) :

$$Ef(X(t_{k+1})) - Ef(\bar{X}_{k+1}) \tag{2.2.11}$$

$$= \sum_{i=0}^{k-1}(EE(f(X_{t_{i+1}, X_{t_i, \bar{X}_i}(t_{i+1})}(t_{k+1}))|X_{t_i, \bar{X}_i}(t_{i+1}))$$

$$- EE(f(X_{t_{i+1}, \bar{X}_{t_i, \bar{X}_i}(t_{i+1})}(t_{k+1}))|\bar{X}_{t_i, \bar{X}_i}(t_{i+1}))$$

$$+ Ef(X_{t_k, \bar{X}_k}(t_{k+1})) - Ef(\bar{X}_{t_k, \bar{X}_k}(t_{k+1})) .$$

According to the definition of $u(s, x)$, (2.2.11) implies

$$|Ef(X(t_{k+1})) - Ef(\bar{X}_{k+1})| \tag{2.2.12}$$

$$= |\sum_{i=0}^{k-1}(Eu(t_{i+1}, X_{t_i, \bar{X}_i}(t_i + h)) - Eu(t_{i+1}, \bar{X}_{t_i, \bar{X}_i}(t_i + h)))$$

$$+ (Ef(X_{t_k, \bar{X}_k}(t_{k+1})) - Ef(\bar{X}_{t_k, \bar{X}_k}(t_{k+1})))|$$

$$\leq \sum_{i=0}^{k-1} E|E(u(t_{i+1}, X_{t_i, \bar{X}_i}(t_i + h)) - u(t_{i+1}, \bar{X}_{t_i, \bar{X}_i}(t_i + h))|\bar{X}_i)|$$

$$+ E|E(f(X_{t_k, \bar{X}_k}(t_{k+1})) - f(\bar{X}_{t_k, \bar{X}_k}(t_{k+1}))|\bar{X}_k)| .$$

We note that the functions $u(s, x)$ and $f(x)$, which belong to $\mathbf{F}$ and so satisfy an inequality of the form (2.2.7), also satisfy the conditional version of this inequality. Suppose that for both $u(s, x)$ and $f(x)$ we have a function $K(x)$ in this inequality with $\kappa = 2m$. Then (2.2.12) implies

$$|Ef(X(t_{k+1})) - Ef(\bar{X}_{k+1})|$$

$$\leq \sum_{i=0}^{k-1} K(1 + E|\bar{X}_i|^{2m})h^{p+1} + K(1 + E|\bar{X}_k|^{2m})h^{p+1}.$$

Assuming that the requirement (d) holds for precisely this $2m$, we arrive at (2.2.5).
□

We will now give a sufficient condition for requirement (d) in Theorem 2.2.1 which is convenient in practice. In Sect. 2.2.5 we will discuss how this condition can be relaxed.

Lemma 2.2.2 *Suppose that for $h < 1$,*

$$|EA(t_k, x, h; \xi_k)| \leq K(1 + |x|)h , \tag{2.2.13}$$

$$|A(t_k, x, h; \xi_k)| \leq M(\xi_k)(1 + |x|)h^{1/2} , \tag{2.2.14}$$

where $M(\xi_k)$ has moments of all orders.

Then for every even number $2m$ the expectations $E|\bar{X}_k|^{2m}$ exist and are uniformly bounded with respect to N and $k = 1, \ldots, N$, if only $E|\bar{X}_0|^{2m}$ exists.

Proof For the i-th coordinate of the vector $\bar{X}_{k+1}$ we have

$$
\begin{aligned}
(\bar{X}_{k+1}^i)^{2m} &= (\bar{X}_k^i + A^i(t_k, \bar{X}_k, h; \xi_k))^{2m} \qquad\qquad\qquad (2.2.15) \\
&= (\bar{X}_k^i)^{2m} + C_{2m}^1 (\bar{X}_k^i)^{2m-1} A^i(t_k, \bar{X}_k, h; \xi_k) \\
&\quad + \sum_{j=2}^{2m} C_{2m}^j (\bar{X}_k^i)^{2m-j} (A^i(t_k, \bar{X}_k, h; \xi_k))^j .
\end{aligned}
$$

Using (2.2.13), we obtain

$$
\begin{aligned}
&|E(\bar{X}_k^i)^{2m-1} A^i(t_k, \bar{X}_k, h; \xi_k)| \qquad\qquad\qquad\qquad (2.2.16) \\
&= |E((\bar{X}_k^i)^{2m-1} E(A^i(t_k, \bar{X}_k, h; \xi_k)|\bar{X}_k))| \\
&\le |E|\bar{X}_k^i|^{2m-1} K(1 + |\bar{X}_k|)h \le K(1 + E|\bar{X}_k|^{2m})h .
\end{aligned}
$$

By (2.2.14), we obtain for $h < 1$ and $j = 2, \ldots, 2m$:

$$
\begin{aligned}
&|E|\bar{X}_k^i|^{2m-j} (A^i(t_k, \bar{X}_k, h; \xi_k))^j| \qquad\qquad\qquad\qquad (2.2.17) \\
&\le E(|\bar{X}_k^i|^{2m-j} (M(\xi_k))^j (1 + |\bar{X}_k|)^j h^{j/2}) \le K(1 + E|\bar{X}_k|^{2m})h .
\end{aligned}
$$

Because of (2.2.15)–(2.2.17) and the inequality $|x|^{2m} \le K \sum_{i=1}^d (x^i)^{2m}$, where the constant K depends on d and m only, we obtain

$$
E \sum_{i=1}^d (\bar{X}_{k+1}^i)^{2m} \le E \sum_{i=1}^d (\bar{X}_k^i)^{2m} + K(1 + E \sum_{i=1}^d (\bar{X}_k^i)^{2m})h .
$$

Using Lemma 1.1.6, this concludes the proof of the lemma. □

Theorem 2.2.1 and Lemma 2.2.2 imply a theorem on the order of accuracy of the method

$$
\begin{aligned}
X_{k+1} &= X_k + \sum_{r=1}^q \sigma_{rk} \xi_{rk} h^{1/2} + a_k h + \sum_{r=1}^q \sum_{i=1}^q (\Lambda_i \sigma_r)_k \xi_{irk} h \qquad (2.2.18) \\
&\quad + \frac{1}{2} \sum_{r=1}^q (\Lambda_r a + L\sigma_r)_k \xi_{rk} h^{3/2} + (La)_k \frac{h^2}{2} ,
\end{aligned}
$$

which is constructed according to (2.1.31).

In (2.2.18) the coefficients σ_{rk}, a_k, $(\Lambda_i \sigma_r)_k$, etc. are calculated at the point (t_k, X_k), and the sets of random variables ξ_{rk}, ξ_{irk} are independent and can be modeled for each k as in (2.1.31).

Theorem 2.2.3 *Suppose the conditions of Lemma 2.1.2 hold. Suppose also that the functions $\Lambda_i \sigma_r$, $\Lambda_r a$, $L \sigma_r$, and La grow at most as a linear function in $|x|$ as $|x|$ grows (the functions a and σ_r satisfy this requirement thanks to the Lipschitz condition (2.1.2)), i.e., (2.2.13)–(2.2.14) hold for (2.2.18). Then the method (2.2.18) has order of accuracy 2 in the sense of weak approximation, i.e., for a sufficiently large class of functions f we have (2.2.5) with $p = 2$ (under the conditions of this theorem, this class of functions contains the functions that belong, together with their partial derivatives with respect to x up to order 6, inclusively, to* **F***).*

The proof of this theorem clearly follows from the properties of the one-step approximation (2.1.31) proved in Sect. 2.1, Lemma 2.2.2, and Theorem 2.2.1.

Example 2.2.4 Consider the one-dimensional equation (2.1.1) with a single noise, i.e. $q = 1$. In this case

$$\xi_{11} = \frac{1}{2}(\xi^2 - 1),$$

where ξ is, e.g., $\mathcal{N}(0, 1)$-distributed or distributed by the law $P(\xi = 0) = 2/3$, $P(\xi = -\sqrt{3}) = P(\xi = \sqrt{3}) = 1/6$. The formula (2.2.18) takes the form

$$X_{k+1} = X_k + \sigma_k \xi_k h^{1/2} + a_k h + \frac{1}{2}(\sigma \frac{\partial \sigma}{\partial x})_k (\xi_k^2 - 1)h \qquad (2.2.19)$$

$$+ \frac{1}{2}(\frac{\partial \sigma}{\partial t} + a \frac{\partial \sigma}{\partial x} + \frac{1}{2}\sigma^2 \frac{\partial^2 \sigma}{\partial x^2} + \sigma \frac{\partial a}{\partial x})_k \xi_k h^{3/2}$$

$$+ (\frac{\partial a}{\partial t} + a \frac{\partial a}{\partial x} + \frac{1}{2}\sigma^2 \frac{\partial^2 a}{\partial x^2})_k \frac{h^2}{2}.$$

This formula was derived in [270] using Taylor expansions of the characteristic functions of the variables $\Delta = X_{t,x}(t + h) - x$ and $\bar{\Delta} = \bar{X}_{t,x}(t + h) - x$.

2.2.2 Runge–Kutta Type Methods

The method (2.2.19) may present some difficulties because of the necessity of computing the derivatives of the coefficients a and σ at each step. Using the idea of Runge–Kutta methods, one can propose a number of ways in which by recalculation one can obtain a method not including all the derivatives participating in (2.2.19). We give here a specific, sufficiently simple method of this kind (it was proposed in [270]):

$$X_{k+1} = X_k + \frac{1}{2}\sigma_k\xi_k h^{1/2} + \frac{1}{2}(a - \sigma\frac{\partial\sigma}{\partial x})_k h + \frac{1}{2}(\sigma\frac{\partial\sigma}{\partial x})_k \xi_k^2 h \quad (2.2.20)$$

$$+\frac{1}{2}a(t_k + h, X_k + \sigma_k\xi_k h^{1/2} + a_k h)h$$

$$+\frac{1}{4}\sigma(t_k + h, X_k + \sigma_k\xi_k(\frac{h}{3})^{1/2} + a_k h)\xi_k h^{1/2}$$

$$+\frac{1}{4}\sigma(t_k + h, X_k - \sigma_k\xi_k(\frac{h}{3})^{1/2} + a_k h)\xi_k h^{1/2},$$

where ξ_k are the same variables as in (2.2.19).

To get convinced of the fact that the method (2.2.20) is a method of order two, we note that

$$\frac{1}{2}a(t + h, x + \sigma\xi h^{1/2} + ah)h \quad\quad\quad (2.2.21)$$

$$= \frac{1}{2}(a + \frac{\partial a}{\partial t}h + \frac{\partial a}{\partial x}ah + \frac{\partial a}{\partial x}\sigma\xi h^{1/2} + \frac{1}{2}\frac{\partial^2 a}{\partial x^2}\sigma^2\xi^2 h)h$$

$$+\frac{1}{2}\frac{\partial^2 a}{\partial x^2}a\sigma\xi h^{5/2} + \frac{1}{2}\frac{\partial^2 a}{\partial t\partial x}\sigma\xi h^{5/2} + O(h^3),$$

$$\frac{1}{4}\sigma(t + h, x + \sigma\xi(\frac{h}{3})^{1/2} + ah)\xi h^{1/2} \quad\quad (2.2.22)$$

$$+\frac{1}{4}\sigma(t + h, x - \sigma\xi(\frac{h}{3})^{1/2} + ah)\xi h^{1/2}$$

$$= \frac{1}{4}(2\sigma + 2\frac{\partial\sigma}{\partial t}h + 2\frac{\partial\sigma}{\partial x}ah + \frac{1}{3}\frac{\partial^2\sigma}{\partial x^2}\sigma^2\xi^2 h)\xi h^{1/2}$$

$$+\frac{1}{4}\frac{\partial^2\sigma}{\partial t^2}\xi h^{5/2} + \frac{1}{2}\frac{\partial^2\sigma}{\partial t\partial x}a\xi h^{5/2} + \frac{1}{12}\frac{\partial^3\sigma}{\partial t\partial x^2}\sigma^2\xi^3 h^{5/2}$$

$$+\frac{1}{4}\frac{\partial^2\sigma}{\partial x^2}a^2\xi h^{5/2} + \frac{1}{12}\frac{\partial^3\sigma}{\partial x^3}a\sigma^2\xi^3 h^{5/2} + \frac{1}{432}\frac{\partial^4\sigma}{\partial x^4}\sigma^4\xi^5 h^{5/2} + O(h^3).$$

Substituting (2.2.21) and (2.2.22) in (2.2.20), we observe that X_{k+1} in (2.2.20) differs from X_{k+1} in (2.2.19), first by the sum

$$s_1 = (\frac{1}{2}\frac{\partial^2 a}{\partial x^2}a\sigma\xi + \frac{1}{2}\frac{\partial^2 a}{\partial t\partial x}\sigma\xi_k + \frac{1}{4}\frac{\partial^2\sigma}{\partial t^2}\xi_k + \frac{1}{2}\frac{\partial^2\sigma}{\partial t\partial x}a\xi_k + \frac{1}{12}\frac{\partial^3\sigma}{\partial t\partial x^2}\sigma^2\xi_k^3$$

$$+\frac{1}{4}\frac{\partial^2\sigma}{\partial x^2}a^2\xi_k + \frac{1}{12}\frac{\partial^3\sigma}{\partial x^3}a\sigma^2\xi_k^3 + \frac{1}{432}\frac{\partial^4\sigma}{\partial x^4}\sigma^4\xi_k^5)h^{5/2} + O(h^3),$$

secondly, the term $\frac{1}{4}\sigma^2\frac{\partial^2\sigma}{\partial x^2}\xi_k h^{3/2}$ in (2.2.19) is replaced by $\frac{1}{12}\sigma^2\frac{\partial^2\sigma}{\partial x^2}\xi_k^3 h^{3/2}$, and, thirdly, the term $\frac{1}{4}\sigma^2\frac{\partial^2 a}{\partial x^2}h^2$ in (2.2.19) is replaced by $\frac{1}{4}\sigma^2\frac{\partial^2 a}{\partial x^2}\xi_k^2 h^2$. It is easy to see that these differences have no influence on the fulfillment of the conditions of the main Theorem 2.2.1 with $p = 2$.

Thus, the method (2.2.20) has order of accuracy two. At each step it requires two recalculations of the function a, three recalculations of the function σ, one calculation of the function $\partial\sigma/\partial x$, and modeling of the single random variable ξ.

Other Runge–Kutta type methods can be found in [3, 245, 351, 359, 386, 387, 428, 440] and also references therein.

Consider: the system in the sense of Stratonovich

$$dX = a(t, X)dt + \sum_{r=1}^{q} \sigma_r(t, X) \circ dw_r(t). \qquad (2.2.23)$$

It can be re-written in the form of Ito (see Sect. 1.1.3) to which the methods of this chapter can be then applied. However, this Ito SDEs contain derivatives of $\sigma_r(t, x)$ and even first weak-order methods (e.g. the Euler scheme (2.0.21)) applied to these Ito SDEs are not derivative-free. Consider the following explicit numerical scheme for (2.2.23) (it is a version of so-called Heun's scheme):

$$X_{k+1} = X_k + ha(t_k, X_k) \qquad (2.2.24)$$
$$+h^{1/2} \sum_{r=1}^{q} (\sigma_r(t_k, X_k) + \sigma_r(t_k, X_k + h^{1/2}\sigma_r(t_k, X_k)\alpha_{rk}))/2\, \alpha_{rk},$$

where α_{rk} are independent random variables taking the values $+1$ and -1 with probabilities $1/2$. Using the standard one-step analysis like in Sect. 2.1 and Theorem 2.2.1, it is not difficult to show that the scheme (2.2.24) for (2.2.23) is of weak order one. Higher order Runge–Kutta type methods for (2.2.23) can be found e.g. in [386].

2.2.3   The Talay–Tubaro Extrapolation Method

Talay and Tubaro proved in [433] that it is possible to expand the global errors of weak methods for stochastic systems in powers of time increment h. Their approach is analogous to the Richardson–Runge extrapolation method for ordinary differential equations and allows us to estimate the global error as well as to improve the accuracy of the method. In particular, we can construct a method of order two applying the Euler method twice with different time steps.

Here we suppose that the coefficients of (2.1.1) are sufficiently smooth and all their derivatives up to a sufficiently large order are bounded.

Theorem 2.2.5 *Let a one-step weak approximation* $\bar{X}_{t,x}(t + h)$ *of the solution* $X_{t,x}(t + h)$ *of (2.1.1) generate a method of order p. Then the global error*

$$R := Ef(X_{t_0,X_0}(T)) - Ef(\bar{X}_{t_0,X_0}(T))$$

of the method has the following expansion

$$R = C_0 h^p + \cdots + C_n h^{p+n} + O(h^{p+n+1}),$$ (2.2.25)

where the constants $C_0, \ldots, C_n$ are independent of h, and n is an integer, $n \geq 0$ (n can be anyhow large if the coefficients of (2.1.1) belong to C^∞ and their derivatives of any order are bounded).

In particular, for the Euler method

$$R = C_0 h + \cdots + C_n h^{1+n} + O(h^{n+2}).$$ (2.2.26)

Proof We begin with the proof of the formula

$$R = Ch + O(h^2)$$ (2.2.27)

for the Euler method

$$X_{k+1} = X_k + \sum_{r=1}^{q} \sigma_{rk} \xi_{rk} h^{1/2} + a_k h,$$ (2.2.28)

where all the ξ_{rk} are independent and are modeled by the law $P(\xi = \pm 1) = 1/2$. As in the proof of the convergence Theorem 2.2.1, we obtain the equality (see (2.2.11) and (2.2.12))

$$R = \sum_{i=0}^{N-1} E(u(t_{i+1}, X_{t_i,\bar{X}_i}(t_{i+1})) - u(t_{i+1}, \bar{X}_{t_i,\bar{X}_i}(t_{i+1})))$$ (2.2.29)

$$= E \sum_{i=0}^{N-1} E(u(t_{i+1}, X_{t_i,\bar{X}_i}(t_{i+1})) - u(t_{i+1}, \bar{X}_{t_i,\bar{X}_i}(t_{i+1}))|\bar{X}_i),$$

where

$$u(s, x) = Ef(X_{s,x}(T)).$$

By Lemma 2.1.9 we get

$$Eu(t + h, X_{t,x}(t + h)) = u(t, x) + hLu(t, x) + \frac{1}{2}h^2 L^2 u(t, x) + O(h^3).$$ (2.2.30)

Expanding $Eu(t + h, \bar{X}_{t,x}(t + h))$ in powers of h by the usual Taylor formula, we obtain

$$Eu(t + h, \bar{X}_{t,x}(t + h)) = u(t, x) + hLu(t, x) + \frac{1}{2}h^2 A(t, x) + O(h^3).$$ (2.2.31)

Two first terms in the right-hand sides of (2.2.30) and (2.2.31) coincide since the one-step order for the Euler method is equal to two. The direct computation gives

$$
\begin{aligned}
A(t, x) = {} & \frac{\partial^2 u}{\partial t^2} + 2 \sum_{i=1}^{d} a^i \frac{\partial^2 u}{\partial t \partial x^i} + \sum_{r=1}^{q} \sum_{i,j=1}^{d} \sigma_r^i \sigma_r^j \frac{\partial^3 u}{\partial t \partial x^i \partial x^j} \\
& + \sum_{i,j=1}^{d} a^i a^j \frac{\partial^2 u}{\partial x^i \partial x^j} + \sum_{r=1}^{q} \sum_{i,j,l=1}^{d} a^i \sigma_r^j \sigma_r^l \frac{\partial^3 u}{\partial x^i \partial x^j \partial x^l} \\
& + \frac{1}{12} \sum_{r,s=1}^{q} \sum_{i,j,l,m=1}^{d} \sigma_r^i \sigma_r^j \sigma_s^l \sigma_s^m \frac{\partial^4 u}{\partial x^i \partial x^j \partial x^l \partial x^m} ,
\end{aligned}
$$

where all the coefficients and the derivatives of the function u are calculated at the point (t, x). We should underline that in fact we do not need the explicit form of the function $A(t, x)$. We have given it for definiteness only. A little later we make use of this observation in the proof of the general assertion.

It follows from (2.2.30) and (2.2.31) that

$$
Eu(t + h, X_{t,x}(t + h)) - Eu(t + h, \bar{X}_{t,x}(t + h)) = h^2 B(t, x) + O(h^3) ,
$$
$$(2.2.32)$$

where $B(t, x) = (L^2 u(t, x) - A(t, x))/2$. The formula (2.2.29) can be rewritten in the form

$$
R = E \sum_{i=0}^{N-1} h^2 B(t, \bar{X}_i) + O(h^2) .
$$
$$(2.2.33)$$

Consider now the $(d + 1)$-dimensional system

$$
dX = a(t, X)dt + \sum_{r=1}^{q} \sigma_r(t, X)dw_r(t), \quad X(t_0) = X_0 ,
$$
$$(2.2.34)$$
$$
dY = B(t, X)dt, \quad Y(t_0) = 0 .
$$

Solving (2.2.34) by the Euler method, we get

$$
E \sum_{i=0}^{N-1} B(t, \bar{X}_i)h = E\bar{Y}(T) = EY(T) + O(h) = C + O(h) ,
$$
$$(2.2.35)$$

where the constant C is equal to

$$
C = EY(T) = E \int_{t_0}^{T} B(\theta, X(\theta))d\theta .
$$
$$(2.2.36)$$

The formula (2.2.27) follows from (2.2.33) and (2.2.35).

Let us prove (2.2.26). Now, instead of (2.2.32), we use the formula

$$Eu(t + h, X_{t,x}(t + h)) - Eu(t + h, \bar{X}_{t,x}(t + h))$$
$$= h^2 B_0(t, x) + \cdots + h^{n+2} B_n(t, x) + O(h^{n+3}),$$

with

$$B_j(t, x) = \frac{1}{(j + 2)!}(L^{2+j}u(t, x) - A_j(t, x)),$$

where $A_j(t, x)$ are coefficients of the corresponding expansion of $Eu(t + h, \bar{X}_{t,x}(t + h))$. As a result, we get instead of (2.2.33):

$$R = E \sum_{i=0}^{N-1} \sum_{j=0}^{n} h^{2+j} B_j(t, \bar{X}_i) + O(h^{n+2}). \tag{2.2.37}$$

For each $j = 0, \ldots, n$ consider the $(d + 1)$-dimensional system like to (2.2.34). Due to (2.2.35) we can write

$$R = \sum_{j=0}^{n} K_j h^{1+j} + hR_1 + O(h^{n+2}), \tag{2.2.38}$$

where $K_j = EY_j(T)$ and

$$R_1 = E \sum_{j=0}^{n} h^j (\bar{Y}_j - Y_j).$$

Clearly, R_1 has a representation analogous to (2.2.38) and we obtain (2.2.26) in a finite number of such steps. The formula (2.2.25) is proved analogously. □

Due to Theorem 2.2.5, we obtain an extension of the well known in the case of deterministic differential equations extrapolation methods to the stochastic case. For example, simulating $u = Ef(X_{t_0, X_0}(T))$ twice by the Euler scheme but with varying time steps $h_1 = h$, $h_2 = \alpha h$, $\alpha > 0$, $\alpha \neq 1$, we obtain $\bar{u}^{h_1} = Ef(\bar{X}_{t_0, X_0}^{h_1}(T))$ and $\bar{u}^{h_2} = Ef(\bar{X}_{t_0, X_0}^{h_2}(T))$. We can expand (see (2.2.27))

$$u = \bar{u}^{h_1} + Ch_1 + O(h^2), \tag{2.2.39}$$
$$u = \bar{u}^{h_2} + Ch_2 + O(h^2),$$

whence

$$C = -\frac{\bar{u}^{h_2} - \bar{u}^{h_1}}{h_2 - h_1} + O(h). \tag{2.2.40}$$

By (2.2.39) and (2.2.40) we get the improved value with error $O(h^2)$:

$$\bar{u}_{imp} = \bar{u}^{h_1} \frac{h_2}{h_2 - h_1} - \bar{u}^{h_2} \frac{h_1}{h_2 - h_1}, \quad u = \bar{u}_{imp} + O(h^2). \tag{2.2.41}$$

Thus, the obtained method has an accuracy of order two. In the same spirit, using three recalculations of $u = E f(X_{t_0, X_0}(T))$ by the Euler method with varying time-steps, one can find C_0 and C_1 from (2.2.26) and, as a consequence, a method of order three can be constructed, and so on. The formula (2.2.25) can be used in the same way. We see that extrapolation procedures make it possible to construct methods of higher orders much simpler to implement than procedures based on one-step approximations of high order. But the extrapolation procedures have certain deficiencies, they are not general-purpose. For example, they do not allow us to use schemes with variable step. So that, similarly to deterministic case, the one-step high-order approximations are of independent significance.

2.2.4 Implicit Method

The following formula holds (cf. Sect. 1.2.2):

$$a(t + h, X(t + h)) = a + \sum_{r=1}^{q} \Lambda_r a \int_t^{t+h} dw_r(\theta) + La \cdot h + \rho_1, \tag{2.2.42}$$

where

$$\rho_1 = \sum_{r=1}^{q} \sum_{i=1}^{q} \Lambda_i \Lambda_r a \int_t^{t+h} (\int_t^{\theta} dw_i(\theta_1)) dw_r(\theta) \tag{2.2.43}$$

$$+ \sum_{r=1}^{q} \sum_{i=1}^{q} \sum_{s=1}^{q} \int_t^{t+h} (\int_t^{\theta} (\int_t^{\theta_1} \Lambda_s \Lambda_i \Lambda_r a(\theta_2, X(\theta_2)) dw_s(\theta_2)) dw_i(\theta_1)) dw_r(\theta)$$

$$+ \sum_{r=1}^{q} \sum_{i=1}^{q} \int_t^{t+h} (\int_t^{\theta} (\int_t^{\theta_1} L\Lambda_i \Lambda_r a(\theta_2, X(\theta_2)) d\theta_2) dw_i(\theta_1)) dw_r(\theta)$$

$$+ \sum_{r=1}^{q} \int_t^{t+h} (\int_t^{\theta} L\Lambda_r a(\theta_1, X(\theta_1)) d\theta_1) dw_r(\theta)$$

$$+ \sum_{r=1}^{q} \int_t^{t+h} (\int_t^{\theta} \Lambda_r La(\theta_1, X(\theta_1)) dw_r(\theta_1)) d\theta.$$

As in Lemma 2.1.2, we can show that

$$|E\rho_1| \leq K(x)h^2, \quad K(x) \in \mathbf{F}, \tag{2.2.44}$$

$$E|\rho_1|^2 \leq K(x)h^2, \quad K(x) \in \mathbf{F}, \tag{2.2.45}$$

$$\left|E\rho_1 \int_t^{t+h} dw_r(\theta)\right| \leq K(x)h^2, \quad K(x) \in \mathbf{F}. \tag{2.2.46}$$

Further,

$$La(t+h, X(t+h)) = La + \rho_2, \tag{2.2.47}$$

where

$$\rho_2 = \sum_{r=1}^{q} \int_t^{t+h} \Lambda_r La(\theta, X(\theta))dw_r(\theta) + \int_t^{t+h} L^2 a(\theta, X(\theta))d\theta. \tag{2.2.48}$$

We can readily prove that

$$|E\rho_2| \leq K(x)h, \quad K(x) \in \mathbf{F}, \tag{2.2.49}$$

$$E|\rho_2|^2 \leq K(x)h, \quad K(x) \in \mathbf{F}, \tag{2.2.50}$$

$$\left|E\rho_2 \int_t^{t+h} dw_r(\theta)\right| \leq K(x)h, \quad K(x) \in \mathbf{F}. \tag{2.2.51}$$

Using the relations (2.1.3), (2.2.42), and (2.2.47), it is easy to obtain the following formula, involving the arbitrary constants α, β :

$$X(t+h) = x + \sum_{r=1}^{q} \sigma_r I_r + \alpha a h + (1-\alpha)a(t+h, X(t+h))h \tag{2.2.52}$$

$$+ \sum_{r=1}^{q}\sum_{i=1}^{q} \Lambda_i \sigma_r I_{ir} + \sum_{r=1}^{q}(L\sigma_r - (1-\alpha)\Lambda_r a)I_r h$$

$$+ \sum_{r=1}^{q}(\Lambda_r a - L\sigma_r)J_r$$

$$+ \beta(2\alpha - 1)La\frac{h^2}{2} + (1-\beta)(2\alpha - 1)La(t+h, X(t+h))\frac{h^2}{2}$$

$$+ \sum_{r=1}^{q}\sum_{i=1}^{q}\sum_{s=1}^{q} \Lambda_s \Lambda_i \sigma_r I_{sir} + \rho - (1-\alpha)\rho_1 h$$

$$- (1-\beta)(2\alpha - 1)\rho_2 h^2.$$

Introduce the two-parameter family of implicit methods

$$\bar{X} = x + \sum_{r=1}^{q} \sigma_r \xi_r h^{1/2} + \alpha a h + (1 - \alpha) a(t + h, \bar{X}) h \tag{2.2.53}$$

$$+ \sum_{r=1}^{q} \sum_{i=1}^{q} \Lambda_i \sigma_r \xi_{ir} h + \sum_{r=1}^{q} (\frac{2\alpha - 1}{2} \Lambda_r a + \frac{1}{2} L\sigma_r) \xi_r h^{3/2}$$

$$+ \beta(2\alpha - 1) L a \frac{h^2}{2} + (1 - \beta)(2\alpha - 1) L a(t + h, \bar{X}) \frac{h^2}{2},$$

where ξ_r, ξ_{ir} can be modeled as in Sect. 2.1 (see (2.1.31)).

We show that (under certain natural assumptions) the method (2.2.53) has order of accuracy two. To this end we consider the equation

$$X - (1 - \alpha) a(t + h, X) h - (1 - \beta)(2\alpha - 1) L a(t + h, X) \frac{h^2}{2} = Z. \tag{2.2.54}$$

We assume that for sufficiently small h and all Z this equation can be solved for X:

$$X = \varphi(t + h, Z). \tag{2.2.55}$$

Introduce vectors Y, $\bar{Y}$:

$$Y = x + \sum_{r=1}^{q} \sigma_r I_r + \alpha a h + \sum_{r=1}^{q} \sum_{i=1}^{q} \Lambda_i \sigma_r I_{ir} \tag{2.2.56}$$

$$+ \sum_{r=1}^{q} (L\sigma_r - (1 - \alpha) \Lambda_r a) I_r h + \sum_{r=1}^{q} (\Lambda_r a - L\sigma_r) J_r + \beta(2\alpha - 1) L a \frac{h^2}{2}$$

$$+ \sum_{r=1}^{q} \sum_{i=1}^{q} \sum_{s=1}^{q} \Lambda_s \Lambda_i \sigma_r I_{sir} + \rho - (1 - \alpha) \rho_1 h - (1 - \beta)(2\alpha - 1) \rho_2 h^2,$$

$$\bar{Y} = x + \sum_{r=1}^{q} \sigma_r \xi_r h^{1/2} + \alpha a h + \sum_{r=1}^{q} \sum_{i=1}^{q} \Lambda_i \sigma_r \xi_{ir} h \tag{2.2.57}$$

$$+ \sum_{r=1}^{q} (\frac{2\alpha - 1}{2} \Lambda_r a + \frac{1}{2} L\sigma_r) \xi_r h^{3/2} + \beta(2\alpha - 1) L a \frac{h^2}{2}.$$

Then by (2.2.52), (2.2.54), and (2.2.56):

$$X(t + h) = \varphi(t + h, Y),$$

and by (2.2.53), (2.2.54), and (2.2.57):

$$\bar{X} = \bar{X}(t+h) = \varphi(t+h, \bar{Y}).$$

Assume that the function $\varphi(t+h, y)$ has partial derivatives with respect to y up to order six, inclusively, and that they together with φ belong to $\mathbf{F}$. For $s = 1, \ldots, 5$ we write

$$|E(\prod_{j=1}^{s} \Delta^{i_j} - \prod_{j=1}^{s} \bar{\Delta}^{i_j})| \tag{2.2.58}$$

$$= |E(\prod_{j=1}^{s} (X^{i_j} - x^{i_j}) - \prod_{j=1}^{s} (\bar{X}^{i_j} - x^{i_j}))|$$

$$= |E(\prod_{j=1}^{s} (\varphi^{i_j}(t+h, Y) - x^{i_j}) - \prod_{j=1}^{s} (\varphi^{i_j}(t+h, \bar{Y}) - x^{i_j}))|.$$

The right-hand side of (2.2.58) is $O(h^3)$ if (see the proof of Theorem 2.1.6):

$$|E(\prod_{j=1}^{s} (Y^{r_j} - x^{r_j}) - \prod_{j=1}^{s} (\bar{Y}^{r_j} - x^{r_j}))| \le K(x)h^3, \ s = 1, \ldots, 5, \tag{2.2.59}$$

and if

$$E \prod_{j=1}^{s} |Y^{r_j} - x^{r_j}| \le K(x)h^3, \ E \prod_{j=1}^{s} |\bar{Y}^{r_j} - x^{r_j}| \le K(x)h^3, \ s = 6. \tag{2.2.60}$$

Taking into account the properties (2.2.44)–(2.2.46) and (2.2.49)–(2.2.51) of the remainders ρ_1 and ρ_2, the relations (2.2.59)–(2.2.60) can be proved as in Sect. 2.1 we proved the analogous relations for the differences $X^{i_j} - x^{i_j}$ and $\bar{X}^{i_j} - x^{i_j}$. So, we have proved that $\bar{X}$, which is implicitly defined by (2.2.53), satisfies

$$|E(\prod_{j=1}^{s} \Delta^{i_j} - \prod_{j=1}^{s} \bar{\Delta}^{i_j})| \le K(x)h^3, \ s = 1, \ldots, 5, \ K(x) \in \mathbf{F}. \tag{2.2.61}$$

We will prove the inequality

$$E \prod_{j=1}^{s} |\bar{\Delta}^{i_j}| \le K(x)h^3, \ s = 6, \ K(x) \in \mathbf{F}. \tag{2.2.62}$$

In fact, the solvability of (2.2.53) for $\bar{X}$ in the form $\bar{X} = \varphi(t+h, \bar{Y})$ with $\varphi \in \mathbf{F}$, implies existence of all sufficiently high moments of $\bar{X}$ if only ξ_r and ξ_{ir} (which participate in the formula for $\bar{Y}$) have sufficiently high moments. Further, since $a \in \mathbf{F}$ and $La \in \mathbf{F}$, moments (up to order six, inclusively) for $a(t + h, \bar{X})$ and $La(t + h, \bar{X})$ exist. Now (2.2.62) immediately follows from (2.2.53).

Finally, assume that $\varphi(t + h, \bar{Y})$ grows at most linearly as $|x|$ goes to infinity. Then the subsequent application of Lemma 2.2.2 and Theorem 2.2.1 leads to the result which we state as the theorem.

Theorem 2.2.6 *Suppose that for sufficiently small h the relation (2.2.54) is solvable for Z : $X = \varphi(t + h, Z)$. Suppose that the function $\varphi(t + h, y)$ has partial deriva- tives with respect to y up to order six, inclusively, that together with φ belong to* **F**. *Finally, assume that the superposition $\varphi(t + h, \bar{Y})$, with $\bar{Y}$ defined by (2.2.57), grows at most linearly as $|x| \to \infty$. Then the implicit method based on (2.2.53) has order of accuracy two in the sense of weak approximation.*

2.2.5 The Concept of Rejecting Exploding Trajectories for SDEs with Nonglobally Lipschitz Coefficients

SDEs with nonglobally Lipschitz coefficients possessing unique solutions make up a very important class in applications. For instance, Langevin-type equations and gradient systems with noise belong to this class (see Chap. 5). At the same time, most numerical methods for SDEs are derived under the global Lipschitz condition which we have used in this chapter so far. If this condition is violated, the behavior of many standard numerical methods in the whole space can potentially lead to incorrect conclusions. This situation is very alarming since we could be forced to refuse many effective methods and/or to resort to some comparatively complicated and inefficient numerical procedures. In this section we consider the concept of rejecting explod- ing trajectories, which allows us to apply any method of weak approximation to a very broad class of SDEs with nonglobally Lipschitz coefficients. Roughly speaking, we require for SDEs from this class just to have regular solutions on a time interval $[t_0, T]$ and to have sufficiently smooth coefficients, i.e., the assumptions made (which are given in terms of Lyapunov functions in Sect. 2.2.5.1) usually hold for SDEs of applicable interest. Following the concept, we discard the approximate trajectories which leave a sufficiently large sphere $S_R := \{x : |x| < R\}$. The theoretical justifi- cation of the concept is given in Sect. 2.2.5.1. We prove that accuracy of any method of weak order p is estimated by $\varepsilon + O(h^p)$, where ε can be made arbitrarily small with growing R, and $|O(h^p)| \leq Kh^p$ can be made arbitrarily small with decreasing h (of course, K may depend on R). Thus, we obtain that the violation of the global Lipschitz condition is not fatal for applying any method of numerical integration. Since the Monte Carlo technique is used for simulation of a mean $Ef(X(T))$, the error estimation $\varepsilon + O(h^p)$ should be increased by the Monte Carlo error. It turns out that in practice the error given by ε is much smaller than the joint numerical integration and Monte Carlo error. Furthermore, in principle, due to the concept of rejecting "bad" trajectories, we can choose a suitable method for solving a system of SDEs with nonglobally Lipschitz coefficients, taking into account all the known methods of numerical integration presented in this book. The application of the concept is discussed in Sect. 2.2.5.2 , where some numerical experiments are presented. See further details in [316, 319]. Also, see Sect. 1.4 where mean-square convergence of numerical methods for SDEs with nonglobally Lipschitz coefficients is discussed.

2.2.5.1   Integration via Paths in a Bounded Domain

Consider the system of Ito SDEs

$$dX = a(t, X)dt + \sum_{l=1}^{q} \sigma_l(t, X)dw_l(t), \qquad X(t_0) = x, \qquad (2.2.63)$$

where X, a, σ_l are d-dimensional column-vectors and $w_l(t)$, $l = 1, \ldots, q$, are independent standard Wiener processes.

We assume that the coefficients of (2.2.63) are sufficiently smooth functions in $[t_0, T] \times \mathbf{R}^d$, and any solution $X(t; t_0, x)$ of (2.2.63) is regular on $[t_0, T]$. We recall that a process is called regular if it is defined for all $t_0 \leq t \leq T$. Denote by $\mathbf{C}^2$ the class of functions defined on $[t_0, T] \times \mathbf{R}^d$ and twice continuously differentiable with respect to x and once with respect to t. A sufficient condition of regularity (see [194]) consists in the existence of a Lyapunov function $V \in \mathbf{C}^2$, $V \geq 0$, which satisfies the inequality

$$LV(t, x) \leq c_0 V(t, x) + c_1, \qquad (t, x) \in [t_0, T] \times \mathbf{R}^d, \qquad (2.2.64)$$

and

$$V_R := \min_{t_0 \leq t \leq T, \ |x| \geq R} V(t, x), \qquad \lim_{R \to \infty} V_R = \infty, \qquad (2.2.65)$$

where c_0 and c_1 are some constants and L is the generating operator:

$$LV(t, x) = \frac{\partial V}{\partial t}(t, x) + \sum_{i=1}^{d} a^i(t, x)\frac{\partial V}{\partial x^i}(t, x) + \frac{1}{2} \sum_{i,j=1}^{d} a^{ij}(t, x)\frac{\partial^2 V}{\partial x^i \partial x^j}(t, x),$$

$$(2.2.66)$$

$$a^{ij} := \sum_{l=1}^{q} \sigma_l^i \sigma_l^j.$$

Moreover, if (2.2.64) and (2.2.65) are fulfilled and if an initial distribution for x (x can be random) is such that $EV(t_0, x)$ exists, then $EV(t, X(t; t_0, x))$ exists for all $t_0 \leq t \leq T$. For instance, if V has an m-polynomial growth at infinity, then there exist the moments of order m for X. We note that it is not required for c_0 to be negative. For definiteness, we consider $c_0 > 0$ and $c_1 > 0$.

Let $S_R := \{x : |x| < R\}$ be an open sphere in $\mathbf{R}^d$, $Q_R = [t_0, T) \times S_R$ be a cylinder in $\mathbf{R}^{d+1}$, $\Gamma_R = \bar{Q}_R \backslash Q_R$, where $\bar{Q}_R$ is the closure of Q_R. The set Γ_R is a part of the boundary of cylinder Q_R consisting of the upper base and the lateral surface. Let τ_R be the first-passage time of the process $(t, X(t; t_0, x))$, $t_0 \le t \le T$, to Γ_R. Clearly, $t_0 \le \tau_R \le T$. We introduce the following events:

$$\Omega_R := \{\omega : |X(s; t_0, x)| < R, \; t_0 \le s < T\} = \{\omega : \tau_R = T\}, \qquad (2.2.67)$$
$$\Lambda_R := \{\omega : \exists \, s \in [t_0, T) \text{ such that } |X(s; t_0, x)| \ge R\} = \{\omega : \tau_R < T\}.$$

Let us obtain an upper bound for the probability

$$p_R := P(\tau_R < T) = P(\Lambda_R), \qquad (2.2.68)$$

assuming (2.2.64) and (2.2.65) (see [194]). Introduce the nonnegative function

$$U(t, x) = (c_1 + 1)(T - t) + \exp(c_0(t_0 - t))V(t, x), \qquad (2.2.69)$$

where $V(t, x)$ is a function satisfying (2.2.64)–(2.2.65). We get

$$LU(t, x) = -(c_1 + 1) - c_0 \exp(c_0(t_0 - t))V(t, x) \qquad (2.2.70)$$
$$+ \exp(c_0(t_0 - t))LV(t, x) \le -1.$$

Due to the Ito formula, we have

$$dU(t, X(t; t_0, x)) = LU(t, X(t; t_0, x))dt \qquad (2.2.71)$$
$$+ \exp(c_0(t_0 - t)) \sum_{i=1}^{d} \frac{\partial V}{\partial x^i}(t, X(t; t_0, x)) \sum_{l=1}^{q} \sigma_l^i(X(t; t_0, x))dw_l(t).$$

Hence

$$U(\tau_R, X(\tau_R; t_0, x)) - U(t_0, x) = \int_0^{\tau_R} LU \, dt \qquad (2.2.72)$$
$$+ \int_0^{\tau_R} \exp(c_0(t_0 - t)) \sum_{i=1}^{d} \frac{\partial V}{\partial x^i} \sum_{l=1}^{q} \sigma_l^i \, dw_l(t).$$

Expectation of the second integral on the right-hand side of (2.2.72) is equal to zero according to the martingale property. Therefore, due to (2.2.70), we get

$$EU(\tau_R, X(\tau_R; t_0, x)) \le U(t_0, x) = (c_1 + 1)(T - t_0) + V(t_0, x). \qquad (2.2.73)$$

By Chebyshev's inequality, we obtain from (2.2.73):

$$p_R \exp(c_0(t_0 - T)) \min_{t_0 \leq t \leq T, |x| \geq R} V(t, x) \leq p_R \min_{t_0 \leq t \leq T, |x| \geq R} U(t, x) \qquad (2.2.74)$$
$$\leq (c_1 + 1)(T - t_0) + V(t_0, x),$$

whence

$$p_R \leq \exp(c_0(T - t_0)) \frac{(c_1 + 1)(T - t_0) + V(t_0, x)}{\min_{t_0 \leq t \leq T, |x| \geq R} V(t, x)}, \qquad (2.2.75)$$

and therefore

$$\lim_{R \to \infty} p_R = 0.$$

Proposition 2.2.7 *Let (2.2.64) and (2.2.65) be fulfilled. Let $f(x)$ be a function such that*

$$|f(x)| \leq V(t, x), \qquad t_0 \leq t \leq T, \ x \in \mathbf{R}^d. \qquad (2.2.76)$$

Then for any $x \in R^d$ and $\varepsilon > 0$ there exists $R(x, \varepsilon) > 0$ such that for any $R > R(x, \varepsilon)$

$$|Ef(X(T; t_0, x)) - E[f(X(T; t_0, x))\chi_{\Omega_R}(\omega)]| < \varepsilon. \qquad (2.2.77)$$

Proof Clearly,

$$\lim_{R \to \infty} f(X(T; t_0, x))\chi_{\Omega_R}(\omega) = f(X(T; t_0, x)), \quad \text{a.s.}$$

Now the conclusion of this proposition follows from the existence of $EV(T, X(T; t_0, x))$ and the Lebesgue theorem on majorized convergence. □

The significance of this proposition consists of the capability to disregard the trajectories running off too far. We are about to show that when systems under consideration are numerically integrated, the approximating trajectories running off too far can also be discarded. Due to this possibility, we are able, in principle, to use any known method of numerical integration for calculating averages.

In this respect we shall rest on the theory of weak approximation for SDEs with globally Lipschitz coefficients developed earlier in this chapter. Let us introduce an auxiliary system with globally Lipschitz coefficients, which coincides with the original system in a sphere $S_{R'}$ somewhat wider than S_R: $S_{R'} \supset S_R$, where $R' = R + r$ and $r > 0$ is a constant. Let the coefficients a^i, σ_l^i and the function f have continuous derivatives up to some order. The requirement on smoothness depends on a particular numerical method used; in general, the higher the order of the method, the more derivatives are needed. We construct coefficients a_R^i, $(\sigma_l^i)_R$ and function f_R so that in $[t_0, T] \times S_{R'}$ they coincide with a^i, σ_l^i and f, respectively, and, in addition, they are bounded in $[t_0, T] \times \mathbf{R}^d$ together with their derivatives up to the same order. This can be done in the following way. Introduce the function $\varphi(z)$ of one variable z:

$$\varphi(z) = \begin{cases} z, & -R' \le z \le R', \\[2mm] R' + \displaystyle\int_{R'}^{z} \frac{dz'}{1 + (z' - R')^k}, & z > R', \\[4mm] -R' - \displaystyle\int_{z}^{-R'} \frac{dz'}{1 + (-R' - z')^k}, & z < -R', \end{cases} \qquad (2.2.78)$$

where $k \ge 2$ is a natural number. Clearly, $\varphi(z)$ is bounded on $\mathbf{R}^1$ together with its derivatives up to order k. Let $g(t, x^1, \ldots, x^d)$ be a function with some continuous derivatives defined in $[t_0, T] \times \mathbf{R}^d$. It is easily seen that

$$g_R(t, x^1, \ldots, x^d) := g(t, \varphi(x^1), \ldots, \varphi(x^d))$$

satisfies the above-mentioned conditions. Moreover, there exists a constant $\rho > r$ (which does not depend on R) such that for any $x = (x^1, \ldots, x^d) \in \mathbf{R}^d$ the point $(\varphi(x^1), \ldots, \varphi(x^d)) \in S_{R+\rho}$. Therefore

$$\sup_{x \in \mathbf{R}^d} |f_R(x)| \le \max_{|x| \le R+\rho} |f(x)|. \qquad (2.2.79)$$

Introduce the auxiliary system of SDEs

$$dX_R = a_R(X_R)dt + \sum_{l=1}^{q} (\sigma_l^i)_R(X_R)dw_l(t). \qquad (2.2.80)$$

We emphasize that this system is used in our theoretical proofs only; it is not used in simulation.

Proposition 2.2.8 *Assume that $V(t, x)$ satisfies (2.2.64), (2.2.65), and*

$$\frac{\min_{t_0 \le t \le T, \, |x| \ge R+\rho} V(t, x)}{\min_{t_0 \le t \le T, \, |x| \ge R} V(t, x)} \le c, \qquad (2.2.81)$$

where c is a constant which is independent of R. Let $f(x)$ be a function such that

$$\lim_{R \to \infty} \frac{\max_{|x| \le R} |f(x)|}{\min_{t_0 \le t \le T, \, |x| \ge R} V(t, x)} = 0. \qquad (2.2.82)$$

Then for any $x \in R^d$ and $\varepsilon > 0$ there exists $R(x, \varepsilon) > 0$ such that for any $R > R(x, \varepsilon)$

$$|Ef_R(X_R(T; t_0, x)) - Ef(X(T; t_0, x))| < \varepsilon. \qquad (2.2.83)$$

Proof Since the solutions $X(t; t_0, x)$ and $X_R(t; t_0, x)$ and also the functions $f(X(t; t_0, x))$ and $f_R(X_R(t; t_0, x))$ coincide on the interval $t \in [t_0, \tau_R]$, we have

$$E[f_R(X_R(T; t_0, x))\chi_{\Omega_R}(\omega)] = E[f(X(T; t_0, x))\chi_{\Omega_R}(\omega)]. \tag{2.2.84}$$

Hence

$$|Ef_R(X_R(T; t_0, x)) - Ef(X(T; t_0, x))| \tag{2.2.85}$$
$$\leq |E[f_R(X_R(T; t_0, x))\chi_{\Lambda_R}(\omega)]| + |E[f(X(T; t_0, x))\chi_{\Lambda_R}(\omega)]|.$$

Proposition 2.2.7 implies

$$\lim_{R \to \infty} |E[f(X(T; t_0, x))\chi_{\Lambda_R}(\omega)]| = 0. \tag{2.2.86}$$

Further, due to (2.2.79) and (2.2.75), we obtain

$$E|f_R(X_R(T; t_0, x))\chi_{\tau_R < T}(\omega)| \leq \sup_{x \in \mathbf{R}^d} |f_R(x)| p_R \leq \max_{|x| \leq R + \rho} |f(x)| p_R \tag{2.2.87}$$
$$\leq \frac{\max_{|x| \leq R + \rho} |f(x)|}{\min_{t_0 \leq t \leq T, \, |x| \geq R + \rho} V(t, x)} \times \frac{\min_{t_0 \leq t \leq T, \, |x| \geq R + \rho} V(t, x)}{\min_{t_0 \leq t \leq T, \, |x| \geq R} V(t, x)}$$
$$\times \exp(c_0(T - t_0))[(c_1 + 1)(T - t_0) + V(t_0, x)].$$

Now, by using the conditions (2.2.81) and (2.2.82), we complete the proof. $\square$

Our next step is to show that approximating paths obtained by a numerical method applied to the system (2.2.63) belong to the bounded domain with a large probability and that averaging via these paths gives a good approximation for the mean $Ef(X(T; t_0, x))$.

Let us start with some necessary auxiliary knowledge of the Markov chains generated by numerical methods. Consider the system of SDEs in the sense of Ito:

$$dY = b(t, Y)dt + \sum_{l=1}^{q} \gamma_l(t, Y)dw_l(t). \tag{2.2.88}$$

We assume that the functions $b(t, y)$ and $\gamma_l(t, y)$, $(t, y) \in [t_0, T] \times \mathbf{R}^d$, have bounded derivatives with respect to t and y up to some order. In particular, the system (2.2.80) satisfies this assumption. As we saw in Chapter 1 and earlier in this chapter, in most cases a method (both mean-square and weak) can be defined by a one-step approximation of the form (see (2.2.1))

$$\bar{Y}(t + h; t, y) = y + A(t, y, h; \xi), \qquad t_0 \leq t < t + h \leq T, \tag{2.2.89}$$

where ξ is a random vector having moments of a sufficiently high order and A is a vector function of dimension d.

Partition the interval $[t_0, T]$ into N equal parts with the step $h = (T - t_0)/N$: $t_0 < t_1 < \cdots < t_N = T$, $t_{k+1} - t_k = h$. According to (2.2.89), we construct the sequence (see (2.2.2))

$$\bar{Y}_0 = Y(t_0) = y, \quad \bar{Y}_{k+1} = \bar{Y}_k + A(t_k, \bar{Y}_k, h; \xi_k), \qquad k = 0, \ldots, N-1, \quad (2.2.90)$$

where ξ_0 is independent of $\bar{Y}_0$, while ξ_k for $k > 0$ are independent of $\bar{Y}_0, \ldots, \bar{Y}_k$, $\xi_0, \ldots, \xi_{k-1}$. The sequence $\bar{Y}_k$ is a Markov chain. Its transition probability function is defined by

$$P(t, y, s, D) = P(\bar{Y}(s; t, y) \in D), \qquad s \geq t, \ t, s = t_0, t_0 + h, \ldots, T, \quad (2.2.91)$$

where $\bar{Y}(s; t, y)$ is the process with discrete time starting at the moment t from y and defined by (2.2.90). The generating operator of the Markov chain is defined by

$$L_h U(t, y) = \frac{1}{h} \int P(t, y, t + h, dz)[U(t + h, z) - U(t, y)] \quad (2.2.92)$$

$$= \frac{1}{h}[EU(t + h, \bar{Y}(t + h; t, y)) - U(t, y)].$$

Denote by $\bar{\tau}_R$ the first exit time of the process $(t_k, \bar{Y}(t_k; t_0, y))$, $k = 0, \ldots, N$, from $[t_0, T) \times S_R$. Due to (2.2.92), we have (see [196])

$$EU(\bar{\tau}_R, \bar{Y}(\bar{\tau}_R; t_0, y)) - U(t_0, y) = E \sum_{k=0}^{\kappa-1} L_h U(t_k, \bar{Y}(t_k; t_0, y))h, \quad (2.2.93)$$

where κ is defined by $t_\kappa = \bar{\tau}_R$.

In what follows we use the following assumption.

Assumption 2.2.9 The one-step approximation (2.2.89) is at least of order two in the weak sense, and the method defined by this approximation converges at least with order one.

By the definition of weak approximation (see (2.0.20))

$$|EU(t + h, \bar{Y}(t + h; t, y)) - EU(t + h, Y(t + h; t, y))| \leq Kh^2, \quad (2.2.94)$$

where K is a positive constant, provided that y belongs to a compact set. At the same time (see Lemma 2.1.9)

$$EU(t + h, Y(t + h; t, y)) = U(t, y) + hLU(t, y) + O(h^2), \quad (2.2.95)$$

where $|O(h^2)| \leq Kh^2$ and K is again independent of y belonging to a compact set. Using (2.2.94) and (2.2.95), we get

$$|EU(t + h, \bar{Y}(t + h; t, y)) - U(t, y) - hLU(t, y)| \leq Kh^2,$$

and then we obtain from (2.2.92)

$$|L_h U(t, y) - LU(t, y)| \leq Kh. \tag{2.2.96}$$

Let us proceed to the system (2.2.80). Apply a numerical method of weak order $p \geq 1$ to the systems (2.2.63) and (2.2.80). As a result, we obtain two Markov chains $\bar{X}_k$ and $(\bar{X}_R)_k$. For the Markov chain $(\bar{X}_R)_k$ (but not for $\bar{X}_k$) we have for $(t, x) \in [t_0, T] \times S_R$ (see (2.2.5))

$$|Ef(X_R(T; t_0, x)) - Ef(\bar{X}_R(T; t_0, x))| \leq Kh^p. \tag{2.2.97}$$

Let L_R be the generating operator for (2.2.80), and $(L_R)_h$ for $\bar{X}_R$. According to (2.2.96), we get

$$|(L_R)_h U(t, x) - L_R U(t, x)| \leq Kh, \qquad (t, x) \in [t_0, T] \times S_R. \tag{2.2.98}$$

If $(t, x) \in [t_0, T] \times S_R$, then

$$L_R U(t, x) = LU(t, x). \tag{2.2.99}$$

Due to (2.2.70), we obtain that $LU \leq -1$. It follows from this inequality together with (2.2.98) and (2.2.99) that for all h small enough

$$(L_R)_h U(t, x) \leq 0, \qquad (t, x) \in [t_0, T] \times S_R. \tag{2.2.100}$$

In future we need the following assumption.

Assumption 2.2.10 If $(\bar{X}_R)_i \in S_R$, $i = 0, \ldots, k$, then $\bar{X}_k = (\bar{X}_R)_k$, $k \leq N$. (Of course, the approximating trajectories are starting from the same point: $\bar{X}_0 = (\bar{X}_R)_0 = x$.)

The assumption is evidently true, for instance, for the explicit Euler method (2.0.21). Moreover, for this method even $\bar{X}_{k+1} = (\bar{X}_R)_{k+1}$ if only $\bar{X}_k = (\bar{X}_R)_k \in S_R$, although $\bar{X}_{k+1}$ may not belong to S_R. The definition of φ (see (2.2.78)) ensures that the coefficients of (2.2.63) and (2.2.80) coincide in the wider domain S_{R+r}. Due to this fact, Assumption 2.2.10 is satisfied for h small enough if ξ_k are bounded. This is the most typical case for weak methods, while, applying mean-square methods, we can use random variables such that $\xi_k h^{1/2}$ are small if h is small. Thus, Assumption 2.2.10 holds for typical mean-square methods as well. Nevertheless, we should pay attention that a method of the type (2.2.89) with initial data from $[t_0, T] \times S_R$ may depend on behavior of a system's coefficients not only in $[t_0, T] \times S_{R+r}$ but, generally speaking, in $[t_0, T] \times \mathbf{R}^d$.

Let us take h ensuring (2.2.100). Denote by $\bar{\tau}_R$ the first exit time of the chain $(t_k, \bar{X}_R(t_k; t_0, x))$ from $[t_0, T) \times S_R$, i.e., $\bar{X}_R(t_k; t_0, x) \in S_R$, $k = 0, 1, \ldots, \kappa - 1$, and either $\bar{X}_R(t_\kappa; t_0, x) = \bar{X}_R(\bar{\tau}_R; t_0, x) \notin S_R$, where $\bar{\tau}_R = t_\kappa$, or $t_\kappa = T$. Applying (2.2.93) to $\bar{X}_R$ and using (2.2.100), we obtain

$$EU(\bar{\tau}_R, \bar{X}_R(\bar{\tau}_R; t_0, x)) \leq U(t_0, x) = (c_1 + 1)(T - t_0) + V(t_0, x). \qquad (2.2.101)$$

Introduce the events

$$\begin{aligned}
\tilde{\Omega}_R &:= \{\omega : |\bar{X}_R(t_k; t_0, x)| < R, \ k = 0, \ldots, N - 1, \\
&\qquad \text{and } |\bar{X}_R(T; t_0, x)| \leq R\} \\
&= \{\omega : (\bar{\tau}_R = T) \setminus ((\bar{\tau}_R = T) \cap (|\bar{X}_R(T; t_0, x)| > R))\}, \\
\tilde{\Lambda}_R &:= \{\omega : (\exists \, t_k, \ k = 0, \ldots, N - 1, \ \text{such that} \\
&\qquad |\bar{X}_R(t_k; t_0, x)| \geq R) \cup (|\bar{X}_R(T; t_0, x)| > R)\} \\
&= \{\omega : (\bar{\tau}_R < T) \cup ((\bar{\tau}_R = T) \cap (|\bar{X}_R(T; t_0, x)| > R))\}.
\end{aligned} \qquad (2.2.102)$$

The event $\tilde{\Lambda}_R$ consists of leaving S_R by $\bar{X}_R$ at one of the moments $t_0, \ldots, t_{N-1}$ or leaving $\bar{S}_R$ at $t_N = T$. We note that in the continuous case (see (2.2.67)) the set $(\tau_R = T) \cap (|X_R(T; t_0, x)| > R)$ is empty.

Let

$$\bar{p}_R = P(\tilde{\Lambda}_R).$$

Analogously to (2.2.74) and (2.2.75), we apply Chebyshev's inequality and obtain from (2.2.101)

$$\bar{p}_R \leq \exp(c_0(T - t_0)) \frac{(c_1 + 1)(T - t_0) + V(t_0, x)}{\min_{t_0 \leq t \leq T, \ |x| \geq R} V(t, x)}. \qquad (2.2.103)$$

Further, analogously to (2.2.87), we obtain

$$E|f_R(\bar{X}_R(T; t_0, x))\chi_{\tilde{\Lambda}_R}(\omega)| \leq \max_{|x| \leq R + \rho} |f(x)|\bar{p}_R. \qquad (2.2.104)$$

We see from the two last inequalities that the expectation $E|f_R(\bar{X}_R(T; t_0, x))\chi_{\tilde{\Lambda}_R}(\omega)|$ is as small as $E|f_R(X_R(T; t_0, x))\chi_{\tilde{\Lambda}_R}(\omega)|$ (cf. (2.2.87)) if only h ensures (2.2.100).

Theorem 2.2.11 *Consider any method satisfying Assumption 2.2.9 which is weakly convergent with order p for systems with sufficiently smooth and bounded derivatives up to some order. Let the conditions of Propositions 2.2.7 and 2.2.8 and Assumption 2.2.10 be fulfilled. Then for any $x \in \mathbf{R}^d$ and $\varepsilon > 0$ there exists $R(x, \varepsilon) > 0$ such that for all $R \geq R(x, \varepsilon)$ and sufficiently small h*

$$|Ef(X(T; t_0, x)) - E[f(\bar{X}(T; t_0, x))\chi_{\tilde{\Omega}_R}(\omega)]| \leq Kh^p + \varepsilon, \qquad (2.2.105)$$

where $K > 0$ depends on x and R.

Proof It has been proved (see Proposition 2.2.8) that for any $\varepsilon > 0$

$$|Ef(X(T; t_0, x)) - Ef_R(X_R(T; t_0, x))| \leq \frac{\varepsilon}{2} \qquad (2.2.106)$$

if R is sufficiently large.

Since the coefficients of system (2.2.80) and the function f_R can be taken so that they have bounded derivatives up to a sufficiently high order, the mentioned method gives for sufficiently small h

$$|Ef_R(X_R(T; t_0, x)) - Ef_R(\bar{X}_R(T; t_0, x))| \leq Kh^p. \qquad (2.2.107)$$

Let us choose $R(x, \varepsilon)$ so that for $R \geq R(x, \varepsilon)$ and sufficiently small h both inequality (2.2.106) and inequality

$$E|f_R(\bar{X}_R(T; t_0, x))\chi_{\tilde{A}_R}(\omega)| \leq \frac{\varepsilon}{2} \qquad (2.2.108)$$

(see (2.2.103) and (2.2.104)) are fulfilled.

Since Assumption 2.2.10 holds, $f_R(x) = f(x)$ for $x \in S_R$, and $\bar{X}_R = \bar{X}$ for $\omega \in \tilde{\Omega}_R$, we get

$$\begin{aligned} Ef_R&(\bar{X}_R(T; t_0, x)) \\ &= E[f_R(\bar{X}_R(T; t_0, x))\chi_{\tilde{\Omega}_R}(\omega)] + E[f_R(\bar{X}_R(T; t_0, x))\chi_{\tilde{A}_R}(\omega)] \quad (2.2.109) \\ &= E[f(\bar{X}(T; t_0, x))\chi_{\tilde{\Omega}_R}(\omega)] + E[f_R(\bar{X}_R(T; t_0, x))\chi_{\tilde{A}_R}(\omega)]. \end{aligned}$$

Inequality (2.2.105) follows from (2.2.106)–(2.2.109). Theorem 2.2.11 is proved. $\square$

Remark 2.2.12 If a method for a particular stochastic system converges, then K in (2.2.105) is bounded for all R (and ε). However, as was discussed in the introduction to this section, a method applied to SDEs with nonglobally Lipschitz coefficients can be divergent. It is obvious that in this case K goes to infinity as $R \to \infty$ ($\varepsilon \to 0$). In practice (see, e.g., our experiments and also a comment on the choice of R in Sect. 2.2.5.2), for a not too big R (and, consequently, not large K) the ε is negligibly small since the divergence is usually due to rare exploding approximate trajectories which have to be discarded. This concept of rejecting exploding trajectories is very practical; it allows us, in particular, to guarantee the accuracy of numerical results obtained even by "divergent" methods. We emphasize that the value of K depends on the choice of a numerical method, as is usual in the global Lipschitz case. Thanks to the above concept, we can exploit the whole arsenal of methods and choose an appropriate scheme depending on the system we are solving.

Remark 2.2.13 It is possible to prove that the proposed concept is also applicable in the case of the Talay–Tubaro extrapolation (see Sect. 2.2.3), i.e., for a sufficiently large R and all sufficiently small h the error can be expanded in powers of h (see (2.2.25)):

$$Ef(X(T; t_0, x)) - E[f(\bar{X}(T; t_0, x))\chi_{\bar{\Omega}_R}(\omega)]$$
$$= \rho(R, h) + C_0 h^p + \cdots + C_n h^{p+n} + O(h^{p+n+1}) \tag{2.2.110}$$

where the constants $C_0, \ldots, C_n$ are independent of h and $\rho(R, h) \to 0$ as $R \to \infty$ uniformly with respect to h.

Due to Remark 2.2.12, ρ is negligibly small for a fixed R in comparison with the term $O(h^{p+n+1})$ (for realistic, not too small h, of course). Therefore it can be supposed that $|\rho| \leq Ch^{p+n+1}$, where C is a positive constant. Then we can use (2.2.110) in practice to estimate the global error as well as to improve the accuracy of the method (see Sect. 2.2.3). For example, simulating $u = Ef(X(T; t_0, x))$ twice by a first-order scheme (i.e., $p = 1$) with two different time steps $h_1 = h$, $h_2 = \alpha h$, $\alpha > 0$, $\alpha \neq 1$, we obtain $\bar{u}^{h_1} = E[f(\bar{X}^{h_1}(T; t_0, x))\chi_{\bar{\Omega}_R}(\omega)]$ and $\bar{u}^{h_2} = E[f(\bar{X}^{h_2}(T; t_0, x))\chi_{\bar{\Omega}_R}(\omega)]$, respectively. We can expand (see (2.2.110) with $p = 1$, $n = 0$):

$$u = \bar{u}^{h_1} + C_0 h_1 + \delta_1, \quad u = \bar{u}^{h_2} + C_0 h_2 + \delta_2,$$

where $|\delta_i| \leq Ch^2$, $i = 1, 2$. Hence C_0 can be estimated as $C_0 \simeq -\frac{\bar{u}^{h_2} - \bar{u}^{h_1}}{h_2 - h_1}$, and we get the improved value

$$\bar{u}_{imp} = \bar{u}^{h_1} \frac{h_2}{h_2 - h_1} - \bar{u}^{h_2} \frac{h_1}{h_2 - h_1}, \quad u = \bar{u}_{imp} + \delta,$$

where $|\delta| \leq Ch^2$.

Example 2.2.14 Consider the following system:

$$dP = -\nabla F(Q)dt - \nu P dt + \sum_{l=1}^{n} \sigma_l dw_l(t), \tag{2.2.111}$$
$$dQ = P dt,$$

where ν is a positive constant and σ_l, $l = 1, \ldots, n$, are n-dimensional constant linearly independent vectors. The authors of [262] proved exponential ergodicity of (2.2.111) assuming that $F \in C^\infty(\mathbf{R}^n)$, $F(q) \geq 0$ for all $q \in \mathbf{R}^n$ and that there exist an $\alpha > 0$ and $0 < \beta < 1$ such that

$$\frac{1}{2}(\nabla F(q), q) \geq \beta F(q) + \nu^2 \frac{\beta(2 - \beta)}{8(1 - \beta)}|q|^2 - \alpha. \tag{2.2.112}$$

The Lyapunov function

$$V(x) = V(p, q) = \frac{1}{2}|p|^2 + F(q) + \frac{\nu}{2}(p, q) + \frac{\nu^2}{4}|q|^2 + 1 \tag{2.2.113}$$
$$\geq 1 + \frac{1}{8}|p|^2 + \frac{\nu^2}{12}|q|^2$$

is used to prove that for any $m \geq 1$ there exist positive c_m, d_m such that

$$L[V(x)]^m \leq -c_m[V(x)]^m + d_m. \tag{2.2.114}$$

The exponential ergodicity means that for any function f with a polynomial growth the following inequality holds:

$$\left| Ef(X(t; 0, x)) - \int f(z)d\mu(z) \right| \leq Ce^{-\lambda t}, \tag{2.2.115}$$

where $C > 0$ and $\lambda > 0$ are some constants. In (2.2.113)–(2.2.115), $x := (p, q)$, $X := (P, Q)$, and μ is an invariant measure for the Markov process defined by (2.2.111).

Resting on (2.2.114), it is not difficult to verify all the assumptions of Theorem 2.2.11 that concern the system under consideration. Due to (2.2.115), application of this theorem to calculation of $Ef(X(t; 0, x))$ gives an approximate value of the ergodic limit.

2.2.5.2 Numerical Experiments

Theorem 2.2.11 has the following practical implication for evaluating expectations of functionals of solutions to SDEs. We pick up a numerical method suitable for a stochastic system under consideration. We choose $R > 0$ such that the solution of the stochastic system equipped with some initial data leaves the sphere S_R of the radius R during a fixed time interval with a relatively small probability. In a lot of cases interesting from the applicable point of view (e.g., Langevin-type equations and gradient systems with noise) it is usually not difficult to guess this value of R by physical reasoning. Anyway, we can test the choice of R in practice as explained below. For the chosen R, we select a time step h for the numerical method, which ensures an accuracy appropriate for our purposes. As usual, the choice of time step h_* is appropriate if, by further decrease of the time step, we obtain a result which is close enough to the one obtained with h_*. The expectation of a functional which we are aiming to find is evaluated according to the Monte Carlo technique by running M independent realizations of the numerical solution to the considered system. According to the concept considered in this section, the value of the functional corresponding to sample trajectories that left the sphere S_R is set to be zero when counted to the expectation. Finally, we say that the choice of R is appropriate if its increase does not essentially affect the result. We also note that there is the Monte Carlo error in this procedure, which is controlled in the standard way by choosing an appropriate M. In practice, the procedure can be modified by assigning a certain value (not zero as we do here) for the trajectories which leave the sphere S_R. This value can be chosen/adjusted in response to experimental results or by physical reasoning.

As we will see in the numerical experiments presented below, the numerical integration and Monte Carlo errors affect accuracy of simulation much more than the error due to canceling "bad" trajectories (ε in (2.2.105)), which is usually negligibly small. Further illustrations see in [316, 319].

Example 2.2.15 Consider the stochastic differential equation

$$dX = -X^3 dt + \sigma dw(t), \quad X(0) = X_0. \tag{2.2.116}$$

It is demonstrated in [262] (see also [379, 431]) that the explicit Euler method for (2.2.116),

$$X_{k+1} = X_k - X_k^3 h + \sigma \Delta_k w, \qquad \Delta_k w := w(t_{k+1}) - w(t_k), \tag{2.2.117}$$

can explode.

For test purposes, we evaluate the functional

$$F = \frac{1}{2} E X^2(T) + E \int_0^T X^4(t) dt. \tag{2.2.118}$$

It can be shown that

$$F = \frac{1}{2} \sigma^2 T.$$

To simulate this functional, we introduce the additional equation

$$dZ = X^4(t) dt, \qquad Z(0) = 0. \tag{2.2.119}$$

Then

$$F = E \left(\frac{1}{2} X^2(T) + Z(T) \right). \tag{2.2.120}$$

The solution of (2.2.119) is approximated as

$$Z_{k+1} = Z_k + X_k^4 h. \tag{2.2.121}$$

By taking $V(x, z) = x^6 + z^2$, it is not difficult to check that the conditions of Propositions 2.2.7 and 2.2.8 are satisfied for the system (2.2.116), (2.2.119). Also, Assumption 2.2.10 holds for the explicit Euler method (2.2.117). Then Theorem 2.2.11 is applicable here; i.e., we can evaluate F from (2.2.118) by using approximate trajectories $(\bar{X}(t), \bar{Z}(t))$, $0 \le t \le T$, which belong to the ball $\{(x, z) : x^2 + z^2 < R^2\}$. In fact, in the case of functionals like that in (2.2.118) it is enough to control the paths $\bar{X}(t)$ only; i.e., the following estimate takes place (cf. (2.2.105)):

$$\left| E\left(\frac{1}{2}X^2(T) + Z(T)\right) - E\left(\frac{1}{2}\bar{X}^2(T) + \bar{Z}(T)\right) \chi_{\tilde{\Omega}_R}(\omega) \right| \le Kh + \varepsilon, \quad (2.2.122)$$

where $\tilde{\Omega}_R$ is defined by $\bar{\tau}_R$ being the first exit time of $(t, \bar{X}(t))$ from the rectangle $[0, T) \times (-R, R)$. This result is valid thanks to the fact that the right-hand sides of (2.2.116), (2.2.119) do not depend on Z. The proof is almost a word-by-word repetition of the proof of Theorem 2.2.11.

We also consider the weak Euler method:

$$X_{k+1} = X_k - X_k^3 h + \sigma \xi_k \sqrt{h}, \quad (2.2.123)$$

where ξ_k are i.i.d. random variables with the law $P(\xi = \pm 1) = 1/2$.

Let us choose a time step $h > 0$ for (2.2.123) such that

$$|X_0| \le \frac{1}{\sqrt{h}} \text{ and } h < \frac{1}{\sigma}\left(1 - \frac{2}{3\sqrt{3}}\right).$$

Then one can directly show that $|X_1| \le \frac{1}{\sqrt{h}}$ and therefore $|X_k| \le \frac{1}{\sqrt{h}}$ for all k. Thus, trajectories of (2.2.123) do not explode, provided that the above conditions on the step h hold. The authors do not exclude a possibility that methods using bounded random variables (like the weak Euler method (2.2.123)) can weakly converge in some nonglobally Lipschitz cases. In general this question concerning convergence is rather complicated (see, e.g., the third example below) and requires further investigation. We should stress that convergence of methods does not undermine the concept of rejecting exploding trajectories. Indeed, suppose that a weak method converges but for a not very small time step it may have exploding trajectories; then results obtained with this time step should be disregarded unless this concept is applied. This is well illustrated in our examples. Further, the concept is universal. It allows us to use any numerical method in the nonglobally Lipschitz case straightaway for a very broad class of SDEs, without any additional analysis at all.

In our experiments we simulate F from (2.2.120) as follows:

$$\bar{F} = \frac{1}{M} \sum_{m=1}^{M} \left(\frac{1}{2}\left[\bar{X}^{(m)}(T)\right]^2 + \bar{Z}^{(m)}(T)\right) \chi_{\tilde{\Omega}_R}(\omega) + \rho_{mc}, \quad (2.2.124)$$

where M is the number of independent realizations $\bar{X}^{(m)}(T)$, $\bar{Z}^{(m)}(T)$ of $\bar{X}(T)$, $\bar{Z}(T)$ that are found due to a numerical method of our choice. The Monte Carlo error ρ_{mc} has zero bias, and its variance equals

$$Var(\rho_{mc}) = \frac{Var\left(\left(\frac{1}{2}\bar{X}^2(T) + \bar{Z}(T)\right)\chi_{\tilde{\Omega}_R}(\omega)\right)}{M}; \quad (2.2.125)$$

i.e., the simulated

Table 2.2.1 Simulation of (2.2.116), (2.2.119) by the Euler methods (2.2.117), (2.2.121) and (2.2.123), (2.2.121) with various time steps h. See the other parameters in the text. The exact value $F = 5$. The "$\pm$" reflects the Monte Carlo error only; it does not reflect the error of the methods

h	Eqs. (2.2.117), (2.2.121)		Eqs. (2.2.123), (2.2.121)
	$\bar{F}$	Trajectories left $(-R, R)$	$\bar{F}$
0.25	6.640 ± 0.010	0.03%	5.962 ± 0.006
0.1	5.409 ± 0.007	0%	5.371 ± 0.007
0.02	5.069 ± 0.007	0%	5.073 ± 0.007
0.01	5.032 ± 0.007	0%	5.037 ± 0.007

$$\hat{F} := \frac{1}{M} \sum_{m=1}^{M} \left(\frac{1}{2} \left[\bar{X}^{(m)}(T) \right]^2 + \bar{Z}^{(m)}(T) \right) \chi_{\tilde{\Omega}_R}(\omega)$$

belongs to the confidence interval

$$\hat{F} \in (E\bar{F} - c\sqrt{Var(\rho_{mc})}, \ E\bar{F} + c\sqrt{Var(\rho_{mc})}) \tag{2.2.126}$$

with the probability 0.95 for $c = 2$ used here.

In Table 2.2.1 some results of our numerical experiments are presented. We take $\sigma = 1$, $X_0 = 0$, $M = 400000$, $T = 10$, and $R = 50$. The "$\pm$" reflects the Monte Carlo error only; it gives the confidence interval with $c = 2$ (see (2.2.126)). If our concept were not applied in the case of the Euler method (2.2.117), (2.2.121) with $h = 0.25$, then there would be an overflow in computer calculations. We also note in passing that both Euler methods produce quite similar results. Of course, the Euler method (2.2.117) is computationally more expensive than (2.2.123) due to the need to simulate Gaussian random variables instead of very simple random variables for (2.2.123).

Example 2.2.16 Consider the oscillator with cubic restoring force and additive noise

$$dQ = Pdt, \tag{2.2.127}$$
$$dP = (Q - Q^3) dt - \nu Pdt + \sigma dw(t),$$

where $w(t)$ is a standard Wiener process and ν and σ are positive constants. This system is exponentially ergodic, and the second moment EQ^2 evaluated with respect to the invariant measure is equal to 2.435 up to 3 d.p.

Fig. 2.2.1 Result of simulations of (2.2.127) by the quasi-symplectic method (2.2.128) with various time steps h. We take $R = 50$; see the other parameters in the text

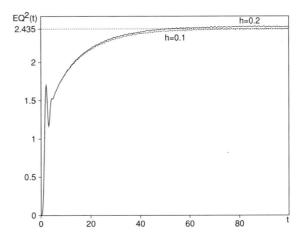

The system (2.2.127) belongs to the class of Langevin equations, for which quasi-symplectic methods are the most effective (see Chap. 5 and also [312]). We apply an explicit quasi-symplectic method of weak order one to (2.2.127):

$$P_{k+1} = (1 - \nu h)\left(P_k + h\left(Q_k - Q_k^3\right) + h^{1/2}\sigma\xi_k\right), \qquad (2.2.128)$$
$$Q_{k+1} = Q_k + h\left(P_k + h\left(Q_k - Q_k^3\right)\right),$$

where ξ_k are i.i.d. random variables with the law $P(\xi = \pm 1) = 1/2$.

The results of simulating $EQ^2(t)$ by this method are presented in Figure 2.2.1. We take the parameters of (2.2.127) as follows: $Q(0) = P(0) = 0$, $\nu = 0.05$, and $\sigma = 1$. For realization of the proposed concept, we choose $R = 50$. The number of independent realizations M used to produce the picture is equal to 400000, which ensures the Monte Carlo error $2\sqrt{Var(\rho_{mc})} \leq 0.008$. For $h = 0.2$, there are 28 trajectories (i.e., 0.007%) that leave the ball of radius 50. We see that, applying the proposed concept (i.e., not taking into account these 28 "bad" trajectories), we obtain a quite accurate approximation of the ergodic limit. If one did not exploit the concept here, then the results obtained with $h = 0.2$ could not be used, since the exploding trajectories lead to numerical overflow in computing the average. For $h = 0.1$, there are no trajectories out of 400000 that leave the ball of radius 50, but this does not exclude the possibility of having exploding trajectories in another series of Monte Carlo runs. Such uncertainty made it uncomfortable to use the results of such experiments without the concept considered in this section. This concept gives us a rigorous basis for making use of any numerical method to solve nonlinear SDEs and for interpreting experiments in which occurrence of exploding trajectories is not excluded.

2.3 Variance Reduction

If we compute $Ef(X(T))$ by the Monte Carlo method, using an approximate method for integrating the system

$$dX = a(t, X)dt + \sum_{r=1}^{q} \sigma_r(t, X)dw_r(t) \tag{2.3.1}$$

to find $X(T)$, two errors arise. One of them is the numerical integration error:

$$Ef(X(T)) = Ef(\bar{X}(T)) + O(h^p).$$

The other is the error of the Monte Carlo method:

$$Ef(\bar{X}(T)) = \frac{1}{M} \sum_{i=1}^{M} f(\bar{X}^{(i)}(T)) \pm c \frac{(Varf(\bar{X}(T)))^{1/2}}{M^{1/2}}, \tag{2.3.2}$$

where, e.g. the values $c = 1, 2, 3$ correspond to the probabilities 0.68, 0.95, 0.997, respectively, that $Ef(\bar{X}(T))$ falls in the corresponding credible interval.

Since $Varf(\bar{X}(T))$ is close to $Varf(X(T))$, we may assume that the error of the Monte Carlo method can be estimated by $(Varf(X(T))/M)^{1/2}$. If $Varf(X(T))$ is large, then to achieve a satisfactory accuracy we have to simulate a very large number of trajectories. If it were possible to change $f(X(T))$ by a variable Z such that $EZ = Ef(X(T))$ but with $VarZ$ substantially smaller than $Varf(X(T))$, then the modeling of Z instead of $f(X(T))$ would make it possible to obtain more accurate results for the same computational costs.

Two variance reduction methods are known: the method of important sampling (see [128, 274, 335, 336, 456]) and the method of control variates (see [335, 336]). A combining method is given in [293, 296].

2.3.1 The Method of Important Sampling

Along with (2.3.1), we consider the system

$$dX = a(t, X)dt - \sum_{r=1}^{q} \mu_r(t, X)\sigma_r(t, X)dt + \sum_{r=1}^{q} \sigma_r(t, X)dw_r(t), \tag{2.3.3}$$

$$dY = \sum_{r=1}^{q} \mu_r(t, X)Ydw_r(t),$$

where μ_r and Y are scalars, μ_r are rather arbitrary functions, however, with good analytical properties (for example, they are sufficiently smooth and have bounded derivatives).

By Girsanov's theorem, we have for any μ_r :

$$y E f(X_{s,x}(T))|_{(2.3.1)} = E Y_{s,x,y}(T) f(X_{s,x}(T))|_{(2.3.3)} .$$

Putting $Z = Y_{s,x,y}(T) f(X_{s,x}(T))$, we see that EZ does not depend on the choice of the μ_r, while for $y = 1$ it equals to the desired quantity. At the same time, $Var Z$ does depend on the μ_r. Then it is natural to regard $\mu_1, \ldots, \mu_q$ as controls and to choose them by the condition that the variance $Var Z = EZ^2 - (EZ)^2$ is minimal. Since EZ is independent of $\mu_1, \ldots, \mu_q$, this choice reduces to solving the following problem from the optimal control theory: it is required to choose the controls $\mu_1, \ldots, \mu_q$ constituting a minimum of the functional

$$I = E Y^2_{s,x,y}(T) f^2(X_{s,x}(T))$$

with respect to (2.3.3).

The function $u(s, x) = E f(X_{s,x}(T))|_{(2.3.1)}$ satisfies the equation

$$Lu \equiv \frac{\partial u}{\partial s} + \sum_{i=1}^{d} a^i \frac{\partial u}{\partial x^i} + \frac{1}{2} \sum_{r=1}^{q} \sum_{i=1}^{d} \sum_{j=1}^{d} \sigma_r^i \sigma_r^j \frac{\partial^2 u}{\partial x^i \partial x^j} = 0 \qquad (2.3.4)$$

with the condition

$$u(T, x) = f(x) \qquad (2.3.5)$$

at the end of the time interval.

Introduce the function

$$v(s, x) y^2 = \min_{\mu_1, \ldots, \mu_q} I = \min_{\mu_1, \ldots, \mu_q} E Y^2_{s,x,y}(T) f^2(X_{s,x}(T))$$

(it is clearly homogeneous of order two in y, which is already reflected in the notation). We write the Bellman equation for this function:

$$\min_{\mu_1, \ldots, \mu_q} (Lvy^2 + \sum_{r=1}^{q} (\sigma_r, \frac{\partial v}{\partial x}) \mu_r y^2 + v \sum_{r=1}^{q} \mu_r^2 y^2) = 0 . \qquad (2.3.6)$$

The minimization condition in (2.3.6) implies (if $v \neq 0$) :

$$\mu_r = -\frac{1}{2v} (\sigma_r, \frac{\partial v}{\partial x}) . \qquad (2.3.7)$$

Thus, v satisfies the equation

$$Lv - \frac{1}{4v}\sum_{r=1}^{q}(\sigma_r, \frac{\partial v}{\partial x})^2 = 0.\tag{2.3.8}$$

Moreover, it is clear that

$$v(T, x) = f^2(x).\tag{2.3.9}$$

Let $f > 0$. Then $v > 0$. By some simple computations, we are readily verify that $\sqrt{v}$ is a solution of the problem (2.3.4)–(2.3.5). Thus, $v = u^2$. By (2.3.7), this implies

$$\mu_r = -\frac{1}{u}(\sigma_r, \frac{\partial u}{\partial x}).\tag{2.3.10}$$

Further, if we write the relation $v = u^2$ in the form

$$EZ^2 = (EZ)^2,$$

then we find that $Var Z = 0$ for μ_r from (2.3.10), i.e., the variable $Y_{s,x,y}(T)\times f(X_{s,x}(T))$ with X and Y from (2.3.3), (2.3.10) is deterministic.

Of course, the controls μ_r, $r = 1, \ldots, q$, cannot be constructed without knowing the function u. Nevertheless, the result obtained establishes that, in principle, it is possible to arbitrarily reduce the variance $Var Z$ by conveniently choosing the functions μ_r.

Note that the reasoning above is not completely rigorous. However, using its results, it is not difficult to prove the following theorem.

Theorem 2.3.1 *Let $f > 0$ and suppose there is a solution $u > 0$ of the problem (2.3.4)–(2.3.5). Suppose there is a solution of the system (2.3.3), (2.3.10) for $t_0 \leq s < T$ and $x \in \mathbf{R}^d$. Then $Z = Y_{s,x,y}(T)f(X_{s,x}(T))$ computed according to (2.3.3), (2.3.10) is a deterministic variable.*

Proof Let $u > 0$ be a solution of (2.3.4)–(2.3.5) and μ_r in (2.3.3) be such that there is a solution of the system (2.3.3). Using Ito's formula, we obtain (taking into account that $Lu = 0$) :

$$d(u(t, X_{s,x}(t)) \cdot Y_{s,x,y}(t))$$

$$= Lu \cdot Y dt - \sum_{r=1}^{q}\mu_r(\sigma_r, \frac{\partial u}{\partial x})Y dt + \sum_{r=1}^{q}(\sigma_r, \frac{\partial u}{\partial x})Y dw_r(t)$$

$$+u\sum_{r=1}^{q}\mu_r Y dw_r(t) + \sum_{r=1}^{q}(\sigma_r, \frac{\partial u}{\partial x})\mu_r Y dt$$

$$= \sum_{r=1}^{q}((\sigma_r, \frac{\partial u}{\partial x}) + \mu_r u)Y dw_r(t),$$

whence

$$u(t, X_{s,x}(t)) \cdot Y_{s,x,y}(t) = u(s, x)y + \int_t^s \sum_{r=1}^q ((\sigma_r, \frac{\partial u}{\partial x}) + \mu_r u) Y dw_r . \quad (2.3.11)$$

For the μ_r from (2.3.10), the relation (2.3.11) reduces to

$$u(t, X_{s,x}(t)) \cdot Y_{s,x,y}(t) = u(s, x)y ,$$

i.e., for each t (so, in particular, for $t = T$) the quantity $u(t, X_{s,x}(t)) \times Y_{s,x,y}(t)$ is deterministic. By (2.3.5), this quantity for $t = T$ is equal to $Y_{s,x,y}(T) f(X_{s,x}(T))$. $\square$

The results obtained can be used in, e.g., the following situation. Let f be a function close to a function f_0, and let the solution of the problem (2.3.4)–(2.3.5) for $f = f_0$ be known and be equal to u_0. If we take μ_r in (2.3.3) equal to

$$\mu_r = -\frac{1}{u_0}(\sigma_r, \frac{\partial u_0}{\partial x}) ,$$

the variance $Var(Y_{s,x,y}(T) f(X_{s,x}(T))$, although not zero, is small.

An illustration of the method of important sampling is given in Example 3.5.3 from Subsection 3.5. See another example in Section 4.7.

Remark 2.3.2 If the condition $f > 0$ in Theorem 2.3.1 is not satisfied, but if, e.g., $f > -C$, $C > 0$, then for $f + C$ the solution of the problem (2.3.4)–(2.3.5) is $u + C$, and the dependence

$$\mu_r = -\frac{1}{u + C}(\sigma_r, \frac{\partial u}{\partial x})$$

in (2.3.3) leads to $Z = Y_{s,x,y}(T)(f(X_{s,x}(T) + C)$ being a deterministic variable (as in Theorem 2.3.1). If f is neither bounded from below nor above but $f = g - h$ with $g > 0$ and $h > 0$ and for each of the functions g, h the conditions of Theorem 2.3.1 hold, then Theorem 2.3.1 can be used for g and h separately to compute Ef.

In the next subsection we consider a more general method of variance reduction which combines the important sampling method considered above with the control variates method.

2.3.2   A Family of Probabilistic Representations for the Cauchy Problem for Parabolic PDEs and Variance Reduction by Control Variates and Combining Method

Consider the Cauchy problem for linear parabolic equation

$$
\frac{\partial u}{\partial s} + \frac{1}{2} \sum_{i,j=1}^{d} a^{ij}(s,x) \frac{\partial^2 u}{\partial x^i \partial x^j} + \sum_{i=1}^{d} b^i(s,x) \frac{\partial u}{\partial x^i} + c(s,x)u + g(s,x) = 0,
$$

(2.3.12)

$$
t_0 \le s < T, \; x \in \mathbf{R}^d,
$$

with the initial condition

$$
u(T,x) = f(x). \tag{2.3.13}
$$

The matrix $a(s,x) = \{a^{ij}(s,x)\}$ is supposed to be symmetric and positive semidefinite.

Let $\sigma(s,x)$ be a matrix obtained from the equation

$$
a(s,x) = \sigma(s,x)\sigma^{\mathsf{T}}(s,x) .
$$

This equation is solvable with respect to σ (for instance, by a lower triangular matrix) at least for a positively definite a.

The solution to the problem (2.3.12)–(2.3.13) has the following probabilistic representation (see [86]):

$$
u(s,x) = E(f(X_{s,x}(T))Y_{s,x,1}(T) + Z_{s,x,1,0}(T)) , \tag{2.3.14}
$$
$$
s \le T, \; x \in \mathbf{R}^d,
$$

where $X_{s,x}(t)$, $Y_{s,x,y}(t)$, $Z_{s,x,y,z}(t)$, $t \ge s$, is the solution of the Cauchy problem for the system of SDEs

$$
dX = b(t,X)dt + \sigma(t,X)dw(t), \; X(s) = x, \tag{2.3.15}
$$
$$
dY = c(t,X)Y dt, \, Y(s) = y,
$$
$$
dZ = g(t,X)Y dt, \; Z(s) = z.
$$

Here $w(t) = (w^1(t), \ldots, w^d(t))^{\mathsf{T}}$ is a d-dimensional standard Wiener process, Y and Z are scalars. If $y = 1$, $z = 0$, we shall use the notation $Y_{s,x}(t) := Y_{s,x,1}(t)$, $Z_{s,x}(t) := Z_{s,x,1,0}(t)$ (analogous notation will be used later for some other variables). In particular, in this simplified notation we write (2.3.14) as

$$
u(s,x) = E[f(X_{s,x}(T))Y_{s,x}(T) + Z_{s,x}(T)]. \tag{2.3.16}
$$

There are various sets of sufficient conditions ensuring connection between the solutions of the Cauchy problem (2.3.12)–(2.3.13) and their probabilistic representations (2.3.14)–(2.3.15). For definiteness, we shall keep the following assumptions.

We assume that the coefficients b, σ, c, and g have bounded derivatives up to some order and additionally c and g are bounded on $[t_0, T] \times \mathbf{R}^d$. Further, we assume that the matrix $a(t, x)$ is positive definite and, moreover, the uniform ellipticity condition holds: there exists $\sigma_0 > 0$ such that

$$\| a^{-1}(t, x) \| = \| (\sigma(t, x)\sigma^{\mathsf{T}}(t, x))^{-1} \| \leq \sigma_0^{-1}, \quad t_0 \leq t \leq T, \ x \in \mathbf{R}^d.$$

As for function $f(x)$, it is assumed to grow at infinity not faster than a polynomial function. It can be both smooth and nonsmooth.

We note that the results of this section can be used under other sets of conditions. For instance, one can consider situations with nonglobally Lipschitz coefficients as in Sect. 2.2.5 (see also [316]) or with matrix $a(t, x)$ which is positive semidefinite.

The value $u(s, x)$ from (2.3.14) can be evaluated using the weak-sense numerical integration of the system (2.3.15) together with the Monte Carlo technique. More specifically, we have

$$u(s, x) \approx E[f(\bar{X}_{s,x}(T))\bar{Y}_{s,x}(T) + \bar{Z}_{s,x}(T)] \tag{2.3.17}$$

$$\approx \frac{1}{M} \sum_{m=1}^{M} [f({}_m\bar{X}_{s,x}(T)){}_m\bar{Y}_{s,x}(T) + {}_m\bar{Z}_{s,x}(T)],$$

where the first approximate equality involves an error due to replacing X, Y, Z by $\bar{X}$, $\bar{Y}$, $\bar{Z}$ (the error is related to the approximate integration of (2.3.15)) and the error in the second approximate equality comes from the Monte Carlo technique; ${}_m\bar{X}_{s,x}(T)$, ${}_m\bar{Y}_{s,x}(T)$, ${}_m\bar{Z}_{s,x}(T)$, $m = 1, \ldots, M$, are independent realizations of $\bar{X}_{s,x}(T)$, $\bar{Y}_{s,x}(T)$, $\bar{Z}_{s,x}(T)$. The weak-sense integration of SDEs is developed sufficiently well and a lot of different effective weak-sense numerical methods have been constructed as we saw earlier in this chapter and will see more in the next Chaps. 3–5. But methods of reducing the second error in (2.3.17) are more intricate.

The error of the Monte Carlo method is evaluated by (see the beginning of Sect. 2.3):

$$\bar{\rho} = c \frac{(Var[f(\bar{X}_{s,x}(T))\bar{Y}_{s,x}(T) + \bar{Z}_{s,x}(T)])^{1/2}}{M^{1/2}},$$

where, e.g., the values $c = 1, 2, 3$ correspond to the probabilities 0.68, 0.95, 0.997, respectively. Introduce

$$\Gamma = \Gamma_{s,x} := f(X_{s,x}(T))Y_{s,x}(T) + Z_{s,x}(T), \tag{2.3.18}$$

$$\bar{\Gamma} = \bar{\Gamma}_{s,x} := f(\bar{X}_{s,x}(T))\bar{Y}_{s,x}(T) + \bar{Z}_{s,x}(T). \tag{2.3.19}$$

Since $Var\,\Gamma_{s,x}$ is close to $Var\,\bar{\Gamma}_{s,x}$, we can assume that the error of the Monte Carlo method is estimated by

$$\rho = c \, \frac{(Var\, \Gamma_{s,x})^{1/2}}{M^{1/2}}. \tag{2.3.20}$$

If $Var\, \Gamma_{s,x}$ is large then to achieve a satisfactory accuracy we have to simulate a very large number of independent trajectories. Clearly, variance reduction is of crucial importance for effectiveness of any Monte Carlo procedure. To reduce the Monte Carlo error, one usually exploits some other probabilistic representations of solutions to considered problems. To obtain various probabilistic representations of the solution to the problem (2.3.12)–(2.3.13), we introduce the system

$$dX = b(t, X)dt - \sigma(t, X)\mu(t, X)dt + \sigma(t, X)dw(t), \quad X(s) = x, \tag{2.3.21}$$

$$dY = c(t, X)Y dt + \mu^{\mathsf{T}}(t, X)Y dw(t), \quad Y(s) = 1, \tag{2.3.22}$$

$$dZ = g(t, X)Y dt + F^{\mathsf{T}}(t, X)Y dw(t), \quad Z(s) = 0, \tag{2.3.23}$$

where μ and F are column-vector functions of dimension d satisfying some regularity conditions (e.g., they have bounded derivatives with respect to x^i up to some order). We should note that $X, \ Y, \ Z$ in (2.3.21)–(2.3.23) differ from $X, \ Y, \ Z$ in (2.3.15), however, this does not lead to any ambiguity. The formula (2.3.16), i.e.

$$u(s, x) = E\Gamma_{s,x}, \tag{2.3.24}$$

remains valid under the new X, Y, Z which is proved below. While the mean $E\Gamma$ does not depend on the choice of μ and F, the variance $Var\, \Gamma = E\Gamma^2 - (E\Gamma)^2$ does. Thus, μ and F can be used to decrease the variance $Var\, \Gamma$ and, consequently, the Monte Carlo error can be reduced.

Let $u(s, x)$ be a solution of the problem (2.3.12)–(2.3.13). Introduce the function (cf. (2.3.18))

$$\Gamma_{s,x}(t) := u(t, X_{s,x}(t))Y_{s,x,1}(t) + Z_{s,x,1,0}(t). \tag{2.3.25}$$

By the Ito formula, we obtain

$$d\Gamma_{s,x}(t) = \frac{\partial u}{\partial t}dt \cdot Y + \frac{1}{2}\sum_{i,j=1}^{d} a^{ij}\frac{\partial^2 u}{\partial x^i \partial x^j}dt \cdot Y + \sum_{i=1}^{d} b^i \frac{\partial u}{\partial x^i}dt \cdot Y$$

$$- \sum_{i=1}^{d}\frac{\partial u}{\partial x^i}\sum_{j=1}^{d}\sigma^{ij}\mu^j dt \cdot Y + \sum_{i=1}^{d}\frac{\partial u}{\partial x^i}\sum_{j=1}^{d}\sigma^{ij}dw^j(t) \cdot Y$$

$$+ ucY dt + \sum_{i=1}^{d}\frac{\partial u}{\partial x^i}\sum_{j=1}^{d}\sigma^{ij}\mu^j Y dt + u\sum_{j=1}^{d}\mu^j Y dw^j(t) + gY dt$$

$$+ \sum_{j=1}^{d} F^j Y dw^j(t).$$

Taking into account that $u(s, x)$ is the solution of (2.3.12)–(2.3.13), we get

$$d\Gamma_{s,x}(t) = Y \sum_{j=1}^{d} \left(\sum_{i=1}^{d} \sigma^{ij} \frac{\partial u}{\partial x^i} + u\mu^j + F^j \right) dw^j(t), \qquad (2.3.26)$$

where σ^{ij}, $\dfrac{\partial u}{\partial x^i}$, u, μ^j, and F^j have $(t, X_{s,x}(t))$ as their arguments, and $X_{s,x}(t)$, $Y_{s,x,1}(t)$ is the solution of (2.3.21)–(2.3.22).

From (2.3.25) and (2.3.13), we get

$$\Gamma_{s,x}(s) = u(s, x), \quad \Gamma_{s,x}(T) = \Gamma_{s,x} = f(X_{s,x}(T))Y_{s,x,1}(T) + Z_{s,x,1,0}(T)$$

and integrating (2.3.26) from s to T, we arrive at

$$f(X_{s,x}(T))Y_{s,x,1}(T) + Z_{s,x,1,0}(T) = u(s, x) \qquad (2.3.27)$$

$$+ \int_s^T Y(t) \sum_{j=1}^{d} \left(\sum_{i=1}^{d} \sigma^{ij} \frac{\partial u}{\partial x^i} + u\mu^j + F^j \right) dw^j(t).$$

Taking expectations on the left and the right of the above formula, we obtain

$$u(s, x) = E\left[f(X_{s,x}(T))Y_{s,x,1}(T) + Z_{s,x,1,0}(T) \right] = E\Gamma_{s,x}(T) = E\Gamma. \qquad (2.3.28)$$

From (2.3.27) and (2.3.28), we get

$$Var\,\Gamma = Var\,\Gamma_{s,x} = E \int_s^T Y_{s,x}^2(t) \sum_{j=1}^{d} \left(\sum_{i=1}^{d} \sigma^{ij} \frac{\partial u}{\partial x^i} + u\mu^j + F^j \right)^2 dt. \qquad (2.3.29)$$

Thus we have proved the following theorem.

Theorem 2.3.3 *Let μ and F be such that for any $x \in \mathbf{R}^d$ there exists a solution to the system (2.3.21)–(2.3.23) on the interval $[s, T]$. Then the probabilistic representation of the solution of the PDE problem (2.3.12)–(2.3.13) is given by (2.3.28) and variance of the representation is given by (2.3.29) provided that the expectation in (2.3.29) exists. In (2.3.29) all the functions σ^{ij}, μ^j, F^j, u, $\partial u/\partial x^i$ have $(t, X_{s,x}(t))$ as their argument.*

In particular, if μ and F are such that

$$\sum_{i=1}^{d} \sigma^{ij} \frac{\partial u}{\partial x^i} + u\mu^j + F^j = 0, \quad j = 1, \dots, d, \qquad (2.3.30)$$

then $Var\,\Gamma = 0$, i.e., Γ is deterministic.

We recall that if we put here $F = 0$ then we obtain the method of important sampling (first considered in [128, 274, 456], see Sect. 2.3.1 above). If we put $\mu = 0$ then we obtain the method of control variates (first considered in [335], see also [336]). Its application is illustrated in Examples 3.5.6 and 3.5.7 of Sect. 3.5. Theorem 2.3.3 establishes the combining method of variance reduction proved in [293] (see also [296]).

Obviously, μ and F satisfying (2.3.30) cannot be constructed without knowing $u(t, x)$, $s \leq t \leq T$, $x \in \mathbf{R}^d$. Nevertheless, the theorem claims a general possibility of variance reduction by a proper choice of the functions μ^j and F^j, $j = 1, \ldots, d$. Theorem 2.3.3 can be used, for example, if we know a function $\hat{u}(t, x)$ related to an approximating problem and being close to $u(t, x)$. In this case we take any $\hat{\mu}^j$, $\hat{F}^j$, $j = 1, \ldots, d$, satisfying

$$\sum_{i=1}^{d} \sigma^{ij} \frac{\partial \hat{u}}{\partial x^i} + \hat{u}\hat{\mu}^j + \hat{F}^j = 0, \tag{2.3.31}$$

and then the variance $Var\,\Gamma$ is although not zero but small.

Let us emphasize that (2.3.30) serves only as a guidance for getting suitable μ and F (recall, that the mean $E\Gamma$ does not depend on the choice of μ and F). In particular, the derivative estimate $\widehat{\partial u/\partial x^i}$ can differ from $\partial \hat{u}/\partial x^i$. In such cases, instead of (2.3.31) we use

$$\sum_{i=1}^{d} \sigma^{ij} \widehat{\frac{\partial u}{\partial x^i}} + \hat{u}\hat{\mu}^j + \hat{F}^j = 0. \tag{2.3.32}$$

In Sect. 2.4 we will consider a systematic method of approximating the functions u and $\partial u/\partial x^i$, $i = 1, \ldots, d$, with relatively low computational cost, and hence to obtain systematic methods of variance reduction. To this end, we will exploit the regression method of evaluating $u(t_k, x)$ and $\partial u/\partial x^i(t_k, x)$ which allows us to use only one set of approximate trajectories starting from the initial position (t_0, x_0).

2.3.3 Pathwise Approach for Derivatives $\partial u/\partial x^i(s, x)$

The probabilistic representation for the derivatives

$$\partial^i(s, x) := \frac{\partial u(s, x)}{\partial x^i}, \quad i = 1, \ldots, d,$$

can be obtained by the straightforward differentiation of (2.3.24) (see, e.g., [129, 293]):

$$\partial^i(s, x) \tag{2.3.33}$$

$$= E\left(\sum_{j=1}^{d} \frac{\partial f(X_{s,x}(T))}{\partial x^j} \delta^i_{s,x} X^j(T) Y_{s,x}(T) + f(X_{s,x}(T)) \delta^i_{s,x} Y(T) + \delta^i_{s,x} Z(T)\right)$$

where

$$\delta^i X^j(t) := \delta^i_{s,x} X^j(t) := \frac{\partial X^j_{s,x}(t)}{\partial x^i}, \quad \delta^i Y(t) := \delta^i_{s,x} Y(t) := \frac{\partial Y_{s,x}(t)}{\partial x^i},$$

$$\delta^i Z(t) := \delta^i_{s,x} Z(t) := \frac{\partial Z_{s,x}(t)}{\partial x^i}, \ s \le t \le T, \ i, j = 1, \ldots, d,$$

satisfy the system of variational equations associated with (2.3.21)–(2.3.23):

$$d\delta^i X = \sum_{j=1}^{d} \frac{\partial(b(t, X) - \sigma(t, X)\mu(t, X))}{\partial x^j} \delta^i X^j dt + \sum_{j=1}^{d} \frac{\partial \sigma(t, X)}{\partial x^j} \delta^i X^j \, dw(t),$$

$$\tag{2.3.34}$$

$$\delta^i X^j(s) = 0 \ \text{if} \ j \ne i, \ \text{and} \ \delta^i X^i(s) = 1,$$

$$d\delta^i Y = \sum_{j=1}^{d} Y \frac{\partial c(t, X)}{\partial x^j} \delta^i X^j dt + c(t, X)\delta^i Y dt \tag{2.3.35}$$

$$+ \sum_{j=1}^{d} Y \frac{\partial \mu^\top(t, X)}{\partial x^j} \delta^i X^j dw(t) + \mu^\top(t, X)\delta^i Y dw(t), \ \delta^i Y(s) = 0,$$

$$d\delta^i Z = \sum_{j=1}^{d} Y \frac{\partial g(t, X)}{\partial x^j} \delta^i X^j dt + g(t, X)\delta^i Y dt \tag{2.3.36}$$

$$+ \sum_{j=1}^{d} Y \frac{\partial F^\top(t, X)}{\partial x^j} \delta^i X^j dw(t) + F^\top(t, X)\delta^i Y dw(t), \ \delta^i Z(s) = 0.$$

Introduce a partition of the time interval $[t_0, T]$, for simplicity the equidistant one: $t_0 < t_1 < \cdots < t_N = T$ with step size $h = (T - t_0)/N$. Let us apply a weak scheme to the systems of SDEs (2.3.21)–(2.3.23), (2.3.34)–(2.3.36) to obtain independent approximate trajectories $(t_k, {}_m\bar{X}(t_k))$, $m = 1, \ldots, M$, all starting from the point (t_0, x), and ${}_m\bar{Y}(t_k)$, ${}_m\bar{Z}(t_k)$, ${}_m\bar{\delta}^i X(t_k)$, ${}_m\bar{\delta}^i Y(t_k)$, ${}_m\bar{\delta}^i Z(t_k)$ with ${}_m\bar{Y}(t_0) = 1$, ${}_m\bar{Z}(t_0) = 0$, ${}_m\bar{\delta}^i X^j(t_0) = 0$ if $j \ne i$, and ${}_m\bar{\delta}^i X^i(t_0) = 1$, ${}_m\bar{\delta}^i Y(t_0) = 0$, ${}_m\bar{\delta}^i Z(t_0) = 0$. Then we obtain the following Monte Carlo estimates of the derivatives $\partial u/\partial x^i(t_0, x)$ from (2.3.33) with $(s, x) = (t_0, x)$:

$$\hat{\partial}^i(t_0, x) = \frac{1}{M} \sum_{m=1}^{M} \left[\sum_{j=1}^{d} \frac{\partial f(_m\bar{X}(T))}{\partial x^j} \,_m\bar{\delta}^i X^j(T) \,_m\bar{Y}(T) \right.$$

$$\left. + f(_m\bar{X}(T)) \,_m\bar{\delta}^i Y(T) + \,_m\bar{\delta}^i Z(T) \right].$$ (2.3.37)

Clearly, the estimates $\hat{\partial}^i(t_k, x)$ for derivatives $\partial u/\partial x^i(t_k, x)$ can be obtained analogously.

Theorem 2.3.3 asserts that the variance in evaluating u by (2.3.24) can reach zero value for some μ and F. In [293] it is proved that for the same μ and F the variance in evaluating ∂^i by (2.3.33) is equal to zero as well (we pay attention that not only μ and F but also their derivatives are present in (2.3.35) and (2.3.36)).

2.3.4 Variance Reduction for Boundary Value Problems

Let G be a bounded domain in $\mathbf{R}^d$ and $Q = [t_0, T) \times G$ be a cylinder in $\mathbf{R}^{d+1}$, $\Gamma = \bar{Q} \backslash Q$ be the part of the cylinder's boundary consisting of the: upper base and lateral surface. Consider the Dirichlet problem for the parabolic equation

$$Lu + g = 0, \quad (s, x) \in Q,$$ (2.3.38)

$$u \mid_\Gamma = f(s, x),$$ (2.3.39)

where

$$Lu := \frac{\partial u}{\partial s} + \frac{1}{2} \sum_{i,j=1}^{d} a^{ij}(s, x) \frac{\partial^2 u}{\partial x^i \partial x^j} + \sum_{i=1}^{d} b^i(s, x) \frac{\partial u}{\partial x^i} + c(s, x)u.$$

We assume that conditions hold which guarantee existence of a sufficiently smooth classical solution $u(s, x)$ of the problem (2.3.38)–(2.3.39).

The solution of the problem (2.3.38)–(2.3.39) has various probabilistic representations:

$$u(s, x) = E\left[f(\tau, X_{s,x}(\tau))Y_{s,x,1}(\tau) + Z_{s,x,1,0}(\tau) \right],$$ (2.3.40)

where $X_{s,x}(t), Y_{s,x,y}(t), Z_{s,x,y,z}(t)$, $s \leq t \leq \tau$, is the solution of the Cauchy problem for the system of SDEs

$$dX = \chi_{\{\tau > t\}}[(b(t, X) - \sigma(t, X)\mu(t, X)) \, dt + \sigma(t, X) \, dw(t)],$$ (2.3.41)
$$dY = \chi_{\{\tau > t\}}[c(t, X)Y \, dt + \mu^\mathsf{T}(t, X)Y \, dw(t)],$$
$$dZ = \chi_{\{\tau > t\}}[g(t, X)Y \, dt + F^\mathsf{T}(t, X)Y \, dw(t)],$$
$$X(s) = x, \quad Y(s) = y, \quad Z(s) = z,$$

$(s, x) \in Q$, and $\tau = \tau_{s,x}$ is the first exit time of the trajectory $(t, X_{s,x}(t))$ to the boundary Γ :

$$\tau = \tau_{s,x} = T \wedge \inf\{t : X_{s,x}(t) \in \Gamma\}.$$

To prove (2.3.40), we take advantage of constructions proposed and justified by Dynkin [86, Ch. 5 and 11] (see also [194, Ch. 3]).

The solutions of (2.3.41) coincide with the solutions of (2.3.21)–(2.3.23) on the interval $[s, \tau)$. The Ito formula (which is of local nature) has the same form for these solutions. Introduce

$$\gamma_{s,x}(t) = u(t, X_{s,x}(t))Y_{s,x,1}(t) + Z_{s,x,1,0}(t), \quad s \le t < \tau.$$

For the differential $d\gamma_{s,x}(t)$, we get an expression of the same form as for $d\Gamma_{s,x}(t)$ in (2.3.26), but now we integrate $d\gamma$ from s to τ. We obtain

$$\gamma_{s,x}(\tau) - \gamma_{s,x}(s) = \int_s^\tau Y_{s,x,1}(t) \sum_{j=1}^d \left(\sum_{i=1}^d \sigma^{ij} \frac{\partial u}{\partial x^i} + u\mu^j + F^j \right) dw^j(t)$$

which implies the probabilistic representation (2.3.40) and the formula for the variance

$$Var\, \gamma_{s,x}(\tau) = E \int_s^\tau Y_{s,x,1}^2(t) \sum_{j=1}^d (\sum_{i=1}^d \sigma^{ij} \frac{\partial u}{\partial x^i} + u\mu^j + F^j)^2 dt. \qquad (2.3.42)$$

In particular, if μ and F are such that

$$\sum_{i=1}^d \sigma^{ij} \frac{\partial u}{\partial x^i} + u\mu^j + F^j = 0, \quad j = 1, \ldots, d, \qquad (2.3.43)$$

then $Var\, \gamma_{s,x}(\tau) = 0$ and $\gamma_{s,x}(\tau) \equiv u(s, x), s \le t \le \tau$, i.e., $\gamma_{s,x}(\tau)$ is deterministic.

We note that an analogous result is true for elliptic boundary value problems as well.

2.4 Conditional Probabilistic Representations and Practical Variance Reduction via Regression

Starting a Monte Carlo simulation, first of all we have to estimate the number of trajectories required to reach a prescribed accuracy. Fortunately, we can easily do this because a reliable estimate of the variance can be obtained by a preliminary numerical experiment using a relatively small set of trajectories. If the required number of trajectories is too large, we run inevitably into the problem of variance reduction. The variance reduction methods of the previous Sect. 2.3 (the method of important sampling, the method of control variates, and the combining method) are based on the assumption that approximations of the solution $u(t, x)$ of the considered problem and its spatial derivatives $\partial u(t, x)/\partial x^i$ are known. However, in general even rough approximations of the desired solution $u(t, x)$ and its derivatives $\partial u/\partial x^i(t, x)$, $i = 1, \ldots, d$, are unknown beforehand. At first sight, it seems that approximating them roughly is not difficult since they can be found by the Monte Carlo technique using a comparatively small number of independent trajectories. But this presupposes evaluating them at many points (t_k, x_k). Computing $u(t_k, x_k)$ and $\partial u/\partial x^i(t_k, x_k)$ by the Monte Carlo technique requires different auxiliary sets of approximate trajectories because of the different starting points (t_k, x_k). This is too expensive, i.e., as a rule, such a procedure is more expensive than simple increase of the number of trajectories starting from the initial position (t_0, x_0), at which we aim to find the value of the solution u.

To make variance reduction practical, a suitable method of constructing $u(t_k, x_k)$ and $\partial u/\partial x^i(t_k, x_k)$ should be comparatively inexpensive. Therefore, we cannot require high accuracy of the estimates for $u(t_k, x_k)$ and $\partial u/\partial x^i(t_k, x_k)$, because there is a trade-off between accuracy and computational expenses. In [322] it was proposed to exploit conditional probabilistic representations. Their employment together with the regression method allows us to evaluate $u(t_k, x)$ and $\partial u/\partial x^i(t_k, x)$ using the single auxiliary set of approximate trajectories starting from the initial position (t_0, x_0) only. This plays a crucial role in obtaining sufficiently inexpensive (but at the same time useful for variance reduction) estimates $\hat{u}(t_k, x)$ and $\widehat{\partial u/\partial x^i}(t_k, x)$. The construction of $\hat{u}$ and $\widehat{\partial u/\partial x^i}$ is accompanied by a number of errors of different nature. Although it is problematic to estimate these errors with sufficient accuracy and confidence, the suitability of $\hat{u}(t_k, x)$ and $\widehat{\partial u/\partial x^i}(t_k, x)$ for variance reduction can be directly verified during computations since the Monte Carlo error can always be estimated very cheaply. We emphasize that the obtained (even rather rough) estimates can effectively be used for accurate evaluating the function u not only at the position (t_0, x_0) but at many other positions as well.

In Sect. 2.4.1 we recall the general scheme of regression method for estimating conditional expectations. Section 2.4.2 is devoted to conditional probabilistic representations of solutions of parabolic equations and their derivatives. These representations together with regression approach play a decisive role in economical estimating of u and $\partial u/\partial x^i$ at all points (t, x), given the only set of trajectories starting from the initial point (t_0, x_0). In Sect. 2.4.3 we obtain the estimate $\hat{u}(s, x)$ and propose to estimate the derivatives $\partial u/\partial x^i(s, x)$ by $\partial \hat{u}/\partial x^i(s, x)$. This estimation of derivatives is inexpensive from the computational point of view, however rather rough. Section 2.4.4 is devoted to the more accurate way of estimating derivatives using linear regression method directly to find $\widehat{\partial u/\partial x^i}(t_k, x)$. In Sect. 2.4.5, we obtain $\widehat{\partial u/\partial x^i}(t_k, x)$ in the case of nonsmooth initial data exploiting probabilistic representations for $\partial u/\partial x^i(s, x)$ which rest on the Malliavin integration by parts. To this aim, we derive a conditional version of the Malliavin integration by parts formula adapted to our context. It should be noted that if the dimension d is large, the procedures of Sects. 2.4.4 and 2.4.5 are computationally very demanding since they require integration of the d^2-dimensional system of first-order variation equations which solution is present in the probabilistic representations for $\partial u/\partial x^i(s, x)$. Therefore, in practice, the inexpensive procedure of Sect. 2.4.3 is preferable if d is large. In Sect. 2.4.7 we give a simple, analytically tractable example to illustrate the benefits of the proposed variance reduction procedure and we also test it on a one-dimensional array of stochastic oscillators. Further illustration can be found in [318, 322]. Although here we illustrate variance reduction methods for the Cauchy problems for parabolic equations, the approach is straightforwardly applicable to the boundary value problems without any additional ideas needed.

We also note that the technique of conditional probabilistic representations together with regression can be used for evaluating various Greeks for American and Bermudan type options [24].

2.4.1   Regression Method of Estimating Conditional Expectation

Let us recall the general scheme of the linear regression method (see, e.g., [152]). Consider a sample $(_mX, \ _mV)$, $m = 1, \ldots, M_r$, from a generic member (X, V) of the sample, where X is a d-dimensional and V is a one-dimensional random variables. We pay attention that we denote by M_r the size of the sample used in the regression while M is the number of realizations used for computing the required quantity $u(t_0, x_0)$ (see (2.3.17)). Let the values of X belong to a domain $\mathbf{D} \subset \mathbf{R}^d$. It is of interest to estimate the regression function

$$c(x) = E(V|X = x). \tag{2.4.1}$$

Let $\{\varphi_l(x)\}_{l=1}^L$ be a set of basis functions each mapping $\mathbf{D}$ to $\mathbf{R}$. As an estimate $\hat{c}(x)$ of $c(x)$, we choose the function of the form $\sum_{l=1}^L \alpha_l \varphi_l(x)$ that minimizes the empirical risk:

$$\hat{\alpha} = \arg \min_{\alpha \in \mathbf{R}^L} \frac{1}{M_r} \sum_{m=1}^{M_r} \left({}_mV - \sum_{l=1}^L \alpha_l \varphi_l(_mX) \right)^2. \tag{2.4.2}$$

So

$$\hat{c}(x) = \sum_{l=1}^L \hat{\alpha}_l \varphi_l(x), \tag{2.4.3}$$

where $\hat{\alpha}_l$ satisfy the system of linear algebraic equations

$$a_{11}\alpha_1 + a_{12}\alpha_2 + \cdots + a_{1L}\alpha_L = b_1 \tag{2.4.4}$$

$$\cdots \cdots \cdots \cdots$$

$$a_{L1}\alpha_1 + a_{L2}\alpha_2 + \cdots + a_{LL}\alpha_L = b_L$$

with

$$a_{ln} = \frac{1}{M_r} \sum_{m=1}^{M_r} \varphi_l(_m X) \varphi_n(_m X), \; b_l = \frac{1}{M_r} \sum_{m=1}^{M_r} \varphi_l(_m X) \, _m V, \; l, n = 1, \ldots, L.$$

(2.4.5)

Thus, the usual base material in the field of regression is a sample $(_m X, \, _m V)$, $m = 1, \ldots, M_r$, from a generic member (X, V) of the sample.

Remark 2.4.1 Although in this paper we use linear regression, other regression methods (see, e.g., [97, 152]) including those from deep learning [136] can be exploited as well.

2.4.2 Conditional Probabilistic Representations for $u(s, x)$ and $\partial u / \partial x^i (s, x)$

The routine (unconditional) probabilistic representations are ideal for the Monte Carlo evaluation of $u(t_0, x_0)$ by using a set of trajectories starting from the point (t_0, x_0). To find $u(s, x)$ by this approach, we need to construct another set of trajectories which starts from (s, x). However, we can use the previous set starting from (t_0, x_0) to compute $u(s, x)$, $s > t_0$, if we make use of conditional probabilistic representations. In this section we introduce the conditional probabilistic representations for solutions of parabolic equations and for derivatives of the solutions.

Along with the unconditional probabilistic representation (2.3.24), (2.3.18), (2.3.21) for $u(s, x)$, we have the following conditional one

$$u(s, x) = E \left(f(X_{s,x}(T)) Y_{s,x}(T) + Z_{s,x}(T) \right)$$
$$= E \left(f(X_{s,x}(T)) Y_{s,x}(T) + Z_{s,x}(T) \text{ with } X := X_{t_0,x_0}(s) | X_{t_0,x_0}(s) = x \right).$$

(2.4.6)

This formula can be considered as the conditional version of the Feynman-Kac formula.

Analogously to (2.4.6), we get for $\partial^i (s, x) = \partial u / \partial x^i (s, x)$ (see (2.3.33)) :

$$\partial^i (s, x) = E \left(\sum_{j=1}^d \frac{\partial f(X_{s,x}(T))}{\partial x^j} \delta_{s,x}^i X^j (T) Y_{s,x}(T) \right.$$
$$+ f(X_{s,x}(T)) \delta_{s,x}^i Y(T) + \delta_{s,x}^i Z(T) \right)$$
$$= E \left(\sum_{j=1}^d \frac{\partial f(X_{s,x}(T))}{\partial x^j} \delta_{s,X}^i X^j (T) Y_{s,x}(T) \right.$$
$$+ f(X_{s,X}(T)) \delta_{s,X}^i Y(T) + \delta_{s,X}^i Z(T) | X := X_{t_0,x_0}(s) = x \right).$$

(2.4.7)

So, we have two different probabilistic representations both for $u(s, x)$ and $\partial^i (s, x)$: the first one is in the form of unconditional expectation (see Sect. 2.3) and the second one (i.e., (2.4.6) and (2.4.7)) is in the form of conditional expectation. The first form is natural to be realized by the Monte Carlo approach and the second one - by a regression method. As we discussed before, it is too expensive to run sets of trajectories starting from various initial points (s, x) and we do have the set of trajectories $(t, {}_m X_{t_0, x_0}(t))$. Taking this into account, the second way (which relies on the conditional probabilistic representations and regression) is more preferable although it is less accurate.

A proof of (2.4.6) and (2.4.7) relies on the following assertion: if ζ is $\tilde{\mathcal{F}}$-measurable, $f(x, \omega)$ is independent of $\tilde{\mathcal{F}}$, and $E f(x, \omega) = \phi(x)$, then (see, e.g., [213])

$$E(f(\zeta, \omega) | \tilde{\mathcal{F}}) = \phi(\zeta).$$

From this assertion, for any measurable g it holds (with $\zeta = X_{t_0, x_0}(s)$, $\tilde{\mathcal{F}} = \sigma\{X_{t_0, x_0}(s)\}$, $f(x, \omega) = g(X_{s, x}(T))$):

$$E(g(X_{s, X}(T)) | X_{t_0, x_0}(s) = x) = Eg(X_{s, x}(T)) \text{ with } X := X_{t_0, x_0}(s),$$

hence (2.4.6) and (2.4.7).

2.4.3 Evaluating $u(s, x)$

In evaluating $u(s, x)$ by regression, the pairs (X, V) and $({}_m X, {}_m V)$ have the form

$$(X, V) \sim \left(X_{t_0, x_0}(s), \ f(X_{s, X}(T)) Y_{s, X}(T) + Z_{s, X}(T) \right), \tag{2.4.8}$$
$$({}_m X, {}_m V) \sim \left({}_m X_{t_0, x_0}(s), \ f({}_m X_{s, {}_m X}(T)) {}_m Y_{s, {}_m X}(T) + {}_m Z_{s, {}_m X}(T) \right).$$

To realize a regression algorithm, we construct the set of trajectories $(t, {}_m X_{t_0, x_0}(t))$. Of course, we construct them approximately at the time moments $s = t_k$ and store the obtained values. So, in reality we have $(t_k, {}_m \tilde{X}_{t_0, x_0}(t_k))$. The time s in (2.4.8) is equal to one of t_k. We note that

$$X_{s, X}(t) = X_{s, X_{t_0, x_0}(s)}(t) = X_{t_0, x_0}(t), \ t \geq s, \tag{2.4.9}$$

i.e., $X_{s, X}(t)$ is a continuation of the base solution starting at the moment t_0 and $X_{s, X}(T)$ in (2.4.8) is equal to $X_{t_0, x_0}(T)$. It is not so for Y :

$$Y_{s, X}(T) \neq Y_{t_0, x_0}(T).$$

Let us recall that $Y_{s, X}(t)$ is the solution of the equation (see (2.3.22)):

$$dY_{s, X} = c(t, X_{s, X}(t)) Y_{s, X} dt + \mu^\mathsf{T}(t, X_{s, X}(t)) Y_{s, X} dw(t), \ Y(s) = 1. \tag{2.4.10}$$

Clearly,

$$Y_{s,X}(t) = \frac{Y_{t_0,x_0}(t)}{Y_{t_0,x_0}(s)}, \quad s \le t \le T, \tag{2.4.11}$$

hence storing $Y_{t_0,x_0}(t)$, we can get $Y_{s,X}(T)$ in (2.4.8).

Analogously, $Z_{s,X}(T) \ne Z_{t_0,x_0}(T)$. It is not difficult to find that

$$Z_{s,X}(t) = \frac{1}{Y_{t_0,x_0}(s)}(Z_{t_0,x_0}(t) - Z_{t_0,x_0}(s)), \quad Z_{s,X}(T) \tag{2.4.12}$$

$$= \frac{1}{Y_{t_0,x_0}(s)}(Z_{t_0,x_0}(T) - Z_{t_0,x_0}(s)).$$

Therefore

$$u(s,x) = E\left(f(X_{t_0,x_0}(T))\frac{Y_{t_0,x_0}(T)}{Y_{t_0,x_0}(s)} \right.$$
$$\left. + \frac{1}{Y_{t_0,x_0}(s)}(Z_{t_0,x_0}(T) - Z_{t_0,x_0}(s)) | X_{t_0,x_0}(s) = x \right).$$

Thus, storing $_m X_{t_0,x_0}(t)$, $_m Y_{t_0,x_0}(t)$, $_m Z_{t_0,x_0}(t)$, $t_0 \le t \le T$ (in fact, storing $_m \bar{X}$, $_m \bar{Y}$, $_m \bar{Z}$ at t_k), we get the pairs $(_m X, _m V)$ from

$$(X,V) \sim \left(X_{t_0,x_0}(s), \ f(X_{t_0,x_0}(T))\frac{Y_{t_0,x_0}(T)}{Y_{t_0,x_0}(s)} \right.$$
$$\left. + \frac{1}{Y_{t_0,x_0}(s)}(Z_{t_0,x_0}(T) - Z_{t_0,x_0}(s)) \right).$$

Having this sample, one can obtain $\hat{u}(s,x)$ by linear regression method (see Sect. 2.4.1):

$$\hat{u}(s,x) = \sum_{l=1}^{L} \hat{\alpha}_l \varphi_l(x). \tag{2.4.13}$$

A very simple estimate $\hat{\partial}^i(s,x)$ for $\partial^i(s,x) = \partial u/\partial x^i(s,x)$ is obtained from (2.4.13) straightforwardly:

$$\hat{\partial}^i(s,x) = \frac{\partial \hat{u}(s,x)}{\partial x^i} = \sum_{l=1}^{L} \hat{\alpha}_l \frac{\partial \varphi_l(x)}{\partial x^i}. \tag{2.4.14}$$

Then from (2.3.31) we find some $\hat{\mu}(s,x)$, $\hat{F}(s,x)$ for any $t_0 < s < T$ (in reality for any t_k) and construct the variate $\hat{\Gamma}(t_0,x_0)$ (see (2.3.16) and (2.3.18)) for $u(t_0,x_0)$ due to the system (2.3.21)–(2.3.23) with $\mu = \hat{\mu}$ and $F = \hat{F}$. We repeat that the variate $\hat{\Gamma}(t_0,x_0)$ is unbiased for any $\hat{\mu}$ and $\hat{F}$. We note that it is sufficient to have rather rough (in comparison with the required accuracy in evaluating $u(t_0,x_0)$) approximations $\hat{\mu}(s,x)$ and $\hat{F}(s,x)$ of some optimal μ and F from (2.3.30). Therefore, it is natural to use a coarser discretization and less Monte Carlo runs in the regression part of evaluating $\hat{u}(s, x)$ due to (2.4.13), i.e., to take M_r in (2.4.2) smaller than M and to construct samples $_mX$ in (2.4.5) with a comparatively rough discretization. Then in computing $u(t_0, x_0)$ with a finer discretization the necessary values of $\hat{\mu}$ and $\hat{F}$ at the intermediate points can be obtained after, e.g., linear interpolation of $\hat{u}$ with respect to time. The success of any regression-based approach clearly depends on the choice of basis functions. This is known to be a rather complicated problem, both in practice and theory. In fact, it is necessary to use special basis tailored to each particular problem. Fortunately, the variance can easily be evaluated during simulation. Therefore, it is not very expensive from the computational point of view to check quality of a given basis if we take coarse discretizations both in the regression part and in the main part of evaluating $u(t_0, x_0)$ and if we take not too large numbers M_r and M of Monte Carlo runs. This can help in choosing a proper basis.

Remark 2.4.2 Clearly, $\hat{\alpha}_l$ depend on s (on t_k). Let us note that the number L and the set $\{\varphi_l(x)\}_{l=1}^L$ may depend on t_k as well.

Remark 2.4.3 It is obvious that in practice we use (2.3.21) with different μ and F in the implementation of the regression and in computing the required quantity $u(t_0, x_0)$. Indeed, in the regression part of the procedure we can take arbitrary μ and F (e.g., both zero) while in computing $u(t_0, x_0)$ we choose μ and F according to (2.3.31) with $\hat{u}$ obtained via the regression or according to (2.3.32) with $\hat{u}$ and $\widehat{\partial u/\partial x^i}$ obtained via the regression.

Remark 2.4.4 At $s = t_0$ the system (2.4.4) degenerates into the only equation (we suppose that not all of $\varphi_l(x_0)$ are equal to zero) :

$$\varphi_1(x_0)\alpha_1 + \cdots + \varphi_L(x_0)\alpha_L \qquad (2.4.15)$$

$$= \frac{1}{M_r}\sum_{m=1}^{M_r}[f(_m\bar{X}_{t_0,x_0}(T))\,_m\bar{Y}_{t_0,x_0}(T) + \,_m\bar{Z}_{t_0,x_0}(T)].$$

Therefore, the coefficients $\alpha_1(t_0), \ldots, \alpha_L(t_0)$ cannot be found from (2.4.15) uniquely. At the same time, the linear combination $\alpha_1(t_0)\varphi_1(x_0) + \cdots + \alpha_L(t_0)\varphi_L(x_0)$, i.e., the estimate

$$\hat{u}(t_0, x_0) = \frac{1}{M_r}\sum_{m=1}^{M_r}[f(_m\bar{X}_{t_0,x_0}(T))\,_m\bar{Y}_{t_0,x_0}(T) + \,_m\bar{Z}_{t_0,x_0}(T)]$$

is defined uniquely. Clearly, when t_k is close to t_0 (for instance, at t_1), the system (2.4.4) is although not degenerate but ill-conditioned. Nevertheless, for such t_k and for x close to x_0, the estimate

$$\hat{u}(t_k, x) = \alpha_1(t_k)\varphi_1(x) + \cdots + \alpha_L(t_k)\varphi_L(x)$$

can be found sufficiently accurately. However, since it is not possible to satisfactory determine the coefficients $\alpha_1(t_k), \ldots, \alpha_L(t_k)$, we cannot get the derivatives $\partial \hat{u}(t_k, x)/\partial x^i$ by direct differentiation as $\alpha_1(t_k)\partial\varphi_1(x)/\partial x^i + \cdots + \alpha_L(t_k)\partial\varphi_L(x)/\partial x^i$. In addition, let us emphasize that such difficulties are not essential for the whole procedure of variance reduction because the variance is equal to the integral (2.3.29) and unsatisfactory knowledge of u and $\partial u/\partial x^i$ on short parts of the interval $[t_0, T]$ does not significantly affect the value of the integral.

2.4.4 Evaluating $\partial u/\partial x^i (s, x)$

The problem of evaluating $\partial u/\partial x^i (s, x)$ is of independent importance due to its connection with numerical computation of Greeks in Finance. Many articles are devoted to pathwise methods of estimating Greeks (see [129] and references therein; see also [293]). In [315] the finite-difference based method is developed and in [108, 109] it is suggested to use Malliavin calculus for computing Greeks. Several pathwise and finite-difference based methods for calculating sensitivities of Bermudan options using regression methods and Monte Carlo simulations are considered in [24] (see also references therein). In this section we propose a conditional version of the pathwise method and in Sect. 2.4.5 – a conditional version of the approach rested on the Malliavin integration by parts for evaluating $\partial u/\partial x^i (s, x)$.

As it was already written, differentiating the equality (2.4.13) we get an estimate for $\partial^i (s, x) = \partial u/\partial x^i (s, x)$ (see (2.4.14)), however, in general, it is rather rough. The more accurate way is to use linear regression method directly.

In evaluating $\partial^i (s, x)$ by regression, the pair (X, V^i) has the form (see (2.4.7)):

$$X = X_{t_0,x_0}(s), \tag{2.4.16}$$

$$V^i = \sum_{j=1}^{d} \frac{\partial f(X_{s,X}(T))}{\partial x^j} \delta^i_{s,X} X^j(T) Y_{s,X}(T) + f(X_{s,X}(T))\delta^i_{s,X} Y(T)$$

$$+ \delta^i_{s,X} Z(T).$$

We already have expressions for $X_{s,X}(T)$, $Y_{s,X}(T)$, $Z_{s,X}(T)$ via $X_{t_0,x_0}(t)$, $Y_{t_0,x_0}(t)$, $Z_{t_0,x_0}(t)$ with t being equal to s and T (see the formulas (2.4.9), (2.4.11), (2.4.12)). Our nearest aim is to express $\delta^i_{s,X} X^j(T)$, $\delta^i_{s,X} Y(T)$, $\delta^i_{s,X} Z(T)$ via $X_{t_0,x_0}(t)$, $Y_{t_0,x_0}(t)$, $Z_{t_0,x_0}(t)$, $\delta^i_{t_0,x_0} X^j(t)$, $\delta^i_{t_0,x_0} Y(t)$, $\delta^i_{t_0,x_0} Z(t)$.

We begin with $\delta^i_{s,X} X^j(t)$. The column-vector $\delta^i_{s,X} X(t)$ is the solution of the linear homogeneous stochastic system (2.3.34) which coefficients depend on $X_{s,X}(t) = X_{t_0,x_0}(t)$. Let the matrix

$$\Phi_{s,X}(t) := \{\delta^i_{s,X} X^j(t)\}$$

be the fundamental matrix of solutions of (2.3.34) normalized at the time s, i.e., $\Phi_{s,X}(s) = I$, where I is the identity matrix. Its element on j-th row and i-th column is equal to $\delta^i_{s,X} X^j(t)$. Clearly,

$$\Phi_{s,X}(t) = \Phi_{t_0,x_0}(t)\Phi^{-1}_{t_0,x_0}(s). \tag{2.4.17}$$

Now let us turn to the column-vector $\delta_{s,X} Y(t)$, consisting of components $\delta^i_{s,X} Y(t)$. We have (see (2.3.35))

$$d\delta_{s,X} Y = Y_{s,X}(t)\Phi^{\mathsf{T}}_{s,X}(t)\, \nabla c(t, X_{s,X}(t))dt \quad +c(t, X_{s,X}(t))\delta_{s,X} Y dt \tag{2.4.18}$$
$$+Y_{s,X}(t)\Phi^{\mathsf{T}}_{s,X}(t)\, \nabla[\mu^{\mathsf{T}}(t, X_{s,X}(t))dw(t)]+ \delta_{s,X} Y \mu^{\mathsf{T}}(t, X_{s,X}(t))dw(t),$$
$$\delta_{s,X} Y(s) = 0.$$

Due to the equality $X_{s,X}(t) = X_{t_0,x_0}(t)$ and (2.4.11) and (2.4.17), we get from (2.4.18)

$$d\delta_{s,X} Y = \frac{Y_{t_0,x_0}(t)}{Y_{t_0,x_0}(s)}[\Phi^{-1}_{t_0,x_0}(s)]^{\mathsf{T}}\Phi^{\mathsf{T}}_{t_0,x_0}(t)\, \nabla c(t, X_{t_0,x_0}(t))dt$$
$$+c(t, X_{t_0,x_0}(t))\delta_{s,X} Y dt$$
$$+\frac{Y_{t_0,x_0}(t)}{Y_{t_0,x_0}(s)}[\Phi^{-1}_{t_0,x_0}(s)]^{\mathsf{T}}\Phi^{\mathsf{T}}_{t_0,x_0}(t)\, \nabla[\mu^{\mathsf{T}}(t, X_{t_0,x_0}(t))dw(t)] \tag{2.4.19}$$
$$+\delta_{s,X} Y \mu^{\mathsf{T}}(t, X_{t_0,x_0}(t))dw(t),\ \delta_{s,X} Y(s) = 0.$$

Taking into account the equality

$$d\delta_{t_0,x_0} Y(t) = Y_{t_0,x_0}(t)\Phi^{\mathsf{T}}_{t_0,x_0}(t)\, \nabla c(t, X_{t_0,x_0}(t))dt + c(t, X_{t_0,x_0}(t))\delta_{t_0,x_0} Y(t)dt$$
$$+ Y_{t_0,x_0}(t)\Phi^{\mathsf{T}}_{t_0,x_0}(t)\, \nabla[\mu^{\mathsf{T}}(t, X_{t_0,x_0}(t))dw(t)] + \delta_{t_0,x_0} Y(t)\mu^{\mathsf{T}}(t, X_{t_0,x_0}(t))dw(t),$$

it is not difficult to verify that

$$\delta_{s,X} Y(t) = \frac{1}{Y_{t_0,x_0}(s)}[\Phi^{-1}_{t_0,x_0}(s)]^{\mathsf{T}}\left(\delta_{t_0,x_0} Y(t) - \frac{Y_{t_0,x_0}(t)}{Y_{t_0,x_0}(s)}\delta_{t_0,x_0} Y(s)\right). \tag{2.4.20}$$

In the similar way we obtain

$$\delta_{s,X} Z(t) = \frac{1}{Y_{t_0,x_0}(s)}[\Phi^{-1}_{t_0,x_0}(s)]^{\mathsf{T}}\left(\delta_{t_0,x_0} Z(t) - \delta_{t_0,x_0} Z(s)\right) \tag{2.4.21}$$
$$-\frac{1}{Y^2_{t_0,x_0}(s)}[\Phi^{-1}_{t_0,x_0}(s)]^{\mathsf{T}}\delta_{t_0,x_0} Y(s)\left(Z_{t_0,x_0}(t) - Z_{t_0,x_0}(s)\right).$$

Hence the column-vector $\partial(s, x)$ with the components $\partial^i(s, x)$ is equal to

$$\partial(s, x) = E\left(\frac{Y_{t_0,x_0}(T)}{Y_{t_0,x_0}(s)}[\Phi_{t_0,x_0}^{-1}(s)]^\mathsf{T}\,\Phi_{t_0,x_0}^\mathsf{T}(T)\,\nabla f(X_{t_0,x_0}(T))\right. \tag{2.4.22}$$

$$+ f(X_{t_0,x_0}(T))\delta_{s,X}Y(T) + \delta_{s,X}Z(T)\big|X_{t_0,x_0}(s) = x\bigg),$$

where $\delta_{s,X}Y(T)$ and $\delta_{s,X}Z(T)$ are from (2.4.20) and (2.4.21).

Thus, storing $_mX_{t_0,x_0}(t)$, $_mY_{t_0,x_0}(t)$, $_mZ_{t_0,x_0}(t)$, $_m\Phi_{t_0,x_0}(t)$, $_m\delta_{t_0,x_0}Y(t)$, $_m\delta_{t_0,x_0}Z(t)$, $t_0 \le t \le T$, we get the corresponding samples

$$(_mX,\ _mV^i) \tag{2.4.23}$$

$$= \left(_mX_{t_0,x_0}(s),\ \left(\frac{_mY_{t_0,x_0}(T)}{_mY_{t_0,x_0}(s)}[\,_m\Phi_{t_0,x_0}^{-1}(s)]^\mathsf{T}\,_m\Phi_{t_0,x_0}^\mathsf{T}(T)\,\nabla f(_mX_{t_0,x_0}(T))\right.\right.$$

$$\left.\left.+ f(_mX_{t_0,x_0}(T))\,_m\delta_{s,m}{}_XY(T) + \,_m\delta_{s,m}{}_XZ(T)\right)^i\right),$$

where $_m\Phi_{t_0,x_0}(s)$ is a realization of the fundamental matrix $\Phi_{t_0,x_0}(s)$ which corresponds to the same elementary event $\omega \in \Omega$ as the realization $_mX_{t_0,x_0}(t)$. We use $(_mX,\ _mV^i)$ for evaluating $\partial^i(s, x), i = 1, \dots, d$, by linear regression method:

$$\hat{\partial}^i(s, x) = \sum_{l=1}^{L} \hat{\beta}_l^i \psi_l(x). \tag{2.4.24}$$

2.4.5 Evaluating $\partial u/\partial x^i(s, x)$ Using the Malliavin Integration by Parts

If $f(x)$ is an irregular function, one can use the procedure recommended in Sect. 2.4.3, where we do not need direct calculations of derivatives $\partial u/\partial x^i$. Another way consists in approximating f by a smooth function with the consequent use of the procedure from Sect. 2.4.4. Because we do not pursue a high accuracy in estimating u and $\partial u/\partial x^i$, such approximation of f can be quite satisfactory. For direct calculation of derivatives $\partial u/\partial x^i$ without smoothing f, we can use the conditional version of the integration-by-parts (Bismut-Elworthy-Li) formula. This formula is successfully applied for evaluating derivatives $\partial u/\partial x^i$ in the case of an irregular f (see, e.g., [93, 108, 109, 341]), in this section we use it for variance reduction purposes.

Let us return to the Cauchy problem for linear parabolic PDE (2.3.12)–(2.3.13) and the probabilistic representation (2.3.24), (2.3.21)–(2.3.23) for its solution from Sect. 2.3.2. We assume here that the matrix $a(t, x) = \sigma(s, x)\sigma^\mathsf{T}(s, x)$ is symmetric and positive definite. We recall the notation (see (2.3.25))

$$\Gamma_{s,x}(t) = u(t, X_{s,x}(t))Y_{s,x,1}(t) + Z_{s,x,1,0}(t). \tag{2.4.25}$$

Clearly, $\Gamma_{s,x}(s) = u(s, x) = E\Gamma_{s,x}(T)$. It follows from (2.3.26) that

$$\Gamma_{s,x}(t) = u(s, x) + \int_s^t Y_{s,x,1}(s') \sum_{j=1}^d \left(\sum_{i=1}^d \sigma^{ij} \frac{\partial u}{\partial x^i} + u\mu^j + F^j \right) dw^j(s'),$$

(2.4.26)

and $\Gamma_{s,x}(t)$, $t \geq s$, is a martingale.

By (2.4.25) and (2.4.26), we get

$$u(s, x) = u(t, X_{s,x}(t))Y_{s,x,1}(t) + Z_{s,x,1,0}(t) \tag{2.4.27}$$

$$- \int_s^t Y_{s,x,1}(s') \sum_{j=1}^d \left(\sum_{i=1}^d \sigma^{ij} \frac{\partial u}{\partial x^i} + u\mu^j + F^j \right) dw^j(s').$$

Since the left-hand side of (2.4.27) does not depend on t, so is the right-hand side. From (2.4.27) we have

$$u(s, x) = E\left[u(t, X_{s,x}(t))Y_{s,x,1}(t) + Z_{s,x,1,0}(t) \right]. \tag{2.4.28}$$

Differentiating (2.4.28) with respect to x^i, we obtain for $i = 1, \ldots, d$,

$$\frac{\partial}{\partial x^i} u(s, x) = E\left[Y_{s,x,1}(t) \sum_{k=1}^d \frac{\partial}{\partial x^k} u(t, X_{s,x}(t)) \frac{\partial}{\partial x^i} X_{s,x}^k(t) \right. \tag{2.4.29}$$

$$\left. + u(t, X_{s,x}(t)) \frac{\partial}{\partial x^i} Y_{s,x,1}(t) + \frac{\partial}{\partial x^i} Z_{s,x,1,0}(t) \right].$$

We again see that the right-hand side of (2.4.29) is independent of t.

It follows from (2.4.27) (see also (2.3.27)) that

$$f(X_{s,x}(T))Y_{s,x,1}(T) + Z_{s,x,1,0}(T) \tag{2.4.30}$$

$$= u(s, x) + \int_s^T Y_{s,x,1}(t) \left(\left[\frac{\partial u}{\partial x} \right]^\mathsf{T} \sigma + u(t, X_{s,x}(t))\mu^\mathsf{T} + F^\mathsf{T} \right) dw(t).$$

Multiplying both parts of (2.4.30) by

$$\int_s^T [dw(t)]^\mathsf{T} \sigma^{-1}(t, X_{s,x}(t)) \frac{\partial}{\partial x^i} X_{s,x}(t),$$

adding and subtracting

$$\int_s^T \left[u(t, X_{s,x}(t)) \frac{\partial}{\partial x^i} Y_{s,x,1}(t) + \frac{\partial}{\partial x^i} Z_{s,x,1,0}(t) \right] dt$$

in the left-hand side, and taking expectation, we obtain for $i = 1, \ldots, d$,

$$E\Bigg\{ \big(f(X_{s,x}(T)) Y_{s,x,1}(T) + Z_{s,x,1,0}(T) \big) \tag{2.4.31}$$

$$\times \int_s^T [dw(t)]^\mathsf{T} \sigma^{-1} \frac{\partial}{\partial x^i} X_{s,x}(t) \Bigg\}$$

$$= \int_s^T E\left[Y_{s,x,1}(t) \left[\frac{\partial u}{\partial x} \right]^\mathsf{T} \frac{\partial}{\partial x^i} X_{s,x}(t) + u \frac{\partial}{\partial x^i} Y_{s,x,1}(t) + \frac{\partial}{\partial x^i} Z_{s,x,1,0}(t) \right] dt$$

$$+ E\int_s^T Y_{s,x,1}(t) u \mu^\mathsf{T} \sigma^{-1} \frac{\partial}{\partial x^i} X_{s,x}(t) dt$$

$$+ E\int_s^T Y_{s,x,1}(t) F^\mathsf{T} \sigma^{-1} \frac{\partial}{\partial x^i} X_{s,x}(t) dt$$

$$- E\int_s^T \left[u \frac{\partial}{\partial x^i} Y_{s,x,1}(t) + \frac{\partial}{\partial x^i} Z_{s,x,1,0}(t) \right] dt.$$

Due to (2.4.29), the integrand in the third line of (2.4.31) does not depend on t and it is equal to $\frac{\partial}{\partial x^i} u(s, x)$, hence this line becomes

$$(T - s) \frac{\partial}{\partial x^i} u(s, x).$$

Thus, we can re-write (2.4.31) in the form

$$(T - s)\frac{\partial}{\partial x^i}u(s, x) \tag{2.4.32}$$

$$= E\left\{ \left(f(X_{s,x}(T))Y_{s,x,1}(T) + Z_{s,x,1,0}(T)\right) \int_s^T [dw(t)]^\mathsf{T} \sigma^{-1}\frac{\partial}{\partial x^i}X_{s,x}(t) \right\}$$

$$- E\int_s^T Y_{s,x,1}(t)u\mu^\mathsf{T}\sigma^{-1}\frac{\partial}{\partial x^i}X_{s,x}(t)dt$$

$$- E\int_s^T Y_{s,x,1}(t)F^\mathsf{T}\sigma^{-1}\frac{\partial}{\partial x^i}X_{s,x}(t)dt$$

$$+ E\int_s^T \left[u\frac{\partial}{\partial x^i}Y_{s,x,1}(t) + \frac{\partial}{\partial x^i}Z_{s,x,1,0}(t) \right] dt.$$

Remark 2.4.5 We note that the derivatives $\frac{\partial}{\partial x^i}u(s, x)$ in (2.4.32) are found without using derivatives of the function $f(t, x)$, i.e., differentiability of $f(t, x)$ is not necessary for spatial differentiability of $u(t, x)$. To gain some intuition, consider the following representation

$$u(s, x) = Ef(X_{s,x}(T)) = \int p(s, x; T, y)f(y)dy,$$

where p is the transitional density of the process X. Therefore, assuming that the density is differentiable, we have

$$\frac{\partial}{\partial x^i}u(s, x) = \int \frac{\partial}{\partial x^i}p(s, x; T, y)f(y)dy$$

$$= \int f(y)p(s, x; T, y)\frac{\partial}{\partial x^i}\ln p(s, x; T, y)dy$$

$$= E\left[f(X_{s,x}(T))\frac{\partial}{\partial x^i}\ln p(s, x; T, X_{s,x}(T)) \right].$$

To complete derivation of the formula for the derivatives $\frac{\partial}{\partial x^i}u(s, x)$, we need to exclude the function $u = (t, X_{s,x}(t))$ from the right-hand side of (2.4.32), where it appears twice: in the term $-E\int_s^T Y_{s,x,1}(t)u\mu^\mathsf{T}\sigma^{-1}\frac{\partial}{\partial x^i}X_{s,x}(t)dt$ and in $E\int_s^T \left[u\frac{\partial}{\partial x^i}Y_{s,x,1}(t) + \frac{\partial}{\partial x^i}Z_{s,x,1,0}(t)\right]dt$. We have

$$u(t, X_{s,x}(t)) = E\left[\left. f(X_{s,x}(T))\frac{Y_{s,x,1}(T)}{Y_{s,x,1}(t)} + \frac{\left(Z_{s,x,1,0}(T) - Z_{s,x,1,0}(t)\right)}{Y_{s,x,1}(t)} \right| \mathcal{F}_t\right].$$

Hence,

$$Y_{s,x,1}(t)u(t, X_{s,x}(t))\mu^\mathsf{T}\sigma^{-1}\frac{\partial}{\partial x^i}X_{s,x}(t) \tag{2.4.33}$$

$$= E\left[\left(f(X_{s,x}(T))Y_{s,x,1}(T) + Z_{s,x,1,0}(T) - Z_{s,x,1,0}(t)\right)\right.$$

$$\left.\left.\mu^\mathsf{T}\sigma^{-1}\frac{\partial}{\partial x^i}X_{s,x}(t)\right| \mathcal{F}_t\right].$$

Substituting (2.4.33) in the third line of (2.4.32), we get

$$-E\int_s^T Y_{s,x,1}(t)u\mu^\mathsf{T}\sigma^{-1}\frac{\partial}{\partial x^i}X_{s,x}(t)dt \tag{2.4.34}$$

$$= -E\int_s^T \left(f(X_{s,x}(T))Y_{s,x,1}(T) + Z_{s,x,1,0}(T) - Z_{s,x,1,0}(t)\right)$$

$$\cdot\mu^\mathsf{T}\sigma^{-1}\frac{\partial}{\partial x^i}X_{s,x}(t)dt$$

$$= -E\int_s^T \left(f(X_{s,x}(T))Y_{s,x,1}(T) + Z_{s,x,1,0}(T)\right)\mu^\mathsf{T}\sigma^{-1}\frac{\partial}{\partial x^i}X_{s,x}(t)dt$$

$$+E\int_s^T Z_{s,x,1,0}(t)\mu^\mathsf{T}\sigma^{-1}\frac{\partial}{\partial x^i}X_{s,x}(t)dt$$

$$= -E\left[\Gamma_{s,x}\int_s^T \mu^\mathsf{T}\sigma^{-1}\frac{\partial}{\partial x^i}X_{s,x}(t)dt\right]$$

$$+E\int_s^T Z_{s,x,1,0}(t)\mu^\mathsf{T}\sigma^{-1}\frac{\partial}{\partial x^i}X_{s,x}(t)dt.$$

For the fifth line of (2.4.32), we obtain

$$E \int_s^T u(t, X_{s,x}(t)) \frac{\partial}{\partial x^i} Y_{s,x,1}(t) dt \tag{2.4.35}$$

$$= E \int_s^T \left[f(X_{s,x}(T)) \frac{Y_{s,x,1}(T)}{Y_{s,x,1}(t)} + \frac{Z_{s,x,1,0}(T) - Z_{s,x,1,0}(t)}{Y_{s,x,1}(t)} \right] \frac{\partial}{\partial x^i} Y_{s,x,1}(t) dt.$$

Substituting (2.4.35) and (2.4.34) in (2.4.32), we arrive at the following variant of the integration-by-parts formula

$$\frac{\partial}{\partial x^i} u(s, x) = \frac{1}{T - s} E \left\{ \Gamma_{s,x} \int_s^T \left[\sigma^{-1} \frac{\partial X_{s,x}(t)}{\partial x^i} \right]^\mathsf{T} dw(t) \right\} \tag{2.4.36}$$

$$- \frac{1}{T - s} E \left[\Gamma_{s,x} \int_s^T \mu^\mathsf{T} \sigma^{-1} \frac{\partial X_{s,x}(t)}{\partial x^i} dt \right]$$

$$+ \frac{1}{T - s} E \int_s^T Z_{s,x}(t) \mu^\mathsf{T} \sigma^{-1} \frac{\partial X_{s,x}(t)}{\partial x^i} dt$$

$$- \frac{1}{T - s} E \int_s^T Y_{s,x}(t) F^\mathsf{T} \sigma^{-1} \frac{\partial X_{s,x}(t)}{\partial x^i} dt$$

$$+ \frac{1}{T - s} E \left[\Gamma_{s,x} \int_s^T \frac{1}{Y_{s,x}(t)} \frac{\partial Y_{s,x}(t)}{\partial x^i} dt \right] - \frac{1}{T - s} E \int_s^T \frac{Z_{s,x}(t)}{Y_{s,x}(t)} \frac{\partial Y_{s,x}(t)}{\partial x^i} dt$$

$$+ \frac{1}{T - s} E \int_s^T \frac{\partial Z_{s,x}(t)}{\partial x^i} dt := \Delta^i(s, x),$$

where $Y_{s,x} = Y_{s,x,1}$ and $Z_{s,x} = Z_{s,x,1,0}$, and μ^T, σ^{-1}, and F^T have $(t, X_{s,x}(t))$ as their arguments. In particular, if $c = 0$, $g = 0$, $\mu = 0$, $F = 0$, we get the well-known integration-by-parts (Bismut-Elworthy-Li) formula (see, e.g. [93, 341]):

$$\frac{\partial}{\partial x^i} u(s, x) = \frac{1}{T - s} E f(X_{s,x}(T)) \int_s^T \left[\sigma^{-1}(t, X_{s,x}(t)) \frac{\partial X_{s,x}(t)}{\partial x^i} \right]^\mathsf{T} dw(t). \tag{2.4.37}$$

We note that here similarly to [93] we derive the integration-by-parts formula using very simple tools, in contrast to e.g. [341, p. 330] and references therein where analysis on the Wiener space is exploited for this purpose.

As in Sect. 2.4.2, together with the unconditional probabilistic representation (2.4.36) for $\frac{\partial}{\partial x^i} u(s, x)$, we have the following conditional one

$$\frac{\partial}{\partial x^i} u(s, x) = E(\Delta^i(s, X)|X := X_{t_0, x_0}(s) = x). \qquad (2.4.38)$$

And again, the formula (2.4.36) is natural for the Monte Carlo approach and (2.4.38) – for a regression method. An implementation of the regression method is rested on the corresponding approximation $(_m X, \ _m V^i)$ of the pair $(X, V^i) = (X_{t_0, x_0}(s), \ \Delta^i(s, X_{t_0, x_0}(s)))$ following the ideas of Sect. 2.4.4.

Remark 2.4.6 Let us comment on how to simulate the terms of (2.4.36). E.g., consider the first term in (2.4.36) and denote it as S. Introduce the system of SDEs consisting of (2.3.21)–(2.3.23), (2.3.34), and

$$dI^i = \left[\sigma^{-1}\frac{\partial X_{s,x}(t)}{\partial x^i}\right]^\top dw(t), \ I^i(s) = 0.$$

Then

$$S = E\left[\left(f(X_{s,x}(T))Y_{s,x,1}(T) + Z_{s,x,1,0}(T)\right) \cdot I^i(T)\right],$$

which can be approximated using weak-sense numerical integration of the introduced SDEs system. The other terms in (2.4.36) can be evaluated analogously.

2.4.6 Two-Run Procedure

The straightforward implementation of evaluating $u(s, x)$ and $\partial u/\partial x^i(s, x)$ by regression as described in Sects. 2.4.3 and 2.4.4 requires storing

$$_m \Lambda(t_k) := (_m X_{t_0, x_0}(t_k), \ _m Y_{t_0, x_0}(t_k), \ _m Z_{t_0, x_0}(t_k),$$
$$_m \Phi_{t_0, x_0}(t_k), \ _m \delta_{t_0, x_0} Y(t_k), \ _m \delta_{t_0, x_0} Z(t_k))$$

(or, more precisely, their approximations $_m \bar{\Lambda}(t_k)$) at all t_k, $k = 1, \ldots, N$, in the main computer memory (RAM) until the end of simulation. This puts a too demanding requirement on the RAM size and limits practicality of the proposed approach since in almost any practical problem a relatively large number of time steps is needed. However, this difficulty can be overcome and we can avoid storing $_m \bar{\Lambda}(t_k)$ at all t_k by implementing the two-run procedure described below.

First, we recall that, as a rule, pseudorandom number generators used for Monte Carlo simulations have the property that the sequence of random numbers obtained by them are easily reproducible (see, e.g., Sect. 2.6 below and references therein). Let us fix a sequence of pseudorandom numbers. The two-run procedure can schematically be presented as follows.

First run:

- simulate M_r number of independent trajectories $_m \bar{\Lambda}(t_k)$, $k = 1, \ldots, N$, with an arbitrary choice of μ and F (e.g., $\mu = 0$ and $F = 0$);

- compute and store the values $_m\bar{\varGamma}$ to form the component V needed for the regression in the second run and compute and store the values

$$_m\bar{Y}(T)\,_m\bar{\varPhi}^\mathsf{T}_{t_0,x_0}(T)\,\nabla f(_m\bar{X}(T)) + f(_m\bar{X}(T))\,_m\overline{\delta Y}(T) +_m\overline{\delta Z}(T)$$

and $_m\bar{Y}(T)$ to form the components V^i in the second run.

Second run:

- reinitialize the random number generator so that it produces the same sequence as for the first run;
- for $k = 1, \ldots, N$

    – simulate the same $_m\bar{\varLambda}(t_k)$, $m = 1, \ldots, M_r$ as in the first run (i.e., they correspond to the same sequence of pseudorandom numbers as in the first run), keeping only the current $_m\bar{\varLambda}(t_k)$ in RAM;
    – use the values stored in RAM during the first run and $_m\bar{\varLambda}(t_k)$ from this run to find $\bar{u}(t_k, x)$ and $\overline{\partial u/\partial x^i}(t_k, x)$ by regression ($_m\bar{\varLambda}(t_k)$ and $_m\bar{\varLambda}(T)$ form the pairs $(_mX, \,_mV)$ and $(_mX, \,_mV^i)$ needed for the regression);
    – use the found $\bar{u}(t_k, x)$ and $\overline{\partial u/\partial x^i}(t_k, x)$ to obtain $\bar{\mu}(t_k, x)$ and $\bar{F}(t_k, x)$ required for variance reduction (see (2.3.31));
    – simulate (2.3.21) with $\mu = \bar{\mu}$ and $F = \bar{F}$ on this step and thus obtain M independent triples
    $$(_m\tilde{X}_{t_0,x_0}(t_k), \,_m\tilde{Y}_{t_0,x_0}(t_k), \,_m\tilde{Z}_{t_0,x_0}(t_k)) = (_m\tilde{X}_{t_{k-1},_m\tilde{X}(t_{k-1})}(t_k),$$
    $$_m\tilde{Y}_{t_{k-1},_m\tilde{X}(t_{k-1}),_m\tilde{Y}(t_{k-1})}(t_k), \,_m\tilde{Z}_{t_{k-1},_m\tilde{X}(t_{k-1}),_m\tilde{Y}(t_{k-1}),_m\tilde{Z}(t_{k-1})}(t_k)),$$
    which we keep in RAM until the next step;

- use the obtained $(_m\tilde{X}_{t_0,x_0}(T), \,_m\tilde{Y}_{t_0,x_0}(T), \,_m\tilde{Z}_{t_0,x_0}(T))$ to get the required $u(t_0, x_0)$ (see (2.3.17)).

We emphasize that in the two-run procedure at each time moment $s = t_k$ we need to keep in memory only the pre-computed values stored at the end of the first run and the values $_m\bar{\varLambda}(t_k)$ and $(_m\tilde{X}_{t_0,x_0}(t_k), \,_m\tilde{Y}_{t_0,x_0}(t_k), \,_m\tilde{Z}_{t_0,x_0}(t_k))$ (only at the current time step k) which is well within RAM limits of a PC.

We note that the two-run realization of the procedure from Sect. 2.4.3 based on using regression for estimating u only is less computationally demanded (both on processor time and RAM and especially for problems of large dimension d) than the procedures of Sects. 2.4.4 and 2.4.5 which estimate the derivatives of u via regression. The two-run procedure was used in the numerical experiments of Sect. 2.4.7.2.

2.4.7 Examples

The first example is partly illustrative and partly theoretical. The second example is numerical. See further illustrations in [322].

2.4.7.1   Heat Equation

Consider the Cauchy problem

$$\frac{\partial u}{\partial t} + \frac{\sigma^2}{2} \frac{\partial^2 u}{\partial x^2} = 0, \quad t_0 \leq t < T, \ x \in \mathbf{R}, \tag{2.4.39}$$

$$u(T, x) = x^2.$$

Its solution is

$$u(t, x) = \sigma^2(T - t) + x^2. \tag{2.4.40}$$

The probabilistic representation (2.3.21), (2.3.24) with $\mu = 0$ takes the form

$$u(s, x) = E\left[X_{s,x}^2(T) + Z_{s,x}(T)\right] = E\Gamma_{s,x}, \tag{2.4.41}$$

$$dX = \sigma dw(t), \quad X(s) = x, \tag{2.4.42}$$

$$dZ = F(t, X)dw(t), \quad Z(s) = 0. \tag{2.4.43}$$

Due to Theorem 2.3.3, we have $Var\Gamma_{s,x} = Var\left[X_{s,x}^2(T) + Z_{s,x}(T)\right] = 0$ for the optimal choice of the function $F(t, x) = -\sigma \partial u/\partial x = -2\sigma x$. We note that in this example $\partial u/\partial x$ and the optimal F do not depend on time t.

For the purpose of this illustrative example, we evaluate $u(0, 0) = E\Gamma_{0,0}$. Let us simulate (2.4.42) exactly (i.e., we have no error of numerical integration):

$$X_0 = x, \quad X_{k+1} = X_k + \sigma \Delta_k w, \quad k = 0, \dots, N - 1, \tag{2.4.44}$$

$$\Delta_k w := w(t_{k+1}) - w(t_k).$$

For $F \equiv 0$, we have $u(0, 0) = E\Gamma_{0,0} \approx \hat{u}(0, 0) = \frac{1}{M} \sum_{m=1}^{M} {}_m X_N^2$, where ${}_m X_N$ are independent realizations of X_N obtained by (2.4.44). Further, $Var\Gamma_{0,0} = 2\sigma^4 T^2$, hence the Monte Carlo error is equal to (see (2.3.20))

$$\rho = c \frac{\sqrt{2\sigma^2 T}}{\sqrt{M}}. \tag{2.4.45}$$

For instance, to achieve the accuracy $\rho = 0.0001$ for $c = 3$ (recall that there is no error of numerical integration here) in the case of $\sigma = 1$ and $T = 10$, one needs to perform $M = 18 \times 10^{10}$ Monte Carlo runs.

To reduce the Monte Carlo error, we estimate $\partial u/\partial x$ by regression to get $\hat{F}(t_k, x)$ close to the optimal $F = -2\sigma x$. As the basis functions for the regression, we take the first two Hermite polynomials:

$$\psi_1(x) = 1, \quad \psi_2(x) = 2x. \tag{2.4.46}$$

We note that in this example the required derivative $\partial u/\partial x$ can be expanded in the basis (2.4.46), i.e., here we do not have any error due to the cut-off of a set of basis functions. In the construction of the estimate for $\partial u/\partial x$, we put $F = 0$ in (2.4.43).

The variational equation associated with (2.4.42) has the form (see (2.3.34)): $d\delta X = 0$, $\delta X(s) = 1$, hence $\delta X(t) = 1$, $t \geq s$. Thus, the sample from (2.4.23) takes the form $({}_m X, {}_m V) = ({}_m X_{t_0,x_0}(s), 2\, {}_m X_{t_0,x_0}(T))$ and the estimator $\hat{\partial}(t_k, x)$ for $\partial u/\partial x(t_k, x)$ is constructed as

$$\hat{\partial}(t_k, x) = \hat{\alpha}_1(t_k) + 2\hat{\alpha}_2(t_k)x, \quad k = 1, \dots, N, \qquad (2.4.47)$$

where $\hat{\alpha}_1(t_k)$ and $\hat{\alpha}_2(t_k)$ satisfy the system of linear algebraic equations (see (2.4.4)–(2.4.5)):

$$a_{11}\alpha_1 + a_{12}\alpha_2 = b_1 \qquad (2.4.48)$$
$$a_{21}\alpha_1 + a_{22}\alpha_2 = b_2,$$

$$a_{11} = 1, \quad a_{12} = a_{21} := a_{12}(t_k) = \frac{1}{M_r}\sum_{m=1}^{M_r} 2 \times {}_m X(t_k), \qquad (2.4.49)$$

$$a_{22} := a_{22}(t_k) = \frac{1}{M_r}\sum_{m=1}^{M_r} 4 \times ({}_m X(t_k))^2,$$

$$b_1 := b_1(t_k) = \frac{1}{M_r}\sum_{m=1}^{M_r} 2 \times {}_m X(T),$$

$$b_2 := b_2(t_k) = \frac{1}{M_r}\sum_{m=1}^{M_r} 4 \times {}_m X(t_k) \times {}_m X(T).$$

Here ${}_m X(t_k)$, $m = 1, \dots, M_r$, are independent realizations of $X(t_k)$, $k = 1, \dots, N$, obtained by (2.4.44). Hence

$$\hat{\alpha}_1(t_k) = \frac{b_1 a_{22} - b_2\, a_{12}}{a_{22} - (a_{12})^2}, \quad \hat{\alpha}_2(t_k) = \frac{b_2 - b_1\, a_{12}}{a_{22} - (a_{12})^2}. \qquad (2.4.50)$$

We define

$$\hat{F}(0, x) = -\frac{\sigma}{M_r}\sum_{m=1}^{M_r} 2 \times {}_m X(T), \qquad (2.4.51)$$

$$\hat{F}(t, x) = -\sigma\left(\hat{\alpha}_1(t_k) + 2\hat{\alpha}_2(t_k)x\right) \quad \text{for } t \in (t_{k-1}, t_k], \quad k = 1, \dots, N.$$

We simulate (2.4.43) with $F = \hat{F}(t, x)$ exactly (i.e., again we have no error of numerical integration):

$$Z_0 = 0,$$
$$Z_{k+1} = Z_k - \sigma\hat{\alpha}_1(t_{k+1})\Delta_k w - 2\sigma^2\hat{\alpha}_2(t_{k+1})w(t_k)\Delta_k w \qquad (2.4.52)$$
$$-\sigma^2\hat{\alpha}_2(t_{k+1})\left[(\Delta_k w)^2 - h\right],$$

The increments $\Delta_k w$ are the same both in (2.4.44) and in (2.4.52) and they are independent of the ones used to estimate $\hat{\alpha}_1$ and $\hat{\alpha}_2$.

We simulate

$$u(0,0) = E\Gamma_{0,0} = E\left(X_N^2 + Z_N\right) \qquad (2.4.53)$$

$$\approx \hat{u}(0,0) = \frac{1}{M_r}\sum_{m=1}^{M_r}\left(_m X_N^2 +_m Z_N\right),$$

where $_m X_N$ and $_m Z_N$ are independent realizations of X_N and Z_N obtained according to (2.4.44) and (2.4.52). We note that the approximation (2.4.53) does not have the numerical integration error and the error due to the cut-off of the basis, it has the Monte Carlo error only.

Using Theorem 2.3.3, one can evaluate $Var\,\Gamma_{0,0}$ in the case of $F = \hat{F}$ defined in (2.4.51) and obtain $Var\,\Gamma_{0,0} \approx 4\sigma^4 T^2/M_r$. Then the Monte Carlo error ρ in this case is equal to (compare with (2.4.45))

$$\rho \approx c\frac{2\sigma^2 T}{\sqrt{MM_r}}. \qquad (2.4.54)$$

This example illustrates that in the absence of the error due to the cut-off of a set of basis functions used in regression and of the numerical integration error the Monte Carlo error is reduced $\sim 1/\sqrt{M_r}$ times by the proposed variance reduction technique. This is, of course, a significant improvement. Indeed, let us return to the example discussed after (2.4.45). The estimate (2.4.54) implies that to achieve the accuracy $\rho = 0.0001$ for $c = 3$ in the case of $\sigma = 1$ and $T = 10$, one can take, e.g., $M = M_r = 6 \times 10^5$, i.e., one can run about 10^5 times less trajectories than when the variance reduction was not used (see the discussion after (2.4.45)). The gain of computational efficiency is significant in spite of the fact that there is an overhead cost of solving the linear system (2.4.48) in the "regression's runs".

Remark 2.4.7 In the above analysis we assumed that "regression's runs" and the Monte Carlo runs for computing the desired value $u(0,0)$ are independent. In practice, this assumption can be dropped and we can use the same paths $X(t)$ for both the "regression's runs" and the Monte Carlo runs. Then, as a rule, we choose $M_r \leq M$.

Remark 2.4.8 We are expecting (see also experiments in Sect. 2.4.7.2) that in the general case the Monte Carlo error after application of this variance reduction technique has the form

$$\rho = O\left(\frac{1}{\sqrt{MM_r}} + \frac{h^{p/2}}{\sqrt{M}} + \frac{err_B}{\sqrt{M}}\right), \qquad (2.4.55)$$

where the first term has the same nature as in this illustrative example (see (2.4.54)); the second term is due to the error of numerical integration (it is assumed that a method of weak order p is used); and the third one arises as a result of the use of a finite set of functions as the basis in the regression while the solution $u(t, x)$ is usually expandable in a basis consisting of infinite number of functions (i.e., this error is due to the cut-off of the basis). We note that finding an appropriate basis for regression in applying this variance reduction approach to a particular problem can be a difficult task and it requires some knowledge about the solution $u(t, x)$ of the considered problem. Roughly speaking, in the proposed implementation of the variance reduction methods (the method of important sampling, the method of control variates or the combining method) we substitute the task of finding an approximate solution to the problem of interest by the task of finding an appropriate basis for the regression.

For complicated systems of SDEs, it is preferable to use regression to approximate the solution $u(t, x)$ and then differentiate this approximation to approximate the derivatives $\partial u / \partial x^i$. In the case of this illustrative example we take the first three Hermite polynomials:

$$\psi_1(x) = 1, \quad \psi_2(x) = 2x, \quad \psi_3(x) = 4x^2 - 2 \tag{2.4.56}$$

as the basis functions for the regression. In this example the required function $u(t, x)$ can be expanded in the basis (2.4.56). We construct the estimator $\hat{u}(t_k, x)$ for $u(t_k, x)$:

$$\hat{u}(t_k, x) = \hat{\alpha}_1(t_k) + 2\hat{\alpha}_2(t_k)x + \hat{\alpha}_3(t_k) \cdot \left(4x^2 - 2\right), \quad k = 1, \ldots, N, \tag{2.4.57}$$

where $\hat{\alpha}_1(t_k), \hat{\alpha}_2(t_k), \hat{\alpha}_3(t_k)$ satisfy the system of linear algebraic equations (2.4.4) with the corresponding coefficients. Further, we approximate the derivative $\partial u / \partial x(t_k, x)$:

$$\frac{\partial u}{\partial x}(t_k, x) \approx 2\hat{\alpha}_2(t_k) + 8\hat{\alpha}_3(t_k)x \tag{2.4.58}$$

with $\hat{\alpha}_2(t_k)$ and $\hat{\alpha}_3(t_k)$ from (2.4.57), and we define

$$\hat{F}(t, x) := -\sigma \left(2\hat{\alpha}_2(t_k) + 8\hat{\alpha}_3(t_k)x\right), \text{ for } t \in [t_{k-1}, t_k), \quad k = 1, \ldots, N, \tag{2.4.59}$$

which we use for variance reduction by putting $F = \hat{F}$ in (2.4.43). In the experiments we simulate (2.4.43) with $F = \hat{F}(t, x)$ exactly (see (2.4.52)). The new estimator for $u(0, 0)$ has the form (2.4.53) again but with the new Z_N corresponding to the choice of $\hat{F}(t, x)$ from (2.4.59).

Table 2.4.1 gives some results of simulating $u(0, 0)$ by (2.4.53) with $F = 0$, $F = \hat{F}$ from (2.4.51), and $F = \hat{F}$ from (2.4.59). We see that for $F = 0$ the Monte Carlo error is consistent with (2.4.45), i.e., it decreases $\sim 1/\sqrt{M}$. When the variance reduction is used, the results in Table 2.4.1 approve the Monte Carlo error estimate (2.4.54). It is quite obvious that $\hat{F}$ from (2.4.51) is a more accurate estimator for the

Table 2.4.1 *Heat equation.* Simulation of $u(0, 0)$ for $\sigma = 1$ and $T = 10$ by (2.4.53) with the corresponding choice of the function F and for various M. The time step $h = 0.1$ and $M_r = M$. The exact value is $u(0, 0) = 10$. The value after "$\pm$"equals to two standard deviations of the corresponding estimator and gives the confidence interval for the corresponding value with probability 0.95 (i.e., $c = 2$)

M	$F = 0$	$F = \hat{F}$ from (2.4.51)	$F = \hat{F}$ from (2.4.59)
10^3	9.67 ± 0.85	9.993 ± 0.045	9.999 ± 0.101
10^4	9.92 ± 0.28	9.9970 ± 0.0058	9.999 ± 0.012
10^5	9.970 ± 0.089	10.0000 ± 0.0003	10.0014 ± 0.0014

exact $F = -2\sigma x$ than $\hat{F}$ from (2.4.59) and then the Monte Carlo error in the first case should usually be less than in the second one which is observed in the experiments as well.

We also did similar experiments in the case of the terminal condition $u(T, x) = x^4$ in (2.4.39). To estimate $\partial u/\partial x$ by regression, we took the basis consisting of the first four Hermit polynomials. The results were analogous to those given above for the case x^2.

2.4.7.2 Ergodic Limit for One-Dimensional Array of Stochastic Oscillators

Consider the one-dimensional array of oscillators [319, 374]:

$$dP^i = -V'(Q^i)\, dt - \lambda \cdot (2Q^i - Q^{i+1} - Q^{i-1})\, dt - \nu P^i\, dt + \sigma\, dw_i(t),$$
$$(2.4.60)$$

$$P^i(0) = p^i,$$
$$dQ^i = P^i\, dt, \quad Q^i(0) = q^i, \quad i = 1, \ldots, n,$$

where periodic boundary conditions are assumed, i.e., $Q^0 := Q^n$ and $Q^{n+1} := Q^1$; $w_i(t)$, $i = 1, \ldots, n$, are independent standard Wiener processes; $\nu > 0$ is a dissipation parameter; $\lambda \geq 0$ is a coupling constant; σ is the noise intensity; and $V(z)$, $z \in \mathbf{R}$, is a potential.

The SDEs (2.4.60) are ergodic with the Gibbs invariant measure μ. We are interested in computing the average of the potential energy with respect to the invariant measure associated with (2.4.60):

$$E_\mu U(Q) = E_\mu \sum_{i=1}^n \left(V(Q^i) + \frac{\lambda}{2} \cdot (Q^i - Q^{i+1})^2 \right).$$

To this end (see further details in Chap. 5 and also [319]), we simulate the system (2.4.60) on a long time interval and approximate the ergodic limit $E_\mu U(Q)$ by $EU(Q(T))$ for a large T. To illustrate variance reduction via regression, we simulate

$$u(0, p, q) = EU(Q_{p,q}(T)) = E\left[U(Q_{p,q}(T)) + Z_{p,q}(T)\right], \qquad (2.4.61)$$

where $Z(t)$, $0 \le t \le T$, satisfies

$$dZ = F^\mathsf{T}(t, P, Q)dw(t), \quad Z(0) = 0. \qquad (2.4.62)$$

We choose the n-dimensional vector function $F(t, p, q)$ to be equal to (see (2.3.31)):

$$F^i(t, p, q) = -\sigma\frac{\partial \hat{u}}{\partial p^i}, \quad i = 1, \ldots, n, \qquad (2.4.63)$$

where $\hat{u} = \hat{u}(t, p, q)$ is an approximation of the function

$$u(t, p, q) := EU(Q_{t,p,q}(T)).$$

We simulate (2.4.60) using the second-order weak quasi-simplectic integrator from Sect. 5.8:

$$P_0 = p, \quad Q_0 = q, \qquad (2.4.64)$$

$$\mathcal{P}^i_{1,k} = e^{-\nu h/2}P^i_k, \quad \mathcal{Q}^i_{1,k} = Q^i_k + \frac{h}{2}\mathcal{P}^i_{1,k},$$

$$\mathcal{P}^i_{2,k} = \mathcal{P}^i_{1,k} + h\left\{-V'(\mathcal{Q}^i_{1,k}) - \lambda \cdot (2\mathcal{Q}^i_{1,k} - \mathcal{Q}^{i+1}_{1,k} - \mathcal{Q}^{i-1}_{1,k})\right\} + h^{1/2}\sigma\xi_{ik},$$

$$P^i_{k+1} = e^{-\nu h/2}\mathcal{P}^i_{2,k}, \quad Q^i_{k+1} = \mathcal{Q}^i_{1,k} + \frac{h}{2}\mathcal{P}^i_{2,k}, \quad i = 1, \ldots, n, \quad k = 0, \ldots, N-1,$$

where ξ_{ik} are i.i.d. random variables with the law

$$P(\xi = 0) = 2/3, \quad P(\xi = \pm\sqrt{3}) = 1/6. \qquad (2.4.65)$$

And we approximate (2.4.62) by the standard second-order weak method (2.2.18) from Sect. 2.2.1:

$$Z_0 = 0, \qquad (2.4.66)$$

$$Z_{k+1} = Z_k + h^{1/2}\sum_{i=1}^n F^i(t_k, P_k, Q_k)\xi_{ik} + \sigma h\sum_{r=1}^n\sum_{i=1}^n \frac{\partial}{\partial p^i}F^r(t_k, P_k, Q_k)\xi_{irk}$$

$$+ \frac{1}{2}h^{3/2}\sum_{i=1}^n \mathcal{L}F^i(t_k, P_k, Q_k)\xi_{ik},$$

$$\xi_{irk} = \frac{1}{2}\xi_{ik}\xi_{rk} - \frac{1}{2}\gamma_{ir}\zeta_{ik}\zeta_{rk}, \quad \gamma_{ir} = \begin{cases} -1, & i < r, \\ 1, & i \geq r, \end{cases}$$

$$\mathcal{L} := \frac{\partial}{\partial t} + \frac{1}{2}\sum_{i=1}^{n}\sum_{j=1}^{n}\frac{\partial^2}{\partial p^i \partial p^j}$$

$$+ \sum_{i=1}^{n}\left(-V'(q^i) - \lambda \cdot (2q^i - q^{i+1} - q^{i-1}) - \nu p^i\right)\frac{\partial}{\partial p^i}$$

$$+ \sum_{i=1}^{n} p^i \frac{\partial}{\partial q^i},$$

where ξ_{ik} and ζ_{jk} are mutually independent random variables, ξ_{ik} are distributed by the law (2.4.65), and the ζ_{ik} – by the law $P(\zeta = \pm 1) = 1/2$.

We consider two potentials: the harmonic potential

$$V(z) = \frac{1}{2}z^2, \quad z \in \mathbf{R}, \tag{2.4.67}$$

and the hard anharmonic potential

$$V(z) = \frac{1}{2}z^2 + \frac{1}{2}z^4, \quad z \in \mathbf{R}. \tag{2.4.68}$$

We define the approximation $\hat{u}(t, p, q)$ used in (2.4.63) at $t = t_k$, $k = 0, \ldots, N - 1$, as follows. First, it is reasonable to put $\partial\hat{u}/\partial p^i(t, p, q) = 0$ for $0 \leq t \leq T_0$ with some relatively small T_0 since for large T the function $u(t, p, q)$, $0 \leq t \leq T_0$, is almost constant due to the ergodicity (the expectation in (2.4.61) is almost independent of the initial condition).

Further, let T_0, T, h, N, and a nonnegative integer $\varkappa$ be such that $T_0 = N_0 h$, $T = Nh$, $N - N_0 = \varkappa N'$, where N_0 and N' are integers. Introduce $\theta_{k'} = t_{N_0 + k'\varkappa}$, $k' = 1, \ldots, N'$.

In the case of harmonic potential the required function $u(t, p, q)$ can be expanded in the basis consisting of the finite number of functions

$$\varphi_l \in \{1, \ p^i, \ q^i, \ p^i p^j, \ q^i q^j, \ p^i q^j, \quad i, j = 1, \ldots, n\}. \tag{2.4.69}$$

In our experiments we deal with three oscillators ($n = 3$), the basis (2.4.69) in this case has 28 functions.

We use the set of functions (2.4.69) as a set of basis functions for regression in both cases of harmonic and hard anharmonic potentials. Namely, using regression as described in Sect. 2.4.3, we construct the estimator $\hat{u}(\theta_{k'}, p, q)$ for $u(\theta_{k'}, p, q)$ as

$$\hat{u}(\theta_{k'}, p, q) = \sum_{l=1}^{L}\hat{\alpha}_l(\theta_{k'})\varphi_l(p, q), \tag{2.4.70}$$

where φ_l are defined in (2.4.69) and $\hat{\alpha}_l(\theta_{k'})$ satisfy the system of linear algebraic equations (2.4.4). The matrix formed from $\hat{\alpha}_l(\theta_{k'})$ is positive definite and we solve the system of linear algebraic equations by Cholesky decomposition. To find the estimator $\hat{u}$, we use M_r independent trajectories.

Then for $T_0 < t_k < T$ we put $\hat{u}(t_k, p, q) = \hat{u}(\theta_{k'}, p, q)$ with $\theta_{k'} \leq t_k < \theta_{k'+1}$. The recalculation of the estimator $\hat{u}$ once per a few number of steps $\varkappa$ reduces the cost of the procedure.

We note that for the basis (2.4.69) the corresponding function F from (2.4.63) is such that some terms in the scheme (2.4.66) are cancelled, in particular, it is not required to simulate the ζ_{ik} in this case.

We compute $u(0, p, q)$ in the usual way:

$$u(0, p, q) = E\left[U(Q_{p,q}(T)) + Z_{p,q}(T)\right] \approx E\left[U(Q_N) + Z_N\right] \qquad (2.4.71)$$

$$\approx \frac{1}{M} \sum_{m=1}^{M} [U(_m Q_N) + {}_m Z_N],$$

by simulating M independent realizations of Q_N, Z_N from (2.4.64), (2.4.66). In these experiments the two-run procedure described in Sect. 2.4.6 was used.

Suppose we would like to compute $u(0, p, q)$ for the particular set of parameters: $n = 3$, $\lambda = 1$, $\nu = 1$, $\sigma = 1$, $T = 10$ and the potentials (2.4.67) and (2.4.68) with accuracy of order 10^{-3}. Since we are using the scheme of order two, we can take $h = 0.02$.

Let us first consider the case of harmonic potential (2.4.67). Without variance reduction (i.e., for $F = 0$), we obtain 0.7500 ± 0.0010 with probability 95% by simulating $M = 1.4 \times 10^6$ trajectories and it takes ~ 541 sec on a PC. When we use the variance reduction technique as described above, it is sufficient to take $T_0 = 2$, $\varkappa = 2$, $M_r = 2 \times 10^4$, $M = 3 \times 10^4$ to get 0.7496 ± 0.0010 in ~ 64 s. In this example the procedure with variance reduction requires 8 time less computational time. All the expenses are taking into account including the time required for the first run of the two-run procedure which is less than 10% of the total time. We recall that in this case the required function $u(t, p, q)$ can be expanded in the finite basis (2.4.69) unlike the case of hard anharmonic potential when such a basis is infinite.

Now consider the case of hard anharmonic potential (2.4.68). Without variance reduction (i.e., for $F = 0$), we obtain 0.6491 ± 0.0011 with probability 95% by simulating $M = 10^6$ trajectories and it takes ~ 403 sec on a PC. With variance reduction, we reach the same level of accuracy 0.6491 ± 0.0011 in ~ 98 sec by choosing, e.g., $T_0 = 2$, $\varkappa = 2$, $M_r = 2.5 \times 10^4$, $M = 5.5 \times 10^4$. Thus, the procedure with variance reduction requires 4 time less computational time.

Some other results of our numerical experiments are presented in Tables 2.4.2 and 2.4.3. They show dependence of the Monte Carlo error on M and M_r. The numerical integration error is relatively small here and it does not essentially affect the results. The case $M_r = 0$ means that the simulation was done without variance reduction. We observe that in both tables for a fixed M_r the Monte Carlo error decreases $\sim 1/\sqrt{M}$. Further, we see from Table 2.4.2 that the Monte Carlo error is

Table 2.4.2 *Harmonic potential.* Two standard deviations of the estimator (2.4.71) in the case of potential (2.4.67) for different M and M_r. $M_r = 0$ means that variance reduction was not used. The other parameters are $n = 3$, $\lambda = 1$, $\nu = 1$, $\sigma = 1$, $T = 10$ and $h = 0.01$, $T_0 = 2$, $\varkappa = 1$

	$M_r = 0$	$M_r = 10^3$	$M_r = 10^4$	$M_r = 10^5$
$M = 10^3$	4.0×10^{-2}	2.6×10^{-2}	–	–
$M = 10^4$	1.2×10^{-2}	7.8×10^{-3}	2.3×10^{-3}	–
$M = 10^5$	3.9×10^{-3}	2.3×10^{-3}	7.9×10^{-4}	2.5×10^{-4}
$M = 10^6$	1.2×10^{-3}	8.2×10^{-4}	2.4×10^{-4}	7×10^{-5}

Table 2.4.3 *Hard anharmonic potential.* Two standard deviations of the estimator (2.4.71) in the case of potential (2.4.68) for different M and M_r. The other parameters are the same as in Table 2.4.2

	$M_r = 0$	$M_r = 10^3$	$M_r = 10^4$	$M_r = 10^5$
$M = 10^3$	3.3×10^{-2}	2.3×10^{-2}	–	–
$M = 10^4$	1.1×10^{-2}	7.4×10^{-3}	3.0×10^{-3}	–
$M = 10^5$	3.5×10^{-3}	2.4×10^{-3}	9.5×10^{-4}	6.7×10^{-4}
$M = 10^6$	1.1×10^{-3}	7.4×10^{-4}	2.9×10^{-4}	2.2×10^{-4}

$\sim 1/\sqrt{M_r}$ for fixed M (for $M_r > 0$, of course) and, consequently, it is $\sim 1/\sqrt{M M_r}$ when the variance reduction is used (we recall that the time step is relatively small here). As it has been noted before, the basis used in the variance reduction is such that the function $u(t, x)$ can be expanded in it in the case of harmonic potential, i.e., err_B in (2.4.55) is equal to 0. These observations are consistent with the Monte Carlo error estimate (2.4.55). For the anharmonic potential, err_B is not equal to zero and we see in Table 2.4.3 that the increase of M_r has less impact on the Monte Carlo error in this case.

2.5 Quasi-Monte Carlo and Multi-level Monte Carlo Methods

As we noted in Sect. 2.3, convergence of the Monte Carlo method is of order $1/\sqrt{M}$ with M being the number of independent simulations, which is rather slow. In previous Sects. 2.3 and 2.4 we considered how the constant at $1/\sqrt{M}$ in the Monte Carlo error can be reduced (i.e., we considered variance reduction methods) in order to make the Monte Carlo method more efficient. There are other approaches to reduce computational cost in computing expectations. In Sect. 2.5.1 we will briefly discuss the quasi-Monte Carlo (QMC) methods [80, 218, 337, 411]. In QMC methods realizations of random numbers used in Monte Carlo simulations are replaced by deterministic points chosen according to some specific rules. It is possible for QMC methods to have error of order $1/M$ up to logarithmic terms which can be a considerable speed up in comparison with the standard Monte Carlo method. The other approach (Sect. 2.5.2) consists in reducing computational complexity of the Monte Carlo method itself via specifically designed control variates, the corresponding technique is known as the multi-level Monte Carlo (MLMC) method [125–127, 159, 190]. To achieve the overall tolerance level ϵ, the computational complexity of the usual Monte Carlo method is $O(\epsilon^{-3})$, while the MLMC method can have a reduced complexity of $O(\epsilon^{-2})$. We do not consider here the other alternative – Smolyak's sparse grids [409] (see also [122, 340, 460]), which is a deterministic method and it can be viewed as a sparse approximation of multi-dimensional integrals. Sparse grid quadrature is a reduction of product quadrature rules which decreases the number of quadrature nodes and allows effective integration in moderately high dimensions. It has a narrow class of applications, e.g. in the case of very specific problems from financial engineering [143] (see also discussion in [467, 469]).

Recall that in this chapter we aim at computing

$$u(t_0, x_0) = Ef(X_{t_0,x_0}(T)), \tag{2.5.1}$$

where $X(t)$, $t \in [t_0, T]$, satisfies the SDEs

$$dX = a(t, X)dt + \sum_{r=1}^{q} \sigma_r(t, X)dw_r(t), \; X(t_0) = x_0. \tag{2.5.2}$$

2.5.1 Quasi-Monte Carlo Methods

Let us start with considering approximation of the integral over a p-dimensional cube $[0, 1]^p$:

$$I_p\varphi = \int_{[0,1]^p} \varphi(y)dy \tag{2.5.3}$$

by the integration rule of the form

$$I_p\varphi \approx \bar{I}_{p,M}\varphi := \frac{1}{M} \sum_{m=1}^{M} \varphi(y_m), \tag{2.5.4}$$

where y_m are some points from $[0, 1]^p$. We see that (2.5.4) resembles the Monte Carlo method (cf. (2.3.2)), and the name quasi-Monte Carlo comes from this analogy with the Monte Carlo method. The difference between the usual Monte Carlo method and QMC methods is that in the former y_m, $m = 1, \ldots, M$, are a random sample from uniform distribution on $[0, 1]^p$, while in the (classical) QMC methods the points y_m are chosen according to a deterministic rule so that the points have small 'discrepancy'.

Similar to random and pseudo-random numbers the law of large numbers must hold for sequences of y_m used in QMC. A deterministic version of the law of large numbers is given by the Koksma–Hlawka inequality (see e.g. [80]), based on which one can obtain the error estimate for (2.5.4):

$$\left| \bar{I}_{p,M}\varphi - I_p\varphi \right| \leq D_M^* V_{HK}(\varphi). \tag{2.5.5}$$

Here $V_{HK}(\varphi)$ is a bounded variation of φ in the sense of Hardy and Krause (see details in [218, 338]). The definition of $V_{HK}(\varphi)$ contains higher order derivatives of φ and it is small for smooth functions. The factor D_M^* is called the discrete star discrepancy and it measures how uniformly the points y_m, $m = 1, \ldots, M$, are distributed on $[0, 1]^p$:

$$D_M^* = D_M^*(y_1, \ldots, y_M)$$
$$:= \sup_{z \in [0,1]^p} \left| \frac{1}{M} \sum_{m=1}^M I\left(y_m \in \prod_{i=1}^p [0, z^i]\right) - \mathrm{vol}_p(z) \right|,$$

where I is the indicator function and $\mathrm{vol}_p(z) = \prod_{i=1}^p z^i$ measures the volume of the p-dimensional domain $\prod_{i=1}^p [0, z^i]$. A sequence with a small D_M^* is called a low-discrepancy sequence and the key objective of QMC methods is to construct sequences with low D_M^*.

There are two types of QMC methods (i.e., two types of construction of points y_m): (i) the open type, which uses the first M points from an infinite sequence and, to increase M, one can take next points from the sequence; and (ii) the closed type, which uses a finite set of points dependent on M and, to increase M, one needs to generate a new set of points. There are several possibilities how a set of low discrepancy points can be constructed with the two main families being (a) digital sequences (the open type) and digital nets (the closed type) and (b) lattice rules (both open and closed types). For specific algorithms for generating such points, see e.g. [80, 218, 337, 411] and references therein. For sufficiently smooth φ, the QMC error (2.5.5) can be of order $(\log M)^p / M$ or better (recall that p here is the dimension of the random space). In comparison, the error of the Monte Carlo method is of order $1/\sqrt{M}$ and it is independent of the dimension. We see that for moderate p, QMC can considerably outperform the usual Monte Carlo method.

In contrast to the Monte Carlo method (cf. (2.3.2)), a fully deterministic QMC method is biased and it lacks a practical error estimate (see e.g. [80]). To address this deficiency of QMC methods, one randomizes QMC methods. A randomized QMC method has the following advantages: it gives an unbiased estimator like the Monte Carlo method, it has a practical error estimate, and it can have a faster rate of convergence that the classical QMC method. At the same time, randomization increases the computational cost of QMC methods. There are many different randomization methods which can be found in [80, 338] and references therein.

Recall that the Euler scheme (1.0.5) for (2.5.2) can be realized in practice by replacing the increments $\Delta_k w_r$ with Gaussian random variables:

$$X_{k+1} = X_k + a(t_k, X_k)h + \sum_{r=1}^{q} \sigma_r(t_k, X_k)\sqrt{h}\xi_{r,k+1}, \qquad (2.5.6)$$

where $\xi_{r,k+1}$ are i.i.d. $\mathcal{N}(0, 1)$ random variables. Introducing the function $\varphi(y)$, $y \in \mathbf{R}^{qN}$, so that

$$\varphi(\xi_{1,1}, \ldots, \xi_{q,1}, \ldots, \xi_{1,N}, \ldots, \xi_{q,N}) = f(X_N), \qquad (2.5.7)$$

we have

$$u(t_0, x_0) \approx \bar{u}(t_0, x_0) := Ef(X_N) \qquad (2.5.8)$$
$$= E\varphi(\xi_{1,1}, \ldots, \xi_{q,1}, \ldots, \xi_{q,N}, \ldots, \xi_{q,N})$$
$$= \frac{1}{(2\pi)^{qN/2}} \int_{\mathbf{R}^{qN}} \varphi(y_{1,1}, \ldots, y_{q,N}) \exp\left(-\frac{1}{2}\sum_{i=1}^{qN} y_i^2\right) dy.$$

To apply a QMC method to the integral from (2.5.8), one needs to transform the integral over $\mathbf{R}^{qN}$ to a one over $[0, 1]^{qN}$ (note that here the dimension of the random space $p = qN$ and it grows both with increase of the number of Wiener processes q and with the number of time steps N of a numerical scheme). There are a number of ways how it can be done, see e.g. the corresponding discussion in [80]. However, in the SDEs context we can just modify the weak-sense method. To this end, we can replace Gaussian random variables in (2.5.6) by uniformly distributed ones:

$$\tilde{X}_{k+1} = \tilde{X}_k + ha(t_k, \tilde{X}_k) + \sqrt{h} \sum_{r=1}^{q} \sigma_r(t_k, \tilde{X}_k)\sqrt{3}(1 - 2\zeta_{r,k+1}), \qquad (2.5.9)$$

where $\zeta_{r,k+1}$ are i.i.d. random variables with the uniform distribution on $[0, 1]$. The scheme (2.5.9) is obviously of weak order 1 and QMC methods are directly applicable to

$$\tilde{u}(t_0, x_0) := Ef(\tilde{X}_N) = E\varphi(\zeta_{1,1}, \ldots, \zeta_{q,N}) = \int_{[0,1]^{qN}} \varphi(y)dy$$

with the appropriately defined φ.

For use of QMC methods in the SDEs context, see e.g. [166, 169, 467]. QMC methods can be efficient and outperform the usual Monte Carlo method when the number of random variables $p = qN$ required for simulating SDEs is relatively small/moderate. When very long simulations are needed, QMC methods are typically not applicable.

2.5.2 Multi-level Monte Carlo Method

In this section we consider how a special choice of control variates can lead to a decrease of computational complexity of the Monte Carlo method.

Let $\bar{X}^h(T)$ be the result of an approximation of the solution $X(T)$ to (2.5.2), which weak order is p and which is computed using a time step h. Assume that a time step h_L is sufficient to reach a desirable accuracy of $Ef(\bar{X}^{h_L}(T))$. Now let $h_L < \cdots < h_0$ be a set of time steps for discretization of the time interval $[t_0, T]$ and $N_l = (T - t_0)/h_l$, $l = 0, \ldots, L$. Note that

$$Ef(\bar{X}^{h_L}(T)) = Ef(\bar{X}^{h_0}(T)) + \sum_{l=1}^{L} E\left[f(\bar{X}^{h_l}(T)) - f(\bar{X}^{h_{l-1}}(T))\right]. \quad (2.5.10)$$

Based on (2.5.10), we can construct the following statistical estimator $\hat{u}$ for $Ef(\bar{X}^{h_L}(T))$:

$$\bar{u} := Ef(\bar{X}^{h_L}(T)) \doteq \hat{u} \quad (2.5.11)$$
$$:= \frac{1}{M_0} \sum_{i=1}^{M_0} f(\bar{X}^{h_0,(0,i)}(T))$$
$$+ \sum_{l=1}^{L} \frac{1}{M_l} \sum_{i=1}^{M_l} \left[f(\bar{X}^{h_l,(l,i)}(T)) - f(\bar{X}^{h_{l-1},(l,i)}(T))\right],$$

where the pairs $(\bar{X}^{h_l,(l,i)}(T), \bar{X}^{h_{l-1},(l,i)}(T))$ for all l and i are independent but the two elements of the same pair are dependent, and M_l are the number of independent runs of the pairs at each level l. We note that in the index (l, i), l stands for the level and i stands for the independent sample. At each level l and for each sample i, we simulate two dependent approximations $\bar{X}^{h_l,(l,i)}(T)$ and $\bar{X}^{h_{l-1},(l,i)}(T)$ with the time steps h_l and h_{l-1}, respectively. Their dependence is typically achieved by simulating both values along the same underlying Wiener process path.

It is clear that $E\hat{u} = \bar{u}$, i.e., the estimator is unbiased and one can also see that this estimator is of the control variates type. Since the pairs $(\bar{X}^{h_l,(l,i)}(T), \bar{X}^{h_{l-1},(l,i)}(T))$ are independent, we get from (2.5.11):

$$Var(\hat{u}) = \frac{1}{M_0} Var\left(f(\bar{X}^{h_0}(T))\right) + \sum_{l=1}^{L} \frac{1}{M_l} Var\left[f(\bar{X}^{h_l}(T)) - f(\bar{X}^{h_{l-1}}(T))\right],$$
$$(2.5.12)$$

where $\bar{X}^{h_l}(T)$ and $\bar{X}^{h_{l-1}}(T)$ are dependent approximations of $X(t)$ computed with the time steps h_l and h_{l-1}, respectively.

The cost of computing $\hat{u}$ from (2.5.11) in terms of the number of steps in the numerical scheme is equal to

$$\text{Cost} = \sum_{l=0}^{L} N_l M_l \quad (2.5.13)$$

(note that we neglect here that the computational cost per step depends on a method).

The main idea of MLMC is as follows. Due to dependence of $\bar{X}^{h_l}(T)$ and $\bar{X}^{h_{l-1}}(T)$, the variance

$$Var\left[f(\bar{X}^{h_l}(T)) - f(\bar{X}^{h_{l-1}}(T))\right]$$

is small, moreover the smaller h_l the smaller is the variance. Consequently, to achieve a certain tolerance level of $Var(\hat{u})$ we can run relatively small number of expensive trajectories corresponding to small h_l keeping M_0 large where simulation of trajectories with h_0 is cheap. This way the computational complexity of the Monte Carlo method can be decreased as we show below.

The usual Monte Carlo method (2.3.2) combined with a method of weak order p using a time step h_L and M_L number of independent trajectories has Cost $= N_L M_L$ and its total (mean-square) error can be expressed as

$$\text{Err} := \sqrt{E\left[\frac{1}{M_L}\sum_{i=1}^{M_L} f(\bar{X}^{h_L,(i)}(T)) - Ef(X(T))\right]^2} = \sqrt{\frac{C_1}{M_L} + \frac{C_2}{N_L^{2p}}}. \quad (2.5.14)$$

For a fixed computational cost, the minimum total error is achieved at

$$N_L = \left[\frac{2pC_2}{C_1}\text{Cost}\right]^{1/(2p+1)} \quad \text{and} \quad M_L = \left[\frac{C_1}{2pC_2}\text{Cost}^{2p}\right]^{1/(2p+1)}.$$

For this choice of N_L and M_L, the numerical integration error and the statistical error are of the same order of smallness, and, to achieve Err $= O(\epsilon)$ for some fixed tolerance $\epsilon > 0$ of the overall approximation (which includes both the numerical integration and statistical errors), we need

$$N_L = O(\epsilon^{-1/p}) \quad \text{and} \quad M_L = O(\epsilon^{-2})$$

with the total cost

$$\text{Cost} = O(\epsilon^{-2-1/p}).$$

E.g., in the case of a scheme of weak order 1, the computational cost of the usual Monte Carlo method is equal to Cost $= O(\epsilon^{-3})$.

Now let us return to the MLMC estimator (2.5.11) and assume that for some $\beta > 0$

$$Var\left[f(\bar{X}^{h_l}(T)) - f(\bar{X}^{h_{l-1}}(T))\right] = O(h_l^{2\beta}). \quad (2.5.15)$$

E.g., if $f(x)$ is globally Lipschitz, the method $\bar{X}$ is of mean-square order β and the pair $(\bar{X}^{h_l}(T), \bar{X}^{h_{l-1}}(T))$ is simulated along the same underlying Wiener path, then (2.5.15) holds. See other examples of numerical schemes and conditions on $f(x)$ when (2.5.15) is valid with some β in e.g. [126, 127] and references therein.

The following complexity result can be proved for (2.5.11) [125, 126]: there are values L and N_l so that

$$\text{Cost} = \begin{cases} O(\epsilon^{-2}), & 2\beta > p, \\ O(\epsilon^{-2}(\log \epsilon)^2), & 2\beta = p, \\ O(\epsilon^{-3+2\beta/p}), & 0 < 2\beta < p. \end{cases} \tag{2.5.16}$$

To illustrate (2.5.16), consider the first-order mean-square scheme (1.0.5), for which $p = \beta = 1$. Then for $\hat{u}$ from (2.5.12) we have for some constants $C_1, C_2 > 0$

$$\text{Err}^2 := E\left[\hat{u} - Ef(X(T))\right]^2 \leq \frac{C_1}{M_0} + C_1 \sum_{l=1}^{L} \frac{1}{M_l N_l^2} + \frac{C_2}{N_L^2}, \tag{2.5.17}$$

where the first two terms in the right-hand side correspond to the statistical error of $\hat{u}$ and the last one – to the error of numerical integration. For the Err to be $O(\epsilon)$, we need all three terms in (2.5.17) to be of order $O(\epsilon^2)$. Hence $M_0 = O(\epsilon^{-2})$ and $N_L = O(\epsilon^{-1})$, i.e. $h_L = O(\epsilon)$. Let $N_l = \tilde{N}2^l$ with some positive integer $\tilde{N}$, then $L = -\log_2 \epsilon + O(1)$. Under fixed Cost from (2.5.13), the optimal choice of $M_l = O(\epsilon^{-2}N_l^{-3/2})$ to ensure that the second term in (2.5.17) is of order $O(\epsilon^2)$:

$$\sum_{l=1}^{L} \frac{1}{M_l N_l^2} = O(\epsilon^2) \sum_{l=1}^{L} \frac{1}{N_l^{1/2}} = O(\epsilon^2) \frac{1}{\tilde{N}^{1/2}} \sum_{l=1}^{L} \frac{1}{2^{l/2}} = O(\epsilon^2).$$

The corresponding computational cost is equal to

$$\text{Cost} = \sum_{l=0}^{L} N_l M_l = O(\epsilon^{-2}) \frac{1}{\tilde{N}^{1/2}} \sum_{l=0}^{L} N_l^{-1/2} = O(\epsilon^{-2}).$$

Hence we observe the considerable reduction of computational cost to achieve the overall tolerance level ϵ : from $O(\epsilon^{-3})$ in the case of the usual Monte Carlo method together with a scheme of weak order one to $O(\epsilon^{-2})$ in the case of the MLMC estimator (2.5.11) together with a mean-square method of order one. Note that if we use the mean-square Euler scheme (recall that its mean-square order is $1/2$ and weak order is 1) in (2.5.11), then (see (2.5.16)) the cost is $O(\epsilon^{-2}(\log \epsilon)^2)$ which is still substantially smaller than $O(\epsilon^{-3})$. However, MLMC relies on numerical methods to be stable (in some sense) for a range of time steps h larger than the h_L required for achieving a desirable accuracy. This is not always possible in practical applications and it then limits applicability of MLMC.

There is a broad literature on MLMC covering both theoretical and implementational considerations, including e.g. how MLMC and QMC methods can be combined. For further reading on MLMC, see the survey paper [126] and references therein.

2.6 Random Number Generators

Implementation of a numerical method for integrating SDEs requires a source of random numbers. For computer simulation purposes, random numbers are typically approximated by pseudorandom ones which are generated via iterative deterministic algorithms. These algorithms produce a sequence of numbers which are in fact not random at all. The program that produces a sequence of pseudorandom numbers is called a pseudorandom number generator which for brevity we will call just a random number generator (RNG).

A RNG should satisfy the following requirements. First, the sequence of random numbers must have appropriate statistical properties, i.e., a RNG is constructed in such a way that it produces the deterministic sequence of numbers which shares enough of the statistical properties of a true random sequence. In particular, RNGs should pass statistical tests on independence and distribution (e.g., uniformity in the case of RNGs for uniform random variables). Further, as a rule, RNGs produce periodic sequences of numbers with a period ρ. Weak-sense numerical integration of SDEs usually requires a large amount M of random numbers. Moreover, due to common practice in Monte Carlo simulations, the period ρ of a used RNG should be significantly larger than M (e.g., there are recommendations not to use more than 5% of a RNG's period in a single calculation [368]). Hence, the period of the generated sequence has to be very large. Second, a RNG has to be as fast as possible to minimize computational time spent on generating random numbers during simulation of SDEs. Third, the sequence of random numbers should be reproducible. This is helpful for testing numerical algorithms and debugging computer codes. Further, some Monte Carlo algorithms themselves can benefit from the possibility of running the same sequence of random numbers twice or more (see e.g. the two-run procedure in Sect. 2.4.6). Fourth, it is preferable for a RNG to be portable, i.e., the same on any operational system or hardware. A parallel implementation puts additional requirements on RNGs.

Random variables with any distribution can be obtained by transforming random variables distributed uniformly over the interval $[0, 1]$, and the primary source of randomness in simulations is RNGs for independent uniformly distributed random numbers. There are several types of uniform RNGs including linear congruential generators (LCG), additive lagged Fibonacci generators, linear feedback shift register generators (LFSR), multiple Fibonacci generators, generalized feedback shift register generators (GFSRs) and their modifications, multiple recursive generators, nonlinear generators, etc. (see [121, 206, 337] and references therein for description of these generators). One can find a large variety of uniform RNGs in literature including their theoretical backgrounds, empirical testing, implementations, recommendations for their use, etc. (see, e.g., [62, 121, 160, 206, 225, 337, 438] and references therein). It is known that in addition to standard statistical tests of RNGs, it is useful to apply application-specific tests which are more relevant to a particular application. A comparative analysis of the RNGs in the context of SDEs integration was considered in [314, Sect. 2.6].

Historically, the most popular and well studied uniform RNGs are linear congruential generators [121, 206, 225, 228]. There are combined multiple LCGs with good statistical properties and long periods [121, 225]. Faster alternatives with good statistical properties include additive lagged Fibonacci generators [62, 256] and GFSRs [121, 225]. Currently, the most popular uniform RNG is the Mersenne twister algorithm (MT19937 and MT19937-64) proposed in [260]. At present it is the default RNG for a number of software systems. This algorithm belongs to the class of twisted GFSRs. It has a very long period $\rho = 2^{19937} - 1$, passed a large set of statistical tests, and it is computationally very efficient.

Most of numerical schemes in this book require a source of either Gaussian or discrete random variables. In the experiments presented in this book we simulated discrete random variables as follows. Let a discrete random variable ξ take values $x_1, \ldots, x_n$ with probabilities $p_1, \ldots, p_n$, and U be a generated uniform number. Then simulate ξ as

$$
\xi = \begin{cases}
x_1 \text{ if } 0 \le U < p_1; \\
x_2 \text{ if } p_1 \le U < p_1 + p_2; \\
\ldots\ldots\ldots\ldots \\
x_n \text{ if } p_1 + \cdots + p_{n-1} \le U < 1.
\end{cases}
$$

Weak methods of this chapter use discrete random variables for which n is small: 2, 3, 4, or 5. Some of these random variables can be simulated much more effectively than in the way described above. For instance, consider i.i.d. discrete random variables ξ_i taking two values ± 1 with probability $1/2$ which are needed for Euler-type methods. These random variables can be effectively generated using "random bit" generators (see, e.g. [206]). At the same time, uniform random numbers can be exploited for our purposes in a very efficient way as it is described below (see also [206, p. 101]). Let an ideal uniform RNG produce 32-bit numbers. Then the sequence of bits in these numbers (with the change of 0 by -1) gives us a sequence ξ_i required for Euler-type methods. These bits should be extracted from the most-significant (left-hand) part of the computer word (e.g., 24 first bits of 32) since in existing RNGs the least significant bits can be not sufficiently random. As a result, producing just one number by a uniform 32-bit RNG, we get 24 numbers ξ_i, i.e., in this way we simulate the i.i.d. discrete random variables ξ_i with the law $P(\xi = \pm 1) = 1/2$ very economically. Moreover, such an approach can be applied to generating some other discrete random variables. Consider, for example, the random variable with the law

$$
P\left(\xi = \pm\sqrt{1 - \sqrt{6}/3}\right) = \frac{3}{8}, \quad P\left(\xi = \pm\sqrt{1 + \sqrt{6}}\right) = \frac{1}{8}, \tag{2.6.1}
$$

which is used in the weak second order methods (see (2.2.18) and (2.1.31)). It is clear that such a number can be generated by three bits of a uniform random number, i.e., we again have the very economical way of generating the needed discrete random numbers.

A source of Gaussian random numbers is needed for all the methods of Chapter 1. There are many choices of algorithms for generation of Gaussian random numbers based on applying some transformations to uniform random numbers (see, e.g. [121, 206, 385] and references therein). The (statistical) quality of Gaussian random numbers generated ultimately depends on the quality of the underlying uniform generator. A particular method for generating Gaussian random numbers, however, may exacerbate some fault in the uniform generator (see [121] and references therein for further discussion).

There are "universal" methods of transforming uniform random variables, which are applicable to almost any distribution and, in particular, to the Gaussian one. The inverse transform algorithm and the general rejection method [121, 206, 385] are among the "universal" methods. The inverse transform algorithm is based on the following assertion. If U is a uniform $(0, 1)$ random variable, then for any continuous distribution function F the random variable X defined by $X = F^{-1}(U)$ has the distribution F. Thus the problem reduces to evaluating the inverse function. Specialized algorithms for inverting normal distribution function are considered, e.g. in [255]. The "universal" methods are usually rather slow and special algorithms are preferable in general.

Polar methods are popular algorithms for generating Gaussian random numbers. The well-known Box-Muller transformation [40] (see also [121, 385]) belongs to this type of methods. The Box-Muller transformation itself is computationally not very efficient since it requires to compute the sine and cosine trigonometric functions. But it is possible to implement the Box-Muller transformation via rejection method and avoid costly evaluations of sine and cosine [252] (see also [121, 206, 385]). The rejection polar method can be described as follows. First, generate two independent uniform $(-1, 1)$ random numbers U_1 and U_2 and set $R^2 = U_1^2 + U_2^2$. Second, if $R^2 \geq 1$ then repeat the first step, otherwise

$$X_1 = U_1 \sqrt{-2 \log R^2 / R^2}, \quad X_2 = U_2 \sqrt{-2 \log R^2 / R^2}. \tag{2.6.2}$$

The obtained X_1 and X_2 are independent normally distributed $\mathcal{N}(0, 1)$ random numbers. On average, the rejection polar method requires 2.546 uniform random numbers, 1 logarithm, 1 square-root, 1 division, and 4.546 multiplications to generate two independent Gaussian random numbers. This method is still very popular and reasonably fast.

Currently, the most popular algorithm for transforming uniform random numbers into Gaussian ones is the so-called ziggurat method which is a special case of the rejection method [254]. It is 3 times or more faster than the rejection polar method and it is a default method for generating Gaussian random numbers in many software systems. The method is based on covering the target Gaussian density $N(0, 1)$ with a set of horizontal equal-area rectangles. The name 'ziggurat' came from the fact that the density of the proposal distribution used in the rejection method is of a form of layered rectangles which resembles the silhouette of a ziggurat.

Monte Carlo simulations are well suited to parallel computers. A common way of parallelizing Monte Carlo simulations is to run identical procedures but with different random number sequences on a set of processors. A communication between the processors is needed only to start the simulation and to make final averaging and output. Then the use of p independent processors should reduce computational costs of the simulation in p times. Of course, this is true only if the results obtained on each processor are statistically independent, i.e., the random number sequences generated in the processors have to be independent. Thus, the problem of parallelization of Monte Carlo simulation consists in finding a good parallel random number generator (PRNG) (see, e.g. [43, 62, 258] and references therein). Note that running a RNG (a centralized RNG) on a particular processor to supply all the other processors with random numbers is not an appropriate solution for parallelizing Monte Carlo simulations because inter-processor communication is very expensive and the desirable speed-up of calculations cannot be achieved in this way. Besides, in this case the requirement of reproducibility of calculations is often difficult to satisfy since different processors may request random numbers in different orders in different runs of the program depending on the network traffic and implementation of the communication software. Thus, each processor has to have its own RNG which produces a sequence of random numbers independent of the sequences on the other processors (see the review [226] and references therein). To parallelize RNGs, the following main techniques are used:

(i) producing parallel streams of random numbers by taking subsequences from a single, long-period RNG [43, 62, 121];
(ii) parametrization of RNGs [258].

The first one is subdivided into (ia) leapfrog method and (ib) sequence splitting. In the leapfrog method the sequence is partitioned among the processors in a cyclic fashion, like a desk of cards dealt to card players. In the sequence splitting, the sequence is partitioned among processors in a block fashion, by splitting it into non-overlapping contiguous sections. The well known problem with this type of PRNG is that although the subsequences are disjoint, this does not necessarily mean that they are uncorrelated. See further discussion in [43, 62] and references therein.

The parametrization method (for a review see [258]) identifies a parameter in the underlying recursion of a serial RNG that can be varied. Each valid value of this parameter leads to a recursion that produces a unique, full-period steam of random numbers. The exact meaning of parametrization depends on the type of RNG. For instance, the additive lagged Fibonacci generator can be parameterized through its initial values [62, 258, 259]. This parametrization relies on the fact [259] that the lagged Fibonacci generator has many disjoint full-period cycles. As a result, different seed tables may produce completely different non-overlapping periodic sequences of numbers. This type of PRNG is quite popular and has been implemented for a number of parallel computers and parallel languages. See also parametrization of LCGs, linear matrix generators, LFSRs, and inverse congruential generators in [258].

Remark 2.6.1 If the number of processors is different in different runs of the program then to ensure the reproducibility in parallel Monte Carlo simulation it is necessary to write the program in terms of "virtual processors", each virtual processor having its own RNG.

A discussion of RNG implementations for computers with vector processors can be found in, e.g. [6, 43] (see also [314]).

Chapter 3
Weak Approximation for Stochastic Differential Equations: Special Cases

In this chapter we apply the general theory built in the previous chapter to three important particular cases of SDEs. First, we construct weak approximations for systems with additive noise (Sect. 3.1), i.e. SDEs with diffusion coefficients independent of the phase variable. Then we propose weak schemes for systems with colored noise (Sect. 3.2), which are differential equations perturbed by the Ornstein–Uhlenbeck-type process. In Sects. 3.3 and 3.4 we consider Wiener integrals, which can be expressed as functionals of solutions of specific SDEs. For Wiener integrals of functionals of integral type, we derive methods of second order of accuracy. For integrals of a more particular, but often encountered, form (integrals of functionals of exponential type), we succeed in constructing efficient methods of fourth order of accuracy. Both unconditional (Sect. 3.3) and conditional (Sect. 3.4) Wiener integrals are considered. In the case of conditional Wiener integrals we exploit a Markovian representation of the Brownian bridge. In Sect. 3.4.5, we consider a generalization to the case of path integrals with respect to nonlinear diffusion bridges (with additive noise). Finally (Sect. 3.5), we present some numerical experiments for Wiener integrals that confirm the theoretical results. Section 3.6 illustrates the use of stochastic numerics in financial mathematics.

3.1 Weak Methods for Systems with Additive Noise

Consider the system of SDEs with additive noise

$$dX = a(t, X)dt + \sum_{r=1}^{q} \sigma_r(t)dw_r(t). \tag{3.1.1}$$

© Springer Nature Switzerland AG 2021
G. N. Milstein and M. V. Tretyakov, *Stochastic Numerics for Mathematical Physics*,
Scientific Computation, https://doi.org/10.1007/978-3-030-82040-4_3

Since the σ_r do not depend on x, numerical methods are substantially simpler for such systems.

3.1.1 Second Order Methods

Since in the case of (3.1.1) the $\Lambda_i \sigma_r$ vanish, the terms ξ_{irk} in the method (2.2.18) (which has order of accuracy two) are absent, and consequently, we only have to model the random variables ξ_{rk} at each step. The method (2.2.18) takes the following form for the system (3.1.1):

$$X_{k+1} = X_k + \sum_{r=1}^{q} \sigma_r(t_k)\xi_{rk}h^{1/2} + a_k h \tag{3.1.2}$$

$$+\frac{1}{2}\sum_{r=1}^{q}\left(\sigma_r' + \left(\sigma_r, \frac{\partial}{\partial x}\right)a\right)_k \xi_{rk}h^{3/2} + (La)_k\frac{h^2}{2}.$$

Moreover, it is not difficult to obtain the fully Runge–Kutta method of order two for (3.1.1):

$$X_{k+1} = X_k + \sum_{r=1}^{q} \sigma_r\left(t_k + \frac{h}{2}\right)\xi_{rk}h^{1/2} \tag{3.1.3}$$

$$+\frac{h}{2}\left[a_k + a\left(t_{k+1}, X_k + \sum_{r=1}^{q}\sigma_r(t_k)\xi_{rk}h^{1/2} + a_k h\right)\right].$$

While for systems of a general form the attempt to arrive at a method of order of accuracy three meets with very awkward constructions, for systems with additive noise the problem of constructing such a method can be solved relatively simply. This problem is considered in the next two subsections.

3.1.2 Main Lemmas for Third Order Methods

To construct a method of order of accuracy three, we write down the following formula for the solution $X_{t,x}(\theta) = X(\theta)$ of (3.1.1):

$$X(t+h) = x + \sum_{r=1}^{q} \sigma_r \int_{t}^{t+h} dw_r(\theta) + ah \tag{3.1.4}$$

$$+ \sum_{r=1}^{q} \Lambda_r a \int_{t}^{t+h} (w_r(\theta) - w_r(t)) d\theta + \sum_{r=1}^{q} \sigma_r' \int_{t}^{t+h} (\theta - t) dw_r(\theta)$$

$$+ La\frac{h^2}{2} + \sum_{r=1}^{q} \sum_{i=1}^{q} \Lambda_i \Lambda_r a \int_{t}^{t+h} \left(\int_{t}^{\theta} (w_i(\theta_1) - w_i(t)) dw_r(\theta_1) \right) d\theta$$

$$+ \sum_{r=1}^{q} \sum_{i=1}^{q} \sum_{s=1}^{q} \Lambda_s \Lambda_i \Lambda_r a \int_{t}^{t+h} \left(\int_{t}^{\theta} \left(\int_{t}^{\theta_1} (w_s(\theta_2) - w_s(t)) dw_i(\theta_2) \right) dw_r(\theta_1) \right) d\theta$$

$$+ \sum_{r=1}^{q} \sigma_r'' \int_{t}^{t+h} \left(\int_{t}^{\theta} (\theta_1 - t) d\theta_1 \right) dw_r(\theta) + \sum_{r=1}^{q} L\Lambda_r a \int_{t}^{t+h} \left(\int_{t}^{\theta} (\theta_1 - t) dw_r(\theta_1) \right) d\theta$$

$$+ \sum_{r=1}^{q} \Lambda_r La \int_{t}^{t+h} \left(\int_{t}^{\theta} (w_r(\theta_1) - w_r(t)) d\theta_1 \right) d\theta + L^2 a\frac{h^3}{6} + \rho,$$

where all coefficients σ_r, $\Lambda_r a$, σ_r', La, $\Lambda_i \Lambda_r a$, $\Lambda_s \Lambda_i \Lambda_r a$, σ_r'', $L\Lambda_r a$, $\Lambda_r La$, $L^2 a$ are calculated at the point (t, x).

Lemma 3.1.1 *The remainder ρ in (3.1.4) satisfies the relations*

$$|E\rho| = O(h^4), \tag{3.1.5}$$

$$E|\rho|^2 = O(h^6), \tag{3.1.6}$$

$$\left| E\rho \int_{t}^{t+h} dw_r(\theta) \right| = O(h^4). \tag{3.1.7}$$

This lemma can be proved similarly to Lemma 2.1.2. To shorten the exposition, we have not listed all the assumptions on the coefficients a and σ_r in detail, they are similar to those in Lemma 2.2.1.2. For brevity here and below we write error estimates in the form as, e.g., in (3.1.5) instead of how we wrote them e.g. in (2.1.5). We introduce the notation

$$I_r = \int\limits_t^{t+h} dw_r(\theta)\,, \quad J_r = \int\limits_t^{t+h} (w_r(\theta) - w_r(t))d\theta\,,$$

$$G_r = \int\limits_t^{t+h} (w_r(\theta) - w_r(t))(\theta - t)d\theta\,,$$

$$J_{ir} = \int\limits_t^{t+h} \left(\int\limits_t^\theta (w_i(\theta_1) - w_i(t))dw_r(\theta_1) \right) d\theta\,,$$

$$J_{sir} = \int\limits_t^{t+h} \left(\int\limits_t^\theta \left(\int\limits_t^{\theta_1} (w_s(\theta_2) - w_s(t))dw_i(\theta_2) \right) dw_r(\theta_1) \right) d\theta\,.$$

Lemma 3.1.2 *The following identities hold:*

$$\int\limits_t^{t+h} (\theta - t)dw_r(\theta) = hI_r - J_r\,, \tag{3.1.8}$$

$$\int\limits_t^{t+h} \left(\int\limits_t^\theta (\theta_1 - t)dw_r(\theta_1) \right) d\theta = 2G_r - hJ_r\,, \tag{3.1.9}$$

$$\int\limits_t^{t+h} \left(\int\limits_t^\theta (w_r(\theta_1) - w_r(t))d\theta_1 \right) d\theta = hJ_r - G_r\,, \tag{3.1.10}$$

$$\int\limits_t^{t+h} \left(\int\limits_t^\theta (\theta_1 - t)d\theta_1 \right) dw_r(\theta) = \frac{1}{2}h^2 I_r - G_r\,. \tag{3.1.11}$$

Proof We prove (3.1.9). We have:

$$d\left(\int\limits_t^\theta (\theta_1 - t)dw_r(\theta_1) \cdot (\theta - t) \right) = \int\limits_t^\theta (\theta_1 - t)dw_r(\theta_1) \cdot d\theta + (\theta - t)^2 dw_r(\theta)\,.$$

Integration of this identity from t to $t + h$ gives

$$\int\limits_t^{t+h} \left(\int\limits_t^\theta (\theta_1 - t)dw_r(\theta_1) \right) d\theta = h \int\limits_t^{t+h} (\theta - t)dw_r(\theta) - \int\limits_t^{t+h} (\theta - t)^2 dw_r(\theta)\,.$$

$$\tag{3.1.12}$$

Further,

$$d((\theta - t)^2(w_r(\theta) - w_r(t))) = 2(\theta - t)(w_r(\theta) - w_r(t))d\theta + (\theta - t)^2 dw_r(\theta),$$

whence

$$\int_t^{t+h} (\theta - t)^2 dw_r(\theta) = h^2(w_r(t + h) - w_r(t)) \tag{3.1.13}$$

$$-2 \int_t^{t+h} (w_r(\theta) - w_r(t))(\theta - t)d\theta = h^2 I_r - 2G_r.$$

The formula (3.1.8) for the integral $\int_t^{t+h}(\theta - t)dw_r(\theta)$ on the right-hand side of (3.1.12) can be obtained in a similar way. Substituting (3.1.8) and (3.1.13) in (3.1.12), we obtain (3.1.9). Thus, (3.1.9) has been proved. The derivation of (3.1.10) and (3.1.11) is even simpler. □

By Lemma 3.1.2, formula (3.1.4) can be written as

$$X(t + h) = x + \sum_{r=1}^q \sigma_r I_r + ah + \sum_{r=1}^q \Lambda_r a J_r + \sum_{r=1}^q \sigma_r'(hI_r - J_r) \tag{3.1.14}$$

$$+ La\frac{h^2}{2} + \sum_{r=1}^q \sum_{i=1}^q \Lambda_i \Lambda_r a J_{ir} + \sum_{r=1}^q L\Lambda_r a(2G_r - hJ_r)$$

$$+ \sum_{r=1}^q \Lambda_r La(hJ_r - G_r) + \sum_{r=1}^q \sigma_r'' \left(\frac{1}{2}h^2 I_r - G_r\right)$$

$$+ L^2 a\frac{h^3}{6} + \sum_{r=1}^q \sum_{i=1}^q \sum_{s=1}^q \Lambda_s \Lambda_i \Lambda_r a J_{sir} + \rho.$$

Lemma 3.1.3 *We have*

$$EJ_{sir} = 0, \quad EJ_{sir}I_j = 0, \quad EJ_{sir}I_jI_l = 0. \tag{3.1.15}$$

Proof The first and last identities in (3.1.15) can be proved using oddness. The second identity can be proved using arguments similar to the ones in the proofs of Lemmas 2.1.3 and 2.1.5. □

As in Sect. 2.1, we introduce an auxiliary vector $\tilde{X}$ equal to the right-hand side of (3.1.14) without the last two terms:

$$\tilde{X} = x + \sum_{r=1}^{q} \sigma_r I_r + ah + \sum_{r=1}^{q} \Lambda_r a J_r + \sum_{r=1}^{q} \sigma'_r (h I_r - J_r) \tag{3.1.16}$$

$$+ La \frac{h^2}{2} + \sum_{r=1}^{q} \sum_{i=1}^{q} \Lambda_i \Lambda_r a J_{ir} + \sum_{r=1}^{q} L \Lambda_r a (2 G_r - h J_r)$$

$$+ \sum_{r=1}^{q} \Lambda_r La (h J_r - G_r) + \sum_{r=1}^{q} \sigma''_r \left(\frac{1}{2} h^2 I_r - G_r \right) + L^2 a \frac{h^3}{6}.$$

Lemma 3.1.4 *The following relations hold:*

$$\left| E \left(\prod_{j=1}^{s} \Delta^{i_j} - \prod_{j=1}^{s} \tilde{\Delta}^{i_j} \right) \right| = O(h^4), \ s = 1, \dots, 7. \tag{3.1.17}$$

The proof of this lemma is based on Lemmas 3.1.1 and 3.1.3 and differs not essentially from the proof of Lemma 2.1.4.

3.1.3 Construction of a Method of Order Three

We construct the one-step approximation $\bar{X}$ on the basis of $\tilde{X}$ as follows:

$$\bar{X} = x + \sum_{r=1}^{q} \sigma_r \xi_r h^{1/2} + ah + \sum_{r=1}^{q} \Lambda_r a \eta_r h^{3/2} + \sum_{r=1}^{q} \sigma'_r (\xi_r - \eta_r) h^{3/2} \tag{3.1.18}$$

$$+ La \frac{h^2}{2} + \sum_{r=1}^{q} \sum_{i=1}^{q} \Lambda_i \Lambda_r a \eta_{ir} h^2 + \sum_{r=1}^{q} L \Lambda_r a (2 \mu_r - \eta_r) h^{5/2}$$

$$+ \sum_{r=1}^{q} \Lambda_r La (\eta_r - \mu_r) h^{5/2} + \sum_{r=1}^{q} \sigma''_r \left(\frac{1}{2} \xi_r - \mu_r \right) h^{5/2} + L^2 a \frac{h^3}{6}.$$

To construct a method of order of accuracy three (the one-step order of accuracy of such a method equals 4), we need fulfillment of the relations

$$\left| E \left(\prod_{j=1}^{s} \bar{\Delta}^{i_j} - \prod_{j=1}^{s} \tilde{\Delta}^{i_j} \right) \right| = O(h^4), \ s = 1, \dots, 7. \tag{3.1.19}$$

Indeed, Lemma 3.1.4 in this case implies

$$\left| E \left(\prod_{j=1}^{s} \Delta^{i_j} - \prod_{j=1}^{s} \bar{\Delta}^{i_j} \right) \right| = O(h^4), \ s = 1, \dots, 7. \tag{3.1.20}$$

Then according to Theorem 2.2.1, a method based on the one-step approximation $\bar{X}$ with the properties (3.1.20) and (3.1.21) will have order of accuracy three. Of course, the above said is valid under the standard assumptions on the coefficients of the system (3.1.1) which, in particular for the approximation (3.1.18), ensure the relation

$$E \prod_{j=1}^{s} |\bar{\Delta}^{i_j}| = O(h^4), \ s = 8. \tag{3.1.21}$$

Thus, we should consider the relations (3.1.19) only.

Lemma 3.1.5 *The following seven groups of identities ensure the relations (3.1.19):*

$$E\xi_r h^{1/2} = E I_r = 0, \ E\eta_r h^{3/2} = E J_r = 0, \tag{3.1.22}$$
$$E\eta_{ir} h^2 = E J_{ir} = 0, \ E\mu_r h^{5/2} = E G_r = 0;$$

$$E\xi_i\xi_r h = E I_i I_r = \delta_{ir} h, \ E\xi_r\eta_j h^2 = E I_r J_j = \delta_{rj}\frac{h^2}{2}, \tag{3.1.23}$$

$$E\xi_i\eta_{jr} h^{5/2} = E I_i J_{jr} = 0, \ E\xi_i\mu_r h^3 = E I_i G_r = \delta_{ir}\frac{h^3}{3},$$

$$E\eta_i\eta_j h^3 = E J_i J_j = \delta_{ij}\frac{h^3}{3}, \ E\eta_i\eta_{jr} h^{7/2} = E J_i J_{jr} = 0;$$

$$E\xi_i\xi_r\eta_{js} h^3 = E I_i I_r J_{js} \tag{3.1.24}$$
$$= \begin{cases} h^3/6 & \text{if } j \neq s \text{ and either } i = j, \ r = s \text{ or } i = s, \ r = j, \\ h^3/3 & \text{if } i = r = j = s, \\ 0 & \text{otherwise,} \end{cases}$$
$$E\xi_i\xi_r\mu_j h^{7/2} = E I_i I_r G_j = 0, \ E\xi_i\eta_j\eta_r h^{7/2} = E I_i J_j J_r = 0;$$

$$E\xi_i\xi_r\xi_j\xi_s h^2 = E I_i I_r I_j I_s \tag{3.1.25}$$
$$= \begin{cases} h^2 & \text{if } \{i, r, j, s\} \text{ consists of two pairs of equal numbers,} \\ 3h^2 & \text{if } i = r = j = s, \\ 0 & \text{otherwise,} \end{cases}$$
$$E\xi_i\xi_r\xi_j\eta_s h^3 = E I_i I_r I_j J_s$$
$$= \begin{cases} h^3/2 & \text{if } \{i, r, j, s\} \text{ consists of two pairs of equal numbers,} \\ 3h^3/2 & \text{if } i = r = j = s, \\ 0 & \text{otherwise,} \end{cases}$$
$$E\xi_i\xi_r\xi_j\eta_{sl} h^{7/2} = E I_i I_r I_j J_{sl} = 0;$$

$$E\xi_i\xi_r\xi_j\xi_s\xi_l h^{5/2} = E I_i I_r I_j I_s I_l = 0, \tag{3.1.26}$$
$$E\xi_i\xi_r\xi_j\xi_s\eta_l h^{7/2} = E I_i I_r I_j I_s J_l = 0;$$

in the following identities we assume, without loss of generality, that $i_1 \leq i_2 \leq i_3 \leq i_4 \leq i_5 \leq i_6$:

$$E \prod_{j=1}^{6} \xi_{i_j} h^3 = E \prod_{j=1}^{6} I_{i_j} \tag{3.1.27}$$

$$= \begin{cases} h^3, & i_1 = i_2 < i_3 = i_4 < i_5 = i_6, \\ 3h^3, & i_1 = i_2 < i_3 = i_4 = i_5 = i_6, \\ 3h^3, & i_1 = i_2 = i_3 = i_4 < i_5 = i_6, \\ 15h^3, & i_1 = i_2 = i_3 = i_4 = i_5 = i_6, \\ 0, & otherwise; \end{cases}$$

$$E \prod_{j=1}^{7} \xi_{i_j} h^{7/2} = E \prod_{j=1}^{7} I_{i_j} = 0. \tag{3.1.28}$$

Proof The proof of this lemma repeats the proof of Lemma 2.1.5 in many respects. Here we will consider the proof of the identity for $I_i I_r J_{js}$ in (3.1.24) only. Without loss of generality, we may put $t = 0$. We introduce the equations

$$dx = w_j dw_s(\theta), \ x(0) = 0,$$
$$dy = xd\theta, \ y(0) = 0.$$

Then

$$y(h) = \int_0^h x(\theta)d\theta = \int_0^h \left(\int_0^\theta w_j(\theta_1)dw_s(\theta_1) \right) d\theta = J_{js}(h),$$
$$E I_i I_r J_{js} = E(w_i(h)w_r(h)y(h)).$$

We have

$$d(w_i w_r y) = w_r y dw_i + w_i y dw_r + w_i w_r x d\theta + y \delta_{ir} d\theta.$$

Hence

$$dE(w_i w_r y) = E(w_i w_r x)d\theta, \tag{3.1.29}$$

since, obviously, Ey vanishes. We turn to evaluation of $E(w_i w_r x)$. We have

$$d(w_i w_r x) = w_r x dw_i + w_i x dw_r + w_i w_r w_j dw_s$$
$$+ x \delta_{ir} d\theta + w_i w_j \delta_{rs} d\theta + w_r w_j \delta_{is} d\theta,$$

whence

$$dE(w_i w_r x) = \delta_{ir} Ex d\theta + E(w_i w_j)\delta_{rs} d\theta + E(w_r w_j)\delta_{is} d\theta. \tag{3.1.30}$$

Since $Ex(t) \equiv 0$, the right-hand side of (3.1.30) does not vanish in three cases only.

The first case: $i = j \neq s = r$. We have

$$dE(w_i w_r x) = \theta d\theta, \quad E(w_i w_r x) = \frac{\theta^2}{2},$$

and (3.1.29) implies

$$E I_i I_r J_{js} = E(w_i(h) w_r(h) y(h)) = \frac{h^3}{6}.$$

The second case: $r = j \neq s = i$. This can be considered in a similar way and leads to the same result.

The third case: $i = j = r = s$. This gives

$$dE(w_i w_r x) = 2\theta d\theta, \quad E I_i I_r J_{js} = \frac{h^3}{3}.$$

So, the identity for $E I_i I_r J_{js}$ in (3.1.24) has been proved. Proofs of the remaining identities in (3.1.22)–(3.1.28) are not more complicated. This proves Lemma 3.1.5. $\square$

To finish construction of the one-step approximation $\bar{X}$ (see (3.1.18)), it remains to choose the random variables ξ_i, η_r, η_{ir}, μ_r so that the relations (3.1.22)–(3.1.28) hold. This can be done by modeling these random variables in various ways. Here we can choose them to be even simpler than in Sect. 2.1: although we are constructing a method of higher order of accuracy, we are doing it for systems of a less general form.

We will look for these variables in the following way. Consider symmetric random variables ξ_i, ν_j, ζ_r, $i, j, r = 1, \ldots, q$, that are all mutually independent (the condition of symmetry can be replaced by the weaker condition of vanishing of the corresponding odd moments), and put

$$\eta_i = \frac{\xi_i}{2} + \nu_i, \quad \eta_{ij} = \frac{1}{6}(\xi_i \xi_j - \zeta_i \zeta_j), \quad \mu_i = \frac{\xi_i}{3}. \tag{3.1.31}$$

Let ξ_i, ν_j, ζ_r have the following moments:

$$E\xi_i = E\xi_i^3 = E\xi_i^5 = E\xi_i^7 = 0, \quad E\nu_j = 0, \quad E\zeta_r = 0, \tag{3.1.32}$$

$$E\xi_i^2 = 1, \quad E\xi_i^4 = 3, \quad E\xi_i^6 = 15, \quad E\nu_j^2 = \frac{1}{12}, \quad E\zeta_i^2 = 1. \tag{3.1.33}$$

Lemma 3.1.6 *Suppose that ξ_i, ν_j, ζ_r are independent random variables with moments satisfying (3.1.32)–(3.1.33). Then the variables (3.1.31) satisfy the relations (3.1.22)–(3.1.28).*

The proof of this lemma consists of a simple verification of the relations (3.1.22)–(3.1.28).

For the identities (3.1.32)–(3.1.33) to be satisfied, the simplest modeling of the random variables ν_j and ζ_r is by the laws $P(\nu = \pm 1/\sqrt{12}) = 1/2$, $P(\zeta = \pm 1) = 1/2$, while ξ_i can be modeled by the law $\mathcal{N}(0, 1)$. However, for ξ_i we can also choose a simpler law. For example, $P(\xi = 0) = 1/3$, $P(\xi = \pm 1) = 3/10$, $P(\xi = \pm\sqrt{6}) = 1/30$.

Since (3.1.20)–(3.1.21) hold, the one-step approximation (3.1.18) with random variables (3.1.31) has order of accuracy four. We summarize the obtained result in the following theorem.

Theorem 3.1.7 *Suppose the coefficient $a(t, x)$ in the system (3.1.1) satisfies the Lipschitz condition*

$$|a(t, x) - a(t, y)| \le K|x - y|.$$

Suppose that $a(t, x)$ together with its partial derivatives up to a sufficiently high order (at least up to order seven, inclusively) belongs to **F**, *and suppose that the coefficients $\sigma_r(t)$ are three time continuously differentiable with respect to $t \in [t_0, T]$. Assume that the functions a, $\Lambda_r a$, La, $\Lambda_i \Lambda_r a$, $L\Lambda_r a$, $\Lambda_r La$, and $L^2 a$ grow at most linearly as $|x|$ goes to infinity. Let random variables ξ_{ik}, ν_{ik}, ζ_{ik} be independent and such that the relations (3.1.32)–(3.1.33) hold. Then the method*

$$X_{k+1} = X_k + \sum_{r=1}^{q} \sigma_r(t_k)\xi_{rk}h^{1/2} + a_k h \tag{3.1.34}$$

$$+ \sum_{r=1}^{q}(\Lambda_r a)_k \left(\frac{\xi_{rk}}{2} + \nu_{rk}\right) h^{3/2} + \sum_{r=1}^{q}\sigma_r'(t_k)\left(\frac{\xi_{rk}}{2} - \nu_{rk}\right)h^{3/2} + (La)_k \frac{h^2}{2}$$

$$+ \frac{1}{6}\sum_{r=1}^{q}\sum_{i=1}^{q}(\Lambda_i \Lambda_r a)_k(\xi_{ik}\xi_{rk} - \zeta_{ik}\zeta_{rk})h^2 + \sum_{r=1}^{q}(L\Lambda_r a)_k \left(\frac{\xi_{rk}}{6} - \nu_{rk}\right) h^{5/2}$$

$$+ \sum_{r=1}^{q}(\Lambda_r La)_k \left(\frac{\xi_{rk}}{6} + \nu_{rk}\right) h^{5/2} + \sum_{r=1}^{q}\sigma_r''(t_k)\xi_{rk}h^{5/2} + (L^2 a)_k \frac{h^3}{6},$$

has order of accuracy three in the sense of weak approximation (i.e., the relation (2.2.5) with $p = 3$ holds for all functions f belonging together with their partial derivatives up to order eight, inclusively, to the class **F**).

Thanks to the lemmas proved in this section, the proof of this theorem follows immediately from Lemma 2.2.2 and Theorem 2.2.1.

Remark 3.1.8 It is not difficult to see that omission of the random variables ν_{rk} in the terms at $h^{5/2}$ does not change the order of the method (3.1.34).

3.2 Weak Schemes for Systems with Colored Noise

In this section we present several weak methods for differential equations with colored noise (see Sect. 1.7):

$$dY = f(Y)dt + G(Y)Zdt \tag{3.2.1}$$

$$dZ = AZdt + \sum_{r=1}^{q} b_r dw_r(t),$$

where Y and f are l-dimensional vectors, Z and b_r are m-dimensional vectors, A is an $m \times m$ matrix, G is an $l \times m$ matrix, and $w_r(t)$ are independent standard Wiener processes. Recall that in the one-dimensional case (see Sect. 1.7) $Z(t)$ is the well-known Ornstein–Uhlenbeck process.

Here we restrict ourselves to presentation of some explicit Runge–Kutta (RK) methods and implicit schemes. See other methods for (3.2.1) as well as some numerical experiments in [298]. The methods can be derived using the corresponding mean-square schemes of Sect. 1.7 and the results of the previous section. We note that the methods can easily be carried over to a non-autonomous system with colored noise.

Obviously, the first order weak method coincides with the Euler scheme. Applying the RK method (3.1.3)–(3.2.1), we obtain the second-order RK method:

$$Y_{k+1} = Y_k + \frac{h}{2}[f(Y_k) + G(Y_k)Z_k] + \frac{h}{2}[f(\tilde{Y}_k) + G(\tilde{Y}_k)\tilde{Z}_k] \tag{3.2.2}$$

$$Z_{k+1} = Z_k + \sum_{r=1}^{q} b_r \xi_{rk} h^{1/2} + \frac{h}{2} A(Z_k + \tilde{Z}_k),$$

where

$$\tilde{Y}_k = Y_k + h[f(Y_k) + G(Y_k)Z_k] \tag{3.2.3}$$

$$\tilde{Z}_k = Z_k + \sum_{r=1}^{q} b_r \xi_{rk} h^{1/2} + AZ_k h,$$

and ξ_{rk} are independent random variables with standard normal distribution $\mathcal{N}(0, 1)$ or distributed according to the law $P(\xi = 0) = 2/3$, $P(\xi = \pm\sqrt{3}) = 1/6$.

Due to specific features of the system (3.2.1), we succeeded in construction of the fully third-order RK scheme:

$$Y_{k+1} = Y_k + \frac{1}{6}(k_1 + 4k_2 + k_3) \tag{3.2.4}$$

$$Z_{k+1} = Z_k + \sum_{r=1}^{q} b_r \xi_{rk} h^{1/2} + \frac{1}{6}(l_1 + 4l_2 + l_3),$$

where

$$\mathbf{f}(y, z) := f(y) + G(y)z \,,$$

$$k_1 = h\mathbf{f}_k, \quad k_2 = h\mathbf{f}\left(Y_k + \frac{k_1}{2}, Z_k + \frac{l_1}{2} + \frac{1}{2}\sum_{r=1}^{q} b_r \xi_{rk} h^{1/2}\right),$$

$$k_3 = h\mathbf{f}\left(Y_k - k_1 + 2k_2, Z_k - l_1 + 2l_2 + \sum_{r=1}^{q} b_r (\xi_{rk} + 6\nu_{rk}) h^{1/2}\right),$$

$$l_1 = hAZ_k, \quad l_2 = hA\left(Z_k + \frac{l_1}{2} + \frac{1}{2}\sum_{r=1}^{q} b_r \xi_{rk} h^{1/2}\right),$$

$$l_3 = hA\left(Z_k - l_1 + 2l_2 + \sum_{r=1}^{q} b_r (\xi_{rk} + 6\nu_{rk}) h^{1/2}\right)$$

and the random variables ξ_{rk} and ν_{rk} are independent and can be simulated either as $\mathcal{N}(0, 1)$ and $\mathcal{N}(0, 1/12)$, respectively, or by the laws

$$P(\xi = 0) = 1/3, \ P(\xi = \pm 1) = 3/10, \ P(\xi = \pm\sqrt{6}) = 1/30,$$
$$P(\nu = \pm 1/\sqrt{12}) = 1/2 \,.$$

Now we present weak implicit schemes. The first-order implicit weak methods coincide with the Euler mean-square schemes (1.7.10), but independent random variables ξ_{rk} can be simulated as $P(\xi = \pm 1) = 1/2$.

The two-parameter family of second-order implicit weak schemes has the form

$$Y_{k+1} = Y_k + \alpha h(f + Gz)_k + (1 - \alpha)h(f + Gz)_{k+1} \tag{3.2.5}$$

$$+ h^{3/2}\sum_{r=1}^{q} G(Y_k)b_r(2\alpha - 1)\xi_{rk}/2$$

$$+ \beta(2\alpha - 1)h^2[L(f + Gz)]_k/2 + (1 - \beta)(2\alpha - 1)h^2[L(f + Gz)]_{k+1}/2 \,,$$

$$Z_{k+1} = Z_k + \sum_{r=1}^{q} b_r \xi_{rk} h^{1/2} + \alpha h A Z_k + (1 - \alpha)h A Z_{k+1}$$

$$+ h^{3/2}\sum_{r=1}^{q} A b_r(2\alpha - 1)\xi_{rk}/2$$

$$+ \beta(2\alpha - 1)h^2 A^2 Z_k/2 + (1 - \beta)(2\alpha - 1)h^2 A^2 Z_{k+1}/2 \,.$$

The random variables ξ_{rk} here are the same as in the scheme (3.2.2)–(3.2.3), and $0 \le \alpha, \ \beta \le 1$.

If the parameter α in (3.2.5) is equal to 1/2, we obtain the trapezoidal weak method which is the simplest one among the family (3.2.5).

3.3 Application of Weak Methods to Computation of Wiener Integrals

Consider Wiener integrals

$$I = \int_{C_{0,0}^d} F(x(\cdot)) \, d\mu_{0,0}(x), \tag{3.3.1}$$

where $\mu_{0,0}(x)$ is a Wiener measure corresponding to Brownian paths with the fixed initial point $(0, 0)$ and

$$F(x(\cdot)) = \varphi(x(T), \int_0^T a(t, x(t))dt). \tag{3.3.2}$$

The integral (3.3.1) is understood in the sense of Lebesgue integral with respect to the measure $\mu_{0,0}(x)$ and is taken over the set $C_{0,0}^d$ of all d-dimensional continuous vector-functions $x(t)$ satisfying the condition $x(0) = 0$ (see, e.g. [120]). A relation of such integrals with quantum physics and some equations of mathematical physics can be found, e.g., in [88, 102, 120, 200, 381].

Numerical evaluation of Wiener integrals is an important and difficult task. Many approaches are proposed for solving this problem (see, e.g. [88, 91, 457] and references therein). As a rule, the known numerical methods reduce a path integral to a high dimensional integral which is then approximated using either classical or Monte Carlo methods. The high order of these integrals makes calculation of the Wiener integrals extremely difficult.

In this section we consider Monte Carlo methods for computing Wiener integrals of functionals of integral type (3.3.1)–(3.3.2) based on the relation between such integrals and stochastic differential equations (see [128, 451] where this approach was proposed).

Let $w(t) = (w^1(t), \ldots, w^d(t))^\mathsf{T}$ be a d-dimensional Wiener process. We introduce the system of SDEs

$$dX^1(t) = dw^1(t) \tag{3.3.3}$$

$$\cdots\cdots\cdots$$

$$dX^d(t) = dw^d(t)$$

$$dZ(t) = a(t, X^1(t), \ldots, X^d(t))dt , \ t \geq s, \tag{3.3.4}$$

with initial conditions

$$X^1(s) = x^1, \ldots, X^d(s) = x^d, \ Z(s) = z . \tag{3.3.5}$$

We will denote the solution of the system (3.3.3)–(3.3.5) by either $X_{s,x}(t)$, $Z_{s,x,z}(t)$, or, if this does not lead to confusion, simply by $X(t)$, $Z(t)$.

The Wiener integral (3.3.1) of the functional (3.3.2) is equal to

$$I = E\varphi(X_{0,0}(T), Z_{0,0,0}(T)).\tag{3.3.6}$$

According to the Monte Carlo method, expectation $E\varphi$ can be estimated by the sum

$$I_M = \frac{1}{M}\sum_{m=1}^{M}\varphi(X_{0,0}^{(m)}(T), Z_{0,0,0}^{(m)}(T)),\tag{3.3.7}$$

where $X^{(m)}(T)$, $Z^{(m)}(T)$, $m = 1, \ldots, M$, are independent realizations of the random variables $X(t)$, $Z(t)$.

An efficiency of this approach is due to the fact that the system (3.3.3)–(3.3.5) has the fixed dimension d and the corresponding accuracy is reached by means of a choice of a method for (3.3.3)–(3.3.4) and a step of numerical integration h and a number M of Monte Carlo simulations. Thus, the problem of calculating the infinite-dimensional Wiener integral I is reduced to the Cauchy problem (3.3.3)–(3.3.5). This problem can naturally be regarded as one-dimensional since it contains one independent variable only. We underline that in other methods the path integral is reduced to a high dimensional Riemann integral and the accuracy is reached on account of increasing its dimension. Of course, for the numerical computation of I it suffices to construct weak approximations of the solution of the system (3.3.3)–(3.3.4) and to use general methods developed in the previous and this chapters. However, in view of the specificity of the problem under consideration we can construct more effective methods. The effectiveness of the constructed algorithms allows us to evaluate integrals (3.3.1)–(3.3.2) for a large dimension d.

Since (3.3.3)–(3.3.5) implies

$$X_{0,0}(t) = w(t), \quad Z_{0,0,0}(t) = \int_0^t a(\theta, X(\theta))d\theta,\tag{3.3.8}$$

and $w(t)$ can be exactly modeled at any moment t, finding an approximation $\bar{Z}_{0,0,0}(t)$ reduces to approximately computing the integral (3.3.8) for $t = T$. Because there are many efficient quadrature formulas, at first glance this problem does not seem difficult. At the same time we have to keep in mind that quadrature formulas have a high order of accuracy only for integrands that are sufficiently smooth with respect to t. In view of nonregularity of $w(t)$, the integrand $a(t, w(t))$ does not satisfy the usual conditions of smoothness. Below we show that the trapezoidal formula applied to the integral (3.3.8) has the second weak order. In view of above said, this requires a separate proof, of course. For Wiener integrals of functionals of exponential type

$$F(x(\cdot)) = \exp\left[\int_0^T f(t, x(t))\, dt\right] \qquad (3.3.9)$$

we derive a method of order four (see Sect. 3.3.2).

3.3.1 The Trapezoidal, Rectangle, and Other Methods of Second Order

We introduce the one-step approximation for the system (3.3.3)–(3.3.4):

$$\bar{X}_{t,x}(t+h) = x + \xi h^{1/2}, \qquad (3.3.10)$$

$$\bar{Z}_{t,x,z}(t+h) = z + \frac{h}{2}(a(t,x) + a(t+h, \bar{X}_{t,x}(t+h))),$$

where $\xi = (\xi^1, \ldots, \xi^d)^\mathsf{T}$ is a d-dimensional random variable with independent coordinates ξ^i, $i = 1, \ldots, d$, such that $E\xi^i = E(\xi^i)^3 = E(\xi^i)^5 = 0$, $E(\xi^i)^2 = 1$, $E(\xi^i)^4 = 3$. We divide the interval $[0, T]$ into N equal parts with step $h = T/N$: $0 = t_0 < t_1 < \cdots < t_N = T$, $t_{k+1} - t_k = h$. Using (3.3.10) we construct the approximate solution

$$X_0 = 0, \quad X_{k+1} = X_k + \xi_k h^{1/2}, \quad Z_0 = 0, \qquad (3.3.11)$$

$$Z_{k+1} = Z_k + \frac{h}{2}(a(t_k, X_k) + a(t_{k+1}, X_{k+1})), \quad k = 0, \ldots, N-1,$$

where ξ_k, $k = 0, \ldots, N-1$, are independent d-dimensional random variables distributed like ξ. We can take, e.g., normally distributed random variables as such ξ_k. However, as in the previous sections, in (3.3.11) we may also use random variables that are more convenient for computing purposes, e.g. with the coordinates taking the values 0, $\sqrt{3}$, $-\sqrt{3}$ with probabilities $2/3$, $1/6$, $1/6$.

Theorem 3.3.1 *Suppose the function a satisfies a global Lipschitz condition. Suppose also that a and φ together with their partial derivatives of order up to six, inclusively, belong to* **F**. *Then*

$$E\varphi(\bar{X}_{0,0}(T), \bar{Z}_{0,0,0}(T)) - E\varphi(X_{0,0}(T), Z_{0,0,0}(T)) = O(h^2). \qquad (3.3.12)$$

Proof The proof rests on Theorem 2.2.1. The conditions (a), (c), (d) of that theorem (see also Lemma 2.2.2) are obviously fulfilled. Therefore it remains to verify the condition (b). It is not difficult to see that in our case the inequality (2.2.4) holds. It can be shown (see details in [274]) that $E\prod_{j=1}^{s} \bar{\Delta}^{i_j}$ for $s = 1, \ldots, 5$ coincides with $E\prod_{j=1}^{s} \Delta^{i_j}$ up to $O(h^3)$. This proves the inequality (2.2.3). $\qquad\square$

We can prove that not only the trapezoidal formula, but any interpolation formula of third order of accuracy with respect to h applied to the integral in (3.3.8) leads to a method of second order of accuracy for Wiener integrals. In particular, an application of the midpoint formula gives

$$I = E\varphi(X_{0,0}(T), Z_{0,0,0}(T)) = E\varphi\left(w(T), \frac{1}{N}\sum_{k=1}^{N} a(t_{k-1/2}, w(t_{k-1/2}))\right) + O(h^2),$$
(3.3.13)

where $t_{k-1/2} = t_k - h/2$.

It is clear that $E\varphi(w(T), \frac{1}{N}\sum_{k=1}^{N} a(t_{k-1/2}, w(t_{k-1/2})))$ can be realized in the form $E\varphi(\bar{X}_{0,0}(T), \bar{Z}_{0,0,0}(T)) = E\varphi(X_N, Z_N)$, where

$$X_0 = 0, \quad X_{k+1} = X_k + (\xi_k + \eta_k)\left(\frac{h}{2}\right)^{1/2},$$
(3.3.14)

$$Z_0 = 0, \quad Z_{k+1} = Z_k + ha\left(t_{k+1/2}, X_k + \eta_k\left(\frac{h}{2}\right)^{1/2}\right),$$

$$k = 0, \ldots, N - 1,$$

and ξ_k and η_k are independent d-dimensional random variables whose coordinates, in turn, are independent $\mathcal{N}(0, 1)$-distributed random variables. As in the trapezoidal method, we can also use simpler random variables.

Remark 3.3.2 We stress that because of the nonregularity of Brownian trajectories separate proofs are required for the result that the considered quadrature formulas for (3.3.8) give the same accuracy as in the deterministic case. To confirm this, we consider the following system of two equations:

$$dX(s) = dw(s), \quad X(0) = 0,$$
$$dZ(s) = X(s)ds, \quad Z(0) = 0,$$

as well as the function $\varphi(z) = z^2$. Here, $EZ^2(h)$ can readily be evaluated exactly: $EZ^2(h) = h^3/3$. We compute the approximation $\bar{Z}(h)$ by Simpson's formula:

$$\bar{Z}(h) = \frac{h}{6}\left(X(0) + 4X\left(\frac{h}{2}\right) + X(h)\right) = \frac{h}{6}\left(4X\left(\frac{h}{2}\right) + X(h)\right).$$

We are immediately convinced that $E\bar{Z}^2(h) = 13h^3/36$, and hence that $E\bar{Z}^2(h) - EZ^2(h) = O(h^3)$ instead of the expected $O(h^5)$, since to compute the integral (3.3.8) for $t = h$ we have used Simpson's formula, which has order of accuracy five in the deterministic case. Moreover, in general, we can prove that there is no "natural" method of third order of accuracy (see details in [274]).

Remark 3.3.3 Define the step random process $w^h(t)$, $0 \leq t \leq T$, to be equal to $w(t_k)$ for $t \in [t_k - h/2, t_k + h/2] \cap [0, T]$. The approximation of the integral $\int_0^T a(w(t))dt$ by the trapezoidal method coincides with $\int_0^T a(w^h(t))dt$. In this subsection we have proved that deviation of $E\varphi \left(\int_0^T a(w(t))dt \right)$ from the expectation of the same expression with the integrand replaced by its approximation has order $O(h^2)$ if only φ and a are sufficiently regular functions. Then one can conjecture that this might also be true for functionals $F(x(\cdot))$ of a more general form than $F(x(\cdot)) = \varphi \left(\int_0^T a(x(t))dt \right)$. In [451] (see also [274]), it is shown that this conjecture is true for a very large class of functionals.

Remark 3.3.4 Note that the piecewise linear approximation $w_h(t)$ of a Wiener process is defined as

$$w_h(t) = w(t_k)\frac{t_{k+1} - t}{h} + w(t_{k+1})\frac{t - t_k}{h}, \quad t_k \leq t \leq t_{k+1}, \tag{3.3.15}$$

and that it differs from the piecewise constant approximation $w^h(t)$ used in the trapezoidal method. The approximation (3.3.15) gives an error of order $O(h)$ only. The simplest way to confirm this is to compute

$$E\int_0^1 w^2(t)dt = \frac{1}{2}, \quad E\int_0^1 (w^h(t))^2 dt = \frac{1}{2}, \quad E\int_0^1 w_h^2(t)dt = \frac{1}{2} - \frac{h}{6}.$$

3.3.2 A Fourth-Order Runge–Kutta Method for Computing Wiener Integrals of Functionals of Exponential Type

This subsection is devoted to the computation of Wiener integrals (3.3.1) of often encountered functionals of exponential type

$$F(x(\cdot)) = \exp\left(\int_0^T a(x(t))dt\right), \quad x(t) \in \mathbf{R}^d. \tag{3.3.16}$$

The functional (3.3.16) is a particular case of the functional (3.3.2), and therefore the results of the previous subsection can be applied here too. In Remark 3.3.2 we have noted that there is no "natural" method of order of accuracy exceeding two for integrating the system (3.3.3)–(3.3.4). However, thanks to the special form of the functional (3.3.16), the computation of the integral (3.3.1) with (3.3.16) can be done by using another system, for which we can successfully develop a method of order four. This system has the form

$$dX^1(t) = dw^1(t), \ X^1(0) = x^1, \tag{3.3.17}$$

$$\cdots\cdots\cdots$$

$$dX^d(t) = dw^d(t), \ X^d(0) = x^d,$$
$$dY(t) = Y(t)a(t, X^1(t), \ldots, X^d(t))dt, \ Y(0) = 1.$$

It can readily be seen that the Wiener integral of the functional (3.3.16) is equal to

$$I = EY(T),$$

where $x^1 = \cdots = x^d = 0$.

Thus, evaluation of the Wiener integral (3.3.1), (3.3.16) leads to the problem of numerical integration of the system (3.3.17). For our purposes, the approximate solution is better, if the difference $EY(t) - E\bar{Y}(t)$ is smaller. As in the previous subsection, despite the facts that the $X(t)$ in (3.3.17) can be found exactly and that the equation for $Y(t)$ does not have stochastic components, we need special proofs for using methods that are well known in the deterministic case.

Let $h = T/N$, $t_k = kh$, $k = 0, \ldots, N$, $t_{k-1/2} = (k - 1/2)h$, $k = 1, \ldots, N$. Consider the following approximation:

$$X_0 = 0, \ X_{k-1/2} = X_{k-1} + \frac{1}{\sqrt{2}}\xi_{k-1/2}h^{1/2}, \tag{3.3.18}$$

$$X_k = X_{k-1/2} + \frac{1}{\sqrt{2}}\xi_k h^{1/2},$$

$$Y_0 = 1, \ Y_k = Y_{k-1} + \frac{1}{6}(k_1 + 2k_2 + 2k_3 + k_4),$$

where the $\xi_{k-1/2}$, ξ_k are mutually independent d-vectors of independent $\mathcal{N}(0, 1)$-distributed components and

$$k_1 = ha(t_{k-1}, X_{k-1})Y_{k-1}, \tag{3.3.19}$$
$$k_2 = ha(t_{k-1/2}, X_{k-1/2})\left(Y_{k-1} + \frac{k_1}{2}\right),$$
$$k_3 = ha(t_{k-1/2}, X_{k-1/2})\left(Y_{k-1} + \frac{k_2}{2}\right),$$
$$k_4 = ha(t_k, X_k)(Y_{k-1} + k_3).$$

Clearly, the method (3.3.18)–(3.3.19) is a Runge–Kutta method of fourth order of accuracy for integrating the last equation in the system (3.3.17) if we consider $a(t, w^1(t), \ldots, w^d(t))$ to be a sufficiently smooth function of t. Since this function is random and nonsmooth in t, we need a separate proof. This proof is rather long. It rests on Theorem 2.2.1 and Lemma 2.1.9 and can be found in [128, 274]. The following result is true under some natural assumptions on the function $a(t, x)$ (e.g., a is non-positive and sufficiently smooth).

Theorem 3.3.5 *The method (3.3.18)–(3.3.19) for the Wiener integral (3.3.1), (3.3.16) is of order four, i.e.,*

$$EY̅(t) - EY(t) = O(h^4).$$ (3.3.20)

Remark 3.3.6 We can also obtain Runge–Kutta methods of second and third order of accuracy. Note that the method (3.3.18)–(3.3.19) requires only a double modeling of a normally-distributed random vector and two evaluations of the function a at each step. The method of third order of accuracy requires the same amount of computations as does the method (3.3.18)–(3.3.19), while the method of second order of accuracy, which is already quite inferior as regards accuracy, requires only a somewhat less amount of computations. Therefore, the Runge–Kutta methods of third and second order of accuracy are of limited interest in this case.

3.4 Conditional Wiener Integrals

Let $C^d_{0,a;T,b}$ be the set of all d-dimensional continuous vector-functions $x(t)$ over $[0, T]$ satisfying the conditions $x(0) = a$, $x(T) = b$. Consider the conditional Wiener integral

$$\mathcal{J} = \int_{C^d_{0,a;T,b}} F(x(\cdot)) \, d\mu^{T,b}_{0,a}(x),$$ (3.4.1)

where F is a functional on $C^d_{0,a;T,b}$ and $\mu^{T,b}_{0,a}(x)$ is the conditional Wiener measure, corresponding to the Brownian paths $X^{T,b}_{0,a}(t)$ with fixed initial and final points, i.e., it corresponds to the d-dimensional Brownian bridge from a at the time $t = 0$ to b at the time $t = T$. The integral (3.4.1) is to be understood in the sense of Lebesgue integral with respect to the measure $\mu^{T,b}_{0,a}(x)$ and is taken over the set $C^d_{0,a;T,b}$ (see, e.g. [120, 403]).

The importance of path integrals (3.4.1) for computing various quantities in quantum statistical mechanics is well known [88, 102, 120, 200, 381, 403]. For instance, the Feynman path integral of the form

$$\mathcal{J} = < a|e^{-TH}|b >$$ (3.4.2)

$$= \int \exp\left(\int_0^T \left[\frac{m\dot{x}^2(t)}{2} - V(x(t))\right] dt\right) \mathcal{D}x(t), \quad H = -\frac{1}{2}\Delta + V,$$

is equivalent to the conditional Wiener integral (3.4.1) with the exponential-type functional

$$F(x(\cdot)) = \exp\left[-\int_0^T V(x(t)) \, dt\right].$$ (3.4.3)

Such quantities as the free energy of the system, the ground state energy, wavefunction, etc. can be written in terms of the integral (3.4.1), (3.4.3) [102, 120, 200, 240, 381, 403]. A wider class of functionals than (3.4.3) is also of interest. For example, correlation functions are expressed via the conditional Wiener integral (3.4.1) with a more general functional than (3.4.3) (see, e.g., [200, 240, 381] and references therein). They are written as the functional averages of products of path positions at different times. For instance, for $d = 1$, a two-point correlation function $\Gamma(\theta)$, $0 \leq \theta \leq T$, has the form

$$\Gamma(\theta) = < x(0)x(\theta) > \tag{3.4.4}$$

$$= \frac{1}{\mathcal{Z}(T)} \int_{-\infty}^{\infty} \int_{C_{0,y;T,y}^1} x(0)\varphi\left(x(\theta), \int_0^T V(t, x(t))\, dt\right) d\mu_{0,y}^{T,y}(x)dy$$

$$= \frac{1}{\mathcal{Z}(T)} \int_{-\infty}^{\infty} \int_{C_{0,y;T,y}^1} y\varphi\left(x(\theta), \int_0^T V(t, x(t))\, dt\right) d\mu_{0,y}^{T,y}(x)dy,$$

where the function

$$\varphi(x, z) = x \exp(-z),$$

and the partition function

$$\mathcal{Z}(T) = \operatorname{Tr} e^{-TH} = \int_{-\infty}^{\infty} \int_{C_{0,y;T,y}^1} \exp\left[-\int_0^T V(x(t))\, dt\right] d\mu_{0,y}^{T,y}(x)dy.$$

Correlation functions contain important information about quantum-mechanical systems and they are observable in scattering experiments (see, e.g. [200]). Other important examples of more general functionals than (3.4.3) are those corresponding to internal and kinetic energies (see, e.g. [53, 101, 425]).

As it is known [177, 188], the d-dimensional Brownian bridge $X(t) = X_{0,a}^{T,b}(t)$, $0 \leq t \leq T$, from $(0, a)$ to (T, b) can be characterized as the pathwise unique solution of the system of SDEs

$$dX = \frac{b - X}{T - t}dt + dw(t), \ 0 \leq t < T, \ X(0) = a, \tag{3.4.5}$$

with

$$X(T) = b. \tag{3.4.6}$$

Clearly, the conditional Wiener integral $\mathcal{J}$ from (3.4.1) is equal to the expectation of the functional taken over all realizations of $X(t)$, $0 \le t \le T$:

$$\mathcal{J} = EF(X). \tag{3.4.7}$$

The solution of (3.4.5) is

$$X(t) = a\frac{T-t}{T} + b\frac{t}{T} + (T-t)\int_0^t \frac{dw(s)}{T-s}. \tag{3.4.8}$$

Hence for any $0 \le \Delta < T - t$

$$X(t + \Delta) = X(t) + \Delta\frac{b - X(t)}{T - t} + (T - t - \Delta)\int_t^{t+\Delta}\frac{dw(s)}{T-s}. \tag{3.4.9}$$

We have

$$E\left[(T - t - h)\int_t^{t+h}\frac{dw(s)}{T-s}\,\middle|\,X(t)\right] = 0, \tag{3.4.10}$$

$$E\left[\left((T - t - h)\int_t^{t+h}\frac{dw(s)}{T-s}\right)^2\,\middle|\,X(t)\right] = \left(1 - \frac{h}{T-t}\right)h.$$

We can exactly simulate the solution of (3.4.5) by a simple recurrent procedure based on the formula

$$X(t + h) = X(t) + h\frac{b - X(t)}{T - t} + h^{1/2}\sqrt{\frac{T - t - h}{T - t}}\,\xi, \quad t + h \le T, \tag{3.4.11}$$

where ξ is a random vector which components are Gaussian random variables with zero mean and unit variance and they are independent of $X(t)$.

The structure of this section is as follows. In Sects. 3.4.1 and 3.4.2 we deal with numerical schemes for conditional Wiener integrals of exponential-type functionals. Functionals of a general form are considered in Sect. 3.4.3 and of integral-type – in Sect. 3.4.4. The case of pinned diffusions is treated in Sect. 3.4.5, where we exploit the results of [60, 78] to express path integrals of integral-type functionals over pinned diffusions as expectations with respect to a Markovian process which solves a system of SDEs. In this case we propose an Euler-type method which is of weak order one. Numerical experiments for methods from the previous and this sections are presented in the next section (Sect. 3.5).

3.4.1 Explicit Runge–Kutta Method of Order Four for Conditional Wiener Integrals of Exponential-Type Functionals

In this section we consider conditional Wiener integrals (3.4.1) of the exponential-type functionals

$$F(x(\cdot)) = \exp\left[\int_0^T f(t, x(t))\,dt\right].\tag{3.4.12}$$

Let us introduce the scalar equation

$$dY = f(t, X(t))\,Y\,dt,\ \ 0 \le t \le T,\ \ Y(0) = 1,\tag{3.4.13}$$

where $X(t)$ is defined by (3.4.5)–(3.4.6) and $f(t, x)$ is the same as in (3.4.12). Then the Wiener integral (3.4.1), (3.4.12) is equal to

$$\mathcal{J} = EY(T).\tag{3.4.14}$$

Thus, evaluation of the Wiener integral (3.4.1), (3.4.12) is reduced to the problem of numerical integration of the system (3.4.5)–(3.4.6), (3.4.13).

Introduce a discretization of the time interval $[0, T]$, for definiteness the equidistant one with a time step $h > 0$:

$$t_k = kh,\ \ k = 0, \ldots, N,\ \ t_N = T,$$

and let $t_{k+1/2} := t_k + h/2$.

Let us formally apply a standard deterministic explicit fourth-order Runge–Kutta method to Eq. (3.4.13) assuming that $X(t)$ is a known function. Then, taking into account (3.4.11), we obtain the following algorithm for integrating the system (3.4.5)–(3.4.6), (3.4.13):

$$X(0) = a,\tag{3.4.15}$$

$$X(t_{k+1/2}) = X(t_k) + \frac{h}{2}\frac{b - X(t_k)}{T - t_k} + \frac{h^{1/2}}{\sqrt{2}}\sqrt{\frac{T - t_{k+1/2}}{T - t_k}}\xi_{k+1/2},$$

$$k = 0, \ldots, N - 1,$$

$$X(t_{k+1}) = X(t_{k+1/2}) + \frac{h}{2}\frac{b - X(t_{k+1/2})}{T - t_{k+1/2}} + \frac{h^{1/2}}{\sqrt{2}}\sqrt{\frac{T - t_{k+1}}{T - t_{k+1/2}}}\xi_{k+1},$$

$$k = 0, \ldots, N - 2,\ \ X(t_N) = b,$$

$$Y_0 = 1, \tag{3.4.16}$$

$$k_1 = f(t_k, X(t_k))Y_k, \quad k_2 = f(t_{k+1/2}, X(t_{k+1/2}))\,[Y_k + hk_1/2]\,,$$

$$k_3 = f(t_{k+1/2}, X(t_{k+1/2}))\,[Y_k + hk_2/2]\,, \quad k_4 = f(t_{k+1}, X(t_{k+1}))\,[Y_k + hk_3]\,,$$

$$Y_{k+1} = Y_k + \frac{h}{6}\,(k_1 + 2k_2 + 2k_3 + k_4)\,, \quad k = 0, \ldots, N-1\,,$$

where $\xi_{k+1/2}$, ξ_{k+1} are d-dimensional random vectors which components are mutually independent random variables with standard normal distribution $\mathcal{N}(0, 1)$.

Since the function $X(t)$ is non-smooth, the deterministic result on the accuracy order of the involved Runge–Kutta method is not applicable here and a separate convergence theorem is needed.

Introduce the operator

$$L = \frac{\partial}{\partial t} + \sum_{i=1}^{d} \frac{b^i - x^i}{T - t} \frac{\partial}{\partial x^i} + \frac{1}{2} \sum_{i=1}^{d} \frac{\partial^2}{(\partial x^i)^2}\,, \quad 0 \le t < T\,. \tag{3.4.17}$$

We observe that this operator contains singularity since the denominator $T - t$ tends to zero as t goes to T.

Consider the function

$$u(t, x) = EY_{t,x,1}(T)\,. \tag{3.4.18}$$

It satisfies the Cauchy problem

$$Lu + fu = 0, \quad 0 \le t < T, \; x \in \mathbf{R}^d\,, \tag{3.4.19}$$

$$u(T, x) = 1\,.$$

We assume that the function $f(t, x)$ is sufficiently smooth, belongs to the class $\mathbf{F}$ (see Definition 2.1.1) together with its partial derivatives of a sufficiently high order and is such that the problem (3.4.19) has a unique solution which is sufficiently smooth and belongs to the class $\mathbf{F}$ together with its partial derivatives of a sufficiently high order. In addition, we suppose that $EY^2(t)$ exists and bounded on $[0, T]$ and that for all sufficiently small h the second moments EY_k^2 are uniformly bounded with respect to h. For instance, the latter conditions are satisfied when the function $f(t, x)$ is bounded. The following theorem is proved under these assumptions on the function $f(t, x)$.

Theorem 3.4.1 *The method (3.4.15)–(3.4.16) applied to evaluation of the conditional Wiener integral (3.4.14) is of fourth order of accuracy, i.e.,*

$$|\mathcal{J} - EY_N| = |EY(T) - EY_N| \le Kh^4, \tag{3.4.20}$$

where the constant K is independent of h.

The proof of Theorem 3.4.1 is based on a thorough analysis of the one-step error which is made in the next section. Complete proofs are available in [313].

3.4.1.1   Theorem on One-Step Error

In this subsection we consider a one-step error of the method (3.4.15)–(3.4.16).

It is convenient to introduce the additional notation for the approximation defined by (3.4.16): $\bar{Y}_{0,a,1}(t_k) = Y_k$ and also $\bar{Y}_{t_k,x,y}(t_i)$, $t_i \geq t_k$, by which we mean the approximation of (3.4.13) started from y at $t = t_k$ with $X(t_k) = x$.

It is not difficult to see that

$$Y_{t,x,y}(t+t') = yY_{t,x,1}(t+t'), \quad \bar{Y}_{t_k,x,y}(t_{k+k'}) = y\bar{Y}_{t_k,x,1}(t_{k+k'}), \tag{3.4.21}$$
$$EY_{t,x,y}(T) = yEY_{t,x,1}(T) = yu(t,x),$$

where $u(t,x)$ is the solution of the problem (3.4.19).

Recall that $t_0 = 0$, $X_0 = a$, $Y_0 = 1$. Using (3.4.18) and (3.4.21) and the fact that we simulate $X_k = X(t_k)$ exactly, we can represent the *global error* of the method (3.4.15)–(3.4.16) (cf. (3.4.20)) in the form

$$\left| EY_{0,a,1}(T) - E\bar{Y}_{0,a,1}(T) \right| = \left| EY_{t_0,X_0,Y_0}(T) - EY_N \right| \tag{3.4.22}$$
$$= |u(t_0,X_0)Y_0 - Eu(t_N,X_N)Y_N|$$

$$= \left| \sum_{k=0}^{N-1} \left[Eu(t_k,X(t_k))Y_k - Eu(t_{k+1},X(t_{k+1}))\bar{Y}_{t_k,X_k,Y_k}(t_{k+1}) \right] \right|$$

$$= \left| \sum_{k=0}^{N-1} EY_k \left[u(t_k,X(t_k)) - u(t_{k+1},X(t_{k+1}))\bar{Y}_{t_k,X_k,1}(t_{k+1}) \right] \right|$$

$$\leq \sum_{k=0}^{N-1} \left| EY_k \left[u(t_k,X(t_k)) - u(t_{k+1},X_{t_k,X_k}(t_{k+1}))\bar{Y}_{t_k,X_k,1}(t_{k+1}) \right] \right| .$$

We have

$$R_k := \left| EY_k \left[u(t_k,X(t_k)) - u(t_{k+1},X_{t_k,X_k}(t_{k+1}))\bar{Y}_{t_k,X_k,1}(t_{k+1}) \right] \right| \tag{3.4.23}$$
$$= \left| EY_k E \left[u(t_k,X_k) - u(t_{k+1},X_{t_k,X_k}(t_{k+1}))\bar{Y}_{t_k,X_k,1}(t_{k+1})|\mathcal{F}_{t_k} \right] \right| .$$

First, we analyze R_k for $k = 0, \ldots, N-2$. To this end, we consider the *one-step error* for $0 \leq t < T - h$:

$$r(t,x) := Eu(t+h,X_{t,x}(t+h))\bar{Y}_{t,x,1}(t+h) - u(t,x). \tag{3.4.24}$$

We rewrite (3.4.16) on a single step in the form:

$$\bar{Y}_{t,x,1}(t+h) = 1 + \frac{h}{6}\left(f_0 + 4f_{1/2} + f_1\right) \tag{3.4.25}$$

$$+\frac{h^2}{6}\left(f_0 f_{1/2} + f_{1/2}^2 + f_{1/2} f_1\right) + \frac{h^3}{12}\left(f_0 f_{1/2}^2 + f_{1/2}^2 f_1\right) + \frac{h^4}{24} f_0 f_{1/2}^2 f_1,$$

where $f_0 := f(t,x)$, $f_{1/2} := f(t+h/2, X_{t,x}(t+h/2))$,   and $f_1 := f(t+h, X_{t,x}(t+h))$.

Using (2.1.35), we get

$$Eu(t+h, X_{t,x}(t+h)) = u(t,x) + hLu(t,x) + \frac{h^2}{2}L^2u(t,x) + \frac{h^3}{6}L^3u(t,x)$$

$$\tag{3.4.26}$$

$$+\frac{h^4}{24}L^4u(t,x) + \int_t^{t+h} \frac{(t+h-\theta)^4}{24} EL^5u(\theta, X_{t,x}(\theta))\, d\theta,$$

$$E f_0 u(t+h, X_{t,x}(t+h)) = f_0 Eu(t+h, X_{t,x}(t+h)) \tag{3.4.27}$$

$$= f_0\left[u(t,x) + hLu(t,x) + \frac{h^2}{2}L^2u(t,x) + \frac{h^3}{6}L^3u(t,x)\right.$$

$$\left. + \int_t^{t+h} \frac{(t+h-\theta)^3}{6} EL^4u(\theta, X_{t,x}(\theta))\, d\theta\right],$$

$$E f_1 u(t+h, X_{t,x}(t+h)) = f_0 u(t,x) + hL\,(fu)\,(t,x) + \frac{h^2}{2}L^2\,(fu)\,(t,x)$$

$$\tag{3.4.28}$$

$$+\frac{h^3}{6}L^3\,(fu)\,(t,x) + \int_t^{t+h} \frac{(t+h-\theta)^3}{6} EL^4\,(fu)\,(\theta, X_{t,x}(\theta))\, d\theta.$$

Further,

$$E f_{1/2} u(t+h, X_{t,x}(t+h)) = E\left(f_{1/2} E\left[u(t+h, X_{t,x}(t+h))|\mathcal{F}_{t+h/2}\right]\right),$$

and by (2.1.34) we obtain

$$E\left[u(t+h, X_{t,x}(t+h))|\mathcal{F}_{t+h/2}\right] = u(t+h/2, X_{t,x}(t+h/2))$$

$$+\frac{h}{2}Lu(t+h/2, X_{t,x}(t+h/2)) + \frac{h^2}{8}L^2u(t+h/2, X_{t,x}(t+h/2))$$

$$+\frac{h^3}{48}L^3u(t+h/2, X_{t,x}(t+h/2))$$

$$+\int_{t+h/2}^{t+h} \frac{(t+h-\theta)^3}{6} E\left[L^4u(\theta, X_{t,x}(\theta))|\mathcal{F}_{t+h/2}\right] d\theta,$$

then

$$Ef_{1/2}u(t+h, X_{t,x}(t+h)) = f_0u(t, x) + \frac{h}{2}L(fu)(t, x) \tag{3.4.29}$$

$$+\frac{h^2}{8}L^2(fu)(t, x) + \frac{h^3}{48}L^3(fu)(t, x)$$

$$+\int_t^{t+h/2} \frac{(t+h/2-\theta)^3}{6} EL^4(fu)(\theta, X_{t,x}(\theta))d\theta + \frac{h}{2}f_0Lu(t, x)$$

$$+\frac{h^2}{4}L(fLu)(t, x) + \frac{h^3}{16}L^2(fLu)(t, x)$$

$$+\frac{h}{2}\int_t^{t+h/2} \frac{(t+h/2-\theta)^2}{2} EL^3(fLu)(\theta, X_{t,x}(\theta))d\theta$$

$$+\frac{h^2}{8}f_0L^2u(t, x) + \frac{h^3}{16}L(fL^2u)(t, x)$$

$$+\frac{h^2}{8}\int_t^{t+h/2} (t+h/2-\theta)EL^2(fL^2u)(\theta, X_{t,x}(\theta))d\theta$$

$$+\frac{h^3}{48}f_0L^3u(t, x) + \frac{h^3}{48}\int_t^{t+h/2} EL(fL^3u)(\theta, X_{t,x}(\theta))d\theta$$

$$+Ef_{1/2}\int_{t+h/2}^{t+h} \frac{(t+h-\theta)^3}{6}L^4u(\theta, X_{t,x}(\theta))d\theta.$$

Analogously, we get

$$Ef_0 f_{1/2} u(t+h, X_{t,x}(t+h)) = f_0^2 u(t,x) + \frac{h}{2} f_0 L\, (f u)\, (t,x) \qquad (3.4.30)$$

$$+ \frac{h^2}{8} f_0 L^2\, (f u)\, (t,x) + f_0 \int\limits_{t}^{t+h/2} \frac{(t+h/2-\theta)^2}{2} EL^3\, (f u)\, (\theta, X_{t,x}(\theta)) d\theta$$

$$+ \frac{h}{2} f_0^2 Lu(t,x) + \frac{h^2}{4} f_0 L\, (f Lu)\, (t,x)$$

$$+ \frac{h}{2} f_0 \int\limits_{t}^{t+h/2} (t+h/2-\theta) EL^2\, (f Lu)\, (\theta, X_{t,x}(\theta)) d\theta + \frac{h^2}{8} f_0^2 L^2 u(t,x)$$

$$+ \frac{h^2}{8} f_0 \int\limits_{t}^{t+h/2} EL\, \big(f L^2 u\big)\, (\theta, X_{t,x}(\theta)) d\theta$$

$$+ f_0 Ef_{1/2} \int\limits_{t+h/2}^{t+h} \frac{(t+h-\theta)^2}{2} L^3 u(\theta, X_{t,x}(\theta))\, d\theta \,,$$

$$Ef_{1/2}^2 u(t+h, X_{t,x}(t+h)) = f_0^2 u(t,x) + \frac{h}{2} L\, \big(f^2 u\big)\, (t,x) \qquad (3.4.31)$$

$$+ \frac{h^2}{8} L^2\, \big(f^2 u\big)\, (t,x) + \int\limits_{t}^{t+h/2} \frac{(t+h/2-\theta)^2}{2} EL^3\, \big(f^2 u\big)\, (\theta, X_{t,x}(\theta)) d\theta$$

$$+ \frac{h}{2} f_0^2 Lu(t,x) + \frac{h^2}{4} L\, \big(f^2 Lu\big)\, (t,x)$$

$$+ \frac{h}{2} \int\limits_{t}^{t+h/2} (t+h/2-\theta) EL^2\, \big(f^2 Lu\big)\, (\theta, X_{t,x}(\theta)) d\theta + \frac{h^2}{8} f_0^2 L^2 u(t,x)$$

$$+ \frac{h^2}{8} \int\limits_{t}^{t+h/2} EL\, \big(f^2 L^2 u\big)\, (\theta, X_{t,x}(\theta)) d\theta$$

$$+ Ef_{1/2}^2 \int\limits_{t+h/2}^{t+h} \frac{(t+h-\theta)^2}{2} L^3 u(\theta, X_{t,x}(\theta))\, d\theta \,,$$

$$Ef_{1/2}f_1u(t+h, X_{t,x}(t+h)) = f_0^2u(t, x) + \frac{h}{2}L\left(f^2u\right)(t, x) \qquad (3.4.32)$$

$$+ \frac{h^2}{8}L^2\left(f^2u\right)(t, x) + \int_t^{t+h/2} \frac{(t+h/2 - \theta)^2}{2}EL^3\left(f^2u\right)(\theta, X_{t,x}(\theta))d\theta$$

$$+ \frac{h}{2}f_0L\left(fu\right)(t, x) + \frac{h^2}{4}L\left(fL\left(fu\right)\right)(t, x)$$

$$+ \frac{h}{2}\int_t^{t+h/2}(t+h/2 - \theta)EL^2\left(fL\left(fu\right)\right)(\theta, X_{t,x}(\theta))d\theta + \frac{h^2}{8}f_0L^2\left(fu\right)(t, x)$$

$$+ \frac{h^2}{8}\int_t^{t+h/2}EL\left(fL^2\left(fu\right)\right)(\theta, X_{t,x}(\theta))d\theta$$

$$+ Ef_{1/2}\int_{t+h/2}^{t+h}\frac{(t+h - \theta)^2}{2}L^3\left(fu\right)(\theta, X_{t,x}(\theta))d\theta,$$

$$Ef_0f_{1/2}^2u(t+h, X_{t,x}(t+h)) = f_0^3u(t, x) + \frac{h}{2}f_0L\left(f^2u\right)(t, x) \qquad (3.4.33)$$

$$+ f_0\int_t^{t+h/2}(t+h/2 - \theta)EL^2\left(f^2u\right)(\theta, X_{t,x}(\theta))d\theta + \frac{h}{2}f_0^3Lu(t, x)$$

$$+ \frac{h}{2}f_0\int_t^{t+h/2}EL\left(f^2Lu\right)(\theta, X_{t,x}(\theta))d\theta$$

$$+ f_0Ef_{1/2}^2\int_{t+h/2}^{t+h}(t+h - \theta)L^2u(\theta, X_{t,x}(\theta))d\theta,$$

$$Ef_{1/2}^2f_1u(t+h, X_{t,x}(t+h)) = f_0^3u(t, x) + \frac{h}{2}L\left(f^3u\right)(t, x) \qquad (3.4.34)$$

$$+ \int_t^{t+h/2}(t+h/2 - \theta)EL^2\left(f^3u\right)(\theta, X_{t,x}(\theta))d\theta + \frac{h}{2}f_0^2L(fu)(t, x)$$

$$+ \frac{h}{2}\int_t^{t+h/2}EL\left(f^2L(fu)\right)(\theta, X_{t,x}(\theta))d\theta$$

$$+ Ef_{1/2}^2\int_{t+h/2}^{t+h}(t+h - \theta)L^2\left(fu\right)(\theta, X_{t,x}(\theta))d\theta,$$

$$Ef_0 f_{1/2}^2 f_1 u(t+h, X_{t,x}(t+h)) = f_0^4 u(t, x) \tag{3.4.35}$$

$$+ f_0 \int_t^{t+h/2} EL\left(f^3 u\right)(\theta, X_{t,x}(\theta)) d\theta + f_0 E f_{1/2}^2 \int_{t+h/2}^{t+h} L\left(fu\right)(\theta, X_{t,x}(\theta)) d\theta.$$

Substituting (3.4.25)–(3.4.35) in (3.4.24), we obtain

$$r = h\left[Lu + fu\right] + \frac{h^2}{2}\left[L^2 u + L(fu) + fLu + f^2 u\right] \tag{3.4.36}$$

$$+ \frac{h^3}{6}[L^3 u + L^2(fu) + fL^2 u + fL\left(fu\right) + L\left(fLu\right) + L\left(f^2 u\right)$$

$$+ f^2 Lu + f^3 u] + \frac{h^4}{24}[L^4 u + L^3(fu) + fL^3 u + fL^2(fu)$$

$$+ f^2 L^2 u + f^2 L\left(fu\right) + f^3 Lu + f^4 u + L^2(fLu) + L^2(f^2 u) + fL(fLu)$$

$$+ fL(f^2 u) + L(fL^2 u) + L(fL(fu)) + L(f^2 Lu) + L(f^3 u)] + \tilde{r},$$

where all the operators and functions are evaluated at the point (t, x) and $\tilde{r}$ accumulates all the integrals present in (3.4.26)–(3.4.35) multiplied by h to the corresponding power. Taking into account that $u(t, x)$ satisfies the equation from (3.4.19), we get

$$r(t, x) = \tilde{r}(t, x). \tag{3.4.37}$$

If the terms in the one-step error $r(t, x)$ of the method (3.4.15)–(3.4.16) (i.e., the terms in $\tilde{r}$) were bounded by $K(x)h^5$, $K(x) \in \mathbf{F}$, for all $t \leq T - h$, the relations (3.4.22)–(3.4.24) would imply that $\sum_{k=0}^{N-2} R_k \leq Ch^4$, where C is independent of h. But we see that the one-step error consists of integrals with integrands containing terms of the form $A(t, x) = L^n\left(q_1 L^l q_2\right)(t, x)$, where $q_1(t, x)$ and $q_2(t, x)$ are some functions from the class $\mathbf{F}$. The functions $A(t, x)$ belong to the class $\mathbf{F}$ for $t \in [0, T_*]$, where $T_* < T$ is a fixed (independent of h) time moment. Then $|r(t, x)| \leq K(x)h^5$, $K(x) \in \mathbf{F}$, $t \in [0, T_*]$, with $K(x)$ depending on T_*. However, the functions $A(t, x)$ do not belong to the class $\mathbf{F}$ for $t \in [0, T)$ due to the singularity in L (see (3.4.17)). Consequently, $r(t, x)$ can not be bounded by $K(x)h^5$, $K(x) \in \mathbf{F}$, for all $t < T$, and a more detailed analysis of the one-step error is required to prove the convergence theorem. In particular, we need to consider the structure of the functions $A(t, x)$ in detail. We always assume that L^0 is an identity operator. The following lemma is proved by induction.

Lemma 3.4.2 *Let $q_1(t, x)$ and $q_2(t, x)$ be sufficiently smooth functions belonging to the class $\mathbf{F}$ together with their partial derivatives of a sufficiently high order. Then for $0 \leq t < T$:*

$$L^n(q_1 L^l q_2)(t, x) = g_0(t, x) + \sum_{j=1}^{m} \sum_{\alpha_j} g_{\alpha_j}(t, x)\psi^{\alpha_j}(t, x), \qquad (3.4.38)$$

$$l, n = 0, 1, \ldots, \quad m = l + n,$$

where α_j is a multi-index such that $\alpha_j = (i_1, \ldots, i_j)$ and each i_k is from $\{1, \ldots, d\}$, the summation in (3.4.38) is over all possible values of α_j, g_0 and g_{α_j} are some functions from the class **F**, and

$$\psi^r = \frac{b^r - x^r}{T - t}, \quad r = 1, \ldots, d,$$

$$\psi^{\alpha_{j+1}} = \frac{b^{i_{j+1}} - x^{i_{j+1}}}{T - t}\psi^{\alpha_j} + \frac{\partial}{\partial x^r}\psi^{\alpha_j}, \quad \alpha_j = (i_1, \ldots, i_j),$$

$$\alpha_{j+1} = (i_1, \ldots, i_j, i_{j+1}), \quad j = 1, 2, \ldots,$$

and for all α_j

$$L\psi^{\alpha_j} = 0.$$

Using specific properties of the functions ψ^{α_j}, the following theorem on one-step error is proved in [313].

Theorem 3.4.3 *The one-step error of the method (3.4.15)–(3.4.16) can be written in the form*

$$r(t, x) = \tilde{r}(t, x) = h^5 S(t, x) + E\rho(t, x; h), \qquad (3.4.39)$$

where $S(t, x)$ is a linear combination of the functions $\psi^{\alpha_2}(t, x)$, $\psi^{\alpha_3}(t, x)$, $\psi^{\alpha_4}(t, x)$, $(T - t)\psi^{\alpha_4}(t, x)$, $h\psi^{\alpha_4}(t, x)$, $(T - t)\psi^{\alpha_5}(t, x)$, $h\psi^{\alpha_5}(t, x)$, $(T - t)^2\psi^{\alpha_6}(t, x)$, $(T - t)h\psi^{\alpha_6}(t, x)$, $h^2\psi^{\alpha_6}(t, x)$, coefficients in this linear combination are independent of t, x, and h; $\rho(t, x; h)$ is such that

$$\left(E\left[\rho(t, X_{0,a}(t); h)\right]^{2n}\right)^{1/2n} \le \frac{Ch^5}{\sqrt{T - t - h}}, \quad t + h < T,$$

with a constant C independent of t and h.

We should emphasize that the most important part of this theorem consists in the equality $r(t, x) = \tilde{r}(t, x)$ which is due to Eqs. (3.4.24)–(3.4.36). Theorem 3.4.3 is a basis for the proof of Theorem 3.4.1 on the global error of the method (3.4.15)–(3.4.16) (see [313]).

3.4.2 Implicit Runge–Kutta Methods for Conditional Wiener Integrals of Exponential-Type Functionals

From the point of view of possible applications, the most interesting case is when the function f is bounded from above, for example, when f is negative (see e.g. (3.4.2)). In this case the explicit Runge–Kutta method from Sect. 3.4.1 may cause some computational problems since, for instance, Y_{k+1} in (3.4.16) can become a large negative number while the exact $Y(t)$ is always positive. Apparently, this may occasionally lead to some instabilities and require a very small time step to achieve a reasonable accuracy. In such a situation an implicit method can behave better.

Let us formally apply the deterministic midpoint method to (3.4.13) provided $X(t)$ is a known function. As a result, we obtain

$$X(h/2) = a + \frac{h}{2}\frac{b-a}{T} + \sqrt{\frac{h}{2}}\sqrt{\frac{T-h/2}{T}}\,\xi_{1/2}, \tag{3.4.40}$$

$$X(t_{k+1/2}) = X(t_{k-1/2}) + h\frac{b - X(t_{k-1/2})}{T - t_{k-1/2}} + \sqrt{h}\sqrt{\frac{T - t_{k+1/2}}{T - t_{k-1/2}}}\,\xi_{k+1/2},$$

$$k = 1, \ldots, N-1,$$

$$Y_0 = 1, \tag{3.4.41}$$

$$Y_{k+1} = Y_k + hf(t_{k+1/2}, X(t_{k+1/2}))\frac{Y_k + Y_{k+1}}{2}, \quad k = 0, \ldots, N-1,$$

where $\xi_{k+1/2}$, $k = 0, \ldots, N-1$, are d-dimensional random vectors which components are mutually independent random variables with standard normal distribution $\mathcal{N}(0, 1)$.

Resolving the implicitness in (3.4.41), we get

$$Y_{k+1} = Y_k \frac{1 + \dfrac{h}{2}f(t_{k+1/2}, X(t_{k+1/2}))}{1 - \dfrac{h}{2}f(t_{k+1/2}, X(t_{k+1/2}))}. \tag{3.4.42}$$

To ensure that the denominator in (3.4.42) does not vanish for all sufficiently small h, we should require that the function $f(t, x)$ is bounded from above, i.e., that $f(t, x) \le c$ for all (t, x), c is a constant. In this case for all sufficiently small h the denominator in (3.4.42) is positive. If $f(t, x) \le 0$, then $-1 \le Y_k \le 1$ for all k.

We prove the convergence theorem for the method (3.4.40)–(3.4.41) under the same assumptions as in Sect. 3.4.1 (see them before Theorem 3.4.1). Note that in the case of $f(t, x) \le 0$, the condition $EY_k^2 \le C$ is satisfied due to the uniform boundedness of the random variables Y_k.

Theorem 3.4.4 *The method (3.4.40)–(3.4.41) applied to evaluation of the conditional Wiener integral (3.4.14) is of second accuracy order, i.e.,*

$$|\mathcal{J} - EY_N| = |EY(T) - EY_N| \le Kh^2, \tag{3.4.43}$$

where the constant K is independent of h.

The proof of this theorem is given in [313].

If we formally apply the deterministic Gauss method of order four (see, e.g., [156, p. 71]) to (3.4.13), assuming that $X(t)$ is a known function, we obtain

$$X(\gamma h) = a + \gamma h \frac{b-a}{T} + \sqrt{\gamma h}\sqrt{\frac{T-\gamma h}{T}}\,\xi_\gamma, \tag{3.4.44}$$

$$X((1-\gamma)h) = X(\gamma h) + (1-2\gamma)h\frac{b - X(\gamma h)}{T - \gamma h}$$

$$+ \sqrt{(1-2\gamma)h}\sqrt{\frac{T-(1-\gamma)h}{T-\gamma h}}\,\xi_{1-\gamma},$$

$$X(t_k + \gamma h) = X(t_{k-1}+(1-\gamma)h) + 2\gamma h\frac{b - X(t_{k-1}+(1-\gamma)h)}{T - t_k + \gamma h}$$

$$+ \sqrt{2\gamma h}\sqrt{\frac{T-t_k-\gamma h}{T-t_k+\gamma h}}\,\xi_{k+\gamma},$$

$$X(t_k + (1-\gamma)h) = X(t_k + \gamma h) + (1-2\gamma)h\frac{b - X(t_k + \gamma h)}{T - t_k - \gamma h}$$

$$+ \sqrt{(1-2\gamma)h}\sqrt{\frac{T-t_{k+1}+\gamma h}{T-t_k-\gamma h}}\,\xi_{k+1-\gamma}, \qquad k = 1,\ldots,N-1,$$

$$Y_0 = 1, \tag{3.4.45}$$

$$k_1 = f(t_k + \gamma h, X(t_k + \gamma h))\left[Y_k + \frac{h}{4}k_1 + \left(\frac{1}{4} - \frac{\sqrt{3}}{6}\right)hk_2\right],$$

$$k_2 = f(t_k + (1-\gamma)h, X(t_k + (1-\gamma)h))\left[Y_k + \left(\frac{1}{4} + \frac{\sqrt{3}}{6}\right)hk_1 + \frac{h}{4}k_2\right],$$

$$Y_{k+1} = Y_k + \frac{h}{2}(k_1 + k_2), \quad k = 0,\ldots,N-1,$$

where $\gamma = \dfrac{1}{2} - \dfrac{\sqrt{3}}{6}$ and $\xi_{k+\gamma}$, $\xi_{k+1-\gamma}$, $k = 0,\ldots,N-1$, are d-dimensional random vectors which components are mutually independent random variables with standard normal distribution $\mathcal{N}(0,1)$.

Resolving (3.4.45) with respect to k_1 and k_2, we get

$$Y_{k+1} = Y_k \frac{1 + \dfrac{h}{4}(f_1 + f_2) + \dfrac{h^2}{12}f_1 f_2}{1 - \dfrac{h}{4}(f_1 + f_2) + \dfrac{h^2}{12}f_1 f_2}, \qquad (3.4.46)$$

where $\quad f_1 := f(t_k + \gamma h, X(t_k + \gamma h)) \quad$ and $\quad f_2 := f(t_k + (1-\gamma)h, X(t_k + (1-\gamma)h))$.

The denominator in (3.4.46) does not vanish for all sufficiently small h for functions $f(t, x)$ being bounded from above. And if $f(t, x) \le 0$, then $-1 \le Y_k \le 1$ for all k.

The intuition built on the previous analysis of the methods (3.4.15)–(3.4.16) and (3.4.40)–(3.4.41) tells us that the method (3.4.44)–(3.4.45) should be of order four. But this assertion turned out to be wrong, the method is of order two only, just as the method (3.4.40)–(3.4.41). We have not found an implicit method for (3.4.14) that satisfies the condition $|Y_k| \le 1$ for $f(t, x) \le 0$ and has the fourth order of accuracy. In this search it was natural to restrict ourselves to standard fourth-order deterministic implicit methods for ordinary differential equations as a basis for potentially higher-order implicit methods for (3.4.14).

The following convergence theorem is valid.

Theorem 3.4.5 *The method (3.4.44)–(3.4.45) applied to evaluation of the conditional Wiener integral (3.4.14) is of second order of accuracy, i.e.,*

$$|\mathcal{J} - EY_N| = |EY(T) - EY_N| \le Kh^2, \qquad (3.4.47)$$

where the constant K is independent of h.

Although the methods (3.4.40)–(3.4.41) and (3.4.44)–(3.4.45) are of the same order of convergence, in our numerical tests (see Sect. 3.5) the method (3.4.44)–(3.4.45) gives more accurate results. Apparently, this is due to the fact that the constant K in (3.4.47) is, in general, less than its counterpart in (3.4.43). At the same time, the method (3.4.40)–(3.4.41) requires one evaluation of f per step, while (3.4.44)–(3.4.45) requires two evaluations of f per step.

3.4.3 Functionals of a General Form

We start this section by specifying the class of functionals for which the corresponding convergence theorem shall be proved. This is done via the formal assumptions listed below. Then in Sect. 3.4.3.1, we give some examples from this class of functionals.

Let us consider functionals $F(x)$ defined on the space $A[0, T]$ of right-continuous d-dimensional vector-functions $x(t)$ on the interval $[0, T]$ without discontinuities of the second kind, i.e., consider functionals on a larger space than $C^d_{0,a;T,b}$. We impose the following three assumptions on F.

Assumption 3.4.6 1. Let $0 < \theta_1 < \cdots < \theta_i < \cdots < \theta_n < T$. Introduce the measure ν_r on $[0, T]^r$ which is the sum of r-dimensional Lebesgue measures on $[0, T]^r$, $(r - 1)$-dimensional Lebesgue measure on the hyperplanes $\{(s_1, \ldots, s_r) \in [0, T]^r : s_j = \theta_i\}$, $i = 1, \ldots, n$, $j = 1, \ldots, r$, and on the diagonal hyperplanes $\{(s_1, \ldots, s_r) \in [0, T]^r : s_i = s_j\}$, $(r - 2)$-dimensional Lebesgue measure on $(r - 2)$-dimensional hyperplanes $\{(s_1, \ldots, s_r) \in [0, T]^r : s_k = \theta_i$ and $s_l = \theta_j$, $k \neq l\}$ and $\{(s_1, \ldots, s_r) \in [0, T]^r : s_i = s_j$ and $s_k = s_l\}$, and so on, including the one-dimensional Lebesgue measure on the lines $\{s_1 = \theta_{i_1}, \ldots, s_{r-1} = \theta_{i_{r-1}}\}$, $i_j \in \{1, \ldots, n\}$, and on the diagonal $\{s_1 = s_2 = \cdots = s_r\}$ plus the unit measures concentrated on the points $(\theta_{i_1}, \ldots, \theta_{i_r})$, $i_j \in \{1, \ldots, n\}$.

2. We assume that the functional $F(x)$ is six times Fréchet differentiable and that its r-th derivative has the following form:

$$F^{(r)}(x)(\delta_1, \ldots, \delta_r) = \int\limits_{[0,T]^r} v^{(r)}(x; s_1, \ldots, s_r)\delta_1(s_1) \cdots \delta_r(s_r)\nu_r(ds_1 \cdots ds_r),$$

$$(3.4.48)$$

$$r = 1, \ldots, 6,$$

where $\delta_i \in A[0, T]$ and the vector-functions $v^{(r)}(x; s_1, \ldots, s_r)$ are symmetric in the arguments $s_1, \ldots, s_r$ and uniformly bounded for $x \in A[0, T]$, $s_i \in [0, T]$.

3. For any function $x \in A[0, T]$ constant on a semi-interval $[c_0, c^0) \subset [0, T]$, there are continuous derivatives

$$\frac{d}{ds}v^{(1)}(x; s); \quad \frac{\partial}{\partial s_1}v^{(2)}(x; s_1, s_2), \; s_1 \neq s_2, \; s_j \neq \theta_i; \quad \frac{d}{ds}v^{(2)}(x; s, s);$$

$$\frac{d}{ds}v^{(2)}(x; s, \theta_i), \quad i = 1, \ldots, n;$$

which are bounded by a constant independent of $[c_0, c^0)$ and $x \in A[0, T]$.

We recall that $F^{(r)}(x)(\delta_1, \ldots, \delta_r)$ are r-linear functionals. Under Assumption 3.4.6 a convergence theorem (Theorem 3.4.9) for the method proposed in Sect. 3.4.3.2 is proved in [84]. We emphasize that the method is applicable much more widely.

Roughly speaking, one might say that we consider functionals of the general form on $A[0, T]$ which satisfy some conditions on smoothness and boundedness. As is usual for any numerical methods, if we weaken the assumptions about the smoothness then, as a rule, the convergence order of the considered method becomes lower. In physical applications, the smoothness part of Assumption 3.4.6 is not particularly restrictive since it is usually satisfied. The assumption on boundedness of derivatives of functionals can be, to some extent, weakened without loss of convergence order but this would significantly complicate the proof of the convergence theorem. At the same time, the common computational practice in quantum statistical mechanics is to curtail potentials so that they and their derivatives remain bounded which usually implies boundedness of derivatives of functionals. Alternatively, the concept of rejecting exploding trajectories from Sect. 2.2.5 (see also [316]) could be exploited here. That is we might choose not to take into account those trajectories which leave a bounded domain S during the time T. The domain S is chosen so that the boundedness condition is satisfied when $x(\cdot) \in S$.

3.4.3.1 Examples of Functionals

To illustrate the class of functionals satisfying Assumption 3.4.6, we give two particular examples here, although many more can be immediately constructed.

Example 3.4.7 We start with the integral-type functionals (see the functional needed to compute the correlation function (3.4.4)):

$$F(x(\cdot)) = \varphi \left(x(\theta), \int_0^T f(t, x(t)) \, dt \right), \quad 0 \le \theta \le T, \ x \in C_{0,a;T,b}^d . \quad (3.4.49)$$

One can check that if the functions $f(t, x)$ and $\varphi(x, z)$ have continuous and bounded derivatives up to a sufficiently high order then Assumption 3.4.6 holds. In particular, the Fréchet derivatives (3.4.48) have the form here:

$$F^{(1)}(x)(\delta_1) = \int_{[0,T]} v^{(1)}(x; s_1) \delta_1(s_1) \nu_1(ds_1)$$

with

$$v^{(1)}(x; s_1) \delta_1(s_1) = \frac{\partial \varphi}{\partial z} \nabla_x f(s_1, x(s_1)) \cdot \delta_1(s_1), \ s_1 \ne \theta;$$

$$v^{(1)}(x; \theta) \delta_1(\theta) = \nabla_x \varphi \cdot \delta_1(\theta);$$

and the measure ν_1 being the sum of the Lebesgue measure on $[0, T]$ and the unit measure concentrated at the point θ;

$$F^{(2)}(x)(\delta_1, \delta_2) = \int_{[0,T]^2} v^{(2)}(x; s_1, s_2) \delta_1(s_1) \delta_2(s_2) \nu_2(ds_1 ds_2)$$

with

$$v^{(2)}(x; s_1, s_2)\delta_1(s_1)\delta_2(s_2) = \frac{\partial^2 \varphi}{\partial z^2} \nabla_x f(s_1, x(s_1)) \cdot \delta_1(s_1) \nabla_x f(s_2, x(s_2)) \cdot \delta_2(s_2),$$

$$s_1 \neq s_2, \ s_i \neq \theta;$$

$$v^{(2)}(x; s, \theta)\delta_1(s)\delta_2(\theta) = \sum_{i=1}^{d} \frac{\partial^2 \varphi}{\partial z \partial x^i} \nabla_x f(s, x(s)) \cdot \delta_1(s)\delta_2^i(\theta), \ s \neq \theta;$$

$$v^{(2)}(x; s, s)\delta_1(s)\delta_2(s) = \frac{\partial \varphi}{\partial z} \sum_{i,j=1}^{d} \frac{\partial^2 f}{\partial x^i \partial x^j}(s, x(s))\delta_1^i(s)\delta_2^j(s), \ s \neq \theta;$$

$$v^{(2)}(x; \theta, \theta)\delta_1(\theta)\delta_2(\theta) = \sum_{i,j=1}^{d} \frac{\partial^2 \varphi}{\partial x^i \partial x^j}\delta_1^i(\theta)\delta_2^j(\theta);$$

and the measure ν_2 being the sum of the two-dimensional Lebesgue measure on $[0, T]^2$, the one-dimensional Lebesgue measures on the lines $\{s_1 = \theta\}$ and $\{s_2 = \theta\}$ and on the diagonal $\{s_1 = s_2\}$, and the unit measure concentrated at the point (θ, θ); the other derivatives can be written analogously. In the above formulas the derivatives of the function φ are taken at the point $\left(x(\theta), \int_0^T f(t, x(t)) \, dt\right)$ and the dot $\cdot$ means the usual scalar product of vectors.

Example 3.4.8 Let functions $f(t, x)$, $g(t, x)$, and $\varphi(z)$ have continuous and bounded derivatives up to a sufficiently high order. Then the functional

$$F(x(\cdot)) = \varphi\left(\int_0^T \int_0^t f(s, x(s)) \, g(t, x(t)) \, ds \, dt\right)$$

satisfies Assumption 3.4.6.

3.4.3.2   Numerical Method

We introduce a discretization of the time interval $[0, T]$

$$0 = t_0 < t_1 < \cdots < t_N = T$$

so that the points θ_i, $i = 1, \ldots, n$, belong to the set $\{t_0, t_1, \ldots, t_N\}$. Let

$$h := \max_{0 \leq k \leq N-1} (t_{k+1} - t_k).$$

and $t_{k+1/2} := (t_{k+1} + t_k)/2$, $k = 0, \ldots, N - 1$.

We also introduce a piecewise constant function $X^h(t)$, $t \in [0, T]$, given by:

$$X^h(t) := a, \quad t \in [0, t_{1/2}); \tag{3.4.50}$$
$$X^h(t) := X(t_k), \quad t \in [t_{k-1/2}, t_{k+1/2}), \quad k = 1, \ldots, N-1;$$
$$X^h(t) := b, \quad t \in [t_{N-1/2}, T].$$

Clearly, trajectories $X^h(t)$ belong to the space $A[0, T]$.

We define the approximation of the conditional Wiener integral $\mathcal{J}$ as follows:

$$\mathcal{J} = EF(X) \approx \bar{\mathcal{J}} = EF(X^h). \tag{3.4.51}$$

This method is analogous to the one used in the case of the usual (unconditional) Wiener measure in Sect. 3.3.1 (see also [451]).

Theorem 3.4.9 *Assume that Assumption 3.4.6 hold. The method (3.4.51), (3.4.50) applied to evaluation of the Wiener integral (3.4.1) is of second order of accuracy, i.e.,*

$$\left| \mathcal{J} - \bar{\mathcal{J}} \right| = \left| EF(X) - EF(X^h) \right| \leq Kh^2, \tag{3.4.52}$$

where the constant K is independent of h.

The proof of the theorem is given in [84].

Remark 3.4.10 The method (3.4.51), (3.4.50) is exact (i.e., there is no integration error) on the class of functionals which depend only on the value of the function $x(t)$ at a finite number of points θ_i, $i = 1, \ldots, n$.

The method (3.4.51), (3.4.50) together with the Monte Carlo technique gives an effective algorithm for computing conditional Wiener integrals, which is very simple to realize in practice. The method (3.4.51), (3.4.50) can be interpreted as a trapezoidal scheme. This interpretation becomes obvious in the case of integral-type functionals (see (3.4.58), (3.4.59)).

Now consider the Euler method, i.e., introduce the piecewise constant function $X_E^h(t)$, $t \in [0, T]$:

$$X_E^h(t) := X(t_k), \quad t \in [t_k, t_{k+1}), \quad k = 0, \ldots, N-1; \quad X_E^h(T) := b. \tag{3.4.53}$$

Theorem 3.4.11 *Assume that Assumption 3.4.6 holds with $r = 1, 2, 3, 4$ in (3.4.48). Then*

$$\tilde{\mathcal{J}} = EF(X_E^h) \tag{3.4.54}$$

approximates $\mathcal{J}$ with the first order of accuracy.

3.4.4 Integral-Type Functionals

In this section we consider conditional Wiener integrals of integral-type functionals:

$$F(x(\cdot)) = \varphi \left(x(\theta), \int_0^T f(t, x(t))\, dt \right), \quad 0 < \theta < T, \ x \in C_{0,a;T,b}^d. \tag{3.4.55}$$

Introduce the scalar process $Z(t)$ satisfying the equation

$$dZ = f(t, X(t))dt, \quad Z(0) = 0, \tag{3.4.56}$$

where $X(t)$ is the solution of (3.4.5)–(3.4.6). Clearly, the conditional Wiener integral $\mathcal{J}$ from (3.4.1) of the functional (3.4.55) is equal to the expectation

$$\mathcal{J} = E\varphi\left(X(\theta), Z(T)\right). \tag{3.4.57}$$

The approximation (3.4.51), (3.4.50) applied to (3.4.1), (3.4.55) results in the trapezoidal method for Z :

$$\mathcal{J} \approx \bar{\mathcal{J}} = E\varphi\left(X(\theta), Z_N\right), \tag{3.4.58}$$

where

$$Z_0 = 0, \tag{3.4.59}$$

$$Z_{k+1} = Z_k + \frac{t_{k+1} - t_k}{2}\left[f(t_k, X(t_k)) + f(t_{k+1}, X(t_{k+1})) \right], \quad k = 0, \ldots, N-1.$$

Recall that the time discretization used here is so that $\theta \in \{t_0, t_1, \ldots, t_N\}$.

If we assume that $\varphi(x, z)$ and $f(t, x)$ have bounded derivatives up to a sufficiently high order, it follows from the general Theorem 3.4.9 that the method (3.4.58), (3.4.59) for (3.4.1), (3.4.55) has the second order of accuracy; i.e., the estimate (3.4.52) is valid for it. The other set of assumptions under which the theorem is valid are that $f(t, x)$ and its derivatives up to a sufficiently high order are bounded and $\varphi(x, z)$ is sufficiently smooth. It is interesting that no method of the form

$$Z_{k+1} = Z_k + (t_{k+1} - t_k)\sum_{i=1}^{3} \alpha_i f(t_k + \beta_i, X(t_k + \beta_i)),$$

$$\alpha_i \in \mathbf{R}, \ \beta_i \in \left[0, t_{k+1} - t_k\right],$$

has order of accuracy higher than two (in the case of usual Wiener integrals, see a similar comment in Remark 3.3.2). At the same time, in the case of integral-type functionals of a particular form – the exponential-type functionals $F(x(\cdot)) =$

$\exp[\int_0^T f(t, x(t)) \, dt]$, the fourth-order Runge–Kutta method (3.3.18)–(3.3.19) was constructed in the previous section.

We made a computational comparison between (3.4.59) and the fourth-order Runge–Kutta method in computing the ground state energy of one particle in a 1D harmonic oscillator. Despite being of lower order, the method (3.4.59) turns out to be preferable due to its stability properties. These follow from preservation by (3.4.59) of such structural properties of exponential-type functionals as positivity and monotonicity, which can be broken down in the case of the fourth-order Runge–Kutta method (3.3.18)–(3.3.19) of Sect. 3.3.2 (see similar observations although in a different context in [321]). Further, instead of the trapezoidal rule (3.4.59), we can use the Simpson rule:

$$Z_0 = 0, \tag{3.4.60}$$

$$Z_{k+1} = Z_k$$
$$+ \frac{t_{k+1} - t_k}{6} \left[f(t_k, X(t_k)) + 4f(t_{k+1/2}, X(t_{k+1/2})) + f(t_{k+1}, X(t_{k+1})) \right],$$
$$k = 0, \ldots, N - 1.$$

Although both methods (3.4.59) and (3.4.60) are of order two, the method (3.4.60) had much smaller bias in our experiments than the method (3.4.59) and thus was computationally more effective.

3.4.5 Extension to the Case of Pinned Diffusions

In this section we extend the Euler method (3.4.54), (3.4.53) to the case of paths of $\mathbf{R}^d$-diffusions

$$d\mathbb{X} = \alpha(t, \mathbb{X})dt + dw(t), \quad \mathbb{X}(t_0) = a, \tag{3.4.61}$$

that are conditioned to pass through a point $b \in \mathbf{R}^d$ at time T, $t_0 \leq t \leq T$. Conditioned diffusions are used, e.g. in parameter estimation problems (see e.g. [78]). We note that the Brownian bridge case considered in the previous sections corresponds to (3.4.61) with $\alpha = 0$.

Analogously to Sect. 3.4.4, we will be interested here in simulating the expectations of integral-type functionals

$$F(x(\cdot)) = \varphi \left(\int_{t_0}^T f(t, x(t)) \, dt \right), \quad x \in C_{t_0,a;T,b}^d, \tag{3.4.62}$$

but now with respect to the measure on paths corresponding to the conditioned diffusion (3.4.61). It is clear that this expectation is equal to

$$J = E\varphi\left(\mathbb{Z}(T)\right),$$

(3.4.63)

where the scalar process $\mathbb{Z}(t)$ satisfies the equation

$$d\mathbb{Z} = f(t, \mathbb{X}(t))dt, \quad \mathbb{Z}(t_0) = 0,$$

(3.4.64)

and $\mathbb{X}(t)$ is the conditional diffusion (3.4.61). In what follows we assume that the functions $\alpha(t, x)$ and $f(t, x)$ are bounded and have bounded derivatives up to a sufficiently high order, and that $\varphi(z)$ is sufficiently smooth.

It is proved in [78] (see also [60]) that the expectation (3.4.63) can be re-written as

$$J = \frac{E\left[\varphi(Z(T))Y(T)\right]}{EY(T)},$$

(3.4.65)

where $Z_{t_0,a,0}(t)$, $Y_{t_0,a,1}(t)$, $t \geq t_0$, satisfy the equations

$$dZ = f(t, X(t))dt, \quad Z(t_0) = 0,$$

(3.4.66)

$$dY = \alpha^\top(t, X)\frac{b - X}{T - t}Ydt + \alpha^\top(t, X)Ydw(t), \quad Y(t_0) = 1,$$

(3.4.67)

with $X(t) = X_{t_0,a}(t)$ being the Brownian bridge from a at the time $t = t_0$ to b at the time $t = T$ (cf. (3.4.5)–(3.4.6)):

$$dX = \frac{b - X}{T - t}dt + dw(t), \quad X(t_0) = a.$$

(3.4.68)

We note that

$$Y(T) = \exp(Q(T)),$$

(3.4.69)

where

$$dQ = \left[\alpha^\top(t, X)\frac{b - X}{T - t} - \frac{1}{2}\alpha^2(t, X)\right]dt + \alpha^\top(t, X)dw(t), \quad Q(t_0) = 0.$$

(3.4.70)

We remark that for $\alpha = 0$ (the Brownian bridge case), J from (3.4.65) coincides with J from (3.4.57).

We introduce a discretization of the time interval $[t_0, T] : t_0 < t_1 < \cdots < t_N = T$, for simplicity equidistant with the time step $h = t_{k+1} - t_k$. To construct the numerical method, we simulate the Brownian bridge $X(t)$ at the nodes t_k exactly (see (3.4.11)):

$$X_{k+1} = X_k + h\frac{b - X_k}{T - t_k} + \sqrt{h}\sqrt{\frac{T - t_{k+1}}{T - t_k}}\xi_{k+1}, \quad X_0 = a,$$

(3.4.71)

and we approximate (3.4.66) and (3.4.70) as follows

$$Z_{k+1} = Z_k + hf(t_k, X_k), \quad Z_0 = 0, \tag{3.4.72}$$

$$Q_{k+1} = Q_k + h \left[\alpha^\top (t_k, X_k) \frac{b - X_k}{T - t_k} - \frac{\alpha^2(t_k, X_k)}{2} \right] \tag{3.4.73}$$

$$+ \sqrt{h} \sqrt{\frac{T - t_{k+1}}{T - t_k}} \alpha^\top (t_k, X_k) \xi_{k+1},$$

$$Q_0 = 0,$$

where ξ_{k+1}, $k = 0, \ldots, N - 1$, are d-dimensional random vectors of which the components are mutually independent random variables with standard normal distribution $\mathcal{N}(0, 1)$.

We remark that we choose to approximate (3.4.69), (3.4.70) rather than (3.4.67) since it was observed (see, e.g. [321] and also Sect. 3.4.4 here) that positivity preservation automatically guaranteed by (3.4.73) has computational advantages while an explicit scheme applied directly to (3.4.73) does not possess this property. We also emphasize that $X(t_k) = X_k$, i.e., there is no numerical error introduced in (3.4.71).

Now we define the approximation of the path integral $\mathcal{J}$ from (3.4.63):

$$\mathcal{J} = E\varphi(\mathbb{Z}(T)) = \frac{E\left[\varphi(Z(T))Y(T)\right]}{EY(T)} \approx \bar{\mathcal{J}} = \frac{E\left[\varphi(Z_N)\exp(Q_N)\right]}{E\exp(Q_N)}. \tag{3.4.74}$$

Introduce the function

$$u(t, x, z)y = E\left[\varphi(Z_{t,x,z}(T))Y_{t,x,y}(T)\right] \tag{3.4.75}$$

and let

$$Y_k = \exp(Q_k). \tag{3.4.76}$$

Under the assumptions we imposed on the coefficients at the beginning of this section, the function $u(t, x, z)$ is smooth in x and z, and sufficiently high moments of $u(t_0, X_{t_0,a}(t), Z_{t_0,a,0}(t)))Y_{t_0,a,1}(t)$ and $u(t_k, X_k, Z_k)Y_k$ and their derivatives with respect to x and z are bounded [123]. The method itself is applicable more widely and the assumptions can be relaxed in the spirit of the comment after Assumption 3.4.6 in Sect. 3.4.3. The following theorem is proved in [84].

Theorem 3.4.12 *The method (3.4.71)–(3.4.73) is of first order of accuracy, i.e.,*

$$\left| E\left[\varphi(Z(T))Y(T)\right] - E\left[\varphi(Z_N)\exp(Q_N)\right] \right| \le Kh, \tag{3.4.77}$$

where the constant K is independent of h.

This theorem has the evident corollary.

Corollary 3.4.13 *The method (3.4.74), (3.4.71)–(3.4.73) for evaluating the path integral (3.4.63) is of first order of accuracy, i.e.,*

$$|\mathcal{J} - \bar{\mathcal{J}}| \le Kh, \tag{3.4.78}$$

where the constant K is independent of h.

Remark 3.4.14 We note that if in (3.4.74) we substitute Q_N simulated by the standard Euler scheme for (3.4.70):

$$Q_{k+1} = Q_k + h\left[\alpha^{\top}(t_k, X_k)\frac{b - X_k}{T - t_k} - \frac{\alpha^2(t_k, X_k)}{2}\right] + \sqrt{h}\,\alpha^{\top}(t_k, X_k)\,\xi_{k+1}, \tag{3.4.79}$$

$$Q_0 = 0,$$

then the method (3.4.74), (3.4.71), (3.4.72), (3.4.79) is of order $O(h \ln h)$ instead of $O(h)$ for (3.4.74), (3.4.71)–(3.4.73).

The Monte Carlo estimator for the path integral (3.4.63) based on the method (3.4.74), (3.4.71)–(3.4.73) has the form

$$\mathcal{J} \approx \bar{\mathcal{J}} = \frac{E\varphi(Z_N)\exp(Q_N)}{E\exp(Q_N)} \approx \hat{\mathcal{J}} = \frac{\sum_{m=1}^{M}\varphi(_mZ_N)\exp(_mQ_N)}{\sum_{m=1}^{M}\exp(_mQ_N)}, \tag{3.4.80}$$

where $_mZ_N, _mQ_N, m = 1, \ldots, M$, are independent realizations of the corresponding random variables. Note that the second approximate equality in (3.4.80) is related to the statistical error.

3.5 Numerical Experiments

The first part of this section (Examples 3.5.1–3.5.4) deals with testing the proposed methods for Wiener integrals (3.3.1) with respect to the "usual" Wiener measure while in the second part (Examples 3.5.5–3.5.9) methods for conditional Wiener integrals (3.4.1) are tested.

In Examples 3.5.1–3.5.4 numerical experiments are mainly related to the computation of the Wiener integral

$$I = \int_{C_{0,0}^{d}} \exp\left(\alpha \int_{0}^{T} x^2(s)ds\right) d\mu_{0,0}(x). \tag{3.5.1}$$

In this case the system (3.3.17) with initial conditions at $t \in [0, T)$ is written as the system of two equations for $t \leq s \leq T$:

$$dX(s) = dw(s), \quad X(t) = x,$$ (3.5.2)
$$dY(s) = \alpha Y(s)X^2(s)ds, \quad Y(t) = y.$$

Recall that

$$I = EY_{0,0,1}(T) = E \exp\left(\alpha \int_0^T X_{0,0}^2(s)ds\right),$$ (3.5.3)

where $X_{0,0}(s)$, $Y_{0,0,1}(s)$ is the solution of the system (3.5.2) for $t = 0$, $x = 0$, $y = 1$. Introduce the function

$$u(t, x) = EY_{t,x,1}(T) = E \exp\left(\alpha \int_t^T X_{t,x}^2(s)ds\right).$$ (3.5.4)

This function satisfies the Cauchy problem

$$\frac{\partial u}{\partial t} + \frac{1}{2}\frac{\partial^2 u}{\partial x^2} + \alpha x^2 u = 0, \quad u(T, x) = 1.$$ (3.5.5)

The solution of (3.5.5) for $\alpha = \lambda^2/2$, $0 \leq \lambda < \pi/2T$, has the form

$$u(t, x) = \exp\left[\frac{1}{2}\lambda x^2 \tan \lambda(T - t) - \frac{1}{2}\ln\cos(\lambda(T - t))\right].$$ (3.5.6)

Since $I = u(0, 0)$, this leads to the well-known result

$$I = (\cos \lambda T)^{-1/2}, \quad \alpha = \frac{\lambda^2}{2}, \quad 0 \leq \lambda < \frac{\pi}{2T}.$$ (3.5.7)

Further, the solution of (3.5.5) for $\alpha = -\lambda^2/2$ has the form

$$u(t, x) = \exp\left[\frac{1}{2}\lambda\frac{1 - \exp 2\lambda(T - t)}{1 + \exp 2\lambda(T - t)}x^2\right.$$ (3.5.8)
$$\left. + \frac{\lambda(T - t)}{2} + \frac{1}{2}\ln\frac{2}{1 + \exp 2\lambda(T - t)}\right],$$

and, consequently,

$$I = \left(\frac{2\exp \lambda T}{1 + \exp 2\lambda T}\right)^{1/2}, \quad \alpha = -\frac{\lambda^2}{2}.$$ (3.5.9)

Now we look for the variance of the variable $Y_{0,0,1}(T)$. For $\alpha = \lambda^2/2$:

$$Var\,Y_{0,0,1}(T) = EY^2_{0,0,1}(T) - (EY_{0,0,1}(T))^2 \tag{3.5.10}$$

$$= E \exp\left(2\alpha \int_0^T X^2_{0,0}(s)ds\right) - (\cos\lambda T)^{-1},\ 0 \leq \lambda < \frac{\pi}{2\sqrt{2T}}.$$

For $\pi/2\sqrt{2T} \leq \lambda < \pi/2T$, the variance is equal to infinity.
For $\alpha = -\lambda^2/2$:

$$Var\,Y_{0,0,1}(T) = \left(\frac{2\exp\sqrt{2}\lambda T}{1 + \exp 2\sqrt{2}\lambda T}\right)^{1/2} - \frac{2\exp\lambda T}{1 + \exp 2\lambda T}. \tag{3.5.11}$$

To reduce variance, let us use Theorem 2.3.1 with $s = 0$, $x = 0$. Note that $u(T, x) = 1$ and take $\alpha = \lambda^2/2$. Then, if (see (2.3.10))

$$\mu(t, x) = -\frac{1}{u}\frac{\partial u}{\partial x} = -\lambda x \tan\lambda(T - t),$$

the variable $Y_{0,0,1}(T)$ computed along the system

$$dX = \lambda(\tan\lambda(T - s))Xds + dw(s),\ X(0) = 0, \tag{3.5.12}$$

$$dY = \frac{\lambda^2}{2}X^2Yds - \lambda(\tan\lambda(T - s))XYdw(s),\ Y(0) = 1,$$

is deterministic.

For $\alpha = -\lambda^2/2$, the variable $Y_{0,0,1}(T)$ becomes deterministic as the solution of the following system:

$$dX = \lambda\frac{1 - \exp 2\lambda(T - s)}{1 + \exp 2\lambda(T - s)}X + dw(s),\ X(0) = 0, \tag{3.5.13}$$

$$dY = -\frac{\lambda^2}{2}X^2Yds - \lambda\frac{1 - \exp 2\lambda(T - s)}{1 + \exp 2\lambda(T - s)}XYdw(s),\ Y(0) = 1.$$

For completeness of exposition, we give the derivation of, e.g., (3.5.8). To this end we change variables in (3.5.5) for $\alpha = -\lambda^2/2$:

$$u = \exp v.$$

We get

$$\frac{\partial v}{\partial t} + \frac{1}{2}\left(\frac{\partial v}{\partial x}\right)^2 + \frac{1}{2}\frac{\partial^2 v}{\partial x^2} - \frac{\lambda^2}{2}x^2 = 0,\ v(T, x) = 0. \tag{3.5.14}$$

We look for a solution of the problem (3.5.14) in the form

$$v(t, x) = \frac{1}{2}p(t)x^2 + r(t), \quad p(T) = r(T) = 0.$$

For $p(t)$ and $r(t)$ we obtain the Cauchy problem

$$p' + p^2 - \lambda^2 = 0, \quad p(T) = 0,$$
$$r' + \frac{1}{2}p = 0, \quad r(T) = 0,$$

which solution leads to formula (3.5.8).

Example 3.5.1 In Table 3.5.1 we give the results of integrating the system (3.5.2) over the interval [0, 1] with initial data $X(0) = 0$, $Y(0) = 1$ for the α's indicated, by the method of first (M_I), second (M_{II}), and third (M_{III}) orders of accuracy constructed in Sect. 3.1 for systems with additive noise. The system (3.5.2) is a system with single noise. Therefore the methods of first (Euler's method) and second (see (3.1.2)) orders of accuracy require the modeling of one random variable per step. In our case, the method of third order of accuracy (see (3.1.34)) requires the modeling of two random variables per step. This is related to the fact that i and r in (3.1.34) can take only the unit value (since $q = 1$) and ζ_{1k}^2 can therefore be replaced by unit (see (3.1.33)). To obtain the results, the variable ξ is modeled (in all three methods) by the $\mathcal{N}(0, 1)$-distribution, and the variable ν is modeled by the law $P(\nu = \pm 1/\sqrt{12}) = 1/2$ (see (3.1.32)–(3.1.33)).

The numbers presented in Table 3.5.1 are approximations of $E\bar{Y}(1)$ computed by

$$E\bar{Y}(1) \simeq \frac{1}{M}\sum_{m=1}^{M} \bar{Y}^{(m)}(1) \tag{3.5.15}$$

$$\pm \frac{2}{M^{1/2}}\left[\frac{1}{M}\sum_{m=1}^{M}(\bar{Y}^{(m)}(1))^2 - \left(\frac{1}{M}\sum_{m=1}^{M}\bar{Y}^{(m)}(1)\right)^2\right]^{1/2},$$

i.e., under the natural assumption that the sample variance is sufficiently close to $Var\bar{Y}(1)$, the quantity $E\bar{Y}(1)$ lies between the given limits with probability 0.95. It is obvious that the true value of the required Wiener integral (3.5.1) at $T = 1$, which is equal to $EY(1)$, differs from $E\bar{Y}(1)$ by $O(h)$ for Euler's method, by $O(h^2)$ for the method (3.1.2), and by $O(h^3)$ for the method (3.1.34). It is also obvious that with increasing M the error of the Monte Carlo method reduces, and if $M \to \infty$, the difference between the tabulated values and the true value of $EY(1)$ tends to the error of the numerical integration. For $\alpha = -1$ and $h = 0.1, 0.2$, it can be seen from the table that this error for Euler's method is one-two units of the second position after the point, for the method of second order it is several units of the third position while for the method of third order it is even less. As α increases, the efficiency of the methods of higher order becomes even more evident.

Table 3.5.1 Computation of the Wiener integral I by methods from Sect. 3.1. The first part of the table: $\alpha = -1$, $T = 1$, $I = EY(1) \doteq 0.6776$; the second part: $\alpha = -0.5$, $T = 1$, $I = EY(1) \doteq 0.8050$; the third part: $\alpha = 0.5$, $T = 1$, $I = EY(1) \doteq 1.3604$; $\gamma := 2\sqrt{Var Y(1)/M}$

h	M	M_I	M_{II}	M_{III}	γ
0.2	100	0.6576 ± 0.0547	0.6316 ± 0.0494	0.6330 ± 0.0494	0.0475
0.1	100	0.6810 ± 0.0532	0.6688 ± 0.0510	0.6695 ± 0.0510	0.0475
0.01	100	0.6650 ± 0.0442	0.6634 ± 0.0441	0.6640 ± 0.0441	0.0475
0.2	10^4	0.6955 ± 0.0052	0.6713 ± 0.0049	0.6749 ± 0.0048	0.0048
0.1	10^4	0.6841 ± 0.0049	0.6743 ± 0.0048	0.6749 ± 0.0048	0.0048
0.2	10^5	0.6973 ± 0.0016	0.6733 ± 0.0015	0.6769 ± 0.0015	0.0015
0.2	100	0.8008 ± 0.0376	0.7734 ± 0.0369	0.7749 ± 0.0365	0.0344
0.1	100	0.8086 ± 0.0371	0.7961 ± 0.0375	0.7967 ± 0.0373	0.0344
0.01	100	0.8018 ± 0.0302	0.8007 ± 0.0329	0.8007 ± 0.0303	0.0344
0.2	10^4	0.8254 ± 0.0034	0.8011 ± 0.0035	0.8030 ± 0.0035	0.0034
0.1	10^4	0.8135 ± 0.0034	0.8028 ± 0.0035	0.8032 ± 0.0034	0.0034
0.2	100	1.2759 ± 0.0763	1.4222 ± 0.1455	1.4443 ± 0.1663	0.1651
0.1	100	1.3093 ± 0.0971	1.3928 ± 0.1352	1.3999 ± 0.1397	0.1651
0.01	100	1.3048 ± 0.0654	1.3105 ± 0.1342	1.3111 ± 0.1344	0.1651
0.2	10^4	1.2356 ± 0.0068	1.3453 ± 0.0124	1.3598 ± 0.0143	0.0165
0.1	10^4	1.2865 ± 0.0087	1.3524 ± 0.0126	1.3572 ± 0.0132	0.0165

The value $\gamma = (2/\sqrt{M})(Var Y(1))^{1/2}$ differs from the sample values of $(2/\sqrt{M}) \times (Var \bar{Y}(1))^{1/2}$, i.e., from the component in (3.5.15):

$$\frac{2}{M^{1/2}} \left[\frac{1}{M} \sum_{m=1}^{M} \left(\bar{Y}^{(m)}(1) \right)^2 - \left(\frac{1}{M} \sum_{m=1}^{M} \bar{Y}^{(m)}(1) \right)^2 \right]^{1/2},$$

first because of the estimation error, and secondly because of the numerical integration error. The components in the columns M_I, M_{II}, and M_{III} cannot become smaller as the order of accuracy increases. With an increase of the order of accuracy these components can become close to γ, because the numerical error increases. Moreover, they may also increase (see, e.g., the data for $\alpha = 0.5$).

We should also stress that the numbers given in the table include the numerical integration error, and therefore the indicated region of variation of them need not cover the true value, especially for methods of low order (see, e.g., the data M_I for $h = 0.2$, $N = 10000$).

Example 3.5.2 In Table 3.5.2 we give the results of integrating the system (3.5.2) by Euler's method (M_E) and the Runge–Kutta method of fourth order of accuracy (M_{R-K}). It can be seen from this table that, e.g., for $\alpha = 0.5$ the numerical integration error of Euler's method is more than one unit of the first position after the point, while in the Runge–Kutta method it moves to the unit of the third position. We draw

Table 3.5.2 Computation of the Wiener integral I by the Euler method and the Runge–Kutta method of fourth order

α	h	M	$EY(1)$	M_E	M_{R-K}	γ
-1	0.2	100	0.6776	0.7299 ± 0.0441	0.7025 ± 0.0436	0.0475
-1	0.2	10^4	0.6776	0.6987 ± 0.0051	0.6770 ± 0.0048	0.0048
-1	0.1	100	0.6776	0.7035 ± 0.0483	0.6883 ± 0.0477	0.0475
-1	0.1	10^4	0.6776	0.6876 ± 0.0048	0.6779 ± 0.0047	0.0048
-0.5	0.2	100	0.8050	0.8502 ± 0.0273	0.8253 ± 0.0295	0.0344
-0.5	0.2	10^4	0.8050	0.8279 ± 0.0034	0.8050 ± 0.0034	0.0034
-0.5	0.1	100	0.8050	0.8259 ± 0.0231	0.8122 ± 0.0339	0.0344
-0.5	0.1	10^4	0.8050	0.8160 ± 0.0034	0.8055 ± 0.0034	0.0034
0.5	0.2	100	1.3604	1.1864 ± 0.0429	1.2685 ± 0.0688	0.1651
0.5	0.2	10^4	1.3604	1.2297 ± 0.0065	1.3536 ± 0.0151	0.0165
0.5	0.1	100	1.3604	1.2611 ± 0.0762	1.3322 ± 0.1126	0.1651
0.5	0.1	10^4	1.3604	1.2824 ± 0.0089	1.3540 ± 0.0134	0.0165

attention to the fact that the sample variance for $\alpha = 0.5$ is less for the Euler method than for the Runge–Kutta method. As already noted in the previous example, the "large" variance of the Runge–Kutta method can in this case be explained by its greater accuracy.

Example 3.5.3 For relatively small ε, we compute the integral

$$I(\lambda, \varepsilon) = I = \int_{C_{0,0}^1} \exp\left[\frac{\lambda^2}{2} \int_0^T (x^2(s) + \varepsilon g(x(s)))ds\right] d\mu_{0,0}(x). \qquad (3.5.16)$$

To reduce the error of the Monte Carlo method, it is natural to integrate the system (see (3.5.12))

$$dX = p(s)Xds + dw(s), \quad X(0) = 0, \qquad (3.5.17)$$

$$dY = Y\frac{\lambda^2}{2}(X^2 + \varepsilon g(X))ds - p(s)XYdw(s), \quad Y(0) = 1,$$

where $p(s) = \lambda \tan \lambda(T - s)$, $T = 1$.

By Euler's method, we integrate the system (3.5.17) and the system

$$dX = dw(s), \quad X(0) = 0, \qquad (3.5.18)$$

$$dY = Y\frac{\lambda^2}{2}(X^2 + \varepsilon g(X))ds, \quad Y(0) = 1.$$

We obtain for $\lambda = 1$, $\varepsilon g(x) = 0.1x^2 \cos x$, $h = 0.0005$, $N = 100$:

for (3.5.17) we have $E\bar{Y}(1) = 1.3599 \pm 0.0053$,

for (3.5.18) we have $E\bar{Y}(1) = 1.3747 \pm 0.0630$.

We see that the error of the Monte Carlo method for the system (3.5.17) is approximately 10 times smaller than the error of the Monte Carlo method for the system (3.5.18).

Example 3.5.4 By the Cameron–Martin formula (see [238]), the value of the Wiener integral of the functional

$$F(x(\cdot)) = \exp\left(-\int_0^1 (x_1^2 + 2x_2^2 + 2x_3^2 + x_4^2 + x_1x_2 + x_2x_3 + x_3x_4)ds\right) \quad (3.5.19)$$

is equal to

$$I = \int_{C_{0,0}^4} F(x(\cdot))\mu_{0,0}(x) = \exp\left(\frac{1}{2}\int_0^1 tr\Gamma(s)ds\right),$$

where the matrix $\Gamma(s)$ can be found from the Cauchy problem

$$\frac{d\Gamma(s)}{ds} = 2Q - \Gamma^2(s), \quad Q = \begin{bmatrix} 1 & 1/2 & 0 & 0 \\ 1/2 & 2 & 1/2 & 0 \\ 0 & 1/2 & 2 & 1/2 \\ 0 & 0 & 1/2 & 1 \end{bmatrix}, \quad \Gamma(1) = 0.$$

Numerical integration of this system gives the above-mentioned Wiener integral: $I \simeq 0.1285$.

Computations of this integral via integration of the system (3.3.17) by the Euler and Runge–Kutta methods are presented in Table 3.5.3. As in Tables 3.5.1 and 3.5.2, along with the Monte Carlo error there arises the error $O(h)$ for the Euler method and $O(h^4)$ for the Runge–Kutta method. As can be seen from the table, this error is essential for the Euler method (see, e.g., the last row of Table 3.5.3, where the exact value 0.1285 of this integral is not covered by the values 0.1285 ± 0.0027), but it is not essential in the case of the Runge–Kutta method for the taken number of trajectories. Recall that the volume of calculations for the considered problems required by the Runge–Kutta method for given h and M is only twice as much as in the case of the Euler method.

Example 3.5.5 We consider the conditional Wiener integral (3.4.1), (3.4.12) with the function $f(t, x)$ of the form

$$f(t, x) = (A(t)x, x) + (a_1(t), x) + a_0(t), \quad (3.5.20)$$

Table 3.5.3 Computation of the multi-dimensional Wiener integral of the functional (3.5.19)

h	M	M_E	M_{R-K}
0.2	25	0.1142 ± 0.0790	0.1410 ± 0.0536
0.2	100	0.0616 ± 0.0503	0.1230 ± 0.0231
0.2	1000	0.0848 ± 0.0179	0.1785 ± 0.0711
0.1	100	0.0997 ± 0.0262	0.1094 ± 0.0239
0.1	1000	0.1126 ± 0.0089	0.1220 ± 0.0081
0.1	10^4	0.1166 ± 0.0029	0.1270 ± 0.0026
0.05	10^4	0.1215 ± 0.0027	0.1264 ± 0.0026

where $A(t)$ is a $d \times d$ symmetric matrix, $a_1(t)$ is a d-dimensional vector, and $a_0(t)$ is a scalar function.

Let $u(t, x)$ be the solution of (3.4.19) with f from (3.5.20). Introduce the function $P(t, x)$:

$$u(t, x) = \exp(P(t, x)). \qquad (3.5.21)$$

This function satisfies the problem

$$LP + (A(t)x, x) + (a_1(t), x) + a_0(t) + \frac{1}{2} \sum_{i=1}^{d} \left(\frac{\partial P}{\partial x^i}\right)^2 = 0, \quad x \in \mathbf{R}^d, \ t < T, \qquad (3.5.22)$$

$$P(T, x) = 0.$$

We look for a solution of (3.5.22) in the form

$$P(t, x) = \frac{1}{2}(P(t)x, x) + (p(t), x) + q(t), \qquad (3.5.23)$$

where $P(t)$ is a $d \times d$ symmetric matrix, $p(t)$ is a d-dimensional vector, and $q(t)$ is a scalar function.

Substituting (3.5.23) in (3.5.22) and collecting terms $(\cdot x, x)$, $(\cdot, x)$ and terms independent of x separately, we arrive at the system for $P(t)$, $p(t)$, and $q(t)$:

$$P'(t) - \frac{2}{T - t}P + 2A(t) + P^2(t) = 0, \quad P(T) = 0, \qquad (3.5.24)$$

$$p'(t) - \frac{1}{T - t}p + \frac{1}{T - t}P(t)b + P(t)p + a_1(t) = 0, \quad p(T) = 0, \qquad (3.5.25)$$

$$q'(t) + \frac{1}{T - t}(p(t), b) + \frac{1}{2}tr\, P(t) + \frac{1}{2}(p(t), p(t)) + a_0(t) = 0, \quad q(T) = 0. \qquad (3.5.26)$$

Note that if $a_1(t) \equiv 0$ and $b = 0$, then $p(t) \equiv 0$. And if in addition $a_0(t) \equiv 0$, then

$$q(t) = \frac{1}{2} \int_t^T tr P(s) \, ds \, .$$

The solution of (3.5.24) can be expanded in (positive) powers of $T - t$. If $A(t)$ is a constant matrix A, then this formal expansion starts with the terms:

$$P(t) = \frac{2}{3} A \cdot (T - t) + \frac{4}{45} A^2 \cdot (T - t)^3 + \cdots . \tag{3.5.27}$$

For test purposes, it is convenient to have an exact solution of (3.5.24)–(3.5.26) in a closed analytical form. To this end, we choose a variable matrix $A(t)$ such that

$$A(t) = A - \frac{2}{9} A^2 \cdot (T - t)^2, \tag{3.5.28}$$

where A is a constant symmetric matrix. Then the exact solution of the system (3.5.24)–(3.5.26) with $b = 0$, $a_0(t) \equiv 0$, and $a_1(t) \equiv 0$ has the form

$$P(t) = \frac{2}{3}(T - t)A, \quad p(t) = 0, \quad q(t) = \frac{(T - t)^2}{6} tr A \, . \tag{3.5.29}$$

Consequently, the solution of (3.5.23) is

$$P(t, x) = \frac{T - t}{3}(Ax, x) + \frac{(T - t)^2}{6} tr A \, . \tag{3.5.30}$$

Then the conditional Wiener integral (3.4.1), (3.4.12) for f from (3.5.20) with $a_0 = 0$, $a_1 = 0$, $A(t)$ from (3.5.28) and for $a = b = 0$ is equal to

$$\mathcal{J} = u(0, 0) = \exp\left(\frac{T^2}{6} tr A\right) \, .$$

In our experiments we take the dimension $d = 4$ and the following matrix A :

$$A = \begin{bmatrix} -1 & -0.5 & 0 & 0 \\ -0.5 & 2 & -0.5 & 0 \\ 0 & -0.5 & -2 & -0.5 \\ 0 & 0 & -0.5 & 1 \end{bmatrix}, \tag{3.5.31}$$

for which $tr A = 0$.

In Table 3.5.4 we give results of simulation of the conditional Wiener integral (3.4.1), (3.4.12) for f from (3.5.20) with $a_0 = 0$, $a_1 = 0$, $A(t)$ from (3.5.28), (3.5.31) and for $a = b = 0$, $T = 1$ by the explicit Runge–Kutta method (3.4.15)–(3.4.16) and

Table 3.5.4 The results of simulation of the conditional Wiener integral (3.4.1)–(3.4.12) for f from (3.5.20) with $a_0 = 0$, $a_1 = 0$, $A(t)$ from (3.5.28), (3.5.31) and for $a = b = 0$, $T = 1$ by the explicit Runge–Kutta method (3.4.15)–(3.4.16) and the implicit Runge–Kutta methods (3.4.40), (3.4.42) and (3.4.44), (3.4.46). The exact solution is 1

h	M	Eqs. (3.4.15)–(3.4.16)	Eqs. (3.4.40), (3.4.42)	Eqs. (3.4.44), (3.4.46)
0.2	10^6	0.9994 ± 0.0013	1.0176 ± 0.0044	1.0040 ± 0.0013
0.1	10^8	1.00002 ± 0.00013	1.00361 ± 0.00015	1.00093 ± 0.00013
0.05	10^8	0.99996 ± 0.00013	1.00089 ± 0.00013	1.00019 ± 0.00013

Table 3.5.5 The results of simulation of the conditional Wiener integral (3.4.1)–(3.4.12) for f from (3.5.20) with $a_0 = 0$, $a_1 = 0$, $A(t)$ from (3.5.28), (3.5.31) and for $a = b = 0$, $T = 1$ by the explicit Runge–Kutta method (3.4.15)–(3.4.16) and the implicit Runge–Kutta methods (3.4.40), (3.4.42) and (3.4.44), (3.4.46) using the variance reduction technique. The exact solution is 1

h	M	Eqs. (3.4.15)–(3.4.16)	Eqs. (3.4.40), (3.4.42)	Eqs. (3.4.44), (3.4.46)
0.1	10^7	0.99977 ± 0.00024	1.00396 ± 0.00050	1.00103 ± 0.00023
0.05	10^7	0.99992 ± 0.00017	1.00098 ± 0.00017	1.00023 ± 0.00016
0.05	10^8	0.99999 ± 0.00005	1.00088 ± 0.00005	1.00027 ± 0.00005
0.01	10^7	1.00003 ± 0.00007	1.00001 ± 0.00007	1.00003 ± 0.00007

the implicit Runge–Kutta methods (3.4.40), (3.4.42) and (3.4.44), (3.4.46). We have two types of errors in numerical simulations here: the error of a method used and the Monte Carlo error. The results in the table are approximations of $E\bar{Y}(1)$ calculated as in (3.5.15). Note that the "$\pm$" reflects the Monte Carlo error only and it does not reflect the error of a method. The results obtained are in agreement with the proved convergence theorems (see also Table 3.5.5). Recall that the implicit methods (3.4.40)–(3.4.41) and (3.4.44)–(3.4.45) are both of order two. In our tests the method (3.4.44)–(3.4.45) performs better. Apparently, this is due to the fact that the constant K in (3.4.47) is, in general, less than its counterpart in (3.4.43).

We also note that for the considered test problem we do not have any numerical instabilities and the explicit method is computationally effective. As has been discussed at the beginning of Sect. 3.4.2, implicit methods should be used in practice when explicit methods are affected by instabilities.

Example 3.5.6 To reduce the Monte Carlo error, variance reduction techniques from Sect. 2.3 can be used. In the case of evaluating Wiener integrals (3.4.1)–(3.4.12) application of method of important sampling changes the linear system (3.4.5) for X to a system with, in general, a nonlinear drift. As a result, we lose the advantage of simulating $X(t)$ exactly and of approximating the conditional Wiener integral by higher-order numerical integrators from Sects. 3.4.1 and 3.4.2. This shortcoming does not arise in the case of the method of control variates from Sect. 2.3.2. That is why, we restrict ourselves here to this method only.

In connection with the evaluation of the Wiener integral (3.4.1), (3.4.12) consider the following system of Ito SDEs (cf. (3.4.5)–(3.4.13)):

$$dX = \frac{b - X}{T - t} dt + dw(t), \quad X(s) = x, \qquad (3.5.32)$$

$$dY = f(t, X(t)) Y \, dt, \quad Y(s) = y, \qquad (3.5.33)$$

$$dZ = G^{\top}(t, X) Y \, dw(t), \quad Z(s) = z. \qquad (3.5.34)$$

Here Z is a scalar and $G(t, x)$ is a column-vector of dimension d with good analytical properties, the other notation is the same as it was in this section before.

It is clear that

$$u(s, x) = EY_{s,x,1}(T) = E\left[Y_{s,x,1}(T) + Z_{s,x,1,0}(T)\right].$$

Theorem 2.3.3 implies that by choosing $G(t, x)$ as

$$G^i = -\frac{\partial u}{\partial x^i}, \quad i = 1, \ldots, d, \qquad (3.5.35)$$

we obtain that the variance of $Y_{s,x,1}(T) + Z_{s,x,1,0}(T)$ is equal to zero.

Applying a numerical method to (3.5.32)–(3.5.34), we get the approximate $\bar{Y}_{s,x,1}(T)$ and $\bar{Z}_{s,x,1,0}(T)$. The variance $Var\left[\bar{Y}_{s,x,1}(T) + \bar{Z}_{s,x,1,0}(T)\right]$ is close to $Var\left[Y_{s,x,1}(T) + Z_{s,x,1,0}(T)\right]$, i.e., it is small in the case of G from (3.5.35), and, consequently, a smaller number of independent realizations M is needed to have a satisfactory accuracy.

For f from (3.5.20) with $a_0 = 0$, $a_1 = 0$, $A(t)$ from (3.5.28), (3.5.31) and for $b = 0$, the solution $u(t, x)$ of (3.4.19) has the form (3.5.21), (3.5.30). Therefore, in this case the vector function G defined in (3.5.35) is equal to

$$G^i(t, x) = -\frac{2}{3}(T - t) \exp(P(t, x)) \sum_{j=1}^{d} A^{ij} x^j, \quad i = 1, \ldots, d, \qquad (3.5.36)$$

where $P(t, x)$ is from (3.5.30) and A is from (3.5.31).

Applying the Euler method to Eq. (3.5.34), we get

$$Z_0 = 0, \qquad (3.5.37)$$
$$Z_{k+1} = Z_k + G^{\top}(t_k, X) Y_k \, \Delta w_k, \quad k = 1, \ldots, N - 1.$$

If we approximate (3.5.32)–(3.5.33) using the explicit fourth-order Runge–Kutta method (3.4.15)–(3.4.16), then Y_k in (3.5.37) is from (3.4.16) and the Wiener increment is

$$\Delta w_k := w(t_{k+1}) - w(t_k) = \frac{h^{1/2}}{\sqrt{2}} \left(\xi_{k+1/2} + \xi_{k+1}\right),$$

where $\xi_{k+1/2}$ and ξ_{k+1} are the same as in (3.4.15)–(3.4.16).

It is clear that $EZ_{k+1} = 0$. This implies that the method (3.4.15)–(3.4.16), (3.5.37) applying to (3.5.32)–(3.5.34) to approximate the Wiener integral $\mathcal{J} = EY(T)$ is of order four, i.e., the above realization of the variance reduction technique does not affect the accuracy of the numerical method. The variance $Var\ Y(T)$ is approximated with accuracy $O(h)$. Consequently, for a fixed number of realizations M the Monte Carlo error in simulations using the variance reduction technique is $\sim 1/\sqrt{h}$ times less than in simulations without variance reduction. In other words, in the case of variance reduction the Monte Carlo error is proportional to $\sqrt{h}/\sqrt{M}$. This is illustrated in Table 3.5.5. In particular, we see for $h = 0.05$ that to produce results of the same quality we need $M = 10^8$ independent trajectories without variance reduction and $M = 10^7$ independent realizations in the variance reduction case (compare Tables 3.5.4 and 3.5.5).

Example 3.5.7 Of course, in practice the solution $u(t, x)$ is not known. However, an approximate solution $\tilde{u}$ to the problem (3.4.19) can be known. In this case we can take $G(t, x)$ in the form of (3.5.35) with $\tilde{u}$ instead of u and we may expect a variance reduction. To illustrate this assertion, we take the function $f(t, x)$ in the form (3.5.20) with the constant matrix $A(t) \equiv A$ from (3.5.31) and $a_0 = 0$, $a_1 = 0$. We also put $b = 0$. In this case we do not know the exact solution $u(t, x)$ of (3.4.19). But for the variance reduction we can use an approximation $\tilde{u}(t, x)$ of the solution based on the formal expansion (3.5.27):

$$\tilde{u}(t, x) = \exp\left(\frac{1}{2}\left(\tilde{P}(t)x, x\right)\right),\tag{3.5.38}$$

where

$$\tilde{P}(t) = \frac{2}{3}A \cdot (T - t).$$

Deriving (3.5.38), we take into account that $tr\ \tilde{P}(t) = 0$ because of the specific choice of the matrix A which is from (3.5.31).

Then we take the function G in (3.5.34) of the form

$$G^i(t, x) = -\frac{\partial \tilde{u}}{\partial x^i}, \quad i = 1, \dots, d.$$

Putting $a = 0$ and $T = 1$, we evaluate the corresponding conditional Wiener integral (3.4.1)–(3.4.12) by the fourth-order explicit Runge–Kutta method (3.4.15)–(3.4.16) with time step $h = 0.01$ and we simulate $M = 10^5$ independent realizations. Without variance reduction, we get: $\mathcal{J} \doteq 1.1536 \pm 0.0093$, while applying the variance reduction technique (i.e., using the method (3.4.15)–(3.4.16), (3.5.37) for (3.5.32)–(3.5.34)) we obtain $\mathcal{J} \doteq 1.1482 \pm 0.0018$. We see that the Monte Carlo error is 5 times less when we use the variance reduction technique.

Table 3.5.6 *Square integral of the Brownian bridge.* Errors in evaluating the conditional Wiener integral (3.4.1), (3.5.39) with $p = 1$ and $p = 4$ and various time steps h. M is the number of Monte Carlo runs

h	M	$p = 1$	$p = 4$
0.20	1×10^9	$6.66 \times 10^{-3} \pm 0.01 \times 10^{-3}$	$-1.3 \times 10^{-3} \pm 0.012 \times 10^{-3}$
0.10	1×10^9	$1.67 \times 10^{-3} \pm 0.009 \times 10^{-3}$	$-0.32 \times 10^{-3} \pm 0.011 \times 10^{-3}$
0.05	5×10^{10}	$0.417 \times 10^{-3} \pm 0.001 \times 10^{-3}$	$-0.080 \times 10^{-3} \pm 0.002 \times 10^{-3}$
0.02	5×10^{10}	$0.067 \times 10^{-3} \pm 0.001 \times 10^{-3}$	$-0.015 \times 10^{-3} \pm 0.002 \times 10^{-3}$

Example 3.5.8 We consider the square integral of the Brownian bridge which has applications in statistics. To test the proposed method, we compute moments of this integral; i.e., we deal with the functionals

$$F(x(\cdot)) = \left(\int_0^1 x^2(t) \, dt \right)^p , \quad p \geq 0, \ x \in C_{0,0;1,0} . \tag{3.5.39}$$

The results of our simulation are presented in Table 3.5.6. The values before "±" are the differences between the exact value of the Wiener integral $\mathcal{J}$ (see (3.4.1)) with $F(x(\cdot))$ from (3.5.39) and its sampled approximations. The reference values for $\mathcal{J}$ are $1/6$ for $p = 1$ and 0.0166799 for $p = 4$ [441]. The values after "±" reflect the Monte Carlo error only; they correspond to the confidence interval for the corresponding estimator with probability 0.95. One can observe convergence with order two that is in good agreement with our theoretical results (Theorem 3.4.9).

Example 3.5.9 Consider the correlation function $\Gamma(\theta)$, $0 \leq \theta \leq T$ (see (3.4.4)):

$$\Gamma(\theta) = < x(0)x(\theta) > \tag{3.5.40}$$

$$= \frac{1}{\mathcal{Z}(T)} \int_{-\infty}^{\infty} \int_{C_{0,y;T,y}} x(0) \, x(\theta) \exp\left(-\int_0^T V(t, x(t)) \, dt \right) d\mu_{0,y}^{T,y}(x) dy$$

$$= \frac{\int_{-\infty}^{\infty} y \mathcal{J}_1(y) dy}{\int_{-\infty}^{\infty} \mathcal{J}_2(y) dy},$$

where

$$\mathcal{J}_1(y) = \int\limits_{C_{0,y;T,y}} x(\theta) \exp\left(-\int\limits_0^T V(t, x(t))\, dt\right) d\mu_{0,y}^{T,y}(x),$$ (3.5.41)

$$\mathcal{J}_2(y) = \int\limits_{C_{0,y;T,y}} \exp\left[-\int\limits_0^T V(x(t))\, dt\right] d\mu_{0,y}^{T,y}(x).$$ (3.5.42)

We evaluate (3.5.40)–(3.5.42) for the harmonic potential

$$V(x) = \frac{\omega^2}{2} x^2$$ (3.5.43)

and for the anharmonic potential

$$V(x) = \frac{\omega^2}{2} x^4.$$ (3.5.44)

We recall (see, e.g. [200, 240]) that T has the meaning of inverse temperature here. In the case of the harmonic potential (3.5.43), the correlation function is equal to [200, Chap. 3]:

$$\Gamma(\theta) = \frac{1}{2\omega} \frac{\cosh \omega(\theta - T/2)}{\sinh(\omega T/2)}, \quad 0 \le \theta \le T.$$ (3.5.45)

We rewrite the integrals in (3.5.40) as

$$\mathcal{G} = \int\limits_{-\infty}^{\infty} y \mathcal{J}_1(y)\, dy = \sqrt{2\pi\sigma_1^2}\, E\left[\eta_1 \mathcal{J}_1(\eta_1) \exp\left(\frac{\eta_1}{2\sigma_1^2}\right)\right],$$ (3.5.46)

$$\mathcal{Z} = \int\limits_{-\infty}^{\infty} \mathcal{J}_2(y)\, dy = \sqrt{2\pi\sigma_2^2}\, E\left[\mathcal{J}_2(\eta_2) \exp\left(\frac{\eta_2}{2\sigma_2^2}\right)\right],$$

where η_1 and η_2 are Gaussian random variables, $\mathcal{N}(0, \sigma_1^2)$ and $\mathcal{N}(0, \sigma_2^2)$, with zero mean and variances σ_1^2 and σ_2^2, respectively. The parameters σ_1^2 and σ_2^2 are chosen so that the variances of the random variables under the expectations in (3.5.46) are small.

The following estimators for $\mathcal{G}$ and $\mathcal{Z}$ are used in our simulation

$$\widehat{\mathcal{G}} = \frac{\sqrt{2\pi\sigma_1^2}}{M} \sum_{m=1}^M \left[{}_m\eta_1 \, {}_m\bar{\mathcal{J}}_1({}_m\eta_1) \exp\left(\frac{{}_m\eta_1}{2\sigma_1^2}\right)\right],$$ (3.5.47)

$$\widehat{\mathcal{Z}} = \frac{\sqrt{2\pi\sigma_2^2}}{M} \sum_{m=1}^M \left[{}_m\bar{\mathcal{J}}_2({}_m\eta_2) \exp\left(\frac{{}_m\eta_2}{2\sigma_2^2}\right)\right],$$

where $_m\eta_1$ and $_m\eta_2$ are sampled from $\mathcal{N}(0, \sigma_1^2)$ and $\mathcal{N}(0, \sigma_2^2)$, respectively, so that the pairs $(_m\eta_1, \; _m\eta_2)$ are independent while $_m\eta_1$ and $_m\eta_2$ in the same pair are dependent: $_m\eta_2 = \sigma_2 \, _m\eta_1/\sigma_1$; $_m\bar{\mathcal{J}}_1(_m\eta_1)$ and $_m\bar{\mathcal{J}}_2(_m\eta_2)$ are values of the corresponding functionals evaluated along a path according to the method (3.4.50); the pairs $(_m\bar{\mathcal{J}}_1(_m\eta_1), \; _m\bar{\mathcal{J}}_2(_m\eta_2))$ are simulated along independent paths while $_m\bar{\mathcal{J}}_1(_m\eta_1)$, $_m\bar{\mathcal{J}}_2(_m\eta_2)$ in the same pair are evaluated along the same path. Recall (see Sect. 3.4.3.2) that a discretization of the time interval $[0, T]$ should be so that the point θ belongs to the set of discretization points $\{t_0, t_1, \ldots, t_N\}$.

The results of the experiment are presented in Tables 3.5.7 and 3.5.8 and on Fig. 3.5.1. The parameters σ_1 and σ_2 are taken 1.2 and 0.8, respectively. As before, in these tables the values before "$\pm$" are estimates of the bias, computed as the difference between the exact $\Gamma(\theta)$ and its sampled approximations, while the values after "$\pm$" give half of the size of the confidence interval for the corresponding estimator with probability 0.95. To compute the bias, the exact values $\Gamma(1) \doteq 0.1840098$ and $\Gamma(8) \doteq 0.0678385$ obtained from (3.5.45) were used. The number of Monte Carlo runs M is chosen here so that the Monte Carlo error is small in comparison with the bias. It is not difficult to see that the experiment illustrates second-order convergence of the method. We note that fitting Ch^2 to, e.g., the data of Table 3.5.7 yields $C \doteq 0.015$, with the maximum absolute value of the residuals being equal to 3×10^{-5}.

In Fig. 3.5.1 (left) the results of simulation of $\Gamma(\theta)$ with $h = 0.2$ are compared with the exact curve from (3.5.45). Thanks to the second-order of accuracy of the proposed numerical method, these curves visually coincide even for this relatively large time step. Figure 3.5.1 (right) demonstrates behavior of the correlation function in the case of the anharmonic potential (3.5.44). The presented curve is obtained with the time step $h = 0.2$ and it visually coincides with the one simulated with $h = 0.05$. These experiments give further confirmation of our theoretical results.

We note that in Examples 3.5.8 and 3.5.9 the second-order method (3.4.51), (3.4.50) and the Euler method (3.4.54), (3.4.53) coincide since in these examples the starting and ending points of Brownian bridge paths coincide.

A further numerical experiment on simulation of the kinetic energy of a bosonic system can be found in [84]. In that example the advantage of the method (3.4.51), (3.4.50) in comparison with the Euler method (which is in general of order one – see Theorem 3.4.11) is clearly seen.

Table 3.5.7 *Correlation function.* The error in evaluating the correlation function $\Gamma(\theta)$ from (3.5.40) in the case of the harmonic potential (3.5.43) with $\omega = 1$, $T = 10$ and $\theta = 1$

h	M	Error
0.250	10^9	$9.78 \times 10^{-4} \pm 0.72 \times 10^{-4}$
0.200	10^9	$6.18 \times 10^{-4} \pm 0.72 \times 10^{-4}$
0.125	10^{10}	$2.45 \times 10^{-4} \pm 0.23 \times 10^{-4}$
0.100	5×10^{10}	$1.46 \times 10^{-4} \pm 0.10 \times 10^{-4}$

Table 3.5.8 *Correlation function.* The error in evaluating the correlation function $\Gamma(\theta)$ from (3.5.40) in the case of the harmonic potential (3.5.43) with $\omega = 1$, $T = 10$, $\theta = 8$, and the number of Monte Carlo runs $M = 10^{11}$

h	Error
0.250	$1.688 \times 10^{-4} \pm 0.079 \times 10^{-4}$
0.200	$1.134 \times 10^{-4} \pm 0.079 \times 10^{-4}$
0.125	$0.331 \times 10^{-4} \pm 0.080 \times 10^{-4}$
0.100	$0.231 \times 10^{-4} \pm 0.080 \times 10^{-4}$

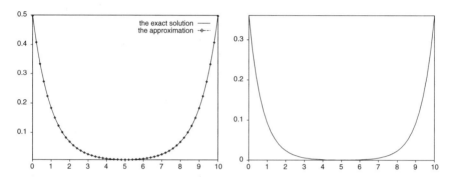

Fig. 3.5.1 *Correlation function.* The dependence of the correlation function $\Gamma(\theta)$ from (3.5.40) on θ simulated with $h = 0.2$ and $M = 10^8$ for $T = 10$. The left figure corresponds to the harmonic potential (3.5.43) and the right figure to the anharmonic potential (3.5.44), both with $\omega = 1$

3.6 Illustration from Financial Mathematics

From the applicable angle, especially beyond academia, SDEs are currently widely used in the three areas: financial engineering, molecular dynamics and Bayesian statistics. Molecular dynamics applications are considered in Chap. 5, which together with Chaps. 2 and 3 also serves as the stochastic numerics foundation for the SDEs' use in statistical applications. In this section we discuss financial engineering/financial mathematics applications. Financial mathematics puts different requirements on numerical approximation of SDEs than molecular dynamics and statistics applications. In the latter one needs to numerically solve very large systems (e.g. consisting of $10^5 - 10^6$ SDEs) over very long time intervals (see details in Chap. 5) and, hence, the need for stochastic geometric integrators arises. These applications typically do not require to do simulation in real time. In contrast financial mathematics applications usually deal with simulating relatively small dimensional systems and over relatively short time intervals, but they need to be delivered essentially in real time (e.g., calibration of models might need to be performed 50 or more times per day, large portfolios of various complex financial instruments need to be marked-to-market daily, etc.). Consequently, models routinely used in the financial sector are relatively simple (especially, in comparison with the ones used in molecular dynamics) and numerical methods for SDEs need to be very fast, including making use of variance reduction techniques (see Sects. 2.3 and 2.4) and MLMC and QMC methods (see Sect. 2.5).

The main aim of financial mathematics is to quantify and hedge risks in the financial world [31, 44, 45, 129, 189, 223, 248, 444]. In the case of pricing and hedging financial derivatives a generic SDEs setting is as follows.

Let $W_i(t)$, $i = 1, \ldots, q$, $0 \le t \le T$, be independent one-dimensional standard Wiener processes on a filtered probability space $(\Omega, \mathcal{F}, \{\mathcal{F}_t\}, Q)$. Here $\{\mathcal{F}_t\}$ is a filtration which is possibly larger than the natural filtration for the q-dimensional standard Wiener process $W(t) = (W_1(t), \ldots, W_q(t))^\mathsf{T}$, and Q is a risk-neutral measure. The market consists of d risky assets and one money market account. The model for the money market account (the current account) is the following differential equation

$$dB = r(t)Bdt, \quad B(0) = 1, \tag{3.6.1}$$

where $r(t)$ is a short rate (or in other words an instantaneous spot rate), which is assumed to be $\mathcal{F}_t$-adapted process with some good properties. The stock prices $S^i(t)$, $i = 1, \ldots, d$, $0 \le t \le T$, are modelled by

$$dS^i = r(t) S_i dt + S^i \sum_{j=1}^{q} \sigma^{ij}(t) \, dW_j(t), \quad i = 1, \ldots, d, \quad S(0) = S_0, \tag{3.6.2}$$

where the elements of volatility matrix $\sigma(t) = \{\sigma^{ij}(t), i = 1, \ldots, d, j = 1, \ldots, q\}$ are $\mathcal{F}_t$-adapted stochastic processes with some good properties.

For simplicity, let us illustrate option pricing in the case of European-type options which gives their holder the right to exercise an option at a fixed maturity time T and receive a payment $f(S(T))$. The arbitrage price of such an option at time $t = 0$ is equal to [31, 189, 223]:

$$u(0, S_0) = E\left[\frac{f(S(T))}{B(T)}\right]. \tag{3.6.3}$$

It is clear that weak-sense numerical methods for SDEs considered in Chap. 2 and in this chapter suit the task of computing (3.6.3).

We note that European-type barrier options can be exercised either at the maturity time T or when the asset prices $S(t)$, $0 \le t \le T$, reach a boundary of a domain (i.e., hit a barrier). In this situation the methods of Chap. 7 for solving Dirichlet PDE problems can be used for option pricing.

Examples of basic models from Financial Engineering are listed below for illustration.

Geometric Brownian motion:

$$dS = r S dt + \sigma S dW(t). \tag{3.6.4}$$

Ornstein–Uhlenbeck process/Vasicek model (see e.g. [31, 223]) for short rates:

$$dr = a [b - r] dt + \sigma d W(t), \ r(0) = r_0. \tag{3.6.5}$$

Cox–Ingersoll–Ross (CIR) model (see e.g. [31, 223]) for short rates:

$$dr = a [b - r] dt + \sigma \sqrt{r} d W(t), \ r(0) = r_0 > 0. \tag{3.6.6}$$

Heston stochastic volatility model [161, 236, 389]:

$$dS = r S dt + \sqrt{v} S d W_1(t), \tag{3.6.7}$$
$$dv = \kappa(\theta - v)dt + \sigma \sqrt{v} \left(\rho d W_1(t) + \sqrt{1 - \rho^2} d W_2(t) \right).$$

SABR stochastic volatility model [8, 153]:

$$dS = \sigma(t) S^\beta \left(\rho d W_1(t) + \sqrt{1 - \rho^2} d W_2(t) \right), \tag{3.6.8}$$
$$d\sigma = \alpha \sigma d W_1(t).$$

In (3.6.4)–(3.6.6) $W(t)$ is a scalar standard Wiener process. In (3.6.4) r and $\sigma > 0$ are constant; in (3.6.5)–(3.6.6) a, b and $\sigma > 0$ are constant. In (3.6.7) r, κ, θ, $\sigma > 0$ and $\rho \in [-1, 1]$ are constant and $W_1(t)$ and $W_2(t)$ are two independent Wiener processes. In (3.6.8) α, $\beta > 0$, and $\rho \in [-1, 1]$ are constant and $W_1(t)$ and $W_2(t)$ are two independent Wiener processes, here either it is assumed that the model is written under a forward measure or the short rate $r = 0$.

LIBOR market model: the LIBOR rates $L(t, T_{i-1})$, $1 \le i \le n, 0 \le t \le T_{i-1} \wedge T_k$ satisfy the SDEs:

$$dL(t, T_{i-1}) = L(t, T_{i-1})\sigma^\top(t, T_{i-1}) \tag{3.6.9}$$

$$\times \begin{cases} -\sum_{l=i}^{k-1} \frac{\delta L(t, T_l)}{\delta L(t, T_l)+1}\sigma(t, T_l)dt + dW(t), \ i < k, \\ dW(t), \ i = k, \\ \sum_{l=k}^{i-1} \frac{\delta L(t, T_l)}{\delta L(t, T_l)+1}\sigma(t, T_l)dt + dW(t), \ i > k, \end{cases}$$

where $W(t)$ is a q-dimensional standard Wiener process written under a corresponding forward measure and the elements of volatility matrices $\sigma(t, T_l) = \{\sigma^{ij}(t, T_l)$, $i = 1, \ldots, n, \ j = 1, \ldots, q\}, l = 0, \ldots, n - 1$, are stochastic processes with some good properties (see details e.g. in [31, 44]).

From the risk management prospective, it is interesting not only to compute prices of options but also their sensitivities or Greeks, which is the topic of the rest of this section that is mainly based on [315].

3.6.1   Evaluation of Greeks

Let us consider a version of the model (3.6.1)–(3.6.2). Here $w(t) = (w_1(t), \dots, w_d(t))^\mathsf{T}$ is a d-dimensional standard Wiener processes on a filtered probability space $(\Omega, \mathcal{F}, \{\mathcal{F}_t\}, Q)$ with $\{\mathcal{F}_t\}$ being the natural filtration for the Wiener process $w(t)$, and Q is a risk-neutral measure. The market model consists of a saving account $B(t)$ (riskless asset) and price processes $S^i(t)$, $i = 1, \dots, d$, (risky assets) satisfying the SDEs

$$dB = r(t, S)B dt, \quad dS^i = b^i(t, S)dt + \sum_{j=1}^{d} \sigma^{ij}(t, S)dw_j(t). \qquad (3.6.10)$$

We note the interpretation of the vector $b(t, x) = (b^1(t, x), \dots, b^d(t, x))^\mathsf{T}$: $b^i(t, S) = r(t, S(t))S^i - r^i(t, S(t))S^i$, where $r^i(t, S(t))$ is a rate of the i-th stock's dividend paid to the share holders. Dependence of the short rate r on S is included for the sake of generality. It is assumed that the interest rate $r(t, x)$, the components of the vector $b(t, x)$ and of the matrix $\sigma(t, x) = \{\sigma^{ij}(t, x)\}$, $t \in [t_0, T]$, $x \in \mathbf{R}_+^d := \{x : x^1 > 0, \dots, x^d > 0\}$, are deterministic functions such that for any $x \in \mathbf{R}_+^d$ the solution $S_{t_0, x}(t)$ of (3.6.10) starting at t_0 from x exists for all $t_0 \le t \le T$ and belongs to $\mathbf{R}_+^d$. Moreover, we assume that the volatility matrix $\sigma(t, x)$ has the full rank for every (t, x), $t \in [t_0, T]$, $x \in \mathbf{R}_+^d$. Under these assumptions the model (B, S) constitutes a complete market [189].

We consider the problem of pricing and hedging a European claim at a maturity time T, specified by a payoff function f which depends on S, by constructing a self-financing portfolio or trading strategy. We allow for a consumption process which is defined by a consumption rate $c(t, S(t))$. The portfolio value $U(t)$ of a trading strategy $(\varphi_t, \psi_t) = (\varphi_t, \psi_t^1, \dots, \psi_t^d)$, where φ_t and ψ_t^j denote the portfolio positions in the saving account $B(t)$ and stocks $S^j(t)$ respectively, is given by

$$U(t) = \varphi_t B(t) + \sum_{j=1}^{d} \psi_t^j S^j(t). \qquad (3.6.11)$$

Further, a trading strategy is self-financing if

$$dU = \varphi_t dB + \sum_{j=1}^{d} \psi_t^j dS^j + \sum_{j=1}^{d} r^j(t, S(t))\psi_t^j S^j(t)dt - c(t, S(t))dt. \qquad (3.6.12)$$

It is well-known (see, e.g., [189]) that in this framework any European claim can be hedged by a uniquely determined self-financing trading strategy. Its value is given by $U(t) = u(t, S(t))$, where $u(t, x)$ satisfies the following Cauchy problem for the PDE:

$$\frac{\partial u}{\partial t} + \frac{1}{2} \sum_{i,j=1}^{d} a^{ij}(t, x)\frac{\partial^2 u}{\partial x^i \partial x^j} + \sum_{i=1}^{d} b^i(t, x)\frac{\partial u}{\partial x^i} - r(t, x)u + c(t, x) = 0,$$

$$(3.6.13)$$

$$t_0 \leq t < T,$$

$$u(T, x) = f(x),$$
$$(3.6.14)$$

where $a^{ij} = \sum_{k=1}^{d} \sigma^{ik}\sigma^{jk}$, $i, j = 1, \ldots, d$.

If $u(t, x)$ is the solution to the problem (3.6.13)–(3.6.14), then the required hedging strategy $(\varphi_t, \psi_t^1, \ldots, \psi_t^d)$ as a function of $(t, S(t))$ is given by

$$\varphi_t = \frac{1}{B(t)}u(t, S(t)) - \sum_{i=1}^{d} \frac{\partial u}{\partial x^i}(t, S(t))S^i(t),$$
$$(3.6.15)$$

$$\psi_t^i = \frac{\partial u}{\partial x^i}(t, S(t)), \quad i = 1, \ldots, d.$$

Thus, to evaluate a hedging strategy for a European claim depending on values of stocks (risky assets) at maturity time, we have to know at every moment the solution of the Cauchy problem for a parabolic equation (the value of the hedging portfolio) and its derivatives (deltas).

For simplicity, we consider the problem (3.6.13)–(3.6.14) in $\mathbf{R}^d$ (not in $\mathbf{R}_+^d$). However, the results obtained for $\mathbf{R}^d$ can be carried over to the case of $\mathbf{R}_+^d$. Besides, we can use a suitable transformation of $\mathbf{R}_+^d$ into $\mathbf{R}^d$, e.g., $X^i = \ln S^i$, $i = 1, \ldots, d$. In the rest of this section we will consider (3.6.13)–(3.6.14) in $\mathbf{R}^d$ and we will write its probabilistic representations with SDEs for $X(t)$ instead of $S(t)$, which should not cause any confusion.

In the rest of this subsection we will restrict ourselves to the particular case of the Cauchy problem (3.6.13)–(3.6.14):

$$\frac{\partial u}{\partial t} + \frac{1}{2} \sum_{i,j=1}^{d} a^{ij}(t, x)\frac{\partial^2 u}{\partial x^i \partial x^j} + \sum_{i=1}^{d} b^i(t, x)\frac{\partial u}{\partial x^i} = 0, \ t_0 \leq t < T, \ x \in \mathbf{R}^d,$$

$$(3.6.16)$$
$$u(T, x) = f(x).$$
$$(3.6.17)$$

The solution of (3.6.16)–(3.6.17) has the following probabilistic representation:

$$u(t, x) = Ef(X_{t,x}(T)),$$
$$(3.6.18)$$

where $X_{t,x}(s)$ is the solution of the Cauchy problem for the system

$$dX = b(s, X)ds + \sigma(s, X)\,dw(s), \ X(t) = x.$$
$$(3.6.19)$$

For differentiable functions f, a probabilistic representation for $\dfrac{\partial u}{\partial x^k}(t, x)$ can be obtained by straightforward differentiation of (3.6.18) (see [46, 293]):

$$\frac{\partial u}{\partial x^k}(t, x) = E\left[\sum_{i=1}^{d} \frac{\partial f}{\partial x^i}(X_{t,x}(T))\, \delta_k X^i(T)\right], \qquad (3.6.20)$$

where $\delta_k X^i(s) := \partial X_{t,x}^i(s)/\partial x^k$, $t \leq s \leq T$, satisfies the system of variational equations associated with (3.6.19):

$$d\delta_k X = \sum_{i=1}^{d} \frac{\partial b(s, X)}{\partial x^i} \delta_k X^i ds + \sum_{i=1}^{d} \frac{\partial \sigma(s, X)}{\partial x^i} \delta_k X^i\, dw(s), \qquad (3.6.21)$$

$$\delta_k X^i(t) = 0, \text{ if } i \neq k, \text{ and } \delta_k X^k(t) = 1.$$

If the problem under consideration depends on a parameter α, i.e., $X = X_{t,x}(s; \alpha)$, $u = u(t, x, \alpha)$, then it is possible to find $\partial u(t, x, \alpha)/\partial \alpha$ in the same way provided the derivative $\delta_\alpha X(s) := \partial X_{t,x}(s; \alpha)/\partial \alpha$ exists. However, if this derivative does not exist, we cannot use the formula like (3.6.20). We are faced with such situation in the case of finding theta $\partial u(t, x)/\partial t$ (e.g., the problem $dX = dw(s)$, $X(t) = x$, $s \geq t$, has the solution $X_{t,x}(s) = x + w(s) - w(t)$ which is evidently nondifferentiable with respect to t). In [293], to evaluate theta, a system of linear parabolic equations is derived which belongs to a class of systems admitting probabilistic representations for their solutions [271]. Similar system can be derived for the problem of evaluation of deltas too. It is shown in [293] that the computational costs of this approach are comparable with those of the straightforward differentiation method (3.6.20)–(3.6.21).

When $f(x)$ is nondifferentiable or a European claim is specified by a more complicated payoff functional than the payoff function $f(X(T))$ at a maturity time T, the authors of [108, 109] proposed to use Malliavin calculus for numerical computation of Greeks. In particular, their approach is based on the integration-by-parts formula which gives

$$(T - t)\frac{\partial u}{\partial x^k}(t, x) = Ef(X_{t,x}(T)) \int_t^T [\sigma^{-1}(s, X_{t,x}(s))\delta_k X(s)]^{\mathsf{T}}\, dw(s). \qquad (3.6.22)$$

Both formulas (3.6.20) and (3.6.22) require computation of $\delta_k X(s)$, i.e., to evaluate deltas by these methods one has to integrate not only the d-dimensional system (3.6.19) but also d additional systems, each of dimension d. This can present severe computational difficulties. At the same time, there is a very simple method which makes use of calculating only the values of u to evaluate deltas. This method rests on the finite difference formula

$$\frac{\partial u}{\partial x^i} = \frac{u(t, x^1, \ldots, x^i + \Delta x, \ldots, x^d) - u(t, x^1, \ldots, x^i - \Delta x, \ldots, x^d)}{2\Delta x} \qquad (3.6.23)$$
$$+ O\left((\Delta x)^2\right).$$

Of course, in (3.6.23) we are forced to use the approximate values $\hat{u}(t, x^1, \ldots, x^i \pm \Delta x, \ldots, x^d)$ instead of $u(t, x^1, \ldots, x^i \pm \Delta x, \ldots, x^d)$:

$$\hat{u}(t, x^1, \ldots, x^i + \Delta x, \ldots, x^d) = \frac{1}{M} \sum_{m=1}^{M} f(\bar{X}_{t, x^1, \ldots, x^i + \Delta x, \ldots, x^d}^{(m)}(T)) \qquad (3.6.24)$$

$$\simeq \bar{u} = Ef(\bar{X}_{t, x^1, \ldots, x^i + \Delta x, \ldots, x^d}(T)) \simeq u = Ef(X_{t, x^1, \ldots, x^i + \Delta x, \ldots, x^d}(T)).$$

The same can be written for $\hat{u}(t, x^1, \ldots, x^i - \Delta x, \ldots, x^d)$. In (3.6.24), $\bar{X}(s)$, $t \le s \le T$, is an approximate solution of (3.6.19) obtained by a scheme of numerical integration, and $\bar{X}^{(m)}(T)$, $m = 1, \ldots, M$, are independent realizations of $\bar{X}(T)$.

There are two errors in (3.6.24): the error of numerical integration, say $O(h^p)$, and the statistical error of the Monte Carlo method which is estimated as $O(1/\sqrt{M})$, i.e.,

$$u \sim \hat{u} + O(h^p) + O(1/\sqrt{M}).$$

Therefore, the error R of the approximation

$$\frac{\partial u}{\partial x^i} \simeq \frac{\hat{u}(t, x^1, \ldots, x^i + \Delta x, \ldots, x^d) - \hat{u}(t, x^1, \ldots, x^i - \Delta x, \ldots, x^d)}{2\Delta x}$$

is evaluated, in general, by

$$R \sim O\left((\Delta x)^2\right) + O\left(\frac{h^p}{\Delta x}\right) + O\left(\frac{1}{\Delta x \sqrt{M}}\right).$$

Due to the presence of small Δx in the denominators, the difference approach seems to be not admissible. Fortunately, more accurate arguments and making use of dependent realizations in simulation of $\hat{u}(t, x^1, \ldots, x^i + \Delta x, \ldots, x^d)$ and $\hat{u}(t, x^1, \ldots, x^i - \Delta x, \ldots, x^d)$ rehabilitate the difference approach. We prove in the case of the weak Euler method ($p = 1$) that the error of numerical integration contributes not $O(h/\Delta x)$ but only $O(h) + O(h^2/\Delta x)$ to the total error of evaluation of the derivatives. This result is due to the Talay–Tubaro expansion of the error of numerical integration (see Sect. 2.2.3). Further, we prove that the method of dependent realizations, which is close to using common random numbers for Monte Carlo estimators (see [41, 129–131, 227]), contributes just $O(1/\sqrt{M})$ to the total error. Thus,

$$R \sim O\left((\Delta x)^2\right) + O(h) + O\left(\frac{h^2}{\Delta x}\right) + O\left(\frac{1}{\sqrt{M}}\right).$$

If we put $\Delta x = \alpha h^{\beta}$, $\alpha > 0$, $1/2 \le \beta \le 1$, then

$$R \sim O(h) + O(1/\sqrt{M}) .$$

Hence, we get the same convergence rate in evaluating deltas as in the evaluation of option prices.

We should note that this inference is obtained here in the case of simple payoff functions which depend on the underlying asset process X at a maturity time T only: $f = f(X(T))$. Most likely, the results obtained can be justified in the case when a contingent claim is defined as a functional of the asset process, for example, $f = f(X(t_1), \ldots, X(t_l))$, $t_0 < \cdots < t_l = T$.

In the one-dimensional case of (3.6.16)–(3.6.17) the estimators $\hat{u}(t, x)$ and $\hat{\partial}u(t, x)$ for the function u and its derivative have the form:

$$\hat{u}(t, x) = \frac{1}{M} \sum_{m=1}^{M} f(\bar{X}_{t,x}^{(m)}(T)), \tag{3.6.25}$$

$$\hat{\partial}u(t, x) = \frac{1}{2\alpha h^{1/2}} \frac{1}{M} \sum_{m=1}^{M} [f(\bar{X}_{t,x+\alpha h^{1/2}}^{(m)}(T)) - f(\bar{X}_{t,x-\alpha h^{1/2}}^{(m)}(T))]. \tag{3.6.26}$$

The properties of estimator $\hat{u}$ are very well investigated, e.g., in Chap. 2 of this book, and in application to evaluating security prices see, e.g., [82]. There are M pairs of approximate trajectories in (3.6.26), each pair consists of a trajectory starting from $x + \alpha h^{1/2}$ and a trajectory starting from $x - \alpha h^{1/2}$. The pairs are independent, but the two trajectories of the same pair are dependent: they correspond to the same realization (as a rule, in the weak sense) of the Wiener process.

In a more general setting estimators for deltas are given in Sect. 3.6.2. Sections 3.6.3 and 3.6.4 are devoted to rigorous proofs. In particular, it is proved in Sect. 3.6.3 that the bias of $\hat{\partial}u$ (together with the numerical differentiation error $O((\Delta x)^2)$) is $O(h)$ and in Sect. 3.6.4 that the variance of $\hat{\partial}u$ is $O(1/M)$. In Sect. 3.6.5 some extensions are considered. While Sects. 3.6.3 and 3.6.4 deal with the weak Euler scheme, the strong (mean-square) Euler scheme is considered in Sect. 3.6.5. Correspondingly, in Sects. 3.6.3 and 3.6.4 the difference approach is rigorously justified for sufficiently smooth payoff functions and in Sect. 3.6.5 it is justified for nonsmooth (and even discontinuous) payoff functions. For evaluation of other price sensitivities (Greeks) by finite differences, see [315]. Results of numerical experiments for estimators of deltas and other Greeks using finite differences including those with Heston stochastic volatility model are presented in [315]. For evaluating various Greeks for American and Bermudan type options, see e.g. [24].

3.6.2 Finite Difference Estimators for Deltas

In high dimensional cases (in practice for $d \geq 3$) it is usually impossible to find $u(t, x)$ for all (t, x) due to the complexity of the problem (3.6.13)–(3.6.14). Since for construction of a hedging strategy one only needs the individual values $u(t, X(t))$ and $\partial u(t, X(t))/\partial x^i$, $i = 1, \ldots, d$, at $(t, X(t))$, the known state of the market, Monte Carlo methods are most appropriate for such problems. The solution to the problem (3.6.13)–(3.6.14) has the following probabilistic representation (the Feynman–Kac formula):

$$u(t, x) = E[f(X_{t,x}(T))Y_{t,x,1}(T) + Z_{t,x,1,0}(T)], \qquad (3.6.27)$$

where $X_{t,x}(s)$, $Y_{t,x,y}(s)$, $Z_{t,x,y,z}(s)$, $s \geq t$, is the solution of the Cauchy problem to the system

$$\begin{aligned}
dX &= b(s, X)ds + \sigma(s, X)\,dw(s), \quad X(t) = x, \qquad (3.6.28)\\
dY &= -r(s, X)Y ds, \quad Y(t) = y,\\
dZ &= c(s, X)Y ds, \quad Z(t) = z.
\end{aligned}$$

In (3.6.28), Y and Z are scalars. We assume that all the coefficients of (3.6.28) have bounded derivatives up to some order.

As it is well-known (see Sects. 2.3 and 2.4), variance reduction is of crucial importance for the effectiveness of any Monte Carlo procedure. Variance reduction methods can be derived from a generalized probabilistic representation for $u(t, x)$ given by the formula (3.6.27) with X, Y, Z satisfying the system (cf. (2.3.21)–(2.3.23) in Sect. 2.3):

$$\begin{aligned}
dX &= b(s, X)ds - \sigma(s, X)\mu(s, X)ds + \sigma(s, X)\,dw(s), \quad X(t) = x, (3.6.29)\\
dY &= -r(s, X)Y ds + \mu^{\mathsf{T}}(s, X)Y\,dw(s), \quad Y(t) = y,\\
dZ &= c(s, X)Y ds + F^{\mathsf{T}}(s, X)Y\,dw(s), \quad Z(t) = z.
\end{aligned}$$

In (3.6.29), $\mu(s, x)$ and $F(s, x)$ are column-vectors of dimension d with good analytical properties (for example, they are bounded and have bounded derivatives up to some order) but arbitrary otherwise.

Along with suitable assumptions on smoothness and boundedness of the coefficients of (3.6.13)–(3.6.14) and (3.6.29), the uniform ellipticity condition is imposed on the matrix σ :

$$\exists\, \lambda > 0, \ y^{\mathsf{T}}\sigma(s, x)\sigma^{\mathsf{T}}(s, x)y \geq \lambda|y|^2, \ \text{for any } s \in [t_0, T], \ x, y \in \mathbf{R}^d. \quad (3.6.30)$$

Further, the function $f(x)$ and its derivatives up to some order are assumed to satisfy inequalities of the form (see Definition 2.1.1)

$$|f(x)| \leq K \cdot (1 + |x|^{\varkappa}), \qquad (3.6.31)$$

where K and $\varkappa$ are positive constants. Under these conditions, the solution $u(t, x)$ of (3.6.13)–(3.6.14) and its derivatives satisfy inequalities of the type (3.6.31) [114].

The aforementioned assumptions look rather restrictive (especially the assumption on smoothness of the payoff function), however the estimators proposed in this work do not directly use any assumed derivatives and can be exploited (at least formally) in a broad fashion. At any rate, if a payoff function is well approximated by a smooth function, the proposed numerical procedures can be applied to the smooth approximating function. Here we restrict ourselves to comparatively not too complicated proofs at the expense of the assumptions made. Some rigorous results are obtained in the case of nonsmooth (and even discontinuous) payoff functions in Sect. 3.6.5.

Let us proceed to construction of estimators for deltas. We approximate the solution $X_{t,x}(s)$, $Y_{t,x,1}(s)$, $Z_{t,x,1,0}(s)$ of (3.6.29) by a weak method, for instance by the weak Euler method (cf. (2.0.21)):

$$\bar{X}_{t,x}(s_{k+1}) := X_{k+1} = X_k + h(b(s_k, X_k) - \sigma(s_k, X_k)\mu(s_k, X_k)) \qquad (3.6.32)$$
$$+ h^{1/2}\sigma(s_k, X_k)\xi_k, \quad k = 0, \dots, N-1,$$
$$\bar{Y}_{t,x,1}(s_{k+1}) := Y_{k+1} = Y_k - hr(s_k, X_k)Y_k + h^{1/2}\mu^{\mathsf{T}}(s_k, X_k)Y_k\xi_k,$$
$$\bar{Z}_{t,x,1,0}(s_{k+1}) := Z_{k+1} = Z_k + hc(s_k, X_k)Y_k + h^{1/2}F^{\mathsf{T}}(s_k, X_k)Y_k\xi_k,$$

$$X_0 = x, \ Y_0 = 1, \ Z_0 = 0,$$

where h is a step of discretization of the time interval $[t, T]$:

$$s_k = t + kh, \quad k = 0, \dots, N, \quad s_N = T ;$$

$\xi_k = (\xi_k^1, \dots, \xi_k^d)^{\mathsf{T}}$ are d-dimensional vectors which components are i.i.d. random variables with the law $P(\xi_k^i = \pm 1) = 1/2$.

The scheme (3.6.32) has the first order of accuracy in the sense of weak approximation. Note that all the results can be carried over on the case of the mean-square Euler method (1.0.2).

For evaluating the price $u(t, x)$, we propose the estimator:

$$\hat{u}(t, x) = \frac{1}{M} \sum_{m=1}^{M} [f(\bar{X}_{t,x}^{(m)}(T))\bar{Y}_{t,x,1}^{(m)}(T) + \bar{Z}_{t,x,1,0}^{(m)}(T)], \qquad (3.6.33)$$

where $\bar{\Lambda}_{t,x}^{(m)}(T) := (\bar{X}_{t,x}^{(m)}(T), \bar{Y}_{t,x,1}^{(m)}(T), \bar{Z}_{t,x,1,0}^{(m)}(T))$, $m = 1, \dots, M$, are independent realizations of $\bar{\Lambda}_{t,x}(T) := (\bar{X}_{t,x}(T), \bar{Y}_{t,x,1}(T), \bar{Z}_{t,x,1,0}(T))$. Let us emphasize once again that the use of the generalized estimator is essential due to the possibility to reduce the Monte Carlo error at the expense of a suitable choice of μ and F.

For evaluating the deltas $\partial u(t, x)/\partial x^i$, $i = 1, \ldots, d$, we propose the estimators:

$$\hat{\partial}_i u(t, x) = \frac{1}{2\alpha h^{1/2}} \frac{1}{M} \sum_{m=1}^{M} \left[f(\bar{X}^{(m)}_{t,x+\alpha h^{1/2}e_i}(T)) \bar{Y}^{(m)}_{t,x+\alpha h^{1/2}e_i,1}(T) \right. \tag{3.6.34}$$

$$+ \bar{Z}^{(m)}_{t,x+\alpha h^{1/2}e_i,1,0}(T) - f(\bar{X}^{(m)}_{t,x-\alpha h^{1/2}e_i}(T)) \bar{Y}^{(m)}_{t,x-\alpha h^{1/2}e_i,1}(T)$$

$$\left. - \bar{Z}^{(m)}_{t,x-\alpha h^{1/2}e_i,1,0}(T) \right],$$

where α is a positive constant, e_i is the unit vector with i-th coordinate equal one and the other coordinates equal zero, the pairs $(\bar{\Lambda}^{(m)}_{t,x+\alpha h^{1/2}e_i}(T), \bar{\Lambda}^{(m)}_{t,x-\alpha h^{1/2}e_i}(T))$, $m = 1, \ldots, M$, are independent but the two realizations $\bar{\Lambda}^{(m)}_{t,x+\alpha h^{1/2}e_i}(T)$ and $\bar{\Lambda}^{(m)}_{t,x-\alpha h^{1/2}e_i}(T)$ of each pair correspond to the same sequence $\xi_k^{(m)}$, i.e., for each m these two realizations are dependent. In the next subsection we prove that the bias of this estimator $\hat{\partial}_i u$ is $O(h)$ and in Sect. 3.6.4 we prove that its variance is $O(1/M)$.

In [315] some alternative estimators were also constructed, the algorithm for which is organized as follows. Introduce the one-step approximation

$$\bar{X}_{t,x}(t + h) = \bar{X} = x + h(b(t, x) - \sigma(t, x)\mu(t, x)) + h^{1/2}\sigma(t, x)\xi. \tag{3.6.35}$$

Let us denote by $\xi_{(\gamma)} = \left(\xi^1_{(\gamma)}, \ldots, \xi^d_{(\gamma)} \right)^\top$, $\gamma = 1, \ldots, 2^d$, all the different values of the vector ξ from (3.6.35) and assign indices to these values so that $\xi_{(2^{d-1}+\iota)} = -\xi_{(\iota)}$, $\iota = 1, \ldots, 2^{d-1}$. For example, if $d = 3$, then $\xi_{(1)} = (1, 1, 1)^\top$, $\xi_{(5)} = -\xi_{(1)} = (-1, -1, -1)^\top$, $\xi_{(2)} = (1, 1, -1)^\top$, $\xi_{(6)} = -\xi_{(2)} = (-1, -1, 1)^\top$, $\xi_{(3)} = (1, -1, 1)^\top$, $\xi_{(7)} = -\xi_{(3)} = (-1, 1, -1)^\top$, $\xi_{(4)} = (-1, 1, 1)^\top$, $\xi_{(8)} = -\xi_{(4)} = (1, -1, -1)^\top$. We denote by $\bar{X}_{(\gamma)}$, $\gamma = 1, \ldots, 2^d$, the value of the vector $\bar{X}$ from (3.6.35) corresponding to $\xi_{(\gamma)}$. Note that $\bar{X}_{(\gamma)}$ and $\xi_{(\gamma)}$ are deterministic (not random) vectors. Let $M = 2^d \times L$. Now we use 2^d runs from the points $(t + h, \bar{X}_{(\gamma)})$, $\gamma = 1, \ldots, 2^d$, and each run consists of L trajectories only. Using these runs, the following estimators were proposed (we denote them as $\check{u} = \check{u}(t, x)$ and $\check{\partial}u = \check{\partial}u(t, x) = (\check{\partial}_1 u, \ldots, \check{\partial}_d u)^\top$) :

$$\check{u} = \frac{1}{2^d} \sum_{\gamma=1}^{2^d} (1 - r(t, x)h + \mu^\top(t, x)\xi_{(\gamma)}h^{1/2}) \tag{3.6.36}$$

$$\cdot \frac{1}{L} \sum_{l=1}^{L} \left[f(\bar{X}^{(l)}_{t+h,\bar{X}_{(\gamma)}}(T)) \bar{Y}^{(l)}_{t+h,\bar{X}_{(\gamma)},1}(T) \right.$$

$$\left. + \bar{Z}^{(l)}_{t+h,\bar{X}_{(\gamma)},1,0}(T) \right] + hc(t, x),$$

$$(\sigma^\mathsf{T}(t,x)\check{\partial}u)^i = \frac{1}{\sqrt{h}}\frac{1}{2^d}\sum_{\iota=1}^{2^{d-1}}\frac{1}{L}\sum_{l=1}^{L}\Big[f(\bar{X}^{(l)}_{t+h,\bar{X}_{(\iota)}}(T))\bar{Y}^{(l)}_{t+h,\bar{X}_{(\iota)},1}(T)$$

$$+\bar{Z}^{(l)}_{t+h,\bar{X}_{(\iota)},1,0}(T) - f(\bar{X}^{(l)}_{t+h,\bar{X}_{(2^{d-1}+\iota)}}(T))\bar{Y}^{(l)}_{t+h,\bar{X}_{(2^{d-1}+\iota)},1}(T)$$

$$-\bar{Z}^{(l)}_{t+h,\bar{X}_{(2^{d-1}+\iota)},1,0}(T)\Big]\xi^i_{(\iota)}, \tag{3.6.37}$$

$$i = 1,\ldots,d.$$

In (3.6.37) the pairs $\left(\bar{A}^{(l)}_{t+h,\bar{X}_{(\iota)}}(T),\bar{A}^{(l)}_{t+h,\bar{X}_{(2^{d-1}+\iota)}}(T)\right)$, $\iota = 1,\ldots,2^{d-1}$, $l = 1,\ldots,L$, are independent but the two realizations from the same pair are dependent, i.e., they correspond to the same realization of the Wiener process. As σ is a nonsingular matrix, the alternative estimators $\check{\partial}_i u(t,x)$, $i = 1,\ldots,d$, can be found from (3.6.37). A detailed analysis of errors for the estimators (3.6.36) and (3.6.37) is given in [315]. In particular, it was shown that biases and variances of these estimators are of the same order that the ones for the estimators (3.6.33) and (3.6.34).

3.6.3   Influence of the Error of Numerical Integration on Evaluating Deltas by Finite Differences

Our aim in this subsection is to answer the question: if in the finite difference (2.3.39) we substitute the approximation $\bar{u}$ from (3.6.38) for u from (3.6.27), how the error of numerical integration influences the error of the evaluation of the derivative.

Theorem 3.6.1 *Let*

$$\bar{u}(t,x) = E\left[f(\bar{X}_{t,x}(T))\bar{Y}_{t,x,1}(T) + \bar{Z}_{t,x,1,0}(T)\right] \tag{3.6.38}$$

with $\bar{X}_{t,x}(T), \bar{Y}_{t,x,1}(T), \bar{Z}_{t,x,1,0}(T)$ obtained by the weak Euler method (3.6.32), and α be a positive number. Then

$$\frac{\bar{u}(t,x+\alpha h^{1/2}v) - \bar{u}(t,x-\alpha h^{1/2}v)}{2\alpha h^{1/2}} - \frac{\partial u(t,x)}{\partial v} = O(h), \tag{3.6.39}$$

where v is a unit vector and

$$\frac{\partial u(t,x)}{\partial v} = \frac{\partial u(t,x+rv)}{\partial r}\bigg|_{r=0}$$

is the derivative of the solution to (3.6.13)–(3.6.14) in the direction of v.

Proof Here we restrict ourselves to the proof for the one-dimensional case, more precisely to the problem (3.6.13)–(3.6.14) with $d = 1, c = g = 0$ (cf. (3.6.16)–(3.6.17)):

$$Lu := \frac{\partial u}{\partial t} + b(t, x)\frac{\partial u}{\partial x} + \frac{\sigma^2(t, x)}{2}\frac{\partial^2 u}{\partial x^2} = 0, \qquad (3.6.40)$$

$$u(T, x) = f(x). \qquad (3.6.41)$$

The solution of this problem has the probabilistic representation (3.6.18)–(3.6.19).
We approximate the solution of (3.6.19) by the weak Euler method:

$$\bar{X}_{t,x}(s_{k+1}) := X_{k+1} = X_k + hb(s_k, X_k) + h^{1/2}\sigma(s_k, X_k)\xi_k, \quad k = 0, \ldots, N - 1, \qquad (3.6.42)$$

where ξ_k are i.i.d. random variables taking the values $+1$ and -1 with probability $1/2$ and $X_0 = x$.

Using (3.6.18) and (3.6.42), we can evaluate the solution $u(t, x)$ of (3.6.40)–(3.6.41) as follows

$$u(t, x) \doteq \bar{u}(t, x) = Ef(\bar{X}_{t,x}(T)). \qquad (3.6.43)$$

As is known (see Sect. 2.2.3), the error of this approximation can be expanded in powers of h :

$$\rho(t, x) := \bar{u}(t, x) - u(t, x) = hE\int_t^T B(\vartheta, X_{t,x}(\vartheta))\, d\vartheta + O(h^2), \qquad (3.6.44)$$

where $B(t, x)$ is determined by the coefficients of the problem (3.6.40)–(3.6.41) and does not depend on h :

$$B(t, x) = (L^2 u(t, x) - A(t, x))/2$$

with

$$A(t, x) = \frac{\partial^2 u}{\partial t^2} + 2b\frac{\partial^2 u}{\partial t \partial x} + \sigma^2\frac{\partial^3 u}{\partial t \partial x^2} + b^2\frac{\partial^2 u}{\partial x^2} + b\sigma^2\frac{\partial^3 u}{\partial x^3} + \frac{1}{12}\sigma^4\frac{\partial^4 u}{\partial x^4}.$$

We approximate the derivative $\dfrac{\partial u}{\partial x}(t, x)$ by the central finite difference:

$$\frac{\partial u}{\partial x}(t, x) = \frac{u(t, x + \Delta x) - u(t, x - \Delta x)}{2\Delta x} + O((\Delta x)^2). \qquad (3.6.45)$$

Replacing u in (3.6.45) by $\bar{u}$ from (3.6.43), we get

$$\frac{\partial u}{\partial x}(t, x) = \frac{u(t, x + \Delta x) - u(t, x - \Delta x)}{2\Delta x} + O((\Delta x)^2)$$

$$= \frac{\bar{u}(t, x + \Delta x) - \bar{u}(t, x - \Delta x)}{2\Delta x} + \frac{h}{2\Delta x}(v(t, x + \Delta x) - v(t, x - \Delta x))$$

$$+ O\left(\frac{h^2}{\Delta x}\right) + O((\Delta x)^2),$$

where

$$v(t, x) = E \int_t^T B(\vartheta, X_{t,x}(\vartheta))\, d\vartheta \tag{3.6.46}$$

which is a smooth function due to the assumptions made in Sect. 3.6.2. Then, expanding $v(t, x \pm \Delta x)$ around (t, x), we obtain

$$\frac{h}{2\Delta x}(v(t, x + \Delta x) - v(t, x - \Delta x)) = O(h) .$$

Therefore

$$\frac{\partial u}{\partial x}(t, x) = \frac{\bar{u}(t, x + \Delta x) - \bar{u}(t, x - \Delta x)}{2\Delta x} + O\left(h + \frac{h^2}{\Delta x} + (\Delta x)^2\right),$$

and if we select $\Delta x = \alpha h^{1/2}$, $\alpha > 0$, then

$$\frac{\partial u}{\partial x}(t, x) = \frac{\bar{u}(t, x + \alpha h^{1/2}) - \bar{u}(t, x - \alpha h^{1/2})}{2\alpha h^{1/2}} + O(h).$$

Thus, the theorem is proved for (3.6.13)–(3.6.14) with $d = 1$, $r = c = 0$. The proof in the general case is not much different from the presented one. □

Remark 3.6.2 Let $\bar{X}_{t,x}(T)$ be obtained by a second-order weak scheme from Sect. 2.2. Then we have the expansion (see Sect. 2.2.3):

$$\bar{u}(t, x) - u(t, x) = h^2 E \int_t^T B(\vartheta, X_{t,x}(\vartheta))\, d\vartheta + O(h^3),$$

(with another function $B(t, x)$, of course) and

$$\frac{\partial u}{\partial x}(t, x) = \frac{\bar{u}(t, x + \Delta x) - \bar{u}(t, x - \Delta x)}{2\Delta x} + O\left(h^2 + \frac{h^3}{\Delta x} + (\Delta x)^2\right) .$$

Taking $\Delta x = \alpha h$, $\alpha > 0$, we get

$$\frac{\partial u}{\partial x}(t, x) = \frac{\bar{u}(t, x + \alpha h) - \bar{u}(t, x - \alpha h)}{2\alpha h} + O(h^2) .$$

An analogous remark can be made for schemes of higher order.

3.6.4 Influence of the Monte Carlo Error on Evaluating Deltas by Finite Differences

It is again enough to restrict ourselves to the one-dimensional case (3.6.40)–(3.6.41). To realize the formula (3.6.43) (or (3.6.38)) in practice, we need to apply the Monte Carlo technique. As a result, in addition to the error of numerical integration considered in the previous section, there is also the Monte Carlo error:

$$\bar{u}(t, x) = E f(\bar{X}_{t,x}(T)) = \hat{u} + r_{\hat{u}} , \qquad (3.6.47)$$

$$\hat{u} = \frac{1}{M} \sum_{m=1}^{M} f(\bar{X}_{t,x}^{(m)}(T)) ,$$

where M is the number of independent realizations $\bar{X}_{t,x}^{(m)}(T)$ of $\bar{X}_{t,x}(T)$ and the Monte Carlo error $r_{\hat{u}}$ has the variance

$$Var(r_{\hat{u}}) = \frac{Var\, f(\bar{X}_{s,x}(T))}{M} = \frac{Var\, f(X_{s,x}(T))}{M} + O\left(\frac{h}{M}\right).$$

Thus, the total error

$$R_{\hat{u}} := \hat{u} - u$$

in the evaluation of u by $\hat{u}$ is estimated as

$$R_{\hat{u}} \sim O(h) + O\left(\frac{1}{\sqrt{M}}\right), \qquad (3.6.48)$$

which is of order $O(h)$ if we choose $M \sim 1/h^2$.

Due to Theorem 3.6.1, we have

$$\frac{\partial u}{\partial x}(t, x) = \frac{1}{2\alpha h^{1/2}} \left[E f(\bar{X}_{t,x+\alpha h^{1/2}}(T)) - E f(\bar{X}_{t,x-\alpha h^{1/2}}(T)) \right] \qquad (3.6.49)$$

$$+ O(h)$$

$$= \frac{1}{2\alpha h^{1/2}} \frac{1}{M} \left[\sum_{m=1}^{M} f(\bar{X}_{t,x+\alpha h^{1/2}}^{(m)}(T)) - \sum_{m=1}^{M} f(\bar{X}_{t,x-\alpha h^{1/2}}^{(m)}(T)) \right]$$

$$+ O(h) + \frac{1}{2\alpha h^{1/2}} \tilde{r}.$$

The factor $\tilde{r}$ at the Monte Carlo error in (3.6.49) has the variance

$$Var(\tilde{r}) = \frac{1}{M}[Var\, f(\bar{X}_{t,x+\alpha h^{1/2}}(T)) + Var\, f(\bar{X}_{t,x-\alpha h^{1/2}}(T))] \quad (3.6.50)$$
$$= \frac{2}{M} Var\, f(X_{t,x}(T)) + O\left(\frac{h}{M}\right).$$

We have assumed here that all the realizations $\bar{X}^{(m)}_{t,x+\alpha h^{1/2}}(T)$ and $\bar{X}^{(m)}_{t,x-\alpha h^{1/2}}(T)$ are independent. The second equality in (3.6.50) is obtained by the following arguments: the values $Var\, f(\bar{X}_{t,x\pm\alpha h^{1/2}}(T))$ differ from $Var\, f(X_{t,x\pm\alpha h^{1/2}}(T))$ by $O(h)$ and $Var\, f(X_{t,x\pm\alpha h^{1/2}}(T))$ differ from $Var\, f(X_{t,x}(T))$ by $O(h)$ too.

Thus, the total error $\tilde{R}$ in the evaluation of $\partial u/\partial x$ due to

$$\frac{\partial u}{\partial x}(t, x) \simeq \frac{1}{2\alpha h^{1/2}} \frac{1}{M} \left[\sum_{m=1}^{M} f(\bar{X}^{(m)}_{t,x+\alpha h^{1/2}}(T)) - \sum_{m=1}^{M} f(\bar{X}^{(m)}_{t,x-\alpha h^{1/2}}(T))\right]$$

is estimated as

$$\tilde{R} \sim O(h) + O\left(\frac{1}{h^{1/2}\sqrt{M}}\right), \quad (3.6.51)$$

which is of order $O(h)$ if $M \sim 1/h^3$ that is $1/h$ times larger than it is required in the evaluation of the solution u itself (see (3.6.48)).

Now, instead of simulating the independent trajectories, let us simulate them in a pairwise dependent way that is similar to the use of common random numbers [41, 129–131, 227]. More precisely, we now simulate M pairs of trajectories, each pair consists of a trajectory starting from $x + \alpha h^{1/2}$ and a trajectory starting from $x - \alpha h^{1/2}$. The pairs are independent but the two trajectories of the same pair are dependent: they correspond to the same realization of the Wiener process in the weak sense. Consider the estimator

$$\hat{\partial}u(t, x) = \frac{1}{2\alpha h^{1/2}} \frac{1}{M} \sum_{m=1}^{M} \left[f(\bar{X}^{(m)}_{t,x+\alpha h^{1/2}}(T)) - f(\bar{X}^{(m)}_{t,x-\alpha h^{1/2}}(T))\right] \quad (3.6.52)$$

introduced in Sect. 3.6.2 (see (3.6.34) and also (3.6.26)).

We have

$$\frac{\partial u}{\partial x}(t, x) = \frac{1}{2\alpha h^{1/2}} \left[Ef(\bar{X}_{t,x+\alpha h^{1/2}}(T)) - Ef(\bar{X}_{t,x-\alpha h^{1/2}}(T))\right] \quad (3.6.53)$$
$$+ O(h)$$
$$= \hat{\partial}u(t, x) + O(h) + \frac{1}{2\alpha h^{1/2}}r_{\hat{\partial}u}.$$

The term $O(h)$ in (3.6.53) is the sum of the error of numerical differentiation equal to $O\left((\Delta x)^2\right) = O(h)$ due to (2.3.39) and the error of numerical integration by the

Euler scheme which is also equal to $O(h)$, this term is deterministic. The variance of the factor $r_{\hat{\partial}u}$ at the Monte Carlo error in (3.6.53) is equal to

$$Var(r_{\hat{\partial}u}) = \frac{Var\left[f(\bar{X}_{t,x+\alpha h^{1/2}}(T)) - f(\bar{X}_{t,x-\alpha h^{1/2}}(T))\right]}{M}. \tag{3.6.54}$$

Since $\bar{X}_{t,x}(T)$ approximates $X_{t,x}(T)$ with the first weak order, we get

$$\begin{aligned}
& Var\left[f(\bar{X}_{t,x+\alpha h^{1/2}}(T)) - f(\bar{X}_{t,x-\alpha h^{1/2}}(T))\right] \tag{3.6.55}\\
&= Var\left[f(X_{t,x+\alpha h^{1/2}}(T)) - f(X_{t,x-\alpha h^{1/2}}(T))\right] + O(h).
\end{aligned}$$

We have

$$\begin{aligned}
& Var\left[f(X_{t,x+\alpha h^{1/2}}(T)) - f(X_{t,x-\alpha h^{1/2}}(T))\right] \\
&= E\left[f(X_{t,x+\alpha h^{1/2}}(T)) - f(X_{t,x-\alpha h^{1/2}}(T))\right]^2 \\
&\quad - \left[E\left(f(X_{t,x+\alpha h^{1/2}}(T)) - f(X_{t,x-\alpha h^{1/2}}(T))\right)\right]^2.
\end{aligned}$$

Due to the conditions imposed on the coefficients of the problem (3.6.40)–(3.6.41) and continuous dependence of solutions to SDEs on the initial data (see [123, Sect. 8]), we obtain

$$E\left[f(X_{t,x+\alpha h^{1/2}}(T)) - f(X_{t,x-\alpha h^{1/2}}(T))\right]^2 \leq Kh \tag{3.6.56}$$

and

$$\left[E\left(f(X_{t,x+\alpha h^{1/2}}(T)) - f(X_{t,x-\alpha h^{1/2}}(T))\right)\right]^2 \leq Kh \tag{3.6.57}$$

(here and in what follows we denote by the same letter K various positive constants independent of h). Hence

$$Var\left[f(X_{t,x+\alpha h^{1/2}}(T)) - f(X_{t,x-\alpha h^{1/2}}(T))\right] \leq Kh \tag{3.6.58}$$

and, consequently,

$$r_{\hat{\partial}u} \sim O(\sqrt{h/M}). \tag{3.6.59}$$

Thus, the total error $R_{\hat{\partial}u}$ in the evaluation of $\partial u/\partial x$ by $\hat{\partial}u$ from (3.6.52) is estimated as

$$R_{\hat{\partial}u} \sim O(h) + O\left(\frac{1}{\sqrt{M}}\right). \tag{3.6.60}$$

We see that to have the total error $\sim h$, it is sufficient to take $M \sim 1/h^2$ like in the evaluation of the solution u itself (cf. (3.6.48)).

In the case of the general problem (3.6.13)–(3.6.14) we consider the estimators (see (3.6.33) and (3.6.34) in Sect. 3.6.2):

$$\hat{u}(t, x) = \frac{1}{M} \sum_{m=1}^{M} [f(\bar{X}_{t,x}^{(m)}(T))\bar{Y}_{t,x,1}^{(m)}(T) + \bar{Z}_{t,x,1,0}^{(m)}(T)] \qquad (3.6.61)$$

and

$$\hat{\partial}_\nu u(t, x) = \frac{1}{2\alpha h^{1/2}} \frac{1}{M} \sum_{m=1}^{M} \Big[f(\bar{X}_{t,x+\alpha h^{1/2}\nu}^{(m)}(T))\bar{Y}_{t,x+\alpha h^{1/2}\nu,1}^{(m)}(T) \qquad (3.6.62)$$
$$+ \bar{Z}_{t,x+\alpha h^{1/2}\nu,1,0}^{(m)}(T)$$
$$- f(\bar{X}_{t,x-\alpha h^{1/2}\nu}^{(m)}(T))\bar{Y}_{t,x-\alpha h^{1/2}\nu,1}^{(m)}(T) - \bar{Z}_{t,x-\alpha h^{1/2}\nu,1,0}^{(m)}(T)\Big],$$

where ν is a unit vector. The above arguments used in the case $d = 1$, $r(t, x) = 0$, $c(t, x) = 0$ can be carried over for the general case. As a result, we obtain the following theorem.

Theorem 3.6.3 Let $\bar{\Lambda}_{t,x}(T) := (\bar{X}_{t,x}(T), \bar{Y}_{t,x,1}(T), \bar{Z}_{t,x,1,0}(T))$ be obtained by the weak Euler method applied to (3.6.29), the pairs $\left(\bar{\Lambda}_{t,x+\alpha h^{1/2}\nu}^{(m)}(T), \bar{\Lambda}_{t,x-\alpha h^{1/2}\nu}^{(m)}(T) \right)$, $m = 1, \ldots, M$, $\alpha > 0$, be independent but the two realizations from the same pair correspond to the same realization of the Wiener process. Then the biases of the estimators $\hat{u}$ from (3.6.61) and $\hat{\partial}_\nu u$ from (3.6.62) are $O(h)$:

$$Bias(\hat{u}(t, x)) = O(h), \quad Bias(\hat{\partial}_\nu u(t, x)) = O(h), \qquad (3.6.63)$$

and their variances are $O(1/M)$:

$$Var(\hat{u}(t, x)) = O(1/M), \quad Var(\hat{\partial}_\nu u(t, x)) = O(1/M). \qquad (3.6.64)$$

Remark 3.6.4 It can be proved that for a weak scheme of order two the bias of the estimator

$$\hat{\partial}_\nu^{(2)} u(t, x) = \frac{1}{2\alpha h} \frac{1}{M} \sum_{m=1}^{M} \Big[f(\bar{X}_{t,x+\alpha h\nu}^{(m)}(T))\bar{Y}_{t,x+\alpha h\nu,1}^{(m)}(T) + \bar{Z}_{t,x+\alpha h\nu,1,0}^{(m)}(T)$$
$$- f(\bar{X}_{t,x-\alpha h\nu}^{(m)}(T))\bar{Y}_{t,x-\alpha h\nu,1}^{(m)}(T) - \bar{Z}_{t,x-\alpha h\nu,1,0}^{(m)}(T)\Big]$$

is $O(h^2)$ and the variance is $O(1/M)$. The bias is the sum of the error of numerical differentiation which is $O(h^2)$ because of $\Delta x = \alpha h$ and the error of numerical integration which is also equal to $O(h^2)$ due to the choice of a weak scheme of order two (see Remark 3.6.2).

3.6.5 Some Extensions

The assumptions of Sect. 2.2.5.1 ensure smoothness of $E[f(X_{t,x}(T))Y_{t,x,1}(T) + Z_{t,x,1,0}(T)]$ with respect to x and the Talay–Tubaro expansion of the error of numerical integration in powers of h. These properties allowed us to prove the main theorem, i.e., Theorem 3.6.3. As it was mentioned in Sect. 3.6.2, the most restrictive assumptions concern the properties of smoothness for payoff functions. Fortunately, the results can be carried over to the case of a much weaker assumption on f if the mean-square Euler scheme is used, i.e., if the solution of (3.6.19) is approximated by (cf. (3.6.42))

$$\bar{X}_{t,x}(s_{k+1}) := X_{k+1} = X_k + hb(s_k, X_k) + h^{1/2}\sigma(s_k, X_k)\zeta_k, \quad k = 0, \ldots, N-1,$$
(3.6.65)

where ζ_k are i.i.d. $\mathcal{N}(0, 1)$ random variables.

For simplicity, we consider the case $d = 1$, $c = g = 0$ in this subsection. Let us represent $Ef(X_{t,x}(T))$ in the form

$$Ef(X_{t,x}(T)) = \int P(t, x, T, dy) f(y),$$
(3.6.66)

where $P(t, x, T, dy)$ is a transition Markov function for the process X. It is clear from (3.6.66) that $Ef(X_{t,x}(T))$ can be smooth for even measurable functions f provided that $P(t, x, T, dy)$ is smooth with respect to x. The assumptions of Sect. 3.6.2 on the coefficients of the considered SDEs and the uniform ellipticity condition (3.6.30) guarantee such properties of $P(t, x, T, dy)$ (to the point, a weaker nondegeneracy condition of Hörmander's type can be taken instead of the uniform ellipticity condition). Using the Malliavin calculus, Bally and Talay showed in [19] that if $\bar{X}$ is simulated by the mean-square Euler scheme (3.6.65), then the error of numerical integration for measurable f satisfying (3.6.31) can be expanded in powers of h. Due to this result, Theorem 3.6.1 can be extended for the wider class of functions f. So, the bias of the estimator (we recall that we consider the case $d = 1$, $c = g = 0$)

$$\hat{\partial}u(t, x) = \frac{1}{2\alpha h^{1/2}} \frac{1}{M} \sum_{m=1}^{M} \left[f(\bar{X}^{(m)}_{t,x+\alpha h^{1/2}}(T)) - f(\bar{X}^{(m)}_{t,x-\alpha h^{1/2}}(T)) \right]$$
(3.6.67)

is $O(h)$ for any measurable function f satisfying (3.6.31). In (3.6.67) the mean-square Euler method is used, the pairs $(\bar{X}^{(m)}_{t,x+\alpha h^{1/2}}(T), \bar{X}^{(m)}_{t,x-\alpha h^{1/2}}(T))$, $m = 1, \ldots, M$, are independent but the two realizations $\bar{X}^{(m)}_{t,x+\alpha h^{1/2}}(T)$ and $\bar{X}^{(m)}_{t,x-\alpha h^{1/2}}(T)$ of each pair correspond to the same sequence $\zeta^{(m)}_k$, $k = 0, \ldots, N-1$, in (3.6.65).

Let us proceed to analysis of the variance of the estimator (3.6.67). Since the inequalities (3.6.56) and (3.6.57) are evidently valid for any globally Lipschitz function f, the variance $Var(\hat{\partial}u(t, x))$ for such functions is $O(1/M)$. So, the total error $R_{\hat{\partial}u}$ of $\hat{\partial}u$ is again (cf. (3.6.60)):

$$R_{\hat{\partial}u} \sim O(h) + O\left(\frac{1}{\sqrt{M}}\right), \tag{3.6.68}$$

i.e., the result of Theorem 3.6.3 remains correct for the class of globally Lipschitz payoff functions.

To evaluate the variance for discontinuous payoff functions, we first consider the step function

$$f(y) = \begin{cases} 0, & y < \lambda \\ 1, & y \geq \lambda \end{cases}.$$

We have

$$\begin{aligned} V &:= Var[f(X_{t,x+\alpha h^{1/2}}(T)) - f(X_{t,x-\alpha h^{1/2}}(T))] \\ &= Ef_1^2 + Ef_2^2 - 2Ef_1f_2 - (Ef_1 - Ef_2)^2, \end{aligned}$$

where $f_1 = f(X_{t,x+\alpha h^{1/2}}(T))$ and $f_2 = f(X_{t,x-\alpha h^{1/2}}(T))$.

Further (see (3.6.66)):

$$Ef_1^2 = Ef_1 = 1 - P(t, x + \alpha h^{1/2}, T, \lambda), \quad Ef_2^2 = Ef_2 = 1 - P(t, x - \alpha h^{1/2}, T, \lambda),$$

$$\begin{aligned} Ef_1f_2 &= P(\min\{X_{t,x+\alpha h^{1/2}}(T), X_{t,x-\alpha h^{1/2}}(T)\} \geq \lambda) \\ &= P(X_{t,x-\alpha h^{1/2}}(T) \geq \lambda) = 1 - P(t, x - \alpha h^{1/2}, T, \lambda) = Ef_2, \end{aligned}$$

where $P(t, x, T, \lambda) = P(X_{t,x}(T) < \lambda) = \int_{-\infty}^{\lambda} P(t, x, T, dy)$.

Hence

$$\begin{aligned} V &= Ef_1 - Ef_2 - (Ef_1 - Ef_2)^2 \tag{3.6.69} \\ &= P(t, x - \alpha h^{1/2}, T, \lambda) - P(t, x + \alpha h^{1/2}, T, \lambda) \\ &\quad - (P(t, x - \alpha h^{1/2}, T, \lambda) - P(t, x - \alpha h^{1/2}, T, \lambda))^2 \\ &= -2P_x'(t, x, T, \lambda) \cdot \alpha h^{1/2} + O(h). \end{aligned}$$

Therefore the variance of the estimator $\hat{\partial}u$ is equal to

$$Var(\hat{\partial}u) = O\left(\frac{1}{h^{1/2}M}\right). \tag{3.6.70}$$

Clearly, the result (3.6.70) is valid for piecewise Lipschitz payoff functions. Thus, for the sufficiently wide class of functions the error R for the estimator $\hat{\partial}u$ is evaluated as

$$R_{\hat{\partial}u} \sim O(h) + O\left(\frac{1}{h^{1/4}\sqrt{M}}\right). \tag{3.6.71}$$

This is worse than (3.6.68) but better than (3.6.51). Although we have not proved that the estimate (3.6.71) is also valid in the case of the weak Euler method (3.6.32), we believe this assertion to be true and, in particular, it is supported by numerical experiments in [315].

Results of this section and, in particular, the use of the method of dependent realizations for Monte Carlo estimators were extended to the case of American options in [23].

Chapter 4
Numerical Methods for SDEs with Small Noise

In the general case many difficulties arise with realizing numerical methods for SDEs. But we know (see, e.g., Chaps. 1–3 where, in particular, such specific systems as systems with additive and colored noises are treated) that numerical methods adapted to specific systems can be more efficient and easier than general methods. An important instance of a stochastic system is given by differential equations with small noise, since often fluctuations, which affect a dynamical system, are sufficiently small.

The system of Ito stochastic differential equations with *small noise* can be written in the form

$$dX = a(t, X)\, dt + \varepsilon^2 b(t, X)\, dt + \varepsilon \sum_{r=1}^{q} \sigma_r(t, X)\, dw_r \,, \quad X(t_0) = X_0 \,, \qquad (4.0.1)$$

$$t \in [t_0, T], 0 \le \varepsilon \le \varepsilon_0 \,,$$

where ε is a small parameter, ε_0 is a positive number, $X = (X^1, \dots, X^d)^\mathsf{T}$, $a(t, x) = (a^1(t, x), \dots, a^d(t, x))^\mathsf{T}$, $b(t, x) = (b^1(t, x), \dots, b^d(t, x))^\mathsf{T}$, $\sigma_r(t, x) = (\sigma_r^1(t, x), \dots, \sigma_r^d(t, x))^\mathsf{T}$, $r = 1, \dots, q$, are d-dimensional vectors, $w_r(t)$, $r = 1, \dots, q$, are independent standard Wiener processes, and X_0 does not depend on $w_r(t) - w_r(t_0)$, $t_0 < t \le T$, $r = 1, \dots, q$.

If the parameter ε is equal to zero, we have a deterministic system for which various effective numerical methods exist. One can imagine that if parameter ε is sufficiently small, i.e., the system (4.0.1) is sufficiently close to the deterministic one, it is also possible to obtain effective methods taking into account that ε is small.

Introduce the equidistant discretization of the interval $[t_0, T]$: $\{t_k : k = 0, 1, \dots, N; t_0 < t_1 < \dots < t_N = T\}$ and the time increment $h = t_{k+1} - t_k$. The errors of the methods proposed in this chapter are estimated in terms of products $h^i \varepsilon^j$, and they are usually of the form $O(h^p + \varepsilon^k h^q)$, $q < p$. The order of such

© Springer Nature Switzerland AG 2021
G. N. Milstein and M. V. Tretyakov, *Stochastic Numerics for Mathematical Physics*,
Scientific Computation, https://doi.org/10.1007/978-3-030-82040-4_4

a method is equal to q which may be low, e.g. $1/2$ or 1. For small ε the product $\varepsilon^k h^q$ also becomes small and, consequently, so does the error. This allows us to construct effective methods with low order but which nevertheless have small errors. For instance, the methods of this chapter are effectively applied to evaluation of the signal-to-noise ratio in systems with stochastic resonance in [443] (see also Chap. 9 in [314]).

In the first part of this chapter (Sects. 4.1–4.3) we derive specific *mean-square* methods for systems with small noise [300]. In these sections, we put $b = 0$ for simplicity, i.e., we consider the system

$$dX = a(t, X)\,dt + \varepsilon \sum_{r=1}^{q} \sigma_r(t, X)\,dw_r\,, \quad X(t_0) = X_0\,. \tag{4.0.2}$$

In Sect. 4.1 we propose our approach to constructing one-step mean-square approximations for solutions of the system (4.0.2) and prove the corresponding theorem on global mean-square error. Various efficient mean-square schemes for systems with small noise are presented in Sect. 4.2. Numerical tests of the proposed mean-square methods are presented in Sect. 4.3.

The second part of this chapter (Sects. 4.4–4.8) is devoted to *weak* approximations of the system with small noise (4.0.1) [301]. In Sect. 4.4 we state the theorem on error estimate of weak methods and illustrate our approach to construction of weak schemes for the system (4.0.1). Various specific weak schemes for systems with small noise are proposed in Sect. 4.5. Section 4.6 deals with the expansion of the global error of weak methods in powers of h and ε. The method of important sampling from Sect. 2.3.1 allows us to effectively reduce the Monte Carlo error in the case of a system with small noise that is demonstrated in Sect. 4.7. Numerical tests of the proposed weak methods are presented in Sect. 4.8.

We note that in the case of the system (4.0.1) the operators L and Λ_r from Chaps. 1 and 2 take the form

$$L = L_1 + \varepsilon^2 L_2, \quad L_1 = \frac{\partial}{\partial t} + \left(a, \frac{\partial}{\partial x}\right) = \frac{\partial}{\partial t} + \sum_{i=1}^{d} a^i \frac{\partial}{\partial x^i}\,,$$

$$L_2 = \left(b, \frac{\partial}{\partial x}\right) + \frac{1}{2}\sum_{r=1}^{q}\left(\sigma_r, \frac{\partial}{\partial x}\right)^2 = \sum_{i=1}^{d} b^i \frac{\partial}{\partial x^i} + \frac{1}{2}\sum_{r=1}^{q}\sum_{i,j=1}^{d}\sigma_r^i \sigma_r^j \frac{\partial^2}{\partial x^i \partial x^j}\,,$$

$$\Lambda_r = \left(\sigma_r, \frac{\partial}{\partial x}\right) = \sum_{i=1}^{d}\sigma_r^i \frac{\partial}{\partial x^i}\,.$$

We will use the following notation for Ito integrals:

$$I_{i_1,\ldots,i_j}(F, t, h)$$
$$= \int_t^{t+h} dw_{i_j}(\theta) \int_t^{\theta} dw_{i_{j-1}}(\theta_1) \int_t^{\theta_1} \cdots \int_t^{\theta_{j-2}} F(\theta_{j-1})\,dw_{i_1}(\theta_{j-1})\,,$$

where $i_1, \ldots, i_j$ are from the set of numbers $\{0, 1, \ldots, q\}$ and $dw_0(\theta_r)$ designates $d\theta_r$, $F(\theta)$ is a deterministic (for simplicity continuous) function; $I_{i_1, i_2, \ldots, i_j}(t, h) \equiv I_{i_1, i_2, \ldots, i_j}(1(\cdot), t, h)$ where $1(\theta)$ is the function which is everywhere equal to one. Properties of Ito integrals were established in Lemma 1.2.1.

4.1 Mean-Square Approximations and Estimation of Their Errors

In Sect. 4.1.1 we illustrate the main idea of this chapter that errors of approximations for the SDEs (4.0.2) can be written in terms of smallness of both time step h and ε. The theorem on global error estimate of mean-square methods for (4.0.2) is stated in Sect. 4.1.2. In Sect. 4.1.3 we discuss how to select the time step h depending on a small but fixed ε. In Sect. 4.1.4 we compare the approach of the previous subsections (where we expand the solution of (4.0.2) in h and then regroup the terms according to their smallness expressed via $\varepsilon^k h^q$) and the approach consisting in expanding the solution first in ε and then in h.

4.1.1 Construction of One-Step Mean-Square Approximation

Consider the one-step approximation of the solution to (4.0.2) with the local order two (cf. (1.2.27)):

$$\bar{X}(t + h) = X(t) + \varepsilon \sum_{r=1}^{q} \sigma_r(t, X(t)) I_r(t, h) + a(t, X(t))h$$

$$+\varepsilon^2 \sum_{i,r=1}^{q} \Lambda_r \sigma_i(t, X(t)) I_{ri}(t, h) + \varepsilon \sum_{r=1}^{q} L_1 \sigma_r(t, X(t)) I_{0r}(t, h)$$

$$+\varepsilon^3 \sum_{r=1}^{q} L_2 \sigma_r(t, X(t)) I_{0r}(t, h) + \varepsilon \sum_{r=1}^{q} \Lambda_r a(t, X(t)) I_{r0}(t, h)$$

$$+\varepsilon^3 \sum_{s,i,r=1}^{q} \Lambda_s \Lambda_i \sigma_r(t, X(t)) I_{sir}(t, h) + L_1 a(t, X(t)) h^2/2$$

$$+\varepsilon^2 L_2 a(t, X(t)) h^2/2. \tag{4.1.1}$$

The remainder $\rho = X(t + h) - \bar{X}(t + h)$ of this approximation has the form (cf. (1.2.22)):

$$\rho = \varepsilon^4 \sum_{r,i,s,j=1}^{q} I_{risj}(\Lambda_r \Lambda_i \Lambda_s \sigma_j, t, h) + \varepsilon^2 \sum_{i,r=1}^{q} I_{0ir}(L_1 \Lambda_i \sigma_r, t, h)$$

$$+\varepsilon^4 \sum_{i,r=1}^{q} I_{0ir}(L_2 \Lambda_i \sigma_r, t, h) + \varepsilon^2 \sum_{i,r=1}^{q} I_{i0r}(\Lambda_i L_1 \sigma_r, t, h)$$

$$+\varepsilon^4 \sum_{i,r=1}^{q} I_{i0r}(\Lambda_i L_2 \sigma_r, t, h) + \varepsilon^2 \sum_{i,r=1}^{q} I_{ir0}(\Lambda_i \Lambda_r a, t, h)$$

$$+\varepsilon^3 \sum_{r,i,s=1}^{q} I_{0sir}(L_1 \Lambda_s \Lambda_i \sigma_r, t, h) + \varepsilon^5 \sum_{r,i,s=1}^{q} I_{0sir}(L_2 \Lambda_s \Lambda_i \sigma_r, t, h)$$

$$+\varepsilon \sum_{r=1}^{q} I_{00r}(L_1^2 \sigma_r, t, h) + \varepsilon^3 \sum_{r=1}^{q} I_{00r}((L_1 L_2 + L_2 L_1)\sigma_r, t, h)$$

$$+\varepsilon^5 \sum_{r=1}^{q} I_{00r}(L_2^2 \sigma_r, t, h) + \varepsilon \sum_{r=1}^{q} I_{0r0}(L_1 \Lambda_r a, t, h)$$

$$+\varepsilon^3 \sum_{r=1}^{q} I_{0r0}(L_2 \Lambda_r a, t, h) + \varepsilon \sum_{r=1}^{q} I_{r00}(\Lambda_r L_1 a, t, h)$$

$$+\varepsilon^3 \sum_{r=1}^{q} I_{r00}(\Lambda_r L_2 a, t, h) + I_{000}(L_1^2 a, t, h)$$

$$+\varepsilon^2 I_{000}((L_1 L_2 + L_2 L_1)a, t, h) + \varepsilon^4 I_{000}(L_2^2 a, t, h). \tag{4.1.2}$$

It is not difficult to obtain

$$E\rho = O(h^3), \ (E\rho^2)^{1/2} = O(h^3 + \varepsilon^2 h^2).$$

The iterated Ito integrals I_{ri} and I_{sir} appearing in the method (4.1.1) cannot be easily simulated. But these integrals are multiplied by ε^2 and ε^3, respectively. That is why, they can be transferred to the remainder and the error of the approximation would still not be large. Further, if we transfer from (4.1.1) not only the terms with complicated Ito integrals but also the terms which are sufficiently small, we obtain the reduced one-step approximation

$$\bar{X}(t+h) = X(t) + \varepsilon \sum_{r=1}^{q} \sigma_r(t, X(t)) I_r(t, h) + a(t, X(t))h$$

$$+\varepsilon \sum_{r=1}^{q} L_1 \sigma_r(t, X(t)) I_{0r}(t, h) + \varepsilon \sum_{r=1}^{q} \Lambda_r a(t, X(t)) I_{r0}(t, h)$$

$$+L_1 a(t, X(t)) h^2/2, \tag{4.1.3}$$

the remainder ρ_1 of which is equal to

$$\rho_1 = \rho + \varepsilon^2 \sum_{i,r=1}^{q} \Lambda_r \sigma_i(t, X(t)) I_{ri}(t, h) + \varepsilon^3 \sum_{r=1}^{q} L_2 \sigma_r(t, X(t)) I_{0r}(t, h) \quad (4.1.4)$$

$$+ \varepsilon^3 \sum_{s,i,r=1}^{q} \Lambda_s \Lambda_i \sigma_r(t, X(t)) I_{sir}(t, h) + \varepsilon^2 L_2 a(t, X(t)) h^2/2,$$

where ρ is taken from (4.1.2).

One can obtain

$$E \rho_1 = O(h^3 + \varepsilon^2 h^2),$$
$$(E\rho_1^2)^{1/2} = O(h^3 + \varepsilon h^{5/2} + \varepsilon^2 h^2 + \varepsilon^3 h^{3/2} + \varepsilon^2 h) = O(h^3 + \varepsilon^2 h). \quad (4.1.5)$$

The terms $\varepsilon h^{5/2}$, $\varepsilon^2 h^2$, and $\varepsilon^3 h^{3/2}$ of the second expression are omitted because they are not greater than $O(h^3 + \varepsilon^2 h)$. Of course, the order of the approximation (4.1.3) is less (the order is equal to one due to the term $\varepsilon^2 \sum_{i,r=1}^{q} \Lambda_r \sigma_i I_{ri}$ in ρ_1) than the order of the approximation (4.1.1), but the error of the approximation (4.1.3) has the small factor ε^2 at h. Thus, we obtain the one-step approximation (4.1.3) which has sufficiently small mean-square local error and is efficient as to simulation of the used random variables.

Using (4.1.2)–(4.1.4), we construct a new approximation by transferring a part of the remainder to the approximation. In this connection we expand the term

$$I_{000}(L_1^2 a, t, h) = \int_t^{t+h} \left(\int_t^{\theta} \left(\int_t^{\theta_1} L_1^2 a(\theta_2, X(\theta_2)) \, d\theta_2 \right) d\theta_1 \right) d\theta$$

$$= L_1^2 a(t, X(t)) h^3/6 + \rho',$$

where

$$E\rho' = O(h^4 + \varepsilon^2 h^3),$$
$$E[(\rho')^2]^{1/2} = O(h^4 + \varepsilon h^{7/2} + \varepsilon^2 h^3) = O(h^4 + \varepsilon^2 h^3).$$

The new approximation has the form

$$\tilde{X}(t + h) = \bar{X}(t + h) + L_1^2 a(t, X(t)) h^3/6, \quad (4.1.6)$$

where $\bar{X}(t + h)$ is taken from (4.1.3). The remainder $\tilde{\rho}$ of the approximation (4.1.6) can be obtained from ρ_1 if we substitute ρ' instead of $h^3 I_{000}(L_1^2 a, t, h)$. It is clear that

$$E\tilde{\rho} = O(h^4 + \varepsilon^2 h^2), \quad E(\tilde{\rho}^2)^{1/2} = O(h^4 + \varepsilon^2 h).$$

Of course, the approximation (4.1.6) can be derived differently to how it has been done above, for instance, from an approximation with local order three, but the suggested way is the simplest.

In this subsection we have demonstrated the main idea of the chapter. In contrast to the general case smallness of terms of an approximation for a system with small noise and of its remainder depends not only on time increment h but also on small parameter ε. This circumstance, as shown above, allows us to construct new numerical methods by excluding complicated terms, for instance, multiple Ito integrals, from a method and including them in its remainder. New methods are efficient as to simulation of the used random variables and have small mean-square errors in the sense of products $\varepsilon^i h^j$. Moreover, the methods contain fewer terms with operators than the corresponding schemes for a general SDEs system.

4.1.2   Theorem on Mean-Square Global Estimate

Before stating the theorem on mean-square global error estimate we note that usually after reducing, estimates of the remainder of a particular one-step approximation are sufficiently simple and often contain only two terms (for instance, see (4.1.5)). However, it is not always the case. For instance, the sum $h^3 + \varepsilon h^{3/2} + \varepsilon^2 h$ cannot be reduced. Detailed analysis of possible errors gives the form of estimates which is used in Theorem 4.1.1 stated below, i.e., the conditions of Theorem 4.1.1 are natural.

Introduce the notation: ρ is a local error of a method, R is a global error (we also call it as method error, mean-square error or error if it does not lead to misunderstanding), r_0 is a fixed natural number, S_1 is either an empty set or a subset of positive integers p which are less than r_0, S_2 is either empty set or a subset of positive integers and semi-integers q which are less than r_0, i.e.,

$$S_1 \subset \{p : 0 < p < r_0, \quad p \text{ is an integer}\},$$

$$S_2 \subset \{q : 0 < q < r_0, \quad q \text{ is either an integer or semi-integer}\}.$$

Below in Theorem 4.1.1 the sum $\sum_{p \in S_1} (\sum_{q \in S_2})$ must be replaced by zero if S_1 (S_2) is an empty set.

Theorem 4.1.1 *Let $\bar{X}_{t,x}(t + h)$ be an approximation of the solution $X_{t,x}(t + h)$ of the system (4.0.2) with initial condition $X(t) = \bar{X}(t) = x$ and $J_1(p)$ and $J_2(q)$ be decreasing functions whose values are natural numbers. If the inequalities*

$$|E\rho| = |E\left(X_{t,x}(t+h) - \bar{X}_{t,x}(t+h)\right)| \tag{4.1.7}$$

$$\leq K\left(1+|x|^2\right)^{1/2}\left(h^{r_0} + \sum_{p\in S_1} h^p \varepsilon^{2J_1(p)}\right),$$

$$\left(E\rho^2\right)^{1/2} = \left(E\left|X_{t,x}(t+h) - \bar{X}_{t,x}(t+h)\right|^2\right)^{1/2} \tag{4.1.8}$$

$$\leq K\left(1+|x|^2\right)^{1/2}\left(h^{r_0} + \sum_{q\in S_2} h^q \varepsilon^{J_2(q)}\right)$$

hold, then

$$\left(E\left|X_{t_0,X_0}(t_k) - \bar{X}_{t_0,X_0}(t_k)\right|^2\right)^{1/2} \tag{4.1.9}$$

$$\leq K\left(1+E|X_0|^2\right)^{1/2}\left(h^{r_0-1} + \sum_{p\in S_1} h^{p-1}\varepsilon^{2J_1(p)} + \sum_{q\in S_2} h^{q-1/2}\varepsilon^{J_2(q)}\right),$$

where the constant $K > 0$ does not depend on discretization step h, parameter ε, $0 \leq \varepsilon \leq \varepsilon_0$, and $k = 1, \ldots, N$.

The proof of Theorem 4.1.1 is similar to the proof of Theorem 1.1.1, and here it is omitted.

Applying Theorem 4.1.1, for example, to the method based on the one-step approximation (4.1.3), we obtain that its mean-square global error is estimated by $O(h^2 + \varepsilon^2 h + \varepsilon^2 h^{1/2}) = O(h^2 + \varepsilon^2 h^{1/2})$. This error is sufficiently small because of the small factor ε^2 at $h^{1/2}$.

It follows from Theorem 4.1.1 that if $r_0 > 1$ and the set S_1 is either empty or every number p of S_1 is greater than one and the set S_2 is either empty or every number q of S_2 is greater than $1/2$, then the corresponding method converges (cf. Theorem 1.1.1). However, the primary meaning of Theorem 4.1.1 is not that it gives convergence order of a method but is that it gives a method's error in terms of h and ε.

4.1.3 Selection of Time Step h Depending on Parameter ε

In practice the parameter ε is small but fixed, and we can usually choose only the step h. Nevertheless, asymptotic behavior of method error under $\varepsilon \to 0$, when h is chosen depending on ε, is interesting in many respects. The inequality (4.1.9) makes such an analysis possible.

Let us choose time increment h so that $h = C\varepsilon^\alpha$, $\alpha > 0$. Then the error of a method can be estimated in powers of small parameter ε

$$\left(E|X_{t_0,X_0}(t_k) - \bar{X}_{t_0,X_0}(t_k)|^2\right)^{1/2} = O(\varepsilon^\beta),$$

where $\beta = \min\{\alpha(r_0 - 1), \min_{p \in S_1}(\alpha(p - 1) + 2J_1(p)), \min_{q \in S_2}(\alpha(q - 1/2) + J_2(q))\}$. The parameter α and a method may be so that a certain term of this method is smaller than $O(\varepsilon^\beta)$. Such a term may be omitted and, in spite of this, the order of the method error does not change with respect to ε.

Let us analyze the method based on the one-step approximation (4.1.3). If $h = C\varepsilon^\alpha$, the mean-square global error of the method (4.1.3) is estimated by $O(\varepsilon^{2\alpha} + \varepsilon^{2+\alpha/2})$. Let us choose α be equal to one. In this case the method error is estimated by $O(\varepsilon^2)$, the order of the terms $\varepsilon L_1 \sigma_r I_{0r}$ and $\varepsilon \Lambda_r a I_{r0}$ is equal to $O(\varepsilon^{5/2})$, and their omission gives $O(\varepsilon^2)$ to the mean-square error. So, in the case of $\alpha = 1$ these terms may be omitted, and that does not lead to substantial increase of the error. Thus, we obtain the new method

$$X_{k+1} = X_k + \varepsilon \sum_{r=1}^{q} (\sigma_r I_r)_k + a_k h + L_1 a_k h^2/2 , \qquad (4.1.10)$$

$$(ER^2)^{1/2} = O(h^2 + \varepsilon h + \varepsilon^2 h^{1/2}) ,$$

where $\sigma_{r_k} = \sigma_r(t_k, X_k)$, $a_k = a(t_k, X_k)$, $(I_r)_k = I_r(t_k, h)$. It is clear that if $h = C\varepsilon^\alpha$, where $\alpha \leq 1$ or $\alpha \geq 2$, the errors of the methods (4.1.3) and (4.1.10) have the same order with respect to ε. But if $h = C\varepsilon^\alpha$, $1 < \alpha < 2$, the method (4.1.10) has the lower order with respect to ε than the method (4.1.3).

4.1.4 (h, ε)-Approach Versus (ε, h)-Approach

In this chapter we construct numerical methods by the (h, ε)-approach, for instance, see the methods (4.1.3) and (4.1.10). According to the (h, ε)-approach, we expand the exact solution $X(t)$ of the system (4.0.2) in powers of time increment h and obtain an expansion which is similar to the stochastic Taylor-type expansion (see Sect. 1.2.2). Then we regroup terms of the expansion with respect to their $h^i \varepsilon^j$ factors and decide which terms must be included in a method. Such a decision depends on the desired mean-square error of the method and on computational complexity of an expansion term, especially on complexity of simulation of the used random variables.

The (ε, h)-approach is based on a different idea. First, the exact solution of the system (4.0.2) is expanded in powers of small parameter ε, for instance,

$$\bar{X}(t) = X^0(t) + \varepsilon X^1(t) , \qquad (4.1.11)$$

$$R = X(t) - \bar{X}(t) = O(\varepsilon^2) ,$$

where $X^0(t)$ and $X^1(t)$ are found as the solutions of the original system under $\varepsilon = 0$ and its system of the first approximation:

$$dX^0 = a(t, X^0)\, dt, \quad X^0(0) = X_0, \tag{4.1.12}$$

$$dX^1 = a'_x(t, X^0)X^1\, dt + \sum_{r=1}^{q} \sigma_r(t, X^0)\, dw_r, \quad X^1(0) = 0. \tag{4.1.13}$$

The system (4.1.12) is the system of deterministic differential equations for which, as is generally known, efficient high-order numerical methods exist, for example,

$$X^0_{k+1} = X^0_k + a_k h + (a\, a'_x + a'_t)_k h^2/2, \tag{4.1.14}$$

$$X^0_0 = X_0, \quad R_0 = O(h^2),$$

where $a_k = a(t_k, X^0_k)$, the $d \times d$-matrix $(a'_x)_k$ is equal to $\partial a(t_k, X^0_k)/\partial x$, d-vector $(a'_t)_k$ is equal to $\partial a(t_k, X^0_k)/\partial t$, and R_0 is the error of the method. The system (4.1.13) is the system of SDEs with additive noise. The Euler method for the system (4.1.13) has the form

$$X^1_{k+1} = X^1_k + \sum_{r=1}^{q} (\sigma_r I_r)_k + (a'_x X^1)_k\, h, \tag{4.1.15}$$

$$X^1_o = 0, \quad (E(R_1)^2)^{1/2} = O(h),$$

where $\sigma_{r_k} = \sigma_r(t_k, X^0_k)$, $(a'_x)_k = \partial a(t_k, X^0_k)/\partial x$, and R_1 is the error of the method. So, we obtain the method (4.1.11), (4.1.14), (4.1.15) for numerical solution of the system (4.0.2) with the error $O(h^2 + \varepsilon^2)$.

One can see that the (h, ε)-approach and the (ε, h)-approach are substantially different. If time increment h tends to zero, a method, constructed by the (ε, h)-approach, does not converge to the exact solution and converges to $X^0(t) + \varepsilon X^1(t)$. In contrast to the (ε, h)-approach, the (h, ε)-approach gives a method which always converges to the exact solution of the system (4.0.2) when $h \to 0$. Our aim is to derive numerical methods for solution of the system (4.0.2) with small but fixed parameter $\varepsilon > 0$. That is why the (h, ε)-approach is more preferable than the (ε, h)-approach.

4.2 Some Specific Mean-Square Methods for Systems with Small Noise

In this section our aim is to construct methods with small mean-square errors (provided that ε is a small parameter) and with simply simulated random variables. Herein we restrict ourselves to the methods which contain the following Ito integrals

$$I_r = h^{1/2}\xi_r, \quad I_{r0} = h^{3/2}(\eta_r/\sqrt{3} + \xi_r)/2, \quad I_{0r} = hI_r - I_{r0}, \tag{4.2.1}$$

$$J_r = \int_0^h \vartheta w_r(\vartheta)\, d\vartheta = h^{5/2}(\xi_r/3 + \eta_r/(4\sqrt{3}) + \zeta_r/(12\sqrt{5})),$$

$$I_{r00} = hI_{r0} - J_r, \quad I_{0r0} = 2J_r - hI_{r0}, \quad I_{00r} = h^2 I_r/2 - J_r,$$

where ξ_r, η_r, ζ_r are independent normally distributed $N(0, 1)$ random variables with zero mean and unit standard deviation. The used random variables (Ito integrals) of all the methods proposed in this section are simulated at each step according to the formulas (4.2.1).

We will restrict ourselves to the set of most common and, in our opinion, useful methods and illustrate the proposed approach to numerical solution of a stochastic system with small noise. A lot of other methods can be derived. First, by adding or omitting some terms one can obtain methods that are similar to the ones given below but have other mean-square errors, for instance, $O(h^5 + \cdots)$, $O(h^6 + \cdots)$. Second, it is possible to derive other types of methods, for instance, implicit Runge-Kutta methods. The presented methods are obtained using the arguments of Sect. 4.1 and further expansion of $X(t + h)$. Their detailed derivation is omitted.

4.2.1   Taylor-Type Numerical Methods

4.2.1.1   Method $O(h + \cdots)$

The simplest numerical method is the Euler one:

$$X_{k+1} = X_k + \varepsilon \sum_{r=1}^{q} (\sigma_r \, I_r)_k + a_k h, \tag{4.2.2}$$

$$E\rho = O(h^2), \quad (E\rho^2)^{1/2} = O(h^2 + \varepsilon^2 h),$$
$$(ER^2)^{1/2} = O(h + \varepsilon^2 h^{1/2}).$$

4.2.1.2   Methods $O(h^2 + \cdots)$ and $O(h^3 + \cdots)$

These methods are based on the one-step approximations which have been derived in Sect. 4.1. The mean-square error of the method (4.1.3) is equal to $O(h^2 + \varepsilon^2 h^{1/2})$. The mean-square error of the method (4.1.10) is estimated by $O(h^2 + \varepsilon h + \varepsilon^2 h^{1/2})$. The method based on the one-step approximation (4.1.6) has the error $(ER^2)^{1/2} = O(h^3 + \varepsilon^2 h^{1/2})$.

4.2.1.3   Methods $O(h^4 + \cdots)$

The following method is obtained:

$$X_{k+1} = X_k + \varepsilon \sum_{r=1}^{q} (\sigma_r I_r)_k + a_k h + \varepsilon \sum_{r=1}^{q} (L_1 \sigma_r I_{0r})_k + \varepsilon \sum_{r=1}^{q} (\Lambda_r a I_{r0})_k \qquad (4.2.3)$$

$$+ L_1 a_k h^2/2 + \varepsilon \sum_{r=1}^{q} (L_1^2 \sigma_r I_{00r})_k + \varepsilon \sum_{r=1}^{q} (L_1 \Lambda_r a I_{0r0})_k$$

$$+ \varepsilon \sum_{r=1}^{q} (\Lambda_r L_1 a I_{r00})_k + L_1^2 a_k h^3/6 + L_1^3 a_k h^4/24 ,$$

$$E\rho = O(h^5 + \varepsilon^2 h^2), \quad (E\rho^2)^{1/2} = O(h^5 + \varepsilon^2 h),$$
$$(ER^2)^{1/2} = O(h^4 + \varepsilon^2 h^{1/2}) .$$

In some cases the derived methods may be improved due to special properties of a particular SDEs system. For instance, let us consider the commutative case, i.e., when $\Lambda_i \sigma_r = \Lambda_r \sigma_i$, or a system with one noise. For such systems, we obtain

$$X_{k+1} = A(t_k, X_k, h; (\xi, \eta, \zeta)_k) + \varepsilon^2 \sum_{i=1}^{q-1} \sum_{r=i+1}^{q} (\Lambda_i \sigma_r I_i I_r)_k \qquad (4.2.4)$$

$$+ \varepsilon^2 \sum_{i=1}^{q} (\Lambda_i \sigma_i (I_i^2 - h)/2)_k + \varepsilon^2 L_2 a_k h^2/2 ,$$

$$E\rho = O(h^5 + \varepsilon^2 h^3), \quad (E\rho^2)^{1/2} = O(h^5 + \varepsilon^2 h^2 + \varepsilon^3 h^{3/2}),$$
$$(ER^2)^{1/2} = O(h^4 + \varepsilon^2 h^{3/2} + \varepsilon^3 h),$$

where $A(t_k, X_k, h; (\xi, \eta, \zeta)_k)$ is equal to the right-hand side of (4.2.3).

Note that for the system with one noise ($q = 1$) the term $\varepsilon^2 \sum_{i=1}^{q-1} \sum_{r=i+1}^{q} (\Lambda_i \sigma_r I_i I_r)_k$ is neglected. One can see that the error of the method (4.2.4) is smaller than the error of the scheme (4.2.3). Moreover, the mean-square order of the method (4.2.4) is equal to one, while the mean-square order of the method (4.2.3) is equal to one-half.

4.2.2   Runge-Kutta-Type Methods

To reduce calculations of derivatives in the methods of Sect. 4.2.1, we propose Runge-Kutta schemes (in fact, they are Runge-Kutta-type methods because they need calculation of a reduced number of derivatives).

4.2.2.1   Method $O(h^2 + \cdots)$

The following method is obtained:

$$X_{k+1} = X_k + \varepsilon \sum_{r=1}^{q} (\sigma_r I_r)_k \tag{4.2.5}$$

$$+ \left[a\!\left(t_k + h, X_k + \varepsilon \sum_{r=1}^{q} (\sigma_r I_r)_k + a_k h\right) + a_k \right] h/2$$

$$+ \varepsilon \sum_{r=1}^{q} (L_1 \sigma_r I_{0r})_k + \varepsilon \sum_{r=1}^{q} [\Lambda_r a (I_{r0} - I_r h/2)]_k \,,$$

$$E\rho = O(h^3), \quad (E\rho^2)^{1/2} = O(h^3 + \varepsilon^2 h),$$
$$(ER^2)^{1/2} = O(h^2 + \varepsilon^2 h^{1/2}).$$

4.2.2.2   Method $O(h^3 + \cdots)$

The following method is obtained:

$$X_{k+1} = X_k + (k_1 + 4k_2 + k_3)/6 + \varepsilon \sum_{r=1}^{q} (\sigma_r I_r)_k \tag{4.2.6}$$

$$+ \varepsilon \sum_{r=1}^{q} (L_1 \sigma_r I_{0r})_k + \varepsilon \sum_{r=1}^{q} (\Lambda_r a I_{r0})_k \,,$$

$$(ER^2)^{1/2} = O(h^3 + \varepsilon^2 h^{1/2}),$$

where

$$k_1 = ha(t_k, X_k), \quad k_2 = ha(t_k + h/2, X_k + k_1/2),$$
$$k_3 = ha(t_{k+1}, X_k - k_1 + 2k_2).$$

4.2.2.3   Methods $O(h^4 + \cdots)$

The following method is obtained:

$$X_{k+1} = X_k + (k_1 + 2k_2 + 2k_3 + k_4)/6 + \varepsilon \sum_{r=1}^{q} (\sigma_r I_r)_k \qquad (4.2.7)$$

$$+\varepsilon \sum_{r=1}^{q} (L_1\sigma_r I_{0r})_k + \varepsilon \sum_{r=1}^{q} (\Lambda_r a I_{r0})_k + \varepsilon \sum_{r=1}^{q} (L_1^2 \sigma_r I_{00r})_k$$

$$+\varepsilon \sum_{r=1}^{q} (L_1 \Lambda_r a I_{0r0})_k + \varepsilon \sum_{r=1}^{q} (\Lambda_r L_1 a I_{r00})_k ,$$

$$(ER^2)^{1/2} = O(h^4 + \varepsilon^2 h^{1/2}) ,$$

where

$$k_1 = ha(t_k, X_k), \quad k_2 = ha(t_k + h/2, X_k + k_1/2),$$
$$k_3 = ha(t_k + h/2, X_k + k_2/2), \quad k_4 = ha(t_{k+1}, X_k + k_3).$$

In the commutative case the method (4.2.7) can be improved as in Sect. 4.2.1.3. The simpler method

$$X_{k+1} = X_k + (k_1 + 2k_2 + 2k_3 + k_4)/6 + \varepsilon \sum_{r=1}^{q} (\sigma_r I_r)_k , \qquad (4.2.8)$$

where k_i are calculated as in (4.2.7), has the larger error in comparison with (4.2.7):

$$(ER^2)^{1/2} = O(h^4 + \varepsilon h + \varepsilon^2 h^{1/2}).$$

In addition to the schemes presented in this subsection, other mean-square Runge-Kutta methods for SDEs with small noise can be found in [49, 300]. Further, mean-square multi-step methods for SDEs with small noise are considered in [50].

4.2.3 Implicit Methods

4.2.3.1 Methods $O(h + \cdots)$

The one-parameter family of implicit Euler schemes has the form

$$X_{k+1} = X_k + \varepsilon \sum_{r=1}^{q} (\sigma_r I_r)_k + \alpha h a_k + (1 - \alpha) h a_{k+1} , \qquad (4.2.9)$$

$$0 \le \alpha \le 1, \quad (ER^2)^{1/2} = O(h + \varepsilon^2 h^{1/2}).$$

4.2.3.2 Methods $O(h^2 + \cdots)$

The two-parameter family of implicit schemes (1.3.7) applied to (4.0.2) has the form

$$X_{k+1} = X_k + \varepsilon \sum_{r=1}^{q} (\sigma_r I_r)_k + \alpha h a_k + (1 - \alpha) h a_{k+1} \qquad (4.2.10)$$

$$+\varepsilon \sum_{r=1}^{q} (\Lambda_r a (I_{r0} - (1 - \alpha) I_r h))_k + \varepsilon \sum_{r=1}^{q} (L_1 \sigma_r I_{0r})_k$$

$$+\beta(2\alpha - 1) L_1 a_k h^2 / 2 + (1 - \beta)(2\alpha - 1) L_1 a_{k+1} h^2 / 2 \,,$$

$$0 \le \alpha \le 1, \ \ 0 \le \beta \le 1 \,,$$

$$E\rho = O(h^3 + \varepsilon^2 h^2), \ \ (E\rho^2)^{1/2} = O(h^3 + \varepsilon^2 h) \,,$$
$$(ER^2)^{1/2} = O(h^2 + \varepsilon^2 h^{1/2}) \,.$$

If $\alpha = 1/2$, we obtain the trapezoidal method which is the simplest of the family (4.2.10):

$$X_{k+1} = X_k + \varepsilon \sum_{r=1}^{q} (\sigma_r I_r)_k + h(a_k + a_{k+1})/2 \qquad (4.2.11)$$

$$+\varepsilon \sum_{r=1}^{q} (L_1 \sigma_r I_{0r})_k + \varepsilon \sum_{r=1}^{q} (\Lambda_r a (I_{r0} - I_r h/2))_k \,,$$

$$(ER^2)^{1/2} = O(h^2 + \varepsilon^2 h^{1/2}) \,.$$

In the commutative case or in the case of one noise the methods (4.2.10)–(4.2.11) can be improved as the method (4.2.3) in Sect. 4.2.1.

4.2.4 Stratonovich SDEs with Small Noise

For some physical applications the Stratonovich interpretation of a stochastic system is preferable. The stochastic system in the Stratonovich sense

$$dX = a(t, X) \, dt + \varepsilon \sum_{r=1}^{q} \sigma_r(t, X) \circ dw_r \,, \ \ X(t_0) = X_0 \,, \qquad (4.2.12)$$

is equivalent to the following system of the Ito SDEs

$$dX = \left[a(t, X) + \frac{\varepsilon^2}{2} \sum_{r=1}^{q} \frac{\partial \sigma_r}{\partial x}(t, X) \sigma_r(t, X) \right] dt + \varepsilon \sum_{r=1}^{q} \sigma_r(t, X) \, dw_r , \quad (4.2.13)$$

$$X(t_0) = X_0 .$$

In the general case ($\varepsilon = 1$) numerical methods constructed for the Ito system are easily rewritten for the Stratonovich system by adding the term $\frac{1}{2} \sum_{r=1}^{q} (\partial \sigma_r / \partial x) \sigma_r$ to the drift (see Sect. 1.1.3). However, in the case of small noise the additional term is multiplied by small factor ε^2. So, the Stratonovich system with small noise (4.2.12) is distinguished from the Ito system $dX = a(t, X)dt + \varepsilon \sum_{r=1}^{q} \sigma_r(t, X)dw_r$ by the small component in the drift, and constructing a numerical method for the system (4.2.13), one must take the magnitude of the additional term into account.

Most of the methods for the system (4.2.12) are obtained from methods for the Ito system (4.0.2) by adding the term $\dfrac{\varepsilon^2}{2} \sum_{r=1}^{q} (\partial \sigma_r / \partial x) \sigma_r h$. Namely, for the Stratonovich system (4.2.12) the methods (4.1.3), (4.1.6), (4.1.10), (4.2.2), (4.2.3), (4.2.5), (4.2.6), (4.2.7) and (4.2.8) acquire the form

$$X_{k+1} = A(t_k, X_k, h; (\xi, \eta, \zeta)_k) + \frac{\varepsilon^2}{2} \sum_{r=1}^{q} \left(\frac{\partial \sigma_r}{\partial x} \sigma_r \right)_k h , \quad (4.2.14)$$

where the expressions $A(t_k, X_k, h; (\xi, \eta, \zeta)_k)$ are calculated according to the same rules as the right-hand sides of the corresponding methods for Ito systems, and the corresponding errors have the same order of smallness.

In the commutative case we obtain

$$X_{k+1} = A(t_k, X_k, h; (\xi, \eta, \zeta)_k) + \frac{\varepsilon^2}{2} \sum_{r=1}^{q} \left(\frac{\partial \sigma_r}{\partial x} \sigma_r \right)_k h \quad (4.2.15)$$

$$+ \varepsilon^2 \sum_{i=1}^{q-1} \sum_{r=i+1}^{q} (\Lambda_i \sigma_r I_i I_r)_k + \varepsilon^2 \sum_{i=1}^{q} \left[\Lambda_i \sigma_i (I_i^2 - h)/2 \right]_k$$

$$+ \varepsilon^2 L_1 \left[\sum_{r=1}^{q} \frac{\partial \sigma_r}{\partial x} \sigma_r \right]_k h^2/4 + \varepsilon^2 (\tilde{L}_2 a)_k h^2/2 ,$$

$$(ER^2)^{1/2} = O(h^4 + \varepsilon^2 h^{3/2} + \varepsilon^3 h) ,$$

where $A(t_k, X_k, h; (\xi, \eta, \zeta)_k)$ is the right-hand side of (4.2.3) and

$$\tilde{L}_2 = \frac{1}{2} \sum_{r=1}^{q} \left(\sigma_r, \frac{\partial}{\partial x} \right)^2 + \frac{1}{2} \sum_{r=1}^{q} \left(\frac{\partial \sigma_r}{\partial x} \sigma_r, \frac{\partial}{\partial x} \right) .$$

Analogously, the method for the Stratonovich system, which corresponds to the method (4.2.7) for the Ito system, can be improved in the commutative case.

The one-parameter family of implicit methods for the Stratonovich system has the form

$$X_{k+1} = A(t_k, X_k, h; (\xi)_k) \tag{4.2.16}$$

$$+ \frac{\varepsilon^2}{2} \sum_{r=1}^{q} \left[\alpha \left(\frac{\partial \sigma_r}{\partial x} \sigma_r \right)_k + (1 - \alpha) \left(\frac{\partial \sigma_r}{\partial x} \sigma_r \right)_{k+1} \right] h,$$

$$(ER^2)^{1/2} = O(h + \varepsilon^2 h^{1/2}),$$

where $A(t_k, X_k, h; (\xi)_k)$ is the right-hand side of (4.2.9).

The two-parameter family of implicit methods for the Stratonovich system has the form

$$X_{k+1} = A(t_k, X_k, h; (\xi, \eta)_k) \tag{4.2.17}$$

$$+ \frac{\varepsilon^2}{2} \sum_{r=1}^{q} \left[\alpha \left(\frac{\partial \sigma_r}{\partial x} \sigma_r \right)_k + (1 - \alpha) \left(\frac{\partial \sigma_r}{\partial x} \sigma_r \right)_{k+1} \right] h,$$

$$(ER^2)^{1/2} = O(h^2 + \varepsilon^2 h^{1/2}),$$

where $A(t_k, X_k, h; (\xi, \eta)_k)$ is the right-hand side of (4.2.10).

4.2.5 Mean-Square Methods for Systems with Small Additive Noise

One of the important particular cases of the system (4.0.2) is the system with additive noise:

$$dX = a(t, X) \, dt + \varepsilon \sum_{r=1}^{q} \sigma_r(t) \, dw_r. \tag{4.2.18}$$

Note that in this case the Stratonovich system coincides with the Ito system.

For the system (4.2.18) we obtain the Taylor-type and Runge-Kutta methods with the errors $O(h^2 + \varepsilon h)$, $O(h^2 + \varepsilon^2 h)$, $O(h^2 + \varepsilon^2 h^{3/2})$, $O(h^3 + \varepsilon^2 h^{3/2})$, $O(h^4 + \varepsilon^2 h^{3/2})$ which are similar to the methods of Sects. 4.2.1–4.2.3. We also obtain the implicit schemes which follow from the schemes of Sect. 4.2.3. Herein we restrict ourselves to the methods with errors $O(h^4 + \cdots)$ which, from our point of view, are the most interesting.

The Taylor-type method with the error $O(h^4 + \varepsilon^2 h^{3/2})$ is

$$X_{k+1} = X_k + \varepsilon \sum_{r=1}^{q} (\sigma_r I_r)_k + a_k h + \varepsilon \sum_{r=1}^{q} \left(\frac{d\sigma_r}{dt} I_{0r} \right)_k \tag{4.2.19}$$

$$+ \varepsilon \sum_{r=1}^{q} (\Lambda_r a I_{r0})_k + (L_1 + \varepsilon^2 L_2) a_k h^2 / 2 + \varepsilon \sum_{r=1}^{q} \left(\frac{d^2 \sigma_r}{dt^2} I_{00r} \right)_k$$

$$+ \varepsilon \sum_{r=1}^{q} (L_1 \Lambda_r a I_{0r0})_k + \varepsilon \sum_{r=1}^{q} (\Lambda_r L_1 a I_{r00})_k + L_1^2 a_k h^3 / 6 + L_1^3 a_k h^4 / 24 ,$$

$$E\rho = O(h^5 + \varepsilon^2 h^3) , \quad (E\rho^2)^{1/2} = O(h^5 + \varepsilon^2 h^2) ,$$
$$(ER^2)^{1/2} = O(h^4 + \varepsilon^2 h^{3/2}) .$$

The Runge-Kutta method with the error $O(h^4 + \varepsilon^2 h^{3/2})$ is

$$X_{k+1} = X_k + (k_1 + 2k_2 + 2k_3 + k_4)/6 + \varepsilon \sum_{r=1}^{q} (\sigma_r I_r)_k \tag{4.2.20}$$

$$+ \varepsilon \sum_{r=1}^{q} \left(\frac{d\sigma_r}{dt} I_{0r} \right)_k + \varepsilon \sum_{r=1}^{q} (\Lambda_r a I_{r0})_k + \varepsilon^2 L_2 a_k h^2 / 2$$

$$+ \varepsilon \sum_{r=1}^{q} \left(\frac{d^2 \sigma_r}{dt^2} I_{00r} \right)_k + \varepsilon \sum_{r=1}^{q} (L_1 \Lambda_r a I_{0r0})_k + \varepsilon \sum_{r=1}^{q} (\Lambda_r L_1 a I_{r00})_k ,$$

$$(ER^2)^{1/2} = O(h^4 + \varepsilon^2 h^{3/2}) ,$$

where

$$k_1 = ha(t_k, X_k) , \quad k_2 = ha(t_k + h/2, X_k + k_1/2) ,$$
$$k_3 = ha(t_k + h/2, X_k + k_2/2) , \quad k_4 = ha(t_{k+1}, X_k + k_3) .$$

It is possible to obtain the simpler method which coincides with the scheme (4.2.8) but its mean-square error in the case of additive noise is equal to $(ER^2)^{1/2} = O(h^4 + \varepsilon h)$.

4.3   Numerical Tests of Mean-Square Methods

In this section we test the methods constructed in the previous sections of this chapter on two examples: evaluation of a Lyapunov exponent (Sect. 4.3.1) and a stochastic laser model (Sect. 4.3.2).

4.3.1 Simulation of the Lyapunov Exponent of a Linear System with Small Noise

It is known [12, 194] that one can investigate stability of a dynamical stochastic system by Lyapunov exponents. The negativeness of upper Lyapunov exponents is an indication of system stability. It is usually impossible to derive analytical expressions for Lyapunov exponents. In this case numerical approaches are useful. For the first time an algorithm of numerical computation of Lyapunov exponents was proposed in [430]. The algorithm is based on weak schemes (see also Sect. 4.8).

Here we calculate the Lyapunov exponent as a convenient example to illustrate the effectiveness (in comparison with ordinary mean-square schemes) of the proposed methods. Although the weak schemes are usually more efficient than mean-square ones, our approach is interesting in itself because we find the exponent together with an approximate trajectory of the SDEs solution.

Let us consider the following two-dimensional linear Ito stochastic system

$$dX = AX dt + \varepsilon \sum_{r=1}^{q} B_r X \, dw_r \,, \tag{4.3.1}$$

where X is a two-dimensional vector, A and B_r are constant 2×2-matrices, $w_r(t)$ are independent standard Wiener processes, $\varepsilon > 0$ is a small parameter. In ergodic case the unique Lyapunov exponent λ of the system (4.3.1) exists [194] and

$$\lambda = \lim_{t\to\infty} \frac{1}{t} E(ln|X(t)|) = \lim_{t\to\infty} \frac{1}{t} ln|X(t)| \,, \tag{4.3.2}$$

where $X(t)$, $t \geq 0$, is a non-trivial solution of the system (4.3.1). The last equality of (4.3.2) holds with probability one. A non-trivial solution of the system (4.3.1) is asymptotically stable with probability one if and only if the Lyapunov exponent λ is negative [194].

In [12, 14] the expansion of Lyapunov exponent of the system (4.3.1) in powers of small parameter ε was obtained. In the case of

$$A = \begin{bmatrix} a & c \\ -c & a \end{bmatrix} \,, \quad B_r = \begin{bmatrix} b_r & d_r \\ -d_r & b_r \end{bmatrix} \,, \quad r = 1, \dots, q \,, \tag{4.3.3}$$

the Lyapunov exponent of the system (4.3.1) is exactly equal to [14]

$$\lambda = a + \frac{\varepsilon^2}{2} \sum_{r=1}^{q} [(d_r)^2 - (b_r)^2] \,. \tag{4.3.4}$$

To test the mean-square methods proposed in this chapter, we choose the case (4.3.3) of the system (4.3.1) with two independent noises. We calculate the function $\bar{\lambda}(t)$:

$$\bar{\lambda}(t) = \frac{1}{t} \ln |\bar{X}(t)| \approx \frac{1}{t} \ln |X(t)| \tag{4.3.5}$$

which in the limit of large time ($t \to \infty$) tends to an approximation of the Lyapunov exponent λ. The approximation $\bar{X}(t)$ of the exact solution $X(t)$ of the system (4.3.1) is simulated by three mean-square schemes: (i) the first order method (1.0.5) with the error $O(h)$ which in our case is efficient as to simulation of the used random variables due to commutativity of the matrices $B_r, r = 1, \ldots, q$, (ii) the simplified version of the Runge-Kutta scheme (4.2.5) with the error $O(h^2 + \varepsilon h + \varepsilon^2 h^{1/2})$, and (iii) the Runge-Kutta scheme (4.2.8) with the error $O(h^4 + \varepsilon h + \varepsilon^2 h^{1/2})$ (see Figs. 4.3.1 and 4.3.2).

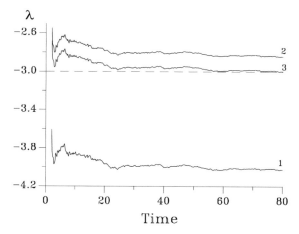

Fig. 4.3.1 Lyapunov exponent. Time dependence of the function $\bar{\lambda}(t)$, $t \geq 2$, for $a = -3, c = 1$, $b_1 = b_2 = 1, d_1 = 1, d_2 = -1, \varepsilon = 0.1, X^1(0) = 0, X^2(0) = 1$ and time step $h = 0.3$. The solution of the system (4.3.1)–(4.3.3) is approximated by (1) the method with the error $O(h)$, (2) the Runge-Kutta method with the error $O(h^2 + \varepsilon h + \varepsilon^2 h^{1/2})$, and (3) the Runge-Kutta method with the error $O(h^4 + \varepsilon h + \varepsilon^2 h^{1/2})$. Dashed line is the exact value of the Lyapunov exponent λ ($\lambda = -3$)

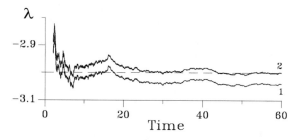

Fig. 4.3.2 Lyapunov exponent. Time dependence of the function $\bar{\lambda}(t)$, $t \geq 2$, for $h = 0.01$, the other parameters are the same as in Fig. 4.3.1. The solution of the system (4.3.1)–(4.3.3) is approximated by (1) the method with the error $O(h)$ and (2) the Runge-Kutta methods with the errors $O(h^2 + \varepsilon h + \varepsilon^2 h^{1/2})$ and $O(h^4 + \varepsilon h + \varepsilon^2 h^{1/2})$. Dashed line is the exact value of the Lyapunov exponent λ ($\lambda = -3$)

4.3.2   Stochastic Model of a Laser

Our second example is devoted to a stochastic model of laser [140, 377] which can be written as the following Stratonovich system

$$dX^1 = (\alpha_0 X^1 - \beta_0 X^2 - (AX^1 - BX^2)XX^*)\,dt \qquad (4.3.6)$$

$$+\varepsilon\left[\sum_{i=1}^{2}(\alpha_i X^1 - \beta_i X^2)\circ dw_i + \sigma\,dw_3\right]$$

$$dX^2 = (\beta_0 X^1 + \alpha_0 X^2 - (BX^1 + AX^2)XX^*)\,dt$$

$$+\varepsilon\left[\sum_{i=1}^{2}(\beta_i X^1 + \alpha_i X^2)\circ dw_i + \sigma\,dw_4\right],$$

where

$$X = X^1 + i\,X^2,\quad X^* = X^1 - i\,X^2.$$

For $\varepsilon = 0$ the system (4.3.6) becomes deterministic. In the case of $\alpha_0/A > 0$ it has asymptotically stable limit cycle $(X^1)^2 + (X^2)^2 = \alpha_0/A$. The radius $\rho = |X|$ under $\varepsilon = 0$ satisfies the equation

$$d\rho/dt = \rho(\alpha_0 - A\rho^2)$$

and does not depend on the detuning parameters β_0 and B. The value ρ^2 for $\varepsilon \neq 0$ satisfies the Stratonovich equation

$$d\rho^2 = 2\rho^2(\alpha_0 - A\rho^2)dt + 2\varepsilon\rho^2(\alpha_1 \circ dw_1 + \alpha_2 \circ dw_2)$$
$$+2\varepsilon\sigma(X_1 \circ dw_3 + X_2 \circ dw_4)$$

and also does not depend on β_0 and B. But the difference equations, which are the result of applying numerical methods to the system (4.3.6), essentially depend not only on the choice of a scheme and time step but also on the detuning parameters, and growing of $|\beta_0 - B|$ leads to vanishing of stable cycle. Therefore, to solve the system (4.3.6), one must use high-order schemes or choose sufficiently small time step. Since the system (4.3.6) contains multiplicative noise and does not belong to the class of systems with commutative noise, the Euler method is the highest order scheme among known mean-square methods with easily simulated random variables (see Chap. 1). The Euler method has the mean-square error $O(h + \varepsilon^2 h^{1/2})$ and in the case of large $|\beta_0 - B|$ too small step h is required. On the other hand, for instance, the method with the mean-square error $O(h^4 + \varepsilon h + \varepsilon^2 h^{1/2})$ allows us to obtain sufficiently accurate approximations of solutions of the system (4.3.6) and, particularly, to simulate phase trajectories.

The radius $\rho = |X_k|$ of a typical trajectory is plotted in Figs. 4.3.3 and 4.3.4. In Fig. 4.3.3 the radius ρ is calculated with the time step $h = 0.005$ by the Euler

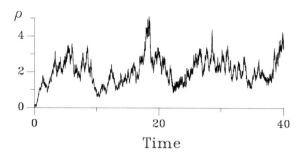

Fig. 4.3.3 Stochastic model of a laser. Time dependence of the radius $\rho = |X_k|$ for $\alpha_0 = 0.5$, $\beta_0 = 1$, $A = 0.1$, $B = 0.4$, $\varepsilon = 0.3$, $\alpha_i = \beta_i = \sigma = 1$, $i = 1, 2$, $X^1(0) = X^2(0) = 0$, and time step $h = 0.005$. The solution X_k of the system (4.3.6) is approximated by the Euler method and by the Runge-Kutta method with the error $O(h^4 + \varepsilon h + \varepsilon^2 h^{1/2})$

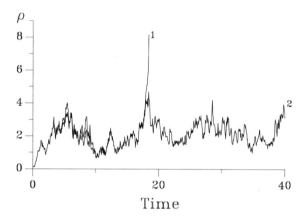

Fig. 4.3.4 Stochastic model of a laser. Time dependence of the radius $\rho = |X_k|$ for time step $h = 0.05$, the other parameters are the same as in Fig. 4.3.3. The solution X_k of the system (4.3.6) is approximated by (1) the Euler method and (2) the Runge-Kutta method with the error $O(h^4 + \varepsilon h + \varepsilon^2 h^{1/2})$

scheme and by the Runge-Kutta scheme with the error $O(h^4 + \varepsilon h + \varepsilon^2 h^{1/2})$ which corresponds to the method (4.2.8) for the Ito system. In this case both methods give the same results, and the trajectories plotted in Fig. 4.3.3 can be considered as exact. As seen in Fig. 4.3.4, if one chooses the larger time step ($h = 0.05$), the Runge-Kutta scheme gives quite well results (compare with Fig. 4.3.3) but the Euler method becomes unstable. In all cases we use the same sample paths for the Wiener processes.

Note that the Runge-Kutta method (4.2.8) and the corresponding method for the Stratonovich system can be improved up to $(ER^2)^{1/2} = O(h^4 + \varepsilon^2 h^{1/2})$ (see the method (4.2.7)).

4.4 The Main Theorem on Error Estimation and General Approach to Construction of Weak Methods

We construct weak schemes for (4.0.1) on the basis of the proposed mean-square methods. See Chap. 2 for definitions and main notions (in particular, Definition 2.1.1 of the class **F**) related to numerical integration of SDEs in the weak sense.

Theorem 4.4.1 *Assume that the following conditions hold.*
(1) The coefficients of the system (4.0.1) are continuous and satisfy a global Lipschitz condition with respect to $x \in \mathbf{R}^d$, *they and their partial derivatives up to a sufficiently high order belong to the class* **F**;
  (2) The error of a one-step approximation $\bar{X}_{t,x}(t+h)$ *of the exact solution* $X_{t,x}(t+h)$ *of the system (4.0.1) with initial condition* $X(t) = \bar{X}(t) = x$ *is estimated by*

$$|Ef(X_{t,x}(t+h)) - Ef(\bar{X}_{t,x}(t+h))| \tag{4.4.1}$$

$$\leq K(x)\left[h^{p+1} + \sum_{l \in S} h^{l+1} \varepsilon^{J(l)}\right], \quad K(x) \in \mathbf{F},$$

where the function $f(x)$ *and its partial derivatives up to a sufficiently high order belong to the class* **F**, *S is a subset of the positive integers* $\{1, 2, \ldots, p\}$, *J is a decreasing function on S with natural values;*
  (3) For a sufficiently large integer m the moments $E|\bar{X}_k|^m$ *exist and are uniformly bounded with respect to* $N, k = 0, 1, \ldots, N,$ *and* $0 \leq \varepsilon \leq \varepsilon_0.$
  Then for any N and $k = 0, 1, \ldots, N$

$$|Ef(X_{t_0,X_0}(t_k)) - Ef(\bar{X}_{t_0,X_0}(t_k))| \leq K\left[h^p + \sum_{l \in S} h^l \varepsilon^{J(l)}\right], \tag{4.4.2}$$

where the constant K does not depend on h, ε, *and* k.

  The proof of Theorem 4.4.1 differs only little from the proof of Theorem 2.2.1 and is therefore omitted.
  According to Theorem 4.4.1, to estimate the global error of a method we need properties of the corresponding one-step approximation, i.e., to prove an error for a weak method we need the estimate (4.4.1). By using the Taylor expansion of the function f, it is possible to obtain the estimate (4.4.1), provided the inequalities (cf. Theorem 2.2.1):

$$\left|E\prod_{j=1}^{s} \Delta^{i_j} - E\prod_{j=1}^{s} \bar{\Delta}^{i_j}\right| \leq K(x)(h^{p+1} + \sum_{l \in S} h^{l+1} \varepsilon^{J(l)}), \ s = 1, \ldots, \bar{s} - 1,$$

$$\tag{4.4.3}$$

$$\Delta^{i_j} = X^{i_j}(t+h) - x^{i_j}, \quad \bar{\Delta}^{i_j} = \bar{X}^{i_j}(t+h) - x^{i_j},$$

$$X(t) = \bar{X}(t) = x, \quad i_j = 1, \ldots, d,$$

$$E \prod_{j=1}^{\bar{s}} |\bar{\Delta}^{i_j}| \leq K(x)(h^{p+1} + \sum_{l \in S} h^{l+1} \varepsilon^{J(l)}), \quad i_j = 1, \ldots, d, \tag{4.4.4}$$

$$E \prod_{j=1}^{\bar{s}} |\Delta^{i_j}| \leq K(x)(h^{p+1} + \sum_{l \in S} h^{l+1} \varepsilon^{J(l)}), \quad i_j = 1, \ldots, d. \tag{4.4.5}$$

hold. The number $\bar{s}$ in (4.4.3)–(4.4.5) must be such that $h^k(\varepsilon h^{1/2})^{\bar{s}-k} = O(h^{p+1} + \sum_{l \in S} h^{l+1} \varepsilon^{J(l)})$ for even $\bar{s} - k$ where $0 \leq k \leq \bar{s}$. The last assertion follows from an analysis of the Taylor expansion of $Ef(X_{t,x}(t+h))$ in powers of h and ε. We underline that $\bar{s}$ is greater than or equal to $(2 + 2\min\{l \in S\})$ and, as a rule, less than $2p + 2$.

In the case of the system (4.0.1) the one-step approximation (2.1.10) can be written in the form

$$\tilde{X}(t+h) = x + \varepsilon \sum_{r=1}^{q} \sigma_r I_r + h(a + \varepsilon^2 b) + \varepsilon^2 \sum_{i,r=1}^{q} \Lambda_i \sigma_r I_{ir} \tag{4.4.6}$$

$$+ \varepsilon \sum_{r=1}^{q} (L_1 + \varepsilon^2 L_2) \sigma_r I_{0r} + \varepsilon \sum_{r=1}^{q} \Lambda_r (a + \varepsilon^2 b) I_{r0}$$

$$+ h^2 (L_1 + \varepsilon^2 L_2)(a + \varepsilon^2 b)/2,$$

$$X(t+h) = \tilde{X}(t+h) + \tilde{\rho}. \tag{4.4.7}$$

The coefficients $\sigma_r, a, b, \Lambda_i \sigma_r$, etc. in (4.4.6) are calculated at the point (t, x), and $\tilde{\rho}$ in (4.4.7) is the remainder.

The weak method based on (4.4.6) has global error $O(h^2)$ (cf. (2.2.18) and Theorem 2.2.3), and for the system (4.0.1) it is written as

$$X_{k+1} = X_k + \varepsilon h^{1/2} \sum_{r=1}^{q} (\sigma_r \xi_r)_k + h(a + \varepsilon^2 b)_k + \varepsilon^2 h \sum_{i,r=1}^{q} (\Lambda_i \sigma_r \xi_{ir})_k \tag{4.4.8}$$

$$+ \varepsilon h^{3/2} \sum_{r=1}^{q} [(L_1 + \varepsilon^2 L_2) \sigma_r \xi_r]_k /2 + \varepsilon h^{3/2} \sum_{r=1}^{q} [\Lambda_r (a + \varepsilon^2 b) \xi_r]_k /2$$

$$+ h^2 [(L_1 + \varepsilon^2 L_2)(a + \varepsilon^2 b)]_k /2,$$

where the random variables are simulated according to

$$\xi_{ir} = (\xi_i \xi_r - \gamma_{ir} \zeta_i \zeta_r)/2, \ \gamma_{ir} = \begin{cases} -1, \ i < j \\ \ \ 1, \ \ i \geq j \end{cases}, \tag{4.4.9}$$

$$P(\xi = 0) = 2/3, \ P(\xi = \pm\sqrt{3}) = 1/6, \ P(\zeta = \pm 1) = 1/2,$$

ξ_r and ζ_i are mutually independent.

A higher-order method based on a one-step approximation of weak order 4 would be too complicated. However, as it will be shown below, a small modification of the one-step approximation (4.4.6) leads to an efficient method with the local error $O(h^4 + \varepsilon^2 h^3)$.

The remainder $\tilde{\rho}$ in (4.4.7) contains terms with factors $\varepsilon^3 h^{3/2}$, $\varepsilon^2 h^2$, $\varepsilon^4 h^2$, $\varepsilon h^{5/2}$, $\varepsilon^3 h^{5/2}$, $\varepsilon^5 h^{5/2}$. Contribution of these terms to the local error of the one-step weak approximation is not worse than $O(\varepsilon^2 h^3)$, which follows from properties of Ito integrals (see Lemmas 1.2.1, 2.1.3 and 2.1.5). Additionally, $\tilde{\rho}$ contains the term $h^3 L_1^2 a/6$, which yields a contribution to the local error of the one-step weak approximation equal to $O(h^3)$, and terms, for example, $\varepsilon^2 h^3$, h^4, etc., which contribute to the local error not more than $O(\varepsilon^2 h^3)$. It is clear that moving the term $h^3 L_1^2 a/6$ from $\tilde{\rho}$ to $\tilde{X}$ leads to a new one-step approximation which is not essentially more complicated, but is at the same time considerably more accurate than the previous one. Of course, this reasoning requires a rigorous proof which can be found in [301].

On this way we get the weak method of order $O(h^3 + \varepsilon^2 h^2)$:

$$X_{k+1} = X_k + \varepsilon h^{1/2} \sum_{r=1}^q (\sigma_r \xi_r)_k + h(a + \varepsilon^2 b)_k + \varepsilon^2 h \sum_{i,r=1}^q (\Lambda_i \sigma_r \xi_{ir})_k \tag{4.4.10}$$

$$+ \varepsilon h^{3/2} \sum_{r=1}^q \left[(L_1 + \varepsilon^2 L_2)\sigma_r \xi_r\right]_k /2 + \varepsilon h^{3/2} \sum_{r=1}^q \left[\Lambda_r (a + \varepsilon^2 b)\xi_r\right]_k /2$$

$$+ h^2 \left[(L_1 + \varepsilon^2 L_2)(a + \varepsilon^2 b)\right]_k /2 + h^3 (L_1^2 a)_k /6$$

with the random variables $(\xi_r)_k$ and $(\xi_{ir})_k$ simulated at each step according to (4.4.9). This method is of second order in h just like the standard second order method (4.4.8), which provides an error $O(h^2)$. But the term h^2 in the error of the method (4.4.10) is multiplied by ε^2. That is why the new method has smaller error than the standard scheme (4.4.8).

Above we shifted a term, which is sufficiently simply to simulate, from the remainder to the method, thereby reducing the error. However, we can also shift some more complicated terms, multiplied by ε^α, from the method to the corresponding remainder. Such a procedure reduces the computational costs (which, of course, is important for applications) while it does not lead to a substantial increase of the error.

For instance, by shifting the complicated (from the computational point of view) terms $\varepsilon^3 h^{3/2} L_2 \sigma_r \xi_r /2$ and $\varepsilon^4 h^2 L_2 b/2$ to the remainder, we obtain a further method to solve the system (4.0.1). It can be seen that such a method has a global error $O(h^3 + \varepsilon^4 h)$. Moreover, if we additionally transfer the terms $\varepsilon^2 h \Lambda_i \sigma_r \xi_{ir}$ and $\varepsilon^3 h^{3/2} \Lambda_r b \xi_r /2$ to the remainder, it can be proved that we do not loose the accuracy of the corresponding method, with respect to both h and ε. We finally arrive at the method

$$X_{k+1} = X_k + \varepsilon h^{1/2} \sum_{r=1}^{q} (\sigma_r \xi_r)_k + h(a + \varepsilon^2 b)_k \qquad (4.4.11)$$

$$+ \varepsilon h^{3/2} \sum_{r=1}^{q} (L_1 \sigma_r \xi_r)_k / 2 + \varepsilon h^{3/2} \sum_{r=1}^{q} (\Lambda_r a \xi_r)_k / 2$$

$$+ h^2 \left[(L_1 + \varepsilon^2 L_2) a \right]_k / 2 + \varepsilon^2 h^2 (L_1 b)_k / 2 + h^3 (L_1^2 a)_k / 6 \,,$$

the global error of which is $O(h^3 + \varepsilon^4 h)$. It is sufficient for its realization to simulate only q independent random variables ξ_r according to the law $P(\xi = \pm 1) = 1/2$. The time-step order of the method (4.4.11) is equal to one, i.e., it is lower than the time-step order of the method (4.4.10) and of the standard method (4.4.8). Nevertheless, for small ε the error behavior of the method (4.4.11) is acceptable. If we, for instance, choose a time-step h with $h = C\varepsilon^{\alpha}$, $0 < \alpha < 4$, the method (4.4.11) even beats the standard method (4.4.8) as far as the degree of smallness with respect to ε is concerned. Furthermore, if we choose a time-step $h = C\varepsilon^{\alpha}$, $0 < \alpha \leq 2$, then the method (4.4.11) is not worse than method (4.4.10) in the same sense. We want to emphasize that, additionally, the method (4.4.11) requires fewer calculations of both the number of simulated random variables and the number of arithmetic operations.

Thus, we have briefly explained how to construct weak methods for a system with small noise. Let us stress that to put these new methods on a sound basis one must thoroughly analyze the remainder and apply Theorem 4.4.1

4.5 Some Specific Weak Methods

In this section we aim at constructing weak methods which have small errors and are sufficiently effective with respect to their computational costs. Below we present the methods without detailed derivation.

4.5.1 Taylor-Type Methods

For the system (4.0.1) we obtain Taylor-type weak methods with errors $O(h^2 + \varepsilon^2 h)$, $O(h^2 + \varepsilon^4 h)$, $O(h^3 + \varepsilon^2 h)$, $O(h^3 + \varepsilon^4 h)$, $O(h^3 + \varepsilon^2 h^2)$, $O(h^3 + \varepsilon^4 h^2)$, $O(h^4 + \varepsilon^2 h)$, $O(h^4 + \varepsilon^2 h^2 + \varepsilon^4 h)$, $O(h^4 + \varepsilon^4 h)$, $O(h^4 + \varepsilon^2 h^2)$, $O(h^4 + \varepsilon^4 h^2)$. In Sect. 4.4 we have derived weak methods with errors $O(h^3 + \varepsilon^4 h)$ and $O(h^3 + \varepsilon^2 h^2)$. More methods can be derived in the same manner. In this subsection we state several methods with errors $O(h^4 + \cdots)$. Others can be found in the preprint [299]. There one also can find some implicit methods.

By the approach stated above it is possible to derive methods with errors $O(h^5 + \cdots)$, $O(h^6 + \cdots)$, etc. But we do not write them because most popular deterministic schemes have orders not higher than 4. Note that it is also possible to derive methods with errors $O(h^3 + \varepsilon^6 h^2)$, $O(h^4 + \varepsilon^6 h^2)$, $O(h^4 + \varepsilon^\alpha h^3)$, $\alpha = 2, \ldots, 8$, however they require large computational efforts.

The method with error $R = O(h^4 + \varepsilon^2 h)$ has the form

$$X_{k+1} = X_k + \varepsilon h^{1/2} \sum_{r=1}^{q} (\sigma_r \xi_r)_k + h(a + \varepsilon^2 b)_k \tag{4.5.1}$$

$$+ h^2 (L_1 a)_k / 2 + h^3 (L_1^2 a)_k / 6 + h^4 (L_1^3 a)_k / 24,$$

where the random variables ξ_r are distributed as

$$P(\xi = \pm 1) = 1/2. \tag{4.5.2}$$

The method with error $R = O(h^4 + \varepsilon^4 h)$ has the form

$$X_{k+1} = X_k + \varepsilon h^{1/2} \sum_{r=1}^{q} (\sigma_r \xi_r)_k + h(a + \varepsilon^2 b)_k \tag{4.5.3}$$

$$+ \varepsilon h^{3/2} \sum_{r=1}^{q} [L_1 \sigma_r (\xi_r / 2 - \eta_r)]_k + \varepsilon h^{3/2} \sum_{r=1}^{q} [\Lambda_r a (\xi_r / 2 + \eta_r)]_k$$

$$+ h^2 [L_1 (a + \varepsilon^2 b)]_k / 2 + \varepsilon^2 h^2 (L_2 a)_k / 2$$

$$+ \varepsilon h^{5/2} \sum_{r=1}^{q} [(L_1^2 \sigma_r + L_1 \Lambda_r a + \Lambda_r L_1 a) \xi_r]_k / 6 + h^3 [L_1^2 (a + \varepsilon^2 b)]_k / 6$$

$$+ \varepsilon^2 h^3 [(L_1 L_2 + L_2 L_1) a]_k / 6 + h^4 (L_1^3 a)_k / 24,$$

where the random variables ξ_r and η_r are distributed as

$$P(\xi = \pm 1) = 1/2, \quad P(\eta = \pm 1/\sqrt{12}) = 1/2. \tag{4.5.4}$$

The method with error $R = O(h^4 + \varepsilon^2 h^2)$ has the form

$$X_{k+1} = X_k + \varepsilon h^{1/2} \sum_{r=1}^{q} (\sigma_r \xi_r)_k + h(a + \varepsilon^2 b)_k + \varepsilon^2 h \sum_{i,r=1}^{q} (\Lambda_i \sigma_r \xi_{ir})_k \tag{4.5.5}$$

$$+ \varepsilon h^{3/2} \sum_{r=1}^{q} [(L_1 + \varepsilon^2 L_2) \sigma_r \xi_r]_k / 2 + \varepsilon h^{3/2} \sum_{r=1}^{q} [\Lambda_r (a + \varepsilon^2 b) \xi_r]_k / 2$$

$$+ h^2 [(L_1 + \varepsilon^2 L_2)(a + \varepsilon^2 b)]_k / 2 + h^3 (L_1^2 a)_k / 6 + h^4 (L_1^3 a)_k / 24,$$

where the random variables ξ_r and ξ_{ir} are simulated according to

$$P(\xi = 0) = 2/3, \ P(\xi = \pm\sqrt{3}) = 1/6, \ \xi_{ir} = (\xi_i\xi_r - \gamma_{ir}\zeta_i\zeta_r)/2, \quad (4.5.6)$$

$$\gamma_{ir} = \begin{cases} -1, & i < r \\ 1, & i \geq r \end{cases},$$

$$P(\zeta = \pm 1) = 1/2,$$

or

$$P(\xi = 0) = 2/3, \ P(\xi = \pm\sqrt{3}) = 1/6, \quad (4.5.7)$$

$$\xi_{ir} = (\xi_i\xi_r - \zeta_{ir})/2, \ \zeta_{ii} = 1, \ \zeta_{ir} = -\zeta_{ri}, \ i \neq r,$$

$$P(\zeta_{ir} = \pm 1) = 1/2, \ i < r.$$

The method with error $R = O(h^4 + \varepsilon^4 h^2)$ has the form

$$X_{k+1} = X_k + \varepsilon h^{1/2} \sum_{r=1}^{q} (\sigma_r\xi_r)_k + h(a + \varepsilon^2 b)_k + \varepsilon^2 h \sum_{i,r=1}^{q} (\Lambda_i\sigma_r\xi_{ir})_k \quad (4.5.8)$$

$$+ \varepsilon h^{3/2} \sum_{r=1}^{q} \left[(L_1 + \varepsilon^2 L_2)\sigma_r(\xi_r - \mu_r) \right]_k + \varepsilon h^{3/2} \sum_{r=1}^{q} \left[\Lambda_r(a + \varepsilon^2 b)\mu_r \right]_k$$

$$+ h^2 \left[(L_1 + \varepsilon^2 L_2)(a + \varepsilon^2 b) \right]_k /2$$

$$+ \varepsilon h^{5/2} \sum_{r=1}^{q} \left[(L_1^2\sigma_r + L_1\Lambda_r a + \Lambda_r L_1 a)\xi_r \right]_k /6$$

$$+ h^3 \left[L_1^2(a + \varepsilon^2 b) \right]_k /6 + \varepsilon^2 h^3 \left[(L_1 L_2 + L_2 L_1)a \right]_k /6 + h^4 (L_1^3 a)_k/24,$$

where ξ_r, ξ_{ir}, and μ_r are simulated, for example, according to

$$\xi_{ir} = (\xi_i\xi_r - \gamma_{ir}\zeta_i\zeta_r)/2, \ \gamma_{ir} = \begin{cases} -1, & i < r \\ 1, & i \geq r \end{cases}, \ P(\zeta = \pm 1) = 1/2, \quad (4.5.9)$$

$$P(\xi = 0) = 2/3, \ P(\xi = \pm\sqrt{3}) = 1/6, \ \mu_r = \xi_r/2 + \zeta_r/\sqrt{12}.$$

Remark 4.5.1 Let us discuss how to choose the increment h given ε, i.e., the inter-dependence of the time increment h and the parameter ε in the methods of this section. We first choose the time increment h to be $h = C\varepsilon^\alpha$. Then the global error of a method can be estimated in powers of the small parameter ε by

$$R = O(\varepsilon^\beta),$$

where

$$\beta = \min\left\{ \alpha p, \min_{l \in S}(\alpha l + J(l)) \right\}.$$

If $h = C\varepsilon^\alpha$, the method (4.5.8) has $R = O(\varepsilon^{4\alpha} + \varepsilon^{2\alpha+4})$, while the method (4.5.1) yields $R = O(\varepsilon^{4\alpha} + \varepsilon^{\alpha+2})$. In the case of $0 < \alpha \leq 2/3$, both errors are bounded by $O(\varepsilon^{4\alpha})$, and so both methods have the same order with respect to ε. However, if $\alpha > 2/3$, the method (4.5.8) has higher order with respect to ε than (4.5.1) (for instance, if $\alpha = 2$, we have $O(\varepsilon^8)$ for (4.5.8) and $O(\varepsilon^4)$ for (4.5.1)). Thus, in the case of a comparatively large time increment h compared to ε (this is of interest mainly if ε is sufficiently small, i.e., when the error estimated by ε^β is not large), complicated methods like (4.5.8) and sufficiently simple methods like (4.5.1) have the same order in ε. In such a situation simple methods are usually preferable because of their considerably lower computational costs. But if one wants to reach an error of high order with respect to ε, complicated methods are preferable.

4.5.2 Runge-Kutta Methods

Below we consider (i) full (derivative free) Runge-Kutta schemes and (ii) Runge-Kutta schemes without derivatives of the coefficients $a(t, x)$ and $b(t, x)$ but with derivatives of the diffusion coefficients $\sigma_r(t, x)$ (semi-Runge-Kutta schemes) which may be useful in the case of simple functions σ_r.

For systems with small noise we obtain full Runge-Kutta methods with errors $O(h^2 + \varepsilon^2 h)$, $O(h^2 + \varepsilon^4 h)$, $O(h^3 + \varepsilon^2 h)$, $O(h^3 + \varepsilon^4 h)$, $O(h^4 + \varepsilon^2 h)$, $O(h^4 + \varepsilon^2 h^2 + \varepsilon^4 h)$, and $O(h^4 + \varepsilon^4 h)$. For higher orders we have succeeded in constructing semi-Runge-Kutta schemes with errors $O(h^3 + \varepsilon^2 h^2)$, $O(h^3 + \varepsilon^4 h^2)$, $O(h^4 + \varepsilon^2 h^2)$ and $O(h^4 + \varepsilon^4 h^2)$.

In this subsection we state several methods with errors $O(h^2 + \cdots)$ and $O(h^4 + \cdots)$. Other Runge-Kutta methods can be found in the preprint [299].

To construct Runge-Kutta methods for system (4.0.1), we use deterministic Runge-Kutta methods as an auxiliary tool. To this end, we select specific deterministic schemes which from our point of view are most appropriate. Obviously, it is possible to derive families of stochastic Runge-Kutta methods which are similar to the proposed ones but use different deterministic Runge-Kutta schemes.

4.5.2.1 Methods $O(h^2 + \cdots)$

The method with error $R = O(h^2 + \varepsilon^2 h)$ has the form

$$X_{k+1} = X_k + \varepsilon h^{1/2} \sum_{r=1}^{q} (\sigma_r \xi_r)_k + \varepsilon^2 h b_k + h(a_k + a(t_{k+1}, X_k + h a_k))/2,$$

$$(4.5.10)$$

where ξ_r are as in (4.5.2).

The method with error $R = O(h^2 + \varepsilon^4 h)$ has the form

$$X_{k+1} = X_k + \varepsilon h^{1/2} \sum_{r=1}^{q} (\sigma_r(t_k, X_k) + \sigma_r(t_{k+1}, X_k + ha_k))\xi_{r_k}/2 \qquad (4.5.11)$$

$$+ h \left[a_k + a(t_{k+1}, X_k + \varepsilon h^{1/2} \sum_{r=1}^{q} (\sigma_r \xi_r)_k + h(a + \varepsilon^2 b)_k) \right]/2$$

$$+ \varepsilon^2 h(b_k + b(t_{k+1}, X_k + ha_k))/2,$$

where ξ_r are as in (4.5.2).

4.5.2.2 Methods $O(h^4 + \cdots)$

The method with error $R = O(h^4 + \varepsilon^2 h)$ has the form

$$X_{k+1} = X_k + \varepsilon h^{1/2} \sum_{r=1}^{q} (\sigma_r \xi_r)_k + \varepsilon^2 h b_k + (k_1 + 2k_2 + 2k_3 + k_4)/6, \quad (4.5.12)$$

where

$$k_1 = ha_k, \ k_2 = ha(t_{k+1/2}, X_k + k_1/2), \ k_3 = ha(t_{k+1/2}, X_k + k_2/2), \quad (4.5.13)$$
$$k_4 = ha(t_{k+1}, X_k + k_3),$$

and ξ_r are as in (4.5.2).

The method with error $R = O(h^4 + \varepsilon^2 h^2 + \varepsilon^4 h)$ has the form

$$X_{k+1} = X_k + \varepsilon h^{1/2} \sum_{r=1}^{q} \left[\sigma_r(t_k, X_k) + \sigma_r(t_{k+1}, X_k + ha_k) \right] \xi_{r_k}/2 \qquad (4.5.14)$$

$$+ (k_1 + 2k_2 + 2k_3 + k_4)/6 + \varepsilon^2 h \left[b_k + b(t_{k+1}, X_k + ha_k) \right]/2,$$

where

$$k_1 = ha_k, \ k_2 = ha(t_{k+1/2}, X_k + k_1/2), \qquad (4.5.15)$$

$$k_3 = ha(t_{k+1/2}, X_k + \varepsilon h^{1/2} \sum_{r=1}^{q} (\sigma_r \xi_r)_k + k_2/2),$$

$$k_4 = ha(t_{k+1}, X_k + \varepsilon h^{1/2} \sum_{r=1}^{q} (\sigma_r \xi_r)_k + k_3 + 3\varepsilon^2 h b_k),$$

and ξ_r are as in (4.5.2).

The method with error $R = O(h^4 + \varepsilon^4 h)$ has the form

$$X_{k+1} = X_k + \varepsilon h^{1/2} \sum_{r=1}^{q} [\sigma_r(t_k, X_k)(\xi_r + 6\eta_r)_k \tag{4.5.16}$$

$$+ 4\sigma_r(t_{k+1/2}, X_k + k_2/2)\xi_{r_k} + \sigma_r(t_{k+1}, X_k + k_1)(\xi_r - 6\eta_r)_k]/6$$

$$+ h[a(t_k, X_k + \varepsilon h^{1/2} \sum_{r=1}^{q}(\sigma_r \eta_r)_k) - a(t_k, X_k - \varepsilon h^{1/2} \sum_{r=1}^{q}(\sigma_r \eta_r)_k)]/2$$

$$+ (k_1 + 2k_2 + 2k_3 + k_4)/6 + \varepsilon^2(l_1 + 3l_2)/4 \,,$$

where

$$k_1 = ha_k \,, \quad k_2 = ha(t_{k+1/2}, X_k + k_1/2) \,, \tag{4.5.17}$$

$$k_3 = ha(t_{k+1/2}, X_k + \varepsilon h^{1/2} \sum_{r=1}^{q}(\sigma_r \xi_r)_k + k_2/2 + \varepsilon^2 l_1/4 + 3\varepsilon^2 l_2/4) \,,$$

$$k_4 = ha(t_{k+1}, X_k + \varepsilon h^{1/2} \sum_{r=1}^{q} \sigma_r(t_{k+1}, X_k + k_1)\xi_{r_k} + k_3 + \varepsilon^2 l_1) \,,$$

$$l_1 = hb_k \,, \quad l_2 = hb(t_k + 2h/3, X_k + 2k_1/9 + 4k_2/9) \,,$$

and ξ_r, η_r are simulated as in (4.5.4). This full Runge-Kutta method requires six recalculations of the function $a(t, x)$, three recalculations of the functions $\sigma_r(t, x)$, and two recalculations of the function $b(t, x)$.

The method with error $R = O(h^4 + \varepsilon^2 h^2)$ has the form

$$X_{k+1} = X_k + \varepsilon h^{1/2} \sum_{r=1}^{q} \left[\sigma_r(t_k, X_k) + \sigma_r(t_{k+1}, X_k) \right] \xi_{r_k}/2 \tag{4.5.18}$$

$$+ \varepsilon^2 h \sum_{i,r=1}^{q}(\Lambda_i \sigma_r \xi_{ir})_k + \varepsilon h^{3/2} \sum_{r=1}^{q} \sum_{i=1}^{n} \left(a^i \frac{\partial \sigma_r}{\partial x^i} \xi_r \right)_k /2$$

$$+ \varepsilon^3 h^{3/2} \sum_{r=1}^{q}(L_2 \sigma_r \xi_r)_k/2 + (k_1 + 2k_2 + 2k_3 + k_4)/6$$

$$+ \varepsilon^2 h[b_k + b(t_{k+1}, X_k + \varepsilon h^{1/2} \sum_{r=1}^{q}(\sigma_r \xi_r)_k + h(a + \varepsilon^2 b)_k)]/2 \,,$$

where k_i, $i = 1, \ldots, 4$, are from (4.5.15) and the used random variables ξ_r, ξ_{ir} are simulated as in the method (4.5.5). The method (4.5.18) contains first and second derivatives of the functions σ_r with respect to x.

Note that in the case of a single noise ($q = 1$) we succeeded in constructing a full Runge-Kutta method with error $O(h^4 + \varepsilon^2 h^2)$ [299].

The method with error $R = O(h^4 + \varepsilon^4 h^2)$ has the form

$$X_{k+1} = X_k + \varepsilon h^{1/2} \sum_{r=1}^{q} [\sigma_r(t_k, X_k)(\xi_r + 6\eta_r)_k \tag{4.5.19}$$

$$+ 4\sigma_r(t_{k+1/2}, X_k + k_2/2)\xi_{r_k} + \sigma_r(t_{k+1}, X_k + k_1)(\xi_r - 6\eta_r)_k]/6$$

$$+ h[a(t_k, X_k + \varepsilon h^{1/2} \sum_{r=1}^{q} (\sigma_r \eta_r)_k) - a(t_k, X_k - \varepsilon h^{1/2} \sum_{r=1}^{q} (\sigma_r \eta_r)_k)]/2$$

$$+ \varepsilon^2 h \sum_{i,r=1}^{q} (\Lambda_i \sigma_r \xi_{ir})_k + \varepsilon^3 h^{3/2} \sum_{r=1}^{q} (L_2 \sigma_r \xi_r)_k/2$$

$$+ (k_1 + 2k_2 + 2k_3 + k_4)/6 + \varepsilon^2 (l_1 + 3l_2)/4 \, ,$$

where

$$k_1 = ha_k \, , \quad k_2 = ha(t_{k+1/2}, X_k + k_1/2) \, , \tag{4.5.20}$$

$$k_3 = ha(t_{k+1/2}, X_k + \varepsilon h^{1/2} \sum_{r=1}^{q} (\sigma_r \xi_r)_k + k_2/2 + \varepsilon^2 l_1/4 + 3\varepsilon^2 l_2/4) \, ,$$

$$k_4 = ha(t_{k+1}, X_k + \varepsilon h^{1/2} \sum_{r=1}^{q} \sigma_r(t_{k+1}, X_k + k_1)\xi_{r_k} + k_3 + \varepsilon^2 l_1) \, ,$$

$$l_1 = hb(t_k, X_k + \varepsilon h^{1/2}(1 + \sqrt{3}) \sum_{r=1}^{q} (\sigma_r \xi_r)_k/2) \, ,$$

$$l_2 = hb(t_k + 2h/3, X_k + 2\varepsilon^2 l_1/3 + 2k_1/9 + 4k_2/9$$

$$+ \varepsilon h^{1/2}(3 - \sqrt{3}) \sum_{r=1}^{q} (\sigma_r \xi_r)_k/6) \, ,$$

and the used random variables are simulated using

$$P(\xi = 0) = 2/3 \, , \quad P(\xi = \pm\sqrt{3}) = 1/6 \, , \quad P(\zeta = \pm 1) = 1/2 \, , \tag{4.5.21}$$

$$\xi_{ir} = (\xi_i \xi_r - \gamma_{ir} \zeta_i \zeta_r)/2 \, , \quad \gamma_{ir} = \begin{cases} -1, \ i < r \\ 1, \ i \geq r \end{cases} \, ,$$

$$\eta_r = \zeta_r/\sqrt{12} \, .$$

Remark 4.5.2 The stochastic system in the Stratonovich sense

$$dX = a(t, X)dt + \varepsilon^2 c(t, X)dt + \varepsilon \sum_{r=1}^{q} \sigma_r(t, X) \circ dw_r, \quad X(t_0) = X_0 \, , \tag{4.5.22}$$

is equivalent to the system in the Ito sense

$$dX = a(t, X)dt + \varepsilon^2 b(t, X)dt + \varepsilon \sum_{r=1}^{q} \sigma_r(t, X)\, dw_r,\qquad (4.5.23)$$

where

$$b(t, x) = c(t, x) + \frac{1}{2}\sum_{r=1}^{q} \frac{\partial \sigma_r}{\partial x}(t, x)\,\sigma_r(t, x).\qquad (4.5.24)$$

In Sects. 4.5.1 and 4.5.2 we have proposed weak methods for the Ito system having the form of (4.5.23). Thus, the methods of Sects. 4.5.1 and 4.5.2 are also appropriate for the Stratonovich system (4.5.22). Note that the full Runge-Kutta methods of Sect. 4.5.2 are no longer full when applied to system (4.5.22), since $b(t, x)$ in (4.5.24) contains derivatives $\partial \sigma_r/\partial x$. However, if the diffusion coefficients σ_r are simple functions, the methods of Sect. 4.5.2 may be efficient and useful for the Stratonovich system (4.5.22). Nevertheless, in some cases we obtained the full Runge-Kutta schemes for (4.5.22) [299].

4.5.3 Weak Methods for Systems with Small Additive Noise

Consider the system with small additive noise

$$dX = a(t, X)dt + \varepsilon \sum_{r=1}^{q} \sigma_r(t)\, dw_r,\quad X(t_0) = X_0.\qquad (4.5.25)$$

For the system (4.5.25) we obtain methods with the errors estimated by $O(h^3 + \varepsilon^6 h^2)$, $O(h^3)$, $O(h^4 + \varepsilon^2 h^3 + \varepsilon^6 h^2)$, $O(h^4 + \varepsilon^6 h^2)$, $O(h^4 + \varepsilon^2 h^3)$, $O(h^4 + \varepsilon^4 h^3)$ and also with the same orders as in Sects. 4.5.1 and 4.5.2. Methods $O(h^4 + \varepsilon^6 h^3)$ and $O(h^4 + \varepsilon^8 h^3)$ are too complicated, and therefore we do not write them. Note that in [299] we also give a few full Runge-Kutta methods for the system with small colored noise, for instance, a scheme with error $O(h^4 + \varepsilon^2 h^3)$.

Methods for the system (4.5.25) with the same orders as in Sects. 4.5.1 and 4.5.2 follow from the corresponding methods for a general system with small noise taking into account that for the system (4.5.25) we have

$$\Lambda_r \sigma_i = 0,\ \ L_2 \sigma_i = 0,\ \ L_1 \sigma_i = \frac{d\sigma_i}{dt},\ \ b = 0.$$

The Runge-Kutta methods $O(h^2 + \cdots)$ and $O(h^4 + \cdots)$ easily follow from the corresponding methods of Sect. 4.5.2. Fortunately, the methods (4.5.18) and

(4.5.19) for the system with additive noise become fully (derivative free) Runge-Kutta schemes. Let us give a more detailed exposition of Taylor-type methods.

Methods $O(h^2 + \cdots)$ easily follow from the corresponding methods of Sect. 4.5.1, and here we do not write them.

The method $O(h^3 + \varepsilon^6 h^2)$ is written as

$$
X_{k+1} = X_k + \varepsilon h^{1/2} \sum_{r=1}^{q} (\sigma_r \xi_r)_k + h a_k \tag{4.5.26}
$$

$$
+ \varepsilon h^{3/2} \sum_{r=1}^{q} \left(\frac{d\sigma_r}{dt} (\xi_r/2 - \eta_r) \right)_k
$$

$$
+ \varepsilon h^{3/2} \sum_{r=1}^{q} \Lambda_r a (\xi_r/2 + \eta_r)_k + h^2 (L_1 + \varepsilon^2 L_2) a_k/2
$$

$$
+ \varepsilon^2 h^2 \sum_{r=1}^{q} \sum_{i=1}^{q} [\Lambda_i \Lambda_r a (\xi_i \xi_r - \zeta_i \zeta_r)]_k /6
$$

$$
+ \varepsilon h^{5/2} \sum_{r=1}^{q} \left[\left(\frac{d^2 \sigma_r}{dt^2} + (L_1 + \varepsilon^2 L_2) \Lambda_r a + \Lambda_r (L_1 + \varepsilon^2 L_2) a \right) \xi_r \right]_k /6
$$

$$
+ h^3 (L_1 + \varepsilon^2 L_2)^2 a_k /6 \,,
$$

where the random variables ξ_r, η_r and ζ_r are simulated as

$$
P(\xi = 0) = 2/3 \,, \quad P(\xi = \pm\sqrt{3}) = 1/6 \,, \quad P(\eta = \pm 1/\sqrt{12}) = 1/2 \,, \tag{4.5.27}
$$

$$
P(\zeta = \pm 1) = 1/2 \,.
$$

The method $O(h^3)$ has the same form (4.5.27) but requires simulation of the needed random variables by the laws

$$
P(\xi = 0) = 1/3 \,, \quad P(\xi = \pm 1) = 3/10 \,, \quad P(\xi = \pm\sqrt{6}) = 1/30 \,, \tag{4.5.28}
$$

$$
P(\eta = \pm 1/\sqrt{12}) = 1/2 \,, \quad P(\zeta = \pm 1) = 1/2 \,.
$$

This method coincides with the third order weak method (3.1.34) for a general system with additive noise ($\varepsilon = 1$).

Methods $O(h^4 + \cdots)$ for the system (4.5.25), except the methods $O(h^4 + \varepsilon^2 h^3 + \varepsilon^6 h^2)$, $O(h^4 + \varepsilon^6 h^2)$, $O(h^4 + \varepsilon^2 h^3)$, $O(h^4 + \varepsilon^4 h^3)$, are obtained from the corresponding methods of Sect. 4.5.1.

The method $O(h^4 + \varepsilon^2 h^3 + \varepsilon^6 h^2)$ is written as

$$X_{k+1} = X_k + \varepsilon h^{1/2} \sum_{r=1}^{q} (\sigma_r \xi_r)_k + h a_k \tag{4.5.29}$$

$$+\varepsilon h^{3/2} \sum_{r=1}^{q} \left(\frac{d\sigma_r}{dt} (\xi_r/2 - \eta_r) \right)_k$$

$$+\varepsilon h^{3/2} \sum_{r=1}^{q} [\Lambda_r a(\xi_r/2 + \eta_r)]_k + h^2 (L_1 + \varepsilon^2 L_2) a_k/2$$

$$+\varepsilon^2 h^2 \sum_{r=1}^{q} \sum_{i=1}^{q} (\Lambda_i \Lambda_r a(\xi_i \xi_r - \zeta_i \zeta_r))_k/6$$

$$+\varepsilon h^{5/2} \sum_{r=1}^{q} \left[\left(\frac{d^2 \sigma_r}{dt^2} + (L_1 + \varepsilon^2 L_2) \Lambda_r a + \Lambda_r (L_1 + \varepsilon^2 L_2) a \right) \xi_r \right]_k/6$$

$$+h^3 (L_1 + \varepsilon^2 L_2)^2 a_k/6 + h^4 L_1^3 a_k/24 \,,$$

where the random variables are as in (4.5.27).

The method $O(h^4 + \varepsilon^6 h^2)$ has the form

$$X_{k+1} = X_k + \varepsilon h^{1/2} \sum_{r=1}^{q} (\sigma_r \xi_r)_k + h a_k \tag{4.5.30}$$

$$+\varepsilon h^{3/2} \sum_{r=1}^{q} \left(\frac{d\sigma_r}{dt} (\xi_r/2 - \eta_r) \right)_k$$

$$+\varepsilon h^{3/2} \sum_{r=1}^{q} [\Lambda_r a(\xi_r/2 + \eta_r)]_k + h^2 (L_1 + \varepsilon^2 L_2) a_k/2$$

$$+\varepsilon^2 h^2 \sum_{r=1}^{q} \sum_{i=1}^{q} [\Lambda_i \Lambda_r a(\xi_i \xi_r - \zeta_i \zeta_r)]_k/6$$

$$+\varepsilon h^{5/2} \sum_{r=1}^{q} \left[(L_1 + \varepsilon^2 L_2) \Lambda_r a \xi_r \right]_k/6$$

$$+\varepsilon h^{5/2} \sum_{r=1}^{q} \left[\Lambda_r (L_1 + \varepsilon^2 L_2) a(\xi_r/6 + \eta_r/2) \right]_k$$

$$+\varepsilon h^{5/2} \sum_{r=1}^{q} \left(\frac{d^2 \sigma_r}{dt^2} (\xi_r/6 - \eta_r/2) \right)_k + h^3 (L_1 + \varepsilon^2 L_2)^2 a_k/6$$

$$+\varepsilon h^{7/2} \sum_{r=1}^{q} \left[\left(\Lambda_r L_1^2 a + L_1 \Lambda_r L_1 a + L_1^2 \Lambda_r a + \frac{d^3 \sigma_r}{dt^3} \right) \xi_r \right]_k/24$$

$$+h^4 L_1^3 a_k/24 + \varepsilon^2 h^4 (L_2 L_1^2 a + L_1^2 L_2 a + L_1 L_2 L_1 a)_k/24 \,,$$

where the random variables are simulated as in (4.5.27).

The method $O(h^4 + \varepsilon^2 h^3)$ *has the form* (4.5.29) but the random variables are simulated as in (4.5.28). Note that this method distinguishes from the method $O(h^3)$ by the additional term $h^4 L_1^3 a_k / 24$ only.

The method $O(h^4 + \varepsilon^4 h^3)$ *has the same form* (4.5.30) as the method $O(h^4 + \varepsilon^6 h^2)$ but the needed random variables are from (4.5.28).

4.6 Expansion of the Global Error in Powers of h and ε

It was shown in Sect. 2.2.3 that it is possible to expand the global errors of methods for stochastic systems in powers of time increment h (see Theorem 2.2.5). Below we expand the global error not only in powers of the time increment h but also in powers of the small parameter ε. The following theorem is an analogue of Theorem 2.2.5.

Theorem 4.6.1 *The global error of the method*

$$X_{k+1} = X_k + \varepsilon h^{1/2} \sum_{r=1}^{q} (\sigma_r \xi_r)_k + h(a + \varepsilon^2 b)_k + h^2 (L_1 a)_k / 2, \qquad (4.6.1)$$

$$P(\xi = \pm 1) = 1/2,$$

is

$$R = O(h^2 + \varepsilon^2 h) = C_1(\varepsilon) h^2 + \varepsilon^2 C_2(\varepsilon) h + O(h^3 + \varepsilon^2 h^2), \qquad (4.6.2)$$

where the functions $C_i(\varepsilon)$, $i = 1, 2$, *do not depend on h and are equal to* $C_i(\varepsilon) = C_i^0 + O(\varepsilon^2)$, *and the constants* C_i^0 *do not depend on both h and ε.*

The proof of this theorem is analogous to the proof of Theorem 2.2.5 and here it is omitted (see details in [301]). The same proof shows that the expansions of the global error for other methods can be obtained in the same way as the expansion (4.6.2) for the method (4.6.1). For instance, for the method (4.5.3) with error $O(h^4 + \varepsilon^4 h)$, we have

$$R = C_1(\varepsilon) h^4 + \varepsilon^2 C_2(\varepsilon) h^3 + \varepsilon^4 C_3(\varepsilon) h^2 + \varepsilon^4 C_4(\varepsilon) h + O(h^5 + \varepsilon^6 h^2).$$

An expansion like (4.6.2) can be used for derivation of extrapolation schemes as follows. Simulate $u^\varepsilon(t_0, X_0) = Ef(X_{t_0, X_0}^\varepsilon(T))$ twice using the method (4.6.1) for given ε with the time steps $h_1 = h$ and $h_2 = \alpha h$, $\alpha > 0$, $\alpha \neq 1$. We obtain $\bar{u}^{\varepsilon, h_1}(t_0, X_0) = Ef(\bar{X}_{t_0, X_0}^{\varepsilon, h_1}(T))$ and $\bar{u}^{\varepsilon, h_2}(t_0, X_0) = Ef(\bar{X}_{t_0, X_0}^{\varepsilon, h_2}(T))$, respectively. We can expand

$$u^\varepsilon = \bar{u}^{\varepsilon, h_1} + C_1(\varepsilon) h_1^2 + \varepsilon^2 C_2(\varepsilon) h_1 + O(h^3 + \varepsilon^2 h^2)$$

and

$$u^\varepsilon = \bar{u}^{\varepsilon, h_2} + C_1(\varepsilon) h_2^2 + \varepsilon^2 C_2(\varepsilon) h_2 + O(h^3 + \varepsilon^2 h^2).$$

This yields

$$\varepsilon^2 C_2(\varepsilon) = \varepsilon^2 \bar{C}_2(\varepsilon) - C_1^0 \times (h_1 + h_2) + O(h^2 + \varepsilon^2 h), \qquad (4.6.3)$$

where $\varepsilon^2 \bar{C}_2(\varepsilon)$ is given by

$$\varepsilon^2 \bar{C}_2(\varepsilon) = (\bar{u}^{\varepsilon,h_1} - \bar{u}^{\varepsilon,h_2})/(h_2 - h_1).$$

On the other hand, setting $\varepsilon = 0$ and using the method (4.6.1) with the time steps h_1 and h_2, we obtain $\bar{u}^{0,h_1}(t_0, X_0) = f(\bar{X}^{0,h_1}_{t_0,X_0}(T))$ and $\bar{u}^{0,h_2}(t_0, X_0) = f(\bar{X}^{0,h_2}_{t_0,X_0}(T))$, where $\bar{X}^{0,h_i}_{t_0,X_0}(t)$ is the corresponding approximation of the solution $X^0_{t_0,X_0}(t)$ of the deterministic system. Then the Runge extrapolation method yields

$$C_1(0) = C_1^0 = \bar{C}_1^0 + O(h), \qquad (4.6.4)$$

where $\bar{C}_1^0$ can be calculated by

$$\bar{C}_1^0 = \left(\bar{u}^{0,h_1} - \bar{u}^{0,h_2}\right)/(h_2^2 - h_1^2),$$

By (4.6.3) and (4.6.4) we obtain an improved value $\bar{u}^{\varepsilon}_{imp}$ with error $O(h^3 + \varepsilon^2 h^2)$ by letting

$$\bar{u}^{\varepsilon}_{imp} = \bar{u}^{\varepsilon,h_1} + \varepsilon^2 \bar{C}_2(\varepsilon)h_1 - \bar{C}_1^0 h_1 h_2. \qquad (4.6.5)$$

In the same spirit, using three recalculations of $u^{\varepsilon}(t_0, X_0) = Ef(X^{\varepsilon}_{t_0,X_0}(T))$ by the method (4.6.1) for given ε and with not equal time steps, one can also find $C_1(\varepsilon)$ and $C_2(\varepsilon)$ from (4.6.2) and obtain yet another improved value.

We conclude that according to our approach to the construction of weak methods for a system with small noise, we can shift some terms, which contribute to the error proportionally to $h^i \varepsilon^j$, from the method to its remainder and vice versa. By calculating the constants $C_i(\varepsilon)$ it is possible to estimate the proper weights of the terms in the sums above and select the most appropriate scheme for solving a given system with small noise, both keeping computational costs low and accuracy high.

4.7   Reduction of the Monte Carlo Error

Here we apply the method of important sampling from Sect. 2.3.1 in the case of systems with small noise.

Together with the system (4.0.1), consider the following (cf. (2.3.3)):

$$dX = a(t, X)dt + \varepsilon^2 b(t, X)dt - \varepsilon \sum_{r=1}^{q} \mu_r(t, X)\sigma_r(t, X) dt \qquad (4.7.1)$$

$$+\varepsilon \sum_{r=1}^{q} \sigma_r(t, X) dw_r,$$

$$dY = \sum_{r=1}^{q} \mu_r(t, X)Y dw_r,$$

where μ_r and Y are scalars.

According to the Girsanov theorem, we have for any μ_r :

$$y E f(X_{s,x}(T))|_{(4.0.1)} = E\left(Y_{s,x,y}(T) f(X_{s,x}(T))\right)|_{(4.7.1)} . \qquad (4.7.2)$$

The function $u(s, x) = E f(X_{s,x}(T))|_{(4.0.1)}$ satisfies the equation

$$Lu \equiv \frac{\partial u}{\partial s} + \sum_{i=1}^{d} a^i \frac{\partial u}{\partial x^i} + \varepsilon^2 \sum_{i=1}^{d} b^i \frac{\partial u}{\partial x^i} + \frac{\varepsilon^2}{2} \sum_{r=1}^{q}\sum_{i=1}^{d}\sum_{j=1}^{d} \sigma_r^i \sigma_r^j \frac{\partial^2}{\partial x^i \partial x^j} = 0$$

$$(4.7.3)$$

subject to the terminal condition at time T :

$$u(T, x) = f(x) . \qquad (4.7.4)$$

Under some mild conditions on the coefficients and on the function f, the solution $u(s, x) = u^\varepsilon(s, x)$ of the problem (4.7.3)–(4.7.4) has the form [113, Chap. 2]:

$$u^\varepsilon(s, x) = u^0(s, x) + \varepsilon^2 u^1(s, x; \varepsilon) . \qquad (4.7.5)$$

The function u^0 satisfies the first-order partial differential equation

$$\frac{\partial u}{\partial s} + \sum_{i=1}^{d} a^i \frac{\partial u}{\partial x^i} = 0 \qquad (4.7.6)$$

under the condition (4.7.4). Obviously, the solution of (4.7.6) has the form

$$u^0(s, x) = f(X_{s,x}^0(T)) , \qquad (4.7.7)$$

where $X_{s,x}^0$ is the solution of the Cauchy problem for the deterministic system of differential equations

$$\frac{dX}{dt} = a(t, X), \quad X(s) = x . \qquad (4.7.8)$$

Applying the Ito formula along the solution of the system (4.7.1), we get the following expression (note that here $Lu = 0$):

$$d\left[u(t, X_{s,x}(t))Y_{s,x,y}(t)\right] = LuY dt - \varepsilon \sum_{r=1}^{q} \mu_r\left(\sigma_r, \frac{\partial u}{\partial x}\right) Y dt$$

$$+\varepsilon \sum_{r=1}^{q}\left(\sigma_r, \frac{\partial u}{\partial x}\right) Y\, dw_r(t) + u \sum_{r=1}^{q} \mu_r Y\, dw_r(t)$$

$$+\varepsilon \sum_{r=1}^{q}\left(\sigma_r, \frac{\partial u}{\partial x}\right)\mu_r Y\, dt$$

$$= \sum_{r=1}^{q}\left(\varepsilon(\sigma_r, \frac{\partial u}{\partial x}) + \mu_r u\right) Y\, dw_r(t).$$

Then

$$u(t, X_{s,x}(t))Y_{s,x,y}(t) = u(s, x)y + \int_{s}^{t} \sum_{r=1}^{q}\left(\varepsilon\left(\sigma_r, \frac{\partial u}{\partial x}\right) + \mu_r u\right) Y dw_r(t).$$

$$(4.7.9)$$

If we suppose that $t = T$, $y = 1$, $\mu_r \equiv 0$, we obtain

$$f(X_{s,x}(T)) = u(s, x) + \int_{s}^{T} \varepsilon \sum_{r=1}^{q}\left(\sigma_r, \frac{\partial u}{\partial x}\right) dw_r(t).$$

Therefore

$$Df(X_{s,x}(T)) = \varepsilon^2 \int_{s}^{T} E\left[\sum_{r=1}^{q}\left(\sigma_r, \frac{\partial u}{\partial x}\right)\right]^2 dt \qquad (4.7.10)$$

because $u(s, x) = Ef(X_{s,x}(T))|_{(4.0.1)}$.

Thus, if we calculate $Ef(X(T))$ by the Monte Carlo technique using a weak method for solving the system (4.0.1), then the Monte Carlo error, evaluated by $c[Df(\bar{X}(T))/N]^{1/2}$ and close to $c[Df(X(T))/N]^{1/2}$, contains a small factor equal to ε.

As can be seen from (4.7.2), the mean value

$$EZ = E\left(Y_{s,x,y}(T)f(X_{s,x}(T))\right)|_{(4.7.1)}$$

does not depend on μ_r, whereas $D\left(Y_{s,x,y}(T)f(X_{s,x}(T))\right)|_{(4.7.1)}$ does depend on μ_r. Thus, below we will select functions μ_r, $r = 1, \ldots, q$, so that the variance DZ becomes less than the variance (4.7.10).

Assume that $f > 0$. Then $u^0 > 0$. Note that if the function f is not positive but there are constants K and C such that $Kf + C > 0$, we can take the function $g = Kf + C$ instead. Then we can simulate Eg and finally obtain Ef.

Setting $t = T$, $y = 1$ in (4.7.9), and

$$\mu_r = -\frac{\varepsilon}{u^0} \left(\sigma_r, \frac{\partial u^0}{\partial x} \right), \quad r = 1, \ldots, q, \tag{4.7.11}$$

we obtain

$$f(X_{s,x}(T))Y = u(s, x) + \int_s^T \varepsilon^3 \sum_{r=1}^q \left[\left(\sigma_r, \frac{\partial u^1}{\partial x} \right) - \left(\sigma_r, \frac{\partial u^0}{\partial x} \right) \frac{u^1}{u^0} \right] dw_r(t).$$

Therefore

$$D\left[f(X_{s,x}(T))Y \right] = \varepsilon^6 \int_s^T E \left(\sum_{r=1}^q \left[\left(\sigma_r, \frac{\partial u^1}{\partial x} \right) - \left(\sigma_r, \frac{\partial u^0}{\partial x} \right) \frac{u^1}{u^0} \right] \right)^2 dt.$$

Hence, the Monte Carlo error for the system (4.7.1) with μ_r from (4.7.11) inherits a small factor equal to ε^3.

The system (4.7.1) with μ_r from (4.7.11) is again a system with small noise, and all the methods proposed above are suitable for finding its solution. We observe that, even if the number M of simulations is small, the Monte Carlo error for this system will be reasonably small. Of course, in order to apply the approach outlined above, we must know the function $u^0(s, x)$.

4.8 Simulation of the Lyapunov Exponent of a Linear System with Small Noise by Weak Methods

To test weak methods proposed in this chapter, we use the two-dimensional linear Ito stochastic system (4.3.1) again:

$$dX = AXdt + \varepsilon \sum_{r=1}^q B_r X \, dw_r, \tag{4.8.1}$$

where X is a two-dimensional vector, A and B_r are constant 2×2-matrices, w_r are independent standard Wiener processes, $\varepsilon > 0$ is a small parameter.

Talay [430] proposed a numerical approach to calculating Lyapunov exponents based on the ergodic property. Using weak methods, Lyapunov exponents are calculated by simulating a single trajectory. This procedure is appealing as it is intuitive and computationally cheap. However, it is difficult to analyze the errors arising from this approach. Below we calculate Lyapunov exponent of (4.8.1) as a convenient example to illustrate efficiency of the proposed methods. We also pay attention to analysis of the errors.

In the ergodic case there exists a unique Lyapunov exponent λ of system (4.8.1) (cf. [194]), with

$$\lambda = \lim_{t \to \infty} \frac{1}{t} E\rho(t) = \lim_{t \to \infty} \frac{1}{t} \rho(t) \ \ a.s.,$$

where $\rho(t) = \ln |X(t)|$, and $X(t), t \geq 0$, is a non-trivial solution of the system (4.8.1). If $D(\rho(t)) \to \infty$ for $t \to \infty$ then [194]

$$E\left(\frac{\rho(t)}{t} - \lambda\right)^2 = D\left(\frac{\rho(t)}{t}\right)\left(1 + \varphi^2(t)\right), \tag{4.8.2}$$

where $\varphi(t) \to 0$ for $t \to \infty$. It is not difficult to show that $D(\rho(t)/t) \to 0$ for $t \to \infty$. From (4.8.2) and the equality

$$D\left(\frac{\rho(t)}{t}\right) = E\left(\frac{\rho(t)}{t} - \lambda\right)^2 - \left[E\left(\frac{\rho(t)}{t}\right) - \lambda\right]^2,$$

we have

$$\left| E\left(\frac{\rho(t)}{t}\right) - \lambda \right| = \varphi(t) \left[D\left(\frac{\rho(t)}{t}\right) \right]^{1/2}. \tag{4.8.3}$$

Herein we consider the system (4.8.1) with matrices A and B_r of the form

$$A = \begin{pmatrix} a & c \\ -c & a \end{pmatrix}, \quad B_r = \begin{pmatrix} b_r & d_r \\ -d_r & b_r \end{pmatrix}, \quad r = 1, 2 \tag{4.8.4}$$

In this case the Lyapunov exponent is given by (4.3.4) (cf. [14]).

By the Monte Carlo technique we numerically calculate the function

$$\lambda(T) = \frac{1}{T} E\rho(T) \approx \bar{\lambda}(T) = \frac{1}{T} E\bar{\rho}(T), \quad \bar{\rho}(T) = \ln |\bar{X}(T)|. \tag{4.8.5}$$

The function $\lambda(t)$ in the limit of large time ($t \to \infty$) tends to the Lyapunov exponent λ. In this case three errors arise: (a) the method error, i.e., $|E\rho(T)/T - E\bar{\rho}(T)/T|$, (b) the Monte Carlo error which is bounded by $c[D(\bar{\rho}(T)/T)]^{1/2}/\sqrt{M}$, and (c) the error with respect to the choice of integration time T (see (4.8.3)).

As can be seen from our computational results, the third error, i.e., $|\lambda(T) - \lambda| = |E(\rho(T)/T) - \lambda|$, is negligibly small, at any rate for $T \geq 2$, as compared to both the method error and the Monte Carlo error.

In our case the function $[D(\bar{\rho}(T)/T)]^{1/2}$ tends to zero with rate $1/\sqrt{T}$. So the Monte Carlo error is proportional to $1/\sqrt{TM}$. Therefore, to reduce the Monte Carlo error we can increase either M or T. As far as the computational costs are concerned, it does not matter whether we increase M or T. In our case Talay's approach requires the same computational costs as the simulation of Lyapunov exponents by the Monte Carlo technique. But using Monte Carlo simulations we find both $E\bar{\rho}(T)/T$ and $D(\bar{\rho}(T)/T)$, which is useful for estimating errors.

We simulate the system (4.8.1) by four different weak schemes: (i) the method (4.6.1) with error $O(h^2 + \varepsilon^2 h)$, which is the simplest method among the weak schemes proposed in this chapter; (ii) the method with error $O(h^2 + \varepsilon^4 h)$:

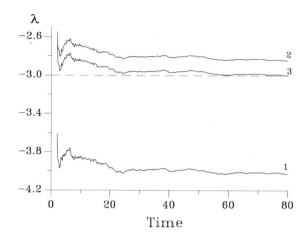

Fig. 4.8.1 Time dependence of the function $\bar{\lambda}(T) = E\bar{\rho}(T)/T$ for time step $h = 0.45$. The other parameters are the same as in Table 4.8.1. The solution of the system (4.8.1), (4.8.4) is approximated by (1) the method (4.6.1), (2) the method (4.8.6), and (3) the standard method (4.4.8). The dashed line shows the exact value of the Lyapunov exponent λ ($\lambda = -2.12$). The number of realizations is $M = 400$ which ensures that the Monte Carlo errors at $T \geq 7$ are not greater than 0.04 for curve 1 and not greater than 0.02 for curves 2, 3 and they are less than the method errors

$$X_{k+1} = X_k + \varepsilon h^{1/2} \sum_{r=1}^{q} (\sigma_r \xi_r)_k + h(a + \varepsilon^2 b)_k \tag{4.8.6}$$

$$+\varepsilon h^{3/2} \sum_{r=1}^{q} (L_1 \sigma_r \xi_r)_k/2 + \varepsilon h^{3/2} \sum_{r=1}^{q} (\Lambda_r a \xi_r)_k/2$$

$$+h^2 \left[L_1 (a + \varepsilon^2 b) \right]_k /2 + \varepsilon^2 h^2 (L_2 a)_k/2 \,,$$

where the random variables ξ_r are distributed as $P(\xi = \pm 1) = 1/2$; (iii) the standard method (4.4.8) with error $O(h^2)$; and (iv) the semi-Runge-Kutta scheme (4.5.19) with error $O(h^4 + \varepsilon^4 h^2)$, which is the most accurate scheme among the weak methods proposed in this chapter for general systems with small noise.

We can infer from Table 4.8.1 and Fig. 4.8.1 that the proposed methods for systems with small noise require less computational effort than the standard ones.

Table 4.8.1 Simulation of $\bar{\lambda}(T)$ for $a = -2$, $c = 1$, $b_1 = b_2 = 2$, $d_1 = 1$, $d_2 = -1$, $\varepsilon = 0.2$, $X^1(0) = 0$, $X^2(0) = 1$, $T = 10$, and for various steps h averaged over M realizations, where $M = 4 \cdot 10^4$ for the methods $O(h^2 + \cdots)$ and $M = 1 \cdot 10^6$ for the method $O(h^4 + \varepsilon^4 h^2)$. The exact solution is $\lambda = -2.12$

h	$O(h^2 + \varepsilon^2 h)$	$O(h^2 + \varepsilon^4 h)$	$O(h^2)$	$O(h^4 + \varepsilon^4 h^2)$
0.3	-2.461 ± 0.004	-2.067 ± 0.002	-2.067 ± 0.002	-2.1228 ± 0.0004
0.2	-2.290 ± 0.003	-2.106 ± 0.002	-2.097 ± 0.002	-2.1195 ± 0.0004
0.1	-2.186 ± 0.002	-2.1198 ± 0.0018	-2.1140 ± 0.0017	-2.1192 ± 0.0004
0.05	-2.150 ± 0.002	-2.1219 ± 0.0018	-2.1186 ± 0.0018	-2.1197 ± 0.0004

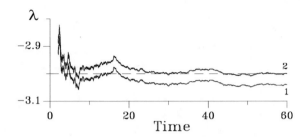

Fig. 4.8.2 Time dependence of the function $\bar{\rho}(T)/T$ computed along a single trajectory using (1) the method (4.8.6) with $h = 0.1$, (2) the standard method (4.4.8) with $h = 0.1$, and (3) the method (4.5.19) with $h = 0.3$. The other parameters are the same as in Table 4.8.1. The dashed line shows the exact value of the Lyapunov exponent λ ($\lambda = -2.12$)

The data of Table 4.8.1 show that the methods $O(h^2 + \varepsilon^2 h)$, $O(h^2 + \varepsilon^4 h)$, and $O(h^2)$ can be improved by using the expansion from Sect. 4.6. For $\varepsilon = 0.2$, for instance, one can calculate $C_1(\varepsilon)$ and $C_2(\varepsilon)$ from the expansion of the global error of the method (4.6.1) (see Theorem 4.6.1) to obtain $C_1(\varepsilon) \approx 2.1$ and $C_2(\varepsilon) \approx 10.2$. Let us emphasize that if some constants in the error expansion have opposite signs then the error will become a non-monotonous function of the time step h and thus may increase while h is decreasing. Such behavior is demonstrated in Table 4.8.1 (see the methods $O(h^2 + \varepsilon^4 h)$ and $O(h^4 + \varepsilon^4 h^2)$).

In Fig. 4.8.2 we show the time dependence of the function $\bar{\rho}(T)/T$ when taking Talay's approach to calculate Lyapunov exponents, i.e., along a single trajectory of a weak scheme. One can see that in this case our methods give accurate results and allow us to reduce computational costs. By the Monte Carlo simulations using the method (4.5.19) with $h = 0.2$, $T = 10$, $M = 10^6$ we achieve an accuracy of $\approx 0.5 \times 10^{-3}$ (see Table 4.8.1).

Remark 4.8.1 Note that the function $\ln|x|$ does not belong to the class **F**. Thus, if $\lambda > 0$, to rigorously deduce conclusions one can consider the function $\ln(1 + |x|)$ instead of $\ln|x|$. The function $\ln(1 + |x|)$ already belongs to the class **F** and $\lim_{t \to \infty} \ln(1 + |X(t)|)/t = \lim_{t \to \infty} \ln(|X(t)|)/t$. As can be seen by carrying out numerical tests, simulations of the function $\ln(1 + |X(t)|)$ yield the same results as simulations of $\ln|X(t)|$. For $\lambda < 0$ one can either switch to the function $\ln(1 + 1/|x|)$ or to the system

$$dX = (\gamma I + AX)dt + \varepsilon \sum_{r=1}^{q} B_r X \, dw_r \qquad (4.8.7)$$

instead of the system (4.8.1). The Lyapunov exponent of the system (4.8.7) is equal to $\gamma + \lambda$, and if we choose γ such that $\gamma + \lambda > 0$, we can use the function $\ln(1 + |x|)$ again.

Chapter 5
Geometric Integrators and Computing Ergodic Limits

In this chapter we construct specific methods for two important classes of stochastic systems which often occur in various applications including molecular dynamics, Bayesian statistics, Physics, Chemistry, Biology.

First we consider *stochastic Hamiltonian systems* which can be written in the form of the Stratonovich SDEs:

$$dP = f(t, P, Q)dt + \sum_{r=1}^{m} \sigma_r(t, P, Q) \circ dw_r(t), \quad P(t_0) = p, \quad (5.0.1)$$

$$dQ = g(t, P, Q)dt + \sum_{r=1}^{m} \gamma_r(t, P, Q) \circ dw_r(t), \quad Q(t_0) = q,$$

where P, Q, f, g, σ_r, γ_r are n-dimensional column-vectors with the components P^i, Q^i, f^i, g^i, σ_r^i, γ_r^i, $i = 1, \ldots, n$, and $w_r(t)$, $r = 1, \ldots, m$, are independent standard Wiener processes.

We denote by $X(t; t_0, x) = (P^{\mathsf{T}}(t; t_0, p, q), Q^{\mathsf{T}}(t; t_0, p, q))^{\mathsf{T}}$, $t_0 \leq t \leq T$, the solution of the problem (5.0.1). A more detailed notation is $X(t; t_0, x; \omega)$, where ω is an elementary event. It is known that $X(t; t_0, x; \omega)$ is a phase flow (diffeomorphism) for almost every ω. See its properties in, e.g. [30, 92, 177].

If there are functions $H_r(t, p, q)$, $r = 0, \ldots, m$, such that (see [30] and Sect. 5.1 below)

$$f^i(t, p, q) = -\partial H_0/\partial q^i, \quad g^i(t, p, q) = \partial H_0/\partial p^i,$$
$$\sigma_r^i(t, p, q) = -\partial H_r/\partial q^i, \quad \gamma_r^i(t, p, q) = \partial H_r/\partial p^i, \quad (5.0.2)$$
$$r = 1, \ldots, m, \quad i = 1, \ldots, n,$$

then the phase flow of (5.0.1) (like the phase flow of a deterministic Hamiltonian system) preserves symplectic structure:

© Springer Nature Switzerland AG 2021   313
G. N. Milstein and M. V. Tretyakov, *Stochastic Numerics for Mathematical Physics*,
Scientific Computation, https://doi.org/10.1007/978-3-030-82040-4_5

$$dP \wedge dQ = dp \wedge dq, \qquad (5.0.3)$$

i.e., the sum of the oriented areas of projections onto the coordinate planes (p^1, q^1), $\ldots$, (p^n, q^n) is an integral invariant [13]. To avoid confusion, we note that the differentials in (5.0.1) and (5.0.3) have different meaning. In (5.0.1) P, Q are treated as functions of time and p, q are fixed parameters, while differentiation in (5.0.3) is made with respect to the initial data p, q.

Let P_k, Q_k, $k = 0, \ldots, N$, $t_{k+1} - t_k = h_{k+1}$, $t_N = T$, be a method for (5.0.1) based on the one-step approximation $\bar{P} = \bar{P}(t + h; t, p, q)$, $\bar{Q} = \bar{Q}(t + h; t, p, q)$. We say that the method preserves symplectic structure if

$$d\bar{P} \wedge d\bar{Q} = dp \wedge dq. \qquad (5.0.4)$$

A lot of attention in deterministic numerical analysis has been paid to symplectic integration of Hamiltonian systems (see, e.g. [154, 155, 398] and references therein). This interest is motivated by the fact that symplectic integrators in comparison with usual numerical schemes allow us to simulate Hamiltonian systems on very long time intervals with high accuracy. As it will be shown in this chapter, symplectic methods for stochastic Hamiltonian systems proposed in Sects. 5.2–5.6 have significant advantages over standard schemes for SDEs.

Sections 5.2–5.4 deal with mean-square symplectic methods [290, 291] while Sect. 5.6 is devoted to weak symplectic integration [312]. We construct symplectic methods for general stochastic Hamiltonian systems (5.0.1)–(5.0.2) as well as higher-order symplectic schemes for Hamiltonian systems with separable Hamiltonians and Hamiltonian systems with additive noise. Symplectic integrators for other specific stochastic Hamiltonian systems such as Hamiltonian systems with colored noise, Hamiltonian systems with small noise, etc. can be found in [289–291, 312]. We also propose volume-preserving methods for stochastic Liouvillian systems (see Sect. 5.5).

It is natural to expect that making use of numerical methods, which are close, in a sense, to symplectic ones, also has some advantages when applying to stochastic systems close to Hamiltonian ones. An important and large class of such systems is *Langevin-type equations* which can be written as the following system of Ito SDEs

$$dP = f(t, Q)dt - \nu\tilde{f}(t, P, Q)dt + \sum_{r=1}^{m}\sigma_r(t, Q)dw_r(t), \quad P(t_0) = p, \quad (5.0.5)$$

$$dQ = g(P)dt, \quad Q(t_0) = q,$$

where P, Q, f, $\tilde{f}$, g, σ_r are n-dimensional column-vectors, ν is a parameter, and $w_r(t)$, $r = 1, \ldots, m$, are independent standard Wiener processes. It is not difficult to verify that this system has the same form in the sense of Stratonovich (see also Example 1.2.8).

The Langevin-type equations (5.0.5) have the widespread occurrence in models from physics, chemistry, and biology. They are used in molecular simulations (see, e.g., Sect. 5.9 here and [69–71, 181, 230, 234, 404] and references therein), in dissipative particle dynamics (see, e.g., [376] and references therein), for studying lattice dynamics in strongly anharmonic crystals [137], for descriptions of noise-induced transport in stochastic ratchets [303], for investigations of the dispersion of passive tracers in turbulent flows (see [367, 380, 399, 439] and references therein), etc. In the second part of this chapter (Sects. 5.7–5.8) we construct special numerical methods (we call them as quasi-symplectic) which preserve some specific properties of the Langevin-type equations [312]. The proposed methods are such that they degenerate to symplectic methods when the system degenerates to a Hamiltonian one and their law of phase volume contractivity is close to the exact one. We further illustrate geometric integration ideas by applying them to such models as Langevin equations and stochastic gradient systems for rigid body dynamics (Sect. 5.9) and the stochastic Landau-Lifshitz equation (Sect. 5.10). Geometric integrators are especially important for computing ergodic limits, which is the topic of Sect. 5.11.

5.1   Preservation of Symplectic Structure

Consider the system (5.0.1). Our nearest aim is to identify a class of stochastic systems, which preserve symplectic structure, i.e., satisfy the condition (5.0.3).

Using the formula of change of variables in differential forms, we obtain

$$
\begin{aligned}
dP \wedge dQ &= dP^1 \wedge dQ^1 + \cdots + dP^n \wedge dQ^n \\
&= \sum_{k=1}^{n} \sum_{l=k+1}^{n} \sum_{i=1}^{n} \left(\frac{\partial P^i}{\partial p^k} \frac{\partial Q^i}{\partial p^l} - \frac{\partial P^i}{\partial p^l} \frac{\partial Q^i}{\partial p^k} \right) dp^k \wedge dp^l \\
&+ \sum_{k=1}^{n} \sum_{l=k+1}^{n} \sum_{i=1}^{n} \left(\frac{\partial P^i}{\partial q^k} \frac{\partial Q^i}{\partial q^l} - \frac{\partial P^i}{\partial q^l} \frac{\partial Q^i}{\partial q^k} \right) dq^k \wedge dq^l \\
&+ \sum_{k=1}^{n} \sum_{l=1}^{n} \sum_{i=1}^{n} \left(\frac{\partial P^i}{\partial p^k} \frac{\partial Q^i}{\partial q^l} - \frac{\partial P^i}{\partial q^l} \frac{\partial Q^i}{\partial p^k} \right) dp^k \wedge dq^l .
\end{aligned}
$$

Hence the phase flow of (5.0.1) preserves symplectic structure if and only if

$$
\sum_{i=1}^{n} \frac{D(P^i, Q^i)}{D(p^k, p^l)} = 0, \quad k \neq l, \tag{5.1.1}
$$

$$
\sum_{i=1}^{n} \frac{D(P^i, Q^i)}{D(q^k, q^l)} = 0, \quad k \neq l, \tag{5.1.2}
$$

and

$$\sum_{i=1}^{n} \frac{D(P^i, Q^i)}{D(p^k, q^l)} = \delta_{kl}, \quad k, l = 1, \ldots, n. \tag{5.1.3}$$

Introduce the notation

$$P_p^{ik} = \frac{\partial P^i}{\partial p^k}, \quad P_q^{ik} = \frac{\partial P^i}{\partial q^k}, \quad Q_p^{ik} = \frac{\partial Q^i}{\partial p^k}, \quad Q_q^{ik} = \frac{\partial Q^i}{\partial q^k}.$$

For a fixed k, we obtain that P_p^{ik}, Q_p^{ik}, $i = 1, \ldots, n$, obey the following system of SDEs

$$\begin{aligned}
dP_p^{ik} &= \sum_{\alpha=1}^{n} \left(\frac{\partial f^i}{\partial p^\alpha} P_p^{\alpha k} + \frac{\partial f^i}{\partial q^\alpha} Q_p^{\alpha k} \right) dt \\
&+ \sum_{r=1}^{m} \sum_{\alpha=1}^{n} \left(\frac{\partial \sigma_r^i}{\partial p^\alpha} P_p^{\alpha k} + \frac{\partial \sigma_r^i}{\partial q^\alpha} Q_p^{\alpha k} \right) \circ dw_r, \\
&\quad P_p^{ik}(t_0) = \delta_{ik},
\end{aligned} \tag{5.1.4}$$

$$\begin{aligned}
dQ_p^{ik} &= \sum_{\alpha=1}^{n} \left(\frac{\partial g^i}{\partial p^\alpha} P_p^{\alpha k} + \frac{\partial g^i}{\partial q^\alpha} Q_p^{\alpha k} \right) dt \\
&+ \sum_{r=1}^{m} \sum_{\alpha=1}^{n} \left(\frac{\partial \gamma_r^i}{\partial p^\alpha} P_p^{\alpha k} + \frac{\partial \gamma_r^i}{\partial q^\alpha} Q_p^{\alpha k} \right) \circ dw_r, \\
&\quad Q_p^{ik}(t_0) = 0.
\end{aligned}$$

Analogously, for a fixed k, P_q^{ik}, Q_q^{ik}, $i = 1, \ldots, n$, satisfy the system

$$\begin{aligned}
dP_q^{ik} &= \sum_{\alpha=1}^{n} \left(\frac{\partial f^i}{\partial p^\alpha} P_q^{\alpha k} + \frac{\partial f^i}{\partial q^\alpha} Q_q^{\alpha k} \right) dt \\
&+ \sum_{r=1}^{m} \sum_{\alpha=1}^{n} \left(\frac{\partial \sigma_r^i}{\partial p^\alpha} P_q^{\alpha k} + \frac{\partial \sigma_r^i}{\partial q^\alpha} Q_q^{\alpha k} \right) \circ dw_r, \\
&\quad P_q^{ik}(t_0) = 0,
\end{aligned} \tag{5.1.5}$$

$$\begin{aligned}
dQ_q^{ik} &= \sum_{\alpha=1}^{n} \left(\frac{\partial g^i}{\partial p^\alpha} P_q^{\alpha k} + \frac{\partial g^i}{\partial q^\alpha} Q_q^{\alpha k} \right) dt \\
&+ \sum_{r=1}^{m} \sum_{\alpha=1}^{n} \left(\frac{\partial \gamma_r^i}{\partial p^\alpha} P_q^{\alpha k} + \frac{\partial \gamma_r^i}{\partial q^\alpha} Q_q^{\alpha k} \right) \circ dw_r, \\
&\quad Q_q^{ik}(t_0) = \delta_{ik}.
\end{aligned}$$

The coefficients in (5.1.4) and (5.1.5) are evaluated at (t, P, Q) with $P = P(t) = [P^1(t; t_0, p, q), \ldots, P^n(t; t_0, p, q)]^\mathsf{T}$, $Q = Q(t) = [Q^1(t; t_0, p, q), \ldots, Q^n(t; t_0, p, q)]^\mathsf{T}$ being a solution to (5.0.1).

Consider the condition (5.1.1). Clearly,

$$\frac{D(P^i(t_0), Q^i(t_0))}{D(p^k, p^l)} = \frac{D(p^i, q^i)}{D(p^k, p^l)} = 0 \,.$$

Therefore, (5.1.1) is fulfilled if and only if

$$\sum_{i=1}^{n} d\,\frac{D(P^i(t), Q^i(t))}{D(p^k, p^l)} = 0 \,. \tag{5.1.6}$$

Due to (5.1.4), we get

$$d\,\frac{\partial P^i}{\partial p^k}\frac{\partial Q^i}{\partial p^l} = d\,P_p^{ik}(t) Q_p^{il}(t)$$

$$= \sum_{\alpha=1}^{n} \left[\left(\frac{\partial f^i}{\partial p^\alpha} P_p^{\alpha k} + \frac{\partial f^i}{\partial q^\alpha} Q_p^{\alpha k} \right) Q_p^{il} \right.$$

$$+ \left. \left(\frac{\partial g^i}{\partial p^\alpha} P_p^{\alpha l} + \frac{\partial g^i}{\partial q^\alpha} Q_p^{\alpha l} \right) P_p^{ik} \right] dt$$

$$+ \sum_{r=1}^{m}\sum_{\alpha=1}^{n} \left[\left(\frac{\partial \sigma_r^i}{\partial p^\alpha} P_p^{\alpha k} + \frac{\partial \sigma_r^i}{\partial q^\alpha} Q_p^{\alpha k} \right) Q_p^{il} \right.$$

$$+ \left. \left(\frac{\partial \gamma_r^i}{\partial p^\alpha} P_p^{\alpha l} + \frac{\partial \gamma_r^i}{\partial q^\alpha} Q_p^{\alpha l} \right) P_p^{ik} \right] \circ dw_r \,.$$

Then (5.1.6) holds if and only if the following equalities take place:

$$\sum_{i=1}^{n}\sum_{\alpha=1}^{n} \left(\frac{\partial f^i}{\partial p^\alpha} P_p^{\alpha k} Q_p^{il} + \frac{\partial f^i}{\partial q^\alpha} Q_p^{\alpha k} Q_p^{il} + \frac{\partial g^i}{\partial p^\alpha} P_p^{\alpha l} P_p^{ik} + \frac{\partial g^i}{\partial q^\alpha} Q_p^{\alpha l} P_p^{ik} \right. \tag{5.1.7}$$

$$\left. - \frac{\partial f^i}{\partial p^\alpha} P_p^{\alpha l} Q_p^{ik} - \frac{\partial f^i}{\partial q^\alpha} Q_p^{\alpha l} Q_p^{ik} - \frac{\partial g^i}{\partial p^\alpha} P_p^{\alpha k} P_p^{il} - \frac{\partial g^i}{\partial q^\alpha} Q_p^{\alpha k} P_p^{il} \right) = 0 \,,$$

$$\sum_{i=1}^{n}\sum_{\alpha=1}^{n} \left(\frac{\partial \sigma_r^i}{\partial p^\alpha} P_p^{\alpha k} Q_p^{il} + \frac{\partial \sigma_r^i}{\partial q^\alpha} Q_p^{\alpha k} Q_p^{il} + \frac{\partial \gamma_r^i}{\partial p^\alpha} P_p^{\alpha l} P_p^{ik} + \frac{\partial \gamma_r^i}{\partial q^\alpha} Q_p^{\alpha l} P_p^{ik} \right. \tag{5.1.8}$$

$$\left. - \frac{\partial \sigma_r^i}{\partial p^\alpha} P_p^{\alpha l} Q_p^{ik} - \frac{\partial \sigma_r^i}{\partial q^\alpha} Q_p^{\alpha l} Q_p^{ik} - \frac{\partial \gamma_r^i}{\partial p^\alpha} P_p^{\alpha k} P_p^{il} - \frac{\partial \gamma_r^i}{\partial q^\alpha} Q_p^{\alpha k} P_p^{il} \right) = 0 \,,$$

$$r = 1, \ldots, m \,.$$

It is not difficult to check that if the functions $f^i(t, p, q)$, $g^i(t, p, q)$ are such that

$$\frac{\partial f^i}{\partial p^\alpha} + \frac{\partial g^\alpha}{\partial q^i} = 0, \quad \frac{\partial f^i}{\partial q^\alpha} = \frac{\partial f^\alpha}{\partial q^i}, \quad \frac{\partial g^i}{\partial p^\alpha} = \frac{\partial g^\alpha}{\partial p^i}, \quad i, \alpha = 1, \ldots, n, \qquad (5.1.9)$$

then (5.1.7) holds, and if the functions $\sigma_r^i(t, p, q)$, $\gamma_r^i(t, p, q)$, $r = 1, \ldots, m$, are such that

$$\frac{\partial \sigma_r^i}{\partial p^\alpha} + \frac{\partial \gamma_r^\alpha}{\partial q^i} = 0, \quad \frac{\partial \sigma_r^i}{\partial q^\alpha} = \frac{\partial \sigma_r^\alpha}{\partial q^i}, \quad \frac{\partial \gamma_r^i}{\partial p^\alpha} = \frac{\partial \gamma_r^\alpha}{\partial p^i}, \quad i, \alpha = 1, \ldots, n, \qquad (5.1.10)$$

then (5.1.8) holds. Thus, if the relations (5.1.9)–(5.1.10) take place, the condition (5.1.1) is fulfilled.

The condition (5.1.2) also holds when (5.1.9)–(5.1.10) are true. This can be proved analogously by using (5.1.5) instead of (5.1.4).

Now consider the condition (5.1.3). Clearly,

$$\sum_{i=1}^{n} \frac{D(P^i(t_0), Q^i(t_0))}{D(p^k, q^l)} = \sum_{i=1}^{n} \frac{D(p^i, q^i)}{D(p^k, q^l)} = \delta_{kl}.$$

Then the condition (5.1.3) is fulfilled if and only if

$$\sum_{i=1}^{n} d \frac{D(P^i(t), Q^i(t))}{D(p^k, q^l)} = 0.$$

Using the same arguments again, we prove that the relations (5.1.9)–(5.1.10) ensure this condition as well.

Finally, noting that the relations (5.0.2) imply (5.1.9)–(5.1.10), we obtain the following proposition.

Theorem 5.1.1 *The phase flow of the system of SDEs*

$$dP^i = -\frac{\partial H}{\partial q^i}(t, P, Q)dt - \sum_{r=1}^{m} \frac{\partial H_r}{\partial q^i}(t, P, Q) \circ dw_r(t)$$

$$dQ^i = \frac{\partial H}{\partial p^i}(t, P, Q)dt + \sum_{r=1}^{m} \frac{\partial H_r}{\partial p^i}(t, P, Q) \circ dw_r(t), \quad i = 1, \ldots, n,$$

with Hamiltonians $H(t, p, q)$, $H_r(t, p, q)$, $r = 1, \ldots, m$, preserves symplectic structure.

Corollary 5.1.2 *The phase flow of a Hamiltonian system with additive noise preserves symplectic structure.*

5.2 Mean-Square Symplectic Methods for Stochastic Hamiltonian Systems

5.2.1 General Stochastic Hamiltonian Systems

As is known [398], in the case of deterministic general Hamiltonian systems symplectic Runge-Kutta (RK) methods are all implicit. Hence it is natural to expect that to construct symplectic methods for general stochastic Hamiltonian systems, fully implicit methods are needed. Here, using fully implicit methods from Section 1.3, we construct symplectic methods for the general Hamiltonian system (5.0.1), (5.0.2). Its Ito form reads

$$dP = f\,dt + \frac{1}{2}\sum_{r=1}^{m}\sum_{j=1}^{n}\frac{\partial\sigma_r}{\partial p^j}\sigma_r^j\,dt + \frac{1}{2}\sum_{r=1}^{m}\sum_{j=1}^{n}\frac{\partial\sigma_r}{\partial q^j}\gamma_r^j\,dt + \sum_{r=1}^{m}\sigma_r\,dw_r(t) \quad (5.2.1)$$

$$dQ = g\,dt + \frac{1}{2}\sum_{r=1}^{m}\sum_{j=1}^{n}\frac{\partial\gamma_r}{\partial p^j}\sigma_r^j\,dt + \frac{1}{2}\sum_{r=1}^{m}\sum_{j=1}^{n}\frac{\partial\gamma_r}{\partial q^j}\gamma_r^j\,dt + \sum_{r=1}^{m}\gamma_r\,dw_r(t).$$

Introduce the following implicit method:

$$P_{k+1} = P_k + fh - \frac{1}{2}\sum_{r=1}^{m}\sum_{j=1}^{n}\left(\frac{\partial\sigma_r}{\partial p^j}\sigma_r^j - \frac{\partial\sigma_r}{\partial q^j}\gamma_r^j\right)h + \sum_{r=1}^{m}\sigma_r\,(\zeta_{rh})_k\sqrt{h} \quad (5.2.2)$$

$$Q_{k+1} = Q_k + gh - \frac{1}{2}\sum_{r=1}^{m}\sum_{j=1}^{n}\left(\frac{\partial\gamma_r}{\partial p^j}\sigma_r^j - \frac{\partial\gamma_r}{\partial q^j}\gamma_r^j\right)h + \sum_{r=1}^{m}\gamma_r\,(\zeta_{rh})_k\sqrt{h},$$

where all the functions have t, P_{k+1}, Q_k as their arguments.

We recall (cf. (1.3.29)) that the random variables ζ_{rh} here are such that

$$\zeta_{rh} = \begin{cases} \xi_r, & |\xi_r| \le A_h, \\ A_h, & \xi_r > A_h, \\ -A_h, & \xi_r < -A_h, \end{cases} \quad (5.2.3)$$

where ξ_r are independent $\mathcal{N}(0, 1)$-distributed random variables and $A_h = \sqrt{2c|\ln h|}$, $c \ge 1$.

Theorem 5.2.1 *The implicit method (5.2.2) for the system (5.2.1), (5.0.2) (or for the system (5.0.1), (5.0.2)) is symplectic and of mean-square order* $1/2$.

Proof The method (5.2.2) belongs to the family (1.3.49) and, consequently, the assertion about its order of convergence follows from Sect. 1.3.5. Let us prove symplecticness of the method. It is convenient to write the one-step approximation corresponding to (5.2.2) in the form

$$\bar{P}^i = p^i - \frac{\partial H_0}{\partial q^i} h - \frac{1}{2} \sum_{r=1}^{m} \sum_{j=1}^{n} \frac{\partial^2 H_r}{\partial q^i \partial p^j} \frac{\partial H_r}{\partial q^j} h \qquad (5.2.4)$$

$$- \frac{1}{2} \sum_{r=1}^{m} \sum_{j=1}^{n} \frac{\partial^2 H_r}{\partial q^i \partial q^j} \frac{\partial H_r}{\partial p^j} h - \sum_{r=1}^{m} \frac{\partial H_r}{\partial q^i} \zeta_{rh} \sqrt{h},$$

$$\bar{Q}^i = q^i + \frac{\partial H_0}{\partial p^i} h + \frac{1}{2} \sum_{r=1}^{m} \sum_{j=1}^{n} \frac{\partial^2 H_r}{\partial p^i \partial p^j} \frac{\partial H_r}{\partial q^j} h$$

$$+ \frac{1}{2} \sum_{r=1}^{m} \sum_{j=1}^{n} \frac{\partial^2 H_r}{\partial p^i \partial q^j} \frac{\partial H_r}{\partial p^j} h + \sum_{r=1}^{m} \frac{\partial H_r}{\partial p^i} \zeta_{rh} \sqrt{h},$$

where $i = 1, \ldots, n$ and all the functions have $t, \bar{P}, q$ as their arguments. Introduce the function $F(t, p, q)$ (h, ζ_{rh} are fixed here):

$$F(t, p, q) = H_0(t, p, q) h + \frac{1}{2} \sum_{r=1}^{m} \sum_{j=1}^{n} \frac{\partial H_r}{\partial q^j} (t, p, q) \frac{\partial H_r}{\partial p^j} (t, p, q) h$$

$$+ \sum_{r=1}^{m} H_r(t, p, q) \zeta_{rh} \sqrt{h}.$$

Then (5.2.4) can be written as

$$\bar{P}^i = p^i - \frac{\partial F}{\partial q^i} (t, \bar{P}, q), \qquad (5.2.5)$$

$$\bar{Q}^i = q^i + \frac{\partial F}{\partial p^i} (t, \bar{P}, q).$$

We have

$$\sum_{i=1}^{n} d\bar{P}^i \wedge d\bar{Q}^i = \sum_{i=1}^{n} d\bar{P}^i \wedge \left(dq^i + \sum_{j=1}^{n} F''_{p^i p^j} d\bar{P}^j + \sum_{j=1}^{n} F''_{p^i q^j} dq^j \right)$$

$$= \sum_{i=1}^{n} d\bar{P}^i \wedge dq^i + \sum_{i=1}^{n} \sum_{j=1}^{n} F''_{p^i p^j} d\bar{P}^i \wedge d\bar{P}^j + \sum_{i=1}^{n} \sum_{j=1}^{n} F''_{p^i q^j} d\bar{P}^i \wedge dq^j.$$

Since $d\bar{P}^i \wedge d\bar{P}^j = -d\bar{P}^j \wedge d\bar{P}^i$, we get

$$\sum_{i=1}^{n} d\bar{P}^i \wedge d\bar{Q}^i = \sum_{i=1}^{n} d\bar{P}^i \wedge dq^i + \sum_{i=1}^{n}\sum_{j=1}^{n} F''_{p^i q^j} d\bar{P}^i \wedge dq^j$$

$$= \sum_{i=1}^{n} d\bar{P}^i \wedge dq^i + \sum_{i=1}^{n}\sum_{j=1}^{n} F''_{q^i p^j} d\bar{P}^j \wedge dq^i. \qquad (5.2.6)$$

Further

$$d\bar{P}^i = dp^i - \sum_{j=1}^{n} F''_{q^i p^j} d\bar{P}^j - \sum_{j=1}^{n} F''_{q^i q^j} dq^j.$$

Substituting $\sum_{j=1}^{n} F''_{q^i p^j} d\bar{P}^j$ from here in (5.2.6), we obtain

$$\sum_{i=1}^{n} d\bar{P}^i \wedge d\bar{Q}^i = \sum_{i=1}^{n} d\bar{P}^i \wedge dq^i + \sum_{i=1}^{n} \left(dp^i - d\bar{P}^i - \sum_{j=1}^{n} F''_{q^i q^j} dq^j \right) \wedge dq^i$$

$$= \sum_{i=1}^{n} dp^i \wedge dq^i - \sum_{i=1}^{n}\sum_{j=1}^{n} F''_{q^i q^j} dq^j \wedge dq^i = \sum_{i=1}^{n} dp^i \wedge dq^i.$$

$$\square$$

A more general symplectic method for the Hamiltonian system (5.0.1), (5.0.2) has the form

$$P_{k+1} = P_k + f(t_k + \beta h, \alpha P_{k+1} + (1-\alpha)P_k, (1-\alpha)Q_{k+1} + \alpha Q_k)h \qquad (5.2.7)$$

$$+ \left(\frac{1}{2} - \alpha\right) \sum_{r=1}^{m}\sum_{j=1}^{n} \left(\frac{\partial\sigma_r}{\partial p^j}\sigma_r^j - \frac{\partial\sigma_r}{\partial q^j}\gamma_r^j\right)h + \sum_{r=1}^{m} \sigma_r \,(\zeta_{rh})_k \sqrt{h},$$

$$Q_{k+1} = Q_k + g(t_k + \beta h, \alpha P_{k+1} + (1-\alpha)P_k, (1-\alpha)Q_{k+1} + \alpha Q_k)h$$

$$+ \left(\frac{1}{2} - \alpha\right) \sum_{r=1}^{m}\sum_{j=1}^{n} \left(\frac{\partial\gamma_r}{\partial p^j}\sigma_r^j - \frac{\partial\gamma_r}{\partial q^j}\gamma_r^j\right)h + \sum_{r=1}^{m} \gamma_r \,(\zeta_{rh})_k \sqrt{h},$$

where $\sigma_r, \gamma_r, r = 1, \ldots, m$, and their derivatives are calculated at $(t_k, \alpha P_{k+1} + (1-\alpha)P_k, (1-\alpha)Q_{k+1} + \alpha Q_k)$, and $\alpha, \beta \in [0, 1]$ are parameters.

Using arguments similar to ones in the proof of Theorem 5.2.1, we obtain the theorem.

Theorem 5.2.2 *The implicit method (5.2.7) for the system (5.0.1), (5.0.2) (or for the system (5.2.1), (5.0.2)) is symplectic and of mean-square order 1/2.*

The method (5.2.2) is a particular case of (5.2.7) when $\alpha = 1$, $\beta = 0$. If $\alpha = \beta = 1/2$ the method (5.2.7) becomes the midpoint method (cf. (1.3.53)):

$$P_{k+1} = P_k + f\left(t_k + \frac{h}{2}, \frac{P_k + P_{k+1}}{2}, \frac{Q_k + Q_{k+1}}{2}\right)h \qquad (5.2.8)$$

$$+ \sum_{r=1}^{m} \sigma_r\left(t_k, \frac{P_k + P_{k+1}}{2}, \frac{Q_k + Q_{k+1}}{2}\right)(\zeta_{rh})_k \sqrt{h},$$

$$Q_{k+1} = Q_k + g\left(t_k + \frac{h}{2}, \frac{P_k + P_{k+1}}{2}, \frac{Q_k + Q_{k+1}}{2}\right)h$$

$$+ \sum_{r=1}^{m} \gamma_r\left(t_k, \frac{P_k + P_{k+1}}{2}, \frac{Q_k + Q_{k+1}}{2}\right)(\zeta_{rh})_k \sqrt{h}.$$

Remark 5.2.3 In the commutative case, i.e., when $\Lambda_i b_r = \Lambda_r b_i$ or in the case of a system with one noise (i.e., $m = 1$) the symplectic method (5.2.8) for (5.0.1), (5.0.2) has the first mean-square order of convergence.

Remark 5.2.4 In the case of Hamiltonians that are separable in the noise part, i.e., when $H_r(t, p, q) = U_r(t, q) + V_r(t, p)$, $r = 1, \ldots, m$, we can obtain symplectic methods for (5.0.1), (5.0.2) which are explicit in stochastic terms and do not need truncated random variables. For instance, (5.2.2) acquires the form

$$P_{k+1} = P_k + f(t_k, P_{k+1}, Q_k)h \qquad (5.2.9)$$

$$+ \frac{h}{2} \sum_{r=1}^{m} \sum_{j=1}^{n} \frac{\partial \sigma_r}{\partial q^j}(t_k, Q_k) \cdot \gamma_r^j(P_{k+1}) + \sum_{r=1}^{m} \sigma_r(t_k, Q_k)\Delta_k w_r,$$

$$Q_{k+1} = Q_k + g(t_k, P_{k+1}, Q_k)h$$

$$- \frac{h}{2} \sum_{r=1}^{m} \sum_{j=1}^{n} \frac{\partial \gamma_r}{\partial p^j}(P_{k+1}) \cdot \sigma_r^j(t_k, Q_k) + \sum_{r=1}^{m} \gamma_r(t_k, P_{k+1})\Delta_k w_r.$$

Note that the method (5.2.9) is implicit in the deterministic terms. See fully explicit symplectic methods for some systems with separable Hamiltonians in the next remark and subsection.

Of course, if it is necessary, fully implicit methods which require truncated random variables can be used in the case of separable Hamiltonians as well.

Remark 5.2.5 It is possible to construct fully explicit symplectic methods for the following partitioned system:

$$dP = f(t, Q)dt + \sum_{r=1}^{m} \sigma_r(t, Q) \circ dw_r(t), \quad P(t_0) = p, \qquad (5.2.10)$$

$$dQ = g(P)dt + \sum_{r=1}^{m} \gamma_r(t)dw_r(t), \quad Q(t_0) = q,$$

with $f^i = -\partial U_0/\partial q^i$, $g^i = \partial V_0/\partial p^i$, $\sigma_r^i = -\partial U_r/\partial q^i$, $r = 1, \ldots, m$, $i = 1, \ldots, n$.

For instance, the explicit partitioned Runge-Kutta (PRK) method (cf. (5.2.17)–(5.2.18))

$$\mathcal{Q}_1 = Q_k + \alpha h g(P_k),$$
$$\mathcal{P}_1 = P_k + h f(t_k + \alpha h, \mathcal{Q}_1) + \frac{h}{2} \sum_{r=1}^{m} \sum_{j=1}^{n} \frac{\partial \sigma_r}{\partial q^j}(t_k, \mathcal{Q}_1) \cdot \gamma_r^j(t_k), \qquad (5.2.11)$$
$$\mathcal{Q}_2 = \mathcal{Q}_1 + (1 - \alpha)h g(\mathcal{P}_1),$$

$$P_{k+1} = \mathcal{P}_1 + \sum_{r=1}^{m} \sigma_r(t_k, \mathcal{Q}_2) \Delta_k w_r, \qquad (5.2.12)$$

$$Q_{k+1} = \mathcal{Q}_2 + \sum_{r=1}^{m} \gamma_r(t_k) \Delta_k w_r, \quad k = 0, \ldots, N-1,$$

with the parameter $0 \leq \alpha \leq 1$ is symplectic and of the mean-square order $1/2$.

A particular case of the system (5.2.10) is considered in the next subsection, where explicit symplectic methods of a higher order are proposed.

5.2.2 Explicit Methods in the Case of Separable Hamiltonians

Consider a special case of the Hamiltonian system (5.0.1), (5.0.2) such that

$$H_0(t, p, q) = V_0(p) + U_0(t, q), \quad H_r(t, p, q) = U_r(t, q), \quad r = 1, \ldots, m.$$
$$(5.2.13)$$

In this case we get the following system in the sense of Stratonovich

$$dP = f(t, Q)dt + \sum_{r=1}^{m} \sigma_r(t, Q) \circ dw_r(t), \quad P(t_0) = p,$$
$$dQ = g(P)dt, \quad Q(t_0) = q, \qquad (5.2.14)$$

with

$$f^i = -\partial U_0/\partial q^i, \ g^i = \partial V_0/\partial p^i, \ \sigma_r^i = -\partial U_r/\partial q^i,$$
$$r = 1, \ldots, m, \ i = 1, \ldots, n. \tag{5.2.15}$$

We note that it is not difficult to consider a slightly more general separable Hamiltonian $H_0(t, p, q) = V_0(t, p) + U_0(t, q)$ but we restrict ourselves to H_0 from (5.2.13). It is obvious that the system (5.2.14) has the same form in the sense of Ito.

For $V_0(p) = \dfrac{1}{2}(M^{-1}p, p)$ with M a constant, symmetric, invertible matrix, the system (5.2.14) takes the form

$$dP = f(t, Q)dt + \sum_{r=1}^{m} \sigma_r(t, Q)dw_r(t), \ P(t_0) = p,$$
$$dQ = M^{-1}Pdt, \ Q(t_0) = q. \tag{5.2.16}$$

This system can be written as a second-order differential equation with multiplicative noise.

Due to specific features of the system (5.2.14), (5.2.15), we have succeeded in construction of explicit partitioned Runge-Kutta (PRK) methods of a higher order.

5.2.2.1 First-Order Methods

A PRK method for (5.2.14) has the form (cf. (5.2.11)–(5.2.12)):

$$\mathcal{Q}_1 = Q_k + \alpha h g(P_k), \ \ \mathcal{P}_1 = P_k + h f(t_k + \alpha h, \mathcal{Q}_1),$$
$$\mathcal{Q}_2 = \mathcal{Q}_1 + (1 - \alpha)h g(\mathcal{P}_1), \tag{5.2.17}$$

$$P_{k+1} = \mathcal{P}_1 + \sum_{r=1}^{m} \sigma_r(t_k, \mathcal{Q}_2)\Delta_k w_r, \ \ Q_{k+1} = \mathcal{Q}_2, \ k = 0, \ldots, N - 1, \tag{5.2.18}$$

where $0 \le \alpha \le 1$ is a parameter.

Theorem 5.2.6 *The explicit method (5.2.17)–(5.2.18) for the system (5.2.14) with (5.2.15) is symplectic and of first mean-square order.*

Proof In the case of the system (5.2.14) the operators Λ_r take the form $\Lambda_r = (\sigma_r, \partial/\partial p)$. Since σ_r do not depend on p, we get $\Lambda_i \sigma_j = 0$. It is known (see Sect. 1.2.3) that in such a case the Euler method has the first mean-square order of accuracy. Comparing the method (5.2.17)–(5.2.18) with the Euler method, it is not difficult to get that the method (5.2.17)–(5.2.18) is of the first mean-square order as well.

Due to (5.2.15), $\partial \sigma_r^i/\partial q^j = \partial \sigma_r^j/\partial q^i$. Using this, we obtain $dP_{k+1} \wedge dQ_{k+1} = dP_1 \wedge dQ_2$. It is easy to prove that $dP_1 \wedge dQ_2 = dP_1 \wedge dQ_1 = dP_k \wedge dQ_k$. Therefore the method (5.2.17)–(5.2.18) is symplectic. $\qquad\square$

Remark 5.2.7 By swapping the roles of p and q, we can propose the following symplectic method of the first mean-square order for the system (5.2.14)–(5.2.15):

$$\mathcal{P} = P_k + \alpha h f(t_k, \mathcal{Q}_k), \quad \mathcal{Q} = Q_k + hg(\mathcal{P}) \tag{5.2.19}$$

$$P_{k+1} = \mathcal{P} + (1 - \alpha)hf(t_{k+1}, \mathcal{Q}) + \sum_{r=1}^{m} \sigma_r(t_k, \mathcal{Q})\Delta_k w_r, \quad Q_{k+1} = \mathcal{Q}. \tag{5.2.20}$$

5.2.2.2 Methods of Order 3/2

Consider the relations

$$\mathcal{P}_i = p + h \sum_{j=1}^{s} \alpha_{ij} f(t + c_j h, \mathcal{Q}_j)$$

$$+ \sum_{j=1}^{s} \sum_{r=1}^{m} \sigma_r(t + d_j h, \mathcal{Q}_j) \left(\lambda_{ij} \varphi_r + \mu_{ij} \psi_r \right), \tag{5.2.21}$$

$$\mathcal{Q}_i = q + h \sum_{j=1}^{s} \hat{\alpha}_{ij} g(\mathcal{P}_j), \quad i = 1, \dots, s,$$

$$\bar{P} = p + h \sum_{i=1}^{s} \beta_i f(t + c_i h, \mathcal{Q}_i)$$

$$+ \sum_{i=1}^{s} \sum_{r=1}^{m} \sigma_r(t + d_i h, \mathcal{Q}_i) (\nu_i \varphi_r + \varkappa_i \psi_r), \tag{5.2.22}$$

$$\bar{Q} = q + h \sum_{i=1}^{s} \hat{\beta}_i g(\mathcal{P}_i)$$

where φ_r, ψ_r do not depend on p and q, the parameters α_{ij}, $\hat{\alpha}_{ij}$, β_i, $\hat{\beta}_i$, λ_{ij}, μ_{ij}, ν_i, $\varkappa_i$ satisfy the conditions

$$\beta_i \hat{\alpha}_{ij} + \hat{\beta}_j \alpha_{ji} - \beta_i \hat{\beta}_j = 0,$$
$$\nu_i \hat{\alpha}_{ij} + \hat{\beta}_j \lambda_{ji} - \nu_i \hat{\beta}_j = 0, \quad \varkappa_i \hat{\alpha}_{ij} + \hat{\beta}_j \mu_{ji} - \varkappa_i \hat{\beta}_j = 0, \quad i, j = 1, \dots, s, \tag{5.2.23}$$

and c_i, d_i are arbitrary parameters.

If $\sigma_r \equiv 0$, the relations (5.2.21)–(5.2.22) coincide with a general form of s-stage PRK methods for deterministic differential equations (see, e.g., [398, p. 34]). It is known [398, 421] that the symplectic condition holds for $\bar{P}$, $\bar{Q}$ from (5.2.21)–(5.2.22) with (5.2.23) in the case of $\sigma_r \equiv 0$. By a generalization of the proof of Theorem 6.2 from [398], we prove the following lemma.

Lemma 5.2.8 *The relations (5.2.21)–(5.2.22) with conditions (5.2.23) preserve symplectic structure, i.e., $d\bar{P} \wedge d\bar{Q} = dp \wedge dq$.*

Proof Denote for a while:

$$f_i = f(t + c_i h, Q_i), \quad g_i = g(\mathcal{P}_i), \quad \sigma_{ri} = \sigma_r(t + d_i h, Q_i).$$

We get

$$d\bar{P} \wedge d\bar{Q} = dp \wedge dq + h \sum_{j=1}^{s} \hat{\beta}_j dp \wedge dg_j + h \sum_{i=1}^{s} \beta_i df_i \wedge dq \qquad (5.2.24)$$

$$+ h^2 \sum_{i=1}^{s} \sum_{j=1}^{s} \beta_i \hat{\beta}_j df_i \wedge dg_j + \sum_{i=1}^{s} \sum_{r=1}^{m} (\nu_i \varphi_r + \varkappa_i \psi_r) d\sigma_{ri} \wedge dq$$

$$+ h \sum_{i=1}^{s} \sum_{j=1}^{s} \sum_{r=1}^{m} (\nu_i \varphi_r + \varkappa_i \psi_r) \hat{\beta}_j d\sigma_{ri} \wedge dg_j.$$

Then we express $dp \wedge dg_j$ from

$$d\mathcal{P}_j \wedge dg_j = dp \wedge dg_j + h \sum_{i=1}^{s} \alpha_{ji} df_i \wedge dg_j$$

$$+ \sum_{i=1}^{s} \sum_{r=1}^{m} (\lambda_{ji} \varphi_r + \mu_{ji} \psi_r) d\sigma_{ri} \wedge dg_j$$

and substitute it in (5.2.24). Analogously, we act with $df_i \wedge dq$ and $d\sigma_{ri} \wedge dq$ finding them from the expressions for $df_i \wedge d\mathcal{Q}_i$ and $d\sigma_{ri} \wedge d\mathcal{Q}_i$. As a result, using (5.2.23), we obtain

$$d\bar{P} \wedge d\bar{Q} = dp \wedge dq + h \sum_{i=1}^{s} \hat{\beta}_i d\mathcal{P}_i \wedge dg_i + h \sum_{i=1}^{s} \beta_i df_i \wedge d\mathcal{Q}_i$$

$$+ \sum_{i=1}^{s} \sum_{r=1}^{m} (\nu_i \varphi_r + \varkappa_i \psi_r) d\sigma_{ri} \wedge d\mathcal{Q}_i.$$

Taking into account that the wedge product is skew-symmetric, the vector-functions f, g, σ_r are gradients, f, σ_r do not depend on p, and g does not depend on q, it is not difficult to see that each of the terms $d\mathcal{P}_i \wedge dg_i$, $df_i \wedge d\mathcal{Q}_i$, $d\sigma_{ri} \wedge d\mathcal{Q}_i$ vanishes. Therefore $d\bar{P} \wedge d\bar{Q} = dp \wedge dq$. $\square$

Introduce the 2-stage explicit PRK method for the system (5.2.14), (5.2.15):

$$\mathcal{Q}_1 = Q_k, \quad \mathcal{P}_1 = P_k + \frac{h}{4} f(t_k, Q_1) + \frac{1}{2} \sum_{r=1}^{m} \sigma_r(t_k, Q_1) \left(3(J_{r0})_k - \Delta_k w_r\right),$$

$$(5.2.25)$$

$$\mathcal{Q}_2 = \mathcal{Q}_1 + \frac{2}{3}hg(\mathcal{P}_1),$$

$$P_2 = \mathcal{P}_1 + \frac{3}{4}hf\left(t_k + \frac{2}{3}h, \mathcal{Q}_2\right) + \frac{3}{2}\sum_{r=1}^{m}\sigma_r\left(t_k + \frac{2}{3}h, \mathcal{Q}_2\right)(-(J_{r0})_k + \Delta_k w_r),$$

$$P_{k+1} = \mathcal{P}_2, \quad \mathcal{Q}_{k+1} = \mathcal{Q}_2 + \frac{h}{3}g(\mathcal{P}_2), \quad k = 0, \dots, N-1, \tag{5.2.26}$$

where

$$J_{r0} := \frac{1}{h}\int_t^{t+h}(w_r(\vartheta) - w_r(t))\,d\vartheta. \tag{5.2.27}$$

We recall (see 1.6.7) that the random variables $\Delta_k w_r(h)$, $(J_{r0})_k$ have a Gaussian joint distribution, and they can be simulated at each step by $2m$ independent $\mathcal{N}(0, 1)$-distributed random variables ξ_{rk} and η_{rk}, $r = 0, \dots, m$:

$$\Delta_k w_r(h) = \xi_{rk}\sqrt{h}, \quad (J_{r0})_k = \left(\xi_{rk}/2 + \eta_{rk}/\sqrt{12}\right)\sqrt{h}.$$

As a result, the method (5.2.25)–(5.2.26) takes the constructive form.

Theorem 5.2.9 *The explicit PRK method (5.2.25)–(5.2.26) for system (5.2.14), (5.2.15) preserves symplectic structure and has the mean-square order 3/2.*

Proof The method (5.2.25)–(5.2.26) has the form of (5.2.21)–(5.2.22) and its parameters satisfy the conditions (5.2.23). Then, Lemma 5.2.8 implies that this method preserves symplectic structure.

The mean-square order of convergence of (5.2.25)–(5.2.26) is proved using the general theory of numerical integration of SDEs of Chap. 1 (see details in [291]). □

Remark 5.2.10 In the case of $\sigma_r = 0$, $r = 1, \dots, m$, the method (5.2.25)–(5.2.26) coincides with the well-known deterministic symplectic PRK method of the second order. Adapting other explicit deterministic second-order PRK methods from [398, 421], it is possible to construct other explicit symplectic methods of the order 3/2 for the system (5.2.14), (5.2.15).

Remark 5.2.11 In the case of a more general system than (5.2.14) methods of the order 3/2 require simulation of repeated Ito integrals which is a laborious problem from the computational point of view.

Lemma 5.2.8 can be generalized for the general separable case, i.e., for the system (5.0.1), (5.0.2) with $H_r = V_r(p) + U_r(t, q)$, $r = 0, 1, \dots, m$, and it can also be generalized for the general stochastic Hamiltonian system (5.0.1), (5.0.2). In the case of systems with one noise repeated Ito integrals can effectively be simulated and generalizations of Lemma 5.2.8 can be used for constructing high-order symplectic methods for Hamiltonian systems with one noise (i.e., when $m = 1$).

5.3 Mean-Square Symplectic Methods for Hamiltonian Systems with Additive Noise

5.3.1 The Case of a General Hamiltonian

In this subsection we consider the general Hamiltonian system with additive noise

$$dP = f(t, P, Q)dt + \sum_{r=1}^{m} \sigma_r(t)dw(t), \quad P(t_0) = p, \qquad (5.3.1)$$

$$dQ = g(t, P, Q)dt + \sum_{r=1}^{m} \gamma_r(t)dw(t), \quad Q(t_0) = q,$$

$$f^i = -\partial H/\partial q^i, \qquad g^i = \partial H/\partial p^i, \quad i = 1, \dots, n, \qquad (5.3.2)$$

where P, Q, f, g, σ_r, γ_r are n-dimensional column-vectors, $w_r(t)$, $r = 1, \dots, m$, are independent standard Wiener processes, and $H(t, p, q)$ is a Hamiltonian.

5.3.1.1 First-Order Methods

Consider the two-parameter family of implicit methods

$$\mathcal{P} = P_k + hf(t_k + \beta h, \alpha\mathcal{P} + (1 - \alpha)P_k, (1 - \alpha)\mathcal{Q} + \alpha Q_k), \qquad (5.3.3)$$

$$\mathcal{Q} = Q_k + hg(t_k + \beta h, \alpha\mathcal{P} + (1 - \alpha)P_k, (1 - \alpha)\mathcal{Q} + \alpha Q_k),$$

$$P_{k+1} = \mathcal{P} + \sum_{r=1}^{m} \sigma_r(t_k)\Delta_k w_r, \quad Q_{k+1} = \mathcal{Q} + \sum_{r=1}^{m} \gamma_r(t_k)\Delta_k w_r, \qquad (5.3.4)$$

$$k = 0, \dots, N - 1,$$

where $\Delta_k w_r(h) := w_r(t_k + h) - w_r(t_k)$ and the parameters $\alpha, \beta \in [0, 1]$.

When $\sigma_r = 0$, $\gamma_r = 0$, $r = 1, \dots, m$, this family coincides with the known family of symplectic methods for deterministic Hamiltonian systems (see [421]).

The following lemma guarantees the unique solvability of (5.3.3) with respect to $\mathcal{P}$, $\mathcal{Q}$ for any P_k, Q_k and sufficiently small h.

Lemma 5.3.1 *Let $F(x; c, s)$ be a continuous d-dimensional vector-function depending on $x \in \mathbf{R}^d$, $c \in \mathbf{R}^d$, and $s \in S$, where S is a set from an R^l. Suppose F has the first partial derivatives $\partial F^i/\partial x^j$, $i, j = 1, \dots, d$, which are uniformly bounded in $\mathbf{R}^d \times \mathbf{R}^d \times S$. Then there is an $h_0 > 0$ such that the equation*

$$x = c + hF(x; c, s) + \nu \tag{5.3.5}$$

is uniquely solvable with respect to x for $0 < h \leq h_0$ *and any* $c \in \mathbf{R}^d$, $\nu \in \mathbf{R}^d$, $s \in S$.
The solution of equation (5.3.5) can be found by the method of simple iteration with an arbitrary initial approximation.

The proof of this lemma is not difficult and it is omitted. The next lemma is true for system (5.3.1) with arbitrary f and g (i.e., f and g may not obey the condition (5.3.2)).

Lemma 5.3.2 *The mean-square order of the methods (5.3.3)–(5.3.4) for the system (5.3.1) is equal to* 1.

The proof is based on comparison of the one-step approximation of the method (5.3.3)–(5.3.4) with the one-step approximation of the Euler method.

The one-step approximation $\tilde{P}$, $\tilde{Q}$ of the method (5.3.3)–(5.3.4) is such that $d\tilde{P} = d\mathcal{P}$, $d\tilde{Q} = d\mathcal{Q}$. Hence $d\tilde{P} \wedge d\tilde{Q} = d\mathcal{P} \wedge d\mathcal{Q}$. The relations for $\mathcal{P}$, $\mathcal{Q}$ coincide with ones for the one-step approximation corresponding to the deterministic symplectic method [421]. Therefore, the method (5.3.3)–(5.3.4) is symplectic as well. From here and Lemma 5.3.2, we get the theorem.

Theorem 5.3.3 *The method (5.3.3)–(5.3.4) for the system (5.3.1)–(5.3.2) preserves symplectic structure and has the first mean-square order of convergence.*

Now consider another family of symplectic methods for system (5.3.1):

$$P_{k+1} = P_k + hf(t_k + \beta h, \alpha P_{k+1} + (1 - \alpha)P_k, (1 - \alpha)Q_{k+1} + \alpha Q_k) \tag{5.3.6}$$
$$+ \sum_{r=1}^{m} \sigma_r(t_k) \Delta_k w_r,$$
$$Q_{k+1} = Q_k + hg(t_k + \beta h, \alpha P_{k+1} + (1 - \alpha)P_k, (1 - \alpha)Q_{k+1} + \alpha Q_k)$$
$$+ \sum_{r=1}^{m} \gamma_r(t_k) \Delta_k w_r, \quad k = 0, \dots, N - 1,$$

with the parameters $\alpha, \beta \in [0, 1]$.

For sufficiently small h, the equations (5.3.6) are uniquely solvable with respect to P_{k+1}, Q_{k+1} according to Lemma 5.3.1.

Theorem 5.3.4 *The method (5.3.6) for the system (5.3.1)–(5.3.2) preserves symplectic structure and has the first mean-square order of convergence.*

Proof Comparing the one-step approximation of the method (5.3.6) with the one-step approximation of the Euler method, one can establish that the mean-square order of the method (5.3.6) is equal to 1.

Now we check symplecticness of the method. Let $\tilde{P}$, $\tilde{Q}$ be the one-step approximation corresponding to the method (5.3.6). Introduce

$$\hat{p} = p + \alpha \sum_{r=1}^{m} \sigma_r(t) \Delta w_r, \quad \hat{q} = q + (1-\alpha) \sum_{r=1}^{m} \gamma_r(t) \Delta w_r,$$

$$\hat{P} = \tilde{P} - (1-\alpha) \sum_{r=1}^{m} \sigma_r(t) \Delta w_r, \quad \hat{Q} = \tilde{Q} - \alpha \sum_{r=1}^{m} \gamma_r(t) \Delta w_r.$$

We have

$$\hat{P} = \hat{p} + hf(t + \beta h, \alpha \hat{P} + (1-\alpha)\hat{p}, (1-\alpha)\hat{Q} + \alpha \hat{q}),$$

$$\hat{Q} = \hat{q} + hg(t + \beta h, \alpha \hat{P} + (1-\alpha)\hat{p}, (1-\alpha)\hat{Q} + \alpha \hat{q}).$$

The relations for $\hat{P}$, $\hat{Q}$ coincide with the one-step approximation corresponding to the symplectic deterministic method. Therefore, $d\hat{P} \wedge d\hat{Q} = d\hat{p} \wedge d\hat{q}$. Further, it is obvious that $d\hat{P} \wedge d\hat{Q} = d\tilde{P} \wedge d\tilde{Q}$ and $d\hat{p} \wedge d\hat{q} = dp \wedge dq$. Consequently, $d\tilde{P} \wedge d\tilde{Q} = dp \wedge dq$, i.e., the method (5.3.6) is symplectic. $\qquad \square$

5.3.1.2 Methods of Order 3/2

For $i = 1, \ldots, s$, consider the relations

$$\mathcal{P}_i = p + h \sum_{j=1}^{s} \alpha_{ij} f(t + c_j h, \mathcal{P}_j, \mathcal{Q}_j) + \varphi_i,$$

$$\mathcal{Q}_i = q + h \sum_{j=1}^{s} \alpha_{ij} g(t + c_j h, \mathcal{P}_j, \mathcal{Q}_j) + \psi_i, \qquad (5.3.7)$$

$$\bar{P} = p + h \sum_{i=1}^{s} \beta_i f(t + c_i h, \mathcal{P}_i, \mathcal{Q}_i) + \eta,$$

$$\bar{Q} = q + h \sum_{i=1}^{s} \beta_i g(t + c_i h, \mathcal{P}_i, \mathcal{Q}_i) + \zeta, \qquad (5.3.8)$$

where φ_i, ψ_i, η, ζ do not depend on p and q, the parameters α_{ij} and β_i satisfy the conditions

$$\beta_i \alpha_{ij} + \beta_j \alpha_{ji} - \beta_i \beta_j = 0, \quad i, j = 1, \ldots, s, \qquad (5.3.9)$$

and c_i are arbitrary parameters.

The equations (5.3.7) are uniquely solvable with respect to $\mathcal{P}_i$, $\mathcal{Q}_i$, $i = 1, \ldots, s$, for any p, q, φ_i, ψ_i, η, ζ and sufficiently small h according to Lemma 5.3.1.

If $\varphi_i = \psi_i = \eta = \zeta = 0$, the relations (5.3.7)–(5.3.8) coincide with a general form of s-stage Runge-Kutta (RK) methods for deterministic differential equations. It is known (see, e.g., Theorem 6.1 in [398]) that the symplectic condition $d\bar{P} \wedge d\bar{Q} = dp \wedge dq$ holds for $\bar{P}$, $\bar{Q}$ from (5.3.7)–(5.3.8) with (5.3.9) and $\varphi_i = \psi_i = \eta = \zeta = 0$. Generalizing this result for arbitrary φ_i, ψ_i, η, ζ, we obtain the following lemma.

Lemma 5.3.5 *The relations (5.3.7)–(5.3.8) with condition (5.3.9) preserve symplectic structure, i.e.,* $d\bar{P} \wedge d\bar{Q} = dp \wedge dq$.

The lemma is proved in [290] by a generalization of the proof of Theorem 6.1 from [398] (see also Lemma 5.2.8 in the previous section).

The next lemma will be used in Theorem 5.3.7 for the Hamiltonian system (5.3.1)–(5.3.2). Consider the general (not necessarily Hamiltonian) system with additive noise

$$dX = a(t, X)dt + \sum_{r=1}^{m} b_r(t)dw_r(t), \quad X(t_0) = X_0, \tag{5.3.10}$$

and introduce the parametric family of one-step approximations for (5.3.10):

$$X_1 = x + \frac{\alpha}{2} ha\left(t + \frac{\alpha}{2}h, X_1\right) + \sum_{r=1}^{m} b_r(t)\left(\lambda_1 J_{r0} + \mu_1 \Delta w_r\right), \tag{5.3.11}$$

$$X_2 = x + \alpha ha\left(t + \frac{\alpha}{2}h, X_1\right) + \frac{1-\alpha}{2} ha\left(t + \frac{1+\alpha}{2}h, X_2\right)$$
$$+ \sum_{r=1}^{m} b_r(t)\left(\lambda_2 J_{r0} + \mu_2 \Delta w_r\right),$$

$$\bar{X} = x + h\left[\alpha a\left(t + \frac{\alpha}{2}h, X_1\right) + (1-\alpha)a\left(t + \frac{1+\alpha}{2}h, X_2\right)\right]$$
$$+ \sum_{r=1}^{m} b_r(t)\Delta w_r + \sum_{r=1}^{m} b'_r(t)I_{0r},$$

where

$$\Delta w_r := w_r(t+h) - w_r(t), \quad I_{0r} := \int_{t}^{t+h} (\vartheta - t) \, dw_r(\vartheta),$$

$$J_{r0} := \frac{1}{h} \int_{t}^{t+h} (w_r(\vartheta) - w_r(t)) \, d\vartheta, \tag{5.3.12}$$

and the parameters α, λ_1, λ_2, μ_1, μ_2 are such that

$$\alpha\lambda_1 + (1-\alpha)\lambda_2 = 1, \quad \alpha\mu_1 + (1-\alpha)\mu_2 = 0, \tag{5.3.13}$$

$$\alpha \left(\frac{\lambda_1^2}{3} + \lambda_1 \mu_1 + \mu_1^2 \right) + (1 - \alpha) \left(\frac{\lambda_2^2}{3} + \lambda_2 \mu_2 + \mu_2^2 \right) = \frac{1}{2}. \tag{5.3.14}$$

For example, the following set of parameters satisfies (5.3.13)–(5.3.14):

$$\alpha = \frac{1}{2}, \; \lambda_1 = \lambda_2 = 1, \; \mu_1 = -\mu_2 = \frac{1}{\sqrt{6}}. \tag{5.3.15}$$

Note that the random variables Δw_r and J_{r0} are of the same mean-square order $O(h^{1/2})$.

Lemma 5.3.6 *The method based on the one-step approximation (5.3.11) with conditions (5.3.13)–(5.3.14) is of the mean-square order 3/2.*

Proof   Due to properties of the Wiener process and Ito integrals, we get

$$E \Delta w_i = 0, \; E \Delta w_i \Delta w_j = \delta_{ij} h, \; E \Delta w_i \Delta w_j \Delta w_k = 0, \tag{5.3.16}$$

$$E \left(\Delta w_i \right)^4 = 3h^2, \; E J_{i0} = 0, \; E J_{i0} J_{j0} = \delta_{ij} \frac{h}{3}, \; E J_{i0} J_{j0} J_{k0} = 0,$$

$$E \left(J_{i0} \right)^4 = \frac{h^2}{3}, \; E \Delta w_i J_{j0} = \delta_{ij} \frac{h}{2}, \; E \Delta w_i \Delta w_j J_{k0} = 0, \; E \Delta w_i J_{j0} J_{k0} = 0.$$

Let $\Delta X_i := X_i - x, \; i = 1, 2$. We have

$$|E \Delta X_i| = O(h), \; E \left(\Delta X_i \right)^{2l} = O(h^l), \; l = 1, 2, 3, 4, \; i = 1, 2,$$
$$\left| E \left(\Delta X_i \right)^3 \right| = O(h^2). \tag{5.3.17}$$

Expand (5.3.11):

$$\Delta X_1 = \frac{\alpha}{2} h a(t, x) + \sum_{r=1}^{m} b_r(t) \left(\lambda_1 J_{r0} + \mu_1 \Delta w_r \right) + \rho_1, \tag{5.3.18}$$

$$\Delta X_2 = \frac{1 + \alpha}{2} h a(t, x) + \sum_{r=1}^{m} b_r(t) \left(\lambda_2 J_{r0} + \mu_2 \Delta w_r \right) + \rho_2, \tag{5.3.19}$$

$$\bar{X} = x + \sum_{r=1}^{m} b_r(t) \Delta w_r + \sum_{r=1}^{m} b_r'(t) I_{0r} + h a(t, x) \tag{5.3.20}$$

$$+ h \sum_{i=1}^{d} \frac{\partial a}{\partial x^i} (t, x) \left(\alpha \Delta X_1^i + (1 - \alpha) \Delta X_2^i \right) + \frac{h^2}{2} \frac{\partial a}{\partial t} (t, x)$$

$$+ \frac{h}{2} \sum_{i,j=1}^{d} \frac{\partial^2 a}{\partial x^i \partial x^j} (t, x) \left(\alpha \Delta X_1^i \Delta X_1^j + (1 - \alpha) \Delta X_2^i \Delta X_2^j \right) + \bar{\rho}.$$

Using (5.3.16)–(5.3.17), one can obtain

$$|E\rho_i| = O(h^2), \quad \left|E\rho_i^l \Delta X_i^k\right| = O(h^2), \quad E\rho_i^2 = O(h^3),$$
(5.3.21)

$$|E\bar\rho| = O(h^3), \quad E\bar\rho^2 = O(h^5).$$
(5.3.22)

Substituting (5.3.18)–(5.3.19) in (5.3.20) and using (5.3.13), we get

$$\bar X = x + \sum_{r=1}^{m} b_r \Delta w_r + \sum_{r=1}^{m} b'_r I_{0r} + ha + \frac{h^2}{2}\frac{\partial a}{\partial t} + \frac{h^2}{2}\sum_{i=1}^{d}\frac{\partial a}{\partial x^i}a^i$$
(5.3.23)

$$+ h\sum_{r=1}^{m}\sum_{i=1}^{d} b_r^i \frac{\partial a}{\partial x^i} J_{r0} + \frac{h^2}{4}\sum_{r=1}^{m}\sum_{i,j=1}^{d}\frac{\partial^2 a}{\partial x^i \partial x^j}b_r^i b_r^j + R,$$

$$R = \frac{h}{2}\sum_{r,l=1}^{m}\sum_{i,j=1}^{d}\frac{\partial^2 a}{\partial x^i \partial x^j}b_r^i b_l^j \left[\alpha\,(\lambda_1 J_{r0} + \mu_1 \Delta w_r)\,(\lambda_1 J_{l0} + \mu_1 \Delta w_l)\right.$$
$$\left. +(1-\alpha)\,(\lambda_2 J_{r0} + \mu_2 \Delta w_r)\,(\lambda_2 J_{l0} + \mu_2 \Delta w_l)\right]$$
$$-\frac{h^2}{4}\sum_{r=1}^{m}\sum_{i,j=1}^{d}\frac{\partial^2 a}{\partial x^i \partial x^j}b_r^i b_r^j + \rho,$$

where the coefficients and their derivatives are calculated at (t, x) and ρ satisfies the same relations as $\bar\rho$ (see (5.3.22)).

The relations (5.3.16) and (5.3.14) imply

$$E[\alpha\,(\lambda_1 J_{r0} + \mu_1 \Delta w_r)\,(\lambda_1 J_{l0} + \mu_1 \Delta w_l)$$
$$+(1-\alpha)\,(\lambda_2 J_{r0} + \mu_2 \Delta w_r)\,(\lambda_2 J_{l0} + \mu_2 \Delta w_l)] = \frac{h}{2}\delta_{rl}.$$
(5.3.24)

Using (5.3.16), (5.3.21)–(5.3.22), and (5.3.24), it is not difficult to get that

$$|ER| = O(h^3), \quad \left(ER^2\right)^{1/2} = O(h^2).$$
(5.3.25)

Comparing (5.3.23) with the one-step approximation of the standard method of mean-square order 3/2 for systems with additive noise (1.6.8), we obtain that the method (5.3.11) is of mean-square order 3/2. $\square$

Now we return to the Hamiltonian system with additive noise (5.3.1). Consider the parametric family of methods:

$$\mathcal{P}_1 = P_k + \frac{\alpha}{2}hf\left(t_k + \frac{\alpha}{2}h, \mathcal{P}_1, \mathcal{Q}_1\right) \tag{5.3.26}$$

$$+ \sum_{r=1}^{m} \sigma_r(t_k)\left(\lambda_1 (J_{r0})_k + \mu_1 \Delta_k w_r\right),$$

$$\mathcal{Q}_1 = Q_k + \frac{\alpha}{2}hg\left(t_k + \frac{\alpha}{2}h, \mathcal{P}_1, \mathcal{Q}_1\right) + \sum_{r=1}^{m} \gamma_r(t_k)\left(\lambda_1 (J_{r0})_k + \mu_1 \Delta_k w_r\right),$$

$$\mathcal{P}_2 = P_k + \alpha hf\left(t_k + \frac{\alpha}{2}h, \mathcal{P}_1, \mathcal{Q}_1\right) + \frac{1-\alpha}{2}hf\left(t_k + \frac{1+\alpha}{2}h, \mathcal{P}_2, \mathcal{Q}_2\right)$$

$$+ \sum_{r=1}^{m} \sigma_r(t_k)\left(\lambda_2 (J_{r0})_k + \mu_2 \Delta_k w_r\right),$$

$$\mathcal{Q}_2 = Q_k + \alpha hg\left(t_k + \frac{\alpha}{2}h, \mathcal{P}_1, \mathcal{Q}_1\right) + \frac{1-\alpha}{2}hg\left(t_k + \frac{1+\alpha}{2}h, \mathcal{P}_2, \mathcal{Q}_2\right)$$

$$+ \sum_{r=1}^{m} \gamma_r(t_k)\left(\lambda_2 (J_{r0})_k + \mu_2 \Delta_k w_r\right),$$

$$P_{k+1} = P_k + h\left[\alpha f\left(t_k + \frac{\alpha}{2}h, \mathcal{P}_1, \mathcal{Q}_1\right) + (1-\alpha)f\left(t_k + \frac{1+\alpha}{2}h, \mathcal{P}_2, \mathcal{Q}_2\right)\right]$$

$$+ \sum_{r=1}^{m} \sigma_r(t_k)\Delta_k w_r + \sum_{r=1}^{m} \sigma_r'(t_k)(I_{0r})_k,$$

$$Q_{k+1} = Q_k + h\left[\alpha g\left(t_k + \frac{\alpha}{2}h, \mathcal{P}_1, \mathcal{Q}_1\right) + (1-\alpha)g\left(t_k + \frac{1+\alpha}{2}h, \mathcal{P}_2, \mathcal{Q}_2\right)\right]$$

$$+ \sum_{r=1}^{m} \gamma_r(t_k)\Delta_k w_r + \sum_{r=1}^{m} \gamma_r'(t_k)(I_{0r})_k,$$

where the parameters α, λ_1, λ_2, μ_1, μ_2 satisfy (5.3.13)–(5.3.14). The formula (5.3.26) contains the random variables $\Delta_k w_r(h)$, $(J_{r0})_k$, $(I_{0r})_k$ whose joint distribution is Gaussian. As usual, they can be simulated at each step by $2m$ independent $\mathcal{N}(0, 1)$-distributed random variables ξ_{rk} and η_{rk}, $r = 0, \ldots, m$:

$$\Delta_k w_r(h) = \sqrt{h}\xi_{rk}, \quad (J_{r0})_k = \sqrt{h}\left(\xi_{rk}/2 + \eta_{rk}/\sqrt{12}\right), \tag{5.3.27}$$

$$(I_{0r})_k = h^{3/2}\left(\xi_{rk}/2 - \eta_{rk}/\sqrt{12}\right).$$

For $\sigma_r \equiv 0$, $\gamma_r \equiv 0$, $r = 1, \ldots, m$, the method (5.3.26) is reduced to the well-known second-order symplectic Runge-Kutta method for deterministic Hamiltonian systems (see, e.g., [398, p. 101]). Let us note that using this deterministic method with $\alpha = 0$ (the midpoint rule), another implicit 3/2-order method for Hamiltonian systems with noise was proposed in [447], however without preserving symplectic structure.

The one-step approximation corresponding to method (5.3.26) is of the form (5.3.11). Therefore, due to Lemma 5.3.6, the method (5.3.26) is of the mean-square order $3/2$. Moreover, this one-step approximation is of the form (5.3.7) with $s = 2$ and

$$\varphi_1 = \sum_{r=1}^{m} \sigma_r \left(\lambda_1 J_{r0} + \mu_1 \Delta w_r \right), \quad \varphi_2 = \sum_{r=1}^{m} \sigma_r \left(\lambda_2 J_{r0} + \mu_2 \Delta w_r \right),$$

$$\psi_1 = \sum_{r=1}^{m} \gamma_r \left(\lambda_1 J_{r0} + \mu_1 \Delta w_r \right), \quad \psi_2 = \sum_{r=1}^{m} \gamma_r \left(\lambda_2 J_{r0} + \mu_2 \Delta w_r \right),$$

$$\eta = \sum_{r=1}^{m} \sigma_r \Delta w_r + \sum_{r=1}^{m} \sigma'_r I_{0r}, \quad \zeta = \sum_{r=1}^{m} \gamma_r \Delta w_r + \sum_{r=1}^{m} \gamma'_r I_{0r},$$

$$\alpha_{11} = \frac{\alpha}{2}, \quad \alpha_{12} = 0, \quad \alpha_{21} = \alpha, \quad \alpha_{22} = \frac{1-\alpha}{2}, \quad \beta_1 = \alpha,$$

$$\beta_2 = 1 - \alpha, \quad c_1 = \frac{\alpha}{2}, \quad c_2 = \frac{1+\alpha}{2}.$$

This set of parameters α_{ij}, β_i, $i, j = 1, 2$, satisfies the conditions (5.3.9). Then due to Lemma 5.3.5, the method (5.3.26) is symplectic. Thus, we have obtained the following theorem.

Theorem 5.3.7 *Under conditions (5.3.13)–(5.3.14) on the parameters, the method (5.3.26) for the system (5.3.1)–(5.3.2) preserves symplectic structure and has the mean-square order* $3/2$.

5.3.2   The Case of Separable Hamiltonians

In this subsection we consider the Hamiltonian system with additive noise (5.3.1), which Hamiltonian has the special structure

$$H(t, p, q) = V(p) + U(t, q). \tag{5.3.28}$$

We note that it is not difficult to consider a slightly more general Hamiltonian $H(t, p, q) = V(t, p) + U(t, q)$ but we restrict ourselves here to (5.3.28). In the case of separable Hamiltonian (5.3.28) the system (5.3.1) takes the partitioned form

$$dP = f(t, Q)dt + \sum_{r=1}^{m} \sigma_r(t)dw_r(t), \quad P(t_0) = p, \tag{5.3.29}$$

$$dQ = g(P)dt + \sum_{r=1}^{m} \gamma_r(t)dw_r(t), \quad Q(t_0) = q,$$

where $f^i = -\partial U/\partial q^i$, $g^i = \partial V/\partial p^i$, $i = 1, \ldots, n$.

Obviously, the implicit symplectic methods from the previous subsection can be applied to the partitioned system (5.3.29), and they take a simpler form in this case (we do not write them down here). We recall that there are no explicit symplectic RK methods for the general system (5.3.1)–(5.3.2). However, for the partitioned system (5.3.29) it is possible to construct explicit symplectic methods just as in the deterministic case.

5.3.2.1   Explicit First-Order Methods

On the basis of the known family of deterministic PRK methods [397, 398, 421], we construct the family of explicit partitioned methods for stochastic system (5.3.29):

$$\mathcal{Q} = Q_k + \alpha h g(P_k), \quad \mathcal{P} = P_k + h f(t_k + \alpha h, \mathcal{Q}), \tag{5.3.30}$$

$$Q_{k+1} = \mathcal{Q} + (1 - \alpha) h g(\mathcal{P}) + \sum_{r=1}^{m} \gamma_r(t_k) \Delta_k w_r,$$

$$P_{k+1} = \mathcal{P} + \sum_{r=1}^{m} \sigma_r(t_k) \Delta_k w_r, \quad k = 0, \dots, N - 1.$$

Since the expressions for dP_{k+1}, dQ_{k+1} coincide with the ones corresponding to the deterministic symplectic method, the method (5.3.30) is symplectic. Further, it is not difficult to show that the method (5.3.30) has the first mean-square order of accuracy. As a result, we obtain the following theorem.

Theorem 5.3.8 *The explicit partitioned method (5.3.30) for the system (5.3.29) preserves symplectic structure and has the first mean-square order of convergence.*

Remark 5.3.9 In the special cases of $\alpha = 0$ and $\alpha = 1$ the method (5.3.30) takes a simpler form. In these cases it requires evaluation of each of the coefficients f, g once per step only.

Remark 5.3.10 It is possible to propose other symplectic first-order methods for (5.3.29) on the basis of the same deterministic PRK methods as above. For instance, the method

$$\mathcal{Q} = Q_k + \alpha h g(P_k) + \sum_{r=1}^{m} \gamma_r(t_k) \Delta_k w_r,$$

$$\mathcal{P} = P_k + h f(t_k + \alpha h, \mathcal{Q}) + \sum_{r=1}^{m} \sigma_r(t_k) \Delta_k w_r, \tag{5.3.31}$$

$$Q_{k+1} = \mathcal{Q} + (1 - \alpha) h g(\mathcal{P}), \quad P_{k+1} = \mathcal{P}, \quad k = 0, \dots, N - 1,$$

is of the first mean-square order and symplectic.

5.3.2.2 Explicit Methods of Order 3/2

Here using specificity of the system (5.3.29), we construct a 3/2-order symplectic *explicit Runge-Kutta* method (other symplectic methods for (5.3.29) are given in [289]).

Introduce the relations (cf. (5.3.7)–(5.3.8)):

$$\mathcal{P}_i = p + h \sum_{j=1}^{s} \alpha_{ij} f(t + c_j h, \mathcal{Q}_j) + \varphi_i, \tag{5.3.32}$$

$$\mathcal{Q}_i = q + h \sum_{j=1}^{s} \hat{\alpha}_{ij} g(\mathcal{P}_j) + \psi_i, \quad i = 1, \ldots, s,$$

$$\bar{P} = p + h \sum_{i=1}^{s} \beta_i f(t + c_i h, \mathcal{Q}_i) + \eta, \quad \bar{Q} = q + h \sum_{i=1}^{s} \hat{\beta}_i g(\mathcal{P}_i) + \zeta, \tag{5.3.33}$$

where φ_i, ψ_i, η, ζ do not depend on p and q, the parameters α_{ij}, $\hat{\alpha}_{ij}$, β_i and $\hat{\beta}_i$ satisfy the conditions

$$\beta_i \hat{\alpha}_{ij} + \hat{\beta}_j \alpha_{ji} - \beta_i \hat{\beta}_j = 0, \quad i, j = 1, \ldots, s, \tag{5.3.34}$$

and c_i are arbitrary parameters.

If $\varphi_i = \psi_i = \eta = \zeta = 0$, the relations (5.3.32)–(5.3.33) coincide with a general form of s-stage PRK methods for deterministic differential equations. By a generalization of the proof of Theorem 6.2 from [398] (see also Lemma 5.3.5 here), it is not difficult to prove the following lemma.

Lemma 5.3.11 *The relations (5.3.32)–(5.3.33) with condition (5.3.34) preserve symplectic structure, i.e., $d\bar{P} \wedge d\bar{Q} = dp \wedge dq$.*

Introduce the parametric family of 2-stage explicit PRK methods for the system (5.3.29):

$$\mathcal{Q}_1 = \mathcal{Q}_k + \sum_{r=1}^{m} \gamma_r(t_k) \left(\hat{\lambda}_1 (J_{r0})_k + \hat{\mu}_1 \Delta_k w_r \right),$$

$$\mathcal{P}_1 = \mathcal{P}_k + h \beta_1 f(t_k + c_1 h, \mathcal{Q}_1) + \sum_{r=1}^{m} \sigma_r(t_k) \left(\lambda_1 (J_{r0})_k + \mu_1 \Delta_k w_r \right),$$

$$\mathcal{Q}_2 = \mathcal{Q}_k + h \hat{\beta}_1 g(\mathcal{P}_1) + \sum_{r=1}^{m} \gamma_r(t_k) \left(\hat{\lambda}_2 (J_{r0})_k + \hat{\mu}_2 \Delta_k w_r \right), \tag{5.3.35}$$

$$\mathcal{P}_2 = \mathcal{P}_k + h \sum_{i=1}^{2} \beta_i f(t_k + c_i h, \mathcal{Q}_i) + \sum_{r=1}^{m} \sigma_r(t_k) \left(\lambda_2 (J_{r0})_k + \mu_2 \Delta_k w_r \right),$$

$$P_{k+1} = P_k + \sum_{r=1}^{m} \sigma_r(t_k)\Delta_k w_r + \sum_{r=1}^{m} \sigma_r'(t_k)(I_{0r})_k + h \sum_{i=1}^{2} \beta_i f(t_k + c_i h, \mathcal{Q}_i),$$

$$Q_{k+1} = Q_k + \sum_{r=1}^{m} \gamma_r(t_k)\Delta_k w_r + \sum_{r=1}^{m} \gamma_r'(t_k)(I_{0r})_k + h \sum_{i=1}^{2} \hat{\beta}_i g(\mathcal{P}_i),$$

(5.3.36)

where the parameters β_i, $\hat{\beta}_i$, c_i, λ_i, $\hat{\lambda}_i$, μ_i, $\hat{\mu}_i$, $i = 1, 2$, satisfy the conditions

$$\beta_1 + \beta_2 = 1, \quad \hat{\beta}_1 + \hat{\beta}_2 = 1, \quad \beta_2\hat{\beta}_1 = 1/2, \quad c_1 = 0, \quad c_2 = \hat{\beta}_1,$$

(5.3.37)

$$\beta_1\hat{\mu}_1 + \beta_2\hat{\mu}_2 = 0, \quad \hat{\beta}_1\mu_1 + \hat{\beta}_2\mu_2 = 0,$$

(5.3.38)

$$\beta_1\hat{\lambda}_1 + \beta_2\hat{\lambda}_2 = 1, \quad \hat{\beta}_1\lambda_1 + \hat{\beta}_2\lambda_2 = 1,$$

$$\beta_1\left(\frac{\hat{\lambda}_1^2}{3} + \hat{\lambda}_1\hat{\mu}_1 + \hat{\mu}_1^2\right) + \beta_2\left(\frac{\hat{\lambda}_2^2}{3} + \hat{\lambda}_2\hat{\mu}_2 + \hat{\mu}_2^2\right) = \frac{1}{2},$$

$$\hat{\beta}_1\left(\frac{\lambda_1^2}{3} + \lambda_1\mu_1 + \mu_1^2\right) + \hat{\beta}_2\left(\frac{\lambda_2^2}{3} + \lambda_2\mu_2 + \mu_2^2\right) = \frac{1}{2},$$

and Δw_r, I_{0r}, J_{r0} are defined in (5.3.12).

For example, the following set of parameters satisfies (5.3.37)–(5.3.38):

$$\beta_1 = \frac{1}{4}, \quad \beta_2 = \frac{3}{4}, \quad \hat{\beta}_1 = \frac{2}{3}, \quad \hat{\beta}_2 = \frac{1}{3}, \quad \lambda_1 = \lambda_2 = \hat{\lambda}_1 = \hat{\lambda}_2 = 1,$$

$$\mu_1 = \frac{1}{2\sqrt{3}}, \quad \mu_2 = -\frac{1}{\sqrt{3}}, \quad \hat{\mu}_1 = \frac{1}{\sqrt{2}}, \quad \hat{\mu}_2 = -\frac{1}{3\sqrt{2}}.$$

(5.3.39)

It is not difficult to see that the method (5.3.35)–(5.3.36) has the form of (5.3.32)–(5.3.33) and its parameters satisfy the conditions (5.3.34). Then, Lemma 5.3.11 implies that this method preserves symplectic structure. Using ideas of the proof of Lemma 5.3.6, we establish that the method (5.3.35)–(5.3.36) with (5.3.37)–(5.3.38) is of mean-square order 3/2. Thus, we have proved the following theorem.

Theorem 5.3.12 *Under conditions (5.3.37)–(5.3.38), the explicit PRK method (5.3.35)–(5.3.36) for system (5.3.29) preserves symplectic structure and has the mean-square order 3/2.*

5.3.3 The Case of Hamiltonian
$$H(t, p, q) = \frac{1}{2} p^{\mathsf{T}} M^{-1} p + U(t, q)$$

Now we propose symplectic methods for the Hamiltonian system (5.3.29), when $\gamma_r(t) = 0$ and the separable Hamiltonian has the special form

$$H(t, p, q) = \frac{1}{2} p^{\mathsf{T}} M^{-1} p + U(t, q), \tag{5.3.40}$$

with M being a constant, symmetric, invertible matrix (i.e., the kinetic energy $V(p)$ in (5.3.28) is equal to $\frac{1}{2} p^{\mathsf{T}} M^{-1} p$). In this case the system (5.3.29) reads

$$dP = f(t, Q)dt + \sum_{r=1}^{m} \sigma_r(t)dw_r(t), \quad P(t_0) = p, \tag{5.3.41}$$
$$dQ = M^{-1}Pdt, \quad Q(t_0) = q,$$

$$f^i = -\partial U/\partial q^i, \quad i = 1, \dots, n. \tag{5.3.42}$$

This system can be written as a second-order differential equation with additive noise

$$\frac{d^2 Q}{dt^2} = M^{-1} f(t, Q) + M^{-1} \sum_{r=1}^{m} \sigma_r(t) \dot{w}_r(t). \tag{5.3.43}$$

Clearly, the symplectic methods from the previous Sects. 5.3.1 and 5.3.2 can be applied to (5.3.41)–(5.3.42). Due to specific features of this system, these methods have a simpler form here. In this subsection we restrict ourselves to explicit methods of orders 2 and 3.

5.3.3.1 Explicit Methods of Order 2

One can prove that the method (5.3.35)–(5.3.36) in application to (5.3.41)–(5.3.42) is of the mean-square order 2. Further, on the basis of the Störmer-Verlet method (the deterministic second-order symplectic method), we construct the method for the system (5.3.41)–(5.3.42):

$$\mathcal{Q} = Q_k + \frac{h}{2} M^{-1} P_k, \tag{5.3.44}$$

$$P_{k+1} = P_k + \sum_{r=1}^{m} \sigma_r(t_k) \Delta_k w_r + hf\left(t_k + \frac{h}{2}, \mathcal{Q}\right) + \sum_{r=1}^{m} \sigma_r'(t_k)(I_{0r})_k$$

$$Q_{k+1} = Q_k + hM^{-1}P_k + \sum_{r=1}^{m} M^{-1}\sigma_r(t_k)(I_{r0})_k + \frac{h^2}{2} M^{-1} f\left(t_k + \frac{h}{2}, \mathcal{Q}\right),$$

$$k = 0, \ldots, N - 1 .$$

Theorem 5.3.13 *The explicit method (5.3.44) for the system (5.3.41)–(5.3.42) is symplectic and of the mean-square order 2.*

Other methods of order 2 are given in [289]. In [402] a symplectic method of mean-square order 1 for (5.3.41)–(5.3.42) is proposed on the basis of the Störmer-Verlet method.

5.3.3.2 Explicit Methods of Order 3

Introduce the integrals

$$(I_{0r})_k = \int_{t_k}^{t_{k+1}} (\vartheta - t_k) \, dw_r(\vartheta), \quad (I_{r0})_k = \int_{t_k}^{t_{k+1}} (w_r(\vartheta) - w_r(t_k)) \, d\vartheta, \qquad (5.3.45)$$

$$(I_{00r})_k := \frac{1}{2} \int_{t_k}^{t_{k+1}} (\vartheta - t_k)^2 \, dw_r(\vartheta), \quad (I_{0r0})_k := \int_{t_k}^{t_{k+1}} \int_{t_k}^{\vartheta_1} (\vartheta_2 - t_k) \, dw_r(\vartheta_2) d\vartheta_1,$$

$$(I_{r00})_k := \int_{t_k}^{t_{k+1}} \int_{t_k}^{\vartheta_1} (w_r(\vartheta_2) - w_r(t_k)) \, d\vartheta_2 d\vartheta_1,$$

$$(J_r)_k = \int_{t_k}^{t_{k+1}} (\vartheta - t_k)(w_r(\vartheta) - w_r(t_k)) \, d\vartheta .$$

The joint distribution of the random variables $\Delta_k w_r(h)$, $(I_{0r})_k$, $(I_{r0})_k$, $(I_{0r0})_k$, $(I_{r00})_k$, $(I_{00r})_k$ is Gaussian. They can be simulated at each step by $3m$ independent $\mathcal{N}(0, 1)$-distributed random variables ξ_{rk}, η_{rk}, and ζ_{rk}, $r = 1, \ldots, m$:

$$\Delta_k w_r = h^{1/2} \xi_{rk}, \quad (I_{r0})_k = h^{3/2} (\eta_{rk}/\sqrt{3} + \xi_{rk})/2, \quad (I_{0r})_k = h \Delta_k w_r - (I_{r0})_k,$$
$$(5.3.46)$$
$$(J_r)_k = h^{5/2} (\xi_{rk}/3 + \eta_{rk}/(4\sqrt{3}) + \zeta_{rk}/(12\sqrt{5})),$$

$$(I_{r00})_k = h(I_{r0})_k - (J_r)_k, \quad (I_{0r0})_k = 2(J_r)_k - h(I_{r0})_k,$$
$$(I_{00r})_k = h^2 \Delta_k w_r/2 - (J_r)_k.$$

Clearly, for $\sigma_r = 0$, $r = 1, \ldots, m$, the stochastic system (5.3.41) is reduced to the deterministic system

$$\frac{dp}{dt} = f(t, q), \quad \frac{dq}{dt} = M^{-1} p . \qquad (5.3.47)$$

The following lemma is true for system (5.3.41) with an arbitrary f (i.e., f may not obey the condition (5.3.42)). Its proof is available in [289].

Lemma 5.3.14 *Let $\bar{q} = q + G(t + h; t, p, q)$, $\bar{p} = p + F(t + h; t, p, q)$ be a one-step approximation of the third-order explicit method for the deterministic system (5.3.47). Suppose an n-dimensional (deterministic) variable $Q = Q(t + h; t, p, q)$ is such that*

$$|Q - q| = O(h).$$

Then the method

$$P_{k+1} = P_k + F(t + h; t, P_k, Q_k) + \sum_{r=1}^{m} \sigma_r(t_k)\Delta_k w_r + \sum_{r=1}^{m} \sigma'_r(t_k)(I_{0r})_k \quad (5.3.48)$$

$$+ \sum_{r=1}^{m} \sigma''_r(t_k)(I_{00r})_k + \sum_{r=1}^{m}\sum_{i=1}^{n}(M^{-1}\sigma_r(t_k))^i \frac{\partial f}{\partial q^i}(t_k, Q_k)(I_{r00})_k,$$

$$Q_{k+1} = Q_k + G(t + h; t, P_k, Q_k) + \sum_{r=1}^{m} M^{-1}\sigma_r(t_k)(I_{r0})_k$$

$$+ \sum_{r=1}^{m} M^{-1}\sigma'_r(t_k)(I_{0r0})_k$$

is of mean-square order 3 for the system (5.3.41) with an arbitrary f.

Using the known deterministic third-order symplectic method (see [398, 419, 421]), we obtain the following method for system (5.3.41)–(5.3.42):

$$Q_1 = Q_k + \frac{7}{24}hM^{-1}P_k, \quad P_1 = P_k + \frac{2}{3}hf\left(t_k + \frac{7h}{24}, Q_1\right), \quad (5.3.49)$$

$$Q_2 = Q_1 + \frac{3}{4}hM^{-1}P_1, \quad P_2 = P_1 - \frac{2}{3}hf\left(t_k + \frac{25h}{24}, Q_2\right),$$

$$Q_3 = Q_2 - \frac{1}{24}hM^{-1}P_2, \quad P_3 = P_2 + hf\left(t_k + h, Q_3\right),$$

$$P_{k+1} = P_3 + \sum_{r=1}^{m} \sigma_r(t_k)\Delta_k w_r + \sum_{r=1}^{m} \sigma'_r(t_k)(I_{0r})_k \quad (5.3.50)$$

$$+ \sum_{r=1}^{m} \sigma''_r(t_k)(I_{00r})_k + \sum_{r=1}^{m}\sum_{i=1}^{n}(M^{-1}\sigma_r(t_k))^i \frac{\partial f}{\partial q^i}(t_k, Q_3)(I_{r00})_k,$$

$$Q_{k+1} = Q_3 + \sum_{r=1}^{m} M^{-1}\sigma_r(t_k)(I_{r0})_k + \sum_{r=1}^{m} M^{-1}\sigma'_r(t_k)(I_{0r0})_k,$$

$$k = 0, \ldots, N - 1.$$

Theorem 5.3.15 *The explicit method (5.3.49)–(5.3.50) for the system (5.3.41)–(5.3.42) is symplectic and of mean-square order 3.*

Proof It is not difficult to check that $dP_{k+1} \wedge dQ_{k+1} = dP_3 \wedge dQ_3$. The expression for $dP_3 \wedge dQ_3$ coincides with the one corresponding to the deterministic third-order symplectic method. This implies that the method (5.3.49)–(5.3.50) is symplectic. By Lemma 5.3.14 we get that the method has the mean-square order 3.   □

5.4   Numerical Tests of Mean-Square Symplectic Methods

5.4.1   Kubo Oscillator

The system of SDEs in the sense of Stratonovich (Kubo oscillator)

$$\begin{aligned} dX^1 &= -aX^2 dt - \sigma X^2 \circ dw(t), \quad X^1(0) = x^1, \\ dX^2 &= aX^1 dt + \sigma X^1 \circ dw(t), \quad X^2(0) = x^2, \end{aligned} \tag{5.4.1}$$

is often used for testing numerical methods. Here a and σ are constants and $w(t)$ is a one-dimensional standard Wiener process.

The phase flow of this system preserves symplectic structure. Moreover, the quantity $\mathcal{H}(x^1, x^2) = (x^1)^2 + (x^2)^2$ is conservative for this system, i.e.,

$$\mathcal{H}(X^1(t), X^2(t)) = \mathcal{H}(x^1, x^2) \text{ for } t \geq 0.$$

This means that a phase trajectory of (5.4.1) belongs to the circle with center at the origin and of radius $\sqrt{\mathcal{H}(x^1, x^2)}$.

We test three methods here. In application to (5.4.1) the symplectic PRK method (5.2.9) takes the form:

$$\begin{aligned} X^1_{k+1} &= X^1_k - aX^2_k h - \frac{\sigma^2}{2} X^1_{k+1} h - \sigma X^2_k \Delta_k w, \\ X^2_{k+1} &= X^2_k + aX^1_{k+1} h + \frac{\sigma^2}{2} X^2_k h + \sigma X^1_{k+1} \Delta_k w. \end{aligned} \tag{5.4.2}$$

This method is implicit in the deterministic part only.

The midpoint method (5.2.8) applied to the system with one noise (5.4.1) reads

$$\begin{aligned} X^1_{k+1} &= X^1_k - a\frac{X^2_k + X^2_{k+1}}{2} h - \sigma \frac{X^2_k + X^2_{k+1}}{2} (\zeta_h)_k \sqrt{h}, \\ X^2_{k+1} &= X^2_k + a\frac{X^1_k + X^1_{k+1}}{2} h + \sigma \frac{X^1_k + X^1_{k+1}}{2} (\zeta_h)_k \sqrt{h}. \end{aligned} \tag{5.4.3}$$

This is a fully implicit method. Note that due to specific features of the system (5.4.1), the formula (5.4.3) is valid (solvable) not only in the case of the truncated random variable ζ_h but also if we put $\Delta_k w$ instead of $(\zeta_h)_k \sqrt{h}$.

The method (5.4.3) is of first mean-square order. The method (5.4.2) is of mean-square order $1/2$ as well as the Euler method:

$$X_{k+1}^1 = X_k^1 - aX_k^2 h - \frac{\sigma^2}{2}X_k^1 h - \sigma X_k^2 \Delta_k w,$$
$$X_{k+1}^2 = X_k^2 + aX_k^1 h - \frac{\sigma^2}{2}X_k^2 h + \sigma X_k^1 \Delta_k w, \qquad (5.4.4)$$

which, of course, is not symplectic.

Figure 5.4.1 gives approximations of a sample phase trajectory of (5.4.1) simulated by the symplectic methods (5.4.2) and (5.4.3) and by the Euler method (5.4.4). The initial condition is $x^1 = 1$, $x^2 = 0$. The corresponding exact phase trajectory belongs to the circle with center at the origin and with the unit radius.

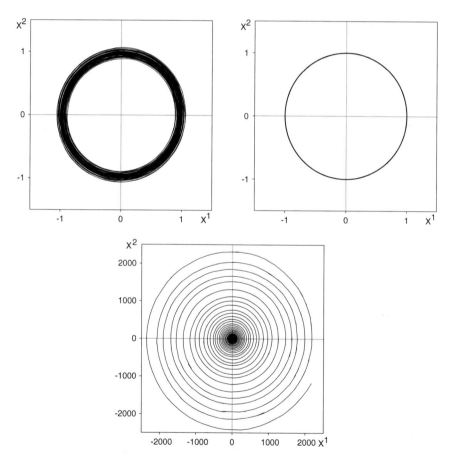

Fig. 5.4.1 A sample phase trajectory of (5.4.1) with $X^1(0) = 1$, $X^2(0) = 0$ obtained by the symplectic method (5.4.2) (top left), the midpoint method (5.4.3) (top right), and by the Euler method (5.4.4) (bottom) for $a = 2$, $\sigma = 0.3$, $h = 0.02$ on the time interval $t \leq 200$

We see that the Euler method is not appropriate for simulation of the oscillator (5.4.1) on long time intervals while the symplectic methods preserve conservative properties of the Kubo oscillator.

These experiments also demonstrate that the midpoint method is much more accurate than the other methods applied. It is not difficult to check that $\mathcal{H}(x^1, x^2)$ is conserved by the midpoint method (5.4.3) but it is not conserved by the symplectic PRK method (5.4.2). This is similar to the deterministic case. Indeed, it is known [397, 398] that symplectic deterministic RK methods (e.g., the midpoint scheme) conserve all quadratic functions that are conserved by the Hamiltonian system being integrated, while deterministic PRK methods do not possess this property.

5.4.2   A Model for Synchrotron Oscillations of Particles in Storage Rings

In [402] a model describing synchrotron oscillations of particles in storage rings under the influence of external fluctuating electromagnetic fields was considered. This model can be written in the following form

$$dP = -\omega^2 \sin(Q)dt - \sigma_1 \cos(Q)dw_1 - \sigma_2 \sin(Q)dw_2, \qquad (5.4.5)$$

$$dQ = Pdt.$$

P and Q are scalars here. The system (5.4.5) is of the form (5.2.14) and therefore its phase flow preserves symplectic structure.

The Euler method for (5.4.5) takes the form

$$P_{k+1} = P_k - h\omega^2 \sin(Q_k) - h^{1/2}(\sigma_1 \cos(Q_k)\Delta_k w_1 + \sigma_2 \sin(Q_k)\Delta_k w_2), \qquad (5.4.6)$$
$$Q_{k+1} = Q_k + hP_k .$$

In application to (5.4.5) the explicit symplectic method (5.2.17)–(5.2.18) with $\alpha = 1$ is written as

$$\mathcal{Q} = Q_k + hP_k , \qquad (5.4.7)$$
$$P_{k+1} = P_k - h\omega^2 \sin(\mathcal{Q}) - h^{1/2}(\sigma_1 \cos(\mathcal{Q})\Delta_k w_1 + \sigma_2 \sin(\mathcal{Q})\Delta_k w_2), \qquad (5.4.8)$$
$$Q_{k+1} = \mathcal{Q}. \qquad (5.4.9)$$

Both methods are of first mean-square order.

Approximations of a sample trajectory of (5.4.5) simulated by the symplectic method (5.4.7) and the Euler method (5.4.6) are plotted on Fig. 5.4.2. The trajectory obtained by the symplectic method with $h = 0.02$ (solid line) visually coincides with the one obtained for a smaller step, e.g. for $h = 0.002$, using the same sample paths for the Wiener processes, i.e., this trajectory visually coincides with the exact solution

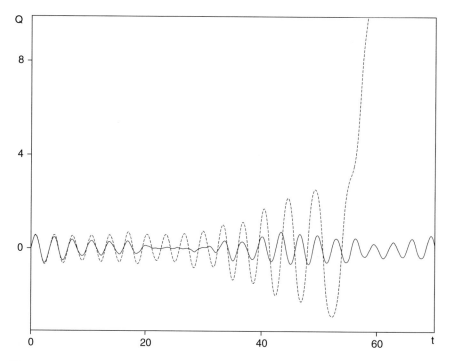

Fig. 5.4.2 A sample trajectory of (5.4.5) for $\omega = 2$, $\sigma_1 = 0.2$, $\sigma_2 = 0.1$, $h = 0.02$. Solid line – the symplectic method (5.4.7), dashed line – the Euler method (5.4.6)

of (5.4.5). This figure clearly demonstrates that the Euler method (dashed line) is unacceptable for simulation of the solution to (5.4.5) on a long time interval while the symplectic method (5.4.7) produces quite accurate results despite both methods have the same mean-square order of accuracy.

In Appendix A.1 of [314] an implementation of the Euler method (5.4.6) and the explicit symplectic method (5.4.7) is considered and the program is given, by which sample trajectories plotted on Fig. 5.4.2 were obtained.

5.4.3 Linear Oscillator with Additive Noise

In this section we consider the following Hamiltonian system with additive noise:

$$dX^1 = X^2 dt + \sigma dw_1(t), \ X^1(0) = X_0^1, \qquad (5.4.10)$$
$$dX^2 = -X^1 dt + \gamma dw_2(t), \ X^2(0) = X_0^2.$$

We have for the solution $X = (X^1, X^2)^\mathsf{T}$ of (5.4.10):

$$X(t_{k+1}) = FX(t_k) + u_k, \ X(0) = X_0, \ k = 0, 1, \ldots, N - 1, \qquad (5.4.11)$$

where

$$F = \begin{bmatrix} \cos h & \sin h \\ -\sin h & \cos h \end{bmatrix},$$

$$u_k = \begin{bmatrix} \sigma \displaystyle\int_{t_k}^{t_{k+1}} \cos(t_{k+1} - s) dw_1(s) + \gamma \displaystyle\int_{t_k}^{t_{k+1}} \sin(t_{k+1} - s) dw_2(s) \\ -\sigma \displaystyle\int_{t_k}^{t_{k+1}} \sin(t_{k+1} - s) dw_1(s) + \gamma \displaystyle\int_{t_k}^{t_{k+1}} \cos(t_{k+1} - s) dw_2(s) \end{bmatrix}.$$

In application to (5.4.10) the explicit symplectic method (5.3.31) with $\alpha = 1$ takes the form

$$X_{k+1}^2 = X_k^2 - hX_k^1 + \gamma \Delta_k w_2, \ X_{k+1}^1 = X_k^1 + hX_{k+1}^2 + \sigma \Delta_k w_1. \qquad (5.4.12)$$

The method (5.4.12) can be written as

$$X_{k+1} = HX_k + v_k, \ k = 0, 1, \ldots, N - 1, \qquad (5.4.13)$$

where $X_k = (X_k^1, X_k^2)^\mathsf{T}$,

$$H = \begin{bmatrix} 1 - h^2 & h \\ -h & 1 \end{bmatrix}, \quad v_k = \begin{bmatrix} \sigma \Delta_k w_1 + \gamma h \Delta_k w_2 \\ \gamma \Delta_k w_2 \end{bmatrix}.$$

Our nearest aim is to analyze propagation of the error $r_k := X_k - X(t_k)$. We get

$$X(t_k) = F^k X_0 + F^{k-1} u_0 + F^{k-2} u_1 + \cdots + u_{k-1}, \qquad (5.4.14)$$

$$X_k = H^k X_0 + H^{k-1} v_0 + H^{k-2} v_1 + \cdots + v_{k-1}. \qquad (5.4.15)$$

Proposition 5.4.1 *Suppose T and h are such that Th^2 is sufficiently small. Then for $k = 0, 1, \ldots, N$, $T = Nh$, the following inequality holds*

$$\|H^k - F^k\| \le \frac{h}{2} + \frac{kh^3}{24} + O(h^2 + Th^3) \le \frac{h}{2} + \frac{Th^2}{24} + O(h^2 + Th^3). \quad (5.4.16)$$

Proof Clearly

$$F^k = \begin{bmatrix} \cos kh & \sin kh \\ -\sin kh & \cos kh \end{bmatrix}.$$

Let us represent H as $H = G\Lambda G^{-1}$ with Λ and G such that $\Lambda = \mathrm{diag}(\lambda_1, \lambda_2)$, $\lambda_{1,2} = 1 - \dfrac{h^2}{2} \pm ih\sqrt{1 - \dfrac{h^2}{4}}$, and the columns of the matrix G are eigenvectors of H corresponding to the eigenvalues λ_1, λ_2. We write the matrices Λ and G in the form

$$\Lambda = \begin{bmatrix} e^{i\varphi} & 0 \\ 0 & e^{-i\varphi} \end{bmatrix}, \quad G = \begin{bmatrix} 1 & 1 \\ e^{i\psi} & e^{-i\psi} \end{bmatrix},$$

where $0 < \varphi, \psi < \dfrac{\pi}{2}$, $\cos\varphi = 1 - \dfrac{h^2}{2}$, $\cos\psi = \dfrac{h}{2}$.

We obtain: $H^k = G\Lambda^k G^{-1}$,

$$H^k - F^k = G(\Lambda^k - G^{-1}F^kG)G^{-1}, \tag{5.4.17}$$

$$\Lambda^k - G^{-1}F^kG \tag{5.4.18}$$

$$= \begin{bmatrix} e^{ki\varphi} - e^{kih} - i\sin kh\dfrac{1 - \sin\psi}{\sin\psi} & -\dfrac{i\sin kh \times e^{-i\psi}}{\sin\psi}\cos\psi \\ \dfrac{i\sin kh \times e^{i\psi}}{\sin\psi}\cos\psi & e^{-ki\varphi} - e^{-kih} + i\sin kh\dfrac{1 - \sin\psi}{\sin\psi} \end{bmatrix}.$$

Let us represent this matrix $\Lambda^k - G^{-1}F^kG$ as the sum $D_1 + D_2$, where $D_2 = \mathrm{diag}(e^{ki\varphi} - e^{kih}, e^{-ki\varphi} - e^{-kih})$. It is not difficult to show that (the norms of matrices are Euclidean)

$$\|G\| = \sqrt{2}(1 + O(h)), \quad \|G^{-1}\| = \frac{\sqrt{2}}{2}(1 + O(h)), \tag{5.4.19}$$

$$\|D_1\| \le \frac{h}{2}(1 + O(h)), \quad \|D_2\| = 2\left|\sin\frac{k\varphi - kh}{2}\right|.$$

Taking into account that $\varphi = \arcsin\left(h\sqrt{1 - \dfrac{h^2}{4}}\right) = h + \dfrac{h^3}{24} + O(h^5)$, $kh \le T$, $k = 0, 1, \ldots, N$, and the assumption on smallness of Th^2, we get

$$\|D_2\| \le \frac{kh^3}{24} + O(h^2) \le \frac{Th^2}{24} + O(h^2), \quad k = 0, 1, \ldots, N. \tag{5.4.20}$$

The inequality (5.4.16) follows from (5.4.17)–(5.4.20). $\square$

Using Proposition 5.4.1, we prove the following assertion.

Proposition 5.4.2 *Let T and h be such that Th^2 is sufficiently small. Suppose $E|X_0|^2 \le C$. Then the mean-square error is estimated as*

$$(E|r_k|^2)^{1/2} \le K \times (T^{1/2}h + T^{3/2}h^2), \quad k = 0, 1, \ldots, N. \tag{5.4.21}$$

In application to (5.4.10) the Euler method can be written in the form

$$\bar{X}_{k+1} = \bar{H}\bar{X}_k + \bar{v}_k = \begin{bmatrix} 1 & h \\ -h & 1 \end{bmatrix} \bar{X}_k + \begin{bmatrix} \sigma \Delta_k w_1 \\ \gamma \Delta_k w_2 \end{bmatrix}. \tag{5.4.22}$$

Analogously to (5.4.17)–(5.4.18), we get $\bar{H}^k - F^k = \bar{G}(\bar{\Lambda}^k - \bar{G}^{-1}F^k\bar{G})\bar{G}^{-1} :=$ $\bar{G}\bar{D}\bar{G}^{-1}$ with

$$\bar{\Lambda} = \begin{bmatrix} 1+ih & 0 \\ 0 & 1-ih \end{bmatrix}, \quad \bar{G} = \begin{bmatrix} 1 & 1 \\ i & -i \end{bmatrix},$$

$$\bar{D} = \begin{bmatrix} (1+ih)^k - e^{ihk} & 0 \\ 0 & (1-ih)^k - e^{-ihk} \end{bmatrix}.$$

Further, $||\bar{G}|| = \sqrt{2}$, $||\bar{G}^{-1}|| = \sqrt{2}/2$, and

$$||\bar{D}|| = [((1+h^2)^{k/2} - 1)^2 + 4(1+h^2)^{k/2} \sin^2 \frac{k(\varphi - h)}{2}]^{1/2}$$

$$\le [(e^{Th/2} - 1)^2 + 4e^{Th/2} \sin^2 \frac{k(\varphi - h)}{2}]^{1/2},$$

where $\varphi = \arcsin \dfrac{h}{\sqrt{1+h^2}} \simeq \dfrac{h}{\sqrt{1+h^2}} + \dfrac{1}{6}\dfrac{h^3}{(1+h^2)^{3/2}}$, $\varphi - h \simeq -\dfrac{h^3}{3}$. Hence if Th is small then

$$||\bar{D}|| \le \left[(e^{Th/2} - 1)^2 + 4e^{Th/2} \sin^2 \frac{k(\varphi - h)}{2} \right]^{1/2} \simeq e^{Th/2} - 1 \simeq Th/2,$$

and it is not difficult to show that the mean-square error of the Euler method is estimated as $O(T^{3/2}h)$.

Consequently, the Euler method can be used on the interval $[0, T_E]$ if $T_E^{3/2}h$ is sufficiently small. Due to Proposition 5.4.2, the error of the symplectic method (5.4.12) on $[0, T_S]$ with $T_S = T_E^2$ is equal to $O(T_E h + T_E^3 h^2)$, i.e., the symplectic method is applicable on substantially longer time intervals than the Euler method. Of course, the Euler method possesses the worse properties than the symplectic method since the absolute values of the eigenvalues of $\bar{H}$ are greater than 1.

Finally, consider the optimal method from [274, p. 61] (the method also uses only the increments $\Delta_k w$ as the information regarding $w(t)$ but it uses this information optimally):

$$\hat{X}_{k+1} = \hat{H}\hat{X}_k + \hat{v}_k \tag{5.4.23}$$

$$= \begin{bmatrix} \cos h & \sin h \\ -\sin h & \cos h \end{bmatrix} \hat{X}_k + \frac{1}{h} \begin{bmatrix} \sigma \sin h \times \Delta_k w_1 + 2\gamma \sin^2 \frac{h}{2} \times \Delta_k w_2 \\ -2\sigma \sin^2 \frac{h}{2} \times \Delta_k w_1 + \gamma \sin h \times \Delta_k w_2 \end{bmatrix}.$$

Evidently, this method is symplectic. And, as $\hat{H} = F$, it has no error in the absence of noise. We get for its error:

$$E|\hat{r}_N|^2 = \sum_{m=0}^{N-1} E|\hat{v}_m - u_m|^2 = N(\sigma^2 + \gamma^2)\frac{h^3}{12} + N \times O(h^5) \simeq \frac{\sigma^2 + \gamma^2}{12}Th^2.$$

Consequently, the error of the optimal method is estimated as $O(T^{1/2}h)$. This implies that the method (5.4.23) is applicable on the longer time interval $[0, T_O] = [0, T_E^3]$ than the symplectic method (5.4.12).

To guarantee the same sample paths for the Wiener processes in realization of the exact, symplectic, and Euler methods, we simulate six independent $\mathcal{N}(0, 1)$-distributed random variables $\xi_{1,k+1}, \eta_{1,k+1}, \zeta_{1,k+1}, \xi_{2,k+1}, \eta_{2,k+1}, \zeta_{2,k+1}$ at every step $k + 1 = 1, \ldots, N - 1$. It is not difficult to show that the needed random variables can be evaluated as

$$\Delta_k w_i = \sqrt{h}\xi_{i,k+1}, \quad \int_{t_k}^{t_{k+1}} \cos(t_{k+1} - s)dw_i(s) = \frac{1}{\sqrt{h}}\sin h \times \xi_{i,k+1} + c_1\eta_{i,k+1},$$

$$\int_{t_k}^{t_{k+1}} \sin(t_{k+1} - s)dw_i(s) = \frac{2}{\sqrt{h}}\sin^2\frac{h}{2} \times \xi_{i,k+1} + c_2\eta_{i,k+1} + c_3\zeta_{i,k+1},$$

$$i = 1, 2,$$

where

$$c_1 = \left(\frac{1}{2}h + \frac{1}{4}\sin 2h - \frac{\sin^2 h}{h}\right)^{1/2}, \quad c_2 = \frac{1}{c_1}\left(\frac{1}{2}\sin^2 h - \frac{2}{h}\sin^2\frac{h}{2}\sin h\right),$$

$$c_3 = \left(\frac{1}{2}h - \frac{1}{4}\sin 2h - \frac{4}{h}\sin^4\frac{h}{2} - c_2^2\right)^{1/2}.$$

In the numerical tests we simulate the system (5.4.10) by (i) the exact formula (5.4.11), (ii) the symplectic method (5.4.12), and (iii) the Euler method (5.4.22). Figure 5.4.3 corresponds to the time interval $[0, 128]$ which approximately contains 20 oscillations of (5.4.10) (note that the period of free oscillations of (5.4.10) is equal to 2π).

The results clearly demonstrate that the Euler method is unacceptable for simulation of the Hamiltonian system (5.4.10) on a long time interval. After 10 oscillations (Fig. 5.4.3) the norm of its error is already half of the norm of the solution, and after

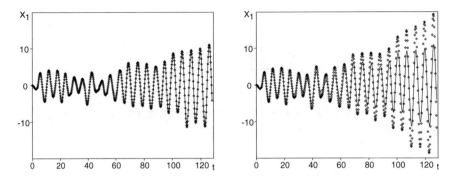

Fig. 5.4.3 A sample trajectory of the solution to (5.4.10) for $\sigma = 0$, $\gamma = 1$, $X_1(0) = X_2(0) = 0$ obtained by the exact formulae (5.4.11) (solid line), the symplectic method (5.4.12) with $h = 0.02$ (points on the left figure), and the Euler method (5.4.22) with $h = 0.02$ (points on the right figure). The points of the symplectic and Euler methods are plotted once per 10 steps, i.e., once per each interval 0.2.

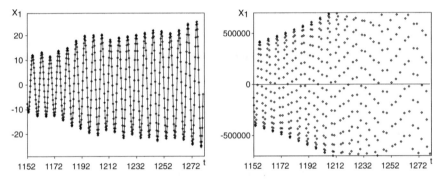

Fig. 5.4.4 Another part of the same sample trajectory as in Fig. 5.4.3. Solid line – the exact solution, points – the symplectic method (left) and the Euler method (right).

200 oscillations (see Fig. 5.4.4) the amplitude of oscillations simulated by the Euler method is greater than the exact amplitude in 50 000 times.

In contrast to the Euler method, the symplectic method reproduces oscillations of the system (5.4.10) quite accurately. After 10 oscillations (Fig. 5.4.3) the norm of its error is approximately 2% of the norm of the solution. But it is more astonishing that after 200 oscillations (see Fig. 5.4.4) the relative error remains the same. The error of the amplitude of oscillations on the considered time interval is also about 2%. As is known, a symplectic method in application to a deterministic oscillator preserves conservative properties of solutions, in particular their boundedness on infinite time interval. One can say that the symplectic method generates a discrete conservative system ("discrete linear oscillator"). It turns out that behavior of this system affected by noise (which is also discrete) is qualitatively identical to the behavior of the continuous Hamiltonian system with noise. For instance, the approximate solution adequately reproduces an increase of the amplitude of the oscillations.

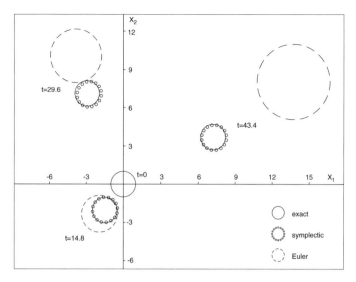

Fig. 5.4.5 The evolution of domains in the phase plane of system (5.4.10) for $\sigma = 0$, $\gamma = 1$. Images of the initial unit circle are obtained at three time moments by the exact mapping, by the mapping in the case of the symplectic method (5.4.12) with $h = 0.05$, and by the mapping in the case of the Euler method (5.4.22) with $h = 0.05$.

Figure 5.4.5 presents evolution of domains in the phase plane of system (5.4.10). The initial domain is the circle with center at the origin and with the unit radius. We plot images of this circle, which are obtained at three time moments by the exact mapping, by the mapping in the case of the symplectic method (5.4.12), and by the mapping in the case of the Euler method (5.4.22). For the considered system (5.4.10), exact images of the unit circle are circles of the unit radius shifted from the origin due to the action of noise. In the case of the Euler method these images are also circles but with increasing radius. In the case of symplectic method (5.4.12) the images of the initial circle are ellipses. In spite of the fact that the symplectic method (5.4.12) and the Euler method (5.4.22) have the same mean-square order of accuracy, these ellipses approximate the exact images much better than the circles obtained by the Euler method.

5.5 Liouvillian Methods for Stochastic Systems Preserving Phase Volume

In the previous sections we considered some Hamiltonian methods for stochastic Hamiltonian systems. These systems (as well as the methods) preserve the symplectic structure and, consequently, preserve the phase volume. In this section we deal with a more general class of systems which preserve the phase volume but may not preserve the symplectic structure.

Let us start with the deterministic d-dimensional system

$$\frac{dX}{dt} = a(t, X), \quad X(t_0) = x, \tag{5.5.1}$$

the phase flow $X(t; t_0, x)$ of which preserves the phase volume. Note that the dimension d may be odd.

Let $D_0 \in \mathbf{R}^d$ be a domain with finite volume. The transformation $X(t; t_0, x)$ maps D_0 into the domain D_t. The volume V_t of the domain D_t is equal to

$$V_t = \int_{D_t} dX^1 \ldots dX^d = \int_{D_0} \left| \frac{D(X^1, \ldots, X^d)}{D(x^1, \ldots, x^d)} \right| dx^1 \ldots dx^d.$$

Then, the volume-preserving condition consists in the equality

$$\left| \frac{D(X^1(t), \ldots, X^d(t))}{D(x^1, \ldots, x^d)} \right| = 1 \tag{5.5.2}$$

or, equivalently, it consists in preservation of the d-form $dX^1 \wedge dX^2 \wedge \cdots \wedge dX^d$.

According to the Liouville theorem, the phase flow of (5.5.1) preserves phase volume if and only if

$$\frac{\partial a^1(t, x)}{\partial x^1} + \cdots + \frac{\partial a^d(t, x)}{\partial x^d} = div\, a = 0. \tag{5.5.3}$$

Numerical methods preserving the phase volume are called *Liouvillian* [419, 423]. Due to our best knowledge, there are no constructive Liouvillian methods for the deterministic system (5.5.1), (5.5.3) of a general form (see [100, 371, 419, 423] and references therein). Some constructive Liouvillian methods for particular cases of (5.5.1), (5.5.3) can be found in [100, 371, 419, 423]. It was shown in [423] that certain methods known to be symplectic are also phase volume preserving. However, it was also demonstrated that in general the relation between these two properties is rather delicate: neither of them implies the other.

Consider the Cauchy problem for the d-dimensional system of SDEs in the sense of Ito:

$$dX = a(t, X)dt + \sum_{r=1}^{m} b_r(t, X)dw_r(t), \quad X(t_0) = x, \tag{5.5.4}$$

the phase flow $X(t; t_0, x; w)$ of which preserves phase volume, i.e., for which the condition (5.5.2) holds.

It is known (see [10, 219] and also [289]) that the phase flow of (5.5.4) preserves phase volume if and only if

$$div\,(a - \frac{1}{2}\sum_{r=1}^{m}\frac{\partial b_r}{\partial x}b_r) = 0, \quad div\,b_r = 0, \; r = 1,\ldots,m. \tag{5.5.5}$$

Let X_k, $k = 0,\ldots,N$, $t_{k+1} - t_k = h_{k+1}$, $t_N = t_0 + T$:

$$X_0 = X(t_0), \quad X_{k+1} = \bar{X}_{t_k,X_k}(t_{k+1}),$$

be a mean-square method for (5.5.4) based on the one-step approximation $\bar{X}_{t,x}(t + h) = \bar{X}(t + h; t, x)$. It is clear that a method preserves phase volume if its one-step approximation satisfies the equality

$$\left| \frac{D(\bar{X}^1,\ldots,\bar{X}^d)}{D(x^1,\ldots,x^d)} \right| = 1 \tag{5.5.6}$$

or equivalently

$$d\bar{X}^1 \wedge \cdots \wedge d\bar{X}^d = dx^1 \wedge \cdots \wedge dx^d. \tag{5.5.7}$$

Taking into account that there are no constructive Liouvillian methods for a general deterministic Liouvillian system, we restrict ourselves here to some particular cases of the stochastic system (5.5.4), (5.5.5).

5.5.1 Liouvillian Methods for Partitioned Systems with Multiplicative Noise

Consider the particular case of (5.5.4):

$$dX = f(t, Y)dt + \sum_{r=1}^{m}\sigma_r(t, Y)\,dw_r(t), \quad X(t_0) = x, \tag{5.5.8}$$

$$dY = g(t, X)dt + \sum_{r=1}^{m}\gamma_r(t)\,dw_r(t), \quad Y(t_0) = y,$$

where X, f, σ_r are l-dimensional column vectors and Y, g, γ_r are n-dimensional column vectors.

It is not difficult to check that the coefficients of (5.5.8) satisfy (5.5.5), i.e., the phase flow of system (5.5.8) preserves phase volume. Note that if $l = n$ and there are U_r, $r = 0,\ldots,m$, and V_0 such that $f^i = -\partial U_0/\partial y^i$, $g^i = \partial V_0/\partial x^i$, and $\sigma_r = -\partial U_r/\partial y^i$, $r = 1,\ldots m$, $i = 1,\ldots,l$, then the system (5.5.8) possesses the symplectic property (cf. (5.2.10), we pay attention that the system (5.2.10) is in the sense of Stratonovich).

Introduce the PRK method for (5.5.8) (cf. (5.2.11)–(5.2.12)):

$$\mathbf{Y}_1 = Y_k + \alpha h g(t_k, X_k), \tag{5.5.9}$$

$$\mathbf{X}_1 = X_k + h f(t_k + \alpha h, \mathbf{Y}_1),$$
$$\mathbf{Y}_2 = \mathbf{Y}_1 + (1 - \alpha) h g(t_{k+1}, \mathbf{X}_1),$$

$$X_{k+1} = \mathbf{X}_1 + \sum_{r=1}^{m} \sigma_r(t_k, \mathbf{Y}_2) \Delta_k w_r, \tag{5.5.10}$$

$$Y_{k+1} = \mathbf{Y}_2 + \sum_{r=1}^{m} \gamma_r(t_k) \Delta_k w_r, \ k = 0, \ldots, N - 1,$$

with the parameter $0 \le \alpha \le 1$.

If $\sigma_r = \gamma_r = 0, r = 1, \ldots, m$, this method coincides with the deterministic Liou-villian method [371, 419, 423].

Theorem 5.5.1   *The method (5.5.9)–(5.5.10) for the system (5.5.8) is Liouvillian and of mean-square order $1/2$.*

Proof   Let us check that the one-step approximation $\bar{X}$, $\bar{Y}$ corresponding to (5.5.9)–(5.5.10) satisfies (5.5.7). Using properties of exterior products, we obtain

$$d\bar{X}^1 \wedge \cdots \wedge d\bar{X}^l \wedge d\bar{Y}^1 \cdots \wedge d\bar{Y}^n = \left(d\mathbf{X}_1^1 + \sum_{r=1}^{m} \sum_{j=1}^{n} \frac{\partial \sigma_r^1}{\partial y^j} d\mathbf{Y}_2^j \right) \wedge \cdots \tag{5.5.11}$$

$$\wedge \left(d\mathbf{X}_1^{l-1} + \sum_{r=1}^{m} \sum_{j=1}^{n} \frac{\partial \sigma_r^{l-1}}{\partial y^j} d\mathbf{Y}_2^j \right) \wedge \left(d\mathbf{X}_1^l + \sum_{r=1}^{m} \sum_{j=1}^{n} \frac{\partial \sigma_r^l}{\partial y^j} d\mathbf{Y}_2^j \right)$$

$$\wedge d\mathbf{Y}_2^1 \wedge \cdots \wedge d\mathbf{Y}_2^n$$

$$= \left(d\mathbf{X}_1^1 + \sum_{r=1}^{m} \sum_{j=1}^{n} \frac{\partial \sigma_r^1}{\partial y^j} d\mathbf{Y}_2^j \right) \wedge \cdots \wedge \left(d\mathbf{X}_1^{l-1} + \sum_{r=1}^{m} \sum_{j=1}^{n} \frac{\partial \sigma_r^{l-1}}{\partial y^j} d\mathbf{Y}_2^j \right)$$

$$\wedge \left(d\mathbf{X}_1^l \wedge d\mathbf{Y}_2^1 \wedge \cdots \wedge d\mathbf{Y}_2^n + \sum_{r=1}^{m} \sum_{j=1}^{n} \frac{\partial \sigma_r^l}{\partial y^j} d\mathbf{Y}_2^j \wedge d\mathbf{Y}_2^1 \wedge \cdots \wedge d\mathbf{Y}_2^n \right)$$

$$= \left(d\mathbf{X}_1^1 + \sum_{r=1}^{m} \sum_{j=1}^{n} \frac{\partial \sigma_r^1}{\partial y^j} d\mathbf{Y}_2^j \right) \wedge \cdots \wedge \left(d\mathbf{X}_1^{l-1} + \sum_{r=1}^{m} \sum_{j=1}^{n} \frac{\partial \sigma_r^{l-1}}{\partial y^j} d\mathbf{Y}_2^j \right)$$

$$\wedge d\mathbf{X}_1^l \wedge d\mathbf{Y}_2^1 \wedge \cdots \wedge d\mathbf{Y}_2^n$$

$$= \cdots = d\mathbf{X}_1^1 \wedge \cdots \wedge d\mathbf{X}_1^l \wedge d\mathbf{Y}_2^1 \wedge d\mathbf{Y}_2^2 \wedge \cdots \wedge d\mathbf{Y}_2^n .$$

Since (5.5.9) corresponds to the deterministic Liouvillian method, it follows from (5.5.11) that the method (5.5.9)–(5.5.10) is Liouvillian.

To prove the mean-square order of (5.5.9)–(5.5.10), we compare it with the Euler method as usual. □

Now put $\gamma_r = 0$, $r = 1, \ldots, m$, in (5.5.8) (cf. (5.2.14)):

$$dX = f(t, Y)dt + \sum_{r=1}^{m} \sigma_r(t, Y)dw_r(t), \quad X(t_0) = x, \tag{5.5.12}$$

$$dY = g(t, X)dt, \quad Y(t_0) = y.$$

The Liouvillian method (5.5.9)–(5.5.10) in application to (5.5.12) is of first mean-square order (cf. Theorem 5.2.6).

Introduce the PRK method for (5.5.12):

$$Y_1 = Y_k, \quad X_1 = X_k + \frac{h}{4} f(t_k, Y_1) + \frac{1}{2} \sum_{r=1}^{m} \sigma_r(t_k, Y_1)(3(J_{r0})_k - \Delta_k w_r), \tag{5.5.13}$$

$$Y_2 = Y_1 + \frac{2}{3} hg\left(t_k + \frac{h}{4}, X_1\right),$$

$$X_2 = X_1 + \frac{3}{4} hf\left(t_k + \frac{2}{3}h, Y_2\right) + \frac{3}{2} \sum_{r=1}^{m} \sigma_r\left(t_k + \frac{2}{3}h, Y_2\right)(-(J_{r0})_k + \Delta_k w_r),$$

$$X_{k+1} = X_2, \quad Y_{k+1} = Y_2 + \frac{h}{3} g(t_{k+1}, X_2), \quad k = 0, \ldots, N - 1. \tag{5.5.14}$$

This method applied to (5.2.14) gives the symplectic method (5.2.25)–(5.2.26).

Theorem 5.5.2 *The method (5.5.13)–(5.5.14) for the system (5.5.12) is Liouvillian and of mean-square order* $3/2$.

Proof By the arguments similar to ones used to obtain (5.5.11) in Theorem 5.5.1, we prove that the one-step approximation corresponding to (5.5.13)–(5.5.14) satisfies the volume-preserving condition (5.5.7). For a proof of the mean-square order, see Theorem 5.2.9. □

5.5.2 Liouvillian Methods for a Volume-Preserving System with Additive Noise

The d-dimensional system with additive noise

$$dX = a(t, X)dt + \sum_{r=1}^{m} b_r(t)dw_r(t), \quad X(t_0) = x, \tag{5.5.15}$$

possesses the volume-preserving property if and only if the condition (5.5.3) holds.

Theorem 5.5.3 *Let* $\bar{X} = X + A(t, X, \bar{X}; h)$ *be a one-step approximation corresponding to the first-order Liouvillian method for the deterministic system* (5.5.1), (5.5.3). *Then the method for the stochastic system* (5.5.15), (5.5.3):

$$X_{k+1} = X_k + A(t_k, X_k, X_{k+1}; h) + \sum_{r=1}^{m} b_r(t_k) \Delta_k w_r \tag{5.5.16}$$

is Liouvillian and of first mean-square order.

Proof We have for the one-step approximation $\bar{X}$ corresponding to (5.5.16): $d\bar{X}^i = dx^i + dA^i$, $i = 1, \ldots, d$. Since these expressions coincide with the ones for the deterministic Liouvillian method, the approximation $\bar{X}$ satisfies (5.5.7) and the method is Liouvillian. The mean-square order of (5.5.16) easily follows from the general theory of Chap. 1.   □

Due to this theorem, construction of first-order Liouvillian methods for Liouvillian systems with additive noise reduces to construction of such methods for deterministic Liouvillian systems. For instance, consider the following Liouvillian system

$$dX^i = a^i(t, X^1, \ldots, X^{i-1}, X^{i+1}, \ldots, X^d)\, dt + \sum_{r=1}^{m} b_r^i(t)\, dw_r(t), \tag{5.5.17}$$
$$X(t_0) = x, \quad i = 1, \ldots, d.$$

In [371] an explicit first-order Liouvillian method for the deterministic system (5.5.1) with $a(t, x)$ as in (5.5.17) was proposed. Using it, we obtain

$$X_{k+1}^i = X_k^i + ha^i(t_k, X_{k+1}^1, \ldots, X_{k+1}^{i-1}, X_k^{i+1}, \ldots, X_k^d) + \sum_{r=1}^{m} b_r^i(t_k) \Delta_k w_r, \tag{5.5.18}$$
$$i = 1, \ldots, d, \quad k = 0, \ldots, N-1.$$

Corollary 5.5.4 *The method* (5.5.18) *for* (5.5.17) *is Liouvillian and of first mean-square order.*

Note that the Liouvillian method (5.5.9)–(5.5.10) for the system (5.5.8) with $\sigma_r(t, y) = \sigma_r(t), r = 1, \ldots, m$, (the partitioned system with additive noise) is of first mean-square order. Further, for the partitioned system (5.5.8) with $\sigma_r(t, y) = \sigma_r(t)$, $r = 1, \ldots, m$, a parametric family of 2-stage explicit Liouvillian PRK methods of mean-square $3/2$ can be easily derived. The form of these methods coincide with the symplectic method (5.3.35)–(5.3.39). Let us also note that for the particular case of system (5.5.12) with $\sigma_r(t, y) = \sigma_r(t)$ and $g(t, x) = M^{-1}x$ where M is a constant, symmetric, invertible matrix, we can construct a Liouvillian method of the third mean-square order. The form of this method coincides with the third-order symplectic method (5.3.49)–(5.3.50).

5.6 Weak Symplectic Methods for Stochastic Hamiltonian Systems

5.6.1 Hamiltonian Systems with Multiplicative Noise

In this subsection weak symplectic methods for Hamiltonian systems with multiplicative noise are constructed. First we consider the general case and then treat the case of separable Hamiltonians.

5.6.1.1 Implicit First-Order Methods for General Stochastic Hamiltonian Systems

Here all the methods are fully implicit (i.e., implicit in both deterministic and stochastic components). Let us recall that in the case of deterministic general Hamiltonian systems symplectic RK methods are all implicit [398].

On the basis of the symplectic method of mean-square order $1/2$ (5.2.7), we propose the weak method:

$$P_{k+1} = P_k + hf(t_k + \beta h, \alpha P_{k+1} + (1 - \alpha)P_k, (1 - \alpha)Q_{k+1} + \alpha Q_k) \quad (5.6.1)$$
$$+ h\left(\frac{1}{2} - \alpha\right) \sum_{r=1}^{m} \sum_{j=1}^{n} \left(\frac{\partial \sigma_r}{\partial p^j}\sigma_r^j - \frac{\partial \sigma_r}{\partial q^j}\gamma_r^j\right) + h^{1/2} \sum_{r=1}^{m} \sigma_r \xi_{rk},$$

$$Q_{k+1} = Q_k + hg(t_k + \beta h, \alpha P_{k+1} + (1 - \alpha)P_k, (1 - \alpha)Q_{k+1} + \alpha Q_k)$$
$$+ h\left(\frac{1}{2} - \alpha\right) \sum_{r=1}^{m} \sum_{j=1}^{n} \left(\frac{\partial \gamma_r}{\partial p^j}\sigma_r^j - \frac{\partial \gamma_r}{\partial q^j}\gamma_r^j\right) + h^{1/2} \sum_{r=1}^{m} \gamma_r \xi_{rk},$$

where $\sigma_r, \gamma_r, r = 1, \ldots, m$, and their derivatives are evaluated at $(t_k, \alpha P_{k+1} + (1 - \alpha)P_k, (1 - \alpha)Q_{k+1} + \alpha Q_k)$, the parameters $\alpha, \beta \in [0, 1]$, and ξ_{rk} are i.i.d. random variables with the law

$$P(\xi = \pm 1) = 1/2. \quad (5.6.2)$$

Note that if $\alpha = \beta = 1/2$ the method (5.6.1) becomes the derivative-free (midpoint) method. The method requires solution of a nonlinear equation at each step (its solvability is proved within the next theorem).

Theorem 5.6.1 *The implicit method (5.6.1) for the system (5.0.1), (5.0.2) is symplectic and of first weak order.*

Proof The symplecticness is proved as in Theorem 5.2.2. Let us prove convergence of the method. Denote by $\bar{X} = \bar{X}(t + h; t, x) = (\bar{P}^\mathsf{T}, \bar{Q}^\mathsf{T})^\mathsf{T}$ the one-step approximation corresponding to the method (5.6.1):

$$\bar{P} = p + hf(t + \beta h, \alpha \bar{P} + (1 - \alpha)p, (1 - \alpha)\bar{Q} + \alpha q) \qquad (5.6.3)$$

$$+ h\left(\frac{1}{2} - \alpha\right) \sum_{r=1}^{m} \sum_{j=1}^{n} \left(\frac{\partial \sigma_r}{\partial p^j} \sigma_r^j - \frac{\partial \sigma_r}{\partial q^j} \gamma_r^j\right) + h^{1/2} \sum_{r=1}^{m} \sigma_r \xi_r,$$

$$\bar{Q} = q + hg(t + \beta h, \alpha \bar{P} + (1 - \alpha)p, (1 - \alpha)\bar{Q} + \alpha q)$$

$$+ h\left(\frac{1}{2} - \alpha\right) \sum_{r=1}^{m} \sum_{j=1}^{n} \left(\frac{\partial \gamma_r}{\partial p^j} \sigma_r^j - \frac{\partial \gamma_r}{\partial q^j} \gamma_r^j\right) + h^{1/2} \sum_{r=1}^{m} \gamma_r \xi_r,$$

where σ_r, γ_r, $r = 1, \ldots, m$, and their derivatives are evaluated at $(t, \alpha \bar{P} + (1 - \alpha)p, (1 - \alpha)\bar{Q} + \alpha q)$.

Using a Lipschitz condition on the coefficients of the system (5.0.1), we prove (cf. Lemma 1.3.6) that there are constants $K > 0$ and $h_0 > 0$ such that for any $h \leq h_0$, $t_0 \leq t \leq t_0 + T$, $x = (p^\mathsf{T}, q^\mathsf{T})^\mathsf{T} \in \mathbf{R}^d$, $d = 2n$, the equation (5.6.3) has a unique solution $\bar{X}$ which satisfies the inequality

$$|\bar{X} - x| \leq K(1 + |x|)\sqrt{h}, \qquad (5.6.4)$$

and this solution can be found by the method of simple iteration with $x = (p^\mathsf{T}, q^\mathsf{T})^\mathsf{T}$ as the initial approximation.

The condition (2.2.4) with $p = 1$ of Theorem 2.2.1 holds for the approximation (5.6.3) due to (5.6.4) (to avoid a confusion, we note that in Theorem 2.2.1 we denote by p the order of a method while in this chapter p means the initial condition for $P(t)$). Let us check fulfillment of the condition (2.2.3) with $p = 1$. To this end, introduce the weak Euler approximation $\hat{X} = (\hat{P}^\mathsf{T}, \hat{Q}^\mathsf{T})^\mathsf{T}$ for the Stratonovich system (5.0.1), (5.0.2):

$$\hat{P} = p + hf + \frac{h}{2} \sum_{r=1}^{m} \sum_{j=1}^{n} \left(\frac{\partial \sigma_r}{\partial p^j} \sigma_r^j + \frac{\partial \sigma_r}{\partial q^j} \gamma_r^j\right) + h^{1/2} \sum_{r=1}^{m} \sigma_r \xi_r,$$

$$\hat{Q} = q + hg + \frac{h}{2} \sum_{r=1}^{m} \sum_{j=1}^{n} \left(\frac{\partial \gamma_r}{\partial p^j} \sigma_r^j + \frac{\partial \gamma_r}{\partial q^j} \gamma_r^j\right) + h^{1/2} \sum_{r=1}^{m} \gamma_r \xi_r, \qquad (5.6.5)$$

where f, g and σ_r, γ_r, $r = 1, \ldots, m$, and their derivatives are evaluated at (t, p, q).

Expanding the terms in the right-hand side of (5.6.3) around (t, p, q) and using (5.6.4) and the corresponding conditions on smoothness and boundedness of the coefficients, it is not difficult to obtain that

$$\left| E\left(\prod_{j=1}^{s} \hat{\Delta}^{i_j} - \prod_{j=1}^{s} \bar{\Delta}^{i_j}\right) \right| \leq K(x)h^2, \ s = 1, 2, 3, \ i_j = 1, \ldots, 2n, \ K(x) \in \mathbf{F},$$

$$(5.6.6)$$

where $\bar{\Delta}^i := \bar{X}^i - x^i$, $\hat{\Delta}^i := \hat{X}^i - x^i$.

Taking into account (5.6.6) and the fact that the Euler approximation (5.6.5) satisfies (2.2.3) with $p = 1$, we get that the approximation (5.6.3) satisfies (2.2.3) with $p = 1$ as well.

Finally, to check the fourth condition of Theorem 2.2.1, we use Lemma 2.2.2 which ensures existence and uniform boundedness of the moments $E|\bar{X}_k|^{\bar{m}}$ under the conditions: (i) $|E\bar{\Delta}| \leq K(1 + |x|)h$ and (ii) $|\bar{\Delta}| \leq M(\xi)(1 + |x|)\sqrt{h}$ with $M(\xi)$ having moments of all orders. The inequalities (5.6.6) and $|E\hat{\Delta}| \leq K(1 + |x|)h$ imply fulfillment of the condition (i), while the condition (ii) holds here due to (5.6.4). $\qquad\square$

Remark 5.6.2 In the case of separable Hamiltonians at noise, i.e., when $H_r(t, p, q) = U_r(t, q) + V_r(t, p)$, $r = 1, \ldots, m$, the method (5.6.1) with $\alpha = 1$, $\beta = 0$ acquires the form

$$P_{k+1} = P_k + f(t_k, P_{k+1}, Q_k)h \tag{5.6.7}$$

$$+ \frac{h}{2} \sum_{r=1}^{m} \sum_{j=1}^{n} \frac{\partial \sigma_r}{\partial q^j}(t_k, Q_k) \cdot \gamma_r^j(t_k, P_{k+1}) + h^{1/2} \sum_{r=1}^{m} \sigma_r(t_k, Q_k)\xi_{rk},$$

$$Q_{k+1} = Q_k + g(t_k, P_{k+1}, Q_k)h$$

$$- \frac{h}{2} \sum_{r=1}^{m} \sum_{j=1}^{n} \frac{\partial \gamma_r}{\partial p^j}(t_k, P_{k+1}) \cdot \sigma_r^j(t_k, Q_k) + h^{1/2} \sum_{r=1}^{m} \gamma_r(t_k, P_{k+1})\xi_{rk}$$

with not too complicated implicitness. Besides, when the Hamiltonians are such that $H_0(t, p, q) = V_0(t, p) + U_0(t, q)$ and $H_r(t, p, q) = \Gamma_r^{\mathsf{T}}(t)p + U_r(t, q)$, $r = 1, \ldots, m$, $\Gamma_r(t)$ are n-dimensional vectors, one obtains fully explicit symplectic methods.

5.6.1.2 Explicit First-Order Methods in the Case of Separable Hamiltonians

Now we consider a special case of the Hamiltonian system (5.0.1), (5.0.2) such that

$$H_0(t, p, q) = V_0(p) + U_0(t, q), \quad H_r(t, p, q) = U_r(t, q), \quad r = 1, \ldots, m. \tag{5.6.8}$$

In this case we get the following system

$$dP = f(t, Q)dt + \sum_{r=1}^{m} \sigma_r(t, Q)dw_r(t), \quad P(t_0) = p,$$
$$dQ = g(P)dt, \quad Q(t_0) = q, \tag{5.6.9}$$

with

$$f^i = -\partial U_0/\partial q^i, \quad g^i = \partial V_0/\partial p^i, \quad \sigma_r^i = -\partial U_r/\partial q^i, \quad r = 1, \ldots, m, \quad i = 1, \ldots, n.$$
$$(5.6.10)$$

Recall that the system (5.6.9) has the same form in the sense of Stratonovich. Due to specific features of the system (5.6.9), (5.6.10) we have succeeded in construction of explicit PRK methods of a higher order.

On the basis of the mean-square PRK method (5.2.17)–(5.2.18) we obtain the weak PRK method for (5.6.9):

$$\mathcal{Q}_1 = \mathcal{Q}_k + \alpha h g(\mathcal{P}_k), \quad \mathcal{P}_1 = \mathcal{P}_k + h f(t_k + \alpha h, \mathcal{Q}_1),$$
$$\mathcal{Q}_2 = \mathcal{Q}_1 + (1 - \alpha) h g(\mathcal{P}_1), \tag{5.6.11}$$

$$P_{k+1} = \mathcal{P}_1 + h^{1/2} \sum_{r=1}^{m} \sigma_r(t_k, \mathcal{Q}_2)\xi_{rk}, \quad Q_{k+1} = \mathcal{Q}_2, \quad k = 0, \ldots, N-1, \tag{5.6.12}$$

where $0 \leq \alpha \leq 1$ is a parameter and ξ_{rk} are i.i.d. random variables with the law (5.6.2).

Theorem 5.6.3 *The explicit method (5.6.11)–(5.6.12) for the system (5.6.9), (5.6.10) is symplectic and of first weak order.*

Proof Due to (5.6.10), $\partial \sigma_r^i/\partial q^j = \partial \sigma_r^j/\partial q^i$. Using this, we obtain $dP_{k+1} \wedge dQ_{k+1} = d\mathcal{P}_1 \wedge d\mathcal{Q}_2$. It is easy to prove that $d\mathcal{P}_1 \wedge d\mathcal{Q}_2 = d\mathcal{P}_1 \wedge d\mathcal{Q}_1 = dP_k \wedge dQ_k$. Therefore, the method (5.6.11)–(5.6.12) is symplectic. The order of convergence is proved as in Theorem 5.6.1 (even simpler). □

Remark 5.6.4 By swapping the roles of p and q, we can propose another symplectic method of first weak order for the system (5.6.9), (5.6.10). Namely, instead of (5.6.11)–(5.6.12) one can propose

$$\mathcal{P}_1 = \mathcal{P}_k + \alpha h f(t_k, \mathcal{Q}_k), \quad \mathcal{Q}_1 = \mathcal{Q}_k + h g(\mathcal{P}_1),$$
$$\mathcal{P}_2 = \mathcal{P}_1 + (1 - \alpha) h f(t_k + h, \mathcal{Q}_1), \tag{5.6.13}$$

$$P_{k+1} = \mathcal{P}_2 + h^{1/2} \sum_{r=1}^{m} \sigma_r(t_k, \mathcal{Q}_1)\xi_{rk}, \quad Q_{k+1} = \mathcal{Q}_1, \quad k = 0, \ldots, N-1. \tag{5.6.14}$$

5.6.1.3 Explicit Second-Order Method in the Case of Separable Hamiltonians

Introduce the explicit PRK method for the system (5.6.9), (5.6.10):

$$\mathcal{Q}_1 = \mathcal{Q}_k + \frac{h}{2} g(\mathcal{P}_k), \tag{5.6.15}$$

$$\mathcal{P}_1 = \mathcal{P}_k + h f\left(t_k + \frac{h}{2}, \mathcal{Q}_1\right) + h^{1/2} \sum_{r=1}^{m} \sigma_r\left(t_k + \frac{h}{2}, \mathcal{Q}_1\right)\xi_{rk},$$

$$P_{k+1} = \mathcal{P}_1, \quad Q_{k+1} = \mathcal{Q}_1 + \frac{h}{2}g(\mathcal{P}_1), \quad k = 0, \ldots, N-1,$$

where ξ_{rk} are i.i.d. random variables with the law

$$P(\xi = 0) = 2/3, \quad P(\xi = \pm\sqrt{3}) = 1/6. \tag{5.6.16}$$

It follows from Lemma 5.3.11 that this method is symplectic. Comparing (5.6.15) with the standard Taylor-type second-order weak method (2.1.31) applied to (5.6.9), we prove that the method (5.6.15) is of weak order 2.

Theorem 5.6.5 *The explicit method (5.6.15) for the system (5.6.9), (5.6.10) is symplectic and of second weak order.*

5.6.2 Hamiltonian Systems with Additive Noise

Consider Hamiltonian systems with additive noise

$$dP = f(t, P, Q)dt + \sum_{r=1}^{m} \sigma_r(t)dw_r(t), \quad P(t_0) = p, \tag{5.6.17}$$

$$dQ = g(t, P, Q)dt + \sum_{r=1}^{m} \gamma_r(t)dw_r(t), \quad Q(t_0) = q,$$

where f and g satisfy (5.0.2).

The first order method for (5.6.17) follows from the method (5.6.1).

5.6.2.1 Implicit Second-Order Methods in the Case of General Hamiltonian System

On the basis of a mean-square symplectic method of order $3/2$ (see (5.3.26)), we construct the weak method:

$$\mathcal{P}_1 = P_k + \frac{\alpha}{2}hf\left(t_k + \frac{\alpha}{2}h, \mathcal{P}_1, \mathcal{Q}_1\right) + \lambda_1 h^{1/2} \sum_{r=1}^{m} \sigma_r\left(t_k + \frac{h}{2}\right)\xi_{rk}, \tag{5.6.18}$$

$$\mathcal{Q}_1 = Q_k + \frac{\alpha}{2}hg\left(t_k + \frac{\alpha}{2}h, \mathcal{P}_1, \mathcal{Q}_1\right) + \lambda_1 h^{1/2} \sum_{r=1}^{m} \gamma_r\left(t_k + \frac{h}{2}\right)\xi_{rk},$$

$$\mathcal{P}_2 = P_k + \alpha h f\left(t_k + \frac{\alpha}{2}h, \mathcal{P}_1, \mathcal{Q}_1\right) + \frac{1-\alpha}{2}hf\left(t_k + \frac{1+\alpha}{2}h, \mathcal{P}_2, \mathcal{Q}_2\right)$$
$$+\lambda_2 h^{1/2}\sum_{r=1}^{m}\sigma_r\left(t_k + \frac{h}{2}\right)\xi_{rk},$$

$$\mathcal{Q}_2 = Q_k + \alpha h g\left(t_k + \frac{\alpha}{2}h, \mathcal{P}_1, \mathcal{Q}_1\right) + \frac{1-\alpha}{2}hg\left(t_k + \frac{1+\alpha}{2}h, \mathcal{P}_2, \mathcal{Q}_2\right)$$
$$+\lambda_2 h^{1/2}\sum_{r=1}^{m}\gamma_r\left(t_k + \frac{h}{2}\right)\xi_{rk},$$

$$P_{k+1} = P_k + h\left[\alpha f\left(t_k + \frac{\alpha}{2}h, \mathcal{P}_1, \mathcal{Q}_1\right) + (1-\alpha)f\left(t_k + \frac{1+\alpha}{2}h, \mathcal{P}_2, \mathcal{Q}_2\right)\right]$$
$$+h^{1/2}\sum_{r=1}^{m}\sigma_r\left(t_k + \frac{h}{2}\right)\xi_{rk},$$

$$Q_{k+1} = Q_k + h\left[\alpha g\left(t_k + \frac{\alpha}{2}h, \mathcal{P}_1, \mathcal{Q}_1\right) + (1-\alpha)g\left(t_k + \frac{1+\alpha}{2}h, \mathcal{P}_2, \mathcal{Q}_2\right)\right]$$
$$+h^{1/2}\sum_{r=1}^{m}\gamma_r\left(t_k + \frac{h}{2}\right)\xi_{rk},$$

where the parameters α, λ_1, λ_2 are such that

$$\alpha\lambda_1 + (1-\alpha)\lambda_2 = \frac{1}{2}, \quad \alpha\lambda_1^2 + (1-\alpha)\lambda_2^2 = \frac{1}{2}, \qquad (5.6.19)$$

and ξ_{rk} are i.i.d. random variables with the law (5.6.16).

For example, the following set of parameters satisfies (5.6.19):

$$\alpha = \frac{1}{2}, \quad \lambda_1 = 0, \quad \lambda_2 = 1. \qquad (5.6.20)$$

The symplecticness follows from Lemma 5.3.5. The order of convergence is proved similarly to the proof of Theorem 5.6.1 comparing (5.6.18) with the standard Taylor-type second-order weak method (2.1.31) applied to (5.6.17).

Theorem 5.6.6 *The implicit method (5.6.18), (5.6.19) for the system (5.6.17) is symplectic and of second weak order.*

5.6.2.2   A Third-Order Method in a Particular Case of Hamiltonian System

Now we propose a symplectic weak method of order 3 for the system with additive noise:

$$dP = f(t, Q)dt + \sum_{r=1}^{m} \sigma_r(t)dw_r(t), \quad f^i(t, Q) = -\frac{\partial U_0}{\partial q^i}, \quad P(t_0) = p, \qquad (5.6.21)$$

$$dQ = M^{-1}Pdt, \quad Q(t_0) = q.$$

On the basis of a symplectic mean-square method of order 3 (see (5.3.49)–(5.3.50)), we construct the weak method:

$$\mathcal{Q}_1 = \mathcal{Q}_k + \frac{7}{24}hM^{-1}\mathcal{P}_k, \quad \mathcal{P}_1 = \mathcal{P}_k + \frac{2}{3}hf\left(t_k + \frac{7h}{24}, \mathcal{Q}_1\right), \qquad (5.6.22)$$

$$\mathcal{Q}_2 = \mathcal{Q}_1 + \frac{3}{4}hM^{-1}\mathcal{P}_1, \quad \mathcal{P}_2 = \mathcal{P}_1 - \frac{2}{3}hf\left(t_k + \frac{25h}{24}, \mathcal{Q}_2\right),$$

$$\mathcal{Q}_3 = \mathcal{Q}_2 - \frac{1}{24}hM^{-1}\mathcal{P}_2, \quad \mathcal{P}_3 = \mathcal{P}_2 + hf(t_k + h, \mathcal{Q}_3),$$

$$P_{k+1} = \mathcal{P}_3 + h^{1/2}\sum_{r=1}^{m}\sigma_r(t_k)\xi_{rk} + h^{3/2}\sum_{r=1}^{m}\sigma_r'(t_k)(\xi_r/2 - \eta_r)_k \qquad (5.6.23)$$

$$+h^{5/2}\sum_{r=1}^{m}\sigma_r''(t_k)\xi_{rk}/6 + h^{5/2}\sum_{r=1}^{m}\sum_{i=1}^{n}(M^{-1}\sigma_r(t_k))^i\frac{\partial f}{\partial q^i}(t_k, \mathcal{Q}_3)\xi_{rk}/6,$$

$$Q_{k+1} = \mathcal{Q}_3 + h^{3/2}\sum_{r=1}^{m}M^{-1}\sigma_r(t_k)(\xi_r/2 + \eta_r)_k + h^{5/2}\sum_{r=1}^{m}M^{-1}\sigma_r'(t_k)\xi_{rk}/6,$$

$$k = 0, \ldots, N - 1,$$

where ξ_{rk}, η_{rk} are mutually independent random variables distributed by the laws

$$P(\xi = 0) = \frac{1}{3}, \quad P(\xi = \pm 1) = \frac{3}{10}, \quad P(\xi = \pm\sqrt{6}) = \frac{1}{30}, \qquad (5.6.24)$$

$$P(\eta = \pm 1/\sqrt{12}) = \frac{1}{2}.$$

The symplecticness of this method follows from Theorem 5.3.15. The order of convergence can be proved by standard arguments using the fact that the corresponding mean-square method (5.3.49)–(5.3.50) has the third order of convergence or by comparing the method (5.6.22)–(5.6.23) with the weak method of order 3 (see (3.1.34)) applied to (5.6.21).

Theorem 5.6.7 *The explicit method (5.6.22)–(5.6.23) for the system (5.6.21) is symplectic and of third weak order.*

5.6.3   Numerical Tests

5.6.3.1   Kubo Oscillator

Consider the Kubo oscillator (cf. (5.4.1))

$$
\begin{aligned}
dX^1 &= -aX^2 dt - \sigma X^2 \circ dw(t), \quad X^1(0) = x^1, \\
dX^2 &= aX^1 dt + \sigma X^1 \circ dw(t), \quad X^2(0) = x^2,
\end{aligned}
\tag{5.6.25}
$$

where a and σ are constants and $w(t)$ is a one-dimensional standard Wiener process. The quantity $\mathcal{H}(x^1, x^2) = \left(x^1\right)^2 + \left(x^2\right)^2$ is conserved by this system:

$$
\mathcal{H}(X^1(t), X^2(t)) = \mathcal{H}(x^1, x^2) \ \text{ for } t \geq 0.
$$

Here we test three specific methods of weak order 1. The weak Euler method in application to (5.6.25) takes the form:

$$
\begin{aligned}
X^1_{k+1} &= X^1_k - ha X^2_k - h\frac{\sigma^2}{2} X^1_k - h^{1/2}\sigma X^2_k \xi_k, \\
X^2_{k+1} &= X^2_k + ha X^1_k - h\frac{\sigma^2}{2} X^2_k + h^{1/2}\sigma X^1_k \xi_k,
\end{aligned}
\tag{5.6.26}
$$

where ξ_k are i.i.d random variables with the law (5.6.2).

The weak midpoint method (the symplectic method (5.6.1) with $\alpha = 1/2$) is written for the autonomous system (5.6.25) as

$$
\begin{aligned}
X^1_{k+1} &= X^1_k - ha\frac{X^2_k + X^2_{k+1}}{2} - h^{1/2}\sigma\frac{X^2_k + X^2_{k+1}}{2}\xi_k, \\
X^2_{k+1} &= X^2_k + ha\frac{X^1_k + X^1_{k+1}}{2} + h^{1/2}\sigma\frac{X^1_k + X^1_{k+1}}{2}\xi_k.
\end{aligned}
\tag{5.6.27}
$$

This is an implicit method in both deterministic and stochastic terms. When applied to (5.6.25), the PRK method (5.6.7) has the form:

$$
\begin{aligned}
X^1_{k+1} &= X^1_k - ha X^2_k - h\frac{\sigma^2}{2} X^1_{k+1} - h^{1/2}\sigma X^2_k \xi_k, \\
X^2_{k+1} &= X^2_k + ha X^1_{k+1} + h\frac{\sigma^2}{2} X^2_k + h^{1/2}\sigma X^1_{k+1}\xi_k.
\end{aligned}
\tag{5.6.28}
$$

This method is symplectic and of first weak order. It is implicit in the deterministic part only.

Let us analyze how accurately these methods approximate $E\mathcal{H}(X^1(t), X^2(t))$. In the case of the Euler method we obtain

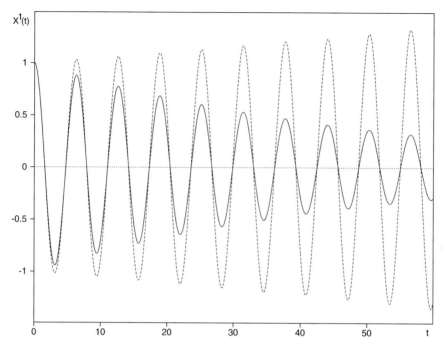

Fig. 5.6.1 The Kubo oscillator (5.6.25). Simulation of $EX^1(t)$ with $X^1(0) = 1$, $X^2(0) = 0$, $a = 1$, $\sigma = 0.2$, $h = 0.05$ on the time interval $t \le 60$. The results obtained by the Euler method (5.6.26) – dashed line. The results obtained by the midpoint method (5.6.27) and the symplectic method (5.6.28) visually coincide with the exact solution (5.6.30) (solid line). The Monte Carlo error is not greater than 0.03 with probability 0.95.

$$E\mathcal{H}(X_k^1, X_k^2) = \left(1 + h^2\left(a^2 + \frac{\sigma^4}{4}\right)\right)^k \times \mathcal{H}(x^1, x^2) \qquad (5.6.29)$$

$$\ge \exp\left(\frac{1}{2}\left(a^2 + \frac{\sigma^4}{4}\right)ht_k\right) \times \mathcal{H}(x^1, x^2),$$

i.e., the quantity grows exponentially fast as t increases.

It is not difficult to check that $\mathcal{H}(x^1, x^2)$ is conserved by the midpoint method (5.6.27). But the PRK method (5.6.28) does not preserve the quantity $\mathcal{H}(x^1, x^2)$. See a similar discussion concerning the mean-square midpoint method in Sect. 5.4.1.

Further, it is not difficult to find the following exact expressions for the Kubo oscillator (5.6.25):

$$EX_{0,x}^1(t) = e^{-\sigma^2 t/2}(x^1 \cos at - x^2 \sin at), \qquad (5.6.30)$$

$$EX_{0,x}^2(t) = e^{-\sigma^2 t/2}(x^2 \cos at + x^1 \sin at).$$

Figure 5.6.1 gives results of Monte Carlo simulation of $EX_{0,x}^1(t)$ by the methods (5.6.26), (5.6.27), and (5.6.28). We see that the Euler method is not appropriate for simulation of the oscillator (5.6.25) on long time intervals while the symplectic methods produce quite accurate results.

5.6.3.2   A Model for Synchrotron Oscillations of Particles in Storage Rings

Consider a model describing synchrotron oscillations of particles in storage rings under the influence of external fluctuating electromagnetic fields from [402] (cf. (5.4.5)):

$$dP = -\omega^2 \sin(Q)dt - \sigma_1 \cos(Q)dw_1 - \sigma_2 \sin(Q)dw_2, \qquad (5.6.31)$$

$$dQ = Pdt,$$

where P and Q are scalars.

Here we test four weak methods: two first-order methods (the Euler method, which is not symplectic, and the symplectic method (5.6.11)–(5.6.12)) and two second-order methods (the standard second-order weak method (2.2.18) and the symplectic method (5.6.15)).

The weak Euler method for (5.6.31) takes the form

$$P_{k+1} = P_k - h\omega^2 \sin(Q_k) - h^{1/2}(\sigma_1 \cos(Q_k)\xi_{1k} + \sigma_2 \sin(Q_k)\xi_{2k}),$$
$$Q_{k+1} = Q_k + hP_k, \qquad (5.6.32)$$

where ξ_{1k}, ξ_{2k} are i.i.d random variables with the law (5.6.2).

In application to (5.6.31) the first-order symplectic method (5.6.11)–(5.6.12) with $\alpha = 1$ is written as

$$\mathcal{Q} = Q_k + hP_k,$$
$$P_{k+1} = P_k - h\omega^2 \sin(\mathcal{Q}) - h^{1/2}(\sigma_1 \cos(\mathcal{Q})\xi_{1k} + \sigma_2 \sin(\mathcal{Q})\xi_{2k}), \quad Q_{k+1} = \mathcal{Q}, \qquad (5.6.33)$$

where ξ_{1k}, ξ_{2k} are i.i.d random variables with the law (5.6.2).

The standard second-order method (2.2.18) applied to (5.6.31) has the form

$$P_{k+1} = P_k - h^{1/2}(\sigma_1 \cos(Q_k)\xi_{1k} + \sigma_2 \sin(Q_k)\xi_{2k}) - h\omega^2 \sin(Q_k) \qquad (5.6.34)$$
$$+ \frac{h^{3/2}}{2}(\sigma_1 \sin(Q_k)\xi_{1k} - \sigma_2 \cos(Q_k)\xi_{2k})P_k - \frac{h^2}{2}\omega^2 \cos(Q_k)P_k,$$
$$Q_{k+1} = Q_k + hP_k - \frac{h^{3/2}}{2}(\sigma_1 \cos(Q_k)\xi_{1k} + \sigma_2 \sin(Q_k)\xi_{2k}) - \frac{h^2}{2}\omega^2 \sin(Q_k),$$

where ξ_{1k}, ξ_{2k} are i.i.d. random variables with the law (5.6.16).

The second-order symplectic method (5.6.15) is written for the system (5.6.31) as

Table 5.6.1 The model (5.6.31). Simulation of $E\mathcal{E}(P(t), Q(t))$ with $P(0) = 1$, $Q(0) = 0$, $\omega = 4$, $\sigma_1 = \sigma_2 = 0.3$, $t = 200$ for various time steps h by the Euler method (5.6.32), the first-order symplectic method (5.6.33), the standard second-order method (5.6.34), and the second-order symplectic method (5.6.35). The exact solution is -6.5. M is a number of independent realizations in the Monte Carlo simulation. Note that the "$\pm$" reflects the Monte Carlo error only (cf. (5.6.37)), it does not reflect the error of a method.

h	M	Eq. (5.6.32)	Eq. (5.6.33)	Eq. (5.6.34)	Eq. (5.6.35)
0.1	10^5	493.3 ± 0.3	-6.268 ± 0.059	462.2 ± 0.6	-6.316 ± 0.059
0.05	10^5	966.1 ± 0.7	-6.397 ± 0.059	0.896 ± 0.094	-6.421 ± 0.058
0.01	$4 \cdot 10^6$	234.5 ± 0.06	-6.503 ± 0.009	-6.456 ± 0.009	-6.502 ± 0.009

$$\mathcal{Q}_1 = Q_k + \frac{h}{2}P_k, \quad \mathcal{P}_1 = P_k - h\omega^2 \sin(\mathcal{Q}_1) - h^{1/2}(\sigma_1 \cos(\mathcal{Q}_1)\xi_{1k} \quad (5.6.35)$$

$$+ \sigma_2 \sin(\mathcal{Q}_1)\xi_{2k}),$$

$$P_{k+1} = \mathcal{P}_1, \quad Q_{k+1} = \mathcal{Q}_1 + \frac{h}{2}\mathcal{P}_1,$$

where ξ_{1k}, ξ_{2k} are i.i.d. random variables with the law (5.6.16).

Consider the quantity

$$\mathcal{E}(p, q) = \frac{p^2}{2} - \omega^2 \cos(q).$$

Its mean value $E\mathcal{E}(P(t), Q(t))$ is treated in physical literature (see, e.g., [402] and references therein) as a mean energy of the system. Under the assumption $\sigma_1 = \sigma_2 = \sigma$ one can obtain that

$$E\mathcal{E}(P_{0,p,q}(t), Q_{0,p,q}(t)) = \mathcal{E}(p, q) + \frac{\sigma^2}{2}t. \quad (5.6.36)$$

In Table 5.6.1 we compare results produced by the four methods given above. We have two types of errors in numerical simulations here: the error of a weak method used and the Monte Carlo error. The results in the table are approximations of $E\mathcal{E}(\bar{P}(t), \bar{Q}(t))$ calculated as

$$E\mathcal{E}(\bar{P}(t), \bar{Q}(t)) \doteq \frac{1}{M}\sum_{m=1}^{M}\mathcal{E}(\bar{P}^{(m)}(t), \bar{Q}^{(m)}(t)) \pm 2\sqrt{\frac{\bar{D}_M}{M}}, \quad (5.6.37)$$

where

$$\bar{D}_M = \frac{1}{M}\sum_{m=1}^{M}[\mathcal{E}(\bar{P}^{(m)}(t), \bar{Q}^{(m)}(t))]^2 - \left[\frac{1}{M}\sum_{m=1}^{M}\mathcal{E}(\bar{P}^{(m)}(t), \bar{Q}^{(m)}(t))\right]^2,$$

i.e., $E\mathcal{E}(\bar{P}(t), \bar{Q}(t))$ belongs to the interval defined in this formula with probability 0.95 (we recall that for sufficiently small h the sampling variance is sufficiently close to the variance of $\mathcal{E}(\bar{P}(t), \bar{Q}(t))$). Note that the "$\pm$" reflects the Monte Carlo error only, it does not reflect the error of a method.

The above experiments with the model (5.6.31) demonstrate once again superiority of symplectic methods in comparison with nonsymplectic ones. We note that the authors of [402] were interested in systems with small noise. Effective symplectic methods in the weak sense for Hamiltonian systems with small noise can be obtained using ideas from Chap. 4.

5.7 Quasi-symplectic Mean-Square Methods for Langevin-Type Equations

In Sect. 5.7.1, we construct mean-square quasi-symplectic methods for Langevin equations which are an important particular case of (5.0.5) when $f(t, q) = f(q)$, $\tilde{f}(t, p, q) = \Gamma p$, Γ is an $n \times n$-dimensional constant matrix, $g(p) = M^{-1} p$, M is a positive definite matrix, and $\sigma_r(t, q) = \sigma_r$, $r = 1, \ldots, m$, are constant vectors. The proposed methods are such that they degenerate to symplectic methods when the system degenerates to a Hamiltonian one and their law of phase volume contractivity is close to the law of the considered system. To construct numerical methods, we use the splitting technique (see, e.g. [398, 414, 464]) and some ideas of [422], where methods for deterministic second-order differential equations with similar properties were obtained. In Sect. 5.7.2, we generalize mean-square methods of Sect. 5.7.1 to the Langevin-type equations (5.0.5) and also to more general systems.

5.7.1 Langevin Equation: Linear Damping and Additive Noise

Consider the Langevin equation

$$dP = f(Q)dt - \nu \Gamma P dt + \sum_{r=1}^{m} \sigma_r dw_r(t), \quad P(t_0) = p, \qquad (5.7.1)$$

$$dQ = M^{-1} P dt, \quad Q(t_0) = q,$$

where P, Q, f are n-dimensional column-vectors, σ_r, $r = 1, \ldots, m$, are n-dimensional constant column-vectors, Γ is an $n \times n$-dimensional constant matrix, $\nu \geq 0$ is a parameter, M is a positive definite matrix, and $w_r(t)$, $r = 1, \ldots, m$, are independent standard Wiener processes. If there is a scalar function $U_0(q)$ such that

$$f^i(q) = -\frac{\partial U_0}{\partial q^i}, \quad i = 1, \ldots, n, \tag{5.7.2}$$

and if $\nu = 0$, then the system (5.7.1) is a Hamiltonian system with additive noise, i.e., its phase flow preserves symplectic structure.

The system (5.7.1) can be written as the second-order differential equation with additive noise:

$$M\ddot{Q} = f(Q) - \nu \Gamma M \dot{Q} + \sum_{r=1}^{m} \sigma_r \dot{w}_r.$$

Let $D_0 \in \mathbf{R}^d$, $d = 2n$, be a domain with finite volume. This domain may be random. We suppose that $D_0 = D_0(\omega)$ is independent of the Wiener processes $w_r(t)$, $t \in [t_0, t_0 + T]$. The transformation $(p, q) \mapsto (P, Q)$ maps D_0 into the domain D_t. The volume V_t of the domain D_t is equal to

$$V_t = \int_{D_t} dP^1 \ldots dP^n dQ^1 \ldots dQ^n \tag{5.7.3}$$

$$= \int_{D_0} \left| \frac{D(P^1, \ldots, P^n, Q^1, \ldots, Q^n)}{D(p^1, \ldots, p^n, q^1, \ldots, q^n)} \right| dp^1 \ldots dp^n dq^1 \ldots dq^n.$$

In the case of the system (5.7.1) the Jacobian J is equal to

$$J = \frac{D(P^1, \ldots, P^n, Q^1, \ldots, Q^n)}{D(p^1, \ldots, p^n, q^1, \ldots, q^n)} = \exp\left(-\nu \, tr\Gamma \cdot (t - t_0)\right). \tag{5.7.4}$$

That is, the system (5.7.1) preserves phase volume for $\nu = 0$. If $\nu > 0$ and $tr\Gamma > 0$ then phase-volume contractivity takes place.

Our aim is to propose mean-square methods based on the one-step approximations

$$\bar{P} = \bar{P}(t + h; t, p, q), \quad \bar{Q} = \bar{Q}(t + h; t, p, q)$$

such that

RL1. *The method applied to (5.7.1)–(5.7.2) degenerates to a symplectic method when $\nu = 0$, i.e., $d\bar{P} \wedge d\bar{Q} = dp \wedge dq$ for $\nu = 0$ and f from (5.7.2);*
RL2. *The Jacobian*

$$\bar{J} = \frac{D(\bar{P}, \bar{Q})}{D(p, q)}$$

does not depend on p, q.

Methods for (5.7.1) satisfying the conditions RL1 and RL2 are called *quasi-symplectic*.

As it is understood, a method is convergent and, consequently, $\bar{J}$ is close to J at any rate. The requirement RL2 is natural since the Jacobian J of the original system (5.7.1) does not depend on p, q. RL2 reflects the structural properties of the system which are connected with the law of phase volume contractivity. It is often possible to reach a stronger property consisting in the equality $\bar{J} = J$. However, such an requirement is too restrictive in general. In the context of deterministic equations the requirement RL2 was introduced in [422].

To construct methods satisfying RL1-RL2, we use ideas of splitting technique (see, e.g. [398, 414]). In connection with (5.7.1), introduce the systems

$$dP_I = f(Q_I)dt + \sum_{r=1}^{m} \sigma_r dw_r(t), \quad P_I(t_0) = p, \tag{5.7.5}$$

$$dQ_I = M^{-1}P_I dt, \quad Q_I(t_0) = q,$$

$$\frac{dP_{II}}{dt} = -\nu \Gamma P_{II}, \quad P_{II}(0) = p, \tag{5.7.6}$$

and denote their solutions as $P_I(t; t_0, p, q)$, $Q_I(t; t_0, p, q)$ and $P_{II}(t; p)$, respectively. The system (5.7.5) with $f(q)$ from (5.7.2) is a Hamiltonian system with additive noise. The system (5.7.6) is a deterministic linear system with constant coefficients, and its solution $P_{II}(t; p)$ can be found explicitly.

5.7.1.1   First-Order Methods

Let $\bar{P}_I = \bar{P}_I(t_0 + h; t_0, p, q)$, $\bar{Q}_I = \bar{Q}_I(t_0 + h; t_0, p, q)$ be a one-step approximation of a symplectic first-order mean-square method for (5.7.5), (5.7.6) (any explicit or implicit method from Sect. 5.3 can be used). Its Jacobian is equal to one, i.e.,

$$\frac{D(\bar{P}_I(t_0 + h; t_0, p, q), \bar{Q}_I(t_0 + h; t_0, p, q))}{D(p, q)} = 1.$$

We construct the one-step approximation $\bar{P}$, $\bar{Q}$ for the solution of (5.7.1)–(5.7.2) as follows

$$\bar{P} = \bar{P}(t_0 + h; t_0, p, q) := P_{II}(h; \bar{P}_I(t_0 + h; t_0, p, q)), \tag{5.7.7}$$

$$\bar{Q} = \bar{Q}(t_0 + h; t_0, p, q) := \bar{Q}_I(t_0 + h; t_0, p, q).$$

We have

$$\bar{J} = \frac{D(\bar{P}, \bar{Q})}{D(p, q)} = \frac{D(P_{II}, \bar{Q}_I)}{D(\bar{P}_I, \bar{Q}_I)} \frac{D(\bar{P}_I, \bar{Q}_I)}{D(p, q)} = J. \tag{5.7.8}$$

Further, if $\nu = 0$, then $\bar{P} = \bar{P}_I$, $\bar{Q} = \bar{Q}_I$, i.e., the approximation (5.7.7) degenerates to the symplectic method for (5.7.1)–(5.7.2) with $\nu = 0$. Thus, the approximation $\bar{P}$, $\bar{Q}$ satisfies both requirements RL1 and RL2.

It is not difficult to prove the following lemma.

Lemma 5.7.1 Let $\bar{P}_I$, $\bar{Q}_I$ be a one-step approximation corresponding to any first-order mean-square method for the system (5.7.5). Then $\bar{P}$, $\bar{Q}$ defined in (5.7.7) is a one-step approximation of the first-order mean-square method for the system (5.7.1).

Thus, due to (5.7.8), we obtain the theorem.

Theorem 5.7.2 Let $\bar{P}_I$, $\bar{Q}_I$ be a one-step approximation corresponding to a symplectic first-order mean-square method for the system (5.7.5), (5.7.2). Then $\bar{P}$, $\bar{Q}$ defined in (5.7.7) is a one-step approximation of the first-order mean-square method for the system (5.7.1) such that (i) it is symplectic when applied to (5.7.1)–(5.7.2) with $\nu = 0$, (ii) its phase volume changes according to the same law as the phase volume of (5.7.1) does, i.e., the Jacobians $\bar{J} = D(\bar{P}, \bar{Q})/D(p, q)$ and $J = D(P, Q)/D(p, q)$ are equal.

Remark 5.7.3 Theorem 5.7.2 also holds for the method based on the following one-step approximation:

$$
\begin{aligned}
\bar{P} &= \bar{P}(t_0 + h; t_0, p, q) := \bar{P}_I(t_0 + h; t_0, P_{II}(h; p), q), \\
\bar{Q} &= \bar{Q}(t_0 + h; t_0, p, q) := \bar{Q}_I(t_0 + h; t_0, P_{II}(h; p), q).
\end{aligned}
\tag{5.7.9}
$$

Remark 5.7.4 In practice, it can be more convenient to use an approximation $\bar{P}_{II}$ of the solution to (5.7.6) instead of the exact solution P_{II} in (5.7.7) (or (5.7.9)). Since (5.7.6) is a deterministic equation, we can exploit a high-order deterministic scheme in order to obtain $\bar{P}_{II}$. In this case the Jacobian $\bar{J}$ approximates the original Jacobian J with the accuracy of the deterministic scheme. Due to the linearity of (5.7.6), this $\bar{J}$ does not depend on the initial data p, q (it depends on $\nu\Gamma$ and h only).

There is another possibility to propose methods for (5.7.1) satisfying RL1-RL2. It consists in direct application of symplectic methods. For instance, the parametric first-order family of implicit methods (5.3.3)–(5.3.4) in application to (5.7.1) takes the form

$$
\begin{aligned}
\bar{P} &= p + hf((1 - \alpha)\bar{Q} + \alpha q) - h\nu\Gamma \cdot (\alpha\bar{P} + (1 - \alpha)p) + \sum_{r=1}^{m} \sigma_r \Delta w_r, \\
\bar{Q} &= q + hM^{-1}(\alpha\bar{P} + (1 - \alpha)p).
\end{aligned}
\tag{5.7.10}
$$

However, it satisfies the requirement RL2 for $\alpha = 0$ and $\alpha = 1$ only. Moreover, due to their specific structure, not all the symplectic methods (see, for example, the explicit method (5.3.30)) can be directly applied to the Langevin equation (5.7.1) itself. Thus, on the way of the direct application of symplectic methods to (5.7.1) we have rather restrictive opportunities. Nevertheless, we can obtain on this way some new methods.

5.7.1.2 Second-Order Methods

In order to construct second-order methods for the Langevin equation (5.7.1) with the properties RL1 and RL2, we use ideas of the method of fractional steps [398, 414, 464]. In the deterministic case (i.e., when $\sigma_r = 0$, $r = 1, \ldots, m$) a second-order method satisfying RL1 and RL2 can be based on the following one-step approximation

$$\bar{P} = \bar{P}(t_0 + h; t_0, p, q) := P_{II}\left(\frac{h}{2}; \bar{P}_I\left(t_0 + h; t_0, P_{II}\left(\frac{h}{2}; p\right), q\right)\right), \quad (5.7.11)$$

$$\bar{Q} = \bar{Q}(t_0 + h; t_0, p, q) := \bar{Q}_I\left(t_0 + h; t_0, P_{II}\left(\frac{h}{2}; p\right), q\right),$$

where $\bar{P}_I$, $\bar{Q}_I$ corresponds to a one-step approximation of a symplectic method for (5.7.5), (5.7.2) with $\sigma_r = 0$.

In the stochastic case the interconnection between terms in (5.7.1) is more complicated and a correction to (5.7.11) is needed to obtain second order accuracy. Consider the following approximation for solution of (5.7.1):

$$\bar{P} = \bar{P}(t_0 + h; t_0, p, q) := P_{II}\left(\frac{h}{2}; \bar{P}_I\left(t_0 + h; t_0, P_{II}\left(\frac{h}{2}; p\right), q\right)\right)$$

$$- \nu \sum_{r=1}^{m} \Gamma \sigma_r (I_{r0} - \frac{h}{2}\Delta w_r), \quad (5.7.12)$$

$$\bar{Q} = \bar{Q}(t_0 + h; t_0, p, q) := \bar{Q}_I\left(t_0 + h; t_0, P_{II}\left(\frac{h}{2}; p\right), q\right),$$

where $\bar{P}_I$, $\bar{Q}_I$ is a one-step approximation corresponding to a symplectic (explicit or implicit) second-order mean-square method for (5.7.5), (5.7.2) (such methods are available in Sect. 5.3),

$$I_{r0} = \int_{t_0}^{t} (w_r(s) - w_r(t_0))\, ds\,.$$

Lemma 5.7.5 *Let $\bar{P}_I$, $\bar{Q}_I$ be a one-step approximation corresponding to any second-order mean-square method for the system (5.7.5). Then $\bar{P}$, $\bar{Q}$ defined in (5.7.12) is a one-step approximation of the second-order mean-square method for the system (5.7.1).*

Proof Due to the assumption, we can write

$$\bar{P}_I(t_0 + h; t_0, p, q) = p + \sum_{r=1}^{m} \sigma_r \Delta w_r + h f(q) \tag{5.7.13}$$

$$+ \frac{h^2}{2} \sum_{i=1}^{n} \left(M^{-1} p \right)^i \frac{\partial f}{\partial q^i} + r_1,$$

$$\bar{Q}_I(t_0 + h; t_0, p, q) = q + h M^{-1} p + \sum_{r=1}^{m} M^{-1} \sigma_r I_{r0} + \frac{h^2}{2} M^{-1} f(q) + r_2,$$

where the remainders r_1 and r_2 are such that

$$|Er_i| = O(h^3), \quad Er_i^2 = O(h^5), \quad i = 1, 2.$$

We also have

$$P_{II}(h; p) = p - h \nu \Gamma p + \frac{h^2}{2} \nu^2 \Gamma^2 p + \rho, \quad \rho = O(h^3). \tag{5.7.14}$$

We obtain from (5.7.12)–(5.7.14) that

$$\bar{P} = p + \sum_{r=1}^{m} \sigma_r \Delta w_r + h \left(f(q) - \nu \Gamma p \right) - \nu \sum_{r=1}^{m} \Gamma \sigma_r I_{r0} \tag{5.7.15}$$

$$+ \frac{h^2}{2} \left[\sum_{i=1}^{n} \left(M^{-1} p \right)^i \frac{\partial f}{\partial q^i} + \nu^2 \Gamma^2 p - \nu \Gamma f(q) \right] + R_1,$$

$$\bar{Q} = q + h M^{-1} p + \sum_{r=1}^{m} M^{-1} \sigma_r I_{r0} + \frac{h^2}{2} M^{-1} [f(q) - \nu \Gamma p] + R_2,$$

where R_1 and R_2 are such that

$$|ER_i| = O(h^3), \quad ER_i^2 = O(h^5), \quad i = 1, 2.$$

It is not difficult to show that the standard Taylor-type mean-square method of order 3/2 for systems with additive noise (see (1.6.8)) has the second order of accuracy when it is applied to (5.7.1). Comparing the one-step approximation of this standard method with (5.7.15), we obtain that the method based on (5.7.12) is of mean-square order 2. □

One can easily check that the approximation (5.7.12) satisfies our requirements RL1 and RL2. The following theorem summarizes the result.

Theorem 5.7.6 *Let* $\bar{P}_I$, $\bar{Q}_I$ *be a one-step approximation corresponding to a symplectic second-order mean-square method for the system (5.7.5), (5.7.2). Then* $\bar{P}$, $\bar{Q}$ *defined in (5.7.12) is a one-step approximation of the second-order mean-square* *method for the system (5.7.1)–(5.7.2) such that (i) it is symplectic when applied to (5.7.1)–(5.7.2) with $\nu = 0$, (ii) its phase volume changes according to the same law as the phase volume of (5.7.1)–(5.7.2) does.*

Let us give a specific example of a method based on (5.7.12):

$$\mathcal{P}_1 = P_{II}\left(\frac{h}{2}; P_k\right), \quad \mathcal{Q}_1 = \mathcal{Q}_k + \frac{h}{2}M^{-1}\mathcal{P}_1, \tag{5.7.16}$$

$$\mathcal{P}_2 = \mathcal{P}_1 + \sum_{r=1}^{m} \sigma_r \Delta_k w_r + h f(\mathcal{Q}_1),$$

$$\mathcal{Q}_2 = \mathcal{Q}_k + h M^{-1} P_k + \sum_{r=1}^{m} M^{-1}\sigma_r (I_{r0})_k + \frac{h^2}{2}M^{-1}f(\mathcal{Q}_1),$$

$$P_{k+1} = P_{II}\left(\frac{h}{2}; \mathcal{P}_2\right) - \nu \sum_{r=1}^{m} \Gamma \sigma_r \left(I_{r0} - \frac{h}{2}\Delta w_r\right), \quad \mathcal{Q}_{k+1} = \mathcal{Q}_2, \quad k = 0, \ldots, N-1.$$

To obtain (5.7.16), we use the explicit symplectic second-order PRK method (5.3.44), which is a generalization of the Störmer-Verlet method.

The random variables $\Delta_k w_r$, $(I_{r0})_k$ have a Gaussian joint distribution, and they can be simulated at each step by $2m$ mutually independent $\mathcal{N}(0, 1)$-distributed random variables ξ_{rk} and η_{rk}, $r = 0, \ldots, m$:

$$\Delta_k w_r = h^{1/2}\xi_{rk}, \quad (I_{r0})_k = h^{3/2}(\xi_{rk} + \eta_{rk}/\sqrt{3})/2. \tag{5.7.17}$$

Note that Remark 5.7.4 is applicable here if one approximates $P_{II}(t)$ using a deterministic method of one-step order not less than 3.

In molecular dynamics several methods based on the deterministic Störmer-Verlet method are used for simulation of the Langevin equation (5.7.1) with diagonal matrix Γ (see [181, 404] and references therein). Effective numerical methods for this type of Langevin equations can be constructed by, for instance, the following splitting

$$dP_I = -\nu\Gamma P_I\, dt + \sum_{r=1}^{m} \sigma_r dw_r(t), \quad dQ_I = M^{-1}P_I\, dt, \quad dP_{II} = f(q)dt.$$

Since P_I, Q_I satisfy the linear system with additive noise, they can be simulated exactly. A number of specific schemes satisfying our requirements RL1-RL2 can be derived using the exact P_I, Q_I and a deterministic symplectic method. Such a second-order method based on the Störmer-Verlet scheme coincides with the method proposed in [404]. In the case of the unit matrix Γ it has the form:

$$\mathcal{P}_1 = P_k + \frac{h}{2} f(Q_k),$$

$$\mathcal{P}_2 = e^{-\nu h} \mathcal{P}_1 + \sum_{r=1}^{m} \sigma_r (\Delta_k w_r - \tilde{I}_{rk}),$$

$$\mathcal{Q}_2 = Q_k + M^{-1} \frac{1 - e^{-\nu h}}{\nu} \mathcal{P}_1 + \frac{M^{-1}}{\nu} \sum_{r=1}^{m} \sigma_r \tilde{I}_{rk},$$

$$P_{k+1} = \mathcal{P}_2 + \frac{h}{2} f(\mathcal{Q}_2), \quad Q_{k+1} = \mathcal{Q}_2,$$

where

$$\tilde{I}_{rk} := \int_{t_k}^{t_{k+1}} \left(1 - e^{-\nu(t_{k+1} - s)}\right) dw_r(s).$$

The random variables $\Delta_k w_r$, $\tilde{I}_{rk}$ have a Gaussian joint distribution. They can be simulated at each step by $2m$ independent $\mathcal{N}(0, 1)$-distributed random variables ξ_{rk} and η_{rk}, $r = 0, \ldots, m$. As a result, the above method can be written in the constructive form.

5.7.1.3 Third-Order Methods

Using ideas of the method of fractional steps, as we did above, it is possible to construct a third-order method for (5.7.1) which satisfies the requirements RL1 and RL2. But such a method contains two fractional steps at which we have to approximate the Hamiltonian system (5.7.5), (5.7.2) using a third-order symplectic method. This makes a method too complicated, and we will use another approach. In [422] a similar problem for deterministic second-order differential equations was solved by a modification of symplectic Runge-Kutta-Nyström (RKN) methods from [420]. Here we modify the symplectic RKN method (5.3.49)–(5.3.50) from Sect. 5.3.3 using some ideas of [422].

As a result, we obtain the method

$$\mathcal{Q}_1 = Q_k + \frac{7}{24} h M^{-1} P_k, \quad \mathcal{P}_1 = P_k + \frac{7}{24} h \left[f(\mathcal{Q}_1) - \nu \Gamma \mathcal{P}_1 \right], \tag{5.7.18}$$

$$\mathcal{Q}_2 = Q_k + \frac{25}{24} h M^{-1} P_k + \frac{h^2}{2} M^{-1} \left[f(\mathcal{Q}_1) - \nu \Gamma \mathcal{P}_1 \right],$$
$$\mathcal{P}_2 = P_k + \frac{2}{3} h \left[f(\mathcal{Q}_1) - \nu \Gamma \mathcal{P}_1 \right] + \frac{3}{8} h \left[f(\mathcal{Q}_2) - \nu \Gamma \mathcal{P}_2 \right]$$

$$\mathcal{Q}_3 = \mathcal{Q}_k + h M^{-1} \mathcal{P}_k + \frac{17}{36} h^2 M^{-1} [f(\mathcal{Q}_1) - \nu \Gamma \mathcal{P}_1]$$

$$+ \frac{1}{36} h^2 M^{-1} [f(\mathcal{Q}_2) - \nu \Gamma \mathcal{P}_2],$$

$$\mathcal{P}_3 = \mathcal{P}_k + \frac{2}{3} h [f(\mathcal{Q}_1) - \nu \Gamma \mathcal{P}_1] - \frac{2}{3} h [f(\mathcal{Q}_2) - \nu \Gamma \mathcal{P}_2] + h [f(\mathcal{Q}_3) - \nu \Gamma \mathcal{P}_3],$$

$$\mathcal{P}_{k+1} = \mathcal{P}_3 + \sum_{r=1}^{m} \sigma_r \Delta_k w_r - \nu \sum_{r=1}^{m} \Gamma \sigma_r \cdot (I_{r0})_k \tag{5.7.19}$$

$$+ \sum_{r=1}^{m} \left[\sum_{i=1}^{n} (M^{-1}\sigma_r)^i \frac{\partial f}{\partial q^i}(\mathcal{Q}_3) + \nu^2 \Gamma^2 \sigma_r \right] (I_{r00})_k,$$

$$\mathcal{Q}_{k+1} = \mathcal{Q}_3 + \sum_{r=1}^{m} M^{-1} \sigma_r \cdot (I_{r0})_k - \nu \sum_{r=1}^{m} M^{-1} \Gamma \sigma_r (I_{r00})_k, \quad k = 0, \dots, N-1,$$

where

$$(I_{r00})_k := \int_{t_k}^{t_k+h} \int_{t_k}^{\vartheta_1} (w_r(\vartheta_2) - w_r(t_k))\, d\vartheta_2 d\vartheta_1 .$$

The joint distribution of the random variables $\Delta_k w_r$, $(I_{r0})_k$, $(I_{r00})_k$ is Gaussian. They can be simulated at each step by $3m$ independent $\mathcal{N}(0, 1)$-distributed random variables ξ_{rk}, η_{rk}, and ζ_{rk}, $r = 0, \dots, m$:

$$\Delta_k w_r = h^{1/2} \xi_{rk}, \quad (I_{r0})_k = h^{3/2}(\xi_{rk} + \eta_{rk}/\sqrt{3})/2, \tag{5.7.20}$$

$$(I_{r00})_k = h^{5/2}(\xi_{rk} + \sqrt{3}\eta_{rk}/2 - \zeta_{rk}/(2\sqrt{5}))/6.$$

The method (5.7.18)–(5.7.19) is implicit in the components $\mathcal{P}_1$, $\mathcal{P}_2$, $\mathcal{P}_3$ and can easily be resolved at each step since the dependence on $\mathcal{P}$ is linear.

For $\nu = 0$ the method (5.7.18)–(5.7.19) coincides with the third-order symplectic method (5.3.49)–(5.3.50) and so it satisfies the requirement RL1. For $\sigma_r = 0$, $r = 1, \dots, m$, (deterministic case), the RKN method (5.7.18)–(5.7.19) satisfies conditions set up in [422]. These conditions ensure that the Jacobian of the deterministic RKN method depends on $\nu \Gamma$ and h only, more precisely:

$$\bar{J}_0 = \bar{J}_0(h, \nu \Gamma) := \frac{D(\mathcal{P}_3, \mathcal{Q}_3)}{D(\mathcal{P}_k, \mathcal{Q}_k)}$$

$$= \frac{\det(I - \frac{3}{8} h \nu \Gamma) \det(I + \frac{25}{24} h \nu \Gamma)}{\det(I + \frac{7}{24} h \nu \Gamma) \det(I + \frac{3}{8} h \nu \Gamma) \det(I + h \nu \Gamma)},$$

where I is the $n \times n$ unit matrix.

We have

$$\bar{J} := \frac{D(P_{k+1}, Q_{k+1})}{D(P_k, Q_k)} = \frac{D(P_{k+1}, Q_{k+1})}{D(\mathcal{P}_3, \mathcal{Q}_3)} \frac{D(\mathcal{P}_3, \mathcal{Q}_3)}{D(P_k, Q_k)} = \bar{J}_0,$$

i.e., the Jacobian $\bar{J}$ does not depend on the initial data P_k, Q_k. Further, it is possible to adapt the proof of Lemma 5.3.14 and prove that the method (5.7.18)–(5.7.19) is of mean-square order 3. Thus, we obtain the theorem.

Theorem 5.7.7 *The method (5.7.18)–(5.7.19) for the system (5.7.1) is of mean-square order 3 and it possesses the following properties: (i) it is symplectic when applied to (5.7.1)–(5.7.2) with $\nu = 0$ and (ii) the Jacobian $D(P_{k+1}, Q_{k+1})$ $/D(P_k, Q_k)$ (i.e., the change of phase volume per step) does not depend on P_k, Q_k.*

5.7.2 Langevin-Type Equation: Nonlinear Damping and Multiplicative Noise

Here we generalize methods of Subsection 5.7.1 to the Langevin-type system (cf. (5.0.5)):

$$dP = f(t, Q)dt - \nu \tilde{f}(t, P, Q)dt + \sum_{r=1}^{m} \sigma_r(t, Q)dw_r(t), \quad P(t_0) = p, \quad (5.7.21)$$

$$dQ = g(P)dt, \quad Q(t_0) = q,$$

where P, Q, f, $\tilde{f}$, g, σ_r are n-dimensional column-vectors, ν is a parameter, and $w_r(t)$, $r = 1, \ldots, m$, are independent standard Wiener processes. Note that the system (5.7.21) has the same form in the sense of Stratonovich.

If there are Hamiltonians $H_0(t, p, q) = V_0(p) + U_0(t, q)$ and $H_r(t, q)$, $r = 1, \ldots, m$, such that

$$f^i = -\partial H_0/\partial q^i, \quad g^i = \partial H_0/\partial p^i, \quad \sigma_r^i = -\partial H_r/\partial q^i, \quad i = 1, \ldots, n, \quad (5.7.22)$$

and if $\nu = 0$, then (5.7.21) is a Hamiltonian system with multiplicative noise (cf. (5.2.14)–(5.2.15)).

Our aim is to construct methods for (5.7.21) such that they inherit the properties RL1-RL2 of the quasi-symplectic methods for the Langevin equation (5.7.1), more precisely we require

RLT1. *The methods become symplectic when the system degenerate to a Hamiltonian one;*

RLT2. *The methods degenerate to those satisfying the requirement RL2 from Sect. 5.7.1 when the system degenerates to the Langevin equation (5.7.1).*

We recall that the Euler method for general systems with multiplicative noise is of order $1/2$. But due to specific features of system (5.7.21), the Euler method (and other usual methods of order $1/2$) applied to (5.7.21) is of order 1. Therefore, we start with methods of order 1.

5.7.2.1   First-Order Methods Based on Splitting

In connection with (5.7.21) introduce the systems (cf. (5.7.5)–(5.7.6)):

$$dP_I = f(t, Q_I)dt + \sum_{r=1}^{m} \sigma_r(t, Q_I)dw_r(t), \quad P_I(t_0) = p, \qquad (5.7.23)$$

$$dQ_I = g(P_I)dt, \quad Q_I(t_0) = q,$$

$$\frac{dP_{II}}{dt} = -\nu \tilde{f}(t, P_{II}, q), \quad P_{II}(t_0) = p, \qquad (5.7.24)$$

and denote their solutions as $P_I(t; t_0, p, q)$, $Q_I(t; t_0, p, q)$ and $P_{II}(t; t_0, p, q)$, respectively.

The system (5.7.23), (5.7.22) is a Hamiltonian system with separable Hamiltonians. Symplectic integrators for such systems are proposed in Sect. 5.2.2. The system (5.7.24) is deterministic.

Let $\bar{P}_I$, $\bar{Q}_I$ be a one-step approximation corresponding to a symplectic method for (5.7.23), (5.7.22) and $\bar{P}_{II}$ be a one-step approximation of a deterministic method for (5.7.24). Introduce the approximation for (5.7.21) as follows

$$\bar{P} = \bar{P}(t_0 + h; t_0, p, q)$$
$$:= \bar{P}_{II}(t_0 + h; t_0, \bar{P}_I(t_0 + h; t_0, p, q), \bar{Q}_I(t_0 + h; t_0, p, q)), \qquad (5.7.25)$$
$$\bar{Q} = \bar{Q}(t_0 + h; t_0, p, q) := \bar{Q}_I(t_0 + h; t_0, p, q).$$

Clearly, the approximation (5.7.25) satisfies the requirements RLT1 and RLT2. Further, using arguments similar to those in the proof of Lemma 5.7.1, we prove the following theorem.

Theorem 5.7.8 *Let $\bar{P}_I$, $\bar{Q}_I$ be a one-step approximation corresponding to a symplectic first-order mean-square method for the system (5.7.23), (5.7.22) and $\bar{P}_{II}$ be a one-step approximation corresponding to a first-order deterministic method for the system (5.7.24). Then $\bar{P}$, $\bar{Q}$ defined in (5.7.25) is a one-step approximation of the first-order mean-square method for the system (5.7.21) such that (i) it is symplectic when applied to (5.7.21)–(5.7.22) with $\nu = 0$, (ii) it satisfies the requirement RL2 from Sect. 5.7.1 when (5.7.21) degenerates to the Langevin equation (5.7.1).*

Let us give a specific example of a first-order splitting method (to this end we use the PRK method (5.2.17)–(5.2.18) from Sect. 5.2.2):

$$\mathcal{Q}_1 = Q_k + \alpha h g(P_k), \quad \mathcal{P}_1 = P_k + h f(t_k + \alpha h, \mathcal{Q}_1), \tag{5.7.26}$$

$$\mathcal{Q}_2 = \mathcal{Q}_1 + (1 - \alpha) h g(\mathcal{P}_1), \quad \mathcal{P}_2 = P_k + h f(t_k + \alpha h, \mathcal{Q}_1) + \sum_{r=1}^{m} \sigma_r(t_k, \mathcal{Q}_2) \Delta_k w_r,$$

$$Q_{k+1} = \mathcal{Q}_2, \quad P_{k+1} = \mathcal{P}_2 - h \nu \tilde{f}(t_k, \mathcal{P}_2, \mathcal{Q}_2).$$

The particular case of system (5.7.21), when $\tilde{f}(t, p, q) = \Gamma(q)p$, Γ is an $m \times m$-dimensional matrix, is of a special interest, in particular due to its application in dissipative particle dynamics (see, e.g. [376] and references therein). In this case the system (5.7.24) becomes deterministic linear system with constant coefficients, which can be solved exactly. If in addition to $f_\nu(t, p, q) = \Gamma(q)p$ the system (5.7.21) is with additive noise (i.e., $\sigma_r(t, q) = \sigma_r(t)$, $r = 1, \ldots, q$) and $g(p) = M^{-1}p$, then the method (5.7.28) (see below) becomes of mean-square order 2. An important example of such systems is the Van der Pol oscillator under external excitations

$$\ddot{Q} = -\omega^2 Q + \varepsilon^2 (1 - Q^2)\dot{Q} + \sigma \dot{w}.$$

Further, our approach can easily be applied to the more general system of Stratonovich SDEs

$$dP = \left(f(t, P, Q) - \nu \tilde{f}(t, P, Q) \right) dt \tag{5.7.27}$$

$$+ \sum_{r=1}^{m} \sigma_r(t, P, Q) \circ dw_r(t), \quad P(t_0) = p,$$

$$dQ = (g(t, P, Q) - \nu \tilde{g}(t, P, Q)) dt + \sum_{r=1}^{m} \gamma_r(t, P, Q) \circ dw_r(t), \quad Q(t_0) = q,$$

where $\nu \geq 0$ is a parameter, P, Q and all the coefficients are n-dimensional column-vectors, and f, g, σ_r, γ_r satisfy (5.0.2). For $\nu = 0$ it coincides with the general Hamiltonian system (5.0.1). As usual, we can split (5.7.27) in two parts: in the Hamiltonian system (5.0.1) and the deterministic system, and then use a relation like (5.7.25) to approximate (5.7.27). In such an approximation we have $\bar{P}_I$, $\bar{Q}_I$ corresponding to a fully implicit symplectic method from Sect. 5.2.1. As a result, we obtain the approximation $\bar{P}$, $\bar{Q}$ for (5.7.27) which satisfies the requirements RLT1-RLT2. Such a method for (5.7.27) based on an approximation of this kind has the mean-square order $1/2$.

5.7.2.2   Methods of Order 3/2

Using the fractional step method, we propose the following approximation for
(5.7.21):

$$\bar{P}(t_0 + h; t_0, p, q)$$
$$:= \bar{P}_{II}\left(t_0 + \frac{h}{2}; t_0, \bar{P}_I\left(t_0 + h; t_0, \bar{P}_{II}\left(t_0 + \frac{h}{2}; t_0, p, q\right), q\right), q\right),$$
$$\bar{Q}_I\left(t_0 + h; t_0, \bar{P}_{II}\left(t_0 + \frac{h}{2}; t_0, p, q\right), q\right)$$
$$-\nu\sum_{r=1}^{m}\sum_{i=1}^{n}\sigma_r^i\frac{\partial \tilde{f}}{\partial p^i}(t_0, p, q)\left[I_{r0} - \frac{h}{2}\Delta w_r\right] - \frac{h^2}{4}\nu\frac{\partial \tilde{f}}{\partial t}(t_0, p, q),$$
$$\bar{Q}(t_0 + h; t_0, p, q) := \bar{Q}_I\left(t_0 + h; t_0, \bar{P}_{II}\left(t_0 + \frac{h}{2}; t_0, p, q\right), q\right),$$

(5.7.28)

where $\bar{P}_I$, $\bar{Q}_I$ is a one-step approximation corresponding to a symplectic method of
order 3/2 for (5.7.23), (5.7.22) (such methods are available in Sect. 5.2.2) and $\bar{P}_{II}$
is a one-step approximation of a second-order deterministic method for (5.7.24).

By arguments similar to those exploited in previous sections, we prove the following theorem.

Theorem 5.7.9 Let $\bar{P}_I$, $\bar{Q}_I$ be a one-step approximation corresponding to a symplectic mean-square method of order 3/2 for the system (5.7.23), (5.7.22), and $\bar{P}_{II}$
be a one-step approximation corresponding to a second-order deterministic method
for the system (5.7.24). Then $\bar{P}$, $\bar{Q}$ defined in (5.7.28) is the one-step approximation of mean-square method of order 3/2 for the system (5.7.21) which satisfies the
requirements $RLT1$-$RLT2$.

As it is also noted before, if $\tilde{f}(t, p, q) = \Gamma(q)p$ then $P_{II}(t)$ can be found explicitly.

5.8   Quasi-symplectic Weak Methods for Langevin-Type Equations

Symplectic methods in the weak sense proposed in Sect. 5.6 together with the ideas
of Sect. 5.7 allow us to derive efficient weak methods for Langevin-type equations.

5.8.1   Langevin Equation: Linear Damping and Additive Noise

In this subsection we propose weak methods for the Langevin equation (5.7.1), which
satisfy the requirements RL1-RL2 from Sect. 5.7.1.

Using the splitting ideas presented in Sect. 5.7.1, we obtain the first-order method.

Theorem 5.8.1 *Let* $\bar{P}_I$, $\bar{Q}_I$ *be a one-step approximation corresponding to a symplectic method of first weak order for the system (5.7.1), (5.7.2). Then* $\bar{P}$, $\bar{Q}$ *defined in (5.7.7) or in (5.7.9) is a one-step approximation of the method of first weak order for the system (5.7.1) which satisfies the requirements RL1-RL2.*

As for $\bar{P}_I$, $\bar{Q}_I$ appearing in the above theorem, one can take the approximation corresponding to the symplectic implicit method (5.6.1) or to the explicit one (5.6.11)–(5.6.12).

Remark 5.8.2 The implicit method (5.6.1) can directly be applied to the Langevin equation (5.7.1). Of course, it satisfies the requirement $RL1$. The method (5.6.1) satisfies the requirement RL2 for $\alpha = 0$ and $\alpha = 1$ only (see also the discussion after (5.7.10) in Sect. 5.7.1).

Now we construct a method of weak order 2. To this end, consider the following approximation for (5.7.1) (cf. (5.7.12)):

$$\bar{P} = \bar{P}(t_0 + h; t_0, p, q) := P_{II}(\frac{h}{2}; \bar{P}_I(t_0 + h; t_0, P_{II}(\frac{h}{2}; p), q)), \qquad (5.8.1)$$

$$\bar{Q} = \bar{Q}(t_0 + h; t_0, p, q) := \bar{Q}_I(t_0 + h; t_0, P_{II}(\frac{h}{2}; p), q),$$

where $\bar{P}_I$, $\bar{Q}_I$ is a one-step approximation corresponding to any symplectic weak second-order method for (5.7.5), (5.7.2) (e.g., one can use the implicit method (5.6.18) or the explicit method (5.6.15)), and $P_{II}(t)$ is the exact solution of (5.7.6).

Theorem 5.8.3 *Let* $\bar{P}_I$, $\bar{Q}_I$ *be a one-step approximation corresponding to a symplectic method of second weak order for the system (5.7.5), (5.7.6). Then* $\bar{P}$, $\bar{Q}$ *defined in (5.8.1) is a one-step approximation of the method of second weak order for the system (5.7.1) which satisfies the requirements RL1-RL2.*

Note that Remark 5.7.4 is applicable for both first and second order methods. Consider the special case of the system (5.7.1):

$$dQ = \frac{P}{m}dt, \quad Q(0) = q, \qquad (5.8.2)$$

$$dP = f(Q)dt - \gamma P dt + \sigma dw(t), \quad P(0) = p,$$

where P, Q, f are n-dimensional column-vectors, $m > 0$, $\sigma > 0$ and $\gamma > 0$ are constants, $w(t)$ is an n-dimensional standard Wiener processes, and there is a potential energy $U_0(q)$ such that

$$f^i(q) = -\frac{\partial U_0}{\partial q^i}, \quad i = 1, \ldots, n.$$

The system of Langevin equations (5.8.2) is a working horse of molecular dynamics [230, 401]. As we discuss later in Example 5.11.1, (5.8.2) is ergodic under some assumptions on $U_0(q)$ and allows us to sample from the Gibbs distribution. In addition to the splitting we have considered above (splitting in the stochastic Hamiltonian system and the ODE for the damping part), we can consider the following splitting

$$dP_I = -\gamma P_I dt + \sigma dw(t) \tag{5.8.3}$$

and

$$dQ_{II} = \frac{P_{II}}{m} dt, \ dP_{II} = f(Q_{II})dt, \tag{5.8.4}$$

i.e., the splitting in the Ornstein-Uhlenbeck process (5.8.3) and the deterministic Hamiltonian system (5.8.4). This splitting is very popular in molecular dynamics (see e.g. [37, 229, 230] and references therein). The SDEs (5.8.3) have the exact solution:

$$P_I(t) = P_I(0)e^{-\gamma t} + \sigma \int\limits_0^t e^{-\gamma(t-s)} dw(s). \tag{5.8.5}$$

Concatenating (5.8.5) and a deterministic symplectic integrator for (5.8.4), one can obtain many quasi-symplectic schemes for the Langevin equations (5.8.2). In [229] (see also [230]) a systematic study of weak second order schemes based on the splitting (5.8.3)–(5.8.4) was conducted and, in particular, it was found that the following method performs better for computing ergodic limits (see Sect. 5.11) for functionals dependent on the position Q only:

$$\mathcal{P}_1 = P_k + \frac{h}{2} f(Q_k), \tag{5.8.6}$$

$$\mathcal{Q}_1 = Q_k + \frac{h}{2m} \mathcal{P}_1,$$

$$\mathcal{P}_2 = \mathcal{P}_1 e^{-\gamma h} + \sqrt{\frac{\sigma^2}{2\gamma}(1 - e^{-2\gamma h})} \xi_k,$$

$$Q_{k+1} = Q_k + \frac{h}{2m} \mathcal{P}_2,$$

$$P_{k+1} = \mathcal{P}_2 + \frac{h}{2} f(Q_{k+1}),$$

where $\xi_k = (\xi_{1,k}, \ldots, \xi_{n,k})^\mathsf{T}$, with their components being i.i.d. random variables either Gaussian $\mathcal{N}(0, 1)$ or with the same law $P(\xi_{i,k} = 0) = 2/3$, $P(\xi_{i,k} = \pm\sqrt{3}) = 1/6$. The scheme (5.8.6) was called 'BAOAB' in [229]. It uses only one evaluation of force $f(q)$ per step.

Theorem 5.8.4 *The numerical scheme (5.8.6) for (5.8.2) is quasi-symplectic and it is of weak order two.*

To get a method of weak order 3 for (5.7.1), we modify the symplectic RKN method (5.6.22)–(5.6.23) as we did in Sect. 5.7.1 in the case of mean-square methods. On this way we obtain the following method

$$Q_1 = Q_k + \frac{7}{24}hM^{-1}P_k, \quad P_1 = P_k + \frac{7}{24}h[f(Q_1) - \nu\Gamma P_1], \tag{5.8.7}$$

$$Q_2 = Q_k + \frac{25}{24}hM^{-1}P_k + \frac{h^2}{2}M^{-1}[f(Q_1) - \nu\Gamma P_1],$$
$$P_2 = P_k + \frac{2}{3}h[f(Q_1) - \nu\Gamma P_1] + \frac{3}{8}h[f(Q_2) - \nu\Gamma P_2],$$

$$Q_3 = Q_k + hM^{-1}P_k + \frac{17}{36}h^2M^{-1}[f(Q_1) - \nu\Gamma P_1]$$
$$+ \frac{1}{36}h^2M^{-1}[f(Q_2) - \nu\Gamma P_2],$$
$$P_3 = P_k + \frac{2}{3}h[f(Q_1) - \nu\Gamma P_1] - \frac{2}{3}h[f(Q_2) - \nu\Gamma P_2] + h[f(Q_3) - \nu\Gamma P_3],$$

$$P_{k+1} = P_3 + h^{1/2}\sum_{r=1}^{m}\sigma_r\xi_{rk} - \nu h^{3/2}\sum_{r=1}^{m}\Gamma\sigma_r \cdot (\xi_r/2 + \eta_r)_k \tag{5.8.8}$$
$$+ h^{5/2}\sum_{r=1}^{m}\left[\sum_{i=1}^{n}(M^{-1}\sigma_r)^i\frac{\partial f}{\partial q^i}(Q_3) + \nu^2\Gamma^2\sigma_r\right]\xi_{rk}/6,$$

$$Q_{k+1} = Q_3 + h^{3/2}\sum_{r=1}^{m}M^{-1}\sigma_r \cdot (\xi_r/2 + \eta_r)_k - \nu h^{5/2}\sum_{r=1}^{m}M^{-1}\Gamma\sigma_r\xi_{rk}/6,$$
$$k = 0, \ldots, N - 1,$$

where ξ_{rk}, η_{rk} are mutually independent random variables distributed by the laws (5.6.24).

The weak order of this method can be proved by standard arguments from Chap. 2 and its phase-volume contractivity properties are proved by the same arguments as those before Theorem 5.7.7.

Theorem 5.8.5 *The method (5.8.7)–(5.8.8) for the system (5.7.1) is quasi-symplectic and it has third weak order.*

5.8.2   Langevin-Type Equation: Nonlinear Damping and Multiplicative Noise

In this subsection we propose weak methods for the Langevin-type equation (5.7.21) which satisfy the requirements RLT1-RLT2 from Sect. 5.7.2. As for first-order methods, we have the theorem.

Theorem 5.8.6 *Let* $\bar{P}_I$, $\bar{Q}_I$ *be a one-step approximation corresponding to a symplectic method of first weak order for the system (5.7.23), (5.7.22), and* $\bar{P}_{II}$ *be a one-step approximation corresponding to a first-order deterministic method for the system (5.7.24). Then* $\bar{P}$, $\bar{Q}$ *defined in (5.7.25) is a one-step approximation of the method of first weak order for the system (5.7.21) which satisfies the requirements RLT1-RLT2.*

A specific method based on $\bar{P}$, $\bar{Q}$ from the above theorem can be written using the implicit symplectic method (5.6.1) or the explicit one (5.6.11)–(5.6.12) for $\bar{P}_I$, $\bar{Q}_I$. Further, as in the case of mean-square methods, the proposed approach can be generalized to a more general system of the form (5.7.27) (see the comment after (5.7.27) in Sect. 5.7.2).

By the method of fractional steps (as in Sect. 5.7) we construct the second order weak method for (5.7.21) on the basis of the symplectic method (5.6.15). The method has the form

$$\mathcal{P}_1 = \bar{P}_{II}\left(t_k + \frac{h}{2}; t_k, P_k, Q_k\right), \quad \mathcal{Q}_1 = Q_k + \frac{h}{2}g(\mathcal{P}_1), \tag{5.8.9}$$

$$\mathcal{P}_2 = \mathcal{P}_1 + hf\left(t_k + \frac{h}{2}, \mathcal{Q}_1\right) + h^{1/2}\sum_{r=1}^{m}\sigma_r\left(t_k + \frac{h}{2}, \mathcal{Q}_1\right)\xi_{rk},$$

$$\mathcal{Q}_2 = \mathcal{Q}_1 + \frac{h}{2}g(\mathcal{P}_2),$$

$$P_{k+1} = \bar{P}_{II}\left(t_k + \frac{h}{2}; t_k, \mathcal{P}_2, \mathcal{Q}_2\right) - \frac{h^2}{4}\nu\frac{\partial \tilde{f}}{\partial t}(t_k, P_k, Q_k), \quad Q_{k+1} = \mathcal{Q}_2,$$
$$k = 0, \ldots, N-1,$$

where ξ_{rk} are i.i.d. random variables with the law (5.6.16) and $\bar{P}_{II}$ is a one-step approximation of any second-order deterministic method for system (5.7.24).

Using a specific approximation instead of $\bar{P}_{II}$, it is possible to modify the method (5.8.9) in such a way that it will become a derivative-free method (i.e., the correction with the derivative $\partial \tilde{f}/\partial t$ can be incorporated in $\bar{P}_{II}$) but we do not consider this here.

The following theorem holds for the method (5.8.9).

Theorem 5.8.7 *The method (5.8.9) for the system (5.7.21) has the second weak order and satisfies the requirements RLT1-RLT2.*

We note that for $\tilde{f}(t, p, q) = \Gamma(q)p$ with Γ being an $m \times m$ dimensional matrix, $P_{II}(t)$ can be found explicitly. Consequently, we can put P_{II} instead of $\bar{P}_{II}$ in (5.8.9).

5.8.3 Numerical Examples

5.8.3.1 Linear Oscillator with Linear Damping under External Random Excitation

Let us consider the linear oscillator with linear damping term and additive noise

$$dX^1 = \omega X^2 dt \qquad (5.8.10)$$

$$dX^2 = (-\omega X^1 - \nu X^2)dt + \frac{\sigma}{\omega}dw(t),$$

where $w(t)$ is a standard Wiener process, ω, ν, σ are positive constants. The system (5.8.10) is dissipative, its invariant measure μ is Gaussian $\mathcal{N}(0, R)$ with the density

$$\rho(x) = (2\pi)^{-1}(\det R)^{-1/2} \exp\left\{-\frac{1}{2}(R^{-1}x, x)\right\}, \qquad (5.8.11)$$

where $R = (\sigma^2/2\nu\omega^2)I$ is the covariance matrix for the two-dimensional process $X = (X^1, X^2)^\mathsf{T}$, I denotes the identity matrix.

The discrete system obtained by the explicit Euler scheme has the form

$$\bar{X}_{k+1}^1 = \bar{X}_k^1 + \omega \bar{X}_k^2 h \qquad (5.8.12)$$

$$\bar{X}_{k+1}^2 = \bar{X}_k^2 - (\omega \bar{X}_k^1 + \nu \bar{X}_k^2)h + \frac{\sigma}{\omega}\Delta_k w.$$

The eigenvalues of the homogeneous part of (5.8.12) are

$$\lambda_{1,2} = 1 - \frac{\nu h}{2} \pm h\sqrt{\frac{\nu^2}{4} - \omega^2}. \qquad (5.8.13)$$

We consider the case when the damping term is small, and that is why we suppose that

$$\frac{\nu}{2} < \omega. \qquad (5.8.14)$$

If (5.8.14) is fulfilled, then $|\lambda_{1,2}|^2 = 1 - \nu h + \omega^2 h^2$, and consequently (5.8.12) is asymptotically stable if and only if

$$h < \frac{\nu}{\omega^2}. \qquad (5.8.15)$$

In this case, the system (5.8.12) possesses a unique invariant measure $\mu_h(x)$ with a Gaussian density $\rho_h(x)$ corresponding to the normal law $\mathcal{N}(0, R_h)$ with zero mean and the covariance matrix

$$R_h = \frac{\sigma^2}{\omega^2 \varkappa} \begin{bmatrix} 1 - \nu h/2 + \omega^2 h^2/2 & -\omega h/2 \\ -\omega h/2 & 1 \end{bmatrix},$$

where

$$\varkappa := 2\nu - 2\omega^2 h - \nu^2 h + \frac{3\nu\omega^2 h^2}{2} - \frac{\omega^4 h^3}{2}.$$

Due to (5.8.14) and (5.8.15), it is possible to prove that $\varkappa > 0$. The elements of R_h can be represented as

$$R_h^{jj} = \frac{\sigma^2}{2\nu\omega^2} \left(1 + \frac{\omega^2 h}{\nu} + O(h\nu) + O\left(\frac{h^2}{\nu^2}\right) \right), \quad j = 1, 2,$$

$$R_h^{ij} = \frac{\sigma^2}{2\nu\omega^2} \left(-\frac{\omega h}{2} - \frac{\omega^3 h^2}{2\nu} + O(h^2 \nu) + O\left(\frac{h^3}{\nu^2}\right) \right), \quad i \neq j,$$

where, for instance, $O\left(\dfrac{h^2}{\nu^2}\right)$ satisfies the inequality $\left| O\left(\dfrac{h^2}{\nu^2}\right) \right| \le C \dfrac{h^2}{\nu^2}$ for all $\nu > 0,\ h > 0$ such that the ratio h/ν is sufficiently small, C is a positive number.

Therefore, if one would like to approximate $\mu(x)$ by $\mu_h(x)$ sufficiently accurately, then the step h must be essentially less than ν/ω^2, i.e., just the fulfillment of the stability condition (5.8.15) is not enough. Suppose our aim is to evaluate

$$\int |x|^2 d\mu(x) = \int |x|^2 \rho(x) dx = \lim_{T \to \infty} E|X_x(T)|^2,$$

where $X_x(t)$ is the solution of (5.8.10) with $X_x(0) = x$.

We can approximate the limit by $E|X_x(T)|^2$ under a sufficiently large T. To evaluate $E|X_x(T)|^2$ by the explicit Euler method, we need to perform $N = T/h$ steps of (5.8.12). If the damping factor ν is small then the time T is rather large and the step h of the Euler method should be very small to satisfy the above condition $h \ll \nu/\omega^2$. Consequently, the number N is huge, and the Euler method is not appropriate for numerical solution of this problem under small ν.

Let us apply the implicit Euler method to system (5.8.10):

$$\bar{X}_{k+1}^1 = \bar{X}_k^1 + \omega \bar{X}_{k+1}^2 h \tag{5.8.16}$$

$$\bar{X}_{k+1}^2 = \bar{X}_k^2 - (\omega \bar{X}_{k+1}^1 + \nu \bar{X}_{k+1}^2)h + \frac{\sigma}{\omega} \Delta_k w.$$

The eigenvalues of the homogeneous part of (5.8.16) are

$$\lambda_{1,2} = 1 - \frac{\nu h + 2\omega^2 h^2}{2(1 + \nu h + \omega^2 h^2)} \pm \frac{\sqrt{\nu^2 h^2 - 4\omega^2 h^2}}{2(1 + \nu h + \omega^2 h^2)} .$$

Under (5.8.14), the eigenvalues are again complex numbers and

$$|\lambda_{1,2}|^2 = 1 - \frac{\nu h + \omega^2 h^2}{1 + \nu h + \omega^2 h^2} .$$

Therefore, in contrast to the explicit Euler method, we need not any restriction on h for asymptotic stability. This can give rise to the illusion about a possibility to choose a comparatively big step h in the implicit Euler scheme. However, the coming evaluations show that such an illusion is very dangerous. Indeed, the system (5.8.16) possesses a unique invariant measure $\mu_h(x)$ corresponding to the normal law $\mathcal{N}(0, R_h)$ with zero mean and the covariance matrix R_h with the elements

$$R_h^{jj} = \frac{\sigma^2}{2\nu\omega^2}\left(1 - \frac{\omega^2 h}{\nu} + O(h\nu) + O\left(\frac{h^2}{\nu^2}\right)\right), \quad j = 1, 2,$$

$$R_h^{ij} = \frac{\sigma^2}{2\nu\omega^2}\left(\frac{\omega h}{2} - \frac{\omega^3 h^2}{2\nu} + O(h^2) + O\left(\frac{h^3}{\nu^2}\right)\right), \quad i \neq j,$$

and we are again forced to take a very small h to reach a satisfactory accuracy.

Now let us use the quasi-symplectic method based on the one-step approximation (5.7.7) with $\bar{P}_I$, $\bar{Q}_I$ from (5.3.30) with $\alpha = 0$. For simplicity we take $\bar{P}_{II} = p - h\nu p$ instead of the exact P_{II} (see Remark 5.7.4). As a result, we get

$$\bar{X}_{k+1}^1 = \bar{X}_k^1 + \omega h(\bar{X}_k^2 - \omega h \bar{X}_k^1) \tag{5.8.17}$$

$$\bar{X}_{k+1}^2 = \left(\bar{X}_k^2 - \omega h \bar{X}_k^1 + \frac{\sigma}{\omega}\Delta_k w\right)(1 - \nu h).$$

In this case, if

$$\frac{\nu}{2} < \omega - \frac{\omega^2 h}{2},$$

the eigenvalues $\lambda_{1,2}$ are complex and

$$|\lambda_{1,2}|^2 = 1 - \nu h.$$

For all not too large h the system (5.8.17) is asymptotically stable and possesses a unique invariant measure with a Gaussian density. The corresponding normal law has zero mean and the covariance matrix with the elements

$$R_h^{11} = \frac{\sigma^2}{2\nu\omega^2}(1 - 2\nu h + O(h^2)), \quad R_h^{22} = \frac{\sigma^2}{2\nu\omega^2}\left(1 - \frac{3}{2}\nu h + O(h^2)\right),$$

$$R_h^{ij} = \frac{\sigma^2}{2\nu\omega^2}\left(\frac{\omega h}{2} - \frac{5}{4}\omega\nu h^2 + O(h^3)\right), \quad i \neq j.$$

We see that the implicit Euler method has advantages in comparison with the explicit Euler method due to its better stability properties. But both of them require too small time-steps to reach sufficient accuracy, in particular, if ν is small. At the same time, the quasi-symplectic method (5.8.17) gives very good results for very large time-steps. This is important, for instance, for the problem of computing a mean due to an invariant law which needs numerical integration over very long time intervals.

As an example, we evaluate $E\left(X^1(T)\right)^2$ for a large T by weak analogues of the implicit Euler method (5.8.16) and the quasi-symplectic method (5.8.17) (i.e., we replace $\Delta_k w$ in these methods by $h^{1/2}\xi_k$, ξ_k are i.i.d. random variables with the law (5.6.2)). Notice that the moments $E\left(X^i(t)X^j(t)\right)$, $i, j = 1, 2$, satisfy a system of linear differential equations and $E\left(X^1(T)\right)^2$ can be found exactly. The results of simulation are presented in Fig. 5.8.1. We see that even for the small step $h = 0.01$ the implicit Euler method tends to a wrong limit with increasing T while the quasi-symplectic method gives quite accurate results, e.g., for $h = 0.1$. The explicit Euler method is unstable for $h = 0.1$ (see (5.8.15)).

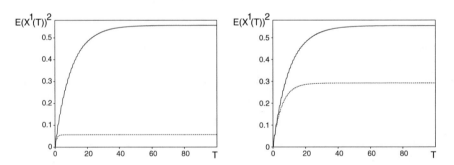

Fig. 5.8.1 The linear oscillator with linear damping (5.8.10). Behavior of $E(X^1(T))^2$ with $X^1(0) = 0$, $X^2(0) = 0$, $\omega = 3$, $\nu = 0.1$, $\sigma = 1$, $h = 0.1$ (left) and $h = 0.01$ (right) on the time interval $t \leq 100$ in the case of the weak implicit Euler method (dashed line) and the weak quasi-symplectic method (solid line, which visually coincides with the exact dependence $E(X^1(T))^2$). The Monte Carlo error is not greater than 0.00005 (left) and 0.0003 (right) for the Euler method and 0.0005 for the quasi-symplectic method with probability 0.95.

5.8.3.2   An Oscillator with Cubic Restoring Force under External random Excitation

Consider the oscillator with cubic restoring force and additive noise

$$\ddot{Q} = Q - Q^3 - \nu \dot{Q} + \sigma \dot{w}, \tag{5.8.18}$$

i.e., (5.7.1) with $U_0(q) = \frac{1}{4}q^4 - \frac{1}{2}q^2$. The dynamical system (5.8.18) is ergodic (see, e.g., [262]) and its invariant measure has the density

$$\rho(p, q) = C \exp\left(-\frac{\nu}{\sigma^2}\left(p^2 + \frac{1}{2}q^4 - q^2\right)\right), \tag{5.8.19}$$

where C is defined by the normalization condition.

Here we compare an implicit quasi-symplectic method and the implicit Euler scheme. We use the implicit quasi-symplectic method based on the one-step approximation (5.7.7) and on the weak implicit symplectic method (5.6.1) with $\alpha = 1/2$. For simplicity we take $\bar{P}_{II} = p - h\nu p$ instead of the exact P_{II} (see Remark 5.7.4). As a result, we get for (5.8.18):

$$\begin{aligned}
\bar{P}_I &= P_k + h\left(\frac{\bar{Q}_I + Q_k}{2} - \frac{(\bar{Q}_I + Q_k)^3}{8}\right) + h^{1/2}\sigma\xi_k, \\
\bar{Q}_I &= Q_k + h(\bar{P}_I + P_k)/2, \\
P_{k+1} &= (1 - \nu h)\bar{P}_I, \quad Q_{k+1} = \bar{Q}_I,
\end{aligned} \tag{5.8.20}$$

where ξ_k are i.i.d. random variables with the law (5.6.2).

In application to (5.8.18) the weak implicit Euler scheme has the form

$$\begin{aligned}
P_{k+1} &= P_k + h\left(Q_{k+1} - Q_{k+1}^3 - \nu P_{k+1}\right) + h^{1/2}\sigma\xi_k \\
Q_{k+1} &= Q_k + h P_{k+1},
\end{aligned} \tag{5.8.21}$$

where ξ_k are i.i.d. random variables with the law (5.6.2).

Figure 5.8.2 gives results of evaluation of $E\left(Q(T)\right)^2$ for a large T by these two methods. We see that even for such a small step as $h = 0.01$ the implicit Euler method tends to a wrong limit with increasing T, while the quasi-symplectic method gives quite accurate results, e.g., for $h = 0.25$.

Now consider the *explicit* quasi-symplectic method based on the one-step approximation (5.7.7) and on the weak explicit symplectic method (5.6.11)–(5.6.12) with $\alpha = 0$. We take $\bar{P}_{II} = p - h\nu p$ instead of the exact P_{II} again. This method for (5.8.18) is written as

$$\begin{aligned}
P_{k+1} &= (1 - \nu h)\left(P_k + h\left(Q_k - Q_k^3\right) + h^{1/2}\sigma\xi_k\right) \tag{5.8.22} \\
Q_{k+1} &= Q_k + h\left(P_k + h\left(Q_k - Q_k^3\right)\right).
\end{aligned}$$

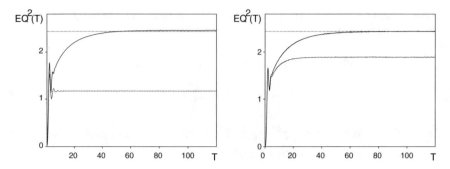

Fig. 5.8.2 The oscillator with cubic restoring force (5.8.18). Behavior of $E(Q(T))^2$ with $P(0) = 0$, $Q(0) = 0$, $\nu = 0.05$, $\sigma = 1$, $h = 0.25$ (left) and $h = 0.01$ (right) on the time interval $t \leq 120$ in the case of the weak implicit Euler method (5.8.21) (dashed line) and the weak quasi-symplectic method (5.8.20) (solid line). The Monte Carlo error is not greater than 0.005 with probability 0.95. The dotted line presents the limit value of $E(Q(T))^2$ as $T \to \infty$ evaluated due to $\int_{-\infty}^{\infty} \int_{-\infty}^{\infty} q^2 \rho(p, q) dp\, dq$ with the invariant measure $\rho(p, q)$ from (5.8.19). This value is equal to 2.435.

Since this quasi-symplectic method is explicit, it is much simpler than (5.8.20). However, for comparatively large h the difference system (5.8.22) has unstable behavior (e.g., for ν, σ as in Fig. 5.8.2 and $h = 0.2$). Most likely, for all sufficiently small h the system (5.8.22) acquires stable behavior (of course, this assertion requires further investigation). For instance, $E\bar{Q}^2(T)$ obtained by (5.8.22) for ν, σ as in Fig. 5.8.2 and $h = 0.1$ visually coincides with the results obtained by the implicit quasi-symplectic method (5.8.20). Thus, even an explicit quasi-symplectic method can effectively be used for solution of Langevin equations on long time intervals in contrast to the implicit Euler method which is more complicated than (5.8.22).

In Appendix A.2 of [314] an implementation of the weak implicit quasi-symplectic method (5.8.20) and the weak implicit Euler method (5.8.21) is considered and the program is given.

5.9 Rigid Body Dynamics: Langevin and Stochastic Gradient Equations

In this section we apply geometric integration ideas to stochastic rigid body dynamics. The presented results are mainly based on [69–71].

Classical molecular dynamics simulation of an isolated system aims at sampling states from a microcanonical (NVE) ensemble with a fixed number of particles N, a fixed volume V, and a fixed total energy E of the system. In practice (e.g., in a laboratory environment) it is often preferable to fix the temperature T of the simulated system and hence to sample from the NVT ensemble (Gibbs measure). In order to sample from the canonical ensemble, the molecular dynamics equations of motion are modified by introducing the interaction of the system with a so-called "thermostat".

There exist a large variety of approaches for introducing such a thermostat, which can be roughly divided into two categories: deterministic (e.g., Nosé-Hoover thermostats) and stochastic (e.g., Brownian dynamics and Langevin equations), depending on whether the resulting equations of motion contain a random component (see [230, 401, 410] and references therein). The use of Brownian dynamics and Langevin equations has the advantage of thermostatting all degrees of freedom of the system independently, without having to rely on efficient energy exchange within the system as in the case of deterministic thermostats. Good energy exchange is particularly difficult to achieve between components of the system which evolve on different time scales (i.e., fast-slow separation of degrees of freedom). Furthermore, for SDE-based thermostats, it is easier to ensure and prove ergodicity (with the Gibbsian invariant measure).

In this section we present Langevin-type equations for the rigid body dynamics in the quaternion representation and effective second-order quasi-symplectic numerical integrators for their simulation [69, 70]. We also consider a Brownian dynamics model (stochastic gradient system) suitable for thermostatted rigid body dynamics together with a first-order geometric integrator.

In Sect. 5.9.1 we recall the Hamiltonian system for rigid body dynamics in the quaternion representation from [268], based on which Langevin (Sect. 5.9.2) and gradient thermostats (Sect. 5.9.3) are introduced. Geometric integrators for these stochastic systems are presented in Sect. 5.9.4. The thermostats and numerical integrators of this section were extensively tested on the TIP4P model of water in [69, 70]. Langevin C integrator of Sect. 5.9.4 is implemented in Large-scale Atomic/Molecular Massively Parallel Simulator (LAMMPS) software library. We also note that in [71] Langevin thermostats of Sect. 5.9.2 were extended to Langevin-type equations that describe the rotational and translational motion of rigid bodies interacting through conservative and non-conservative forces and hydrodynamic coupling and are suitable for modelling e.g. colloidal suspensions or solvated macromolecules [410]. For these Langevin-type equations, a weak second-order geometric integrator was constructed in [71] based on the Langevin C scheme of Sect. 5.9.4.

5.9.1 Hamiltonian System

Consider n rigid three-dimensional bodies with center-of-mass coordinates $\mathbf{r} = ((r^1)^\mathsf{T}, \ldots, (r^n)^\mathsf{T})^\mathsf{T} \in \mathbf{R}^{3n}$, $r^j = (r^j_1, r^j_2, r^j_3)^\mathsf{T} \in \mathbf{R}^3$, and the rotational coordinates in the quaternion representation $\mathbf{q} = ((q^1)^\mathsf{T}, \ldots, (q^n)^\mathsf{T})^\mathsf{T}$, $q^j = (q^j_0, q^j_1, q^j_2, q^j_3)^\mathsf{T}$, such that $|q^j| = 1$, i.e., $q^j \in \mathbb{S}^3$, which is the three-dimensional unit sphere with center at the origin. Following [268] (see also [69–71, 158] and references therein), the Hamiltonian for a system of rigid bodies can be written as

$$H(\mathbf{r}, \mathbf{p}, \mathbf{q}, \pi) = \frac{\mathbf{p}^\mathsf{T}\mathbf{p}}{2m} + \sum_{j=1}^{n}\sum_{l=1}^{3} V_l(q^j, \pi^j) + U(\mathbf{r}, \mathbf{q}), \qquad (5.9.1)$$

where $\mathbf{p} = ((p^1)^\mathsf{T}, \ldots, (p^n)^\mathsf{T})^\mathsf{T} \in \mathbb{R}^{3n}$, $p^j = (p_1^j, p_2^j, p_3^j)^\mathsf{T} \in \mathbb{R}^3$, are the center-of-mass momenta conjugate to $\mathbf{r}$; $\boldsymbol{\pi} = ((\pi^1)^\mathsf{T}, \ldots, (\pi^n)^\mathsf{T})^\mathsf{T}$, $\pi^j = (\pi_0^j, \pi_1^j, \pi_2^j, \pi_3^j)^\mathsf{T}$ are the angular momenta conjugate to $\mathbf{q}$ such that $q^{j\mathsf{T}}\pi^j = 0$, i.e., $\pi^j \in T_{q^j}\mathbb{S}^3$, which is the tangent space of $\mathbb{S}^3$ at q^j; and $U(\mathbf{r}, \mathbf{q})$ is the potential energy. The second term in (5.9.1) represents the rotational kinetic energy with

$$V_l(q, \pi) = \frac{1}{8I_l}\left[\pi^\mathsf{T} S_l q\right]^2, \quad l = 1, 2, 3, \tag{5.9.2}$$

where the three constant 4-by-4 matrices S_l are

$$S_1 = \begin{bmatrix} 0 & -1 & 0 & 0 \\ 1 & 0 & 0 & 0 \\ 0 & 0 & 0 & 1 \\ 0 & 0 & -1 & 0 \end{bmatrix}, \quad S_2 = \begin{bmatrix} 0 & 0 & -1 & 0 \\ 0 & 0 & 0 & -1 \\ 1 & 0 & 0 & 0 \\ 0 & 1 & 0 & 0 \end{bmatrix}, \quad S_3 = \begin{bmatrix} 0 & 0 & 0 & -1 \\ 0 & 0 & 1 & 0 \\ 0 & -1 & 0 & 0 \\ 1 & 0 & 0 & 0 \end{bmatrix},$$

and I_l are the principal moments of inertia of the molecule. We also introduce $S_0 = \mathrm{diag}(1, 1, 1, 1)$, the matrix $D = \mathrm{diag}(0, 1/I_1, 1/I_2, 1/I_3)$, and the orthogonal matrix:

$$S(q) = [S_0 q, S_1 q, S_2 q, S_3 q] = \begin{bmatrix} q_0 & -q_1 & -q_2 & -q_3 \\ q_1 & q_0 & -q_3 & q_2 \\ q_2 & q_3 & q_0 & -q_1 \\ q_3 & -q_2 & q_1 & q_0 \end{bmatrix}.$$

Note that $q^\mathsf{T} S(q) = (1, 0, 0, 0)$ and $q^\mathsf{T} S(q)D = (0, 0, 0, 0)$. The rotational kinetic energy of a molecule can be expressed in terms of the matrices D and S as follows:

$$\sum_{l=1}^{3} V_l(q, \pi) = \frac{1}{8}\pi^\mathsf{T} S(q)DS^\mathsf{T}(q)\pi.$$

We have

$$\sum_{l=1}^{3} \nabla_\pi V_l(q, \pi) = \frac{1}{4}\sum_{l=1}^{3}\frac{1}{I_l} S_l q \left[S_l q\right]^\mathsf{T} \pi \tag{5.9.3}$$

$$= \frac{1}{4}S(q)DS^\mathsf{T}(q)\pi,$$

$$\sum_{l=1}^{3} \nabla_q V_l(q, \pi) = -\frac{1}{4}\sum_{l=1}^{3}\frac{1}{I_l}\left[\pi^\mathsf{T} S_l q\right] S_l \pi.$$

We assume that the potential energy $U(\mathbf{r}, \mathbf{q})$ is a sufficiently smooth function. Let $f^j(\mathbf{r}, \mathbf{q}) = -\nabla_{r^j} U(\mathbf{r}, \mathbf{q}) \in \mathbb{R}^3$, the net force acting on molecule j, and $F^j(\mathbf{r}, \mathbf{q}) = -\tilde{\nabla}_{q^j} U(\mathbf{r}, \mathbf{q}) \in T_{q^j}\mathbb{S}^3$, which is the rotational force. Note that, while ∇_{r^j} is the gradient in the Cartesian coordinates in $\mathbf{R}^3$, $\tilde{\nabla}_{q^j}$ is the directional derivative tangent to the three dimensional sphere $\mathbb{S}^3$ implying that

$$\mathbf{q}^\mathsf{T} \tilde{\nabla}_{q^j} U(\mathbf{r}, \mathbf{q}) = 0. \tag{5.9.4}$$

5.9.2 Langevin Equations

In [70] the following Langevin thermostat for rigid body dynamics was proposed

$$dR^j = \frac{P^j}{m} dt, \quad R^j(0) = r^j, \tag{5.9.5}$$

$$dP^j = f^j(\mathbf{R}, \mathbf{Q}) dt$$

$$- \gamma P^j dt + \sqrt{\frac{2m\gamma}{\beta}} dw^j(t), \quad P^j(0) = p^j,$$

$$dQ^j = \frac{1}{4} S(Q^j) D S^\mathsf{T}(Q^j) \Pi^j dt, \quad Q^j(0) = q^j, \tag{5.9.6}$$

$$|q^j| = 1,$$

$$d\Pi^j = \frac{1}{4} \sum_{l=1}^{3} \frac{1}{I_l} \left((\Pi^j)^\mathsf{T} S_l Q^j \right) S_l \Pi^j dt + F^j(\mathbf{R}, \mathbf{Q}) dt$$

$$- \Gamma J(Q^j) \Pi^j dt + \sqrt{\frac{2M\Gamma}{\beta}} \sum_{l=1}^{3} S_l Q^j d W_l^j(t),$$

$$\Pi^j(0) = \pi^j, \quad (q^j)^\mathsf{T} \pi^j = 0, \quad j = 1, \ldots, n,$$

where $(\mathbf{w}^\mathsf{T}, \mathbf{W}^\mathsf{T})^\mathsf{T} = ((w^1)^\mathsf{T}, \ldots, (w^n)^\mathsf{T}, (W^1)^\mathsf{T}, \ldots, (W^n)^\mathsf{T})^\mathsf{T}$ is a $(3n + 3n)$-dimensional standard Wiener process with $w^j = (w_1^j, w_2^j, w_3^j)^\mathsf{T}$ and $W^j = (W_1^j, W_2^j, W_3^j)^\mathsf{T}$; $\gamma \geq 0$ and $\Gamma \geq 0$ are the damping coefficients for the translational and rotational motions and $\beta = 1/(k_B T) > 0$ is the inverse temperature. In the above equations we also use

$$J(q) = \frac{M}{4} S(q) D S^\mathsf{T}(q), \quad M = \frac{4}{\sum_{l=1}^{3} \frac{1}{I_l}}. \tag{5.9.7}$$

The part corresponding to translational degrees of freedom (5.9.5) coincides with the Langevin equations (5.8.2). Note that we use a different notation here (in this section on rigid bodies) for the particle positions ($\mathbf{R}$ instead of Q) in comparison with how Langevin and Hamilton equations were written in the previous sections of this chapter.

It is not difficult to show that the thermostat (5.9.5)–(5.9.6) has the following properties:

- The Ito interpretation of the SDEs (5.9.5)–(5.9.6) coincides with its Stratonovich interpretation.
- The solution of (5.9.5)–(5.9.6) preserves the quaternion length

$$|Q^j(t)| = 1, \quad j = 1, \ldots, n. \tag{5.9.8}$$

- The solution of (5.9.5)–(5.9.6) automatically preserves the constraint:

$$\mathbf{Q}^\mathsf{T}(t)\boldsymbol{\Pi}(t) = 0, \quad \text{for all } t \geq 0. \tag{5.9.9}$$

Physically, $Q^j(t)$ are constrained to unit spheres; therefore the momenta $\Pi^j(t)$ should also have three degrees of freedom and satisfy the physical requirement (5.9.9), i.e., $\Pi^j(t) \in T_{Q^j}\mathbb{S}^3$.

- Assume that the solution $X(t) = (\mathbf{R}^\mathsf{T}(t), \mathbf{P}^\mathsf{T}(t), \mathbf{Q}^\mathsf{T}(t), \boldsymbol{\Pi}^\mathsf{T}(t))^\mathsf{T}$ of (5.9.5)–(5.9.6) is an ergodic process [194, 412] on

$$\mathbb{D} = \{x = (\mathbf{r}^\mathsf{T}, \mathbf{p}^\mathsf{T}, \mathbf{q}^\mathsf{T}, \boldsymbol{\pi}^\mathsf{T})^\mathsf{T} \in \mathbf{R}^{14n} :$$
$$|q^j| = 1, (q^j)^\mathsf{T} \pi^j = 0, \quad j = 1, \ldots, n\}.$$

Then it can be shown that the invariant measure of $X(t)$ is Gibbsian with the density $\rho(\mathbf{r}, \mathbf{p}, \mathbf{q}, \boldsymbol{\pi})$ on $\mathbb{D}$:

$$\rho(\mathbf{r}, \mathbf{p}, \mathbf{q}, \boldsymbol{\pi}) \propto \exp(-\beta H(\mathbf{r}, \mathbf{p}, \mathbf{q}, \boldsymbol{\pi})), \tag{5.9.10}$$

which corresponds to the NVT ensemble of rigid bodies, as required.

5.9.3   Stochastic Gradient System

Gradient systems are popular in molecular dynamics for thermostatting translational degrees of freedom [230, 319, 401] (see also references therein). In this section we consider Brownian dynamics for thermostatting rigid bodies, i.e., a stochastic gradient system for center-of-mass and rotational coordinates, which was proposed in [70].

It is easy to verify that

$$\int_{\mathbb{D}_{mom}} \exp(-\beta H(\mathbf{r}, \mathbf{p}, \mathbf{q}, \boldsymbol{\pi}))d\mathbf{p}d\boldsymbol{\pi} \tag{5.9.11}$$

$$\propto \exp(-\beta U(\mathbf{r}, \mathbf{q})) =: \tilde{\rho}(\mathbf{r}, \mathbf{q}),$$

where $(\mathbf{r}^\mathsf{T}, \mathbf{q}^\mathsf{T})^\mathsf{T} \in \mathbb{D}' = \{(\mathbf{r}^\mathsf{T}, \mathbf{q}^\mathsf{T})^\mathsf{T} \in \mathbb{R}^{7n} : |q^j| = 1\}$ and the domain of conjugate momenta $\mathbb{D}_{\mathrm{mom}} = \{(\mathbf{p}^\mathsf{T}, \boldsymbol{\pi}^\mathsf{T})^\mathsf{T} \in \mathbb{R}^{7n} : \mathbf{q}^\mathsf{T}\boldsymbol{\pi} = 0\}$.

Introduce the gradient system in the form of Stratonovich SDEs:

$$d\mathbf{R} = \frac{v}{m}\mathbf{f}(\mathbf{R}, \mathbf{Q})dt + \sqrt{\frac{2v}{m\beta}}d\mathbf{w}(t), \quad \mathbf{R}(0) = \mathbf{r}, \tag{5.9.12}$$

$$dQ^j = \frac{\Upsilon}{M}F^j(\mathbf{R}, \mathbf{Q})dt + \sqrt{\frac{2\Upsilon}{M\beta}}\sum_{l=1}^{3} S_l Q^j \circ dW_l^j(t), \tag{5.9.13}$$

$$Q^j(0) = q^j, \quad |q^j| = 1, \quad j = 1, \ldots, n,$$

where "$\circ$" indicates the Stratonovich form of the SDEs, parameters $v > 0$ and $\Upsilon > 0$ control the speed of evolution of the gradient system (5.9.12)–(5.9.13), $\mathbf{f} = ((f^1)^\mathsf{T}, \ldots, (f^n)^\mathsf{T})^\mathsf{T}$ and the rest of the notation is as in (5.9.5)–(5.9.6). Note that, unlike the case of (5.9.5)–(5.9.6), the Stratonovich and Ito interpretations of the SDEs (5.9.12)–(5.9.13) do not coincide.

This gradient thermostat possesses the following properties.

- As in the case of (5.9.5)–(5.9.6), the solution of (5.9.12)–(5.9.13) preserves the quaternion length.
- Assume that the solution $X(t) = (\mathbf{R}^\mathsf{T}(t), \mathbf{Q}^\mathsf{T}(t))^\mathsf{T} \in \mathbb{D}'$ of (5.9.12)–(5.9.13) is an ergodic process [194]. Then, by the usual means of the stationary Fokker-Planck equation, one can show that its invariant measure is Gibbsian with the density $\tilde{\rho}(\mathbf{r}, \mathbf{q})$ from (5.9.11).

If a thermostat is used only to control the temperature of a system, and not to mimic the dynamical effects of an implicit solvent, the Langevin thermostat (5.9.5)–(5.9.6) is suitable for computing both dynamical and static quantities (provided the damping coefficients are relatively small). By contrast, the gradient thermostat (5.9.12)–(5.9.13) can be used to compute only static quantities [319, 401] in such systems. Note that in the part corresponding to translational degrees of freedom (5.9.12) coincides with the usual stochastic gradient system (see (5.11.32) considered later in this section and also [230, 232, 319]). If we put $F^j = 0$, then $Q^j(t)$ from (5.9.13) becomes the Wiener process on a three-dimensional sphere $\mathbb{S}^3$ [56, 92, 177].

5.9.4 Numerical Methods

In this section we present geometric integrators for the Langevin thermostat (5.9.5)–(5.9.6) (Sects. 5.9.4.1–5.9.4.3) and for the gradient thermostat (5.9.12)–(5.9.13) (Sect. 5.9.4.4) which were proposed in [70]. The numerical methods for the Langevin thermostat are based on the splitting technique. It was observed in [229] that numerical schemes based on different splittings might have considerably different properties. Roughly speaking, Langevin thermostat SDEs (5.9.5)–(5.9.6) consist of three components: Hamiltonian + damping + noise. The integrator Langevin A is based on the splitting of (5.9.5)–(5.9.6) into a system close to a stochastic Hamiltonian system (Hamiltonian + noise) and the deterministic system of linear differential equations corresponding to the Langevin damping. It is an analog of (5.8.1). The other two schemes, Langevin B and C, are based on splitting of (5.9.5)–(5.9.6) into the deterministic Hamiltonian system and the Ornstein-Uhlenbeck process (damping + noise) using their different concatenations (cf. the splitting (5.8.3)–(5.8.4) for the Langevin equations (5.8.2) used for translational degrees of freedom). All three schemes are of weak order 2 and use one evaluation of forces per step. The numerical method for the gradient thermostat (5.9.12)–(5.9.13) also uses one force evaluation per step, but it is of weak order 1. To preserve the length of quaternions in the case of numerical integration of (5.9.12)–(5.9.13), ideas of Lie-group type integrators for differential equations on manifolds [154] are used.

In what follows we assume that (5.9.5)–(5.9.6) and (5.9.12)–(5.9.13) have to be solved on a time interval $[0, T]$ and, for simplicity, we use a uniform time discretization with the step $h = T/N$.

In Sects. 5.9.4.1–5.9.4.3 we use the mapping $\Psi_{t,l}(q, \pi): (q, \pi) \mapsto (Q, \Pi)$ defined by (see [69–71, 268]):

$$Q = \cos(\chi_l t)q + \sin(\chi_l t)S_l q ,$$
$$\Pi = \cos(\chi_l t)\pi + \sin(\chi_l t)S_l \pi , \tag{5.9.14}$$

where

$$\chi_l = \frac{1}{4I_l}\pi^\mathsf{T} S_l q .$$

We also introduce a composite map

$$\Psi_t := \Psi_{t/2,3} \circ \Psi_{t/2,2} \circ \Psi_{t,1} \circ \Psi_{t/2,2} \circ \Psi_{t/2,3} , \tag{5.9.15}$$

where "$\circ$" denotes function composition, i.e., $(g \circ f)(x) := g(f(x))$.

We note that Langevin A, B and C degenerate to the deterministic integrator from [268] when the Langevin thermostat (5.9.5)–(5.9.6) degenerates to the deterministic Hamiltonian system (i.e., when the thermostat is "switched off"). We note that the deterministic integrator of [268] is symplectic and it preserves the length of quaternions automatically.

5.9.4.1 Geometric Integrator Langevin A

The geometric integrator of this section is based on splitting the Langevin system (5.9.5)–(5.9.6) into the system (5.9.5)–(5.9.6) without the damping term, which is close to a stochastic Hamiltonian system, and the deterministic system of linear differential equations

$$\dot{\mathbf{p}} = -\gamma \mathbf{p}$$
$$\dot{\pi}^j = -\Gamma J(q^j)\pi^j, \ j = 1, \ldots, n. \tag{5.9.16}$$

The resulting numerical method has the form:

$$\mathbf{P}_0 = \mathbf{p}, \ \mathbf{R}_0 = \mathbf{r}, \tag{5.9.17}$$
$$\mathbf{Q}_0 = \mathbf{q} \text{ with } |q^j| = 1, j = 1, \ldots, n,$$
$$\boldsymbol{\Pi}_0 = \boldsymbol{\pi} \text{ with } \mathbf{q}^\mathsf{T}\boldsymbol{\pi} = 0,$$
$$\mathcal{P}_{1,k} = e^{-\gamma\frac{h}{2}}\mathbf{P}_k,$$
$$\Pi^j_{1,k} = e^{-\Gamma J(Q^j_k)\frac{h}{2}}\Pi^j_k, \ j = 1, \ldots, n,$$

$$\mathcal{P}_{2,k} = \mathcal{P}_{1,k} + \frac{h}{2}\mathbf{f}(\mathbf{R}_k, \mathbf{Q}_k) + \frac{\sqrt{h}}{2}\sqrt{\frac{2m\gamma}{\beta}}\xi_k$$

$$\Pi^j_{2,k} = \Pi^j_{1,k} + \frac{h}{2}F^j(\mathbf{R}_k, \mathbf{Q}_k) + \frac{\sqrt{h}}{2}\sqrt{\frac{2M\Gamma}{\beta}}\sum_{l=1}^{3}S_l Q_k \eta^{j,l}_k,$$

$$j = 1, \ldots, n,$$

$$\mathbf{R}_{k+1} = \mathbf{R}_k + \frac{h}{m}\mathcal{P}_{2,k},$$

$$(Q^j_{k+1}, \Pi^j_{3,k}) = \Psi_h(Q^j_k, \Pi^j_{2,k}), \ j = 1, \ldots, n,$$

$$\Pi^j_{4,k} = \Pi^j_{3,k} + \frac{h}{2}F^j(\mathbf{R}_{k+1}, \mathbf{Q}_{k+1}) + \frac{\sqrt{h}}{2}\sqrt{\frac{2M\Gamma}{\beta}}\sum_{l=1}^{3}S_l Q_{k+1}\eta^{j,l}_k, j = 1, \ldots, n,$$

$$\mathcal{P}_{3,k} = \mathcal{P}_{2,k} + \frac{h}{2}\mathbf{f}(\mathbf{R}_{k+1}, \mathbf{Q}_{k+1}) + \frac{\sqrt{h}}{2}\sqrt{\frac{2m\gamma}{\beta}}\xi_k,$$

$$\mathbf{P}_{k+1} = e^{-\gamma\frac{h}{2}}\mathcal{P}_{3,k},$$
$$\Pi^j_{k+1} = e^{-\Gamma J(Q^j_{k+1})\frac{h}{2}}\Pi^j_{4,k}, \ j = 1, \ldots, n,$$
$$k = 0, \ldots, N-1,$$

where $\xi_k = (\xi_{1,k}, \ldots, \xi_{3n,k})^\mathsf{T}$ and $\eta^{j,l}_k$, $l = 1, 2, 3$, $j = 1, \ldots, n$, with their components being i.i.d. random variables with the same probability distribution

$$P(\theta = 0) = 2/3, \ P(\theta = \pm\sqrt{3}) = 1/6. \tag{5.9.18}$$

Proposition 5.9.1 *The numerical scheme (5.9.17)–(5.9.18) for (5.9.5)–(5.9.6) is quasi-symplectic, it preserves the structural properties (5.9.8) and (5.9.9) and it is of weak order two.*

We note that one can choose ξ_k and $\eta_k^{j,l}$, $l = 1, 2, 3$, $j = 1, \ldots, n$, so that their components are i.i.d. Gaussian random variables with zero mean and unit variance. In this case the weak order of the scheme remains second as for the simple discrete distribution (5.9.18). Let us remark in passing that in the case of Gaussian random variables the above scheme also converges in the mean-square sense with order one. Further, the numerical scheme (5.9.17)–(5.9.18) becomes of the form (5.8.1) (see Theorem 5.8.3) when the Langevin system (5.9.5)–(5.9.6) does not have rotational degrees of freedom, i.e. when it coincides with (5.8.2).

We also note that explicit evaluation of a matrix exponent is not required as we have

$$e^{-\Gamma J(q)\frac{h}{2}} = S(q)e^{-\frac{\Gamma M h}{8}D}S^{\mathsf{T}}(q) = \sum_{l=1}^{3}e^{-\frac{\Gamma M h}{8 I_l}}S_l q\,[S_l q]^{\mathsf{T}}. \tag{5.9.19}$$

5.9.4.2   Geometric Integrator Langevin B

The geometric integrator of this section is based on the following splitting:

$$d\mathbf{P}_I = -\gamma\mathbf{P}_I\,dt + \sqrt{\frac{2m\gamma}{\beta}}d\mathbf{w}(t),$$

$$d\Pi_I^j = -\Gamma J(q)\Pi_I^j dt + \sqrt{\frac{2M\Gamma}{\beta}}\sum_{l=1}^{3}S_l q\,dW_l^j(t); \tag{5.9.20}$$

$$d\mathbf{R}_{II} = \frac{\mathbf{P}_{II}}{m}\,dt$$

$$d\mathbf{P}_{II} = \mathbf{f}(\mathbf{R}_{II}, \mathbf{Q}_{II})dt,$$

$$dQ_{II}^j = \frac{1}{4}S(Q_{II}^j)DS^{\mathsf{T}}(Q_{II}^j)\Pi_{II}^j dt,$$

$$d\Pi_{II}^j = F^j(\mathbf{R}_{II}, \mathbf{Q}_{II})dt + \frac{1}{4}\sum_{l=1}^{3}\frac{1}{I_l}\left[(\Pi_{II}^j)^{\mathsf{T}}S_l Q_{II}^j\right]S_l\Pi_{II}^j dt, \tag{5.9.21}$$

$$j = 1, \ldots, n.$$

The SDEs (5.9.20) have the exact solution:

$$\mathbf{P}_I(t) = \mathbf{P}_I(0)e^{-\gamma t} + \sqrt{\frac{2m\gamma}{\beta}}\int_0^t e^{-\gamma(t-s)}d\mathbf{w}(s),$$

$$\Pi_I^j(t) = e^{-\Gamma J(q)t}\Pi_I^j(0) + \sqrt{\frac{2M\Gamma}{\beta}}\sum_{l=1}^{3}\int_0^t e^{-\Gamma J(q)(t-s)}S_l q\,dW_l^j(s). \tag{5.9.22}$$

To construct a method based on the splitting (5.9.20)–(5.9.21), we take half a step of (5.9.20) using (5.9.22), one step of the symplectic method for (5.9.21) from [268], and again half a step of (5.9.20). Note that Langevin C in the next section uses the same splitting (5.9.20)–(5.9.21) but a different concatenation.

The vector $\int_0^t e^{-\Gamma J(q)(t-s)} S_l q \, dW_l^j(s)$ in (5.9.22) is Gaussian with zero mean and covariance

$$C_l(t; q) = \int\limits_0^t e^{-\Gamma J(q)(t-s)} S_l q (S_l q)^\mathsf{T} e^{-\Gamma J(q)(t-s)} ds.$$

The covariance matrix $C(t; q)$ of the sum $\sum_{l=1}^3 \int_0^t e^{-\Gamma J(q)(t-s)} S_l q \, dW_l^j(s)$ is equal to

$$C(t; q) = \frac{2}{M\Gamma} S(q) \Lambda_C(t; \Gamma) S^\mathsf{T}(q),$$

where

$$\Lambda_C(t; \Gamma) = \mathrm{diag}(0, I_1(1 - \exp(-M\Gamma t/(2I_1))), I_2(1 - \exp(-M\Gamma t/(2I_2))),$$
$$I_3(1 - \exp(-M\Gamma t/(2I_3)))).$$

If we introduce a 4×3-dimensional matrix $\sigma(t, q)$ such that

$$\sigma(t; q) \sigma^\mathsf{T}(t; q) = C(t; q), \tag{5.9.23}$$

e.g., $\sigma(t; q)$ with the columns

$$\sigma_l(t; q) = \sqrt{\frac{2}{M\Gamma} I_l \left(1 - e^{-\frac{M\Gamma t}{2I_l}}\right)} S_l q,$$

$l = 1, 2, 3$, then the expression for $\Pi_l^j(t)$ in (5.9.22) can be written as

$$\Pi_l^j(t) = e^{-\Gamma J(q)t} \Pi_l^j(0) + \sqrt{\frac{2M\Gamma}{\beta}} \sum_{l=1}^3 \sigma_l(t; q) \chi_l^j,$$

where χ_l^j are independent Gaussian random variables with zero mean and unit variance.

Using the above calculations and analogous procedures for the linear momenta, we obtain the following quasi-symplectic scheme for (5.9.5)–(5.9.6):

$$\mathbf{P}_0 = \mathbf{p}, \quad \mathbf{R}_0 = \mathbf{r}, \quad \mathbf{Q}_0 = \mathbf{q}, \quad |q^j| = 1, \quad j = 1, \ldots, n$$
$$\Pi_0 = \pi, \quad \mathbf{q}^\mathsf{T}\pi = 0$$

$$\mathcal{P}_{1,k} = \mathbf{P}_k e^{-\gamma h/2} + \sqrt{\frac{m}{\beta}(1 - e^{-\gamma h})}\xi_k$$

$$\Pi_{1,k}^j = e^{-\Gamma J(Q_k^j)\frac{h}{2}}\Pi_k^j + \sqrt{\frac{4}{\beta}}\sum_{l=1}^{3}\sqrt{I_l\left(1 - e^{-\frac{M\Gamma h}{4I_l}}\right)}S_l Q_k^j \eta_k^{j,l}, \quad j = 1, \ldots, n,$$

$$\tag{5.9.24}$$

$$\mathcal{P}_{2,k} = \mathcal{P}_{1,k} + \frac{h}{2}\mathbf{f}(\mathbf{R}_k, \mathbf{Q}_k),$$

$$\Pi_{2,k}^j = \Pi_{1,k}^j + \frac{h}{2}F^j(\mathbf{R}_k, \mathbf{Q}_k), \quad j = 1, \ldots, n,$$

$$\mathbf{R}_{k+1} = \mathbf{R}_k + \frac{h}{m}\mathcal{P}_{2,k},$$

$$(Q_{k+1}^j, \Pi_{3,k}^j) = \Psi_h(Q_k^j, \Pi_{2,k}^j), \quad j = 1, \ldots, n,$$

$$\Pi_{4,k}^j = \Pi_{3,k}^j + \frac{h}{2}F^j(\mathbf{R}_{k+1}, \mathbf{Q}_{k+1}), \quad j = 1, \ldots, n,$$

$$\mathcal{P}_{3,k} = \mathcal{P}_{2,k} + \frac{h}{2}\mathbf{f}(\mathbf{R}_{k+1}, \mathbf{Q}_{k+1}),$$

$$\mathbf{P}_{k+1} = \mathcal{P}_{3,k}e^{-\gamma h/2} + \sqrt{\frac{m}{\beta}(1 - e^{-\gamma h})}\zeta_k,$$

$$\Pi_{k+1}^j = e^{-\Gamma J(Q_{k+1}^j)\frac{h}{2}}\Pi_{4,k}^j + \sqrt{\frac{4}{\beta}}\sum_{l=1}^{3}\sqrt{I_l\left(1 - e^{-\frac{M\Gamma h}{4I_l}}\right)}S_l Q_{k+1}^j \varsigma_k^{j,l},$$

$$j = 1, \ldots, n,$$
$$k = 0, \ldots, N - 1,$$

where $\xi_k = (\xi_{1,k}, \ldots, \xi_{3n,k})^\mathsf{T}$, $\zeta_k = (\zeta_{1,k}, \ldots, \zeta_{3n,k})^\mathsf{T}$ and $\eta_k^j = (\eta_{1,k}^j, \ldots, \eta_{3,k}^j)^\mathsf{T}$, $\varsigma_k^j = (\varsigma_{1,k}^j, \ldots, \varsigma_{3,k}^j)^\mathsf{T}$, $j = 1, \ldots, n$, with their components being i.i.d. random variables with the same law (5.9.18).

Proposition 5.9.2 *The numerical scheme (5.9.24), (5.9.18) for (5.9.5)–(5.9.6) is quasi-symplectic, it preserves (5.9.8) and (5.9.9) and it is of weak order two.*

We note that in the case of translational degrees of freedom Langevin B coincides with the scheme called 'OBABO' in [229].

5.9.4.3 Geometric Integrator: Langevin C

This integrator is based on the same splitting (5.9.20)–(5.9.21) as Langevin B but using a different concatenation: we take half a step of a symplectic method for (5.9.21), one step of (5.9.20) using (5.9.22), and again half a step of (5.9.21). The resulting method takes the form

$$\mathbf{P}_0 = \mathbf{p}, \quad \mathbf{R}_0 = \mathbf{r}, \quad \mathbf{Q}_0 = \mathbf{q},$$

$$|q^j| = 1, \quad j = 1, \dots, n,$$

$$\mathbf{\Pi}_0 = \boldsymbol{\pi}, \quad \mathbf{q}^\mathsf{T}\boldsymbol{\pi} = 0,$$

$$\mathcal{P}_{1,k} = \mathbf{P}_k + \frac{h}{2}\mathbf{f}(\mathbf{R}_k, \mathbf{Q}_k),$$

$$\Pi_{1,k}^j = \Pi_k^j + \frac{h}{2}F^j(\mathbf{R}_k, \mathbf{Q}_k), \quad j = 1, \dots, n,$$

$$\mathbf{R}_{1,k} = \mathbf{R}_k + \frac{h}{2m}\mathcal{P}_{1,k}, \tag{5.9.25}$$

$$(\mathcal{Q}_{1,k}^j, \Pi_{2,k}^j) = \Psi_{h/2}(\mathcal{Q}_k^j, \Pi_{1,k}^j), \quad j = 1, \dots, n,$$

$$\mathcal{P}_{2,k} = \mathcal{P}_{1,k}e^{-\gamma h} + \sqrt{\frac{m}{\beta}(1 - e^{-2\gamma h})}\xi_k$$

$$\Pi_{3,k}^j = e^{-\Gamma J(\mathcal{Q}_{1,k}^j)h}\Pi_{2,k}^j + \sqrt{\frac{4}{\beta}}\sum_{l=1}^{3}\sqrt{I_l\left(1 - e^{-\frac{M\Gamma h}{2I_l}}\right)}S_l\mathcal{Q}_{1,k}^j\eta_k^{j,l},$$

$$j = 1, \dots, n,$$

$$\mathbf{R}_{k+1} = \mathbf{R}_{1,k} + \frac{h}{2m}\mathcal{P}_{2,k},$$

$$(\mathcal{Q}_{k+1}^j, \Pi_{4,k}^j) = \Psi_{h/2}(\mathcal{Q}_{1,k}^j, \Pi_{3,k}^j), \quad j = 1, \dots, n,$$

$$\mathbf{P}_{k+1} = \mathcal{P}_{2,k} + \frac{h}{2}\mathbf{f}(\mathbf{R}_{k+1}, \mathbf{Q}_{k+1}),$$

$$\Pi_{k+1}^j = \Pi_{4,k}^j + \frac{h}{2}F^j(\mathbf{R}_{k+1}, \mathbf{Q}_{k+1}), \quad j = 1, \dots, n,$$

where $\xi_k = (\xi_{1,k}, \dots, \xi_{3n,k})^\mathsf{T}$ and $\eta_k^j = (\eta_{1,k}^j, \dots, \eta_{3,k}^j)^\mathsf{T}$, $j = 1, \dots, n$, with their components being i.i.d. random variables with the same law (5.9.18).

Proposition 5.9.3 *The numerical scheme (5.9.25) for (5.9.5)–(5.9.6) is quasi-symplectic, it preserves (5.9.8) and (5.9.9) and it is of weak order two.*

We note that in the case of translational degrees of freedom Langevin C coincides with the scheme (5.8.6). In [229] it was shown to be the most efficient scheme among various types of splittings of Langevin equations for systems without rotational degrees of freedom.

5.9.4.4 Numerical Scheme for the Gradient System

To preserve the length of quaternions in the case of numerical integration of the gradient system (5.9.12)–(5.9.13), ideas of Lie-group type integrators for deterministic ordinary differential equations on manifolds are used [52, 154]. The main idea is to rewrite the components Q^j of the solution to (5.9.12)–(5.9.13) in the form $Q^j(t) = \exp(Y^j(t))Q^j(0)$ and then solve numerically the SDEs for the 4×4-matrices $Y^j(t)$. To this end, introduce the 4×4 skew-symmetric matrices:

$$\mathbb{F}_j(\mathbf{r}, \mathbf{q}) = F^j(\mathbf{r}, \mathbf{q})\left(q^j\right)^{\mathsf{T}} - q^j(F^j(\mathbf{r}, \mathbf{q}))^{\mathsf{T}},$$
$$j = 1, \dots, n.$$

Note that $\mathbb{F}_j(\mathbf{r}, \mathbf{q})q^j = F^j(\mathbf{r}, \mathbf{q})$ under $|q^j| = 1$ and the equations (5.9.13) can be written as

$$dQ^j = \frac{\Upsilon}{M}\mathbb{F}_j(\mathbf{R}, \mathbf{Q})Q^j\, dt + \sqrt{\frac{2\Upsilon}{M\beta}}\sum_{l=1}^{3} S_l Q^j \circ dW_l^j(t),$$

$$Q^j(0) = q^j, \quad |q^j| = 1. \tag{5.9.26}$$

We also remark that if $\mathbb{F}_j(\mathbf{r}, \mathbf{q}) = 0$, Q^j are Wiener processes on the three-dimensional sphere [56, 92, 177]. One can show that

$$Y^j(t+h) = h\frac{\Upsilon}{M}\mathbb{F}_j(\mathbf{R}(t), \mathbf{Q}(t)) + \sqrt{\frac{2\Upsilon}{M\beta}}\sum_{l=1}^{3}\left(W_l^j(t+h) - W_l^j(t)\right)S_l$$

$$+ \text{ terms of higher order.}$$

Consequently, we derived the following numerical method for (5.9.12)–(5.9.13):

$$\mathbf{R}_0 = \mathbf{r}, \quad \mathbf{Q}_0 = \mathbf{q}, \quad |q^j| = 1, \quad j = 1, \dots, n, \tag{5.9.27}$$

$$\mathbf{R}_{k+1} = \mathbf{R}_k + h\frac{\upsilon}{m}\mathbf{f}(\mathbf{R}_k, \mathbf{Q}_k) + \sqrt{h}\sqrt{\frac{2\upsilon}{m\beta}}\xi_k,$$

$$Y_k^j = h\frac{\Upsilon}{M}\mathbb{F}_j(\mathbf{R}_k, \mathbf{Q}_k) + \sqrt{h}\sqrt{\frac{2\Upsilon}{M\beta}}\sum_{l=1}^{3}\eta_k^{j,l}S_l,$$

$$Q_{k+1}^j = \exp(Y_k^j)Q_k^j, \quad j = 1, \dots, n,$$

where $\xi_k = (\xi_{1,k}, \dots, \xi_{3n,k})^{\mathsf{T}}$ and $\xi_{i,k}, i = 1, \dots, 3n, \eta_k^{j,l}, l = 1, 2, 3, j = 1, \dots, n,$ are i.i.d. random variables with the same law

$$P(\theta = \pm 1) = 1/2. \tag{5.9.28}$$

Since the matrix Y_k^j is skew symmetric, the exponent $\exp(Y_k^j)$ can be effectively computed using the Rodrigues formula (see Appendix in [70]). The geometric integrator (5.9.27) possesses properties stated in the following proposition.

Proposition 5.9.4 *The numerical scheme (5.9.27) for (5.9.12)–(5.9.13) preserves the length of quaternions, i.e., $|Q_k^j| = 1$, $j = 1, \ldots, n$, for all k, and it is of weak order one.*

Let us remark that in the case of Gaussian random variables the above scheme also converges in the mean-square sense with order $1/2$. It is not difficult to derive a method of mean-square order one for (5.9.12)–(5.9.13), which preserves the length of quaternions.

5.10 Stochastic Landau-Lifshitz Equation

In this section we further illustrate the use of geometric integration ideas for SDEs. Following [265], we consider here the stochastic Landau-Lifshitz equation (SLL) and numerical methods for it which preserve the length of spins. The SLL equation can be used for Atomistic Spin Dynamics (ASD) simulations at constant, finite temperature. As in the previous section, we use here a stochastic thermostat to sample from the canonical ensemble. Dynamics of magnetic materials have been theoretically studied since the work by Landau and Lifshitz [224]. The recent interest in this area has been rapidly growing due to novel applicable fields such as spintronics and laser-induced ultrafast spin dynamics. Consequently, ASD simulations are important from many points of view, and for its success one needs fast and accurate numerical integration of the SSL equations.

In systems of applicable interest the number n of spins is typically of order 10^6. Due to the interactions, one has to solve a system of $3n$ coupled non-linear SDEs. To compute quantities in equilibrium (i.e., with respect to the invariant measure of the SLL equation), this large system should be simulated over long time intervals, usually from 10 fs to 1 ns. This is a challenging computational task. Effective numerical integrators for the SLL equation should be, on the one hand, sufficiently stable and on the other hand very fast. The latter rules out the use of implicit integrators such as the midpoint scheme (1.3.53). Despite its superior stability properties which allows large step sizes, typically 10 fs, the midpoint scheme is slow in practice since the implicitness requires solution of $3n$ non-linear coupled equations at every time step. Langevin spin dynamics simulations could be done using the Heun method (2.2.24), which has the advantage of being fast in terms of the number of operations per time step. However, this method has poor stability properties requiring a relatively small step size, typically ranging from 0.01 fs to 1 fs, depending on whether it is complimented by a projection of spin length to the length of 1 at every time step. We also note that since the accuracy of the first-principle magnetic interaction parameters is limited to 10%, the accuracy of numerical methods is, to some extent, less important than their stability (in the sense of the ability to use larger step sizes for long time simulations). Hence, both the standard implicit and explicit numerical integrators are not optimal for ASD. Also, ASD simulations are often used to study systems with different interactions and/or different symmetries. Therefore, in addition one should require from numerical integrators for ASD to be simple and universal in their implementation.

Geometric integrators for SSL satisfying the above requirements were constructed in [265] and are presented in this section. These methods exploit the idea of semi-implicitness, which was also the basis for constructing fast and accurate numerical schemes for the deterministic Landau-Lifshitz equation (see e.g. [20] and references therein).

In Sect. 5.10.1, we formulate the SLL equation. In Sect. 5.10.2, the two semi-implicit methods (SIA and SIB) are considered and their properties are discussed. SIA and SIB were tested in numerical experiments in [265]. In particular, it was shown that, due to the enhanced stability, SIB allows time steps by a factor of $10 \div 10^3$ larger than the standard Heun method (2.2.24). Both SIA and SIB are included in UppASD software library.

5.10.1   Model

The deterministic Landau-Lifshitz equation in dimensionless variables can be written in the form

$$\frac{dX^i}{dt} = -X^i \times B^i(\mathbf{X}) - \alpha X^i \times [X^i \times B^i(\mathbf{X})], \quad i = 1, \ldots, n, \qquad (5.10.1)$$

where n is the number of spins, $X^i = (X_x^i, X_y^i, X_z^i)^\mathsf{T}$ are three-dimensional column-vectors representing unit spin vectors and $\mathbf{X} = ((X^1)^\mathsf{T}, \ldots, (X^n)^\mathsf{T})^\mathsf{T}$ is a $3n$-dimensional column-vector formed by the X^i; B^i is the effective field acting on spin i; $\alpha \geq 0$ is the damping parameter. In (5.10.1) the time is normalized by the precession frequency $\omega_{\hat{B}} = \gamma \hat{B}$, where $\hat{B}$ is some reference magnetic field strength, and the effective field $B = ((B^1)^\mathsf{T}, \ldots, (B^n)^\mathsf{T})^\mathsf{T}$ is also normalized by $\hat{B}$ and is given by

$$B(\mathbf{x}) = -\nabla H(\mathbf{x}), \qquad (5.10.2)$$

where H is the Hamiltonian of the problem. Then

$$B^i(\mathbf{x}) = -\nabla_i H(\mathbf{x}),$$

where ∇_i is the gradient with respect to the Cartesian components of the effective magnetic field acting on spin i. Note that in this section the multiplication $\times$ means the vector product.

For ASD, the most important contributions to the Hamiltonian are the Heisenberg exchange for the interaction between the spins H_{ex}, the Zeeman energy for the interaction with an external field H_{ext}, and the uniaxial anisotropy H_{ani} defining a preferential direction of the spins:

$$H = H_{ex} + H_{ext} + H_{ani}, \qquad (5.10.3)$$

where

$$H_{ex}(\mathbf{x}) = -\sum_{i \neq j} J_{ij} x^i x^j, \quad H_{ext}(\mathbf{x}) = -B_0 \sum_i x^i, \quad H_{ani}(\mathbf{x}) = K \sum_i (x^i e_K)^2.$$

Here J_{ij} are the exchange parameters, B_0 is the uniform external field, K is the strength of the anisotropy, and e_K is a unit vector that defines the anisotropy axis. Note that with these contributions to the Hamiltonian the effective fields B^i are linear in x. In realistic materials usually $|J_{ij}| \gg |B_0| \gg |K|$. For the exchange parameters themselves, typically $J_{i(i+1)} \gg J_{i(i+j)}$, $j > 1$, i.e., all spins interact with each other but the nearest-neighbor interactions dominate. Since all the spins interact, the equation (5.10.1) involves simultaneous solution of a $3n$ system of non-linear equations. Due to the interactions between the spins, each effective field B^i is time-dependent. The time-dependence of the effective field is usually considered as the main source of instability in the numerical integration.

In order to perform spin dynamics at finite temperature, fluctuations are included according to the Brownian motion approach for spins by adding fluctuating torques to the equation (5.10.1) (see e.g. [216]). The stochastic Landau-Lifshitz (SLL) equation is then given by

$$dX^i = X^i \times a_i(\mathbf{X})dt + X^i \times \sigma(X^i) \circ dW^i(t), \qquad (5.10.4)$$
$$X^i(0) = x_0^i, \quad |x_0^i| = 1, \quad i = 1, \dots, n,$$

where $W^i(t) = (W_x^i(t), W_y^i(t), W_z^i(t))^\mathsf{T}$, $i = 1, \dots, n$; $W_x^i(t)$, $W_y^i(t)$, $W_z^i(t)$, $i = 1, \dots, n$, are independent standard Wiener processes; $a_i(\mathbf{x})$, $\mathbf{x} \in \mathbf{R}^{3n}$, are three-dimensional column-vectors defined by

$$a_i(\mathbf{x}) = -B^i(\mathbf{x}) - \alpha x^i \times B^i(\mathbf{x}) ; \qquad (5.10.5)$$

$\sigma(x)$, $x \in \mathbf{R}^3$, is a 3×3-matrix such that

$$\sigma(x)y = -\sqrt{2D}y - \alpha\sqrt{2D}x \times y \qquad (5.10.6)$$

for any $y \in \mathbf{R}^3$, and

$$D = \frac{\alpha}{(1+\alpha^2)} \frac{k_b T}{\hat{X}\hat{B}}, \qquad (5.10.7)$$

with $\hat{X}$ being the (non-normalized) magnetization of each spin. Note that, as usual, the symbol 'o' in (5.10.4) means that the corresponding stochastic integral is interpreted in the Stratonovich sense.

Let us consider some properties of the solution to (5.10.4)–(5.10.7). Its the solution is an ergodic process. The system (5.10.4)–(5.10.7) has noncommutative noise, which makes higher-order methods complicated. Further, the length of each individual spin is a constant of motion, i.e.,

$$|X^i(t)| = 1, \ i = 1, \ldots, n, \ t \geq 0. \tag{5.10.8}$$

Indeed, we have

$$d\frac{1}{2}|X^i|^2 = X^i dX^i = X^i \left[X^i \times a_i(\mathbf{X})\right] dt + X^i \left[X^i \times \sigma(X^i) \circ dW_i(t)\right] = 0.$$

Other general conservation laws of (5.10.4)–(5.10.7) and also of (5.10.1) do not exist. However, when we restrict ourselves to realistic systems, we have the damping coefficient $\alpha \ll 1$. This means that, in practice, solutions of (5.10.4)–(5.10.7) are, in a sense, close to the deterministic solutions of (5.10.1) with $\alpha = 0$. Hence the precessional motion can usually be considered as dominant. In turn, the largest contribution to the precessional motion is due to the exchange interaction. Therefore, it is relevant to examine the conservation laws for $\alpha = 0$. Since the Hamiltonian has no explicit time-dependence, energy is conserved for this case. Further, when only Heisenberg exchange is included we have for the total spin:

$$\sum_i \frac{dX^i}{dt} = \sum_{i \neq j} J_{ij} X^i \times X^j = \sum_{i>j} J_{ij} (X^i \times X^j + X^j \times X^i) = 0 \tag{5.10.9}$$

since $J_{ij} = J_{ji}$. We recall that the orientation of individual spins is time dependent, which makes the effective field acting on each spin time dependent due to the exchange interaction. However, at the same time, the symmetry of the exchange interaction ensures that the total spin is time-independent. Therefore, the conservation of total spin is an important property for stable numerical integration of the exchange interaction. By the same arguments, when an external field is added, the total spin will precess in the external field:

$$\sum_i \frac{dX^i}{dt} = B_0 \times \sum_i X^i. \tag{5.10.10}$$

For this case, the length of the total spin is a constant of motion, as well as the component of the total spin along B_0. Hence the energy is also conserved but the transversal components of the total spin with respect to B_0 oscillate in time. When anisotropy is included, there are no conservation properties associated with the total spin.

5.10.2 Semi-implicit Numerical Methods

In this subsection we consider numerical integrators for the stochastic Landau-Lifshitz equation (5.10.4)–(5.10.6). Throughout we use (for simplicity) a uniform discretization of a time interval $[0, t_*]$ with step size $h = t_*/N$. The value at the initial step is $X_0^i = x_0^i$, $i = 1, \ldots, n$, and X_k^i, $i = 1, \ldots, n$, denotes the approximate solution $X^i(t_k)$, $i = 1, \ldots, n$, to the SLL equation at time t_k, $k = 1, \ldots, N$.

Here we describe two semi-implicit integration schemes, simply called semi-implicit A (SIA) and semi-implicit B (SIB). They are called semi-implicit since they require only to solve n or, $2n$ in the case of the SIB scheme, linear 3×3 systems at each time-step, which can be done analytically. The starting point for derivation of the semi-implicit methods is the midpoint scheme (1.3.53). To deal with implicitness in (1.3.53), we replace $\mathbf{X}_{k+1}$ in the argument of a_i and σ in the midpoint scheme by a predictor $\mathcal{X}_k$. As a consequence, resolving the implicitness at each time step is simplified (in comparison to the midpoint scheme) to solving a linear 3×3 system per spin that is independent of the interactions between the spins. The difference between SIA and SIB is the choice for $\mathcal{X}_k$. Both semi-implicit methods have effectively the same computational cost as explicit schemes.

Similar to the Heun method (2.2.24), for the SIA scheme we take the Euler approximation for the predictor $\mathcal{X}_k$. The SIA method for the SLL equation reads

$$\mathcal{X}_k^i = X_k^i + h X_k^i \times a_i(\mathbf{X}_k) + h^{1/2} X_k^i \times \sigma(X_k^i)\xi_{k+1}^i, \qquad (5.10.11)$$
$$i = 1, \ldots, n,$$

$$X_{k+1}^i = X_k^i + h \frac{X_k^i + X_{k+1}^i}{2} \times a_i \left(\frac{\mathbf{X}_k + \mathcal{X}_k}{2} \right)$$
$$+ h^{1/2} \frac{X_k^i + X_{k+1}^i}{2} \times \sigma \left(\frac{X_k^i + \mathcal{X}_k^i}{2} \right) \xi_{k+1}^i, \; i = 1, \ldots, n,$$
$$k = 1, \ldots, N,$$

where $\xi_{k+1}^i = \left(\xi_{k+1}^{i,1}, \xi_{k+1}^{i,2}, \xi_{k+1}^{i,3} \right)^{\mathsf{T}}$; $\xi_k^{i,j}$, $j = 1, 2, 3$, $i = 1, \ldots, n$, $k = 1, \ldots, N$, are i.i.d. random variables which can be distributed according to, e.g., $P(\xi_k^{i,j} = \pm 1) = 1/2$. Alternatively, we can choose $\xi_k^{i,j}$ being distributed as the ξ_h defined as in Sect. 1.3.3 (see (1.3.29)): for $\zeta \sim \mathcal{N}(0, 1)$ define

$$\xi_h = \begin{cases} \zeta, & |\zeta| \le A_h, \\ A_h, & \zeta > A_h, \\ -A_h, & \zeta < -A_h, \end{cases} \qquad (5.10.12)$$

where $A_h = \sqrt{2|\ln h|}$.

SIA can be viewed as a second iteration for the implicit equation due to the midpoint scheme (1.3.53). As zero approximation of $\mathbf{X}_{k+1}$, we took $\mathbf{X}_k$ and then the second iteration was constructed so that the length of individual spins is preserved.

One can see that the first iteration (or in other words the prediction step) of SIA does not preserve the spin length. In contrast, the SIB method keeps the spin-length conserving the midpoint scheme's structure at both iterations and, according to numerical tests of [265], this modification is important for the performance of the semi-implicit schemes.

The SIB method for the SLL equation reads

$$\mathcal{X}_k^i = X_k^i + h \frac{X_k^i + \mathcal{X}_k^i}{2} \times a_i(\mathbf{X}_k) + h^{1/2} \frac{X_k^i + \mathcal{X}_k^i}{2} \times \sigma(X_k^i)\xi_{k+1}^i, \ i = 1, \ldots, n,$$

(5.10.13)

$$X_{k+1}^i = X_k^i + h \frac{X_k^i + X_{k+1}^i}{2} \times a_i \left(\frac{\mathbf{X}_k + \mathcal{X}_k}{2} \right)$$

$$+ h^{1/2} \frac{X_k^i + X_{k+1}^i}{2} \times \sigma \left(\frac{X_k^i + \mathcal{X}_k^i}{2} \right) \xi_{k+1}^i, \ i = 1, \ldots, n,$$

$$k = 1, \ldots, N,$$

where $\xi_{k+1}^i = \left(\xi_{k+1}^{i,1}, \xi_{k+1}^{i,2}, \xi_{k+1}^{i,3} \right)^{\mathsf{T}}$; $\xi_l^{i,j}$ are i.i.d. random variables as in SIA.

Remark 5.10.1 One can continue the process and make several iterations for the implicit equation due to the midpoint scheme (1.3.53), e.g., as shown in numerical tests (see [265]) about 10 iterations were sufficient to resolve the implicitness up to the machine accuracy. However, in practice the use of several iterations would be too computationally expensive while SIB already demonstrates stability and accuracy comparable with the midpoint scheme.

As the Heun and mid-point methods, SIA and SIB are of weak order one for both choices of the distributions of $\xi_k^{i,j}$ (discrete and continuous). and they are of mean-square order $1/2$ if $\xi_k^{i,j}$ have the cut-off Gaussian distribution (5.10.12). In the deterministic case (i.e., $D = 0$) they have second-order convergence, again the same as the deterministic versions of the Heun and mid-point methods.

When α is small, the SLL equation (5.10.4)–(5.10.7) is a system with small multiplicative noise. In this case the weak-sense errors of all the methods considered in this section are of order $O(h^2 + \alpha^2 h)$ (see Sect. 4.4). The smallness of noise can be further exploited to construct high accuracy but low order efficient methods following the recipe from Chap. 4.

We now discuss conservation properties of the schemes. The Heun method (2.2.24) with projection (i.e., with normalization of the spin length at every step: $X_{k+1}^i = X_{k+1}^{*i}/|X_{k+1}^{*i}|$, where X_{k+1}^{*i} is the result of application of the usual Heun method) applied to the SLL equation has only one conservation property – norm-preservation which is due to the projection step. The Heun method without the projection step would conserve the total spin in the deterministic case with $\alpha = 0$ and under only the Heisenberg exchange, but it always violates norm-preservation. Omitting the projection step gives very poor results for the interaction with an external magnetic field. In practice the projection step can be exploited for error control.

Energy is not conserved by the Heun method when $\alpha = 0$. It has the advantage of being very flexible, its implementation is independent of the symmetry of the system and types of interactions used. The method is also fast since integration can be done for each spin separately.

Due to the structure of the mid-point scheme applied to (5.10.4)–(5.10.7), the difference $X_{k+1}^i - X_k^i$ is always perpendicular to $X_k^i + X_{k+1}^i$. Therefore $(X_k^i + X_{k+1}^i)(X_{k+1}^i - X_k^i) = 0$ and hence $|X_{k+1}^i|^2 = |X_k^i|^2$, i.e., the length of each spin is exactly preserved by the mid-point scheme without any need of projection. In the deterministic case with $\alpha = 0$ and under only the Heisenberg exchange, the mid-point scheme conserves the total spin. It also preserves the total energy for the case of $\alpha = 0$. Preservation of all the main structural properties of the SLL equation by the mid-point scheme comes at a cost. Since all spins are coupled, a system of $3n$ non-linear algebraic equations has to be solved at each time step. This is a major limitation for application of the mid-point scheme to ASD, where the number of spins is typically of order $n = 10^6$.

The SIA method preserves the constraint $|X^i(t)| = 1$ exactly, without the need of projection. This follows directly from the observation that the norm conservation of each spin is independent of the point at which a_i and σ are evaluated. Let us now look at SIA in the deterministic case with $\alpha = 0$. Regarding total spin, the relevant symmetry property is:

$$\frac{X_k^i + X_{k+1}^i}{2} \times \frac{X_k^j + \mathcal{X}_k^j}{2} + \frac{X_k^j + X_{k+1}^j}{2} \times \frac{X_k^i + \mathcal{X}_k^i}{2} \neq 0 \,,$$

which is violated since the Euler approximation for $\mathcal{X}_k^i$ depends only on the orientation of the spins at the current time step ($\mathbf{X}_k$), but not on $\mathcal{X}_k^i$, whereas X_{k+1}^i is also determined by the value X_{k+1}^i itself. Owing to this difference, for $\alpha = 0$ the total spin cannot be preserved by SIA. Also, the energy is not a conserved quantity by SIA and the scheme introduces numerical damping.

Unlike SIA, SIB has the norm-conserving midpoint structure for both X_k^i and $\mathcal{X}_k^i$. In the case of a two-spin deterministic system with $\alpha = 0$ both energy and total spin are conserved quantities of SIB. Hence for this system SIB has the same conservation properties as the mid-point scheme. At the same time, implementation-wise very little additional computational efforts are required by SIB compared to the Heun scheme and SIA. Numerical experiments of [265] suggest that SIB outperforms SIA while SIA is slightly better than the Heun method with projection. This observation implies, in particular, that the built-in norm conservation alone is not sufficient for obtaining superior numerical integrators for ASD and preservation of other structural properties of the SLL equation should guide one in constructing effective numerical methods.

5.11 Ergodic Limits

For many applications, it is interesting to compute the mean of a given function with respect to the invariant law of the diffusion, i.e. the ergodic limit. To evaluate these mean values, one often has to integrate a system over comparatively long time intervals (even if the convergence to ergodic limits is exponential), especially in the case of Langevin equations with small dissipation. This is one of the most serious difficulties from the computational point of view. Geometric integrators considered in the previous sections of this chapter demonstrate superiority over long time intervals in comparison with standard schemes for SDEs. For instance, in the case of small dissipation the Langevin equations are close to stochastic Hamiltonian systems. In such a rather typical situation numerical methods based on symplectic integrators are the most appropriate.

Consider the system of Ito SDEs

$$dX = b(X)dt + \sum_{l=1}^{r} \sigma_l(X)dw_l(t), \; X(0) = x, \tag{5.11.1}$$

where X, a, σ_l are d-dimensional column-vectors and $w_l(t)$, $l = 1, \ldots, r$, are independent standard Wiener processes. We use the notation $X(t)$ and $X(t; x)$ for solutions of the problem (5.11.1).

In what follows we assume that the following conditions are imposed on (5.11.1).

(A1) The coefficients of (5.11.1) are sufficiently smooth functions in $\mathbf{R}^d$.
(A2) The solution of (5.11.1) is *regular*, i.e., it is defined for all $t \geq 0$.
(A3) The process $X(t)$ is *ergodic*, i.e., there exists a unique invariant measure μ of X and independently of $x \in \mathbf{R}^d$ there exists the limit

$$\lim_{t \to \infty} E\varphi(X(t; x)) = \int \varphi(x) \, d\mu(x) := \varphi^{erg} \tag{5.11.2}$$

for any function $\varphi(x)$ with polynomial growth at infinity.
(A4) The Markov transition function $P(t, x, dy)$ (it is homogeneous for (5.11.1)) and the invariant measure $\mu(dx)$ have sufficiently smooth densities $p(t, x, y)$ and $\rho(x)$, respectively.

In connection with the condition (A2), we recall sufficient conditions for regularity of SDEs' solution. Denote by $\mathbf{C}^2$ the class of functions defined on $[0, \infty) \times \mathbf{R}^d$ and twice continuously differentiable with respect to x and once with respect to t. A sufficient condition of regularity (see [194] and also Section 2.2.5.1) consists in existing a nonnegative function $V \in \mathbf{C}^2$ which satisfies the inequality

$$LV(t, x) \leq c_0 V(t, x) + c_1, \; (t, x) \in [0, \infty) \times \mathbf{R}^d, \tag{5.11.3}$$

and

$$\lim_{R \to \infty} \inf_{|x| \geq R} V(t, x) = \infty, \tag{5.11.4}$$

where c_0 and c_1 are some constants and L is the operator

$$L := \frac{\partial}{\partial t} + \mathcal{L}$$

and $\mathcal{L}$ is the generator for (5.11.1):

$$\mathcal{L} := \sum_{i=1}^{d} b^i(x) \frac{\partial}{\partial x^i} + \frac{1}{2} \sum_{i,j=1}^{d} a^{ij}(x) \frac{\partial^2}{\partial x^i \partial x^j}, \quad a^{ij} := \sum_{l=1}^{r} \sigma_l^i \sigma_l^j.$$

Moreover, if (5.11.3) and (5.11.4) are fulfilled and if an initial distribution for x (x can be random) is such that $EV(0, x)$ exists, then $EV(t, X(t; x))$ exists for all $t \geq 0$. For instance, if V has an m-polynomial growth at infinity, then there exist moments of order m for X.

The transition density $p(t, x, y)$ from the assumption (A4) satisfies the Fokker-Planck (forward Kolmogorov) equation

$$\frac{\partial p}{\partial t}(t, x, y) + \sum_{i=1}^{d} \frac{\partial}{\partial y^i} \left(b^i(y) \, p(t, x, y) \right)$$

$$- \frac{1}{2} \sum_{i,j=1}^{d} \frac{\partial^2}{\partial y^i \partial y^j} \left(a^{ij}(y) \, p(t, x, y) \right) = 0, \quad t > 0, \tag{5.11.5}$$

$$p(0, x, y) = \delta(y - x),$$

and the invariant density $\rho(x)$ satisfies the stationary Fokker-Planck equation

$$\sum_{i=1}^{d} \frac{\partial}{\partial x^i} \left(b^i(x) \, \rho(x) \right) - \frac{1}{2} \sum_{i,j=1}^{d} \frac{\partial^2}{\partial x^i \partial x^j} \left(a^{ij}(x) \, \rho(x) \right) = 0. \tag{5.11.6}$$

We are interested here in systems which solutions satisfy a stronger condition than (A3):

(A3e) The process $X(t)$ is *exponentially ergodic*, i.e., for any $x \in \mathbf{R}^d$ and any function φ with a polynomial growth we have the following strengthening of (5.11.2):

$$\left| E\varphi(X(t; x)) - \varphi^{erg} \right| \leq Ce^{-\lambda t}, \quad t \geq 0, \tag{5.11.7}$$

where $C > 0$ and $\lambda > 0$ are some constants.

Example 5.11.1 Consider the Langevin equations

$$dP = f(Q)dt - \gamma Pdt + \sqrt{\frac{2\gamma}{\beta}}dw(t), \quad P(0) = P_0 = p, \qquad (5.11.8)$$

$$dQ = M^{-1}Pdt, \quad Q(0) = Q_0 = q,$$

where P, Q, $f(q) = -\nabla U(q)$ are n-dimensional column-vectors, $\gamma > 0$ is a damping parameter, M is a non-singular, diagonal matrix, and $w(t) = (w_1(t), \ldots, w_n(t))^\mathsf{T}$ and $w_l(t)$ are independent standard Wiener processes. Let the potential $U(q)$ satisfy the following assumptions: $U(q) \geq 0$ for all $q \in \mathbf{R}^n$ and there exists an $\alpha_1 > 0$ and $0 < \alpha_2 < 1$ such that

$$\frac{1}{2}(\nabla U(q), q) \geq \alpha_2 U(q) + \gamma^2 \frac{\alpha_2(2 - \alpha_2)}{8(1 - \alpha_2)}|q|^2 - \alpha_1. \qquad (5.11.9)$$

Exponential ergodicity of (5.11.8) was proved under these assumptions in [262]. It can directly be verified that the Gibbs distribution

$$\rho(p, q) \propto \exp\left(-\beta\left\{\frac{1}{2}p^\mathsf{T}M^{-1}p + U(q)\right\}\right) \qquad (5.11.10)$$

is a solution of the stationary Fokker-Planck equation (cf. (5.11.6)) corresponding to (5.11.8). The Langevin equations (5.11.8) are very popular in, e.g., molecular simulation [230, 401], when one is interested in constant temperature dynamics (in other words, in sampling from the canonical ensemble).

Example 5.11.2 Assume that $\rho(y) > 0$, $y \in \mathbf{R}^d$, is a sufficiently smooth density function which is known up to a constant of proportionality. Consider the stochastic gradient system

$$dY = \frac{1}{2}\nabla \log \rho(Y)dt + dw(t). \qquad (5.11.11)$$

The solution $Y(t)$ of (5.11.11) is exponentially ergodic under the following conditions [379]: $\rho(y)$ is a bounded density,

$$\log \rho(y) \in C^2(\mathbf{R}^d), \qquad (5.11.12)$$

and there exists $0 < \alpha < 1$ such that

$$\liminf_{|y|\to\infty} \left[(1 - \alpha)|\nabla \log \rho(y)|^2 + \Delta \log \rho(y)\right] > 0, \qquad (5.11.13)$$

where Δ is the Laplace operator. Stochastic gradient systems are widely used in Bayesian statistics, molecular dynamics, and deep learning.

5.11.1 Ensemble-Averaging Estimators

It follows from (5.11.7) (and (5.11.2)) that for any $\varepsilon > 0$ there exists $T_0 > 0$ such that for all $T \geq T_0$

$$\left| E\varphi(X(T; x)) - \varphi^{erg} \right| \leq \varepsilon. \tag{5.11.14}$$

Thus, the problem of computing the ergodic limit is reduced to evaluating the expectation $E\varphi(X(T))$ at a finite (though usually large) time T, which is an ensemble-averaging estimator for φ^{erg}.

Resting on (5.11.14), we can consider the following Monte Carlo (numerical ensemble averaging) estimator for the ergodic limit φ^{erg}:

$$\hat{\varphi}^{erg} = \frac{1}{M} \sum_{m=1}^{M} \varphi\left(\bar{X}^{(m)}(T; x)\right), \tag{5.11.15}$$

where M is the number of independent approximate realizations and $\bar{X}(T; x)$ is a weak approximation of $X(T; x)$ with order $p > 0$. The total error

$$R_{\hat{\varphi}^{erg}} := \hat{\varphi}^{erg} - \varphi^{erg} \tag{5.11.16}$$

consists of three parts: the error ε of the approximation (5.11.14); the error of numerical integration Kh^p; and the Monte Carlo error; i.e.,

$$R_{\hat{\varphi}^{erg}} \sim Kh^p + \varepsilon + O\left(\frac{1}{\sqrt{M}}\right).$$

More precisely,

$$\text{Bias}(\hat{\varphi}^{erg}) = \left| E\hat{\varphi}^{erg} - \varphi^{erg} \right| \leq Kh^p + \varepsilon \tag{5.11.17}$$

and the estimator's variance is

$$\text{Var}(\hat{\varphi}^{erg}) = O(1/M). \tag{5.11.18}$$

Here, in this ensemble average approach to computing ergodic limits, each error is controlled by its own parameter: sufficiently large T ensures smallness of ε; time step h (as well as the choice of a numerical method) controls the numerical integration error; the statistical error is regulated by choosing an appropriate number of independent trajectories M. For more reading on ensemble averaging, see [319].

5.11.2   Time-Averaging Estimators

The other approach for calculating ergodic limits is based on time averaging, which
is very popular in the physical and Bayesian statistics community.

Following [263], let us, for simplicity, restrict our consideration to the case when
the solution $X(t)$ of (5.11.1) is on the torus $\mathbf{T}^d$ and the coefficients $b\colon \mathbf{T}^d \to \mathbf{R}^d$
and $\sigma_l\colon \mathbf{T}^d \to \mathbf{R}^d$. In this setting, given two vector fields g and $\tilde{g}$ on $\mathbf{T}^d$, one defines
the Lie-bracket as $[g, \tilde{g}](x) := (g \cdot \nabla \tilde{g})(x) - (\tilde{g} \cdot \nabla g)(x)$. The vector $[g, \tilde{g}](x)$ may
be thought of as the new direction generated by infinitesimally following g, then $\tilde{g}$,
then $-g$ and finally $-\tilde{g}$. To track how the noise spans the tangent space at all points,
introduce the increasing set of vector fields

$$\Upsilon_0 = \mathrm{span}\{b, \sigma_1, \ldots, \sigma_r\}, \quad \Upsilon_{n+1} = \mathrm{span}\{g, [\bar{g}, g] : g \in \Upsilon_n, \bar{g} \in \Upsilon_0\}.$$

Assumption 5.11.3   Assume that one of the following two assumptions hold:

1. (Elliptic Setting) The matrix-valued function $a(x)$ is uniformly positive definite.
2. (Hypoelliptic setting) The functions b and σ_l are $C^\infty(\mathbf{T}^d, \mathbf{R}^d)$ and such that,
    for some n and all $x \in \mathbf{T}^d$, $\Upsilon_n = \Upsilon_n(x) = \mathbf{R}^d$ and (5.11.1) possesses a unique
    stationary measure.

In averaging, homogenization and ergodic theory if one can solve the relevant
Poisson equation then one can prove results about limits of a time-average (see e.g.
[96, 356]). Recalling that μ is the unique stationary measure of (5.11.1) and given
$\varphi\colon \mathbf{T}^d \to \mathbf{R}$, we have in this setting:

$$\varphi^{erg} = \int_{\mathbf{T}^d} \varphi(x)\, \mu(dx). \tag{5.11.19}$$

Let ψ solve the Poisson equation

$$\mathcal{L}\psi = \varphi - \varphi^{erg}. \tag{5.11.20}$$

Under Assumption 5.11.3, (5.11.1) possesses a unique solution which is at least as
smooth as φ [263]. In what follows, we will assume that φ is sufficiently smooth.

Remark 5.11.4   The main technical issues that one needs to address in order to
extend the results discussed below from the $\mathbf{T}^d$ case to general ergodic SDEs on $\mathbf{R}^d$
is proof of the existence of well-controlled solutions to the above Poisson equation
(cf. [353–355]).

By the Ito formula, we get

$$\psi(X(t)) - \psi(x_0) = \int_0^t \left(\varphi(X(s)) - \varphi^{erg} \right) ds \tag{5.11.21}$$

$$+ \sum_{l=1}^{r} \sum_{i=1}^{d} \int_0^t \frac{\partial}{\partial x^i} \psi(X(s)) \sigma_l^i(X) dw_l(t).$$

Rearranging this, we obtain that

$$\frac{1}{t} \int_0^t \varphi(X(s)) ds - \varphi^{erg} = \frac{\psi(X(t)) - \psi(x)}{t} - \frac{1}{t} M(t), \tag{5.11.22}$$

where $M(t) = \sum_{l=1}^{r} \sum_{i=1}^{d} \int_0^t \frac{\partial}{\partial x^i} \psi(X(s)) \sigma_l^i(X) dw_l(t)$. Since ψ is bounded, the first term on the right-hand side goes to zero as $t \to \infty$. To see that the last term also goes to zero as $t \to \infty$ observe that

$$\frac{1}{t^2} E\left(M(t)^2 \right) = \frac{1}{t^2} E\langle M \rangle(t) \leq \frac{K}{t}$$

for some $K > 0$ independent of time since $\nabla \psi$ and σ_l are both bounded on $\mathbf{T}^d$ (here $\langle M \rangle(t)$ is the quadratic variation of $M(t)$). Consequently, we have shown that for any initial x

$$E\left(\frac{1}{T} \int_0^T \varphi(X(s)) ds - \varphi^{erg} \right)^2 \leq \frac{K}{T}, \tag{5.11.23}$$

which is a version of what is often called the mean ergodic theorem.

We can view $\frac{1}{T} \int_0^T \varphi(X(s)) ds$ as a time-averaging estimator for φ^{erg} and it follows from (5.11.22) that

$$\text{Bias} \left(\frac{1}{T} \int_0^T \varphi(X(s)) ds \right) = E\left(\frac{1}{T} \int_0^T \varphi(X(s)) ds - \varphi^{erg} \right) = O\left(\frac{1}{T} \right). \tag{5.11.24}$$

One can also show that

$$\text{Var} \left(\frac{1}{T} \int_0^T \varphi(X(s)) ds \right) = O\left(\frac{1}{T} \right). \tag{5.11.25}$$

The estimate (5.11.25) easily follows from (5.11.23) and (5.11.24) but can also be obtained directly from (5.11.22) using an additional mixing condition.

To obtain an almost sure (a.s.) result, for any $\varepsilon > 0$ introduce $A(T;\varepsilon) = \{\frac{1}{T}\sup_{t\leq T}$ $|M(t)| > T^{\varepsilon-\frac{1}{2}}\}$. By the Doob inequality for continuous martingales and the fact that $E|M(T)|^2 \leq KT$, we get

$$P(A(T;\varepsilon)) \leq \frac{E|M(T)|^2}{T^{1+2\varepsilon}} \leq \frac{K}{T^{2\varepsilon}} \cdot$$

Hence

$$\sum_{n\in\mathbf{N}} P(A(2^n;\varepsilon)) < \infty$$

and the Borel-Cantelli lemma implies that there exists an a.s. bounded random variable $C(\omega) > 0$ and n_0 so that for every $n \in \mathbf{N}$ with $n \geq n_0$ one has

$$\frac{1}{2^n} \sup_{t\leq 2^n} |M(t)| \leq \frac{C(\omega)}{2^{n(1/2-\varepsilon)}} \cdot$$

. Hence with probability one, for every $t \in [2^n, 2^{n+1})$ with $n \geq n_0$ one has

$$\frac{1}{t}|M(t)| \leq \frac{1}{2^n} \sup_{2^n\leq s\leq 2^{n+1}} |M(s)| \leq \frac{2}{2^{n+1}} \sup_{s\leq 2^{n+1}} |M(s)|$$

$$\leq \frac{2C(\omega)}{2^{(n+1)(1/2-\varepsilon)}} \leq \frac{2C(\omega)}{t^{1/2-\varepsilon}} \cdot$$

By combining this estimate with (5.11.22) and the fact that ψ is bounded, we see that for any $\varepsilon > 0$ and for every $t \geq 2^{n_0}$ one has

$$\left|\frac{1}{T}\int_0^T \varphi(X(s))\, ds - \varphi^{erg}\right| \leq \frac{2\max_{x\in\mathbf{T}^d} |\psi|}{T} + \frac{C(\omega)}{T^{1/2-\varepsilon}} \qquad \text{a.s.} \qquad (5.11.26)$$

for some a.s. bounded $C(\omega) > 0$. We note that (5.11.26) implies that for any initial x

$$\lim_{t\to\infty} \frac{1}{t}\int_0^t \varphi(X(s;x))\, ds = \varphi^{erg} \qquad \text{a.s.} \qquad (5.11.27)$$

In other words, the strong law of large number holds starting from any initial x. Assumptions 5.11.3 are sufficient to ensure that (5.11.1) has a unique stationary measure μ. Hence, it follows from Birkoff's ergodic theorem that (5.11.27) holds for μ-a.e. initial x. The above argument not only shows that the result holds for all x, but also gives quantitative estimates on the rate of convergence.

Based on (5.11.27) and by approximating a single trajectory, one gets for a sufficiently large T:

$$\frac{1}{T} \int_0^T \varphi(X(s;x))ds \approx \check{\varphi}^{erg} = \check{\varphi}_L^{erg} := \frac{1}{L} \sum_{k=1}^L \varphi(\bar{X}(t_k;x)), \qquad (5.11.28)$$

where $Lh = T$, $\bar{X}(t_k;x)$ is a weak approximation of $X(t_k;x)$ with order $p > 0$, and $\check{\varphi}^{erg}$ is a numerical time-averaging estimator. Let us emphasize that T in (5.11.28) is much larger than T in (5.11.14) and (5.11.15) because T in (5.11.28) should be such that it not just ensures the distribution of $X(t)$ to be close to the invariant distribution (like is required from T in (5.11.14)) but it should also guarantee smallness of variance of $\check{\varphi}^{erg}$.

In [429] it was shown that several numerical methods for SDEs, including the Euler scheme, have unique stationary measures which converge at the expected rate to the unique stationary measure of the ergodic SDEs with nondegenerate noise and globally Lipschitz coefficients in $\mathbf{R}^d$. In [433] an expansion of the numerical integration error in powers of time step was obtained which allows one to use the Richardson-Romberg extrapolation to improve the accuracy.

Under some conditions on the coefficients of (5.11.1), the estimator $\check{\varphi}^{erg}$ has the following properties [263, 429]:

$$\text{Bias}(\check{\varphi}_L^{erg}) \leq K_1 h^p + \frac{K_2}{T}; \qquad (5.11.29)$$

and [263]

$$\text{Var}(\check{\varphi}_L^{erg}) \leq \frac{K}{T}. \qquad (5.11.30)$$

Furthermore, there exists a deterministic constant K so that for h sufficiently small, positive $\varepsilon > 0$, and T sufficiently large one has [263]:

$$\left| \check{\varphi}_L^{erg} - \varphi^{erg} \right| \leq K h^p + \frac{C(\omega)}{T^{1/2-\varepsilon}} \quad \text{a.s.,} \qquad (5.11.31)$$

where $C(\omega) > 0$ is an a.s. bounded random variable depending on ε and the particular φ. In [429] the proof of (5.11.29) is based on the backward Kolmogorov equation and its properties. Proofs of (5.11.29)–(5.11.31) in [263] are based on using the Poisson equation (5.11.20).

In practice one usually estimates the statistical error of the time averaging estimator $\check{\varphi}_L^{erg}$ as follows. We run a long trajectory MT split into M blocks of a large length $T = hL$ each. We evaluate the time-averaging estimators $_m\check{\varphi}_L^{erg}$, $m = 1, \ldots, M$, for each block. Since T is large and a time decay of correlations is usually fast, $_m\check{\varphi}_L^{erg}$ can be considered as almost uncorrelated. We compute the sample variance

$$\hat{D} = \frac{1}{M} \sum_{m=1}^M \left(_m\check{\varphi}_L^{erg} \right)^2 - \left(\frac{1}{M} \sum_{m=1m}^M \check{\varphi}_L^{erg} \right)^2.$$

For a sufficiently large T and M, $E\check{\varphi}_L^{erg}$ belongs to the confidence interval

$$E\,\check{\varphi}_L^{erg} \in \left(\check{\varphi}_{LM}^{erg} - c\frac{\sqrt{\hat{D}}}{\sqrt{M}}, \check{\varphi}_{LM}^{erg} + c\frac{\sqrt{\hat{D}}}{\sqrt{M}}\right),$$

with probability, for example 0.95 for $c = 2$ and 0.997 for $c = 3$. Note that $E\,\check{\varphi}_L^{erg}$ contains the two errors forming the bias as explained in (5.11.29). We also pay attention to the fact that $\hat{D} \sim 1/T$, i.e., it is inverse proportional to the product hL.

For more reading on time averaging, see [38, 263, 319, 379, 429] (see also references therein).

5.11.3 An Efficient Second-Order Accurate Scheme in Approximating Ergodic Limits

The bias (5.11.17) and (5.11.29) of both numerical ensemble and time averaging estimators is typically dominated by the numerical integration error. As it was shown in [263, 319, 429] (see also references therein) and discussed in the two previous subsections, if a method of weak order p is used for computing ergodic limits then the corresponding order in the bias for the ergodic limit is at least p. One then can ask the question whether it is possible to construct a method which numerical integration order in approximating ergodic limits is higher than its finite-time weak order.

Consider the stochastic gradient system (cf. Example 5.11.2)

$$dX = a(X)dt + \sigma dw, \ X(0) = X_0, \tag{5.11.32}$$

where the force

$$a(x) := -\nabla V(x), \tag{5.11.33}$$

$V(x)$, $x \in \mathbf{R}^d$, is a potential energy function and $\sigma > 0$ is a constant which characterizes the strength of the additive noise. Under assumptions as in Example 5.11.2, this system is exponentially ergodic with the unique invariant measure which density $\rho \propto \exp(-\beta V)$, where $\beta = 2\sigma^{-2}$.

In [229] the following method for (5.11.32) was proposed

$$X_{k+1} = X_k + ha(X_k) + \sigma\frac{\sqrt{h}}{2}(\xi_k + \xi_{k+1}), \tag{5.11.34}$$

where $\xi_k = (\xi_k^1, \ldots, \xi_k^i)^{\mathsf{T}}$ and ξ_k^i, $i = 1, \ldots, d$, $k = 1, \ldots$, are i.i.d. random variables with the law $\mathcal{N}(0, 1)$. This method is very similar in form to the Euler method and it has the same cost as the Euler method (one evaluation of the force per time step), but in contrast to the Euler method the increments of the scheme (5.11.34) are not independent when conditioned on X_k.

As we will see below, the scheme (5.11.34) has the second order of accuracy in computing ergodic limits. But let us first consider an illustrative example.

Example 5.11.5 Let $a(x) = -\alpha x$ with $\alpha > 0$, then $X(t)$ from (5.11.32) is the Ornstein-Uhlenbeck process, which is Gaussian with $EX_x(t) = xe^{-\alpha t}$ and $Cov(X_x(s), X_x(t)) = \dfrac{\sigma^2}{2\alpha}(e^{-\alpha(t-s)} - e^{-\alpha(t+s)})$ for $s \leq t$. It is not difficult to calculate that for the Euler scheme :

$$EX_N = x_0(1 - \alpha h)^N = x_0 e^{-\alpha T}(1 + O(h)),$$

$$Var(X_N) = \frac{\sigma^2}{2\alpha} \frac{1 - (1 - \alpha h)^{2N}}{1 + \alpha h}$$

$$= \frac{\sigma^2}{2\alpha}(1 - e^{-2\alpha\tau}) - \frac{\sigma^2}{2}h + e^{-2\alpha T}O(h) + O(h^2), \quad \alpha h < 1,$$

where $|O(h^p)| \leq Kh$ with $K > 0$ independent of T, and for the scheme (5.11.34):

$$EX_N = x_0(1 - \alpha h)^N = x_0 e^{-\alpha T}(1 + O(h)),$$

$$Var(X_N) = \frac{\sigma^2}{2\alpha}\left[1 - \frac{(1 - \alpha h)^{2N}}{1 - \alpha h}\right] = \frac{\sigma^2}{2\alpha}(1 - e^{-2\alpha T}) + e^{-2\alpha T}O(h).$$

We see that although both schemes have first order accuracy on finite time intervals, the ergodic limit of the scheme (5.11.34) is exact while the ergodic limit of the Euler scheme approximates the ergodic limit of the Ornstein-Uhlenbeck process with order one which is usually the case for weak schemes of order one.

Under some assumptions on the potential $V(x)$ and $\varphi(x)$, it was proved in [232] that the scheme (5.11.34) is first order weakly convergent and for all sufficiently small $h > 0$ its error has the form

$$E\varphi(X_x(T)) - E\varphi(X_N) = C_0(T, x)h + C(T, x)h^2, \tag{5.11.35}$$

where

$$C_0(T, x) = E\int_0^T B_0(t, X_x(t))dt, \tag{5.11.36}$$

$$\begin{aligned}
B_0(t, x) = \frac{1}{2}\Bigg[& \sum_{i,j=1}^d a^j(x)\frac{\partial a^i}{\partial x^j}(x)\frac{\partial u}{\partial x^i}(t, x) \\
& + \frac{\sigma^2}{2}\sum_{i,j=1}^d \frac{\partial a^i}{\partial x^j}(x) a^i(x)\frac{\partial^2}{\partial x^i \partial x^j}u(t, x) \\
& + \frac{\sigma^2}{2}\sum_{i,j=1}^d \frac{\partial^2 a^i}{(\partial x^j)^2}(x)\frac{\partial}{\partial x^i}u(t, x)\Bigg],
\end{aligned}$$

and
$$|C(T, x)| \le K(1 + |x|^{\varkappa} e^{-\lambda T}),$$

for some $K > 0$, $\varkappa \in \mathbf{N}$ and $\lambda > 0$ independent of h and T. Here $u(t, x)$ is the solution of the corresponding backward Kolmogorov equation. The proof exploits the idea of the proof of Theorem 2.2.5 from Sect. 2.2.3.

Furthermore (see [232]), the coefficient $C_0(T, x)$ from (5.11.36) goes to zero as $T \to \infty$:
$$|C_0(T, x)| \le K(1 + |x|^{\varkappa}) e^{-\lambda T} \qquad (5.11.37)$$

for some constants $K > 0$, $\varkappa \in \mathbf{N}$ and $\lambda > 0$, i.e., over a long integration time the scheme (5.11.34) is of order two up to exponentially small correction. We note that the average of $B_0(t, x)$ with respect to the invariant measure is equal to zero which is the reason why the scheme (5.11.34) is second order accurate in approximating ergodic limits.

In the case of the Euler scheme one can get the same error expansion as (5.11.35) for the scheme (5.11.34) but with a different $B_0(t, x) = B_0^E(t, x)$ (see the proof of Theorem 2.2.5):

$$
\begin{aligned}
B_0^E(t, x) = \frac{1}{2} \Bigg[& \sum_{i,j=1}^{d} a^j \frac{\partial u}{\partial x^j} a^i \frac{\partial u}{\partial x^i} + \frac{\sigma^2}{2} \sum_{i,j} \frac{\partial^2 a^j}{(\partial x^i)^2} \frac{\partial u}{\partial x^j} + \frac{\sigma^2}{2} \sum_{i,j=1}^{d} a^i \frac{\partial^3 u}{\partial x^i (\partial x^j)^2} \\
& + \sigma^2 \sum_{i,j=1}^{d} \frac{\partial a^j}{\partial x^i} \frac{\partial^2 u}{\partial x^j \partial x^i} + \frac{\sigma^4}{6} \sum_{i,j=1}^{d} \frac{\partial^4 u}{(\partial x^i)^2 (\partial x^j)^2} \Bigg].
\end{aligned}
$$

The average of $B_0^E(t, x)$ with respect to the invariant measure is not equal to zero and, consequently, the Euler scheme approximates ergodic limits with order one – the same order as its weak convergence over a finite time interval (see also Example 5.11.5).

A number of numerical tests for the scheme (5.11.34) are presented in [229, 232] confirming superior accuracy of (5.11.34) in comparison with the Euler scheme and even with the more expensive Heun method (2.2.24). Works related to this topic include [4, 37, 230–232, 453].

Remark 5.11.6 In this section we considered ergodic SDEs in the whole space $\mathbf{R}^d$. Making use of reflected SDEs together with the weak-sense numerical integration to compute ergodic limits in a bounded domain G and on its boundary ∂G as well to sample from distributions whose support is a compact set $\bar{G}$ and from distributions whose support is a $d - 1$ dimensional hyper-surface ∂G was considered in [233].

Chapter 6
Simulation of Space and Space-Time Bounded Diffusions

"Ordinary" mean-square methods from Chap. 1, intended to solve SDEs on a finite time interval, are based on a time discretization. The space-time point, corresponding to an "ordinary" one-step approximation constructed at a time point t_k, lies on the d-dimensional plane $t = t_k$, which belongs to the $(d + 1)$-dimensional semi-space $[T_0, \infty) \times \mathbf{R}^d$. The "ordinary" mean-square methods give both time and phase components of the approximate trajectory. They ensure smallness of time increments at each step, but space increments can take arbitrary large values with some probability. To approximate SDEs in a bounded domain, we have to control space increments at each step so that the constructed approximation belongs to the bounded domain. Of course, this cannot be achieved by "ordinary" mean-square methods, and thus special methods are required.

In the first part of this chapter (Sects. 6.1–6.2) we propose a mean-square approximation of an autonomous diffusion process in a space bounded domain [279, 282], while in the second part (Sects. 6.3 –6.5) we construct algorithms for space-time diffusions in space-time bounded domains [304]. In the last section (Sect. 6.6), we give a brief survey on simulation of reflected diffusions.

In Sect. 6.1 we consider the autonomous system of SDEs without drift

$$dX = \chi_{\tau_x > t} \sigma(X) dw(t), \quad X(0) = x, \tag{6.0.1}$$

in a bounded domain $G \subset \mathbf{R}^d$ with boundary ∂G. Here $w(t) = (w^1(t), \ldots, w^d(t))^\mathsf{T}$, $t \geq 0$, is a standard $\mathcal{F}_t$-measurable Wiener process of dimension d defined on a probability space $(\Omega, \mathcal{F}, P)$, where $\mathcal{F}_t$ is a nondecreasing family of σ-subalgebras of $\mathcal{F}$; $X = (X^1, \ldots, X^d)^\mathsf{T}$ is a vector of dimension d, $\sigma(x) = \{\sigma^{ij}(x)\}$ is a matrix of dimension $d \times d$, τ_x is a random time at which the path $X_x(t)$ leaves the region G.

In addition to (6.0.1), we introduce the system with the coefficients frozen at x :

$$d\bar{X} = \sigma(x) dw(t), \quad \bar{X}(0) = x. \tag{6.0.2}$$

© Springer Nature Switzerland AG 2021
G. N. Milstein and M. V. Tretyakov, *Stochastic Numerics for Mathematical Physics*,
Scientific Computation, https://doi.org/10.1007/978-3-030-82040-4_6

Let $U_r \subset \mathbf{R}^d$ be an open sphere of radius r with center at the origin and with the boundary ∂U_r. Denote by $\bar{\theta}$ the first time at which the process $w(t)$ leaves the sphere U_r. Clearly, $w(\bar{\theta})$ has the uniform distribution on ∂U_r. Let $U_r^\sigma(x)$ be an open ellipsoid with the boundary $\partial U_r^\sigma(x)$ obtained from the sphere U_r by the linear transformation $\sigma(x)$ and the shift x. It is assumed that r is small enough to satisfy the inclusion: $U_r^\sigma(x) \subset G$. The solution $\bar{X}_x(t)$ of the problem (6.0.2) at the time $\bar{\theta}$ is equal to

$$\bar{X}_x(\bar{\theta}) = x + \sigma(x)w(\bar{\theta}), \quad \bar{X}_x(\bar{\theta}) \in \partial U_r^\sigma(x), \tag{6.0.3}$$

and $\bar{\theta}$ is the first exit time of the trajectory $\bar{X}_x(t)$ from $U_r^\sigma(x)$. We note that if $\tau_x \le \bar{\theta}$ then $X_x(t) = X_x(\tau_x)$ for $t \ge \tau_x$ due to (6.0.1). It turns out that $\bar{X}_x(\bar{\theta})$ is close to $X_x(\bar{\theta})$ in the mean-square sense. So, the point $\bar{X}_x(\bar{\theta})$ is an approximation of a point which belongs to the phase trajectory starting at x.

Construction of the point $(\bar{\theta}, \bar{X}_x(\bar{\theta}))$ amounts to modeling $\bar{\theta}$ and $\bar{X}_x(\bar{\theta})$ separately because of their independence. It is important to underline that if we are interested in phase trajectories only, it is possible to simulate them without modeling $\bar{\theta}$. To simulate $\bar{X}_x(\bar{\theta})$, we need only $w(\bar{\theta})$ which has uniform distribution on ∂U_r, i.e., modeling the point $\bar{X}_x(\bar{\theta}) \in \partial U_r^\sigma(x)$ is a fairly simple problem. Various algorithms are available to sample from uniform distribution on a d-dimensional sphere, see e.g. [253, 332, 436].

Let $\bar{X}_0 = x$, $\bar{X}_1 = \bar{X}_x(\bar{\theta})$. We find the point $\bar{X}_2$ on the boundary $\partial U_r^\sigma(\bar{X}_1)$ in the same way as we found $\bar{X}_1$ coming from $\bar{X}_0 = x$. Then we construct $\bar{X}_3$ and so on until a point $\bar{X}_{\bar{\nu}}$ with a random subscript $\bar{\nu}$ (see Algorithm 6.1.3). As a result, the sequence $\bar{X}_0, \ldots, \bar{X}_{\bar{\nu}}$ is obtained which can be considered as a mean-square approximation of the phase trajectory of the solution $X_x(t)$. If the point $\bar{X}_{\bar{\nu}}$ is sufficiently close to the boundary ∂G, it is possible to simulate the exit point $X_x(\tau_x)$. In Sect. 6.2, we construct an approximation for an autonomous system with *drift* in a space bounded domain.

Thus, the algorithm of Sects. 6.1–6.2 is based on a space discretization (quantization) using *a random walk over small spheres*. It gives the points which are close in the mean-square sense to the points of the real phase trajectory for SDEs in the space bounded domain. To realize the algorithm, the exit point of the Wiener process from a d-dimensional ball has to be constructed at each step. The algorithm gives only the phase component of the approximate trajectory without modelling the corresponding time component like the algorithm over touching spheres [331]. The space-time point lies on the d-dimensional lateral surface of a semi-cylinder with sphere base in the $(d+1)$-dimensional semi-space $[T_0, \infty) \times \mathbf{R}^d$. The algorithm ensures smallness of the phase increments at each step, but the non-simulated time increments can take arbitrary large values with some probability.

In Sects. 6.3–6.5, the mean-square approximations are considered which control boundedness of both space increments and time increments at each step. In addition they give approximate values for both phase and time components of the space-time diffusion in the space-time bounded domain. The space-time point lies on a bounded d-dimensional manifold. It is possible to solve this problem in a constructive manner by the implementation of a space-time discretization by a random walk over boundaries of small space-time parallelepipeds.

In Sect. 6.4, we consider the system of SDEs

$$dX = \chi_{\tau_{t,x}>s}b(s, X)ds + \chi_{\tau_{t,x}>s}\sigma(s, X)dw(s), \quad X(t) = X_{t,x}(t) = x, \quad (6.0.4)$$

in a space-time bounded domain $Q = [T_0, T_1) \times G \subset \mathbf{R}^{d+1}$. Here X and b are d-dimensional vectors, σ is a $d \times d$-matrix, $(w(s), \mathcal{F}_s)$, $s \geq T_0$, is a d-dimensional standard Wiener process defined on a probability space $(\Omega, \mathcal{F}, P)$, G is a bounded open domain in $\mathbf{R}^d$, and the Markov moment $\tau_{t,x}$ is the first-passage time of the process $(s, X_{t,x}(s))$, $s \geq t$, to $\Gamma = \overline{Q}\backslash Q$. The set Γ is a part of the boundary ∂Q consisting of the lateral surface and the upper base of the cylinder $\overline{Q}$. We put $X_{t,x}(s) = X_{t,x}(\tau_{t,x})$ under $s \geq \tau_{t,x}$, and thus, the process $(s, X_{t,x}(s))$ is defined for all $t \leq s < T_1$. The coefficients $b^i(s, x)$ and $\sigma^{ij}(s, x)$, $(s, x) \in \overline{Q}$, and the boundary ∂G are assumed to be sufficiently smooth, while the strict ellipticity condition is imposed on the matrix $a(s, x) := \sigma(s, x)\sigma^\mathsf{T}(s, x)$.

The mean-square approximations for (6.0.4) are based on a space-time discretization by *a random walk over boundaries of small space-time parallelepipeds*. It turns out that the first exit point $(\bar{\theta}, w(\bar{\theta}))$ of the space-time Brownian motion $(s, w(s))$, $s > 0$, from the space-time parallelepiped $\Pi_{r,l} = [0, lr^2) \times C_r$, where $C_r \subset \mathbf{R}^d$ is a cube with center at the origin and edge length equal to $2r$, can be simulated in a sufficiently easy way (some aspects of the space-time Brownian motion in the one-dimensional case, $d = 1$, are considered in [180]). To construct a one-step approximation, we introduce the system with frozen coefficients (both t, x fixed)

$$d\bar{X} = b(t, x)ds + \sigma(t, x)dw(s), \quad \bar{X}(t) = x. \quad (6.0.5)$$

As an approximation of the point $(t + \bar{\theta}, X_{t,x}(t + \bar{\theta}))$ of the space-time diffusion $(s, X_{t,x}(s))$, $s \geq t$, we take the point $(t + \bar{\theta}, \bar{X}_{t,x}(t + \bar{\theta}))$, where $\bar{X}_{t,x}(t + \bar{\theta})$ is a solution of (6.0.5):

$$\bar{X}_{t,x}(t + \bar{\theta}) = x + b(t, x)\bar{\theta} + \sigma(t, x)(w(t + \bar{\theta}) - w(t)), \quad (6.0.6)$$

and $(\bar{\theta}, w(t + \bar{\theta}) - w(t))$ is the exit point of the space-time Brownian motion $(s - t, w(s) - w(t))$, $s > t$, from the space-time parallelepiped $\Pi_{r,l}$.

The point $(t + \bar{\theta}, \bar{X}_{t,x}(t + \bar{\theta}))$ lies on the lateral surface or on the upper base of a certain parallelepiped obtained from $\Pi_{r,l}$ by a linear transformation, i.e., it is constructed on a bounded d-dimensional manifold in contrast to the "ordinary" mean-square approximations of Chap. 1 and to the approximations of Sects. 6.1–6.2, which are constructed on the d-dimensional unbounded manifolds.

On the basis of the one-step approximation (6.0.6), we form a Markov chain $(\bar{\vartheta}_k, \bar{X}_k)$ which belongs to Q at each step and approximates the points $(\bar{\vartheta}_k, X(\bar{\vartheta}_k))$ of the trajectory $(s, X_{t,x}(s))$, $s \geq t$, in the mean-square sense (see Algorithm 6.4.3). Section 6.3 is devoted to simulation of space-time Brownian motions which is the basis for our algorithms. A global algorithm is proposed and convergence theorems are proved in Sect. 6.4. An approximation for the space-time exit point is also constructed in Sect. 6.4. Using results of Sect. 6.4, we consider algorithms for FBSDEs with random terminal time in Sect. 10.3 (see also [317]). Numerical examples are given in Sect. 6.5.

6.1   Mean-Square Approximation for Autonomous SDEs without Drift in a Space Bounded Domain

In this section we consider the autonomous system of SDEs (see (6.0.1))

$$dX = \chi_{\tau_x > t} \sigma(X) dw(t), \ X(0) = x, \tag{6.1.1}$$

in a bounded domain $G \subset \mathbf{R}^d$ with a boundary ∂G.

The following conditions are assumed to be satisfied:

(i) G is a convex open bounded set with the twice continuously differentiable boundary ∂G;
(ii)   the coefficients $\sigma^{ij}(x)$ belong to the class $C^{(2)}(\bar{G})$;
(iii)   the matrix

$$a(x) = \sigma(x)\sigma^{\mathsf{T}}(x), \ a(x) = \{a^{ij}(x)\},$$

satisfies the strict ellipticity condition, i.e.,

$$\lambda_1^2 = \min_{x \in \bar{G}} \min_{1 \le i \le d} \lambda_i^2(x) > 0,$$

where $\lambda_1^2(x) \le \lambda_2^2(x) \le \cdots \le \lambda_d^2(x)$ are eigenvalues of the matrix $a(x)$.

Also introduce $\lambda_d^2 = \max_{x \in \bar{G}} \lambda_d^2(x)$. Then for any $x \in \bar{G}$, $y \in \mathbf{R}^d$ the following inequality

$$\lambda_1^2 \sum_{i=1}^d (y^i)^2 \le \sum_{i,j=1}^d a^{ij}(x) \, y^i y^j \le \lambda_d^2 \sum_{i=1}^d (y^i)^2 \tag{6.1.2}$$

holds.

Due to (6.1.2), the first exit time τ_x, at which the path $X_x(t)$ leaves the region G, is finite with probability one. We shall consider the process $X_x(t)$ defined on $0 \le t < \infty$ regarding it as the stopped one after τ_x.

A local approximation theorem is given in the next subsection. In Sect. 6.1.2 we prove two convergence theorems. The first one is on approximation properties of the sequence $\bar{X}_0, \ldots, \bar{X}_{\bar{\nu}}$ until leaving an open domain $D \subset G$ with $\rho(\partial D, \partial G) > 0$, which does not depend on r. In the second convergence theorem, the point $\bar{X}_{\bar{\nu}}$ belongs to a boundary layer which decreases in a prescribed way with decreasing r, i.e., $\bar{X}_{\bar{\nu}}$ becomes sufficiently close to ∂G with decreasing r (more precisely, $\rho(\bar{X}_{\bar{\nu}}, \partial G) =$

$O(r^{1-\varepsilon})$ with a sufficiently small $\varepsilon > 0$). In both situations the mean-square order of accuracy is equal to $O(r)$. The second theorem is important for approximation of the exit point $X_x(\tau_x)$. It is shown (Sect. 6.1.3) that this point can be approximated by $\bar{X}_{\bar{\nu}}$ with a mean-square order which is close to $O(\sqrt{r})$.

6.1.1 Local Approximation of Diffusion in a Space Bounded Domain

In Sects. 6.1.1–6.1.3, $X_x(t)$ is the solution of the problem (6.1.1), $X_{t_0,x}(t)$, $t \geq t_0$, is the solution of the system from (6.1.1) with initial data $X(t_0) = x$, and $\bar{X}_x(t)$ is found from (6.0.2).

Let Γ_δ be the interior of a δ-neighborhood of the boundary ∂G belonging to G. Obviously, if $x \in G \backslash \Gamma_{2\lambda_d r}$, then the inclusion $U_r^\sigma(x) \subset U_{2r}^\sigma(x) \subset G$ holds for all sufficiently small r.

Theorem 6.1.1 *For every natural number n there exists a constant $K > 0$ such that for any sufficiently small $r > 0$ and for any $x \in G \backslash \Gamma_{2\lambda_d r}$ the following inequality*

$$E|X_x(\bar{\theta}) - \bar{X}_x(\bar{\theta})|^{2n} \leq K r^{4n} \tag{6.1.3}$$

holds.

Proof Let the Markov moment θ be the first time at which the process $X_x(t)$ leaves the ellipsoid $U_{2r}^\sigma(x)$. First we prove the theorem for $n = 1$. We have

$$E|X_x(\bar{\theta}) - \bar{X}_x(\bar{\theta})|^2 = E\left|\int_0^{\bar{\theta}} (\chi_{\tau_x > s}\sigma(X_x(s)) - \sigma(x))dw(s)\right|^2 \tag{6.1.4}$$

$$= E\int_0^{\bar{\theta}} |\chi_{\tau_x > s}\sigma(X_x(s)) - \sigma(x)|^2 ds$$

$$= E\int_0^{\bar{\theta} \wedge \theta} |\sigma(X_x(s)) - \sigma(x)|^2 ds$$

$$+ E\int_{\bar{\theta} \wedge \theta}^{\bar{\theta}} |\chi_{\tau_x > s}\sigma(X_x(s)) - \sigma(x)|^2 ds$$

$$\leq E\int_0^{\bar{\theta} \wedge \theta} |\sigma(X_x(s)) - \sigma(x)|^2 ds + K \times E(\bar{\theta} - \bar{\theta} \wedge \theta).$$

Here the notation $|x|$ means the Euclidean norm of a vector x and $|\sigma|$ means $(\operatorname{tr}\sigma\sigma^{\mathsf{T}})^{1/2}$ of a matrix σ.

Since $E\bar{\theta} = r^2/d$, then $E(\bar{\theta} \wedge \theta) \le r^2/d$. Further, $X_x(s) \in U^{\sigma}_{2r}(x)$ for $s \in (0, \bar{\theta} \wedge \theta)$. Therefore

$$E|X_x(\bar{\theta} \wedge \theta) - \bar{X}_x(\bar{\theta} \wedge \theta)|^2 = E \int_0^{\bar{\theta}\wedge\theta} |\sigma(X_x(s)) - \sigma(x)|^2 ds \qquad (6.1.5)$$

$$\le Kr^2 \, E(\bar{\theta} \wedge \theta) \le Kr^4.$$

Using (6.1.2), it is easy to show that if $\xi \in \overline{U}^{\sigma}_r(x)$, $\eta \in \partial U^{\sigma}_{2r}(x)$, then $|\xi - \eta| \ge \lambda_1 r$. Because $\bar{X}_x(\bar{\theta} \wedge \theta) \in \overline{U}^{\sigma}_r(x)$, $X_x(\theta) \in \partial U^{\sigma}_{2r}(x)$, we have for every $m > 0$

$$E(\chi_{\theta<\bar{\theta}}|X_x(\bar{\theta} \wedge \theta) - \bar{X}_x(\bar{\theta} \wedge \theta)|^m) = E(\chi_{\theta<\bar{\theta}}|X_x(\theta) - \bar{X}_x(\bar{\theta} \wedge \theta)|^m) \qquad (6.1.6)$$

$$\ge P(\theta < \bar{\theta}) \, \lambda_1^m r^m .$$

On the other hand

$$E(\chi_{\theta<\bar{\theta}}|X_x(\bar{\theta} \wedge \theta) - \bar{X}_x(\bar{\theta} \wedge \theta)|^m) \qquad (6.1.7)$$

$$\le (P(\theta < \bar{\theta}))^{1/2} \times (E|X_x(\bar{\theta} \wedge \theta) - \bar{X}_x(\bar{\theta} \wedge \theta)|^{2m})^{1/2}$$

$$= (P(\theta < \bar{\theta}))^{1/2} \times (E| \int_0^{\bar{\theta}\wedge\theta} (\sigma(X_x(s)) - \sigma(x))dw(s)|^{2m})^{1/2}.$$

Let i be one of the indices $1, \ldots, d$. Introduce the variable

$$Z(t) = X^i_x(\bar{\theta} \wedge \theta \wedge t) - \bar{X}^i_x(\bar{\theta} \wedge \theta \wedge t)$$

$$= \int_0^{\bar{\theta}\wedge\theta\wedge t} \sum_{j=1}^d (\sigma^{ij}(X_x(s)) - \sigma^{ij}(x))dw^j(s)$$

$$= \int_0^t \chi_{\bar{\theta}\wedge\theta\ge s}\varphi(s)dw(s),$$

where $\varphi(s)$ is the i-th row vector of the matrix $\sigma(X_x(s)) - \sigma(x)$. We do not write the index i at Z and φ because this does not lead to any misunderstanding. Clearly, $Z(t)$, $t \ge 0$, is a uniformly bounded scalar, and

$$|\varphi(s)| \le |\sigma(X_x(s)) - \sigma(x)| \le Kr, \; 0 \le s \le \bar{\theta} \wedge \theta .$$

We have for every natural $m \geq 1$:

$$dZ^{2m}(t) = 2mZ^{2m-1}(t)\chi_{\bar{\theta} \wedge \theta \geq t}\varphi(t)dw(t) + m(2m-1)Z^{2m-2}(t)\chi_{\bar{\theta} \wedge \theta \geq t}|\varphi(t)|^2 dt .$$

Hence

$$EZ^{2m}(t) = m(2m-1)E\int_0^t Z^{2m-2}(s)\chi_{\bar{\theta} \wedge \theta \geq s}|\varphi(s)|^2 ds$$

$$\leq Km(2m-1)r^2 \times E(\bar{\theta} \wedge \theta \times \max_{0 \leq s \leq t}|Z(s)|^{2m-2}) .$$

Applying the Hölder inequality with $p = 2m/(2m-2)$ (see such a recipe, e.g., in [123]) and taking into account that (see (7.4.29))

$$E(\bar{\theta} \wedge \theta)^m \leq E\bar{\theta}^m \leq \frac{m!}{d^m}r^{2m},$$

we get

$$E|Z(t)|^{2m} \leq Km(2m-1)r^2 \times (E\max_{0 \leq s \leq t}|Z(s)|^{2m})^{(2m-2)/2m} \qquad (6.1.8)$$

$$\times (E(\bar{\theta} \wedge \theta)^m)^{1/m}$$

$$\leq Km(2m-1)r^4 \times (E\max_{0 \leq s \leq t}|Z(s)|^{2m})^{(2m-2)/2m} .$$

As $Z(t)$ is a martingale, we can use the Doob inequality

$$E\max_{0 \leq s \leq t}|Z(s)|^{2m} \leq (\frac{2m}{2m-1})^{2m}E|Z(t)|^{2m} .$$

Now we obtain from (6.1.8):

$$E|Z(t)|^{2m} \leq Kr^{4m},$$

where K does not depend on t (of course, K depends on m).
Hence

$$E\left|\int_0^{\bar{\theta} \wedge \theta}(\sigma(X_x(s)) - \sigma(x))dw(s)\right|^{2m} \leq Kr^{4m}. \qquad (6.1.9)$$

The inequalities (6.1.6), (6.1.7), and (6.1.9) imply

$$P(\theta < \bar{\theta}) \times \lambda_1^m r^m \leq K \times (P(\theta < \bar{\theta}))^{1/2} \times r^{2m}.$$

Therefore, for every positive m (recall K depends on m)

$$P(\theta < \bar{\theta}) \leq Kr^{2m}. \tag{6.1.10}$$

Further,

$$E(\bar{\theta} - \bar{\theta} \wedge \theta) = E\chi_{\theta < \bar{\theta}}(\bar{\theta} - \bar{\theta} \wedge \theta) \leq (P(\theta < \bar{\theta}))^{1/2} (E(\bar{\theta} - \bar{\theta} \wedge \theta)^2)^{1/2}$$
$$\leq (P(\theta < \bar{\theta}))^{1/2} (E\bar{\theta}^2)^{1/2} \leq K(P(\theta < \bar{\theta}))^{1/2} r^2,$$

whence

$$E(\bar{\theta} - \bar{\theta} \wedge \theta) \leq Kr^{m+2}. \tag{6.1.11}$$

Using this inequality for $m = 2$ together with (6.1.4) and (6.1.5), we arrive at (6.1.3) for $n = 1$. Thus, the theorem is proved for $n = 1$.

For an arbitrary positive integer n, we get

$$E|X_x(\bar{\theta}) - \bar{X}_x(\bar{\theta})|^{2n} \tag{6.1.12}$$

$$= E| \int_0^{\bar{\theta} \wedge \theta} (\sigma(X_x(s)) - \sigma(x))dw(s) + \int_{\bar{\theta} \wedge \theta}^{\bar{\theta}} (\chi_{\tau_x > s}\sigma(X_x(s)) - \sigma(x))dw(s)|^{2n}$$

$$\leq KE| \int_0^{\bar{\theta} \wedge \theta} (\sigma(X_x(s)) - \sigma(x))dw(s)|^{2n}$$

$$+ KE| \int_{\bar{\theta} \wedge \theta}^{\bar{\theta}} (\chi_{\tau_x > s}\sigma(X_x(s)) - \sigma(x))dw(s)|^{2n},$$

where the constant K depends on n only. The first term on the right-hand side is bounded by Kr^{4n} due to (6.1.9). The second term can be bounded as follows (see (6.1.4) and (6.1.11) for $m = 4n - 2$) :

$$E| \int_{\bar{\theta} \wedge \theta}^{\bar{\theta}} [\chi_{\tau_x > s}\sigma(X_x(s)) - \sigma(x)]dw(s)|^{2n}$$

$$= E(| \int_{\bar{\theta} \wedge \theta}^{\bar{\theta}} [\chi_{\tau_x > s}\sigma(X_x(s)) - \sigma(x)]dw(s)|^2$$

$$\times |X_x(\bar{\theta}) - X_x(\bar{\theta} \wedge \theta) - \bar{X}_x(\bar{\theta}) + \bar{X}_x(\bar{\theta} \wedge \theta)|^{2n-2})$$

$$\leq KE| \int_{\bar{\theta} \wedge \theta}^{\bar{\theta}} [\chi_{\tau_x > s}\sigma(X_x(s)) - \sigma(x)]dw(s)|^2 \leq KE(\bar{\theta} - \bar{\theta} \wedge \theta) \leq Kr^{4n}.$$

Now (6.1.12) implies (6.1.3). Theorem 6.1.1 is proved. □

Remark 6.1.2 Clearly, the inequality (6.1.10) remains true if θ is the first time at which the process $X_x(t)$ leaves the ellipsoid $U^\sigma_{(1+\alpha)r}(x)$ for any $\alpha > 0$. Therefore, the condition $x \in G \backslash \Gamma_{2\lambda_d r}$ in Theorem 6.1.1 may be replaced by $x \in G \backslash \Gamma_{(1+\alpha)\lambda_d r}$, $\alpha > 0$. Moreover, it is not difficult to show that the theorem remains true under the condition $x \in G \backslash \Gamma_{(1+r^\beta)\lambda_d r}$ if only $0 \le \beta < 2$. But for definiteness we take here and in what follows the layer $\Gamma_{2\lambda_d r}$.

6.1.2 Global Algorithm for Diffusion in a Space Bounded Domain

Algorithm 6.1.3 *Let $\bar{\theta}_1$ be the first time at which the Wiener process $w(t)$ leaves the sphere U_r, $\bar{\theta}_1 + \bar{\theta}_2$ be the first time at which the process $w(t) - w(\bar{\theta}_1)$, $t \ge \bar{\theta}_1$, leaves the same sphere U_r and so on. Let $x \in G \backslash \Gamma_{2\lambda_d r}$. We construct a recurrence sequence of random vectors $\bar{X}_k$, $k = 0, 1, \ldots, \bar{\nu}$:*

$$\begin{aligned}
\bar{X}_0 &= x \\
\bar{X}_1 &= \bar{X}_0 + \sigma(\bar{X}_0) w(\bar{\theta}_1) \\
&\cdot \cdot \cdot \cdot \cdot \cdot \cdot \cdot \cdot \cdot \cdot \cdot \\
\bar{X}_{k+1} &= \bar{X}_k + \sigma(\bar{X}_k)(w(\bar{\theta}_1 + \cdots + \bar{\theta}_{k+1}) - w(\bar{\theta}_1 + \cdots + \bar{\theta}_k)), \\
&\cdot \cdot \cdot \cdot \cdot \cdot \cdot \cdot \cdot \cdot \cdot \cdot
\end{aligned}$$

where $\bar{\nu} = \bar{\nu}_x$ is the first number for which $\bar{X}_k \in \Gamma_{2\lambda_d r}$.

Of course, the random moment $\bar{\nu}$ also depends on the domain $G \backslash \Gamma_{2\lambda_d r}$ which is left by $\bar{X}_{\bar{\nu}}$. Therefore, the more detailed notation for $\bar{\nu} = \bar{\nu}_x$ is $\bar{\nu} = \bar{\nu}_x(G \backslash \Gamma_{2\lambda_d r})$. Let us set $\bar{\theta}_k = 0$ and $\bar{X}_k = \bar{X}_{\bar{\nu}}$ for $k > \bar{\nu}$.

We have obtained the random walk

$$\bar{X}_0, \ldots, \bar{X}_k, \ldots,$$

which stops at a random step $\bar{\nu}$. It is a Markov chain.

We start consideration of properties of this Markov chain by obtaining some average characteristics of $\bar{\nu} = \bar{\nu}_x$. In connection with the homogeneous Markov chain $\bar{X}_k$, we introduce the one-step transition function

$$P(x, B) = P\left(\bar{X}_1 \in B \mid \bar{X}_0 = x\right),$$

where B is a Borel set belonging to $\bar{G}$.

Define an operator P acting on functions $v(x)$, $x \in \bar{G}$, by the formula

$$Pv(x) = \int_{\bar{G}} P(x, dy) v(y) = Ev(\bar{X}_1), \quad \bar{X}_0 = x,$$

and an operator
$$Av(x) = Pv(x) - v(x)$$

which is called the generator of the chain. The generator gives an average increment of the function v on the trajectory of the considered chain per step.

Consider the boundary value problem in $\bar{G}$:

$$Pv(x) - v(x) = -g(x), \quad x \in G \setminus \Gamma_{2\lambda_d r}, \tag{6.1.13}$$

$$v(x) = 0, \quad x \in \Gamma_{2\lambda_d r}, \tag{6.1.14}$$

which is connected with the chain $\bar{X}_k$.

The solution of the problem is the following function (see [450]):

$$v(x) = E \sum_{k=0}^{\bar{\nu}_x - 1} g(\bar{X}_k), \quad \bar{X}_0 = x. \tag{6.1.15}$$

If $g \equiv 1$ then
$$v(x) = E\bar{\nu}_x.$$

Further, if $v(x)$ is the solution of the boundary value problem (6.1.13)–(6.1.14) with the function $g(x)$ satisfying the inequality

$$g(x) \geq 1$$

in $G \setminus \Gamma_{2\lambda_d r}$, then, thanks to (6.1.15), we obtain

$$E\bar{\nu}_x \leq v(x). \tag{6.1.16}$$

Lemma 6.1.4 *There exists a constant $C > 0$ depending only on a diameter of the domain G such that the inequality*

$$E\bar{\nu}_x \leq \frac{C}{\lambda_1^2 r^2} \tag{6.1.17}$$

holds.

Proof Introduce the function

$$V(x) = \begin{cases} A^2 - x^2, & x \in G \setminus \Gamma_{2\lambda_d r}, \\ 0, & x \in \Gamma_{2\lambda_d r}, \end{cases} \tag{6.1.18}$$

where constant A^2 is such that for all $x \in \bar{G}$ we have

$$A^2 - x^2 \geq 0, \; x \in \bar{G}.$$

This function satisfies the boundary condition (6.1.14).

Let a point x be such that $U_r^\sigma(x) \subset G \setminus \Gamma_{2\lambda_d r}$. Now we evaluate $PV(x) - V(x)$. The measure $P(x, B)$ concentrates on $\partial U_r^\sigma(x)$, and, due to the inclusion $U_r^\sigma(x) \subset G \setminus \Gamma_{2\lambda_d r}$, the function $V(y)$ on $\partial U_r^\sigma(x)$ is equal to $A^2 - y^2$. Let dS be an area element of the surface ∂U_r and S be the area of this surface. We have

$$PV(x) = EV(\bar{X}_1) = EV(x + \sigma(x) w(\bar{\theta})) \qquad (6.1.19)$$

$$= \frac{1}{S} \int_{\partial U_r} \left(A^2 - (x + \sigma(x) z)^2\right) dS$$

$$= A^2 - x^2 - \frac{2}{S} \int_{\partial U_r} (x, \sigma(x) z) \, dS - \frac{1}{S} \int_{\partial U_r} (\sigma(x) z)^2 \, dS.$$

Clearly,

$$\int_{\partial U_r} (x, \sigma(x) z) \, dS = 0,$$

and, due to the strict ellipticity condition

$$(\sigma(x) z)^2 \geq \lambda_1^2 \sum_{i=1}^n (z^i)^2 = \lambda_1^2 r^2,$$

the equality (6.1.19) implies

$$PV(x) - V(x) \leq -\lambda_1^2 r^2. \qquad (6.1.20)$$

Now let $x \in G \setminus \Gamma_{2\lambda_d r}$ but the part of $U_r^\sigma(x)$ can belong to $\Gamma_{2\lambda_d r}$. We temporarily introduce the function $\bar{V}(y)$ which is equal to $A^2 - y^2$ on the entire surface $\partial U_r^\sigma(x)$. Therefore, as in (6.1.19) and (6.1.20), we obtain

$$P\bar{V}(x) = A^2 - x^2 - \frac{1}{S} \int_{\partial U_r} (\sigma(x) z)^2 \, dS \leq A^2 - x^2 - \lambda_1^2 r^2.$$

Since $V(y) \leq \bar{V}(y)$ on $\partial U_r^\sigma(x)$, we have $PV(x) \leq P\bar{V}(x)$ and, consequently, the inequality (6.1.20) is proved for all $x \in G \setminus \Gamma_{2\lambda_d r}$.

It obviously follows from (6.1.20) that the function

$$v(x) = \frac{V(x)}{\lambda_1^2 r^2}$$

satisfies (6.1.13)–(6.1.14) with $g \geq 1$. Hence, in view of (6.1.16), we obtain (6.1.17) with $C = \max_{x \in \bar{G}} V(x)$.   □

Remark 6.1.5 It is clear that the proof of Lemma 6.1.4 remains unchanged if we use the function $V(x)$ of the form:

$$V(x) = \begin{cases} A^2 + (a, x) - x^2, & x \in G \setminus \Gamma_{2\lambda_d r}, \\ 0, & x \in \Gamma_{2\lambda_d r}, \end{cases}$$

where the constant A^2 and the vector a are such that for all $x \in \bar{G}$ the inequality

$$A^2 + (a, x) - x^2 \geq 0$$

holds. On account of the choice of a and A^2, the bound (6.1.17) can be strengthened. For example, let $x^* \in G$ be a point such that

$$rad\ G = \max_{x \in \partial G} (x - x^*)^2 = \min_{y \in \bar{G}} \max_{x \in \partial G} (x - y)^2 .$$

If we take V as

$$V(x) = \begin{cases} rad\ G - (x - x^*)^2, & x \in G \setminus \Gamma_{2\lambda_d r}, \\ 0, & x \in \Gamma_{2\lambda_d r}, \end{cases}$$

we obtain the inequality

$$E\bar{\nu}_x \leq \frac{rad\ G - (x - x^*)^2}{\lambda_1^2 r^2}, \quad x \in G \setminus \Gamma_{2\lambda_d r} .$$

Below we will need the result from [183, p. 297], which we present here in the form convenient for our purposes.

Lemma 6.1.6 *Let*

$$AV_1(x) \leq -\left(1 - e^{-\alpha}\right) V_1(x), \quad x \in G \setminus \Gamma_{2\lambda_d r},$$

where $V_1(x) \geq 1, x \in \bar{G}$. Then the inequality

$$Ee^{\alpha \bar{\nu}_x} \leq V_1(x)$$

holds.

We also need the next lemma.

Lemma 6.1.7 *For all sufficiently small r the following inequality holds*

$$E \left(1 + \frac{\lambda_1^2}{1+C} r^2 \right)^{\bar{\nu}_x} \leq 1 + C, \quad C = \max_{x \in \bar{G}} V(x), \tag{6.1.21}$$

where $V(x)$ is from (6.1.18).

Proof We take $V_1 = V + 1$. From (6.1.20) we get

$$AV_1(x) = PV_1(x) - V_1(x) \leq -\lambda_1^2 r^2, \quad x \in G \setminus \Gamma_{2\lambda_d r} .$$

For all sufficiently small r we obtain

$$AV_1(x) \leq -\lambda_1^2 r^2 \frac{V_1(x)}{1+C} = -\left(1 - e^{\ln(1 - \lambda_1^2 r^2 / (1+C))} \right) V_1(x), \quad x \in G \setminus \Gamma_{2\lambda_d r} .$$

Then it follows from Lemma 6.1.6 that

$$E \exp \left(-\bar{\nu}_x \ln \left(1 - \frac{\lambda_1^2 r^2}{1+C} \right) \right) = E \left(\frac{1}{1 - \lambda_1^2 r^2 / (1+C)} \right)^{\bar{\nu}_x} \leq V_1(x) \leq 1 + C$$

which implies (6.1.21). $\square$

Corollary 6.1.8 *The probability $P(\bar{\nu}_x \geq L/r^2)$ decreases exponentially as L increases. More precisely, for every $L > 0$*

$$P(\bar{\nu}_x \geq L/r^2) \leq (1 + C) e^{-\alpha_r \lambda_1^2 L/(1+C)} \tag{6.1.22}$$

with $\alpha_r \to 1$ as $r \to 0$. The constant C in (6.1.22) is the same as in (6.1.21).

The corollary easily follows from Lemma 6.1.7 by Chebyshev's inequality. To prove convergence theorems, we need the following lemma.

Lemma 6.1.9 *For every natural number n there exists a constant $K > 0$ such that for any sufficiently small $r > 0$ and for any $x, y \in G \setminus \Gamma_{2\lambda_d r}$ the inequality*

$$E \left| \int_0^{\bar{\theta}} (\chi_{\tau_x > s} \sigma(X_x(s)) - \chi_{\tau_y > s} \sigma(X_y(s))) dw(s) \right|^{2n} \tag{6.1.23}$$

$$\leq K |x - y|^{2n} r^{2n} + K r^{4n}$$

holds.

Proof We have

$$
\int_0^{\bar{\theta}} (\chi_{\tau_x>s}\sigma(X_x(s)) - \chi_{\tau_y>s}\sigma(X_y(s)))dw(s)
$$

$$
= \int_0^{\bar{\theta}} (\chi_{\tau_x>s}\sigma(X_x(s)) - \sigma(x))dw(s) - \int_0^{\bar{\theta}} (\chi_{\tau_y>s}\sigma(X_y(s)) - \sigma(y))dw(s)
$$

$$
+ \int_0^{\bar{\theta}} (\sigma(x) - \sigma(y))dw(s)
$$

$$
= (X_x(\bar{\theta}) - \bar{X}_x(\bar{\theta})) - (X_y(\bar{\theta}) - \bar{X}_y(\bar{\theta})) + (\sigma(x) - \sigma(y)) w(\bar{\theta}) .
$$

Hence

$$
|\int_0^{\bar{\theta}} (\chi_{\tau_x>s}\sigma(X_x(s)) - \chi_{\tau_y>s}\sigma(X_y(s)))dw(s)|^{2n}
$$

$$
= |(X_x(\bar{\theta}) - \bar{X}_x(\bar{\theta})) - (X_y(\bar{\theta}) - \bar{X}_y(\bar{\theta})) + (\sigma(x) - \sigma(y)) \times w(\bar{\theta})|^{2n}
$$

$$
\le K|X_x(\bar{\theta}) - \bar{X}_x(\bar{\theta})|^{2n} + K|X_y(\bar{\theta}) - \bar{X}_y(\bar{\theta})|^{2n} + K|\sigma(x) - \sigma(y)|^{2n} |w(\bar{\theta})|^{2n},
$$

where the constant K depends on n only.

Now Theorem 6.1.1 and the relations

$$
|\sigma(x) - \sigma(y)| \le K|x - y|, \quad |w(\bar{\theta})|^{2n} = r^{2n}
$$

imply (6.1.23).   □

Let D be an open domain such that $\bar{D} \subset G$ and $\Delta := \rho(\partial D, \partial G)$. We consider $r < \Delta$ so that $D \subset G\backslash\Gamma_{2\lambda_d r}$. Let $x \in D$ and $\bar{\nu} = \bar{\nu}_x = \bar{\nu}_x(D)$ be the first moment at which $\bar{X}_{\bar{\nu}} \in G\backslash D$. For brevity, we preserve the old notation $\bar{\nu}$ for the new Markov moment $\bar{\nu}_x(D)$ as this does not cause any confusion. As before, we set $\bar{\theta}_k = 0$ and $\bar{X}_k = \bar{X}_{\bar{\nu}}$ for $k > \bar{\nu}$, i.e., we stop the above constructed trajectory $\bar{X}_k$ at the moment $\bar{\nu} = \bar{\nu}_x(D) < \bar{\nu}_x(G\backslash\Gamma_{2\lambda_d r})$. Therefore, the inequality (6.1.17) is fulfilled for the moment $\bar{\nu} = \bar{\nu}_x(D)$ as well.

Now consider the sequence

$$
X_0 = x
$$
$$
X_1 = X_x(\bar{\theta}_1)
$$

$$
. \quad . \quad . \quad . \quad . \quad . \quad . \quad . \quad . \quad . \quad .
$$

$$
X_{k+1} = X_x(\bar{\theta}_1 + \cdots + \bar{\theta}_{k+1}) = X_{\bar{\theta}_1+\cdots+\bar{\theta}_k, X_k}(\bar{\theta}_1 + \cdots + \bar{\theta}_{k+1})
$$

$$
. \quad . \quad . \quad . \quad . \quad . \quad . \quad . \quad . \quad . \quad .
$$

which is connected with the solution of the system (6.0.1).

If $\bar{\theta}_1 + \cdots + \bar{\theta}_k \geq \tau_x$ then, of course, $X_k = X_x(\tau_x)$, and if $k > \bar{\nu} = \bar{\nu}_x(D)$ then $X_k = X_{\bar{\nu}}$ as $\bar{\theta}_{\bar{\nu}+1} = \cdots = \bar{\theta}_k = 0$. Thus, X_k stops at a random step $\bar{\nu} \wedge \kappa$, where $\kappa = \min\{k : \bar{\theta}_1 + \cdots + \bar{\theta}_k > \tau_x\}$ if $\tau_x < \bar{\theta}_1 + \cdots + \bar{\theta}_{\bar{\nu}}$ and $\kappa = \bar{\nu}$ otherwise. The sequence X_k, just as $\bar{X}_k$, is a Markov chain. Furthermore, both $\bar{X}_k$ and X_k are martingales over σ-algebras $\mathcal{F}_0 = \{\emptyset, \Omega\}$, $\mathcal{F}_k = \mathcal{F}_{\bar{\theta}_1 + \cdots + \bar{\theta}_k}$, $k = 1, 2, \ldots$.

Consider the sequences $\bar{X}_k$, X_k for $N = L/r^2$ steps. The closeness of $\bar{X}_k$ to X_k during N steps is established in the following theorem.

Theorem 6.1.10 *Let $\bar{\nu} = \bar{\nu}_x(D)$ be the first exit time of the approximate trajectory $\bar{X}_k$ from the domain D. There exist constants $K > 0$ and $\gamma > 0$ (which do not depend on x, r, L, and Δ) such that for any $x \in D$ and for any sufficiently small $r > 0$ the inequality*

$$(E \max_{1 \leq k \leq \bar{\nu} \wedge N} |X_k - \bar{X}_k|^2)^{1/2} = (E \max_{1 \leq k \leq N} |X_k - \bar{X}_k|^2)^{1/2} \leq \frac{K}{\Delta} e^{\gamma L} r \qquad (6.1.24)$$

holds.

Proof Let ν be the first number at which $X_\nu \in \Gamma_{2\lambda_d r}$. More precisely,

$$\nu = \begin{cases} \min\{k : X_k \in \Gamma_{2\lambda_d r}, \ k \leq \bar{\nu}\}, \\ \infty, \ X_k \notin \Gamma_{2\lambda_d r}, \ k = 1, \ldots, \bar{\nu}. \end{cases} \qquad (6.1.25)$$

Clearly, for a sufficiently small r (if only $D \subset G\backslash\Gamma_{2\lambda_d r}$ and $3\lambda_d r \leq \Delta/2$)

$$|X_\nu - \bar{X}_\nu| \geq \frac{\Delta}{2}, \ \text{if } \nu \leq \bar{\nu}. \qquad (6.1.26)$$

Introduce the sequences $\bar{X}_{\nu \wedge m}$, $X_{\nu \wedge m}$ stopped at ν and the differences

$$d_m = X_{\nu \wedge m} - \bar{X}_{\nu \wedge m}, \ m = 0, 1, \ldots .$$

As ν is a Markov moment with respect to the system of σ-algebras $(\mathcal{F}_m)$, the stopped sequences $(\bar{X}_{\nu \wedge m}, \mathcal{F}_m)$, $(X_{\nu \wedge m}, \mathcal{F}_m)$ and $(d_m, \mathcal{F}_m)$ are martingales. The sequence $\bar{X}_{\nu \wedge m}$ $(X_{\nu \wedge m})$ is the Markov chain $\bar{X}_m$ (X_m) stopped at the moment ν. This is equivalent to the fact that $\bar{\theta}_m = 0$ not only for $m > \bar{\nu}$ but also for $m > \nu$, i.e., we may consider $\bar{\theta}_m = 0$ for $m > \bar{\nu} \wedge \nu$. Consequently, if $\bar{\nu} \wedge \nu = k$ then $d_k = d_{k+1} = \cdots = d_N$. This implies $d_k^2 = d_{k+1}^2 = \cdots = d_N^2$.

We have

$$d_m = d_1 \chi_{\bar{\nu} \wedge \nu = 1} + \cdots + d_{m-1} \chi_{\bar{\nu} \wedge \nu = m-1} + d_m \chi_{\bar{\nu} \wedge \nu \geq m}$$

and

$$d_{m-1} = d_1 \chi_{\bar{\nu} \wedge \nu = 1} + \cdots + d_{m-2} \chi_{\bar{\nu} \wedge \nu = m-2} + d_{m-1} \chi_{\bar{\nu} \wedge \nu = m-1} + d_{m-1} \chi_{\bar{\nu} \wedge \nu \geq m} .$$

Therefore

$$d_m = d_{m-1} + (d_m - d_{m-1})\chi_{\bar{\nu}\wedge\nu\geq m} \ . \tag{6.1.27}$$

Analogously,

$$d_m^2 = d_{m-1}^2 + (d_m^2 - d_{m-1}^2)\chi_{\bar{\nu}\wedge\nu\geq m} \ .$$

We get

$$
\begin{aligned}
d_m &= X_m - \bar{X}_m = X_x(\bar{\theta}_1 + \cdots + \bar{\theta}_m) - \bar{X}_m \tag{6.1.28} \\
&= X_{\bar{\theta}_1 + \cdots + \bar{\theta}_{m-1}, X_{m-1}}(\bar{\theta}_1 + \cdots + \bar{\theta}_m) - \bar{X}_m \\
&= X_{\bar{\theta}_1 + \cdots + \bar{\theta}_{m-1}, X_{m-1}}(\bar{\theta}_1 + \cdots + \bar{\theta}_m) - X_{\bar{\theta}_1 + \cdots + \bar{\theta}_{m-1}, \bar{X}_{m-1}}(\bar{\theta}_1 + \cdots + \bar{\theta}_m) \\
&\quad + X_{\bar{\theta}_1 + \cdots + \bar{\theta}_{m-1}, \bar{X}_{m-1}}(\bar{\theta}_1 + \cdots + \bar{\theta}_m) - \bar{X}_m \ .
\end{aligned}
$$

The first difference at the right-hand side of (6.1.28) is the error of the solution due to the error in the initial data at the time $(\bar{\theta}_1 + \cdots + \bar{\theta}_{m-1})$ accumulated to the $(m-1)$-st step. The second difference is the one-step error at the m-th step.

For $m \leq \bar{\nu} \wedge \nu$ the vectors $\bar{X}_{m-1}$ and X_{m-1} belong to $G\backslash\Gamma_{2\lambda_{dr}}$, and we obtain from the equality (6.1.28):

$$\chi_{\bar{\nu}\wedge\nu\geq m}d_m \tag{6.1.29}$$

$$= \chi_{\bar{\nu}\wedge\nu\geq m}(X_{m-1} + \int_{\bar{\theta}_1 + \cdots + \bar{\theta}_{m-1}}^{\bar{\theta}_1 + \cdots + \bar{\theta}_m} \chi(s) \times \sigma(X_{\bar{\theta}_1 + \cdots + \bar{\theta}_{m-1}, X_{m-1}}(s))dw(s))$$

$$- \chi_{\bar{\nu}\wedge\nu\geq m}(\bar{X}_{m-1} + \int_{\bar{\theta}_1 + \cdots + \bar{\theta}_{m-1}}^{\bar{\theta}_1 + \cdots + \bar{\theta}_m} \bar{\chi}(s) \times \sigma(X_{\bar{\theta}_1 + \cdots + \bar{\theta}_{m-1}, \bar{X}_{m-1}}(s))dw(s))$$

$$+ \chi_{\bar{\nu}\wedge\nu\geq m}(X_{\bar{\theta}_1 + \cdots + \bar{\theta}_{m-1}, \bar{X}_{m-1}}(\bar{\theta}_1 + \cdots + \bar{\theta}_m) - \bar{X}_m) \ .$$

Here

$$\chi(s) := \chi_{\tau(\bar{\theta}_1 + \cdots + \bar{\theta}_{m-1}, X_{m-1}) > s} \ , \quad \bar{\chi}(s) := \chi_{\tau(\bar{\theta}_1 + \cdots + \bar{\theta}_{m-1}, \bar{X}_{m-1}) > s} \ ,$$

where $\tau(\bar{\theta}_1 + \cdots + \bar{\theta}_{m-1}, x)$ is a random time at which the path $X_{\bar{\theta}_1 + \cdots + \bar{\theta}_{m-1}, x}(t)$ leaves the region G.

For brevity, we also introduce the following notation

$$\sigma(s) := \sigma(X_{\bar{\theta}_1 + \cdots + \bar{\theta}_{m-1}, X_{m-1}}(s)), \quad \bar{\sigma}(s) := \sigma(X_{\bar{\theta}_1 + \cdots + \bar{\theta}_{m-1}, \bar{X}_{m-1}}(s)) \ .$$

From (6.1.29) and (6.1.27) we obtain

$$d_m - d_{m-1} = (d_m - d_{m-1})\chi_{\bar{\nu}\wedge\nu\geq m} \tag{6.1.30}$$

$$= \chi_{\bar{\nu}\wedge\nu\geq m} \int\limits_{\bar{\theta}_1+\cdots+\bar{\theta}_{m-1}}^{\bar{\theta}_1+\cdots+\bar{\theta}_m} (\chi(s)\times\sigma(s) - \bar{\chi}(s)\times\bar{\sigma}(s))dw(s)$$

$$+\chi_{\bar{\nu}\wedge\nu\geq m}(X_{\bar{\theta}_1+\cdots+\bar{\theta}_{m-1},\bar{X}_{m-1}}(\bar{\theta}_1+\cdots+\bar{\theta}_m) - \bar{X}_m).$$

Due to $\mathcal{F}_{m-1}$-measurability of the random variable $\chi_{\bar{\nu}\wedge\nu\geq m}$, the equality (6.1.30) implies

$$E(d_m - d_{m-1})^2$$

$$\leq 2E\chi_{\bar{\nu}\wedge\nu\geq m}E(|\int\limits_{\bar{\theta}_1+\cdots+\bar{\theta}_{m-1}}^{\bar{\theta}_1+\cdots+\bar{\theta}_m} (\chi(s)\times\sigma(s) - \bar{\chi}(s)\times\bar{\sigma}(s))dw(s)|^2 \mid \mathcal{F}_{m-1})$$

$$+2E\chi_{\bar{\nu}\wedge\nu\geq m}E(|X_{\bar{\theta}_1+\cdots+\bar{\theta}_{m-1},\bar{X}_{m-1}}(\bar{\theta}_1+\cdots+\bar{\theta}_m) - \bar{X}_m|^2 \mid \mathcal{F}_{m-1}).$$

By the conditional versions of Lemma 6.1.9 and Theorem 6.1.1 with $n = 1$, we obtain

$$E(d_m - d_{m-1})^2 \leq Kr^2 E(\chi_{\bar{\nu}\wedge\nu\geq m}d_{m-1}^2) + Kr^4 \leq Kr^2 Ed_{m-1}^2 + Kr^4, \tag{6.1.31}$$

where the constant K does not depend on x, r, L, and Δ.

Because $(d_m, \mathcal{F}_m)$ is a martingale, we have

$$Ed_m^2 = Ed_{m-1}^2 + E(d_m - d_{m-1})^2. \tag{6.1.32}$$

The relations (6.1.31) and (6.1.32) imply

$$Ed_m^2 \leq Ed_{m-1}^2 + Kr^2 Ed_{m-1}^2 + Kr^4, \quad d_0 = 0.$$

From here we get for $N = L/r^2$:

$$Ed_N^2 = E|X_{\nu\wedge N} - \bar{X}_{\nu\wedge N}|^2 \leq [(1 + Kr^2)^{L/r^2} - 1] \times Kr^2 \leq Ke^{2\gamma L} r^2, \tag{6.1.33}$$

where the constant $\gamma > 0$ does not depend on x, r, L, and Δ.

Further, it is not difficult to obtain that $X_{\bar{\nu}\wedge\nu\wedge N} = X_{\nu\wedge N}$, $\bar{X}_{\bar{\nu}\wedge\nu\wedge N} = \bar{X}_{\nu\wedge N}$. Indeed, this is evident for $\bar{\nu} \geq \nu \wedge N$. And this is valid for $\bar{\nu} < \nu \wedge N$ because both X and $\bar{X}$ stop after the moment $\bar{\nu}$. Hence,

$$E|X_{\bar{\nu}\wedge\nu\wedge N} - \bar{X}_{\bar{\nu}\wedge\nu\wedge N}|^2 \leq Ke^{2\gamma L} r^2. \tag{6.1.34}$$

Let us prove now that

$$P(\nu \leq \bar{\nu} \wedge N) \leq K \frac{e^{2\gamma L}}{\Delta^2} r^2. \tag{6.1.35}$$

In fact, due to (6.1.26), we have

$$E\chi_{\nu \leq \bar{\nu} \wedge N}|X_{\bar{\nu} \wedge \nu \wedge N} - \bar{X}_{\bar{\nu} \wedge \nu \wedge N}| = E\chi_{\nu \leq \bar{\nu} \wedge N}|X_\nu - \bar{X}_\nu| \tag{6.1.36}$$

$$\geq P(\nu \leq \bar{\nu} \wedge N) \times \frac{\Delta}{2}.$$

On the other hand, using (6.1.33), we get

$$E\chi_{\nu \leq \bar{\nu} \wedge N}|X_{\bar{\nu} \wedge \nu \wedge N} - \bar{X}_{\bar{\nu} \wedge \nu \wedge N}| \tag{6.1.37}$$
$$\leq (P(\nu \leq \bar{\nu} \wedge N))^{1/2} \times (E|X_{\bar{\nu} \wedge \nu \wedge N} - \bar{X}_{\bar{\nu} \wedge \nu \wedge N}|^2)^{1/2}$$
$$\leq K(P(\nu \leq \bar{\nu} \wedge N))^{1/2} \times e^{\gamma L} r.$$

The relations (6.1.36) and (6.1.37) imply (6.1.35).

Since $X_{\bar{\nu} \wedge N} = X_N$, $\bar{X}_{\bar{\nu} \wedge N} = \bar{X}_N$, we obtain from (6.1.34) and (6.1.35):

$$E|X_N - \bar{X}_N|^2 = E|X_{\bar{\nu} \wedge N} - \bar{X}_{\bar{\nu} \wedge N}|^2 \tag{6.1.38}$$
$$= E\chi_{\nu \geq \bar{\nu} \wedge N}|X_{\bar{\nu} \wedge N} - \bar{X}_{\bar{\nu} \wedge N}|^2 + E\chi_{\nu < \bar{\nu} \wedge N}|X_{\bar{\nu} \wedge N} - \bar{X}_{\bar{\nu} \wedge N}|^2$$
$$= E\chi_{\nu \geq \bar{\nu} \wedge N}|X_{\bar{\nu} \wedge \nu \wedge N} - \bar{X}_{\bar{\nu} \wedge \nu \wedge N}|^2$$
$$+ E\chi_{\nu < \bar{\nu} \wedge N}|X_{\bar{\nu} \wedge N} - \bar{X}_{\bar{\nu} \wedge N}|^2$$
$$\leq E|X_{\bar{\nu} \wedge \nu \wedge N} - \bar{X}_{\bar{\nu} \wedge \nu \wedge N}|^2 + KP(\nu \leq \bar{\nu} \wedge N) \leq K \frac{e^{2\gamma L}}{\Delta^2} r^2.$$

Using Doob's inequality for the martingale $(X_m - \bar{X}_m, \mathcal{F}_m)$, we arrive at (6.1.24). Theorem 6.1.10 is proved.   □

Remark 6.1.11 It follows from Theorem 6.1.13 that it is possible to avoid the multiplier $1/\Delta$ in (6.1.24), i.e., the following inequality

$$(E \max_{1 \leq k \leq \bar{\nu} \wedge N} |X_k - \bar{X}_k|^2)^{1/2} = (E \max_{1 \leq k \leq N} |X_k - \bar{X}_k|^2)^{1/2} \leq K e^{\gamma L} r \tag{6.1.39}$$

is valid.

Theorem 6.1.12 *Let* $\bar{\nu} = \bar{\nu}_x(D)$. *The inequality*

$$(E \max_{1 \leq k \leq \bar{\nu}} |X_k - \bar{X}_k|^2)^{1/2} \leq K(e^{\gamma L} r/\Delta + e^{-\alpha_r \lambda_1^2 L/2(1+C)}) \tag{6.1.40}$$

holds.

Proof Introduce two sets: $\mathcal{C} = \{\bar{\nu} \leq L/r^2\}$ and $\Omega \setminus \mathcal{C} = \{\bar{\nu} > L/r^2\}$. Due to (6.1.22) and (6.1.24), we have (below l is the diameter of G):

$$\begin{aligned}
E\left|X_{\bar{\nu}} - \bar{X}_{\bar{\nu}}\right|^2 &= E(\left|X_{\bar{\nu}} - \bar{X}_{\bar{\nu}}\right|^2; \mathcal{C}) + E(\left|X_{\bar{\nu}} - \bar{X}_{\bar{\nu}}\right|^2; \Omega \setminus \mathcal{C}) \quad (6.1.41) \\
&= E(\left|X_{\bar{\nu} \wedge N} - \bar{X}_{\bar{\nu} \wedge N}\right|^2; \mathcal{C}) + E(\left|X_{\bar{\nu}} - \bar{X}_{\bar{\nu}}\right|^2; \Omega \setminus \mathcal{C}) \\
&\leq E(\left|X_{\bar{\nu} \wedge N} - \bar{X}_{\bar{\nu} \wedge N}\right|^2) + l^2 \, P(\Omega \setminus \mathcal{C}) \\
&\leq K \frac{e^{2\gamma L}}{\Delta^2} r^2 + l^2 (1 + C) e^{-\alpha_r \lambda_1^2 L/(1+C)},
\end{aligned}$$

whence (6.1.40) follows. □

The domain D in Theorems 6.1.10 and 6.1.12 is not changed with decreasing r. Now consider the domain $G \setminus \Gamma_{cr^{1-1/n}}$, where $c > 0$ is a certain number and $n \geq 2$ is a natural number. Let $x \in G$ and let r be sufficiently small such that $\Gamma_{cr^{1-1/n}} \supset \Gamma_{2\lambda_d r}$ and $x \in G \setminus \Gamma_{cr^{1-1/n}}$. We construct the approximate phase trajectory $\bar{X}_k$ until its exit into the layer $\Gamma_{cr^{1-1/n}}$, i.e., we stop the approximate trajectory, which was constructed at the beginning of this subsection, at the moment $\bar{\nu} = \bar{\nu}_x(G \setminus \Gamma_{cr^{1-1/n}})$. This stopping moment satisfies the inequality

$$\bar{\nu}_x(G \setminus \Gamma_{cr^{1-1/n}}) < \bar{\nu}_x(G \setminus \Gamma_{2\lambda_d r}).$$

As before, we preserve the same notation both for $\bar{X}_k$ with the new stopping moment and for the stopping moment $\bar{\nu} = \bar{\nu}_x(G \setminus \Gamma_{cr^{1-1/n}})$ as there is no risk of ambiguity. And as before $N = L/r^2$. Theorems 6.1.13–6.1.15 are proved in [282].

Theorem 6.1.13 *Let $\bar{\nu} = \bar{\nu}_x(G \setminus \Gamma_{cr^{1-1/n}})$ be the first exit time of the approximate trajectory $\bar{X}_k$ from the domain $G \setminus \Gamma_{cr^{1-1/n}}$. There exist constants $K > 0$ and $\gamma > 0$ (which do not depend on x, r, and L) such that for any sufficiently small $r > 0$ the inequality*

$$(E \max_{1 \leq k \leq \bar{\nu} \wedge N} \left|X_k - \bar{X}_k\right|^2)^{1/2} = (E \max_{1 \leq k \leq N} \left|X_k - \bar{X}_k\right|^2)^{1/2} \leq K e^{\gamma L} r \quad (6.1.42)$$

holds.

Theorem 6.1.14 *Let $\bar{\nu} = \bar{\nu}_x(G \setminus \Gamma_{cr^{1-1/n}})$. The inequality*

$$(E \max_{1 \leq k \leq \bar{\nu}} \left|X_k - \bar{X}_k\right|^2)^{1/2} \leq K(e^{\gamma L} r + e^{-\alpha_r \lambda_1^2 L/2(1+C)})$$

is valid.

Theorem 6.1.15 *Let $n > 1$, $l \geq 1$ be some natural numbers and $\bar{\nu} = \bar{\nu}_x(G \setminus \Gamma_{cr^{1-1/n}})$ be the first exit moment of the approximate trajectory $\bar{X}_k$ from the domain $G \setminus \Gamma_{cr^{1-1/n}}$.*

*There exist constants $K > 0$ and $\gamma > 0$ (which do not depend on x, r, L) such that
for any sufficiently small $r > 0$ the inequality*

$$(E \max_{1 \le k \le \bar{\nu} \wedge N} |X_k - \bar{X}_k|^{2l})^{1/2l} = (E \max_{1 \le k \le N} |X_k - \bar{X}_k|^{2l})^{1/2l} \le K e^{\gamma L} r \qquad (6.1.43)$$

holds.

6.1.3 Simulation of the Exit Point $X_x(\tau_x)$

In the previous subsection we constructed the point $\bar{X}_N = \bar{X}_{\bar{\nu} \wedge N}$, where $N = L/r^2$,
$\bar{\nu} = \bar{\nu}_x(G \backslash \Gamma_{cr^{1-1/n}})$. What is the distance between $\bar{X}_N$ and the exit point $X_x(\tau_x)$?
Which point on ∂G can be taken as an approximation of $X_x(\tau_x)$?
  On the set $\mathcal{C} = \{\bar{\nu} \le L/r^2\}$ we have $\bar{X}_N = \bar{X}_{\bar{\nu}} \in \Gamma_{cr^{1-1/n}}$. Let $\xi_x(\omega)$, $\omega \in \mathcal{C}$, be
a point on ∂G such that

$$|\bar{X}_N - \xi_x| \le cr^{1-1/n}, \ \omega \in \mathcal{C}. \qquad (6.1.44)$$

It is natural to take this point as an approximation of the exit point $X_x(\tau_x)$ if $\bar{X}_N \in$
$\Gamma_{cr^{1-1/n}}$. Due to Theorem 6.1.13 and (6.1.44), we obtain

$$E(|X_N - \xi_x|^2 ; \mathcal{C}) \le K(c^2 + e^{2\gamma L}) \times r^{2-2/n} . \qquad (6.1.45)$$

We need the following auxiliary lemma.

Lemma 6.1.16 *There exists a constant K such that for any $x \in \bar{G}$, $y \in \partial G$ the
inequality*

$$E(X_x(\tau_x) - y)^2 \le K |x - y|$$

holds.

Proof Consider the Dirichlet problem

$$\frac{1}{2} \sum_{i,j=1}^{d} a^{ij}(x) \frac{\partial^2 u}{\partial x^i \partial x^j} = 0, \ x \in G,$$

$$u \mid_{\partial G} = (x - y)^2.$$

The solution of this problem is

$$u_y(x) = E(X_x(\tau_x) - y)^2.$$

Due to the conditions $(i) - (iii)$ from the beginning of Sect. 6.1, $u_y \in C^{(4)}(\bar{G})$ (see [330]). Since $u_y(y) = 0$, we obtain

$$u_y(x) = u_y(x) - u_y(y) \leq K |x - y| .$$

□

We have defined the variable $\xi_x(\omega)$ on $\mathcal{C}$ only. Now we complete the definition by letting $\xi_x(\omega)$ for $\omega \in \Omega \setminus \mathcal{C}$ be, e.g., the point on ∂G nearest to $\bar{X}_N$. According to Lemma 6.1.16, we have

$$E((X_x(\tau_x) - \xi_x)^2 \mid \mathcal{F}_N) = E((X_{X_N}(\tau_{X_N}) - \xi_x)^2 \mid \mathcal{F}_N) \leq K |X_N - \xi_x| .$$

Since $\mathcal{C} \in \mathcal{F}_N$, the above inequality and (6.1.45) imply

$$E((X_x(\tau_x) - \xi_x)^2; \mathcal{C}) \leq K E(|X_N - \xi_x|; \mathcal{C})$$
$$\leq K (E(|X_N - \xi_x|^2; \mathcal{C}))^{1/2} \leq K \left(c + e^{\gamma L}\right) \times r^{1-1/n} .$$

We can also evaluate the expectation $E(X_x(\tau_x) - \xi_x)^2$ analogously to (6.1.41). As a result, we obtain the following theorem.

Theorem 6.1.17 *Let* $\xi_x(\omega) \in \partial G$ *be the nearest point to* $\bar{X}_N$. *Then*

$$(E([X_x(\tau_x) - \xi_x]^2; \mathcal{C}))^{1/2} \leq K e^{\gamma L/2} \times r^{1/2 - 1/2n} ,$$
$$[E(X_x(\tau_x) - \xi_x)^2]^{1/2} \leq K e^{\gamma L/2} \times r^{1/2 - 1/2n} + K e^{-\alpha_r \lambda_1^2 L/2(1+C)} ,$$

where for clarity we reduced some non-essential constants.

6.2 Systems with Drift in a Space Bounded Domain

In this section we construct a local approximation for the autonomous system with drift

$$dX = \chi_{\tau_x > t} b(X) dt + \chi_{\tau_x > t} \sigma(X) dw(t), \quad X(0) = x_0 . \tag{6.2.1}$$

Freezing the coefficients at the point x_0, we obtain the system

$$d\bar{X} = b(x_0) dt + \sigma(x_0) dw(t), \quad \bar{X}(0) = x_0 . \tag{6.2.2}$$

Because of the drift $b(x_0)$, the symmetry is broken down. The distribution of the process $\bar{X}(\bar{\theta})$ on the surface of some ellipsoid, where $\bar{\theta}$ is the first exit time of $\bar{X}$ from the ellipsoid, is already not simple. To seek another surface with a simple distribution of $\bar{X}_{x_0}(\bar{\theta})$ on it is not an easy problem as well.

Instead of (6.2.2), let us take the process satisfying another equation:

$$d\bar{X} = \bar{b}\left(\bar{X}\right) dt + \bar{\sigma}\left(\bar{X}\right) dw\left(t\right), \quad \bar{X}\left(0\right) = x_0, \tag{6.2.3}$$

where

$$\bar{b}\left(x_0\right) = b\left(x_0\right), \ \bar{\sigma}\left(x_0\right) = \sigma\left(x_0\right).$$

It is clear that the solutions of (6.2.1) and (6.2.3) are close in a small neighborhood of the point x_0. If we are able to find $\bar{b}$, $\bar{\sigma}$, and some surface such that it is easy to construct the exit point $\bar{X}_{x_0}\left(\theta\right)$ then the problem of a local approximation of phase trajectories for systems with drift will be solved.

To this end, let us consider a sufficiently smooth one-to-one transformation $g : U_r^{\sigma}\left(x_0\right) \longrightarrow V_r^{\sigma}\left(x_0\right)$ with the inverse transformation $f : V_r^{\sigma}\left(x_0\right) \longrightarrow U_r^{\sigma}\left(x_0\right)$, where $V_r^{\sigma}\left(x_0\right) \subset G$ is a set. We have $f\left(g\left(x\right)\right) = x$ for $x \in U_r^{\sigma}\left(x_0\right)$ and $g\left(f\left(x\right)\right) = x$ for $x \in V_r^{\sigma}\left(x_0\right)$.

Theorem 6.2.1 *Let a transformation g be such that the relations*

$$g\left(x_0\right) = x_0 = f\left(x_0\right), \tag{6.2.4}$$

$$\frac{1}{2} \sum_{m,k=1}^{d} \frac{\partial^2 g}{\partial x^m \partial x^k}\left(x_0\right) \sum_{j=1}^{d} \sigma^{mj}\left(x_0\right) \sigma^{kj}\left(x_0\right) = b\left(x_0\right), \tag{6.2.5}$$

and

$$\left\{\frac{\partial g^i}{\partial x^m}\left(x_0\right)\right\} = I \tag{6.2.6}$$

are fulfilled (I is the identity matrix).
   Then the system (6.2.3) with

$$\bar{b}\left(x\right) = \frac{1}{2} \sum_{m,k=1}^{d} \frac{\partial^2 g}{\partial x^m \partial x^k}\left(f\left(x\right)\right) \sum_{j=1}^{d} \sigma^{mj}\left(x_0\right) \sigma^{kj}\left(x_0\right) \tag{6.2.7}$$

and

$$\bar{\sigma}\left(x\right) = \left\{\sum_{m=1}^{d} \frac{\partial g^i}{\partial x^m}\left(f\left(x\right)\right) \sigma^{mj}\left(x_0\right)\right\} \tag{6.2.8}$$

has the solution

$$\bar{X}\left(t\right) = g\left(x_0 + \sigma\left(x_0\right) w\left(t\right)\right), \ t \in \left[0, \bar{\theta}\right),$$

which belongs to $V_r^{\sigma}\left(x_0\right)$. For this solution, $\bar{\theta}$ is the first time at which $\bar{X}\left(t\right)$ leaves the domain $V_r^{\sigma}\left(x_0\right)$.

Proof It is clear that for $0 \leq t \leq \bar{\theta}$

$$\bar{X}(t) = g(x_0 + \sigma(x_0) w(t)) \in V_r^{\sigma}(x_0), \quad \bar{X}(0) = g(x_0) = x_0,$$
$$\bar{X}(\bar{\theta}) = g(x_0 + \sigma(x_0) w(\bar{\theta})) \in \partial V_r^{\sigma}(x_0),$$

and

$$x_0 + \sigma(x_0) w(t) = f(\bar{X}(t)), \quad 0 \leq t \leq \bar{\theta}. \tag{6.2.9}$$

Using Ito's formula and (6.2.9), we obtain

$$\begin{aligned}
d\bar{X} &= dg(x_0 + \sigma(x_0) w(t)) \\
&= \sum_{m=1}^{d} \frac{\partial g}{\partial x^m}(x_0 + \sigma(x_0) w(t)) \sum_{j=1}^{d} \sigma^{mj}(x_0)\, dw^j(t) \\
&\quad + \frac{1}{2} \sum_{m,k=1}^{d} \frac{\partial^2 g}{\partial x^m \partial x^k}(x_0 + \sigma(x_0) w(t)) \sum_{j=1}^{d} \sigma^{mj}(x_0)\, \sigma^{kj}(x_0)\, dt \\
&= \frac{1}{2} \sum_{m,k=1}^{d} \frac{\partial^2 g}{\partial x^m \partial x^k}(f(\bar{X}(t))) \sum_{j=1}^{d} \sigma^{mj}(x_0)\, \sigma^{kj}(x_0)\, dt \\
&\quad + \sum_{m=1}^{d} \frac{\partial g}{\partial x^m}(f(\bar{X}(t))) \sum_{j=1}^{d} \sigma^{mj}(x_0)\, dw^j(t).
\end{aligned}$$

Due to (6.2.4)–(6.2.6), it is obvious now that $\bar{X}(t)$ is the solution of the system (6.2.3) with the coefficients given by (6.2.7)–(6.2.8). □

To obtain some constructive methods, we form the transformation

$$g(x) = \left(g^1\left(x^1, \ldots, x^d\right), \ldots, g^d\left(x^1, \ldots, x^d\right) \right)^{\mathsf{T}}$$

as

$$g(x) = \left(g^1\left(x^1\right), \ldots, g^d\left(x^d\right) \right)^{\mathsf{T}},$$

where each component is a function of a single argument. Then the inversion of g is reduced to the inversion of functions of a single argument. Let $f^i\left(x^i\right)$ be the inverse function of $g^i\left(x^i\right)$. Then, the system (6.2.3) with (6.2.7)–(6.2.8) takes the form

$$d\bar{X}^i = \frac{1}{2}(g^i)''\left(f^i\left(\bar{X}^i\right)\right) \sum_{j=1}^{d} \left(\sigma^{ij}(x_0)\right)^2 dt \tag{6.2.10}$$

$$+ (g^i)'\left(f^i\left(\bar{X}^i\right)\right) \sum_{j=1}^{d} \sigma^{ij}(x_0)\, dw^j(t), \quad i = 1, \ldots, d.$$

The constraints (6.2.5) and (6.2.6) are written now as

$$\frac{1}{2}(g^i)''\left(x_0^i\right) = \frac{b^i\left(x_0\right)}{\sum_{k=1}^{d}\left(\sigma^{ik}\left(x_0\right)\right)^2} = \beta_0^i, \quad i = 1, \ldots, d,$$

and

$$(g^i)'\left(x_0^i\right) = 1, \quad i = 1, \ldots, d.$$

Let us give some concrete methods of constructing functions g^i. *The first method* is based on the function

$$g^i\left(x^i\right) = x^i + \beta_0^i\left(x^i - x_0^i\right)^2, \quad \beta_0^i = \frac{b^i\left(x_0\right)}{\sum_{k=1}^{d}[\sigma^{ik}\left(x_0\right)]^2}.$$

The corresponding inverse function $f^i\left(x^i\right)$ is equal to

$$f^i\left(x^i\right) = \begin{cases} x_i^0 + \left(-1 + \sqrt{1 + 4\beta_0^i\left(x^i - x_0^i\right)}\right)/2\beta_0^i, & \beta_0^i \neq 0, \\ x^i, & \beta_0^i = 0. \end{cases}$$

The system (6.2.10) acquires the form

$$d\bar{X}^i = b^i\left(x_0\right)dt + \sqrt{1 + 4\beta_0^i\left(\bar{X}^i - x_0^i\right)}\sum_{j=1}^{d}\sigma^{ij}\left(x_0\right)dw^j\left(t\right),$$

$$\bar{X}^i\left(0\right) = x_0^i,$$

and on the interval $\left[0, \bar{\theta}\right]$ it has the solution

$$\bar{X}^i\left(t\right) = x_0^i + \sum_{j=1}^{d}\sigma^{ij}\left(x_0\right)w^j\left(t\right) + \beta_0^i\left(\sum_{j=1}^{d}\sigma^{ij}\left(x_0\right)w^j\left(t\right)\right)^2,$$

which at the moment $\bar{\theta}$ belongs to $\partial V_r^\sigma\left(x_0\right) : \bar{X}\left(\bar{\theta}\right) \in \partial V_r^\sigma\left(x_0\right)$. The equation of the surface $\partial V_r^\sigma\left(x_0\right)$ has the form

$$\left(\sigma^{-1}\left(x_0\right)\lambda, \sigma^{-1}\left(x_0\right)\lambda\right) = r^2, \tag{6.2.11}$$

where λ is a vector with the components $\lambda^i, i = 1, \ldots, d$:

$$\lambda^i = \begin{cases} \left(-1 + \sqrt{1 + 4\beta_0^i\left(x^i - x_0^i\right)}\right)/2\beta_0^i, & \beta_0^i \neq 0, \\ x^i - x_0^i, & \beta_0^i = 0. \end{cases}$$

The second method is based on the functions

$$g^i \left(x^i\right) = x_0^i + \frac{x^i - x_0^i}{1 - \beta_0^i \left(x^i - x_0^i\right)} \, .$$

The inverse function f^i is equal to

$$f^i \left(x^i\right) = x_0^i + \frac{x^i - x_0^i}{1 + \beta_0^i \left(x^i - x_0^i\right)} \, .$$

The system (6.2.10) acquires the form

$$d\bar{X}^i = \left(1 + \beta_0^i \left(\bar{X}^i - x_0^i\right)\right)^3 b^i \left(x_0\right) dt$$

$$+ \left(1 + \beta_0^i \left(\bar{X}^i - x_0^i\right)\right)^2 \sum_{j=1}^{d} \sigma^{ij} \left(x_0\right) dw^j \left(t\right), \quad \bar{X}^i \left(0\right) = x_0^i \, ,$$

and on the interval $[0, \bar{\theta}]$ it has the solution

$$\bar{X}^i \left(t\right) = x_0^i + \frac{\sum_{j=1}^{d} \sigma^{ij} \left(x_0\right) w^j \left(t\right)}{1 - \beta_0^i \sum_{j=1}^{d} \sigma^{ij} \left(x_0\right) w^j \left(t\right)} \, .$$

The equation of the surface $\partial V_r^\sigma \left(x_0\right)$ has the form (6.2.11) with

$$\lambda^i = \frac{x^i - x_0^i}{1 + \beta_0^i \left(x^i - x_0^i\right)}, \quad i = 1, \ldots, d \, .$$

This surface is a slightly deformed ellipsoid.

6.3 Space-Time Brownian Motion

In this section we propose an algorithm for simulation of the exit point of the space-time Brownian motion from a space-time parallelepiped. This algorithm is the basis for the one-step approximation of the solution to (6.0.4) considered in Sect. 6.4. To make the exposition self-contained, we start with some auxiliary knowledge and distributions of a one-dimensional Wiener process.

6.3.1   Auxiliary Knowledge

Let G be a bounded domain in $\mathbf{R}^d$, $Q = [T_0, T_1) \times G$ be a cylinder in $\mathbf{R}^{d+1}$, $\Gamma = \overline{Q} \backslash Q$. The set Γ is a part of the boundary of the cylinder Q consisting of the upper base and the lateral surface.

Consider the first boundary value problem for the equation of parabolic type

$$\frac{\partial u}{\partial t} + \frac{1}{2} \sum_{i,j=1}^d a^{ij}(t, x) \frac{\partial^2 u}{\partial x^i \partial x^j} + \sum_{i=1}^d b^i(t, x) \frac{\partial u}{\partial x^i} + c(t, x)u \qquad (6.3.1)$$

$$+ e(t, x) = 0, \quad (t, x) \in Q,$$

with the initial condition on the upper base

$$u(T_1, x) = f(x), \quad x \in \overline{G}, \qquad (6.3.2)$$

and the boundary condition on the lateral surface

$$u(t, x) = g(t, x), \quad T_0 \le t \le T_1, \ x \in \partial G. \qquad (6.3.3)$$

Introduce the function φ defined on Γ which is equal to $f(x)$ on the upper base and to $g(t, x)$ on the lateral surface. Then the conditions (6.3.2)–(6.3.3) can be rewritten shortly as

$$u \mid \Gamma = \varphi. \qquad (6.3.4)$$

The coefficients $a^{ij} = a^{ji}$ are assumed to satisfy the property of strict ellipticity in $\overline{Q}$, i.e.,

$$\lambda_1^2 = \min_{(t,x) \in \overline{Q}} \min_{1 \le i \le d} \lambda_i^2(t, x) > 0,$$

where $\lambda_1^2(t, x) \le \lambda_2^2(t, x) \le \cdots \le \lambda_d^2(t, x)$ are eigenvalues of the matrix $a(t, x) = \{a^{ij}(t, x)\}$.

Let $\lambda_d^2 = \max_{(t,x) \in \overline{Q}} \lambda_d^2(t, x)$. Then, for any $(t, x) \in \overline{Q}$ and $y \in R^d$ the inequality

$$\lambda_1^2 \sum_{i=1}^d (y^i)^2 \le \sum_{i,j=1}^d a^{ij}(t, x) y^i y^j \le \lambda_d^2 \sum_{i=1}^d (y^i)^2 \qquad (6.3.5)$$

holds.

The solution to the problem (6.3.1), (6.3.4) has the following probabilistic representation [86], [123, p. 299]:

$$u(t, x) = E\left[\varphi(\tau, X_{t,x}(\tau))Y_{t,x,1}(\tau) + Z_{t,x,1,0}(\tau)\right], \qquad (6.3.6)$$

where $X_{t,x}(s)$, $Y_{t,x,y}(s)$, $Z_{t,x,y,z}(s)$, $s \geq t$, is the solution of the Cauchy problem to the following system of stochastic differential equations

$$dX = b(s, X)ds + \sigma(s, X)dw(s), \quad X(t) = x, \quad (6.3.7)$$
$$dY = c(s, X)Yds, \quad Y(t) = y,$$
$$dZ = e(s, X)Yds, \quad Z(t) = z.$$

Here the point (t, x) belongs to Q, $\tau = \tau_{t,x}$ is the first-passage time of the trajectory $(s, X_{t,x}(s))$ to the boundary Γ. In the system (6.3.7), Y and Z are scalars, $w(s) = (w^1(s), \ldots, w^d(s))^\mathsf{T}$ is a d-dimensional standard Wiener process, $b(s, x)$ is a column-vector of dimension d compounded from the coefficients $b^i(s, x)$, $\sigma(s, x)$ is a matrix of dimension $d \times d$ which is obtained from the equation

$$\sigma(s, x)\sigma^\mathsf{T}(s, x) = a(s, x), \quad a(s, x) = \{a^{ij}(s, x)\}. \quad (6.3.8)$$

Setting in (6.3.1), (6.3.7)

$$c = 0, \ e = 0, \ f = 0, \ g = \chi_{(\partial G)_0}(x), \quad (6.3.9)$$

where $(\partial G)_0 \subseteq \partial G$, we get the formula:

$$u(t, x) = P(\tau_{t,x} < T_1, \ X_{t,x}(\tau_{t,x}) \in (\partial G)_0), \ T_0 \leq t < T_1, \quad (6.3.10)$$

where the time $\tau_{t,x}$ is the first-passage time of the trajectory $X_{t,x}(s)$ to the boundary ∂G.

In particular, if

$$c = 0, \ e = 0, \ f = 0, \ g = 1, \quad (6.3.11)$$

then

$$u(t, x) = P(\tau_{t,x} < T_1), \ T_0 \leq t < T_1. \quad (6.3.12)$$

Setting in (6.3.1), (6.3.7)

$$c = 0, \ e = 0, \ f = \chi_{G_0}(x), \ g = 0, \quad (6.3.13)$$

where $G_0 \subset G$, we get the formula:

$$u(t, x) = P(\tau_{t,x} \geq T_1, \ X_{t,x}(T_1) \in G_0). \quad (6.3.14)$$

In autonomous case (i.e., a^{ij}, b^i, c, e, g do not depend on t) we shall consider the first boundary value problem for parabolic equations in the following form:

$$\frac{\partial u}{\partial t} = \frac{1}{2} \sum_{i,j=1}^{d} a^{ij}(x) \frac{\partial^2 u}{\partial x^i \partial x^j} + \sum_{i=1}^{d} b^i(x) \frac{\partial u}{\partial x^i} \qquad (6.3.15)$$

$$+ c(x)u + e(x), \quad t > 0, \ x \in G,$$

$$u(0, x) = f(x), \ x \in \overline{G}, \qquad (6.3.16)$$

$$u(t, x) = g(x), \ t > 0, \ x \in \partial G. \qquad (6.3.17)$$

Using (6.3.9)–(6.3.10) and (6.3.13)–(6.3.14), it is not difficult to obtain that the function

$$u(t, x) = P(\tau_{0,x} < t, \ X_{0,x}(\tau_{0,x}) \in (\partial G)_0), \ t > 0, \qquad (6.3.18)$$

is the solution of the problem (6.3.15)–(6.3.17) under (6.3.9); the function

$$u(t, x) = P(\tau_{0,x} < t), \ t > 0, \qquad (6.3.19)$$

is the solution of the problem (6.3.15)–(6.3.17) under (6.3.11); the function

$$u(t, x) = P(\tau_{0,x} \geq t, \ X_{0,x}(t) \in G_0) \qquad (6.3.20)$$

is the solution of the problem (6.3.15)–(6.3.17) under (6.3.13). Here $X_{0,x}(s)$ is the solution to the Cauchy problem

$$dX = b(X)ds + \sigma(X)dw(s), \ X(0) = x, \qquad (6.3.21)$$

and $\tau_{0,x}$ is the first-passage time of the trajectory $X_{0,x}(s)$ to the boundary ∂G.

6.3.2   Some Distributions for One-Dimensional Wiener Process

Some of the distributions for the Wiener process, which we give in this section (see Sects. 6.3.2 and 6.3.3), can be found in the literature. For instance, in [33, 87, 180] some distributions for the one-dimensional Wiener process are written down in a certain form. For completeness of the exposition, we derive all the distributions here and give them in the forms, which are suitable for practical realization.

Introduce the first-passage time $\tau_x := \tau_{0,x}$ of the one-dimensional Wiener process $x + W(t)$, $-1 \leq x \leq 1$, $t > 0$, to the boundary of the interval $[-1, 1]$. Let us derive analytical formulas for

$$u(t, x) = P(\tau_x < t).$$

From (6.3.15)–(6.3.17) under (6.3.11), we obtain that the function (see (6.3.19))

$$v(t, x) = u(t, x) - 1 = P(\tau_x < t) - 1$$

satisfies the following boundary value problem

$$\frac{\partial v}{\partial t} = \frac{1}{2}\frac{\partial^2 v}{\partial x^2}, \quad t > 0, \ -1 < x < 1, \tag{6.3.22}$$

$$v(0, x) = -1, \quad v(t, -1) = v(t, 1) = 0. \tag{6.3.23}$$

By the method of separation of variables, we get the formula

$$P(\tau_x < t) = 1 - \frac{4}{\pi}\sum_{k=0}^{\infty}\frac{(-1)^k}{2k+1}\cos\frac{\pi(2k+1)x}{2}\exp(-\frac{1}{8}\pi^2(2k+1)^2 t). \tag{6.3.24}$$

Further, extending the initial data in (6.3.22)–(6.3.23) in the odd way on the whole axis and solving the obtained Cauchy problem, we get another form for the same distribution

$$P(\tau_x < t) = 1 - \int_{-1}^{1} G(t, x, y)dy, \tag{6.3.25}$$

where

$$G(t, x, y) = \frac{1}{\sqrt{2\pi t}}\sum_{k=-\infty}^{\infty}[\exp(-\frac{1}{2t}(x - 4k - y)^2) \tag{6.3.26}$$

$$- \exp(-\frac{1}{2t}(x - (4k+2) + y)^2)].$$

We shall use the formulas (6.3.24) and (6.3.25) under $x = 0$. Denote $\tau = \tau_0$,

$$\mathcal{P}(t) := P(\tau < t),$$

and introduce the density $\mathcal{P}'(t)$. From (6.3.24) and (6.3.25) one can obtain the following lemma.

Lemma 6.3.1 *Let τ be the first-passage time of the one-dimensional standard Wiener process $W(t)$ to the boundary of the interval $[-1, 1]$. Then the following formulas for its distribution and density take place*

$$\mathcal{P}(t) = 1 - \frac{4}{\pi}\sum_{k=0}^{\infty}\frac{(-1)^k}{2k+1}\exp(-\frac{1}{8}\pi^2(2k+1)^2 t), \quad t > 0, \tag{6.3.27}$$

and

$$P(t) = 2 \sum_{k=0}^{\infty} (-1)^k \operatorname{erfc} \frac{2k+1}{\sqrt{2t}} \, , \quad t > 0, \tag{6.3.28}$$

$$P'(t) = \frac{\pi}{2} \sum_{k=0}^{\infty} (-1)^k (2k+1) \exp(-\frac{1}{8}\pi^2 (2k+1)^2 t) \, , \quad t > 0, \tag{6.3.29}$$

and

$$P'(t) = \frac{2}{\sqrt{2\pi t^3}} \sum_{k=0}^{\infty} (-1)^k (2k+1) \exp(-\frac{1}{2t}(2k+1)^2) \, , \quad t > 0. \tag{6.3.30}$$

Recall

$$\operatorname{erfc} x = \frac{2}{\sqrt{\pi}} \int_{x}^{\infty} \exp(-s^2) \, ds, \quad \operatorname{erfc} 0 = 1. \tag{6.3.31}$$

The formulas (6.3.27) and (6.3.29) are suitable for calculations in the case of large t, and the formulas (6.3.28) and (6.3.30) are suitable for small t. The remainders of the series (6.3.29) and (6.3.30) are evaluated by the quantities

$$r_k(t) = \frac{\pi}{2}(2k+3) \exp\left(-\frac{1}{8}\pi^2 (2k+3)^2 t\right)$$

and

$$\rho_k(t) = \frac{2}{\sqrt{2\pi t^3}}(2k+3) \exp\left(-\frac{1}{2t}(2k+3)^2\right),$$

respectively.

These quantities coincide for $t = \dfrac{2}{\pi}$ and

$$r_k(t) < r_k\left(\frac{2}{\pi}\right), \quad t > \frac{2}{\pi},$$

$$\rho_k(t) < r_k\left(\frac{2}{\pi}\right), \quad t < \frac{2}{\pi}.$$

If we take k, for example, equal to 2, then

$$r_2\left(\frac{2}{\pi}\right) = \frac{7\pi}{2}e^{-49\pi/4} < 2.13 \times 10^{-16},$$

and consequently,

$$
\bar{P}'(t) = \begin{cases} \dfrac{2}{\sqrt{2\pi t^3}}(e^{-1/2t} - 3e^{-9/2t} + 5e^{-25/2t}) , & 0 < t < \dfrac{2}{\pi}, \\ \dfrac{\pi}{2}(e^{-\pi^2 t/8} - 3e^{-9\pi^2 t/8} + 5e^{-25\pi^2 t/8}) , & t > \dfrac{2}{\pi}, \end{cases} \tag{6.3.32}
$$

differs from $P'(t)$ by a quantity of 2.13×10^{-16} on the whole interval $[0, \infty)$.
It is not difficult to evaluate that

$$
\bar{P}(t) = \int_0^t \bar{P}'(s)ds \tag{6.3.33}
$$

differs from $P(t)$ on the whole interval $[0, \infty)$ by $(8/7\pi)e^{-49\pi/4} < 7.04 \times 10^{-18}$.
This high accuracy is typically sufficient for practical calculations. See the curves of
the distribution $P(t)$ and its density $P'(t)$ in Fig. 6.3.1.

Denote the inverse function to P by P^{-1}, and let γ be a random variable uniformly
distributed on $[0, 1]$. Then the random variable

$$
\tau = P^{-1}(\gamma)
$$

is distributed by the law $P(t)$.

To simulate this law in practice, we have to solve the following equation

$$
\bar{P}(t) = \gamma . \tag{6.3.34}
$$

Let us note that because of the analytical simplicity of the function $\bar{P}(t)$ it is
natural to use the Newton method for solving (6.3.34).

Fig. 6.3.1 The distribution
function $P(t)$ and the density
$P'(t)$.

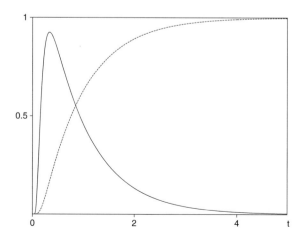

Lemma 6.3.2 *For the conditional probability*

$$Q(\beta; t) := P(W(t) < \beta \mid |W(s)| < 1, \ 0 < s < t) \, ,$$

where $-1 < \beta \leq 1$, *the following equalities hold:*

$$Q(\beta; t) = \frac{P(W(t) < \beta, \ \tau \geq t)}{P(\tau \geq t)} \tag{6.3.35}$$

$$= \frac{1}{1 - P(t)} \frac{2}{\pi} \sum_{k=0}^{\infty} \frac{1}{2k+1} \left((-1)^k + \sin \frac{\pi(2k+1)\beta}{2} \right)$$

$$\times \exp \left(-\frac{1}{8} \pi^2 (2k+1)^2 t \right) ,$$

and

$$Q(\beta; t) = \frac{1}{1 - P(t)} \tag{6.3.36}$$

$$\times \sum_{k=0}^{\infty} \frac{(-1)^k}{2} \left(\operatorname{erfc} \frac{2k-1}{\sqrt{2t}} - \operatorname{erfc} \frac{2k+\beta}{\sqrt{2t}} \right.$$

$$\left. - \operatorname{erfc} \frac{2k+2-\beta}{\sqrt{2t}} + \operatorname{erfc} \frac{2k+3}{\sqrt{2t}} \right) .$$

Proof The first equality in (6.3.35) flows out of equivalence of the events $\{|W(s)| < 1, \ 0 < s < t\}$ and $\{\tau \geq t\}$. Let us prove the second one. To this end consider the probability

$$u(t, x) = P(\tau_x \geq t, \ \alpha \leq x + W(t) < \beta),$$

where $\alpha \geq -1$.

Due to (6.3.15)–(6.3.17), (6.3.20) under (6.3.13), this probability is the solution of the boundary value problem:

$$\frac{\partial u}{\partial t} = \frac{1}{2} \frac{\partial^2 u}{\partial x^2}, \ t > 0, \ -1 < x < 1, \tag{6.3.37}$$

$$u(0, x) = \chi_{[\alpha, \beta)}(x), \ u(t, -1) = u(t, 1) = 0, \ t > 0. \tag{6.3.38}$$

Solving this problem, we get

$$u(t, x) = \frac{2}{\pi} \sum_{k=1}^{\infty} \frac{1}{k} \sin \frac{\pi k(\alpha + \beta)}{2} \sin \frac{\pi k(\beta - \alpha)}{2} \sin \pi kx \, \exp\left(-\frac{1}{2}\pi^2 k^2 t\right)$$

$$+ \frac{4}{\pi} \sum_{k=0}^{\infty} \frac{1}{2k + 1} \sin \frac{\pi(2k + 1)(\beta - \alpha)}{4} \cos \frac{\pi(2k + 1)(\beta + \alpha)}{4}$$

$$\times \cos \frac{\pi(2k + 1)x}{2} \exp\left(-\frac{1}{8}\pi^2(2k + 1)^2 t\right) .$$

As $P(W(t) < \beta, \ \tau \geq t) = u(t, 0)$ under $\alpha = -1$, $x = 0$, we arrive at (6.3.35) from here. The equality (6.3.36) follows from

$$u(t, x) = \frac{1}{\sqrt{2\pi t}} \int_{\alpha}^{\beta} G(t, x, y) dy$$

obtained analogously to (6.3.25). □

Let us note that the series (6.3.35) and (6.3.36) are of the Leibniz type, (6.3.35) is convenient for calculations in the case of large t, and (6.3.36) is convenient for small t. We draw the reader's attention to the denominator $(1 - \mathcal{P}(t))$ in (6.3.35), which is close to zero for $t \gg 1$. But it is not difficult to transform (6.3.35) to a form suitable for calculations. See the curves of the distribution $\mathcal{Q}(\beta; t)$ for some values of t in Fig. 6.3.2.

Let the function $\mathcal{Q}^{-1}(\cdot; t)$ be the inverse function to $\mathcal{Q}(\cdot; t)$ for every fixed t and γ be a random variable uniformly distributed on $[0, 1]$. Then the random variable

$$\xi = \mathcal{Q}^{-1}(\gamma; t)$$

has $\mathcal{Q}(\beta; t)$ as its distribution function.

Fig. 6.3.2 The distribution function $Q(\beta; \cdot)$; for $t \geq 0.5$ the curves visually coincide

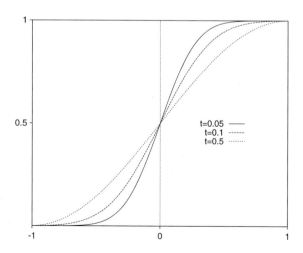

6.3.3 Simulation of the Exit Time and Exit Point of Wiener Process from a Cube

Let $C \subset \mathbf{R}^d$ be a d-dimensional cube with center at the origin and with the edge length equal to 2. We assume that all the edges of the cube are parallel to the coordinate axes, i.e., $C = \{x = (x^1, \ldots, x^d) : |x^i| < 1, \ i = 1, \ldots, d\}$. Let $W(s) = (W^1(s), \ldots, W^d(s))^\mathsf{T}$ be a d-dimensional standard Wiener process, τ be the first-passage time of $W(s)$ to the boundary ∂C of the cube C.

Let us give the following evident result in the form of a lemma.

Lemma 6.3.3 *The distribution function $P_d(t)$ for τ is equal to*

$$P_d(t) = P(\tau < t) = 1 - (1 - \mathcal{P}(t))^d \tag{6.3.39}$$

and the random variable

$$\tau = \mathcal{P}^{-1}(1 - \gamma^{1/d}) \tag{6.3.40}$$

is distributed by the law $P_d(t)$.

Our nearest goal is to construct an algorithm for simulation of the point $(\tau, W(\tau))$. To this end, we obtain some distributions connected with the d-dimensional Wiener process.

Lemma 6.3.4 *Let τ^j be the first-passage time of the component $W^j(t)$ to the boundary of the interval $[-1, 1]$. Then*

$$P \left(\bigcap_{i \neq j} (W^i(\tau^j) < \beta^i \ , \ |W^i(s)| < 1, \ 0 < s < \tau^j) \middle| \tau^j \right) \tag{6.3.41}$$

$$= (1 - \mathcal{P}(\tau^j))^{d-1} \prod_{i \neq j} \mathcal{Q}(\beta^i; \tau^j) .$$

Proof We shall use an assertion of the following kind: if $\zeta \geq 0$ is $\tilde{\mathcal{F}}$-measurable (where $\tilde{\mathcal{F}}$ is a σ-subalgebra of a general σ-algebra $\mathcal{F}$), a random variable $\varphi(t, \omega)$ for every $t \geq 0$ does not depend on $\tilde{\mathcal{F}}$ and $E\varphi(t, \omega) = h(t)$, then $E(\varphi(\zeta, \omega)| \ \tilde{\mathcal{F}}) = h(\zeta)$ (see [123, p. 67], [213, p. 158]).

Due to Lemma 6.3.2 and independence of the processes $W^i(s)$, we get for any $t \geq 0$,

$$P \left(\bigcap_{i \neq j} (W^i(t) < \beta^i, \ |W^i(s)| < 1, \ 0 < s < t) \right) = (1 - \mathcal{P}(t))^{d-1} \prod_{i \neq j} \mathcal{Q}(\beta^i; t) .$$

This equality implies (6.3.41) in accordance with the above-mentioned assertion because the processes $W^i(s)$, $i \neq j$, do not depend on the process $W^j(s)$. □

Introduce the random variable $\varkappa$ which takes the value j for $\omega \in \{\omega : W^j(\tau) = \pm 1\}$. This variable is defined uniquely with probability 1, and $P(\varkappa = j) = 1/d$. Let $\nu := W^\varkappa(\tau)$. Clearly, the distribution law for ν is given by $P(\nu = -1) = P(\nu = 1) = 1/2$.

Lemma 6.3.5 *The following equality takes place:*

$$P(\varkappa = j, \ \tau < \theta, \ \bigcap_{i \neq j}(W^i(\tau) < \beta^i)) \tag{6.3.42}$$

$$= \int_0^\theta (1 - \mathcal{P}(\vartheta))^{d-1} \prod_{i \neq j} \mathcal{Q}(\beta^i; \vartheta) \mathcal{P}'(\vartheta) d\vartheta .$$

Proof We have

$$P(\varkappa = j, \ \tau < \theta, \ \bigcap_{i \neq j}(W^i(\tau) < \beta^i))$$

$$= P(\bigcap_{i \neq j}(W^i(\tau^j) < \beta^i , \ |W^i(s)| < 1, \ 0 < s < \tau^j), \ \tau^j < \theta)$$

$$= \int_0^\theta P\left(\bigcap_{i \neq j}(W^i(\tau^j) < \beta^i , \ |W^i(s)| < 1, \ 0 < s < \tau^j) \ \middle|\ \tau^j = \vartheta \right) d\mathcal{P}_{\tau^j}(\vartheta) ,$$

where $\mathcal{P}_{\tau^j}(\vartheta)$ is the distribution function for τ^j. Clearly $\mathcal{P}_{\tau^j}(\vartheta) = \mathcal{P}(\vartheta)$. Now the assertion (6.3.42) follows from Lemma 6.3.4. □

Lemma 6.3.6 *The following equality holds:*

$$P\left(\bigcap_{i \neq j}(W^i(\tau) < \beta^i) \ \middle|\ \varkappa = j, \ \tau = \theta \right) = \prod_{i \neq j} \mathcal{Q}(\beta^i; \theta) . \tag{6.3.43}$$

Proof The random variables $\varkappa$ and τ are independent. Indeed, $P(\varkappa = 1, \ \tau < \theta) = \cdots = P(\varkappa = d, \ \tau < \theta)$ due to symmetry. Hence $P(\varkappa = i, \ \tau < \theta) = (1/d)P(\tau < \theta) = P(\varkappa = i)P(\tau < \theta)$. Further (see (6.3.39))

$$dP(\varkappa = j, \ \tau < \theta) = \frac{1}{d} d\mathcal{P}_d(\theta) = (1 - \mathcal{P}(\theta))^{d-1} \mathcal{P}'(\theta) d\theta .$$

From here we get

$$P\left(\varkappa = j, \ \tau < \theta, \ \bigcap_{i \neq j}(W^i(\tau) < \beta^i)\right) \qquad (6.3.44)$$

$$= \int_0^\theta P\left(\bigcap_{i \neq j}(W^i(\tau) < \beta^i)\ \middle|\ \varkappa = j, \ \tau = \vartheta\right)(1 - \mathcal{P}(\vartheta))^{d-1}\mathcal{P}'(\vartheta)d\vartheta\ .$$

Comparing (6.3.42) with (6.3.44), we obtain (6.3.43).   □

Let us note that the point $(\tau, \ W(\tau)) \in [0, \infty) \times \partial C$, i.e., this point belongs to the lateral surface of the unbounded semi-cylinder $[0, \infty) \times C$ with cubic base in $(d + 1)$-dimensional space of variables $(t, x^1, \ldots, x^d)$.

Theorem 6.3.7 (*Algorithm for simulating exit point to lateral surface of cylinder with cubic base*). *Let* $\varkappa, \ \nu, \ \gamma, \gamma^1, \ldots, \gamma^{d-1}$ *be independent random variables. Let* $\varkappa$ *and* ν *be simulated by the laws* $P(\varkappa = j) = 1/d, \ j = 1, \ldots, d; \ P(\nu = \pm 1) = 1/2$, *and let* $\gamma, \gamma^1, \ldots, \gamma^{d-1}$ *be uniformly distributed on* $[0, 1]$.
Then the point $(\tau, \ \xi) = (\tau, \xi^1, \ldots, \xi^d)$ *with*

$$\tau = \mathcal{P}^{-1}(1 - \gamma^{1/d}), \ \xi^1 = \mathcal{Q}^{-1}(\gamma^1; \tau), \ldots, \ \xi^{\varkappa-1} = \mathcal{Q}^{-1}(\gamma^{\varkappa-1}; \tau), \qquad (6.3.45)$$
$$\xi^\varkappa = \nu, \ \xi^{\varkappa+1} = \mathcal{Q}^{-1}(\gamma^\varkappa; \tau), \ldots, \ \xi^d = \mathcal{Q}^{-1}(\gamma^{d-1}; \tau)$$

has the same distribution as $(\tau, \ W(\tau))$.

This theorem is a simple consequence of Lemmas 6.3.3 and 6.3.6.

Corollary 6.3.8 *Let* $C_r = \{x = (x^1, \ldots, x^d): \ |x^i| < r, \ i = 1, \ldots, d\} \subset \mathbf{R}^d$ *be a d-dimensional cube with center at the origin and with edge length equal to $2r$. Let $\bar\theta$ be the first-passage time of the d-dimensional standard Wiener process $w(s)$ to the boundary ∂C_r of the cube C_r. Then the point*

$$(\bar\theta, \bar{w}) = (r^2\tau, r\xi),$$

where (τ, ξ) is simulated by the algorithm for simulating exit point to lateral surface of cylinder with the cubic base C, has the same distribution as $(\bar\theta, w(\bar\theta))$.

The proof easily follows from the fact that if $W(t)$ is a Wiener process, then $w(t) = rW(t/r^2)$ is a Wiener process as well.

6.3.4 Simulation of the Exit Point of the Space-Time Brownian Motion from a Space-Time Parallelepiped with Cubic Base

Now let us consider the space-time parallelepiped $\Pi = [0, l) \times C \subset \mathbf{R}^{d+1}$, where the cube $C \subset \mathbf{R}^d$ is defined at the beginning of the previous subsection. We will construct an algorithm for simulating the exit point $(\tau(l), W(\tau(l)))$ from the parallelepiped Π. The random variable $\tau(l)$ is found as $\min(\tau, l)$, where τ is the first-passage time of $W(s)$ to the boundary ∂C as above. The distribution function of $\tau(l)$ is equal to

$$P(\tau(l) < t) = \begin{cases} 1 - (1 - \mathcal{P}(t))^d, & t \leq l, \\ 1, & t > l. \end{cases} \tag{6.3.46}$$

Theorem 6.3.9 (*Algorithm for simulating exit point from a space-time parallelepiped with cubic base*). *Let $\iota, \varkappa, \nu, \gamma, \gamma^1, \ldots, \gamma^{d-1}$ be independent random variables. Let ι be simulated by the law*

$$P(\iota = -1) = 1 - (1 - \mathcal{P}(l))^d, \quad P(\iota = 1) = (1 - \mathcal{P}(l))^d,$$

and the random variables $\varkappa, \nu, \gamma, \gamma^1, \ldots, \gamma^{d-1}$ be simulated as in Theorem 6.3.7.

Then a random point $(\tau(l), \xi)$, distributed as the exit point $(\tau(l), W(\tau(l)))$, is simulated by the following algorithm.

If the simulated value of ι is equal to -1, then the point $(\tau(l), \xi)$ belongs to the lateral surface of Π, and

$$\tau(l) = \mathcal{P}^{-1}(1 - [1 - \gamma(1 - (1 - \mathcal{P}(l))^d]^{1/d}),$$
$$\xi^1 = \mathcal{Q}^{-1}(\gamma^1; \tau(l)), \ldots, \quad \xi^{\varkappa-1} = \mathcal{Q}^{-1}(\gamma^{\varkappa-1}; \tau(l)), \quad \xi^\varkappa = \nu,$$
$$\xi^{\varkappa+1} = \mathcal{Q}^{-1}(\gamma^\varkappa; \tau(l)), \ldots, \quad \xi^d = \mathcal{Q}^{-1}(\gamma^{d-1}; \tau(l));$$

otherwise, when $\iota = 1$, the point $(\tau(l), \xi)$ belongs to the upper base of Π, and

$$\tau(l) = l,$$
$$\xi^1 = \mathcal{Q}^{-1}(\gamma; l), \xi^2 = \mathcal{Q}^{-1}(\gamma^1; l), \ldots, \xi^d = \mathcal{Q}^{-1}(\gamma^{d-1}; l).$$

Proof Using Lemma 6.3.3, we have

$$P(\tau(l) < l) = P(\tau < l) = 1 - (1 - \mathcal{P}(l))^d, \tag{6.3.47}$$
$$P(\tau(l) = l) = P(\tau \geq l) = (1 - \mathcal{P}(l))^d.$$

The conditional probability $P(\tau(l) < t \mid \tau(l) < l)$ is equal to

$$P(\tau(l) < t \mid \tau(l) < l) = \frac{P((\tau(l) < t) \bigcap (\tau(l) < l))}{P(\tau(l) < l)}$$
$$= \chi_{[l,\infty)}(t) + \chi_{[0,l)}(t) \frac{P(\tau < t)}{P(\tau < l)},$$

and the random variable $\mathcal{P}^{-1}(1 - [1 - \gamma(1 - (1 - \mathcal{P}(l))^d]^{1/d})$ is distributed by the law $P(\tau(l) < t \mid \tau(l) < l)$.

Carrying out reasoning similar to Lemmas 6.3.4, 6.3.5, and 6.3.6, we obtain

$$
P\left(\bigcap_{i\neq j}(W^i(\tau(l)) < \beta^i)\ \middle|\ \varkappa = j, \tau(l) = \theta < l\right) = \chi_{[0,l)}(\theta)\prod_{i\neq j}\mathcal{Q}(\beta^i;\theta).
$$

(6.3.48)

Further, the equality

$$
P\left(\bigcap_{i=1}^{d}(W^i(\tau(l)) < \beta^i)\ \middle|\ \tau(l) = l\right)
$$

(6.3.49)

$$
= P\left(\bigcap_{i=1}^{d}(W^i(l) < \beta^i)\ \middle|\ \tau \geq l\right)
$$

$$
= \frac{1}{P(\tau \geq l)}P\left(\bigcap_{i=1}^{d}(W^i(l) < \beta^i,\ |W^i(s)| < 1,\ 0 < s < l)\right)
$$

$$
= \frac{1}{\displaystyle\prod_{i=1}^{d}P(\tau^i \geq l)}P\left(\bigcap_{i=1}^{d}(W^i(l) < \beta^i, \tau^i \geq l)\right) = \prod_{i=1}^{d}\mathcal{Q}(\beta^i; l)
$$

holds due to the mutual independence of the components W^i, $i = 1, \ldots, d$, and Lemma 6.3.2. Now the statement of the theorem easily follows from (6.3.47)–(6.3.49).   □

The following corollary has the same proof as Corollary 6.3.8.

Corollary 6.3.10 *Let* $\Pi_{r,l} = [0, lr^2) \times C_r = \{(t, x) = (t, x^1, \ldots, x^d) : 0 \leq t < lr^2,\ |x^i| < r,\ i = 1, \ldots, d\} \subset \mathbf{R}^{d+1}$ *be a space-time parallelepiped. Let* $\bar{\theta}$ *be the first-passage time of the process* $(s, w(s))$, $s > 0$, *to the boundary* $\partial\Pi_{r,l}$. *Then the point*

$$
(\bar{\theta}, \bar{w}) = (r^2\tau(l), r\xi),
$$

where $(\tau(l), \xi)$ *is simulated by the algorithm for simulating the exit point from the space-time parallelepiped* Π, *has the same distribution as* $(\bar{\theta}, w(\bar{\theta}))$.

6.4 Approximations for SDEs in a Space-Time Bounded Domain

The simulation of the space-time Brownian motion considered in the previous section is the basis for our algorithms. Local approximation theorem is proved in Sect. 6.4.1. A global algorithm is proposed and convergence theorems are proved in Sect. 6.4.2. An approximation for the space-time exit point is constructed in

Sect. 6.4.3. Sect. 6.4.4 briefly describes an algorithm for (6.0.4) which is based on simulation of exit points for the Brownian motion with drift. Numerical examples are given in Sect. 6.5.

6.4.1 Local Mean-Square Approximation in a Space-Time Bounded Domain

Let us consider the system of SDEs (see (6.0.4))

$$dX = \chi_{\tau_{t,x}>s}b(s, X)ds + \chi_{\tau_{t,x}>s}\sigma(s, X)dw(s), \quad X(t) = X_{t,x}(t) = x, \quad (6.4.1)$$

in a space-time bounded domain $Q = [T_0, T_1) \times G \subset \mathbf{R}^{d+1}$. Here and in the next subsections we assume that the coefficients of (6.4.1) belong to the class $C^{1,2}([T_0, T_1] \times \overline{G})$, the boundary ∂G of the domain G is twice continuously differentiable, and the strict ellipticity condition is imposed on the matrix $a(s, x) = \sigma(s, x)\sigma^\mathsf{T}(s, x)$ (see (6.3.5)).

Introduce the space-time parallelepiped $\Pi_{r,l}^{\sigma(t,x)}(x)$:

$$\Pi_{r,l}^{\sigma(t,x)}(x) = \bigcup_{0 \le s < lr^2} \{t + s\} \times C_r^{\sigma(t,x)}(x + b(t, x)s),$$

where $(t, x) \in Q$ and $C_r^{\sigma(t,x)}(x + b(t, x)s)$ is the space parallelepiped in $\mathbf{R}^d$ obtained from the open cube C_r by the linear transformation $\sigma(t, x)$ and the shift $x + b(t, x)s$, and as in the previous section, C_r is the cube with center at the origin and with edges of length $2r$ which are parallel to the coordinate axes. The time size of $\Pi_{r,l}^{\sigma(t,x)}(x)$ is taken lr^2 since the characteristic exit time of a diffusion process from a space cube of linear size r is proportional to r^2.

Let Γ_δ be an intersection of a δ-neighborhood of the set Γ with the domain Q. Recall that the set Γ is a part of the boundary ∂Q consisting of the lateral surface and the upper base of the cylinder $\overline{Q}$. The size δ of the layer Γ_δ may depend on r. The condition of strict ellipticity ensures for any $\beta > 0$ the existence of a constant $\alpha > 0$ such that under all sufficiently small r for every point $(t, x) \in Q \setminus \Gamma_{\alpha r}$ the following relations take place:

$$\Pi_{r,l}^{\sigma(t,x)}(x) \subset Q, \quad \min_{0 \le s \le lr^2} \rho(\partial C_r^{\sigma(t,x)}(x + b(t, x)s), \partial G) \ge \beta r. \quad (6.4.2)$$

Indeed, due to the property of strict ellipticity, we get

$$\max_{0 \le s \le lr^2} \rho(x, \partial C_r^{\sigma(t,x)}(x + b(t, x)s)) \le lr^2 \max_{(s,y) \in \overline{Q}} |b(s, y)| + 2r\sqrt{d}\lambda_d.$$

It is easy to see that if we take

$$\alpha = lr \max_{(s,y)\in Q} |b(s,y)| + 2\sqrt{d}\lambda_d + \beta,$$

then for a sufficiently small r the relations (6.4.2) are fulfilled. The values β, α, and r used below are assumed to ensure (6.4.2).

To construct a one-step approximation for the system (6.4.1), we consider the system with frozen coefficients (see (6.0.5)):

$$d\bar{X} = b(t,x)ds + \sigma(t,x)dw(s), \quad \bar{X}(t) = x, \quad (t,x) \in Q\backslash\Gamma_{\alpha r}. \qquad (6.4.3)$$

Let $\bar{\theta}$ be the first-passage time of the process $(s-t, w(s) - w(t))$, $s > t$, to the boundary $\partial\Pi_{r,l}$ of the space-time parallelepiped $\Pi_{r,l} = [0, lr^2) \times C_r \subset \mathbf{R}^{d+1}$. Clearly, $\bar{\theta} \leq lr^2$. The point $(\bar{\theta}, w(t + \bar{\theta}) - w(t))$ is simulated in accordance with Corollary 6.3.10.

Let us take the point $(t + \bar{\theta}, \bar{X}_{t,x}(t + \bar{\theta}))$ with $\bar{X}_{t,x}(t + \bar{\theta})$ calculated by

$$\bar{X}_{t,x}(t + \bar{\theta}) = x + b(t,x)\bar{\theta} + \sigma(t,x)(w(t + \bar{\theta}) - w(t)) \qquad (6.4.4)$$

as an approximation of the point $(t + \bar{\theta}, X_{t,x}(t + \bar{\theta})))$, $(t,x) \in Q\backslash\Gamma_{\alpha r}$, where $X_{t,x}(s)$ is a solution of the system (6.4.1). Recall that if $t + \bar{\theta} \geq \tau_{t,x}$, then $X_{t,x}(t + \bar{\theta}) = X_{t,x}(\tau_{t,x})$.

The point $(t + \bar{\theta}, \bar{X}_{t,x}(t + \bar{\theta}))$ belongs to the lateral surface or to the upper base of the space-time parallelepiped $\Pi_{r,l}^{\sigma(t,x)}(x) \subset Q$. It follows from (6.4.2) that

$$\rho(\bar{X}_{t,x}(t + s), \partial G) \geq \beta r, \ 0 \leq s \leq lr^2. \qquad (6.4.5)$$

Theorem 6.4.1 For every natural m there exists a constant $K > 0$ such that for any sufficiently small r and for any point $(t,x) \in Q\backslash\Gamma_{\alpha r}$ the inequality

$$E\left|X_{t,x}(t + \bar{\theta}) - \bar{X}_{t,x}(t + \bar{\theta})\right|^{2m} \leq K r^{4m} \qquad (6.4.6)$$

holds.

Proof We have (see (6.4.1)) that $\tau_{t,x} \leq T_1$, $X_{t,x}(s) \in G$ for $s \in [t, \tau_{t,x})$, and $X_{t,x}(s) = X_{t,x}(\tau_{t,x})$ for $s \geq \tau_{t,x}$.

Let us rewrite the local error in the form

$$E \left| X_{t,x}(t + \bar{\theta}) - \bar{X}_{t,x}(t + \bar{\theta}) \right|^{2m} \tag{6.4.7}$$

$$= E \left| \int_t^{t+\bar{\theta}} \left(\chi_{\tau_{t,x}>s} b(s, X_{t,x}(s)) - b(t, x) \right) ds \right.$$

$$\left. + \int_t^{t+\bar{\theta}} \left(\chi_{\tau_{t,x}>s} \sigma(s, X_{t,x}(s)) - \sigma(t, x) \right) dw(s) \right|^{2m}$$

$$\leq K E \left| \int_t^{t+\bar{\theta}} \left(\chi_{\tau_{t,x}>s} b(s, X_{t,x}(s)) - b(t, x) \right) ds \right|^{2m}$$

$$+ K E \left| \int_t^{(t+\bar{\theta}) \wedge \tau_{t,x}} \left(\sigma(s, X_{t,x}(s)) - \sigma(t, x) \right) dw(s) \right|^{2m}$$

$$+ K E \left| \int_{(t+\bar{\theta}) \wedge \tau_{t,x}}^{t+\bar{\theta}} \sigma(t, x) \, dw(s) \right|^{2m}.$$

We obtain for the first term in (6.4.7):

$$K E \left| \int_t^{t+\bar{\theta}} \left(\chi_{\tau_{t,x}>s} b(s, X_{t,x}(s)) - b(t, x) \right) ds \right|^{2m} \leq K E \bar{\theta}^{2m} \leq K r^{4m} \tag{6.4.8}$$

because of boundedness of $b(s, x)$, $(s, x) \in \overline{Q}$, and $\bar{\theta} \leq lr^2$.

Below we need the following inequality for Ito integrals in the case of the scalar Wiener process (see, e.g., [123, p.26]):

$$E \left| \int_t^{t+T} \varphi(s) dw(s) \right|^{2m} \leq (m(2m-1))^{m-1} T^{m-1} \int_t^{t+T} E \varphi^{2m}(s) ds, \tag{6.4.9}$$

$$m = 1, 2, \dots .$$

Clearly, in the case of the d-dimensional Wiener process the inequality (6.4.9) implies

$$E \left| \int_t^{t+T} \varphi(s) dw(s) \right|^{2m} \leq K T^{m-1} \int_t^{t+T} E \sum_{i,j=1}^d (\varphi^{ij}(s))^{2m} ds, \quad m = 1, 2, \dots, \tag{6.4.10}$$

where the constant K depends on m. If φ is bounded, we also have

$$E \left| \int_t^{t+T} \varphi(s) dw(s) \right|^{2m} \leq K T^m, \quad m = 1, 2, \dots . \tag{6.4.11}$$

Due to the inequality (6.4.10), smoothness of $\sigma(s, x)$ for $(s, x) \in \overline{Q}$, and $(t + \bar{\theta}) \wedge \tau_{t,x} \leq t + lr^2$, we obtain for the second term of (6.4.7):

$$K E \left| \int_t^{(t+\bar{\theta}) \wedge \tau_{t,x}} (\sigma(s, X(s)) - \sigma(t, x)) \, dw(s) \right|^{2m} \tag{6.4.12}$$

$$= K E \left| \int_t^{t+lr^2} \chi_{(t+\bar{\theta}) \wedge \tau_{t,x} > s} \left(\sigma(s, X_{t,x}(s)) - \sigma(t, x) \right) dw(s) \right|^{2m}$$

$$\leq K r^{2m-2} \int_t^{t+lr^2} E \left(\chi_{(t+\bar{\theta}) \wedge \tau_{t,x} > s} \sum_{i,j=1}^d \left| \sigma^{ij}(s, X_{t,x}(s)) - \sigma^{ij}(t, x) \right|^{2m} \right) ds$$

$$\leq K r^{2m-2} \int_t^{t+lr^2} \left(E \chi_{\tau_{t,x} > s} \left| X_{t,x}(s) - x \right|^{2m} + (s - t)^{2m} \right) ds$$

$$\leq K r^{2m-2} \int_t^{t+lr^2} E \chi_{\tau_{t,x} > s} \left| X_{t,x}(s) - x \right|^{2m} ds + K r^{6m}.$$

Further,

$$E \chi_{\tau_{t,x} > s} \left| X_{t,x}(s) - x \right|^{2m} = E \chi_{\tau_{t,x} > s} \left| \int_t^s b(s', X_{t,x}(s')) \, ds' \right.$$

$$\left. + \int_t^s \sigma(s', X_{t,x}(s')) dw(s') \right|^{2m},$$

whence, due to (6.4.11),

$$E \chi_{\tau_{t,x} > s} \left| X_{t,x}(s) - x \right|^{2m} \leq K (s - t)^{2m} + K (s - t)^m.$$

Substituting this inequality in (6.4.12), we obtain

$$
K E \left| \int_{t}^{(t+\bar{\theta})\wedge\tau_{t,x}} \left(\sigma(s, X_{t,x}(s)) - \sigma(t, x)\right) dw(s) \right|^{2m} \leq K r^{4m}. \tag{6.4.13}
$$

It follows from the inequalities (6.4.8) and (6.4.13) that

$$
E \left| X_{t,x}(\tau_{t,x} \wedge (t + \bar{\theta})) - \bar{X}_{t,x}(\tau_{t,x} \wedge (t + \bar{\theta})) \right|^{2m} \leq K r^{4m}. \tag{6.4.14}
$$

Now let us estimate the third term in (6.4.7). Due to (6.4.10), we have

$$
K E \left| \int_{(t+\bar{\theta})\wedge\tau_{t,x}}^{t+\bar{\theta}} \sigma(t, x) dw(s) \right|^{2m} \tag{6.4.15}
$$

$$
= K E \left| \int_{t}^{t+lr^2} \left(\chi_{(t+\bar{\theta})>s} - \chi_{(t+\bar{\theta})\wedge\tau_{t,x}>s}\right) \sigma(t, x) dw(s) \right|^{2m}
$$

$$
\leq K r^{2m-2} \int_{t}^{t+lr^2} E \left(\chi_{(t+\bar{\theta})>s} - \chi_{(t+\bar{\theta})\wedge\tau_{t,x}>s}\right) ds
$$

$$
= K r^{2m-2} E \chi_{\tau_{t,x}<(t+\bar{\theta})} \left((t + \bar{\theta}) - -(t + \bar{\theta}) \wedge \tau_{t,x}\right)
$$

$$
\leq K r^{2m} P(\tau_{t,x} < t + \bar{\theta}).
$$

We evaluate the probability $P(\tau_{t,x} < t + \bar{\theta})$ using the recipe from Sect. 6.1.1. If $\tau_{t,x} < t + \bar{\theta}$, then $\tau_{t,x} < T_1$ and, consequently, $X_{t,x}(\tau_{t,x}) \in \partial G$. At the same time, according to (6.4.5),

$$
\rho(\bar{X}_{t,x}(\tau_{t,x} \wedge (t + \bar{\theta})), \partial G) \geq \beta r .
$$

Therefore,

$$
E \left(\chi_{\tau_{t,x}<t+\bar{\theta}} \left| X_{t,x}(\tau_{t,x} \wedge (t + \bar{\theta})) - \bar{X}_{t,x}(\tau_{t,x} \wedge (t + \bar{\theta})) \right|^{m}\right)
$$

$$
\geq P(\tau_{t,x} < t + \bar{\theta}) (\beta r)^{m}, \quad m = 1, 2, \dots .
$$

On the other hand, due to (6.4.14), we have

$$P(\tau_{t,x} < t + \bar{\theta})\,(\beta r)^m$$

$$\leq E\left(\chi_{\tau_{t,x}<t+\bar{\theta}}\left|X_{t,x}(\tau_{t,x} \wedge (t+\bar{\theta})) - \bar{X}_{t,x}(\tau_{t,x} \wedge (t+\bar{\theta}))\right|^m\right)$$

$$\leq \sqrt{P(\tau_{t,x} < t + \bar{\theta})}$$

$$\times \left[E\left|X_{t,x}(\tau_{t,x} \wedge (t+\bar{\theta})) - \bar{X}_{t,x}(\tau_{t,x} \wedge (t+\bar{\theta}))\right|^{2m}\right]^{1/2}$$

$$\leq K r^{2m}\sqrt{P(\tau_{t,x} < t + \bar{\theta})}.$$

Consequently,

$$P(\tau_{t,x} < t + \bar{\theta}) \leq K r^{2m}, \quad m = 1, 2, \ldots. \tag{6.4.16}$$

Now the inequality (6.4.7) together with (6.4.8), (6.4.13), and (6.4.15) gives (6.4.6). Theorem 6.4.1 is proved. $\qquad\square$

Remark 6.4.2 In the case of additive noise, i.e., when $\sigma(t, x) \equiv \sigma(t)$, we prove under the assumptions of Theorem 6.4.1 that

$$E\left|X_{t,x}(t + \bar{\theta}) - \bar{X}_{t,x}(t + \bar{\theta})\right|^{2m} \leq K\, r^{6m}.$$

6.4.2   Global Algorithm in a Space-Time Bounded Domain

In this subsection we construct a random walk over small space-time parallelepipeds based on the one-step approximation (6.4.4).

Algorithm 6.4.3 *Let* $(\bar{\theta}_1, w(t + \bar{\theta}_1) - w(t))$ *be the first exit point of the process* $(s - t, w(s) - w(t))$, $s > t$, *from the parallelepiped* $\Pi_{r,l}$ *simulated in accordance with Corollary 6.3.10,* $(\bar{\theta}_2, w(t + \bar{\theta}_1 + \bar{\theta}_2) - w(t + \bar{\theta}_1))$ *be the exit point of the process* $(s - t - \bar{\theta}_1, w(s) - w(t + \bar{\theta}_1))$, $s > t + \bar{\theta}_1$, *from the parallelepiped* $\Pi_{r,l}$, *and so on.*

Suppose that $(t, x) \in Q\backslash\Gamma_{\alpha r}$. *Then, we construct the recurrence sequence* $(\bar{\vartheta}_k, \bar{X}_k)$, $k = 0, 1, \ldots, \bar{\nu}$:

$$\bar{\vartheta}_0 = t, \ \bar{X}_0 = x,$$
$$\bar{\vartheta}_k = \bar{\vartheta}_{k-1} + \bar{\theta}_k,$$
$$\bar{X}_k = \bar{X}_{k-1} + b(\bar{\vartheta}_{k-1}, \bar{X}_{k-1})\bar{\theta}_k + \sigma(\bar{\vartheta}_{k-1}, \bar{X}_{k-1})(w(\bar{\vartheta}_k) - w(\bar{\vartheta}_{k-1})),$$

$$k = 1, \ldots, \bar{\nu},$$

where the number $\bar{\nu} = \bar{\nu}_{t,x}$ *is the first one for which* $(\bar{\vartheta}_k, \bar{X}_k) \in \Gamma_{\alpha r}$. *If* $(t, x) \in \Gamma_{\alpha r}$, *we put* $\bar{\nu} = 0$.

Let $(\bar{\vartheta}_k, \bar{X}_k) = (\bar{\vartheta}_{\bar{\nu}}, \bar{X}_{\bar{\nu}})$ for $k > \bar{\nu}$. The obtained sequence $(\bar{\vartheta}_k, \bar{X}_k)$, $k = 0, 1, \ldots,$ is a Markov chain stopped at the Markov moment $\bar{\nu}$. It is clear that the random number of steps $\bar{\nu}$ depends on the domain $Q\setminus\Gamma_{\alpha r}$. That is why, the more rigorous notation for $\bar{\nu}$ is $\bar{\nu}_{t,x}(Q\setminus\Gamma_{\alpha r})$.

Using the technique described in Sect. 6.1.2, we prove the theorems on average characteristics of $\bar{\nu}_{t,x} = \bar{\nu}_{t,x}(Q\setminus\Gamma_{\alpha r})$.

Theorem 6.4.4 *The mean number of steps $\bar{\nu}_{t,x}(Q\setminus\Gamma_{\alpha r})$ is estimated as*

$$E\bar{\nu}_{t,x}(Q\setminus\Gamma_{\alpha r}) \leq \frac{K}{r^2}, \qquad (6.4.17)$$

where the positive constant K does not depend on r.

Theorem 6.4.5 *For every $L > 0$, the inequality*

$$P\left\{\bar{\nu}_{t,x}(Q\setminus\Gamma_{\alpha r}) \geq \frac{L}{r^2}\right\} \leq (1 + T_1 - T_0) \exp(-c_r \frac{\gamma}{1 + T_1 - T_0} L), \qquad (6.4.18)$$
$$c_r \to 1 \text{ as } r \to 0,$$

is valid.

Further, we need two auxiliary lemmas. Their proofs are available in [304].

Lemma 6.4.6 *There exists a constant K such that for all r small enough and all $(t, x) \in Q\setminus\Gamma_{\alpha r}$ the inequality*

$$\left|E(X_{t,x}(t + \bar{\theta}_1) - \bar{X}_{t,x}(t + \bar{\theta}_1))\right| \leq K r^4 \qquad (6.4.19)$$

is valid.

Lemma 6.4.7 *Let the random variable Z be defined by the relation*

$$X_{t,x}(t + \bar{\theta}_1) - X_{t,y}(t + \bar{\theta}_1) = x - y + Z.$$

Then for every natural m there exists a positive constant K such that for any r small enough and all $(t, x), (t, y) \in Q\setminus\Gamma_{\alpha r}$ the inequalities

$$E|Z|^m \leq K r^m \left(|x - y|^m + r^m\right), \qquad (6.4.20)$$

$$|EZ| \leq K r^2 \left(|x - y| + r^2\right) \qquad (6.4.21)$$

hold.

Remark 6.4.8 If $\sigma(t, x) \equiv \sigma(t)$, we obtain under the assumptions of Lemma 6.4.7 (cf. Remark 6.4.2):
$$E|Z|^m \leq K r^m \left(|x - y|^m + r^{2m}\right).$$

For every $\varepsilon \in (0, 1]$ and any $\beta > 0$ it is possible to introduce the layer $\Gamma_{\alpha r^{1-\varepsilon}}$ with a constant α such that under a sufficiently small r and for every $(t, x) \in Q \backslash \Gamma_{\alpha r^{1-\varepsilon}}$ the following relations together with the relations (6.4.2) take place:

$$\Pi_{r,l}^{\sigma(t,x)}(x) \subset Q, \quad \min_{0 \le s \le lr^2} \rho(\partial C_r^{\sigma(t,x)}(x + b(t, x)s), \partial G) \ge \beta r^{1-\varepsilon} .$$

Clearly, $\Gamma_{\alpha r} \subset \Gamma_{\alpha r^{1-\varepsilon}}$.

The Markov moment $\bar{\nu}_{t,x}(Q \backslash \Gamma_{\alpha r^{1-\varepsilon}})$, when the chain $(\bar{\vartheta}_k, \bar{X}_k)$ leaves the domain $Q \backslash \Gamma_{\alpha r^{1-\varepsilon}}$, satisfies the inequality

$$\bar{\nu}_{t,x}(Q \backslash \Gamma_{\alpha r^{1-\varepsilon}}) \le \bar{\nu}_{t,x}(Q \backslash \Gamma_{\alpha r}) .$$

We shall use the old notation $(\bar{\vartheta}_k, \bar{X}_k)$ for the new Markov chain, which is constructed by the same rules as above but stops in the layer $\Gamma_{\alpha r^{1-\varepsilon}}$ at the new Markov moment $\bar{\nu} = \bar{\nu}_{t,x}(Q \backslash \Gamma_{\alpha r^{1-\varepsilon}})$. We believe that the use of the same notation $(\bar{\vartheta}_k, \bar{X}_k)$ for various Markov chains and $\bar{\nu}$ for various stopping moments will cause no confusion below.

Consider the sequence $(\bar{\vartheta}_k, X_k)$, $k = 0, 1, \ldots$, with X_k :

$$X_0 = x,$$
$$X_1 = X_{t,x}(\bar{\vartheta}_1)$$

$$\cdot \ \cdot \ \cdot \ \cdot \ \cdot \ \cdot \ \cdot$$

$$X_k = X_{t,x}(\bar{\vartheta}_k) = X_{\bar{\vartheta}_{k-1}, X_{k-1}}(\bar{\vartheta}_k)$$

$$\cdot \ \cdot \ \cdot \ \cdot \ \cdot \ \cdot \ \cdot$$

connected with the system (6.4.1).

The sequence $(\bar{\vartheta}_k, X_k)$ is a Markov chain, which stops at the random moment $\bar{\nu}$ due to $\bar{\vartheta}_k = \bar{\vartheta}_{\bar{\nu}}$ under $k > \bar{\nu}$. The following theorem states the closeness of X_k and $\bar{X}_k$ for $N = L/r^2$ steps (see its proof in [304]).

Theorem 6.4.9 *Let* $\bar{\nu} = \bar{\nu}_{t,x}(Q \backslash \Gamma_{\alpha r^{1-\varepsilon}})$, $0 < \varepsilon \le 1$, *be the first exit moment of the Markov chain* $(\bar{\vartheta}_i, \bar{X}_i)$, $i = 1, 2, \ldots$, *from the domain* $Q \backslash \Gamma_{\alpha r^{1-\varepsilon}}$. *Then, there exist constants* $K > 0$ *and* $\gamma > 0$ *such that for all* r *small enough the inequality*

$$\left(E \left| X_{N \wedge \bar{\nu}} - \bar{X}_{N \wedge \bar{\nu}} \right|^2 \right)^{1/2} = \left(E \left| X_N - \bar{X}_N \right|^2 \right)^{1/2} \le K e^{\gamma L} r$$

holds.

Remark 6.4.10 In the case of additive noise (see Remarks 6.4.2 and 6.4.8) we have faster convergence:

$$\left(E \left| X_{N \wedge \bar{\nu}} - \bar{X}_{N \wedge \bar{\nu}} \right|^2 \right)^{1/2} = \left(E \left| X_N - \bar{X}_N \right|^2 \right)^{1/2} \le K e^{\gamma L} r^2$$

under the assumptions of Theorem 6.4.9.

Theorem 6.4.11 *Let* $\bar{\nu} = \bar{\nu}_{t,x}(Q \backslash \Gamma_{\alpha r^{1-\varepsilon}})$, $0 < \varepsilon \le 1$, *be the first exit moment of the Markov chain* $(\bar{\vartheta}_i, \bar{X}_i)$, $i = 1, 2, \ldots$, *from the domain* $Q \backslash \Gamma_{\alpha r^{1-\varepsilon}}$. *Then, there exist constants* $K > 0$ *and* $\gamma > 0$ *such that for all* r *small enough the inequality*

$$\left(E \left| X_{\bar{\nu}} - \bar{X}_{\bar{\nu}} \right|^2 \right)^{1/2} \le K \left(e^{\gamma L} r + e^{-c_r \gamma L/2} \right)$$

holds.

Proof Introduce two sets $\mathcal{C} = \{\bar{\nu} \le L/r^2\}$ and $\Omega \backslash \mathcal{C} = \{\bar{\nu} > L/r^2\}$. Let l be a diameter of G. Using Theorems 6.4.5 and 6.4.9, we obtain

$$\begin{aligned}
E \left| X_{\bar{\nu}} - \bar{X}_{\bar{\nu}} \right|^2 &= E \left(\left| X_{\bar{\nu}} - \bar{X}_{\bar{\nu}} \right|^2 ; \mathcal{C} \right) + E \left(\left| X_{\bar{\nu}} - \bar{X}_{\bar{\nu}} \right|^2 ; \Omega \backslash \mathcal{C} \right) \\
&= E \left(\left| X_{N \wedge \bar{\nu}} - \bar{X}_{N \wedge \bar{\nu}} \right|^2 ; \mathcal{C} \right) + E \left(\left| X_{\bar{\nu}} - \bar{X}_{\bar{\nu}} \right|^2 ; \Omega \backslash \mathcal{C} \right) \\
&\le E \left| X_N - \bar{X}_N \right|^2 + l^2 P(\Omega \backslash \mathcal{C}) \le K e^{2\gamma L} r^2 + K e^{-c_r \gamma L} .
\end{aligned}$$

$\square$

Remark 6.4.12 Let $\bar{\nu} = \bar{\nu}_{t,x}(Q \backslash \Gamma_{\alpha r^{1-\varepsilon}})$, $0 < \varepsilon \le 1$, be the first exit moment of the Markov chain $(\bar{\vartheta}_i, \bar{X}_i)$, $i = 1, 2, \ldots$, from the domain $Q \backslash \Gamma_{\alpha r^{1-\varepsilon}}$. Then, for every natural m there exist constants $K > 0$ and $\gamma > 0$ such that for all r small enough the following inequalities:

$$E \left| X_N - \bar{X}_N \right|^{2m} \le K e^{2\gamma L} r^{2m}, \tag{6.4.22}$$

$$E \left| X_{\bar{\nu}} - \bar{X}_{\bar{\nu}} \right|^{2m} \le K \left(e^{2\gamma L} r^{2m} + e^{-c_r \gamma L} \right), \tag{6.4.23}$$

and

$$P(\tau_{t,x} < \bar{\vartheta}_N) \le K e^{2\gamma L} r^{2n}, \quad n = 1, 2, \ldots, \tag{6.4.24}$$

hold [304].

6.4.3   Approximation of the Exit Point $(\tau, X(\tau))$

In this subsection we are interested in an approximation of the exit point $\left(\tau_{t,x}, X_{t,x}(\tau_{t,x}) \right)$ of the space-time diffusion $(s, X_{t,x}(s))$, $s \ge t$, from the space-time domain Q. For the sake of simplicity in proofs we restrict ourselves to the case of the convex domain G.

We have $(\bar{\vartheta}_N, \bar{X}_N) = (\bar{\vartheta}_{\bar{\nu}}, \bar{X}_{\bar{\nu}}) \in \Gamma_{\alpha r^{1-\varepsilon}}$ on the set $\mathcal{C} = \{\bar{\nu} \le L/r^2\}$. Let $(\bar{\tau}_{t,x}, \xi_{t,x})(\omega)$, $\omega \in \mathcal{C}$, be a point on Γ defined as follows. If $\bar{\vartheta}_{\bar{\nu}} \ge T_1 - \alpha r^{1-\varepsilon}$ then

$\bar{\tau}_{t,x} = T_1$ and $\xi_{t,x} = \bar{X}_{\bar{\nu}} \in G$, otherwise (i.e., when $\rho(\bar{X}_{\bar{\nu}}, \partial G) \le \alpha r^{1-\varepsilon}$) $\bar{\tau}_{t,x} = \bar{\vartheta}_{\bar{\nu}}$ and a point $\xi_{t,x} \in \partial G$ is such that

$$\left| \bar{X}_{\bar{\nu}} - \xi_{t,x} \right| \le \alpha r^{1-\varepsilon}, \ \omega \in \mathcal{C} . \tag{6.4.25}$$

To complete the definition of $(\bar{\tau}_{t,x}, \xi_{t,x})(\omega)$ on the set $\Omega \backslash \mathcal{C}$, we put $\bar{\tau}_{t,x}$ be equal to $\bar{\vartheta}_N$ and $\xi_{t,x}$ be a point on ∂G nearest to $\bar{X}_N$. It is natural to take the point $(\bar{\tau}_{t,x}, \xi_{t,x})$ as an approximate one to the exit point $(\tau_{t,x}, X_{t,x}(\tau_{t,x}))$.

Below we need the following lemma [304] (it is analogous to Lemma 6.1.16).

Lemma 6.4.13 *There exists a constant $K > 0$ such that for all $(t, x) \in \overline{Q}$ and $y \in \partial G$ the inequalities*

$$E(\tau_{t,x} - t) \le K|x - y|, \tag{6.4.26}$$

$$E\left(X_{t,x}(\tau_{t,x}) - y\right)^2 \le K|x - y| \tag{6.4.27}$$

are valid.

Theorem 6.4.14 *Let $\bar{\nu} = \bar{\nu}_{t,x}(Q \backslash \Gamma_{\alpha r^{1-\varepsilon}})$, $0 < \varepsilon \le 1$, be the first exit moment of the Markov chain $(\bar{\vartheta}_i, \bar{X}_i)$, $i = 1, 2, \ldots$, from the domain $Q \backslash \Gamma_{\alpha r^{1-\varepsilon}}$. Then, there exist positive constants K and γ such that for all r small enough the inequalities*

$$\left[E\left(\left| X_{t,x}(\tau_{t,x}) - \xi_{t,x} \right|^2 ; \mathcal{C} \right) \right]^{1/2} \le K \, r^{(1-\varepsilon)/2} \tag{6.4.28}$$

and

$$\left[E \left| X_{t,x}(\tau_{t,x}) - \xi_{t,x} \right|^2 \right]^{1/2} \le K(r^{(1-\varepsilon)/2} + e^{-c_r \gamma L/2}) \tag{6.4.29}$$

hold.

Proof Consider the distance between $X_{t,x}(\tau_{t,x})$ and $\xi_{t,x}$ on $\mathcal{C}$:

$$E\left(\left| X_{t,x}(\tau_{t,x}) - \xi_{t,x} \right|^2 ; \mathcal{C} \right) \tag{6.4.30}$$

$$= E\left(\chi_{\bar{\vartheta}_N \ge T_1 - \alpha r^{1-\varepsilon}} \left| X_{t,x}(\tau_{t,x}) - \xi_{t,x} \right|^2 ; \mathcal{C} \right)$$

$$+ E\left(\chi_{\bar{\vartheta}_N < T_1 - \alpha r^{1-\varepsilon}} \left| X_{t,x}(\tau_{t,x}) - \xi_{t,x} \right|^2 ; \mathcal{C} \right) .$$

We get for the first term of (6.4.30):

$$E\left(\chi_{\bar{\vartheta}_N \ge T_1 - \alpha r^{1-\varepsilon}} \left| X_{t,x}(\tau_{t,x}) - \xi_{t,x} \right|^2 ; \mathcal{C} \right) \tag{6.4.31}$$

$$= E \chi_{\bar{\vartheta}_N \ge T_1 - \alpha r^{1-\varepsilon}} \left| X_{t,x}(\tau_{t,x}) - \bar{X}_N \right|^2$$

$$\le 2 E \chi_{\bar{\vartheta}_N \ge T_1 - \alpha r^{1-\varepsilon}} \left| X_{t,x}(\tau_{t,x}) - X_N \right|^2 + 2E \left| X_N - \bar{X}_N \right|^2 .$$

Due to Theorem 6.4.9, the second term of (6.4.31) is estimated by $K e^{2\gamma L} r^2$, and we have for the first term of (6.4.31):

$$E \chi_{\bar{\vartheta}_N \geq T_1 - \alpha r^{1-\varepsilon}} \left| X_{t,x}(\tau_{t,x}) - X_N \right|^2$$

$$= E \left| \chi_{\bar{\vartheta}_N \geq T_1 - \alpha r^{1-\varepsilon}} (X_{t,x}(\tau_{t,x}) - X_N) \right|^2$$

$$\leq 2E \left| \int_{\bar{\vartheta}_N \wedge \tau_{t,x}}^{\tau_{t,x}} \chi_{\bar{\vartheta}_N \geq T_1 - \alpha r^{1-\varepsilon}} b(s, X_{t,x}(s)) ds \right|^2$$

$$+ 2E \left| \int_{\bar{\vartheta}_N \wedge \tau_{t,x}}^{\tau_{t,x}} \chi_{\bar{\vartheta}_N \geq T_1 - \alpha r^{1-\varepsilon}} \sigma(s, X_{t,x}(s)) dw(s) \right|^2$$

$$\leq K E \chi_{\bar{\vartheta}_N \geq T_1 - \alpha r^{1-\varepsilon}} \left(\tau_{t,x} - \tau_{t,x} \wedge \bar{\vartheta}_N \right)^2$$

$$+ K E \chi_{\bar{\vartheta}_N \geq T_1 - \alpha r^{1-\varepsilon}} \left(\tau_{t,x} - \tau_{t,x} \wedge \bar{\vartheta}_N \right)$$

$$\leq K E \chi_{\bar{\vartheta}_N \geq T_1 - \alpha r^{1-\varepsilon}} (T_1 - \bar{\vartheta}_N)^2 + K E \chi_{\bar{\vartheta}_N \geq T_1 - \alpha r^{1-\varepsilon}} (T_1 - \bar{\vartheta}_N) \leq K r^{1-\varepsilon},$$

whence it follows that

$$E \left(\chi_{\bar{\vartheta}_N \geq T_1 - \alpha r^{1-\varepsilon}} \left| X_{t,x}(\tau_{t,x}) - \xi_{t,x} \right|^2 ; \mathcal{C} \right) \leq K r^{1-\varepsilon}. \tag{6.4.32}$$

Consider the second term of (6.4.30). Due to its definition, the point $\xi_{t,x}(\omega)$, $\omega \in \mathcal{C}$, belongs to ∂G if $\bar{\vartheta}_N < T_1 - \alpha r^{1-\varepsilon}$. Then by the conditional version of Lemma 6.4.13, we get (note that $\xi_{t,x}$ is measurable with respect to $\mathcal{F}_N$)

$$E \left(\chi_{\bar{\vartheta}_N < T_1 - \alpha r^{1-\varepsilon}} \left| X_{t,x}(\tau_{t,x}) - \xi_{t,x} \right|^2 ; \mathcal{C} \right)$$

$$= E \left(\chi_{\bar{\vartheta}_N < T_1 - \alpha r^{1-\varepsilon}} E \left(\left| X_{\bar{\vartheta}_N, X_N}(\tau_{\bar{\vartheta}_N, X_N}) - \xi_{t,x} \right|^2 / \mathcal{F}_N \right); \mathcal{C} \right)$$

$$\leq K E \left(\chi_{\bar{\vartheta}_N < T_1 - \alpha r^{1-\varepsilon}} \left| X_N - \xi_{t,x} \right|; \mathcal{C} \right).$$

Theorem 6.4.9 and the inequality (6.4.25) imply

$$E \left(\chi_{\bar{\vartheta}_N < T_1 - \alpha r^{1-\varepsilon}} \left| X_N - \xi_{t,x} \right|; \mathcal{C} \right) \tag{6.4.33}$$

$$\leq \left[E \left(\chi_{\bar{\vartheta}_N < T_1 - \alpha r^{1-\varepsilon}} \left| X_N - \xi_{t,x} \right|^2 ; \mathcal{C} \right) \right]^{1/2}$$

$$\leq \left[2E \left| X_N - \bar{X}_N \right|^2 + 2 \left(E \chi_{\bar{\vartheta}_N < T_1 - \alpha r^{1-\varepsilon}} \left| \bar{X}_N - \xi_{t,x} \right|^2 ; \mathcal{C} \right) \right]^{1/2}$$

$$\leq K e^{\gamma L} r + 2\alpha r^{1-\varepsilon} \leq K r^{1-\varepsilon}.$$

Thus,

$$E \left(\chi_{\bar{\vartheta}_N < T_1 - \alpha r^{1-\varepsilon}} \left| X_{t,x}(\tau_{t,x}) - \xi_{t,x} \right|^2 ; \mathcal{C} \right) \leq K r^{1-\varepsilon}.$$

Substituting this inequality and the inequality (6.4.32) in (6.4.30), we get (6.4.28).

The inequality (6.4.29) is obtained by Theorem 6.4.5 analogously to the proof of Theorem 6.4.11. Theorem 6.4.14 is proved. $\qquad\square$

Theorem 6.4.15 *Under the assumptions of Theorem 6.4.14, the inequalities*

$$E\left(|\tau_{t,x} - \bar{\tau}_{t,x}|; \mathcal{C}\right) \le K r^{1-\varepsilon}, \tag{6.4.34}$$

$$E|\tau_{t,x} - \bar{\tau}_{t,x}| \le K(r^{1-\varepsilon} + e^{-\alpha_r \gamma L}) \tag{6.4.35}$$

hold.

Proof Recall that $\tau_{t,x} \le T_1$, $\bar{\vartheta}_N \le T_1$. Further, $\bar{\tau}_{t,x} = T_1$ under $\bar{\vartheta}_N \ge T_1 - \alpha r^{1-\varepsilon}$ and $\bar{\tau}_{t,x} = \bar{\vartheta}_N$ otherwise. Consequently, $\bar{\tau}_{t,x} \ge \bar{\vartheta}_N$. Let below $\tau := \tau_{t,x}$, $\bar{\tau} := \bar{\tau}_{t,x}$.
Consider the difference $|\tau - \bar{\tau}|$ on the set $\mathcal{C}$. We have

$$E\left(|\tau - \bar{\tau}|; \mathcal{C}\right) = E\left((\bar{\tau} - \tau \wedge \bar{\tau}); \mathcal{C}\right) + E\left((\tau - \tau \wedge \bar{\tau}); \mathcal{C}\right). \tag{6.4.36}$$

We get for the first term:

$$\begin{aligned}
E\left((\bar{\tau} - \tau \wedge \bar{\tau}); \mathcal{C}\right) &\le E(\bar{\tau} - \tau \wedge \bar{\tau}) = E\chi_{\tau < \bar{\tau}}(\bar{\tau} - \tau \wedge \bar{\tau}) \\
&= E\chi_{\tau < \bar{\vartheta}_N}(\bar{\tau} - \tau \wedge \bar{\tau}) + E\chi_{\bar{\vartheta}_N \le \tau < \bar{\tau}}(\bar{\tau} - \tau \wedge \bar{\tau}) \\
&\le (T_1 - T_0) P(\tau < \bar{\vartheta}_N) + E\chi_{T_1 - \alpha r^{1-\varepsilon} \le \tau < T_1}(T_1 - \tau).
\end{aligned}$$

Then using (6.4.24) under $n = 1$, we obtain

$$E\left((\bar{\tau} - \tau \wedge \bar{\tau}); \mathcal{C}\right) \le K e^{2\gamma L} \cdot r^2 + \alpha r^{1-\varepsilon} \le K r^{1-\varepsilon}. \tag{6.4.37}$$

Consider the second term of (6.4.36). Due to $\xi_{t,x} \in \partial G$ under $\bar{\vartheta}_N < T_1 - \alpha r^{1-\varepsilon}$, Lemma 6.4.13, and the inequality (6.4.33), we get

$$\begin{aligned}
E\left((\tau - \tau \wedge \bar{\tau}); \mathcal{C}\right) &= E\left(\chi_{\bar{\tau} < \tau}(\tau - \tau \wedge \bar{\tau}); \mathcal{C}\right) \\
&= E\left(\chi_{\bar{\vartheta}_N < \tau}\chi_{\bar{\vartheta}_N < T_1 - \alpha r^{1-\varepsilon}}(\tau - \tau \wedge \bar{\vartheta}_N); \mathcal{C}\right) \\
&= E\left(\chi_{\bar{\vartheta}_N < T_1 - \alpha r^{1-\varepsilon}}(\tau_{\bar{\vartheta}_N, X_N} - \bar{\vartheta}_N); \mathcal{C}\right) \\
&= E\left(\chi_{\bar{\vartheta}_N < T_1 - \alpha r^{1-\varepsilon}} E(\tau_{\bar{\vartheta}_N, X_N} - \bar{\vartheta}_N / \mathcal{F}_N); \mathcal{C}\right) \\
&\le K E\left(\chi_{\bar{\vartheta}_N < T_1 - \alpha r^{1-\varepsilon}} |X_N - \xi_{t,x}|; \mathcal{C}\right) \le K r^{1-\varepsilon}.
\end{aligned}$$

Substituting this inequality and the inequality (6.4.37) in (6.4.36), we get (6.4.34).

The inequality (6.4.35) is obtained by Theorem 6.4.5 analogously to the proof of Theorem 6.4.11. Theorem 6.4.15 is proved. $\qquad\square$

6.4.4 Simulation of Space-Time Brownian Motion with Drift

In this section we have dealt so far with the one-step approximation $(t + \bar{\theta}, \bar{X}_{t,x}(t + \bar{\theta}))$, $(t, x) \in Q \backslash \Gamma_{ar}$ (see (6.4.4)), which is based on the simulation of the exit point $(\bar{\theta}, w(t + \bar{\theta}) - w(t))$ of the process $(s - t, w(s) - w(t))$, $s > t$, from the space-time parallelepiped $\Pi_{r,l} = [0, lr^2) \times C_r$ with the cubic base C_r.

It is possible to derive other constructive one-step approximations. In this subsection we briefly consider a one-step approximation based on a simulation of exit points for the Brownian motion with drift $W_\mu(s)$:

$$W_\mu(s) = \mu s + W(s), \quad W_\mu(0) = 0,$$

where μ is a d-dimensional fixed vector and $W(s)$ is a d-dimensional standard Wiener process.

If $(\bar{\theta}, w_\mu(t + \bar{\theta}) - w_\mu(t))$ is the first exit point of the process $(s - t, w_\mu(s) - w_\mu(t))$, $s > t$, under $\mu = \sigma^{-1}(t, x)b(t, x)$, $(t, x) \in Q \backslash \Gamma_{ar}$, from the space-time parallelepiped $[0, l) \times C_r$, $l \leq T_1 - t$, then it is easy to see that the approximation

$$\bar{X}_{t,x}(t + \bar{\theta}) = x + \sigma(t, x)(w_\mu(t + \bar{\theta}) - w_\mu(t)) \tag{6.4.38}$$

belongs to the space parallelepiped $\overline{C}_r^\sigma(x)$ even under not small l.

Then we are able to ensure belonging of $\bar{X}_{t,x}(t + \bar{\theta})$ to $\bar{G}$, and, consequently, $(t + \bar{\theta}, \bar{X}_{t,x}(t + \bar{\theta}))$ to $\bar{Q}$, but the smallness of time size of the space-time parallelepiped $[0, l) \times C_r$ is already not required in contrast to the approximation (6.4.4).

The approximation (6.4.38) is more universal than the approximation (6.4.4). However, the approximation (6.4.4) is computationally simpler than (6.4.38) and is quite appropriate for the majority of problems.

Algorithms for simulating exit points of the Brownian motion with drift $W_\mu(s)$ are available in [304].

6.5 Numerical Examples

The numerical methods proposed in this chapter are widely applicable. These methods are the first ones which can constructively approximate trajectories of a diffusion process in bounded domains. They can also be applied to solving boundary value problems through a Monte Carlo technique on a level with the weak methods. Let us underline that the methods from this chapter give an estimator for a solution to the Dirichlet problem for parabolic and elliptic equations with constant coefficients which do not contain the error of numerical integration in comparison with the weak methods (see the next chapter).

Here we give three numerical examples. The first and the second examples deal with solving boundary value problems. In the second example an elliptic problem is considered, nevertheless, we need the simulation of the space-time exit points. The third example concerns the stability analysis of linear autonomous system of SDEs and uses simulation of space-time trajectories essentially.

Example 6.5.1 Let us consider an application of random walks over touching space-time parallelepipeds to the Dirichlet problem for parabolic equation (6.3.1)–(6.3.3) in the case when the coefficients are constant. This problem has the probabilistic representation (6.3.6)–(6.3.7), which we use for the Monte Carlo procedure here.

Let $(\bar{\vartheta}_k, \bar{X}_k)$ be a Markov chain which is formed analogously to the one of Sect. 6.4.2 but wandering is realized over touching space-time parallelepipeds (instead of small space-time parallelepipeds in Sect. 6.4.2) and is finished in the layer Γ_δ at a random step $\bar{\nu}$, where $\delta > 0$ is a sufficiently small constant. The equation with frozen coefficients (6.4.4), which we are able to simulate exactly, coincides with (6.3.7) when its coefficients are constant. Consequently, the chain $(\bar{\vartheta}_k, \bar{X}_k)$ coincides with the chain $(\bar{\vartheta}_k, X_k)$. In the considered case, the solution $u(t, x)$ to the Dirichlet problem (6.3.1)–(6.3.3) under $c = 0$ and $e = 0$ is simulated as (see (6.3.6))

$$
u(t, x) \doteq \bar{u}(t, x) = \frac{1}{M} \sum_{m=1}^{M} \varphi(\bar{X}_{\bar{\nu}}^{(m)}) \pm 2[\bar{D}/M]^{1/2} ,
$$

where

$$
\varphi(\bar{X}_{\bar{\nu}}^{(m)}) = \begin{cases} f(\bar{X}_{\bar{\nu}}^{(m)}), & \bar{\vartheta}_{\bar{\nu}}^{(m)} \in (T_1 - \delta, T_1], \\ g(\bar{X}_{\bar{\nu}}^{(m)}), & \bar{\vartheta}_{\bar{\nu}}^{(m)} \notin (T_1 - \delta, T_1], \end{cases}
$$

$$
\bar{D} = \frac{1}{M} \sum_{m=1}^{M} \left[\varphi(\bar{X}_{\bar{\nu}}^{(m)})) \right]^2 - \left[\frac{1}{M} \sum_{m=1}^{M} \varphi(\bar{X}_{\bar{\nu}}^{(m)}) \right]^2 ,
$$

and M is a number of independent Markov chains $\left(\bar{\vartheta}_k^{(m)}, \bar{X}_k^{(m)} \right)$, $m = 1, \ldots, M$.

Because the simulated values $(\bar{\vartheta}_k, \bar{X}_k)$ coincide with the points of exact solution $(\bar{\vartheta}_k, X_k)$ here, the estimator $\bar{u}(t, x)$ does not contain the error of numerical integration (naturally, there are the Monte Carlo error depending on M and the error due to approximation of the boundary conditions depending on δ).

The mean number of steps of the random walk over touching spheres up to the boundary of space domain G is estimated by $C \ln(l/2\delta)$ (see, e.g., [95, 394] and also Theorem 7.4.12 in Sect. 7.4.3), if G is convex and l is its diameter. In our case the value of $\bar{\nu}$ is also estimated by $C \ln(l/2\delta)$.

Another Monte Carlo approach, whereby a random walk is made on a maximum square and the differential Laplace operator is approximated by a difference one, was proposed in [172].

Table 6.5.1 Test results for the boundary value problem (6.5.1)–(6.5.3). The exact solution $u(1, 0.7, 0.4) = 0.4796$ ($\delta = 0.00001$)

M	$\bar{u}(1, 0.7, 0.4) \pm 2[\bar{D}/M]^{1/2}$	$E\bar{\nu}$
1000	0.4460 ± 0.0527	3.142
4000	0.4780 ± 0.0270	3.257
10^5	0.4782 ± 0.0054	3.272

As an illustration, we take the following heat equation in the domain $Q = [0, T_1) \times G$, $G = \{x = (x_1, x_2) : |x_1| < 2, \ |x_2| < 1\}$ (this example is similar to one in [172]):

$$\frac{\partial u}{\partial t} = \frac{1}{2}\Delta u, \ t > 0, \ |x_1| < 2, \ |x_2| < 1, \tag{6.5.1}$$

with the initial and boundary conditions

$$u(0, x) = 2, \tag{6.5.2}$$

$$u(t, x) \mid_{\partial G} = 0, \ t > 0. \tag{6.5.3}$$

By changing time $t = T_1 - s$ in (6.5.1)–(6.5.3), we obtain the corresponding boundary value problem (like (6.3.1)–(6.3.3)) with the initial condition on the upper base. The results of numerical test are presented in Table 6.5.1.

Example 6.5.2 Consider the boundary value problem for biharmonic equation

$$L^2 u + c_1(x)Lu + c_2(x)u = f(x), \ x \in G \subset \mathbf{R}^d, \tag{6.5.4}$$

$$u \mid_{\partial G} = \varphi(x), \quad Lu \mid_{\partial G} = \psi(x), \tag{6.5.5}$$

where L is an operator of elliptic type:

$$L = \frac{1}{2}\sum_{i,j=1}^{d} a^{ij}(x)\frac{\partial^2}{\partial x^i \partial x^j} + \sum_{i=1}^{d} b^i(x)\frac{\partial}{\partial x^i},$$

and $c_1(x)$, $c_2(x)$, $f(x)$, $\varphi(x)$, and $\psi(x)$ are some known functions.

Introducing the function $v = Lu$, we obtain the system of elliptic equations

$$Lu - v = 0, \ x \in G, \ u \mid_{\partial G} = \varphi(x), \tag{6.5.6}$$

$$Lv + c_1(x)v + c_2(x)u = f(x), \ x \in G, \ v \mid_{\partial G} = \psi(x). \tag{6.5.7}$$

Let us give a probabilistic representation of the solution to the problem (6.5.6)–(6.5.7) (the first probabilistic representation for the problem (6.5.6)–(6.5.7) in the case of constant c_1 and c_2 was obtained in [193]). To this end introduce the system of SDEs

$$dX = b(X)\,ds + \sigma(X)\,dw(s), \tag{6.5.8}$$

$$\frac{dY_1}{ds} = c_2(X)Y_2 \tag{6.5.9}$$

$$\frac{dY_2}{ds} = -Y_1 + c_1(X)Y_2,$$

where $w(s)$ is a standard d-dimensional Wiener process, $b(x)$ is the d-dimensional vector with the components $b^i(x)$, Y_1 and Y_2 are scalars, and $\sigma(x)$ is a matrix that is obtained from the equality

$$a(x) = \sigma(x)\sigma^{\mathsf{T}}(x), \quad a(x) = \{a^{ij}(x)\}.$$

Under some conditions on the coefficients of the problem (6.5.6)–(6.5.7), its solution $(u(x), v(x))$ has the following form (see [271]):

$$u(x) = E\left[\varphi(X_x(\tau))Y_1^{(1)}(\tau) + \psi(X_x(\tau))Y_2^{(1)}(\tau)\right] \tag{6.5.10}$$

$$- E\int_0^\tau f(X_x(s))Y_2^{(1)}(s)\,ds,$$

$$v(x) = E\left[\varphi(X_x(\tau))Y_1^{(2)}(\tau) + \psi(X_x(\tau))Y_2^{(2)}(\tau)\right] - E\int_0^\tau f(X_x(s))Y_2^{(2)}(s)\,ds,$$

where τ is the first exit time of the process $X_x(s)$, $X(0) = x$, from the domain G, and $(Y_1^{(1)}, Y_2^{(1)})$ is the solution of the system (6.5.9) with initial data $Y_1^{(1)}(0) = 1$, $Y_2^{(1)}(0) = 0$, and $(Y_1^{(2)}, Y_2^{(2)})$ has the initial data $Y_1^{(2)}(0) = 0$, $Y_2^{(2)}(0) = 1$.

The probabilistic representation (6.5.8)–(6.5.10) for the boundary value problem (6.5.4)–(6.5.5) can be used for solving the problem (6.5.4)–(6.5.5) by implementing the random walk over small space-time parallelepipeds together with the Monte Carlo technique. If the coefficients of the elliptic operator L and the scalars c_1, c_2, f are constant, we can use the random walk over touching space parallelepipeds that gives an estimator which is free from the error of numerical integration. Note that in this case a sufficient condition, under which the representation (6.5.10) is valid, consists in $c_1 \le 0$, $c_2 \ge 0$.

As an illustration, consider the following two-dimensional problem in the square $G = \{x = (x_1, x_2) : |x_1| < 1, \ |x_2| < 1\}$:

$$\frac{1}{4}\Delta^2 u = 1, \ x \in G, \tag{6.5.11}$$

$$u \mid_{\partial G} = \varphi(x), \quad \varphi(x_1, \pm 1) = \frac{1 + x_1^4}{12}, \quad \varphi(\pm 1, x_2) = \frac{1 + x_2^4}{12},$$

$$\frac{1}{2} \Delta u \mid_{\partial G} = \psi(x), \quad \psi(x_1, \pm 1) = \frac{1 + x_1^2}{2}, \quad \psi(\pm 1, x_2) = \frac{1 + x_2^2}{2}. \qquad (6.5.12)$$

Introducing the function $v = \frac{1}{2} \Delta u$ as above, we obtain the system of elliptic equations

$$\frac{1}{2} \Delta u - v = 0, \quad x \in G, \quad u \mid_{\partial G} = \varphi(x) \qquad (6.5.13)$$

$$\frac{1}{2} \Delta v = 1, \quad x \in G, \quad v \mid_{\partial G} = \psi(x). \qquad (6.5.14)$$

Its exact solution is

$$u(x) = \frac{x_1^4 + x_2^4}{12}, \quad v(x) = \frac{x_1^2 + x_2^2}{2}.$$

Of course, one can solve the problem (6.5.13)–(6.5.14) sequentially: first find the function v from the problem (6.5.14) and then u from (6.5.13). But such an approach requires knowing the function v in the whole domain G even if one needs the solution (u, v) only at individual points of the domain G. In the latter case, the Monte Carlo approach is more preferable.

For the system (6.5.13)–(6.5.14), the formulas (6.5.8)–(6.5.10) acquire the form

$$u(x) = E\varphi(x + w(\tau)) - E[\tau\psi(x + w(\tau))] + \frac{1}{2} E\tau^2,$$

$$v(x) = E\psi(x + w(\tau)) - E\tau,$$

where τ is the first exit time of the process $x + w(s)$ from the domain G.

To simulate the point $(\tau, x + w(\tau))$, we use the random walk over touching space squares, which is finished in a δ-neighborhood of the boundary ∂G belonging to G. Recall that we are able to exactly simulate both the exit point and the exit time of the Wiener process from a square in accordance with Theorem 6.3.7. Then for the same reasons as in Example 6.5.1, the corresponding estimator $(\bar{u}, \bar{v})$ does not contain the error of numerical integration. The comment on the mean number of steps $E\bar{\nu}$ from Example 6.5.1 is also valid here. Let us underline that the method of random walk over touching spheres in the space domain G cannot be applied to this problem because we essentially use the simulation of both the exit point $x + w(\tau)$ and the exit time τ. The results of numerical tests are given in Table 6.5.2.

An implementation of the random walk over touching space squares can be found in [314, Appendix A.3].

Table 6.5.2 Test results for the boundary value problem (6.5.11)–(6.5.12) ($\delta = 0.00001$)

M	x_1	x_2	$u(x_1, x_2)$	$\bar{u}(x_1, x_2)$	$v(x_1, x_2)$	$\bar{v}(x_1, x_2)$	$E\bar{v}$
10^4	0.3	0.5	0.00588	0.0051 ± 0.0037	0.17000	0.1656 ± 0.0082	4
10^5				0.0059 ± 0.0012		0.1698 ± 0.0026	4
10^6				0.0058 ± 0.0004		0.1700 ± 0.0008	4
10^4	0.7	0.8	0.05414	0.0539 ± 0.0022	0.56500	0.5598 ± 0.0064	4
10^5				0.0541 ± 0.0006		0.5638 ± 0.0020	4
10^6				0.0542 ± 0.0002		0.5646 ± 0.0006	4
10^4	0.9	0.9	0.10935	0.1090 ± 0.0011	0.81000	0.8067 ± 0.0039	3
10^5				0.1092 ± 0.0003		0.8088 ± 0.0012	3
10^6				0.1093 ± 0.0001		0.8097 ± 0.0004	3

Example 6.5.3 Consider the second-order linear autonomous Ito system of SDEs

$$dX = AX\, dt + \sum_{i=1}^{2} B_i X\, dw_i(t), \tag{6.5.15}$$

where X is a two-dimensional vector, A and B_i, $i = 1, 2$, are constant 2×2-matrices, $w_i(t)$, $i = 1, 2$, are independent standard Wiener processes.

Various characteristics describing asymptotic behavior of solutions of the system (6.5.15), such as the Lyapunov exponent, the moment Lyapunov exponents, the stability index, and some others, are considered in [11, 12, 194] (see also references therein). The Lyapunov exponent λ^* of system (6.5.15) (cf. [194]) is defined as

$$\lambda^* := \lim_{t \to \infty} \frac{1}{t} E \ln |X_x(t)| = \lim_{t \to \infty} \frac{1}{t} \ln |X_x(t)| \ a.s., \tag{6.5.16}$$

and the moment Lyapunov exponent $g(p)$ is defined as

$$g(p) := \lim_{t \to \infty} \frac{1}{t} E \ln |X_x(t)|^p, \ \ p \in \mathbf{R}, \tag{6.5.17}$$

where $X_x(t)$, $t \geq 0$, is a nontrivial solution to system (6.5.15).

The limits λ^* and $g(p)$ exist, and they are independent of x, $x \neq 0$, in the ergodic case. The limit $g(p)$ is a convex analytic function of $p \in \mathbf{R}$, $g(0) = 0$, $g(p)/p$ increases with growing p, and

$$g'(0) = \lambda^*. \tag{6.5.18}$$

If $\lambda^* < 0$ then the trivial solution to system (6.5.15) is a.s. asymptotically stable. It is well known and it follows from (6.5.18) that in this case $g(p)$ is negative for all sufficiently small $p > 0$, i.e., the solution $X = 0$ of (6.5.15) is p-stable for such p. If $g(p) \to +\infty$ as $p \to +\infty$, then the equation

$$g(p) = 0 \tag{6.5.19}$$

has the unique root $\gamma^* > 0$, which is known as the stability index.

It is clear that the solution $X = 0$ of (6.5.15) is p-stable for $0 < p < \gamma^*$ and p-unstable for $p > \gamma^*$. The stability index γ^* is connected with the asymptotic behavior of the probability $V_\delta(x)$ of the exit of $X_x(t)$ from the ball $|x| < \delta$ (see [22]): $V_\delta(x) := P\{\sup_{t \geq 0} |X_x(t)| > \delta\}$, $|x|/\delta \to 0$. It turns out that there exists a constant $K > 0$ such that for all $\delta > 0$ and $|x| < \delta$ the following inequality takes place:

$$\frac{1}{K}(|x|/\delta)^{\gamma^*} \leq V_\delta(x) \leq K\,(|x|/\delta)^{\gamma^*}. \tag{6.5.20}$$

The unstable case, when the equation (6.5.19) has a negative root γ^*, is considered analogously [22].

The stability properties of the system (6.5.15) can also be characterized by the exit time τ of $X_x(t)$ from a certain neighborhood of the origin. In [220] the value of $Ee^{-\mu\tau}$, $\mu > 0$, is simulated. By the algorithms proposed in this chapter, we are able to evaluate the distribution function $P(\tau < t)$, which can be a good characteristic for describing transient behavior related to the system (6.5.15). Naturally, we are also able to evaluate functionals on τ, e.g., $Ee^{-\mu\tau}$.

We take the following particular case of the two-dimensional system (6.5.15) for our numerical tests:

$$dX_1 = (aX_1 + cX_2)\,ds + b_1 X_1\,dw_1(s) + b_2 X_2\,dw_2(s) \tag{6.5.21}$$
$$dX_2 = (-cX_1 + aX_2)\,ds + b_1 X_2\,dw_1(s) - b_2 X_1\,dw_2(s),$$
$$X(0) = X_x(0) = x.$$

The function $g(p)$, the Lyapunov exponent λ^*, and the stability index γ^* for this system are equal to [278]:

$$g(p) = p \times (a + \frac{1}{2}(b_2^2 - b_1^2)) + \frac{1}{2}p^2 b_1^2, \tag{6.5.22}$$
$$\lambda^* = g'(0) = a + \frac{1}{2}(b_2^2 - b_1^2),$$
$$\gamma^* = -\frac{2a + (b_2^2 - b_1^2)}{b_1^2}.$$

Here we evaluate the distribution function $P(\tau < t)$, where τ is the first exit time of $X_x(s)$ under $X(0) = (1, 1)^\mathsf{T}$ from the square $G = \{(x_1, x_2) : |x_i| < 3,\ i = 1, 2\}$. To simulate the system (6.5.21), we use the random walk over boundaries of small space-time parallelepipeds constructed in Sect. 6.4.2. The algorithm allows finding $\bar{\tau}$ (see Sect. 6.4.3), which is close to τ. The sampling distribution function $\bar{P}_M(t)$ is calculated as

Fig. 6.5.1 The distribution
function $\bar{P}(t)$ for $a = -1$,
$c = 1$, $b_2 = 2$,
$X(0) = (1, 1)^{\mathsf{T}}$, $r = 0.02$,
$M = 5000$, and for various
$b_1 : (1)\ b_1 = 0.1$
$(\lambda^* = 0.995, \gamma^* = -199)$,
$(2)\ b_1 = 0.6\ (\lambda^* = 0.82$,
$\gamma^* = -4.556)$, $(3)\ b_1 = \sqrt{5}$
$(\lambda^* = -1.5, \gamma^* = 0.6)$, and
$(4)\ b_1 = 3\ (\lambda^* = -3.5$,
$\gamma^* = 0.778)$

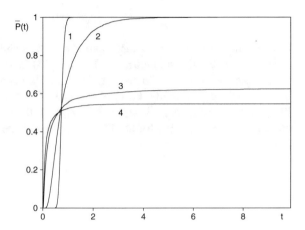

$$\bar{P}_M(t) = \begin{cases} 0, & t \leq \bar{\tau}_1^{(M)}, \\ m/M, & \bar{\tau}_m^{(M)} < t \leq \bar{\tau}_{m+1}^{(M)}, \\ 1, & t > \bar{\tau}_M^{(M)}, \end{cases}$$

where $\{\bar{\tau}_1^{(M)}, \dots, \bar{\tau}_M^{(M)}\}$ is a sample point of size M sorting in the ascending order, it corresponds to the random variable $\bar{\tau}$.

The sampling function $\bar{P}_M(t)$ is close to the distribution function $\bar{P}(t) = P(\bar{\tau} < t)$ under a sufficiently big M, and $\bar{P}(t)$ is close to $P(\tau < t)$ under a sufficiently small r. We control the accuracy of our simulations by increasing M and decreasing r. We select M and r such that the curves $\bar{P}_M(t)$ are visually almost identical under larger values of M and smaller values of r. The parameter l of space-time parallelepipeds $\Pi_{r,l}$ used in the simulations is taken equal to 1.

Figure 6.5.1 presents the behavior $\bar{P}(t) \doteq \bar{P}_M(t)$ under fixed a, c, b_2, and various b_1. Increase of b_1 leads to stabilization (see the formulas (6.5.22)). It is interesting to note (see Fig. 6.5.1) that the probability of the exit of $X_x(s)$ from G at small times t under $\lambda^* > 0$ (unstable case) is lower than the corresponding probability under $\lambda^* < 0$ (stable case). It can be explained in the following way. The radius $\rho(s) = \sqrt{X_1^2(s) + X_2^2(s)}$ satisfies the equation

$$d\rho = \left(a + \frac{b_2^2}{2} \right) \rho\, ds + b_1 \rho\, dw_1(s). \tag{6.5.23}$$

Due to the selection of the parameters, the Lyapunov exponent λ^* is positive (unstable case) under relatively small b_1 and large b_2. In this case, the first term of (6.5.23) plays the main role and the influence of noise is relatively small. So, there is a lag time before the trajectory $X_x(s)$ leaves the domain G. In the stable case our parameters

Fig. 6.5.2 The distribution function $\bar{P}(t)$ for $a = -1$, $c = 1$, $X(0) = (1, 1)^\mathsf{T}$, $\lambda^* = -1.5$, $M = 5000$, and for various γ^* : (1) $\gamma^* = 1/3$ ($b_1 = 3$, $b_2 = 2.828$, $r = 0.02$), (2) $\gamma^* = 0.6$ ($b_1 = \sqrt{5}$, $b_2 = 2$, $r = 0.02$), and (3) $\gamma^* = 2.479$ ($b_1 = 1.1$, $b_2 = 0.4683$, $r = 0.05$)

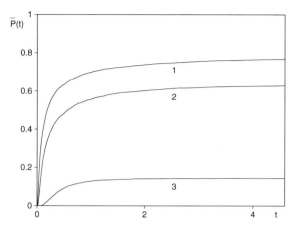

are such that b_1 is large and the second term of (6.5.23) plays an essential role. Then the trajectory $X_x(s)$ can leave the domain G during a small time interval with a rather large probability.

Figure 6.5.2 illustrates the behavior of $\bar{P}(t)$ under fixed a, c, and $\lambda^* = a + (b_2^2 - b_1^2)/2$ for various values of the stability index γ^* (see (6.5.22)). One can see that the probability of the exit of the trajectory $X_x(s)$ from G decreases with increasing of γ^* in accordance with (6.5.20).

Figures 6.5.1 and 6.5.2 also demonstrate that in the unstable case the trajectory leaves the neighborhood of the origin during a finite time interval with the probability equal to 1 (see the curves 1 and 2 on Fig. 6.5.1). However, in the stable case, the probability $P(\tau < \infty)$ of leaving the neighborhood of the origin by the trajectory is less than 1. This probability decreases with decrease of the Lyapunov exponent λ^* (see the curves 3 and 4 on Fig. 6.5.1) and with increase of the stability index γ^* (see Fig. 6.5.2).

6.6 Mean-Square Approximation of Diffusion with Reflection

Consider an autonomous system of SDEs with normal reflection at the boundary

$$dX = b(X)\,ds + \sigma(X)\,dw(s) + \nu(X)I_{\partial G}(X)\,dL(s), \quad X(0) = x, \qquad (6.6.1)$$

in a convex domain G in $\mathbf{R}^d$ with nonempty interior. Here $w(t)$, $t \geq 0$, is a standard d-dimensional Wiener process, $b(x)$ is a d-dimensional vector, $\sigma(x)$ is a $d \times d$-dimensional matrix, $\nu(x)$ is the direction of the inner normal to the surface ∂G at a point $x \in \partial G$, $L(t)$, $t \geq 0$, is the local time of the process X on the boundary ∂G of the domain G, and $I_A(x)$ is the indicator function of a set A. The domain G may be unbounded. For instance, G can be a convex polyhedron, i.e., $G = \cap_{i=1}^{n} G_i$, where G_i are half-spaces; in particular, G can be a half-space.

We recall that the local time $L(t)$ is a continuous, nondecreasing random process which increases on the set $\{t \geq 0, X(t) \in \partial G\}$ only. The Lebesgue measure of this set is zero. For theoretical aspects of SDEs with reflections see, e.g., [111, Section 1.6] and [177, 180, 237, 405].

We briefly observe two types of mean-square methods for (6.6.1): the projection scheme [57, 235, 361, 362, 406, 407] and the penalization scheme [239, 362, 407]. Introduce the projection map onto $\bar{G}$:

$$\Pi(x) = \arg\min_{y \in \bar{G}} |x - y| , \quad x \in \mathbf{R}^d .$$

We note that if $x \in \bar{G}$ then $\Pi(x) = x$.

The projection scheme can be written in the form

$$X_{k+1} = \Pi \left(X_k + h \, b(X_k) + \sigma(X_k) \Delta w_k \right) , \tag{6.6.2}$$

where, as usual, h is a time step and $\Delta w_k := w(t_{k+1}) - w(t_k)$. It is clear that according to this scheme we have the standard mean-square Euler approximations (see Sect. 1.1.5) while a point $\hat{X}_{k+1} = X_k + h \, b(X_k) + \sigma(X_k) \Delta w_k$ is inside the domain $\bar{G}$. If $\hat{X}_{k+1} \notin \bar{G}$ then the next value X_{k+1} is taken as the projection of $\hat{X}_{k+1}$ on the boundary ∂G.

To introduce the penalization scheme, we need an additional notation:

$$\beta(x) := x - \Pi(x), \quad \beta_\lambda(x) := \beta(x)/\lambda, \quad x \in \mathbf{R}^d .$$

We note that $\beta(x) = 0$ for $x \in \bar{G}$. Consider the system of SDEs

$$dX_\lambda = b(X_\lambda)ds + \sigma(X_\lambda) \, dw(s) - \beta_\lambda(X_\lambda) \, ds, \quad X(0) = x . \tag{6.6.3}$$

It is known [264] that for any $0 < \alpha < 1$ the estimate

$$E \sup_{t \in [0,T]} |X(t) - X_\lambda(t)|^2 \leq C \lambda^{1-\alpha} \tag{6.6.4}$$

holds under some conditions on the coefficients of (6.6.1).

Approximating the solution of (6.6.3) by the standard mean-square Euler scheme and choosing $\lambda = h$, one obtains the penalization scheme for (6.6.1):

$$X_{k+1} = X_k + h \, b(X_k) - h \, b_h(X_k) + \sigma(X_k) \, \Delta w_k \tag{6.6.5}$$
$$= \Pi(X_k) + h \, b(X_k) + \sigma(X_k) \, \Delta w_k.$$

The choice $\lambda = h$ in (6.6.5) can be explained in the following way (further details are available in [362]). Suppose we have applied the mean-square Euler method to

(6.6.3). Then the penalization scheme contains the penalty term $h\beta_\lambda(X_k)$. If $\lambda < h$, the penalty term may push X_{k+1} inwards too much for G because when this term is nonzero it is generically of order $h^{3/2}/\lambda$ which, if $\lambda = h^{1+\varepsilon}$, $\varepsilon > 0$, is too large compared with the increments of X which are of order $h^{1/2}$. For small $h \leq \lambda$ the mean-square error of the penalization scheme is estimated as $O([h\ln(1/h)]^{1/4} + \lambda^{1/2-\alpha})$ for any $\alpha > 0$ [362] (cf. (6.6.4) and Theorem 6.6.1). For $\lambda = h^{1/2}$ the error of the method is $O(h^{1/4-\alpha})$ for any $\alpha > 0$ while for $\lambda = h$ the estimates given in Theorem 6.6.1 are valid.

The following convergence theorem was proved in [407] (see also [361, 362]).

Theorem 6.6.1 *Assume that the coefficients of (6.6.1) satisfy a Lipschitz condition. Let $\bar{X}(t) := X_k$ for $t \in [t_k, t_{k+1})$, where X_k are obtained by either the projection scheme (6.6.2) or by the penalization scheme (6.6.5). Then for every $T > 0$ and natural number p the following estimates hold:*

(i) in the case of a general convex domain G:

$$\left(E \sup_{t\in[0,T]} |X(t) - \bar{X}(t)|^{2p} \right)^{1/2p} = O([h\ln(1/h)]^{1/4}),$$

(ii) in the case of a convex polyhedron G:

$$\left(E \sup_{t\in[0,T]} |X(t) - \bar{X}(t)|^{2p} \right)^{1/2p} = O([h\ln(1/h)]^{1/2}).$$

In the case of a convex polyhedron it was possible to obtain the rate of convergence $O([h\ln(1/h)]^{1/2})$ (see [361, 362, 406, 407]) due to the fact from [85] that an estimate of $\sup_{t\in[0,T]} |X(t) - \bar{X}(t)|^{2p}$ does not contain a term with local time. It is probably impossible to obtain the rate of convergence $O([h\ln(1/h)]^{1/2})$ for other convex domains in $\mathbf{R}^d$ (see further details in [361, 406, 407]).

Chapter 7
Random Walks for Linear Boundary Value Problems

Chapters 2 and 3 are devoted to weak numerical methods for SDEs which are suitable for solving the Cauchy problem for linear parabolic equations. The previous chapter deals with mean-square approximations of SDEs in bounded domains, and its results can be applied for solving boundary value problems. However, since solutions of boundary value problems for parabolic and elliptic equations can be represented as expectations of solutions of the corresponding systems of SDEs in bounded domains, one can apply far simpler weak approximations which certainly should be subject to restrictions related to nonexit from the bounded domains. Such weak methods are considered in this chapter.

In Sects. 7.1 and 7.2, we present various weak methods for the Dirichlet problem for parabolic equations obtained in [275, 276, 311]. Section 7.3 deals with random walks for the Dirichlet problem for elliptic equations [281, 292, 311]. In Sects. 7.4 and 7.5, specific weak methods for the Dirichlet problem for elliptic equations are constructed [281]. Estimates of errors of these specific methods contain only low-order derivatives of the solution to the Dirichlet problem. This can be important, in particular, for problems with a boundary layer (i.e., for problems with small parameter at higher derivatives). Section 7.6 is devoted to solution of the Neumann problem for linear parabolic equations [233, 276, 277]. Other random walks for linear boundary value problems are proposed in [35, 64, 132, 134, 239]. See other approaches to the Monte Carlo methods for solving partial differential equations in [95, 220, 394] and references therein. In the case of SDEs driven by Levy processes [9], weak methods for the Dirichlet problem for parabolic integro-differential equation were considered in [77].

© Springer Nature Switzerland AG 2021
G. N. Milstein and M. V. Tretyakov, *Stochastic Numerics for Mathematical Physics*,
Scientific Computation, https://doi.org/10.1007/978-3-030-82040-4_7

7.1   Algorithms for Solving the Dirichlet Problem Based on Time-Step Control

Let G be a bounded domain in $\mathbf{R}^d$ and $Q = [T_0, T_1) \times G$ be a cylinder in $\mathbf{R}^{d+1}$, $\Gamma = \bar{Q} \backslash Q$ be the part of the cylinder's boundary consisting of the upper base and lateral surface. Consider the Dirichlet problem for the parabolic equation:

$$\frac{\partial u}{\partial t} + \frac{1}{2}\sum_{i,j=1}^{d} a^{ij}(t,x)\frac{\partial^2 u}{\partial x^i \partial x^j} + \sum_{i=1}^{d} b^i(t,x)\frac{\partial u}{\partial x^i} + c(t,x)u$$

$$+ g(t,x) = 0, \quad (t,x) \in Q, \tag{7.1.1}$$

$$u\mid_\Gamma = \varphi(t,x). \tag{7.1.2}$$

The form of Eq. (7.1.1) is convenient for a probabilistic approach: the "initial" condition is prescribed at the final time moment $t = T_1$, and the equation is considered for $t < T_1$.

We assume that the coefficients $a^{ij} = a^{ji}$ satisfy the strict ellipticity condition in $\bar{Q}$ and that conditions hold which guarantee existence of the sufficiently smooth in $\bar{Q}$ solution of (7.1.1)–(7.1.2). We recall [114, 222, 241] that these conditions consist in some smoothness requirements imposed on the functions $a^{ij}(t,x)$, $b^i(t,x)$, $c(t,x)$, $g(t,x)$, on the boundary ∂G of the domain G, and on the function φ which is finally assumed to satisfy (7.1.1) on the boundary of the upper base of $\bar{Q}$.

The solution of the problem (7.1.1)–(7.1.2) has the following probabilistic representation:

$$u(t,x) = E\left[\varphi(\tau, X_{t,x}(\tau))Y_{t,x,1}(\tau) + Z_{t,x,1,0}(\tau)\right], \tag{7.1.3}$$

where $X_{t,x}(s)$, $Y_{t,x,y}(s)$, $Z_{t,x,y,z}(s)$, $s \geq t$, is the solution of the Cauchy problem for the system of SDEs:

$$dX = (b(s,X) - \sigma(s,X)\mu(s,X))\,ds + \sigma(s,X)\,dw(s), \quad X(t) = x, \tag{7.1.4}$$

$$dY = c(s,X)Y\,ds + \mu^\mathsf{T}(s,X)Y\,dw(s), \quad Y(t) = y, \tag{7.1.5}$$

$$dZ = g(s,X)Y\,ds + F^\mathsf{T}(s,X)Y\,dw(s), \quad Z(t) = z, \tag{7.1.6}$$

$(t,x) \in Q$, and $\tau = \tau_{t,x}$ is the first exit time of the trajectory $(s, X_{t,x}(s))$ to the boundary Γ. In (7.1.4)–(7.1.6), $w(s) = (w^1(s), \ldots, w^d(s))^\mathsf{T}$ is a standard Wiener process, $b(s,x)$ is a d-dimensional column-vector composed of the coefficients $b^i(s,x)$, the $d \times d$ matrix $\sigma(s,x)$ is obtained from the formula $\sigma(s,x)\sigma^\mathsf{T}(s,x) = a(s,x)$, $a(s,x) = \{a^{ij}(s,x)\}$, $i,j = 1,\ldots,d$, $\mu(s,x)$ and $F(s,x)$ are arbitrary d-dimensional column-vectors sufficiently smooth in $\bar{Q}$, Y and Z are scalars.

For $\mu(s,x) = 0$ and $F(s,x) = 0$, the formula (7.1.3) gives the standard probabilistic representation [86, 123]. For $\mu(s,x) \neq 0$ and $F(s,x) = 0$, (7.1.3) follows from Girsanov's theorem. It is also clear that the term $F(s,x) \neq 0$ does not affect the validity of this formula. Thus, the mean of the random variable appearing under the symbol of expectation in (7.1.3) does not depend on μ and F. At the same time, other properties of this random variable can essentially depend on μ and F. In particular, choosing μ or/and F, it is possible to reach zero variance of this random variable (see Sect. 2.3). This property is of great importance since we use the Monte Carlo technique for solving linear boundary value problems by probabilistic methods. This is a reason for us to construct random walks taking into account dependence on μ and F.

To realize the representation (7.1.3), we need an approximation of the trajectory $(s, X(s))$, and this approximation should satisfy some restrictions related to nonexit from the domain $\bar{Q}$. Such approximations are considered in Sects. 7.1 and 7.2.

Approximations of this section (originally proposed in [275, 276]) are based on controlling a time step of numerical integration of the system (7.1.4). The step is chosen so that (of course, aside of reaching a required accuracy) the next state of a Markov chain approximating in the weak sense the solution of (7.1.4) remains in the domain $\bar{Q}$ with probability one. This leads to a decrease of the time step when the chain is close to the boundary Γ of the domain Q. The chain is stopped in a narrow zone near the boundary so that values of the solution $u(t, x)$ in this zone can be approximated quite accurately by the known values of the function φ on the boundary. Another type of approximations is considered in Sect. 7.2. There the step of numerical integration of the system (7.1.4) is constant for points belonging to a certain time layer $t = t_k$. But when a point is close to the boundary, we make an intermediate (auxiliary) step of the random walk, which preserves the point in the time layer $t = t_k$.

In Sect. 7.1.1, some one-step weak approximations for (7.1.4)–(7.1.6) are given together with their error analysis while convergence theorems are proved in Sect. 7.1.2. In this section we restrict ourselves to the case of $\mu = 0$ and $F = 0$ in (7.1.4)–(7.1.6).

7.1.1 Theorems on One-Step Approximation

Let $(t_0, x_0) \in Q$, ξ be a point with uniform distribution on the surface of an open unit ball $U_1 \subset \mathbf{R}^d$ with center at the origin, and $r > 0$ be a number. Introduce an open ellipsoid $U_r^{\sigma(t_0, x_0)}(x_0)$ obtained from the ball U_1 by the linear transformation $r\sigma(t_0, x_0)$ and the shift $x_0 + b(t_0, x_0)r^2/d$. For any point $(t_0, x_0) \in Q$, there is an $r > 0$ such that $x_0 \in U_r^{\sigma(t_0, x_0)}(x_0)$ and the cylinder $\Pi_r^{\sigma(t_0, x_0)}(t_0, x_0) = [t_0, t_0 + r^2/d] \times U_r^{\sigma(t_0, x_0)}(x_0) \subset Q$. Below r is always assumed to be sufficiently small. Obviously, the point (t_1, X_1) with

$$t_1 = t_0 + \frac{r^2}{d}, \quad X_1 = x_0 + b(t_0, x_0)\frac{r^2}{d} + r\sigma(t_0, x_0)\xi, \qquad (7.1.7)$$

belongs to the boundary of the upper base of $\bar{\Pi}_r^{\sigma(t_0, x_0)}(t_0, x_0)$.

We also introduce

$$Y_1 = y_0 + c(t_0, x_0) y_0 \frac{r^2}{d}, \quad Z_1 = z_0 + g(t_0, x_0) y_0 \frac{r^2}{d}. \tag{7.1.8}$$

Theorem 7.1.1 *Let the function f be the restriction to $\bar{\Pi}_r^{\sigma(t_0,x_0)}(t_0, x_0)$ of the solution of the problem (7.1.1)–(7.1.2) having in $\bar{Q}$ continuous derivatives $D_x^m D_t^k u, 0 \leq m + 2k \leq 4, k = 0, 1$. Let $X_{t_0,x_0}(s), Y_{t_0,x_0,y_0}(s), Z_{t_0,x_0,y_0,z_0}(s)$ be the solution of the system (7.1.4)–(7.1.6) and ϑ be the exit time of the process $(s, X_{t_0,x_0}(s))$ from $\Pi_r^{\sigma(t_0,x_0)}(t_0, x_0)$ (i.e., the point $(\vartheta, X_{t_0,x_0}(\vartheta))$ belongs either to the upper base or to the lateral surface of the cylinder $\bar{\Pi}_r^{\sigma(t_0,x_0)}(t_0, x_0)$). Then*

$$E\left[f(t_1, X_1)Y_1 + Z_1 - f(\vartheta, X_{t_0,x_0}(\vartheta))Y_{t_0,x_0,y_0}(\vartheta) - Z_{t_0,x_0,y_0,z_0}(\vartheta)\right] = O(r^4). \tag{7.1.9}$$

Proof We have

$$E\left[f(\vartheta, X_{t_0,x_0}(\vartheta))Y_{t_0,x_0,y_0}(\vartheta) + Z_{t_0,x_0,y_0,z_0}(\vartheta)\right] = u(t_0, x_0)y_0 + z_0.$$

Since $f(t_1, X_1) = u(t_1, X_1)$, the difference under the expectation symbol in (7.1.9) is equal to

$$u(t_1, X_1)Y_1 + Z_1 - u(t_0, x_0)y_0 - z_0 \tag{7.1.10}$$

$$= u\left(t_0 + \frac{r^2}{d}, x_0 + b(t_0, x_0)\frac{r^2}{d} + r\sigma(t_0, x_0)\xi\right) \times \left(y_0 + c(t_0, x_0)y_0\frac{r^2}{d}\right)$$

$$+ z_0 + g(t_0, x_0)y_0\frac{r^2}{d} - u(t_0, x_0)y_0 - z_0$$

$$= [u(t_0, x_0) + \frac{\partial u}{\partial t}(t_0, x_0)\frac{r^2}{d}$$

$$+ \sum_{i=1}^{d} \frac{\partial u}{\partial x^i}(t_0, x_0)\{b(t_0, x_0)\frac{r^2}{d} + r\sigma(t_0, x_0)\xi\}^i$$

$$+ \frac{1}{2}\sum_{i,j=1}^{d} \frac{\partial^2 u}{\partial x^i \partial x^j}(t_0, x_0)\{b(t_0, x_0)\frac{r^2}{d} + r\sigma(t_0, x_0)\xi\}^i$$

$$\times \{b(t_0, x_0)\frac{r^2}{d} + r\sigma(t_0, x_0)\xi\}^j]$$

$$\times (1 + c(t_0, x_0)\frac{r^2}{d})y_0 + \rho + g(t_0, x_0)y_0\frac{r^2}{d} - u(t_0, x_0)y_0.$$

The remainder ρ contains the terms of the form

$$\frac{\partial^2 u}{\partial t \partial x^i}(t_0, x_0)\frac{r^2}{d}r\{\sigma(t_0, x_0)\xi\}^i,$$

$$\frac{\partial^3 u}{\partial x^i \partial x^j \partial x^k}(t_0, x_0)r^3\{\sigma(t_0, x_0)\xi\}^i\{\sigma(t_0, x_0)\xi\}^j\{\sigma(t_0, x_0)\xi\}^k,$$

having zero expectations, and it also contains the terms of order $O(r^4)$ (uniform in $(t_0, x_0) \in Q$). This is due to the fact that $E\xi^i = 0$, $E\xi^i\xi^j\xi^k = 0$, $i, j, k = 1, \ldots, d$, and due to the assumption that $u(t, x)$ has continuous in $\bar{Q}$ derivatives:

$$\frac{\partial^2 u}{\partial t^2}, \quad \frac{\partial^3 u}{\partial x^i \partial x^j \partial x^k}, \quad \frac{\partial^4 u}{\partial x^i \partial x^j \partial x^k \partial x^l}.$$

Using the equalities $E\xi^i\xi^j = \delta_{ij}/d$, $i, j = 1, \ldots, d$, we obtain from (7.1.10):

$$E[u(t_1, X_1)Y_1 + Z_1 - u(t_0, x_0)y_0 - z_0]$$

$$= \left(\frac{\partial u}{\partial t} + \frac{1}{2}\sum_{i,j=1}^{d} a^{ij}\frac{\partial^2 u}{\partial x^i \partial x^j} + \sum_{i=1}^{d} b^i\frac{\partial u}{\partial x^i} + cu + g\right)\frac{r^2}{d}y_0 + O(r^4) = O(r^4),$$

i.e., (7.1.9) is proved.

For a completeness of the exposition, we derive the relations:

$$E\xi^i = 0, \quad E\xi^i\xi^j = \delta_{ij}/d, \quad E\xi^i\xi^j\xi^k = 0, \quad i, j, k = 1, \ldots, d. \tag{7.1.11}$$

The first and the third groups of equalities in (7.1.11) follow from the symmetry. The equalities $E\xi^i\xi^i = 1/d$, $i = 1, \ldots, d$, follow from the relation $(\xi^1)^2 + \cdots + (\xi^d)^2 = 1$. To prove $E\xi^i\xi^j = 0$, $i \neq j$, we note that ξ has the same distribution as $w(\theta)$, where θ is the exit time of the d-dimensional Wiener process from the unit ball U_1. Further, the function $v = x^i x^j$, $i \neq j$, is the solution of the problem

$$\frac{1}{2}\Delta v = 0, \quad v|_{\partial U_1} = x^i x^j.$$

From here $0 = v(0) = Ew^i(\theta)w^j(\theta) = E\xi^i\xi^j$, and we have obtained the relations (7.1.11). Theorem 7.1.1 is proved. □

Now let $C_1 \subset \mathbf{R}^d$ be the cube with center at the origin and with the coordinates of the vertices equal to ± 1, i.e., its edges are parallel to the coordinate axes and their length is equal to 2. Introduce the parallelepiped $C_r^{\sigma(t_0, x_0)}(x_0)$ obtained from the cube C_1 by the linear transformation $r\sigma(t_0, x_0)$ and the shift $x_0 + b(t_0, x_0)r^2/d$ and also the cylinder $\Pi_r^{\sigma(t_0, x_0)}(t_0, x_0) = [t_0, t_0 + r^2/d] \times C_r^{\sigma(t_0, x_0)}(x_0) \subset Q$. Let ν be the random vector whose coordinates ν^i, $i = 1, \ldots, d$, are mutually independent random variables taking the values ± 1 with probability $1/2$. It is clear that the point (t_1, X_1) with

$$t_1 = t_0 + r^2, \quad X_1 = x_0 + b(t_0, x_0)\, r^2 + \sigma(t_0, x_0)\, r\nu, \tag{7.1.12}$$

finds itself on the vertices of the upper base of $\bar{\Pi}_r^{\sigma(t_0, x_0)}(t_0, x_0)$.

We also introduce

$$Y_1 = y_0 + c(t_0, x_0)\, y_0 r^2, \quad Z_1 = z_0 + g(t_0, x_0)\, y_0 r^2. \tag{7.1.13}$$

Theorem 7.1.2 *Let the function f be the restriction to $\bar{\Pi}_r^{\sigma(t_0, x_0)}(t_0, x_0)$ of the solution of the problem (7.1.1)–(7.1.2) having in $\bar{Q}$ continuous derivatives $D_x^m D_t^k u$, $0 \le m + 2k \le 4$, $k = 0, 1$. Let $X_{t_0, x_0}(s)$, $Y_{t_0, x_0, y_0}(s)$, $Z_{t_0, x_0, y_0, z_0}(s)$ be the solution of the system (7.1.4)–(7.1.6) and ϑ be the exit time of the process $(s, X_{t_0, x_0}(s))$ from $\Pi_r^{\sigma(t_0, x_0)}(t_0, x_0)$. Then (7.1.9) holds.*

Proof Consider the difference

$$d_1 = u(t_1, X_1)Y_1 + Z_1 - u(t_0, x_0)y_0 - z_0. \tag{7.1.14}$$

Doing transformations of d_1 as in the proof of Theorem 7.1.1 and then taking the expectation of d_1, we obtain (7.1.9).   □

Remark 7.1.3 In Theorems 7.1.1 and 7.1.2 we obtain that one-step orders of accuracy of the approximations (7.1.7)–(7.1.8) and (7.1.12)–(7.1.13) are equal to $O(r^4)$. The time step is proportional to r^2 and the proposed methods have one-step order 2 with respect to the time step as the weak Euler method (2.0.21).

Remark 7.1.4 "Ordinary" weak methods (see Chap. 2) usually involve an analysis of the difference

$$E\left[f(\bar{X}_{t_0, x_0}(t_0 + h)) - f(X_{t_0, x_0}(t_0 + h)) \right],$$

i.e., the d-dimensional manifold in $(d + 1)$-dimensional space of variables $(t, x^1, \ldots, x^d)$, on which the function f is defined, has a special form $t = t_1$, and, in particular, it is unbounded. In Theorems 7.1.1 and 7.1.2 this manifold is already bounded, it is a part of the cylinder's boundary. Then, associating t_1 with ϑ, it is natural to have t_1 in more general methods as a random variable.

Introduce the one-step approximation:

$$t_1 = t_0 + \Delta t(t_0, x_0, \xi; r), \quad X_1 = x_0 + \Delta x(t_0, x_0, \xi; r), \tag{7.1.15}$$
$$Y_1 = y_0 + \Delta y(t_0, x_0, \xi; r)\, y_0, \quad Z_1 = z_0 + \Delta z(t_0, x_0, \xi; r)\, y_0,$$

where ξ is a random vector and r is a small parameter, possibly multidimensional, connected with the size of steps in $t, x^1, \ldots, x^d$. Here we restrict ourselves to a one-dimensional r. We assume that for all possible values of ξ increments Δt and Δx are such that the point $(t_1, X_1) \in \bar{Q}$.

Our nearest aim is to obtain sufficient conditions which ensure the order of accuracy $O(r^4)$ for the approximation (7.1.15). To this end, we substitute (7.1.15) in (7.1.14) and expand the difference d_1 in the powers of Δt, Δx^i, Δy, Δz. Assuming that

$$E \Delta t^2 = O(r^4), \quad E \Delta t \Delta x^i = O(r^4), \quad E \Delta x^i \Delta x^j \Delta x^k = O(r^4), \qquad (7.1.16)$$
$$E \Delta t \Delta y = O(r^4), \quad E \Delta x^i \Delta y = O(r^4), \quad i, j, k = 1, \dots, d,$$

we get

$$d_1 = u(t_0, x_0) + \frac{\partial u}{\partial t}(t_0, x_0) \, y_0 \Delta t + \sum_{i=1}^{d} \frac{\partial u}{\partial x^i}(t_0, x_0) \, y_0 \Delta x^i$$

$$+ \frac{1}{2} \sum_{i,j=1}^{d} \frac{\partial^2 u}{\partial x^i \partial x^j}(t_0, x_0) \, y_0 \Delta x^i \Delta x^j + \Delta z + \rho,$$

where $E\rho = O(r^4)$.

The relation $E d_1 = O(r^4)$ holds provided that in addition to (7.1.16) the random increments Δt, Δx^i, Δy, Δz satisfy the conditions:

$$E \Delta t = r^2 + O(r^4), \quad E \Delta x^i = b^i(t_0, x_0) \, r^2 + O(r^4), \qquad (7.1.17)$$
$$E \Delta x^i \Delta x^j = a^{ij}(t_0, x_0) \, r^2 + O(r^4), \quad E \Delta x^i \Delta y = c(t_0, x_0) \, y_0 r^2 + O(r^4),$$
$$E \Delta z = g(t_0, x_0) \, y_0 r^2 + O(r^4), \quad i, j = 1, \dots, d.$$

Thus, we have proved the following theorem.

Theorem 7.1.5 *Assume that the solution of (7.1.1)–(7.1.2) has continuous derivatives $D_x^m D_t^k u$, $0 \leq m + 2k \leq 4$, $k = 0, 1$, in $\bar{Q}$. Let the approximation (7.1.15) give a point (t_1, X_1) such that for all sufficiently small $r > 0$ the line segment connecting the points (t_0, x_0) and (t_1, X_1) entirely belongs to $\bar{Q}$. Then, under the conditions (7.1.16)–(7.1.17) the approximation (7.1.15) has the one-step order of accuracy $O(r^4)$, i.e.,*

$$E [u(t_1, X_1) Y_1 + Z_1 - u(t_0, x_0) \, y_0 - z_0] = O(r^4). \qquad (7.1.18)$$

It is obvious that the conditions (7.1.16)–(7.1.17) are met by both Theorem 7.1.1 (with r^2 replaced by r^2/d) and Theorem 7.1.2.

Example 7.1.6 Let Δt be a random variable taking two values 0 and $2r^2$ with probability $1/2$. Consider the approximation

$$t_1 = t_0 + \Delta t, \qquad (7.1.19)$$

$$X_1 = x_0 + [b(t_0, x_0) \, r^2 + \frac{r}{\sqrt{2}} \sigma(t_0, x_0) \, \nu] \frac{2r^2 - \Delta t}{r^2},$$

$$Y_1 = y_0 + c(t_0, x_0) \, y_0 r^2, \quad Z_1 = z_0 + g(t_0, x_0) \, y_0 r^2,$$

where $\nu = (\nu^1, \ldots, \nu^d)^\mathsf{T}$ is a random vector with mutually independent components taking the values ± 1 with probability $1/2$. In addition, Δt and ν are also independent. The relations (7.1.16)–(7.1.17) are easily verified. Thus, the approximation (7.1.19) has the one-step order of accuracy $O(r^4)$. Note that in the case of the classical heat equation $\partial u/\partial t + (1/2)\partial^2 u/\partial x^2 = 0$ (considered with the reverse time, of course) the configuration of knots in the scheme (7.1.19) is the same as in the simplest implicit finite-difference scheme.

Example 7.1.7 Let Δt be a random variable taking two values 0 and r with probabilities $1 - r$ and r, respectively. Consider the approximation

$$t_1 = t_0 + \Delta t, \quad X_1 = x_0 + b(t_0, x_0) r^2 + r\sigma(t_0, x_0) \nu, \qquad (7.1.20)$$
$$Y_1 = y_0 + c(t_0, x_0) y_0 r^2, \quad Z_1 = z_0 + g(t_0, x_0) y_0 r^2.$$

Here the relations (7.1.17) are fulfilled, but we have $E\Delta t^2 = O(r^3)$ only instead of $E\Delta t^2 = O(r^4)$ in (7.1.16). Therefore, the one-step order of accuracy of (7.1.20) is equal to $O(r^3)$. In this example the point (t, X) jumps from one time layer to another with a larger step (r instead of r^2) but such jumps occur comparatively rarely.

Now we construct methods of a higher order. Here we restrict ourselves to the known weak approximations of solutions of SDEs (see Chap. 2).

Suppose the coefficients of the system (7.1.4)–(7.1.6) are defined in the strip $[T_0, T_1] \times \mathbf{R}^d$. Consider a weak approximation $\bar{X}_{t_0,x_0}(t_0 + h)$, $\bar{Y}_{t_0,x_0,y_0}(t_0 + h)$, $\bar{Z}_{t_0,x_0,y_0,z_0}(t_0 + h)$ of the solution of (7.1.4)–(7.1.6) with one-step error $O(h^{p+1})$. We recall (see Chap. 2) that this means the fulfillment of the relation

$$Ef(\bar{X}_{t_0,x_0}(t_0 + h), \bar{Y}_{t_0,x_0,y_0}(t_0 + h), \bar{Z}_{t_0,x_0,y_0}(t_0 + h)) \qquad (7.1.21)$$
$$-Ef(X_{t_0,x_0}(t_0 + h), Y_{t_0,x_0,y_0}(t_0 + h), Z_{t_0,x_0,y_0}(t_0 + h)) = O(h^{p+1})$$

for a sufficiently wide class of functions $f(x, y, z)$. In particular, (7.1.21) holds for a function of the form $f(x, y, z) = f(x)y + z$.

Theorem 7.1.8 *Assume that the coefficients and the solution of the problem (7.1.1)–(7.1.2) are sufficiently smooth functions in $\bar{Q}$. Let an approximation $\bar{X}_{t_0,x_0}(t_0 + h)$, $\bar{Y}_{t_0,x_0,y_0}(t_0 + h)$, $\bar{Z}_{t_0,x_0,y_0,z_0}(t_0 + h)$ have one-step weak order of accuracy $O(h^{p+1})$. Let for all sufficiently small $h > 0$ the point $(t_0 + h, \bar{X}_{t_0,x_0}(t_0 + h)) \in \bar{Q}$ provided $(t_0, x_0) \in Q$. Then*

$$E[u(t_0 + h, \bar{X}_{t_0,x_0}(t_0 + h))\bar{Y}_{t_0,x_0,y_0}(t_0 + h) + \bar{Z}_{t_0,x_0,y_0,z_0}(t_0 + h)$$
$$- u(t_0, x_0) y_0 - z_0] = O(h^{p+1}), \qquad (7.1.22)$$

i.e., the input approximation has the same order of accuracy in the bounded domain.

Proof We extend the coefficients $a^{ij}(t, x)$, $b^i(t, x)$, $c(t, x)$ and the solution $u(t, x)$ of the problem (7.1.1)–(7.1.2) onto the strip $[T_0, T_1] \times \mathbf{R}^d$ preserving their properties of smoothness and boundedness [267]. This extension can be done in such a way that the property of strict ellipticity is also preserved (with a new constant). As an extension of $g(t, x)$, we take the evaluated value:

$$g(t, x) = -\frac{\partial u}{\partial t}(t, x) - \frac{1}{2} \sum_{i,j=1}^{d} a^{ij}(t, x) \frac{\partial^2 u}{\partial x^i \partial x^j}(t, x) - \sum_{i=1}^{d} b^i(t, x) \frac{\partial u}{\partial x^i}(t, x)$$

$$-c(t, x)u(t, x), \quad (t, x) \in [T_0, T_1] \times \mathbf{R}^d.$$

As a result, we obtain the extension of the boundary value problem (7.1.1)–(7.1.2) to the Cauchy problem, the characteristic system of which is the system (7.1.4)–(7.1.6) with the extended coefficients. Then

$$u(t_0, x_0) y_0 + z_0 = E[u(t_0 + h, X_{t_0,x_0}(t_0 + h))Y_{t_0,x_0,y_0}(t_0 + h)$$
$$+Z_{t_0,x_0,y_0,z_0}(t_0 + h)].$$

Substituting this expression in (7.1.22), we obtain an expression of the form (7.1.21), in which the function $f(x, y, z)$ is equal to $u(t_0 + h, x)y + z$ (here $t_0 + h$ plays the role of a parameter). Theorem 7.1.8 is proved.   □

In Chap. 2 there are weak methods with one-step error $O(h^3)$ for general systems of SDEs, and in Chap. 3 there are weak methods with one-step error $O(h^4)$ for systems with additive noise. Since we can choose bounded random variables in these methods, we can use them for solving the boundary value problem (7.1.1)–(7.1.2).

Example 7.1.9 Let in (7.1.1):

$$\frac{1}{2} \sum_{i,j=1}^{d} a^{ij}(t, x) \frac{\partial^2}{\partial x^i \partial x^j} = \frac{1}{2} \Delta,$$

where Δ is the Laplace operator. Then, the system (7.1.4)–(7.1.6) takes the form (recall that in this section we restrict ourselves to the case $\mu = 0$, $F = 0$):

$$dX = b(s, X)ds + dw(s)$$
$$dY = c(s, X)Y\, ds$$
$$dZ = g(s, X)Y\, ds.$$

As an example, we write the method with one-step error $O(r^6)$ (cf. (2.1.31)):

$$t_1 = t_0 + r^2, \tag{7.1.23}$$

$$X_1 = x_0 + r\xi + r^2 b(t_0, x_0) + \frac{1}{2} r^3 \sum_{i=1}^{d} \frac{\partial b}{\partial x^i}(t_0, x_0)\, \xi^i$$

$$+ \frac{1}{2} r^4 \left(\frac{\partial b}{\partial t} + \sum_{i=1}^{d} b^i \frac{\partial b}{\partial x^i} + \frac{1}{2} \sum_{i=1}^{d} \frac{\partial^2 b}{(\partial x^i)^2} \right) (t_0, x_0),$$

$$Y_1 = y_0 + r^2 c(t_0, x_0)\, y_0 + \frac{1}{2} r^3 y_0 \sum_{i=1}^{d} \frac{\partial c}{\partial x^i}(t_0, x_0)\, \xi^i$$

$$+ \frac{1}{2} r^4 y_0 \left(\frac{\partial c}{\partial t} + \sum_{i=1}^{d} b^i \frac{\partial c}{\partial x^i} + c^2 + \frac{1}{2} \sum_{i=1}^{d} \frac{\partial^2 c}{(\partial x^i)^2} \right) (t_0, x_0),$$

$$Z_1 = z_0 + r^2 g(t_0, x_0)\, y_0 + \frac{1}{2} r^3 y_0 \sum_{i=1}^{d} \frac{\partial g}{\partial x^i}(t_0, x_0)\, \xi^i$$

$$+ \frac{1}{2} y_0 r^4 \left(\frac{\partial g}{\partial t} + \sum_{i=1}^{d} b^i \frac{\partial g}{\partial x^i} + cg + \frac{1}{2} \sum_{i=1}^{d} \frac{\partial^2 g}{(\partial x^i)^2} \right) (t_0, x_0),$$

where $\xi = (\xi^1, \ldots, \xi^d)^{\mathsf{T}}$ is a random vector with mutually independent components distributed as $P(\xi^i = 0) = 2/3$, $P(\xi^i = \pm\sqrt{3}) = 1/6$. Note that X_1 takes a finite number of values.

7.1.2 Numerical Algorithms and Convergence Theorems

Let $r > 0$ be sufficiently small. Denote by Γ_{r^2} the intersection of the r^2-neighborhood of the boundary Γ with the domain Q. We construct the algorithm based on the one-step approximation (7.1.7)–(7.1.8).

Algorithm 7.1.10 *Let*

$$t_{k+1} = t_k + \frac{r_{k+1}^2}{d}, \tag{7.1.24}$$

$$X_{k+1} = X_k + b(t_k, X_k)\frac{r_{k+1}^2}{d} + r_{k+1}\sigma(t_k, X_k)\xi_{k+1}, \quad X_0 = x_0,$$

$$Y_{k+1} = Y_k + c(t_k, X_k)Y_k\frac{r_{k+1}^2}{d}, \quad Y_0 = 1,$$

$$Z_{k+1} = Z_k + g(t_k, X_k)Y_k\frac{r_{k+1}^2}{d}, \quad Z_0 = 0, \quad k = 0, 1, \ldots,$$

where $\xi_1, \ldots, \xi_k, \ldots$ *are mutually independent random variables each of which has the same distribution as* ξ *in (7.1.7)*

The sequence r_k determining the algorithm is constructed as follows. Define the function $\rho_r(t, x)$, $(t, x) \in Q \backslash \Gamma_{r^2}$: if $\Pi_r^{\sigma(t,x)}(t, x) \in Q$, we set $\rho_r(t, x) = r$; otherwise we find $\rho_r(t, x) < r$ such that $\bar{\Pi}_{\rho_r(t,x)}^{\sigma(t,x)}(t, x)$ touches the boundary Γ. In this way the function $\rho_r(t, x)$ is defined everywhere in $Q \backslash \Gamma_{r^2}$.

Let $(t_k, X_k) \in Q \backslash \Gamma_{r^2}$. Set $r_{k+1} = \rho_r(t_k, X_k)$ and then obtain the next point $(t_{k+1}, X_{k+1}, Y_{k+1}, Z_{k+1})$ according to (7.1.24). This walk terminates at a random step $\varkappa$ as soon as $(t_\varkappa, X_\varkappa) \in \Gamma_{r^2}$. Then we find the point $(\bar{t}_\varkappa, \bar{X}_\varkappa)$ on Γ which is the nearest to $(t_\varkappa, X_\varkappa)$.

It is proved (see Theorem 7.1.12 below) that $(\bar{t}_\varkappa, \bar{X}_\varkappa)$ weakly approximates $(\tau, X_{t_0, x_0}(\tau))$ with the accuracy $O(r^2)$, i.e.,

$$E[\varphi(\bar{t}_\varkappa, \bar{X}_\varkappa)Y_\varkappa + Z_\varkappa - u(t_0, x_0)] \tag{7.1.25}$$
$$= E[\varphi(\bar{t}_\varkappa, \bar{X}_\varkappa)Y_\varkappa + Z_\varkappa - \varphi(\tau, X_{t_0, x_0}(\tau))Y_{t_0, x_0, 1}(\tau) - Z_{t_0, x_0, 1, 0}(\tau)] = O(r^2).$$

On the basis of the one-step approximation (7.1.12)–(7.1.13), we construct the following algorithm.

Algorithm 7.1.11 *Let*

$$t_{k+1} = t_k + r_{k+1}^2, \tag{7.1.26}$$
$$X_{k+1} = X_k + b(t_k, X_k)\, r_{k+1}^2 + \sigma(t_k, X_k)\, r_{k+1} \nu_{k+1}, \quad X_0 = x_0,$$
$$Y_{k+1} = Y_k + c(t_k, X_k)\, Y_k r_{k+1}^2, \quad Y_0 = 1,$$
$$Z_{k+1} = Z_k + g(t_k, X_k)\, Y_k r_{k+1}^2, \quad Z_0 = 0, \quad k = 0, 1, \dots ,$$

where $\nu_1, \dots, \nu_k, \dots$ are mutually independent random variables distributed as ν in (7.1.12). Define the function $\rho_r(t, x)$, $(t, x) \in Q \backslash \Gamma_{r^2}$: if $\Pi_r^{\sigma(t,x)}(t, x) \in Q$, then $\rho_r(t, x) = r$, otherwise we find $\rho_r(t, x) < r$ such that $\bar{\Pi}_{\rho_r(t,x)}^{\sigma(t,x)}(t, x)$ touches the boundary Γ. Then we recursively define the sequence $r_{k+1} = \rho_r(t_k, X_k)$ and the random walk in accordance with (7.1.26) up to the moment $\varkappa$ when the point $(t_\varkappa, X_\varkappa)$ reaches Γ_{r^2}.

Note that in the case of convex G the point $X_\varkappa \in \partial G$ (of course, if $t_\varkappa < \tau$), and hence $(t_\varkappa, X_\varkappa) = (\bar{t}_\varkappa, \bar{X}_\varkappa) \in \Gamma$. The relation (7.1.25) also holds here.

The sequences (t_k, X_k, Y_k, Z_k) defined in (7.1.24) and (7.1.26) are Markov chains. The pair (t_k, X_k), which is of primary interest in what follows, is also a Markov chain.

The more general one-step approximation (7.1.15) determines a Markov chain if we take r as a control parameter with a corresponding synthesizing function $r = \rho_r(t, x)$. Let an approximation of the form (7.1.15) have one-step order of accuracy $O(r^{2s+2})$, $s \geq 1$. Then it is natural to introduce a narrower neighborhood $\Gamma_{r^{2s}}$ of the boundary Γ instead of Γ_{r^2}. According to (7.1.15), we construct the Markov chain:

$$t_{k+1} = t_k + \Delta t(t_k, X_k, \xi_{k+1}; \rho_r(t_k, X_k)), \tag{7.1.27}$$
$$X_{k+1} = X_k + \Delta x(t_k, X_k, \xi_{k+1}; \rho_r(t_k, X_k)), \quad X_0 = x_0,$$
$$Y_{k+1} = Y_k + \Delta y(t_k, X_k, \xi_{k+1}; \rho_r(t_k, X_k)) Y_k, \quad Y_0 = 1,$$
$$Z_{k+1} = Z_k + \Delta z(t_k, X_k, \xi_{k+1}; \rho_r(t_k, X_k)) Y_k, \quad Z_0 = 0, \quad k = 0, 1, \dots.$$

Assume the function $\rho_r(t, x)$ to be such that for $(t_k, X_k) \in Q \backslash \Gamma_{r^{2s}}$ the point $(t_{k+1}, X_{k+1}) \in \bar{Q}$. The random walk stops at the moment $\varkappa$ when $(t_\varkappa, X_\varkappa)$ reaches $\Gamma_{r^{2s}}$. As before, $(\bar{t}_\varkappa, \bar{X}_\varkappa)$ is the nearest to $(t_\varkappa, X_\varkappa)$ point on the boundary Γ.

Theorem 7.1.12 *Suppose $E\varkappa = O(1/r^2)$ and the components Y_k, Z_k of the chain (7.1.27) are uniformly bounded. Let the approximation (7.1.15) have the one-step order of accuracy $O(r^{2s+2})$, $s \geq 1$, uniformly in $(t_0, x_0) \in Q$ for uniformly bounded y_0, z_0. Then $(\bar{t}_\varkappa, \bar{X}_\varkappa)$ approximates $(\tau, X_{t_0,x_0}(\tau))$ with the weak order of accuracy $O(r^{2s})$, i.e.,*

$$|E[\varphi(\bar{t}_\varkappa, \bar{X}_\varkappa)Y_\varkappa + Z_\varkappa - \varphi(\tau, X_{t_0,x_0}(\tau))Y_{t_0,x_0,1}(\tau) - Z_{t_0,x_0,1,0}(\tau)]|$$
$$= |E[\varphi(\bar{t}_\varkappa, \bar{X}_\varkappa)Y_\varkappa + Z_\varkappa - u(t_0, x_0)]| \leq Kr^{2s}. \tag{7.1.28}$$

Proof Since $\varkappa$ is finite, we have

$$u(\bar{t}_\varkappa, \bar{X}_\varkappa)Y_\varkappa + Z_\varkappa - u(t_0, x_0) \tag{7.1.29}$$
$$= \big(u(\bar{t}_\varkappa, \bar{X}_\varkappa)Y_\varkappa + Z_\varkappa\big) - (u(t_\varkappa, X_\varkappa)Y_\varkappa + Z_\varkappa)$$
$$+ \sum_{k=0}^{\varkappa-1} \big[u(t_{k+1}, X_{k+1})Y_{k+1} + Z_{k+1} - (u(t_k, X_k)Y_k + Z_k)\big]$$
$$= u(\bar{t}_\varkappa, \bar{X}_\varkappa)Y_\varkappa - u(t_\varkappa, X_\varkappa)Y_\varkappa$$
$$+ \sum_{k=0}^{\infty} \big[u(t_{k+1}, X_{k+1})Y_{k+1} + Z_{k+1} - (u(t_k, X_k)Y_k + Z_k)\big],$$

where $(X_k, Y_k, Z_k) = (X_\varkappa, Y_\varkappa, Z_\varkappa)$ for $k \geq \varkappa$. We also put $r_{k+1} = 0$ for $k \geq \varkappa$.

According to the assumption of the theorem on the order of the approximation (7.1.15), we have for $(t_k, X_k) \in Q \backslash \Gamma_{r^{2s}}$:

$$|E\left(u(t_{k+1}, X_{k+1})Y_{k+1} + Z_{k+1} - u(t_k, X_k)Y_k - Z_k | t_k, X_k\right)| \tag{7.1.30}$$
$$\leq C\rho_r^{2s+2}(t_k, X_k) = Cr_{k+1}^{2s+2}.$$

Since the distance between the points $(t_\varkappa, X_\varkappa)$ and $(\bar{t}_\varkappa, \bar{X}_\varkappa)$ is not larger than r^{2s} and $Y_\varkappa$ is bounded, the absolute value of the first term in (7.1.29) is of order $O(r^{2s})$. Further, due to the condition $E\varkappa < \infty$ and the uniform boundedness of the terms of series in (7.1.29), expectation of the series is equal to the series of expectations of the terms. Using (7.1.29) and (7.1.30), we obtain

$$|E[\varphi(\bar{t}_{\varkappa}, \bar{X}_{\varkappa})Y_{\varkappa} + Z_{\varkappa} - u(t_0, x_0)]|$$

$$\leq Cr^{2s} + \sum_{k=0}^{\infty} |EE\,(u(t_{k+1}, X_{k+1})Y_{k+1} + Z_{k+1} - u(t_k, X_k)Y_k - Z_k | t_k, X_k)|$$

$$\leq Cr^{2s} + Cr^{2s} E \sum_{k=0}^{\varkappa-1} r_{k+1}^2 \leq Cr^{2s} \left(1 + r^2 E\varkappa\right) \leq Kr^{2s}.$$

Theorem 7.1.12 is proved. ☐

The conditions of Theorem 7.1.12 are quite natural. Indeed, Y and Z satisfy the system (7.1.4)–(7.1.6) where the coefficients $c(s, X(s))$ and $g(s, X(s))$ are bounded due to the boundedness of $X(s)$ and finiteness of the integration interval. This implies (recall that $\mu = 0$ and $F = 0$) boundedness of $Y(s)$ and $Z(s)$, and the assumption on uniform boundedness of Y and Z becomes justified. Further, the assumption on uniformity in $(t, x) \in Q$ of the accuracy order of the approximation is justified by the corresponding smoothness of solution u of the problem (7.1.1)–(7.1.2) in $\bar{Q}$. The condition $E\varkappa = O(1/r^2)$ is related to the fact that for a lot of methods the discrete process (t_k, X_k) mainly walks with a large step and the step decreases only when the walk is close to the boundary. Near the boundary this walk is similar to a random walk over touching spheres which terminates rapidly. For instance, the mean number of steps of the random walk over touching spheres before reaching an ε-neighborhood of the boundary is estimated as $|\ln \varepsilon|$ (in our case this value is estimated as $|\ln r^{2s}|$ $\sim |\ln r|$). Of course, trajectories of the Markov chains (t_k, X_k) can be complicated, and a point can spend some time near the boundary and then jump inside the domain. Nevertheless, the majority of steps is spent on approaching the boundary and it is estimated as $O(1/r^2)$. A rigorous justification of the relation $E\varkappa = O(1/r^2)$ involves consideration of a boundary value problem for the corresponding Markov chain. Now we present some results for such a problem in connection with the chain (t_k, X_k).

If we take t, x instead of t_0, x_0 in the first two formulas of (7.1.27), then the obtained relations determine the transition of the chain per step. Let $v(t, x)$ be a function defined in $\bar{Q}$. Introduce the one-step transition function:

$$Pv(t, x) = Ev(t_1, X_1)$$

and consider the boundary value problem in $\bar{Q}$:

$$Pv - v = -1, \quad (t, x) \in Q \setminus \Gamma_{r^{2s}}, \tag{7.1.31}$$

$$v(t, x) = 0, \quad (t, x) \in \Gamma_{r^{2s}}. \tag{7.1.32}$$

As is known [450], the solution of this boundary value problem is $v(t, x) = E\varkappa$, where $\varkappa$ corresponds to the chain (t_k, X_k) starting from the point (t, x).

Further, if we find a function $v(t, x) \geq 0$ in $Q \setminus \Gamma_{r^{2s}}$ and $v(t, x) = 0$ in $\Gamma_{r^{2s}}$ such that

$$Pv - v \leq -c < 0, \tag{7.1.33}$$

then

$$E\varkappa \leq \frac{v(t, x)}{c}. \tag{7.1.34}$$

As an example, let us consider the method (7.1.26) in detail. Let $r > 0$ be a sufficiently small number, $\rho_r(t, x)$ be the function defined above. The formulas

$$t_1 = \begin{cases} t + \Delta t(t, x, \xi; \rho_r(t, x)) = t + \rho_r^2, & (t, x) \in Q \setminus \Gamma_{r^2}, \\ t, & (t, x) \in \Gamma_{r^2}, \end{cases} \tag{7.1.35}$$

$$X_1 = \begin{cases} x + \Delta x(t, x, \xi; \rho_r(t, x)) = x + b(t, x) \rho_r^2 + \sigma(t, x) \rho_r \nu, & (t, x) \in Q \setminus \Gamma_{r^2}, \\ x, & (t, x) \in \Gamma_{r^2}, \end{cases}$$

define the transition of the chain per step.

Theorem 7.1.13 *The method (7.1.26) converges with the weak order of accuracy $O(r^2)$. Mean number of steps is equal to $E\varkappa = O(1/r^2)$.*

Proof To prove the theorem, it is enough to obtain an estimate of $E\varkappa$ because the other conditions of Theorem 7.1.12 obviously hold. To estimate $E\varkappa$, we use (7.1.33)–(7.1.34) with v of the form:

$$v(t, x) = \begin{cases} T_1 - t + v_0, & (t, x) \in Q \setminus \Gamma_{r^2}, \\ 0, & (t, x) \in \Gamma_{r^2}, \end{cases} \tag{7.1.36}$$

where $v_0 > 0$ is a constant. If a point (t, x) is such that $\rho_r(t, x) = r$, then the operator P applied to the function $v = T_1 - t$ gives $T_1 - t - r^2$ when $T_1 - t - r^2 \geq r^2$ and 0 when $T_1 - t - r^2 < r^2$. The operator P applied to the constant v_0 gives a value which is not greater than this constant (speaking here about functions, we consider them in $Q \setminus \Gamma_{r^2}$ only; for instance, the constant v_0 is the function of (t, x) which is equal to v_0 for $(t, x) \in Q \setminus \Gamma_{r^2}$ and equal to 0 for $(t, x) \in \Gamma_{r^2}$). As a result, for such points (t, x) we have

$$Pv - v \leq -r^2 .$$

If a point (t, x) is such that $\rho_r(t, x) < r$, then the boundary of the upper base of the corresponding cylinder (right parallelepiped) either entirely belongs to the upper base of the cylinder $\bar{Q}$ (i.e., when $t_1 = T_1$) or touches the lateral surface of the cylinder $\bar{Q}$. In the first case the operator P gives zero for both $T_1 - t$ and v_0, and thus

$$Pv - v \leq -v_0 .$$

In the second case, for a sufficiently small r the point (t_1, X_1) get into Γ_{r^2} with a probability $p \geq p_0$, where p_0 is independent of r. Hence the operator P applied to the constant v_0 is less than v_0 at least by $p_0 v_0$. Thus

$$Pv - v \leq -p_0 v_0 .$$

It follows from the above estimates that we can take r^2 as c in (7.1.33)–(7.1.34) and $v(t, x) \leq T_1 - t + v_0$. Theorem 7.1.13 is proved. $\qquad \square$

Remark 7.1.14 The same function v allows us to prove convergence theorems for the methods from Examples 7.1.6, 7.1.7, and 7.1.9. We emphasize that the order of accuracy of the method from Example 7.1.9 is $O(r^4)$, but the mean number of steps is still $O(1/r^2)$. The common property of all these methods, due to which the function satisfying the conditions (7.1.33)–(7.1.34) is so easily constructed, is as follows: for random walks over touching cylinders the probability that the point gets into r^{2s}-neighborhood of the boundary is bounded from below by a certain number p_0 which is independent of r. In fact, it would be enough if this probability were bounded from below by Kr^2, $K > 0$. Then, we have the inequality $Pv - v \leq -Kr^2 v_0$ instead of $Pv - v \leq -p_0 v_0$, and, due to (7.1.33)–(7.1.34), the relation $E\varkappa = O(1/r^2)$ remains valid. We note in connection with the method (7.1.24) that if the radius of the d-dimensional sphere is not larger that r, then the area of its segment of the height r^2 is larger than $Kr^{(d-1)/2}$, where $K > 0$ is a constant. Taking this into account, it is possible to prove a convergence theorem like Theorem 7.1.13 for the method (7.1.24) in the case of $d = 1, \ldots, 5$ using the function (7.1.36). If $d > 5$ then the function of a more complicated form is required to prove such a theorem.

Remark 7.1.15 The majority of the results of this section can be carried over to elliptic equations (see also Sect. 7.3).

7.2 The Simplest Random Walk for the Dirichlet Problem for Parabolic Equations

In Sect. 7.1, approximations are based on controlling a time step of numerical integration of the system (7.1.4). In this section the step of numerical integration of the system (7.1.4) is constant for points belonging to a certain time layer $t = t_k$. But when a point is close to the boundary, we make an intermediate (auxiliary) step of the random walk, which preserves the point in the time layer $t = t_k$. The result of this auxiliary step is such that the point, which is close to the boundary, is replaced by two points with some probabilities using an interpolation. One of these new points belongs to the boundary and if it is realized, the walk terminates. The other point is inside the domain so that, starting from it, we can make a new step of numerical integration without leaving the domain $\bar{Q}$. The approach of Sect. 7.1 is probably more universal. However, methods of Sect. 7.2 are of independent interest from both theoretical and applied points of view due to their simplicity. The methods of Sect. 7.2

are the simplest among methods of order $O(h)$ because inside the domain we use the Euler weak approximations and near the boundary we exploit linear interpolation. In addition, these methods have a layer structure and, in particular, a constant-size step of numerical integration can be used. Due to this fact, it becomes possible to apply extrapolation methods.

The main algorithm of random walk is proposed in Sect. 7.2.1. In Sect. 7.2.2, its convergence with weak order of accuracy $O(h)$ is proved. Some similar algorithms of random walks are given in Sect. 7.2.3. One of these algorithms is distinguished by being the most simple but its order of convergence is $O(\sqrt{h})$. Finally, some numerical tests of the proposed methods are presented in Sect. 7.2.4. We note that the simplest random walk was used e.g. for evaluating the price and hedging strategy of American options in [288], for barrier options in [210], and for computing characteristics of a fluid-lubricated bearing with external random forcing in [15]. Numerical tests are presented there as well.

7.2.1   The Algorithm of the Simplest Random Walk

We apply the weak explicit Euler approximation with the simplest simulation of noise to the system (7.1.4)–(7.1.6) (cf. (2.0.21)):

$$X_{t,x}(t+h) \approx X = x + h\,(b(t,x) - \sigma(t,x)\,\mu(t,x)) + h^{1/2}\sigma(t,x)\,\xi, \quad (7.2.1)$$

$$Y_{t,x,y}(t+h) \approx Y = y + hc(t,x)\,y + h^{1/2}\mu^\mathsf{T}(t,x)\,y\,\xi, \quad (7.2.2)$$

$$Z_{t,x,y,z}(t+h) \approx Z = z + hg(t,x)\,y + h^{1/2}F^\mathsf{T}(t,x)\,y\,\xi, \quad (7.2.3)$$

where $h > 0$ is a step of integration (a sufficiently small number), $\xi = (\xi^1, \ldots, \xi^d)^\mathsf{T}$, ξ^i, $i = 1, \ldots, d$, are mutually independent random variables taking the values ± 1 with probability $1/2$. Clearly, the random vector X takes 2^d different values.

Introduce the set of points close to the boundary (a boundary zone) $S_{t,h} \subset \bar{G}$ on the layer t: we say that $x \in S_{t,h}$ if at least one of the 2^d values of the vector X is outside $\bar{G}$. It is not difficult to see that due to compactness of $\bar{Q}$ there is a constant $\lambda > 0$ such that if the distance from $x \in G$ to the boundary ∂G is equal to or greater than $\lambda\sqrt{h}$ then x is outside the boundary zone and, therefore, for such x all the realizations of the random variable X belong to $\bar{G}$.

Since restrictions connected with nonexit from the domain $\bar{G}$ should be imposed on an approximation of the system (7.1.4), the formulas (7.2.1)–(7.2.3) can be used only for the points $x \in \bar{G}\setminus S_{t,h}$ on the layer t, and a special construction is required for points from the boundary zone.

Let $x \in S_{t,h}$. Denote by $x^\pi \in \partial G$ the projection of the point x on the boundary of the domain G (the projection is unique because h is sufficiently small and ∂G is smooth) and by $n(x^\pi)$ the unit vector of internal normal to ∂G at x^π. Introduce the random vector $X^\pi_{x,h}$ taking two values x^π and $x + h^{1/2}\lambda n(x^\pi)$ with probabilities $p = p_{x,h}$ and $q = q_{x,h} = 1 - p_{x,h}$, respectively, where

$$p_{x,h} = \frac{h^{1/2}\lambda}{|x + h^{1/2}\lambda n(x^\pi) - x^\pi|}.$$

If $v(x)$ is a twice continuously differentiable function with the domain of definition $\bar{G}$, then an approximation of $v(x)$ by the expectation $Ev(X_{x,h}^\pi)$ corresponds to linear interpolation and

$$v(x) = Ev(X_{x,h}^\pi) + O(h) = pv(x^\pi) + qv(x + h^{1/2}\lambda n(x^\pi)) + O(h). \quad (7.2.4)$$

We emphasize that the second value $x + h^{1/2}\lambda n(x^\pi)$ does not belong to the boundary zone. We also note that p is always greater than $1/2$ (since the distance from x to ∂G is less than $h^{1/2}\lambda$) and that if $x \in \partial G$ then $p = 1$ (since in this case $x^\pi = x$).

Let a point $(t_0, x_0) \in Q$. We would like to find the value $u(t_0, x_0)$. Introduce a discretization of the interval $[t_0, T]$, for definiteness the equidistant one:

$$t_0 < t_1 < \cdots < t_N = T, \quad h := (T - t_0)/N.$$

To approximate the solution of the system (7.1.4), we construct a Markov chain (t_k, X_k) which stops when it reaches the boundary Γ at a random step $\varkappa$. The number k takes nonnegative integer values not greater than N.

We set $X_0' = x_0$. If $X_0' \notin S_{t_0,h}$ then we take $X_0 = X_0'$. If $X_0' \in S_{t_0,h}$ then the random variable X_0 takes two values: either $X_0'^\pi \in \partial G$ with probability $p_{X_0',h}$ or $X_0' + h^{1/2}\lambda n(X_0'^\pi) \notin S_{t_0,h}$ with probability $q_{X_0',h}$. If $X_0 = X_0'^\pi$ (i.e., $(t_0, X_0) \in \Gamma$), then we put $\varkappa = 0$, $X_\varkappa = X_0'^\pi$, and the random walk is finished. Let X_k, $k < N$, be constructed and either $X_k \in \partial G$ (i.e., the chain is stopped at one of the instances $t_0, \ldots, t_k$ and, consequently, $\varkappa \le k$) or $X_k \notin S_{t_k,h}$. Now our aim is to construct X_{k+1} with the same properties. Suppose that the chain does not stop until t_k inclusively, i.e., $\varkappa > k$. Introduce X_{k+1}' due to (7.2.1) with $t = t_k$, $x = X_k$, $\xi = \xi_k$ (we assume that all ξ_k are mutually independent and distributed as ξ):

$$X_{k+1}' = X_k + h\left(b(t_k, X_k) - \sigma(t_k, X_k)\,\mu(t_k, X_k)\right) + h^{1/2}\sigma(t_k, X_k)\,\xi_k. \quad (7.2.5)$$

If $k + 1 = N$ then we set $X_{k+1} = X_{k+1}'$. If $k + 1 < N$, we obtain X_{k+1} using X_{k+1}' as we got X_0 using X_0'. More precisely, we use the following rule. If $X_{k+1}' \notin S_{t_{k+1},h}$ then we take $X_{k+1} = X_{k+1}'$. If $X_{k+1}' \in S_{t_{k+1},h}$ then the random variable X_{k+1} takes two values: either $X_{k+1}'^\pi \in \partial G$ with probability $p_{X_{k+1}',h}$ or $X_{k+1}' + h^{1/2}\lambda n(X_{k+1}'^\pi) \notin S_{t_{k+1},h}$ with probability $q_{X_{k+1}',h}$. If $X_{k+1} = X_{k+1}'^\pi$ (i.e., $(t_{k+1}, X_{k+1}) \in \Gamma$), then we put $\varkappa = k + 1$, $X_\varkappa = X_{k+1}'^\pi$, and the random walk is finished.

So, the main steps in constructing the chain (t_k, X_k) are as follows. The state X_0 is formed by the rule: if $X_0' := x_0$ does not belong to the boundary zone $S_{t_0,h}$ ($X_0' \notin S_{t_0,h}$), we put $X_0 = X_0'$; if $X_0' \in S_{t_0,h}$, the state X_0 is defined as the random variable which takes two values: either $X_0'^\pi$ on the boundary ∂G or the other value in the internal part of the domain G. It can be said that the first value accounts for a "stopping" behavior of the original process X while the second value corresponds to a "walking" behavior. Further, if the stopping moment $\varkappa$ is larger than zero, i.e., if the realized value of X_0 does not belong to the boundary, X_1' is obtained by the step according to the usual formula. Then we obtain X_1 using X_1' just as X_0 has been got by X_0', and so on.

We also introduce an extended chain (t_k, X_k, Y_k, Z_k). We put $Y_0 = 1$ and $Z_0 = 0$. Let $\varkappa > k$ and X_k, Y_k, Z_k be known. Then Y_{k+1} and Z_{k+1} are evaluated in accordance with the system (7.2.2)–(7.2.3) for $t = t_k$, $x = X_k$, $y = Y_k$, $z = Z_k$, $\xi = \xi_k$. The extended chain stops at the same step $\varkappa$.

Below we write the constructed algorithm formally.

Algorithm 7.2.1 *STEP 0.*　　　$X_0' = x_0$, $Y_0 = 1$, $Z_0 = 0$, $k = 0$.

STEP 1.　If $X_k' \notin S_{t_k,h}$ then $X_k = X_k'$ and go to STEP 3.
　　　　　If $X_k' \in S_{t_k,h}$ then either $X_k = X_k'^{\pi}$ with probability
　　　　　$p_{X_k',h}$ or $X_k = X_k' + h^{1/2}\lambda text with probability $q_{X_k',h}$.
STEP 2.　If $X_k = X_k'^{\pi}$ then STOP and $\varkappa = k$,
　　　　　$X_\varkappa = X_k'^{\pi}$, $Y_\varkappa = Y_k$, $Z_\varkappa = Z_k$.
STEP 3.　Simulate ξ_k and find X_{k+1}', Y_{k+1}, Z_{k+1} according to (7.2.5),
　　　　　(7.2.2)–(7.2.3) for $t = t_k$, $x = X_k$, $y = Y_k$, $z = Z_k$,
　　　　　$\xi = \xi_k$.
STEP 4.　If $k + 1 = N$, STOP and $\varkappa = N$, $X_\varkappa = X_N'$, $Y_\varkappa = Y_N$,
　　　　　$Z_\varkappa = Z_N$, otherwise $k := k + 1$ and return to STEP 1.

It may happen so that it is more rational to choose both h and λ depending on the chain's state: h_k and λ_k. Then, in Theorem 7.2.2 (see below) one should put $h = \max_{0 \le k < N} h_k$. In practice, one can take $\lambda_k = |\sigma(t_k, X_k)|$, possibly with small corrections.

7.2.2 Convergence Theorem

Denote by ν_{t_0,x_0} the number of those t_k at which X_k' gets into the set $S_{t_k,h}$. The event $\{\nu_{t_0,x_0} > n\}$ implies an event such that first n trials in a certain trial scheme are unsuccessful (Γ is not attained by (t_k, X_k)). Moreover, the probability of each unsuccess is less than $1/2$. Therefore, the following estimate takes place:

$$P\{\nu_{t_0,x_0} > n\} \le \frac{1}{2^n}. \tag{7.2.6}$$

But these arguments are not completely rigorous since the probability of attaining the boundary by X_k depends on X_k'. Along with the chain X_k, let us consider a chain $\tilde{X}_k$ which differs from the original one in the following way only: when X_k' gets into the boundary zone $S_{t_k,h}$, each time a coin is tossed so that the new chain hits the boundary at $X_k'^{\pi}$ with probability $1/2$ and it hits the same state as the original chain with probability $1/2$, i.e., the state $X_k' + h^{1/2}\lambda n(X_k'^{\pi}) \notin S_{t_k,h}$. Since the new chain terminates with a smaller probability than the original chain, we have $\nu_{t_0,x_0} \leq \tilde{\nu}_{t_0,x_0}$. Therefore, $P\{\nu_{t_0,x_0} > n\} \leq P\{\tilde{\nu}_{t_0,x_0} > n\}$. But the arguments before (7.2.6) are rigorous for $\tilde{\nu}_{t_0,x_0}$, i.e., $P\{\tilde{\nu}_{t_0,x_0} > n\} \leq 1/2^n$. The last two inequalities imply (7.2.6). As a conclusion, we state the following lemma.

Lemma 7.2.2 *The inequality (7.2.6) holds together with the inequalities*

$$P\{\nu_{t_0,x_0} = n\} \leq \frac{1}{2^{n-1}}, \tag{7.2.7}$$

$$E\nu_{t_0,x_0} \leq C, \quad E\nu^2_{t_0,x_0} \leq C, \tag{7.2.8}$$

where C does not depend on t_0, x_0, h.

We extend the definition of the constructed chain for all k by the rule: if $k > \varkappa$, then $(t_k, X_k, Y_k, Z_k) = (t_\varkappa, X_\varkappa, Y_\varkappa, Z_\varkappa)$.

We assume that the functions $a^{ij}(t,x)$, $b^i(t,x)$, $c(t,x)$, $g(t,x)$ together with their first partial derivatives in t, x^m and second partial derivatives in x^m are continuous in $\bar{Q}$, the domain G has twice continuously differentiable boundary ∂G, the function $\varphi(t,x)$, $(t,x) \in \Gamma$, is at least of the same smoothness as the solution $u(t,x)$, and, finally, the function $\varphi(t,x)$ satisfies (7.1.1) on the boundary of the upper base of the cylinder $\bar{Q}$. Then, the classical solution $u(t,x)$ of the problem (7.1.1)–(7.1.2) has continuous in $\bar{Q}$ derivatives (see, e.g., [114, 222, 241]):

$$\frac{\partial^2 u}{\partial t^2}, \quad \frac{\partial^3 u}{\partial t \partial x^i \partial x^j}, \quad \frac{\partial^3 u}{\partial x^i \partial x^j \partial x^m}, \quad \frac{\partial^4 u}{\partial x^i \partial x^j \partial x^m \partial x^l}.$$

Now we prove a lemma on one-step error. Introduce

$$d_k = u(t_k, X_k) Y_k + Z_k - u(t_k, X'_k) Y_k - Z_k,$$
$$d'_k = u(t_{k+1}, X'_{k+1}) Y_{k+1} + Z_{k+1} - u(t_k, X_k) Y_k - Z_k,$$

$$k = 0, \ldots, N - 1.$$

Clearly, X'_k belongs to the layer $t = t_k$; the variable d_k can be nonzero in the case of $X'_k \in S_{t_k,h}$ only; if $\varkappa > k$, then $X_k \notin S_{t_k,h}$ and all the 2^d realizations of the random variable X'_{k+1} belong to $\bar{G}$; if $\varkappa \leq k$, then $t_k = t_{k+1} = t_\varkappa$, $X'_{k+1} = X_k = X_\varkappa$, $Y_{k+1} = Y_k = Y_\varkappa$, $Z_{k+1} = Z_k = Z_\varkappa$ and, consequently, $d'_k = 0$. We also note that it is not difficult to show that for sufficiently small h the component Y_k is positive.

Lemma 7.2.3 *The following inequalities hold:*

$$|E(d_k \mid X'_k, Y_k, Z_k)| \leq C h Y_k I_{S_{t_k,h}}(X'_k) \chi_{\varkappa \geq k}, \tag{7.2.9}$$

$$|E(d'_k \mid X_k, Y_k, Z_k)| \leq C h^2 Y_k \chi_{\varkappa > k}. \tag{7.2.10}$$

Proof The inequality (7.2.9) follows from the interpolation relation (7.2.4) and from the reminder before the lemma. The inequality (7.2.10) is a consequence of the fact that the one-step accuracy order of the Euler method (7.2.1)–(7.2.3) in the weak sense is $O(h^2)$.   □

Theorem 7.2.4 *Algorithm 7.2.1 has weak order of accuracy $O(h)$, i.e., the inequality*

$$|E(\varphi(t_{\varkappa}, X_{\varkappa}) Y_{\varkappa} + Z_{\varkappa}) - u(t_0, x_0)| \leq Ch \tag{7.2.11}$$

holds with $C > 0$ independent of t_0, x_0, h.

Proof We have

$$R := E(\varphi(t_{\varkappa}, X_{\varkappa}) Y_{\varkappa} + Z_{\varkappa}) - u(t_0, x_0) = E(u(t_{\varkappa}, X_{\varkappa}) Y_{\varkappa} + Z_{\varkappa}) - u(t_0, x_0)$$

$$= E(u(t_0, X_0) Y_0 + Z_0) - u(t_0, x_0)$$

$$+ E \sum_{k=0}^{\varkappa-1} (u(t_{k+1}, X_{k+1}) Y_{k+1} + Z_{k+1} - u(t_k, X_k) Y_k - Z_k)$$

$$= E \sum_{k=0}^{N-1} (d_k + d_k') = \sum_{k=0}^{N-1} E E(d_k' \mid X_k, Y_k, Z_k) + \sum_{k=0}^{N-1} E E(d_k \mid X_k', Y_k, Z_k).$$

Due to (7.2.10), the absolute value of the first sum is estimated by $Ch^2 \sum_{k=0}^{N-1} EY_k$. It is not difficult to show that $EY_k \leq C$, $k = 0, \dots, N - 1$. Thus, the first sum is $O(h)$ uniformly in t_0, x_0, h. Using (7.2.9) and Lemma 7.2.2, we estimate the second sum:

$$\left| \sum_{k=0}^{N-1} E E(d_k \mid X_k', Y_k, Z_k) \right| \leq Ch \, E \sum_{k=0}^{N-1} Y_k I_{S_{t_k,h}}(X_k') \chi_{\varkappa \geq k} \tag{7.2.12}$$

$$\leq Ch \, E \left(\max_{0 \leq k \leq N-1} Y_k \times \sum_{k=0}^{N-1} I_{S_{t_k,h}}(X_k') \right) = Ch \, E \left(\max_{0 \leq k \leq N-1} Y_k \times \nu_{t_0, x_0} \right)$$

$$\leq Ch \left(E \max_{0 \leq k \leq N-1} Y_k^2 \right)^{1/2} \left(E \nu_{t_0, x_0}^2 \right)^{1/2} \leq Ch \left(E \max_{0 \leq k \leq N-1} Y_k^2 \right)^{1/2}.$$

Let $c(t, x) \leq \bar{c}$, $(t, x) \in \bar{Q}$, where $\bar{c}$ is a constant. Introduce the sequence

$$\tilde{Y}_0 = 1, \quad \tilde{Y}_k = \tilde{Y}_{k-1}(1 + h\bar{c} + h^{1/2}\mu(t_{k-1}, X_{k-1}) \xi_{k-1}), \quad k \leq \varkappa;$$
$$\tilde{Y}_k = \tilde{Y}_{\varkappa}, \quad k > \varkappa.$$

It is evident that $Y_k \leq \tilde{Y}_k$. It is not difficult to see that the sequence $V_k = (1 + h\bar{c})^{-k} \tilde{Y}_k$ is a martingale. Hence,

$$E \max_{0 \leq k \leq N-1} V_k^2 \leq 4E V_{N-1}^2 \leq C$$

(as before, here C is independent of h), and this together with (7.2.12) implies (7.2.11). Theorem 7.2.4 is proved.   □

We note that despite some steps of Algorithm 7.2.1 are very rough (their one-step errors are $O(h)$), this algorithm converges, and, moreover, its global order of convergence is $O(h)$. We have this rate of convergence due to the fact that the number of rough steps (on average) is bounded from above by a constant which does not grow as $h \to 0$.

Remark 7.2.5 If we assume that the global error R of Algorithm 7.2.1 can be expanded in powers of time step h:

$$R = C_0 h + O(h^2), \tag{7.2.13}$$

then it becomes possible to use the extrapolation method and to obtain a method of order two applying two times Algorithm 7.2.1 with different time steps. Namely, let $\bar{u}^{h_1}$ and $\bar{u}^{h_2}$ be approximations of $u(t_0, x_0)$ calculated according to Algorithm 7.2.1. Then (see Sect. 2.2.3)

$$\bar{u}_{imp} = \bar{u}^{h_1} \frac{h_2}{h_2 - h_1} - \bar{u}^{h_2} \frac{h_1}{h_2 - h_1}, \quad u(t_0, x_0) = \bar{u}_{imp} + O(h^2). \tag{7.2.14}$$

We do not give here a proof of (7.2.13) (this is apparently more difficult in comparison with the proof of Theorem 2.2.5) however our experiments confirm the natural rule expressed by formula (7.2.14).

7.2.3 Other Random Walks

The next algorithm is obtained by a simplification of Algorithm 7.2.1. Indeed, as soon as X_k gets into the boundary domain $S_{t_k, h}$, the random walk terminates, i.e., $\varkappa = k$, and $\bar{X}_\varkappa = X_k^\pi$, $Y_\varkappa = Y_k$, $Z_\varkappa = Z_k$ is taken as the final state of the Markov chain. Let us write this algorithm formally.

Algorithm 7.2.6 *STEP 0.* $X_0 = x_0$, $Y_0 = 1$, $Z_0 = 0$, $k = 0$.

STEP 1.   If $X_k \notin S_{t_k, h}$ then go to STEP 2.
            If $X_k \in S_{t_k, h}$ then STOP and $\varkappa = k$, $\bar{X}_\varkappa = X_k^\pi$,
            $Y_\varkappa = Y_k$, $Z_\varkappa = Z_k$.
STEP 2.   Simulate ξ_k and find X_{k+1}, Y_{k+1}, Z_{k+1} according to
            (7.2.1)–(7.2.3) for $t = t_k$, $x = X_k$, $y = Y_k$,
            $z = Z_k$, $\xi = \xi_k$.
STEP 3.   If $k + 1 = N$, STOP and $\varkappa = N$, $\bar{X}_\varkappa = X_N$, $Y_\varkappa = Y_N$,
            $Z_\varkappa = Z_N$, otherwise $k := k + 1$ and return to STEP 1.

In this algorithm one-step errors of all steps except the last one is $O(h^2)$. The error on the last step (i.e., on the step $\varkappa$) is estimated as $O(\sqrt{h})$. It is easy to prove the following theorem (we note in passing that the conditions on smoothness of the solution and parameters mentioned before Lemma 7.2.3 can be weakened for this theorem).

Theorem 7.2.7 *Algorithm 7.2.6 has weak order of accuracy* $O(\sqrt{h})$:

$$|E(\varphi(t_\varkappa, \bar{X}_\varkappa) Y_\varkappa + Z_\varkappa) - u(t_0, x_0)| \le C\sqrt{h},$$

where C is independent of t_0, x_0, h.

A similar random walk was proposed in [64]. In contrast to Algorithm 7.2.6, which terminates when the chain gets into the boundary zone $S_{t_k, h} \subset \bar{G}$, the method from [64] terminates when the chain exits from $\bar{G}$ and the projection of the point, having exited from the domain, on the boundary ∂G is taken as the final state of the chain.

Remark 7.2.8 If we assume that the global error R of Algorithm 7.2.6 can be expanded as

$$R = C_0 h^{1/2} + O(h),$$

then analogously to the rule (7.2.14) we obtain

$$\bar{u}_{imp} = \bar{u}^{h_1} \frac{h_2^{1/2}}{h_2^{1/2} - h_1^{1/2}} - \bar{u}^{h_2} \frac{h_1^{1/2}}{h_2^{1/2} - h_1^{1/2}}, \quad u(t_0, x_0) = \bar{u}_{imp} + O(h). \quad (7.2.15)$$

Now we construct a random walk with a more accurate one-step approximation for $x \in S_{t,h}$ than in Algorithm 7.2.1. Let $x \in S_{t,h}$. We denote by $\xi^{(i)} = (\xi^{(i),1}, \ldots, \xi^{(i),d})^{\mathsf{T}}$, $i = 1, \ldots, 2^d$, the values of the vector ξ from (7.2.1), and we assign indices to these values so that $\xi^{(2^{d-1}+i)} = -\xi^{(i)}$, $i = 1, \ldots, 2^{d-1}$. We denote by $X^{(i)}$ the value of the vector X from (7.2.1) corresponding to $\xi^{(i)}$. At least one of the points $X^{(i)} \notin \bar{G}$ since $x \in S_{t,h}$. We connect the point x with those $X^{(i)}$ which are outside $\bar{G}$ by the curves $\eta^{(i)}(\theta)$:

$$\eta^{(i)}(\theta) = (x + \theta h (b(t, x) - \sigma(t, x) \mu(t, x)) + \sqrt{\theta h} \, \sigma(t, x) \xi^{(i)}), \quad \theta \in [0, 1].$$

There is a value $\theta = \theta^{(i)}$, $0 < \theta^{(i)} < 1$, (it is unique because h is sufficiently small and ∂G is smooth) such that the point $\eta^{(i)} := \eta^{(i)}(\theta^{(i)})$ belongs to the boundary ∂G. We put $\theta^{(i)} = 1$ and $\eta^{(i)} = X^{(i)}$ for those points $X^{(i)}$ which belong to $\bar{G}$.

Introduce the pair:

$$(\vartheta^{(i)}, \eta^{(i)}) := (t + \theta^{(i)} h, \eta^{(i)}). \quad (7.2.16)$$

Let α be the random variable taking values $\{1, \ldots, 2^d\}$ and distributed by the law

$$p_i = P\{\alpha = i\} = \frac{\gamma}{\sqrt{\theta^{(i)}}\left(\sqrt{\theta^{(i)}} + \sqrt{\theta^{(2^{d-1}+i)}}\right)}, \quad i = 1, \ldots, 2^{d-1}, \quad (7.2.17)$$

$$p_i = P\{\alpha = i\} = \frac{\gamma}{\sqrt{\theta^{(i)}}\left(\sqrt{\theta^{(i)}} + \sqrt{\theta^{(i-2^{d-1})}}\right)}, \quad i = 2^{d-1} + 1, \ldots, 2^d,$$

where γ is uniquely found from the condition $\sum_{i=1}^{2^d} P\{\alpha = i\} = 1$.

Now we construct a Markov chain (ϑ_k, X_k) which stops reaching the boundary Γ at a random step $\varkappa$. The index k takes nonnegative integer values not greater than N, and $\vartheta_k = t_k$, at least for $k < \varkappa$. We put $\vartheta_0 = t_0$, $X_0 = x_0$. Let $(\vartheta_k, X_k) \in Q$, i.e., $\varkappa > k$, $k \leq N - 1$, $\vartheta_k = t_k$. Let us define $(\vartheta_{k+1}, X_{k+1})$. If $X_k \notin S_{t_k, h}$, then $\vartheta_{k+1} = t_{k+1}$ and X_{k+1} is found in accordance with (7.2.1) for $t = t_k$, $x = X_k$, $\xi = \xi_k$. If $X_k \in S_{t_k, h}$, then

$$(\vartheta_{k+1}, X_{k+1}) = (\vartheta_k^{(\alpha_k)}, \eta_k^{(\alpha_k)}),$$

where $(\vartheta_k^{(\alpha_k)}, \eta_k^{(\alpha_k)})$ is found as $(\vartheta^{(i)}, \eta^{(i)})$ from (7.2.16) for $t = t_k$, $x = X_k$, $i = \alpha_k$ (both ξ_k and α_k are independent of previous history; α_k is distributed as α; it is also clear that at each step we simulate either ξ_k or α_k). In this case, due to the construction of the pair $(\vartheta_k^{(\alpha_k)}, X_k^{(\alpha_k)})$, the point X_{k+1} either belongs to ∂G with probability $p \geq 1/2^d$ and the random walk terminates (in this case $\varkappa = k + 1$, $(\vartheta_\varkappa, X_\varkappa) = (\vartheta_{k+1}, X_{k+1})$) or $X_{k+1} \notin \partial G$ (this means that the realized α_k is such that $\theta_k^{(\alpha_k)} = 1$ and, therefore, $\vartheta_{k+1} = t_{k+1}$) and for $k + 1 < N$ the random walk should be continued. For $k + 1 = N$ the point $(\vartheta_{k+1}, X_{k+1}) \in \Gamma$ and the random walk terminates with $\varkappa = N$, $(\vartheta_\varkappa, X_\varkappa) = (T_1, X_N)$. Thus, the chain (ϑ_k, X_k) has been constructed.

Note that (ϑ_k, X_k) remains in the domain $\bar{Q}$ with probability 1; $\varkappa$ takes nonnegative integer values not greater than N.

Now we introduce an extended Markov chain $(\vartheta_k, X_k, Y_k, Z_k)$. We put $Y_0 = 1$, $Z_0 = 0$. Let $\varkappa > k$ and Y_k, Z_k be known. If $X_k \notin S_{t_k, h}$, then Y_{k+1}, Z_{k+1} are evaluated in accordance with (7.2.1)–(7.2.3) for $t = t_k$, $x = X_k$, $y = Y_k$, $z = Z_k$, $\xi = \xi_k$. If $X_k \in S_{t_k, h}$, then

$$Y_{k+1} = Y_k + \theta_k^{(\alpha_k)} h c(\vartheta_k, X_k) Y_k + \sqrt{\theta_k^{(\alpha_k)} h} \, \mu^\mathsf{T}(\vartheta_k, X_k) Y_k \, \xi^{(\alpha_k)}, \quad (7.2.18)$$

$$Z_{k+1} = Z_k + \theta_k^{(\alpha_k)} h g(\vartheta_k, X_k) Y_k + \sqrt{\theta_k^{(\alpha_k)} h} \, F^\mathsf{T}(\vartheta_k, X_k) Y_k \, \xi^{(\alpha_k)},$$

where $\theta_k^{(i)}$, $i = 1, \ldots, 2^d$, and α_k are the same as in evaluation of $(\vartheta_{k+1}, X_{k+1})$.

The constructed algorithm formally takes the following form.

Algorithm 7.2.9 *STEP 0.* $X_0 = x_0, \; Y_0 = 1, \; Z_0 = 0, \; k = 0.$

STEP 1.   *If $X_k \notin S_{t_k,h}$ then go to STEP 3.*
*If $X_k \in S_{t_k,h}$ then simulate α_k according to the distribution
(7.2.17), find $(\vartheta_{k+1}, X_{k+1}) = (\vartheta_k^{(\alpha_k)}, \eta_k^{(\alpha_k)})$ according to (7.2.16)
for $i = \alpha_k, \; t = \vartheta_k, \; x = X_k$ and find $Y_{k+1}, \; Z_{k+1}$ according
to (7.2.18).*

STEP 2.   *If $X_{k+1} \in \partial G$ then STOP and put $(\vartheta_{\varkappa}, X_{\varkappa}, Y_{\varkappa}, Z_{\varkappa})$
$= (\vartheta_{k+1}, X_{k+1}, Y_{k+1}, Z_{k+1})$; otherwise go to STEP 4.*

STEP 3.   *Put $\vartheta_{k+1} = t_{k+1}$, simulate ξ_k, and find $X_{k+1}, \; Y_{k+1}, \; Z_{k+1}$
according to (7.2.1)–(7.2.3) for $t = \vartheta_k, \; x = X_k,$
$y = Y_k, \; z = Z_k, \; \xi = \xi_k.$*

STEP 4.   *If $k + 1 = N$, STOP and put $(\vartheta_{\varkappa}, X_{\varkappa}, Y_{\varkappa}, Z_{\varkappa}) = (\vartheta_N, X_N,$
$Y_N, Z_N)$; otherwise put $k := k + 1$ and return to STEP 1.*

We prove the following lemma on one-step error of Algorithm 7.2.9.

Lemma 7.2.10 *The following inequality is valid:*

$$|E(u(\vartheta_{k+1}, X_{k+1}) Y_{k+1} + Z_{k+1} - u(\vartheta_k, X_k) Y_k - Z_k \mid X_k, Y_k, Z_k)| \qquad (7.2.19)$$
$$\leq Ch^2 Y_k I_{G \setminus S_{t_k,h}}(X_k) \, \chi_{\varkappa > k} + Ch^{3/2} Y_k I_{S_{t_k,h}}(X_k) \, \chi_{\varkappa > k}.$$

Proof For $X_k \in G \setminus S_{t_k,h}$ the inequality (7.2.19) is a consequence of the fact that the one-step accuracy order of the Euler method (7.2.1)–(7.2.3) in the weak sense is $O(h^2)$.

Let $X_k \in S_{t_k,h}$. Then we have

$$\rho := E(u(\vartheta_{k+1}, X_{k+1}) Y_{k+1} + Z_{k+1} - u(\vartheta_k, X_k) Y_k \qquad (7.2.20)$$
$$- Z_k \mid X_k, Y_k, Z_k)$$
$$= \sum_{i=1}^{2^d} p_i \left(u(\vartheta_{k+1}^{(i)}, X_{k+1}^{(i)}) Y_{k+1}^{(i)} + Z_{k+1}^{(i)} \right) - u(\vartheta_k, X_k) Y_k - Z_k,$$

where p_α is from (7.2.17).

Substituting the expressions for $\vartheta_{k+1}^{(i)}, X_{k+1}^{(i)}, Y_{k+1}^{(i)}, Z_{k+1}^{(i)}$ in (7.2.20) and expanding $u(\vartheta_{k+1}^{(i)}, X_{k+1}^{(i)})$ in a series in powers of h around the point (ϑ_k, X_k), we obtain

$$\rho = \sum_{i=1}^{2^d} p_i Y_k \left\{ \left[u(\vartheta_k, X_k) + \frac{\partial u}{\partial t} \theta_k^{(i)} h \right. \right. \tag{7.2.21}$$

$$+ \sum_{j=1}^{d} \frac{\partial u}{\partial x^j} \left(\theta_k^{(i)} h (b^j - (\sigma\mu)^j) + \sqrt{\theta_k^{(i)} h} (\sigma\xi^{(i)})^j \right)$$

$$+ \frac{1}{2} \sum_{j,l=1}^{d} \frac{\partial^2 u}{\partial x^j \partial x^l} \theta_k^{(i)} h (\sigma\xi^{(i)})^j (\sigma\xi^{(i)})^l + O(h^{3/2}) \right]$$

$$\times \left(1 + \theta_k^{(i)} hc + \sqrt{\theta_k^{(i)} h} \, \mu^{\mathsf{T}} \xi^{(i)} \right) + \left(\theta_k^{(i)} hg + \sqrt{\theta_k^{(i)} h} \, F^{\mathsf{T}} \xi^{(i)} \right) \right\}$$

$$+ \sum_{i=1}^{2^d} p_i Z_k - u(\vartheta_k, X_k) Y_k - Z_k,$$

where the derivatives of u and the coefficients b, σ, μ, c, g, F are evaluated at the point (ϑ_k, X_k).

Rearranging the terms in (7.2.21), we get

$$\rho = Y_k h \tag{7.2.22}$$

$$\times \sum_{i=1}^{2^d} p_i \theta_k^{(i)} \left[\frac{\partial u}{\partial t} + \frac{1}{2} \sum_{j,l=1}^{d} \frac{\partial^2 u}{\partial x^j \partial x^l} \sum_{m=1}^{d} \sigma_{jm} \sigma_{lm} + \sum_{j=1}^{d} \frac{\partial u}{\partial x^j} b^j + cu + g \right]$$

$$+ Y_k \sqrt{h} \sum_{i=1}^{2^d} p_i \sqrt{\theta_k^{(i)}} \left[\sum_{j=1}^{d} \frac{\partial u}{\partial x^j} (\sigma\xi^{(i)})^j + u\mu^{\mathsf{T}} \xi^{(i)} + F^{\mathsf{T}} \xi^{(i)} \right]$$

$$- Y_k h \sum_{j=1}^{d} \frac{\partial u}{\partial x^j} (\sigma\mu)^j \sum_{i=1}^{2^d} p_i \theta_k^{(i)} + Y_k h \sum_{j=1}^{d} \frac{\partial u}{\partial x^j} \sum_{i=1}^{2^d} p_i \theta_k^{(i)} \mu^{\mathsf{T}} \xi^{(i)} (\sigma\xi^{(i)})^j$$

$$+ \frac{1}{2} Y_k h \sum_{j,l=1}^{d} \frac{\partial^2 u}{\partial x^j \partial x^l} \sum_{n,m=1, n\neq m}^{d} \sum_{i=1}^{2^d} p_i \theta_k^{(i)} \sigma_{jn} \xi^{(i),n} \sigma_{lm} \xi^{(i),m} + O(h^{3/2}).$$

The first sum in (7.2.22) is equal to zero since $u(t, x)$ satisfies the equation (7.1.1). We have from (7.2.17): $p_i \sqrt{\theta_k^{(i)}} = p_{2^{d-1}+i} \sqrt{\theta_k^{(2^{d-1}+i)}}, i = 1, \ldots, 2^{d-1}$, and, therefore, for any $r = 1, \ldots, d$ we obtain

$$\sum_{i=1}^{2^d} p_i \sqrt{\theta_k^{(i)}} \xi^{(i),r} = \sum_{i=1}^{2^{d-1}} \xi^{(i),r} \left(p_i \sqrt{\theta_k^{(i)}} - p_{2^{d-1}+i} \sqrt{\theta_k^{(2^{d-1}+i)}} \right) = 0,$$

whence the second sum in (7.2.22) is equal to zero. It follows from (7.2.17) that $p_i \theta_k^{(i)} + p_{2^{d-1}+i} \theta_k^{(2^{d-1}+i)} = \gamma$. Then it is not difficult to check that for $n \neq m$:

$$\sum_{i=1}^{2^d} p_i \theta_k^{(i)} \, \xi^{(i),n} \, \xi^{(i),m} = \gamma \sum_{i=1}^{2^{d-1}} \xi^{(i),n} \xi^{(i),m} = \frac{\gamma}{2} \sum_{i=1}^{2^d} \xi^{(i),n} \xi^{(i),m} = 0.$$

This implies that the last sum in (7.2.22) is equal to zero and the sum before the last one is equal to $Y_k h \sum_{j=1}^d \frac{\partial u}{\partial x^j} (\sigma \mu)^j \sum_{i=1}^{2^d} p_i \theta_k^{(i)}$. Thus, $\rho = O(h^{3/2})$. Lemma 7.2.10 is proved.   □

The next theorem is proved similarly to the proof of Theorem 7.2.4.

Theorem 7.2.11 *Algorithm 7.2.9 has weak order of accuracy $O(h)$:*

$$|E(\varphi(\vartheta_\varkappa, X_\varkappa) Y_\varkappa + Z_\varkappa) - u(t_0, x_0)| \leq Ch,$$

where $C > 0$ is independent of t_0, x_0, h.

We note that the chain $(\vartheta_k, X_k, Y_k, Z_k)$ is more computationally expensive, but due to smaller errors in the boundary domain, it allows us to obtain more accurate results than the chain from Sect. 7.2.1. Further, if instead of the Euler approximation X, Y, Z from (7.2.1)–(7.2.3) we use an approximation X, Y, Z with one-step error $O(h^{5/2})$ or higher, then the corresponding modification of Algorithm 7.2.9 has weak order $O(h^{3/2})$. For instance, we can exploit the second-order weak method (2.2.18) from Sect. 2.2.1 for this purpose.

Remark 7.2.12 Algorithm 7.2.9 is used in Sect. 9.1.3 to construct a layer (deterministic) method for solving the Dirichlet problem for semilinear parabolic equations.

Remark 7.2.13 We note that the algorithms from Sects. 7.2.1 and 7.2.3 can be applied in the case when the domain Q has the form

$$Q = \{(t, x) : T_0 \leq t < T, \ x \in G_t\},$$

i.e., when G depends on t (see [288]).

7.2.4 Numerical Tests

Consider the Dirichlet problem for the heat equation:

$$\frac{\partial u}{\partial t} = \frac{1}{2} (1.21 - x_2^2 - x_3^2) \frac{\partial^2 u}{\partial x_1^2} + \frac{1}{2} \frac{\partial^2 u}{\partial x_2^2} + \frac{1}{2} \frac{\partial^2 u}{\partial x_3^2} + 6(1 - 0.5e^{-t}) \qquad (7.2.23)$$

$$\times \, (x_1^2(1.21 - x_2^2 - x_3^2) + x_2^2) + 0.5e^{-t}(1.21 - x_1^4 - x_2^4), \quad t \in (0, T], \ x \in U_1,$$

$$u(0, x) = \frac{1}{2}(1.21 - x_1^4 - x_2^4), \quad x \in \bar{U}_1, \tag{7.2.24}$$

$$u(t, x) = (1.21 - x_1^4 - x_2^4)(1 - 0.5e^{-t}), \quad t \in [0, T], \quad x \in \partial U_1,$$

where $U_1 \subset \mathbf{R}^3$ is a unit ball with center at the origin. This problem has the solution:

$$u(t, x) = (1.21 - x_1^4 - x_2^4)(1 - 0.5e^{-t}).$$

By changing time $t = T - s$ the problem (7.2.23)–(7.2.24) is rewritten in the form (7.1.1)–(7.1.2) which is suitable for the probabilistic approach.

The results of simulation of $u(0.6, 0, 0, 0)$ and $u(15, 0, 0, 0)$ by Algorithms 7.2.1 and 7.2.6 are given in Table 7.2.1. The values in Table 7.2.1 are approximations of $\bar{u} = E(\varphi(t_\varkappa, X_\varkappa) Y_\varkappa + Z_\varkappa)$ evaluated as

$$\bar{u} \doteq \frac{1}{M} \sum_{m=1}^{M} \left(\varphi\left(t_\varkappa^{(m)}, X_\varkappa^{(m)}\right) Y_\varkappa^{(m)} + Z_\varkappa^{(m)} \right) \pm 2\sqrt{\frac{\bar{D}_M}{M}}, \tag{7.2.25}$$

where

$$\bar{D}_M = \frac{1}{M} \sum_{m=1}^{M} \left[\varphi\left(t_\varkappa^{(m)}, X_\varkappa^{(m)}\right) Y_\varkappa^{(m)} + Z_\varkappa^{(m)} \right]^2$$
$$- \left[\frac{1}{M} \sum_{m=1}^{M} \left(\varphi\left(t_\varkappa^{(m)}, X_\varkappa^{(m)}\right) Y_\varkappa^{(m)} + Z_\varkappa^{(m)} \right) \right]^2.$$

Thus, assuming that the sampling variance is sufficiently close to $\bar{D}_M$, $\bar{u}$ belongs to the interval defined in (7.2.25) with probability 0.95. The Monte Carlo error

Table 7.2.1 Parabolic problem. The results of approximate solution of the problem (7.2.23)–(7.2.24) according to Algorithm 7.2.1 (the upper table) and Algorithm 7.2.6 (the lower table) for $\mu \equiv 0$, $F \equiv 0$. The exact solution $u(0.6, 0, 0, 0) \doteq 0.87797$, $u(15, 0, 0, 0) \doteq 1.21000$.

h	M	$\bar{u}(0.6, 0, 0, 0)$	$\approx E\varkappa$	$\bar{u}(15, 0, 0, 0)$	$\approx E\varkappa$
0.04	$1 \cdot 10^3$	0.7805 ± 0.0127	7	1.0340 ± 0.0189	7
0.01	$4 \cdot 10^4$	0.8449 ± 0.0022	30	1.1520 ± 0.0035	32
0.0016	$1 \cdot 10^6$	0.8716 ± 0.0005	193	1.1997 ± 0.0007	207
0.0001	$4 \cdot 10^6$	0.8776 ± 0.0002	3111	1.2095 ± 0.0004	3326
h	M	$\bar{u}(0.6, 0, 0, 0)$	$\approx E\varkappa$	$\bar{u}(15, 0, 0, 0)$	$\approx E\varkappa$
0.04	$1 \cdot 10^3$	0.7127 ± 0.0115	6	0.9560 ± 0.0167	6
0.01	$4 \cdot 10^4$	0.7941 ± 0.0020	27	1.0719 ± 0.0030	28
0.0016	$1 \cdot 10^6$	0.8440 ± 0.0004	183	1.1526 ± 0.0007	193
0.0001	$4 \cdot 10^6$	0.8695 ± 0.0002	3064	1.1955 ± 0.0004	3262

for values in Table 7.2.1 is not greater than errors of numerical integration. As it follows from this table, results of the experiment are in quite good agreement with theoretical results. Numerical tests of Algorithm 7.2.9 also gave results corresponding to the theory. We have also checked the rules given by Remarks 7.2.5 and 7.2.8. For example, using the results from the lower part of Table 7.2.1 for $h_1 = 0.0001$, $h_2 = 0.0016$, we get due to formula (7.2.15): $\bar{u}(0.6, 0, 0, 0) = 0.8780 \pm 0.0004$, $\bar{u}(15, 0, 0, 0) = 1.2098 \pm 0.0008$.

An implementation of Algorithm 7.2.1 is available in Appendix A4 of [314].

7.3   Random Walks for the Elliptic Dirichlet Problem

Consider the Dirichlet problem for elliptic equation

$$\frac{1}{2} \sum_{i,j=1}^{d} a^{ij}(x) \frac{\partial^2 u}{\partial x^i \partial x^j} + \sum_{i=1}^{d} b^i(x) \frac{\partial u}{\partial x^i} + c(x)u + g(x) = 0, \ x \in G, \quad (7.3.1)$$

$$u\mid_{\partial G} = \varphi(x). \quad (7.3.2)$$

We assume that the coefficients $a^{ij} = a^{ji}$ satisfy the strict ellipticity condition in $\bar{G}$, i.e.,

$$\lambda_1^2 = \min_{x \in \bar{G}} \min_{1 \le i \le d} \lambda_i^2(x) > 0,$$

where $\lambda_1^2(x) \le \lambda_2^2(x) \le \cdots \le \lambda_d^2(x)$ are eigenvalues of the matrix $a(x) = \{a^{ij}(x)\}$. Let $\lambda_d^2 = \max_{x \in \bar{G}} \lambda_d^2(x)$. Then, for any $x \in \bar{G}$ and $y \in \mathbf{R}^d$ the following inequality takes place:

$$\lambda_1^2 \sum_{i=1}^{d} (y^i)^2 \le \sum_{i,j=1}^{d} a^{ij}(x) y^i y^j \le \lambda_d^2 \sum_{i=1}^{d} (y^i)^2. \quad (7.3.3)$$

We also assume that conditions hold which guarantee existence of the unique solution $u(x)$ of the problem (7.3.1)–(7.3.2) from the class $C^4(\bar{G})$. We recall [330] that it is sufficient for the above to require that the functions $a^{ij}(x)$, $b^i(x)$, $c(x)$, $g(x)$ are from the class $C^2(\bar{G})$, G is an open domain with twice continuously differentiable boundary ∂G, $\varphi(x) \in C^4(\bar{G})$, and $c(x) \le 0$, $x \in \bar{G}$.

The solution of the problem (7.3.1)–(7.3.2) has the probabilistic representation:

$$u(x) = E\left[\varphi(X_x(\tau))Y_{x,1}(\tau) + Z_{x,1,0}(\tau)\right], \quad (7.3.4)$$

where $X_x(s)$, $Y_{x,y}(s)$, $Z_{x,y,z}(s)$, $s \ge 0$, is the solution of the Cauchy problem for the system of SDEs:

$$dX = (b(X) - \sigma(X)\mu(X)) \, ds + \sigma(X) \, dw(s), \quad X(0) = x, \tag{7.3.5}$$
$$dY = c(X)Y \, ds + \mu^{\mathsf{T}}(X)Y \, dw(s), \quad Y(0) = y, \tag{7.3.6}$$
$$dZ = g(X)Y \, ds + F^{\mathsf{T}}(X)Y \, dw(s), \quad Z(0) = z, \tag{7.3.7}$$

$x \in G$, and $\tau = \tau_x$ is the first exit time of the trajectory $X_x(s)$ to the boundary ∂G.

The notation here and in what follows is similar to the notation in Sects. 7.1 and 7.2.

7.3.1 The Simplest Random Walk for Elliptic Equations

In this subsection, the results of Sect. 7.2 are carried over to the elliptic problem (7.3.1)–(7.3.2). Here we restrict ourselves to the case $\mu \equiv 0$.

First we construct an algorithm of random walk, which is similar to Algorithm 7.2.1. We apply the weak explicit Euler approximation with the simplest simulation of noise to the system (7.3.5)–(7.3.7):

$$X_{t,x}(t+h) \approx X = x + hb(x) + h^{1/2}\sigma(x)\,\xi, \tag{7.3.8}$$
$$Y_{t,x,y}(t+h) \approx Y = y + hc(x)\,y, \tag{7.3.9}$$
$$Z_{t,x,y,z}(t+h) \approx Z = z + hg(x)\,y + h^{1/2}F^{\mathsf{T}}(x)\,y\,\xi. \tag{7.3.10}$$

Introduce the boundary zone $S_h \subset \bar{G}$: $x \in S_h$ if at least one of the 2^d values of the vector X is outside $\bar{G}$. Let a constant $\lambda > 0$ be such that if the distance from $x \in G$ to the boundary ∂G is equal to or greater than $\lambda\sqrt{h}$, then x is outside the boundary zone and, therefore, for such x all the realizations of the random variable X belong to $\bar{G}$.

Let $x \in S_h$. Introduce the random vector $X_{x,h}^{\pi}$ taking two values x^{π} and $x + h^{1/2}\lambda n(x^{\pi})$ with probabilities $p = p_{x,h}$ and $q = q_{x,h} = 1 - p_{x,h}$, respectively, where

$$p_{x,h} = \frac{h^{1/2}\lambda}{|x + h^{1/2}\lambda n(x^{\pi}) - x^{\pi}|},$$

$x^{\pi} \in \partial G$ is the projection of the point x on the boundary ∂G, and $n(x^{\pi})$ is the unit vector of internal normal to ∂G at x^{π}.

To approximate the solution of the system (7.3.5), we construct a Markov chain X_k which stops when it reaches the boundary ∂G at a random step $\varkappa$.

We set $X_0' = x_0$. If $X_0' \notin S_h$ then we take $X_0 = X_0'$. If $X_0' \in S_h$ then the random variable X_0 takes two values: either $X_0'^{\pi} \in \partial G$ with probability $p_{X_0',h}$ or $X_0' + h^{1/2}\lambda n(X_0'^{\pi}) \notin S_h$ with probability $q_{X_0',h}$. If $X_0 = X_0'^{\pi}$, we put $\varkappa = 0$, $X_{\varkappa} = X_0'^{\pi}$, and the random walk is finished. Let X_k be constructed and either $X_k \in \partial G$ (i.e., the chain is stopped at one of the previous steps and, consequently, $\varkappa \le k$) or $X_k \notin S_h$. Now our aim is to construct X_{k+1} with the same properties. Suppose that the chain does not stop until the step k inclusively, i.e., $\varkappa > k$. Introduce X'_{k+1} due to (7.3.8) with $t = t_k$, $x = X_k$, $\xi = \xi_k$:

$$X'_{k+1} = X_k + hb(X_k) + h^{1/2}\sigma(X_k)\,\xi_k. \tag{7.3.11}$$

Now we obtain X_{k+1} using X'_{k+1} as we got X_0 using X'_0. More precisely, we use the following rule. If $X'_{k+1} \notin S_h$ then we take $X_{k+1} = X'_{k+1}$. If $X'_{k+1} \in S_h$ then the random variable X_{k+1} takes two values: either $X'^{\pi}_{k+1} \in \partial G$ with probability $p_{X'_{k+1},h}$ or $X'_{k+1} + h^{1/2}\lambda n(X'^{\pi}_{k+1}) \notin S_h$ with probability $q_{X'_{k+1},h}$. If $X_{k+1} = X'^{\pi}_{k+1}$, we put $\varkappa = k+1$, $X_\varkappa = X'^{\pi}_{k+1}$, and the random walk is finished. So, the random walk X_k has been constructed.

Clearly, X_k remains in the domain $\bar{G}$ with probability 1.

We also introduce an extended chain (X_k, Y_k, Z_k). We put $Y_0 = 1$ and $Z_0 = 0$. Let $\varkappa \geq k+1$ and Y_k, Z_k be known. Then the values Y_{k+1}, Z_{k+1} are evaluated in accordance with the system (7.3.9)–(7.3.10) for $x = X_k$, $y = Y_k$, $z = Z_k$, $\xi = \xi_k$.

The constructed algorithm can be written as follows.

Algorithm 7.3.1 *STEP 0.* $X'_0 = x_0$, $Y_0 = 1$, $Z_0 = 0$, $k = 0$.

STEP 1.   If $X'_k \notin S_{t_k,h}$ then $X_k = X'_k$ and go to STEP 3.
           If $X'_k \in S_{t_k,h}$ then either $X_k = X'^{\pi}_k$ with probability
           $p_{X'_k,h}$ or $X_k = X'_k + h^{1/2}\lambda n(X'^{\pi}_k)$ with probability $q_{X'_k,h}$.
STEP 2.   If $X_k = X'^{\pi}_k$ then STOP and $\varkappa = k$,
           $X_\varkappa = X'^{\pi}_k$, $Y_\varkappa = Y_k$, $Z_\varkappa = Z_k$.
STEP 3.   Simulate ξ_k and find X'_{k+1}, Y_{k+1}, Z_{k+1} according to (7.3.11),
           (7.3.9)–(7.3.10) for $x = X_k$, $y = Y_k$, $z = Z_k$, $\xi = \xi_k$.
STEP 4.   Put $k := k+1$ and return to STEP 1.

This algorithm is similar to Algorithm 7.2.1 for parabolic equation. At the same time, we emphasize that here $\varkappa$ can take arbitrary large values in contrast to Algorithm 7.2.1.

Denote by ν_{x_0} the number of those k at which X'_k gets into the set S_h.

Lemma 7.3.2 *The following inequalities hold:*

$$P\{\nu_{x_0} = n\} \leq \frac{1}{2^{n-1}}, \quad E\nu_{x_0} \leq C, \tag{7.3.12}$$

$$E\varkappa \leq \frac{C}{h}, \tag{7.3.13}$$

where $C > 0$ does not depend on t_0 and h.

Proof The inequalities (7.3.12) are proved analogously to Lemma 7.2.2. Let us prove the inequality (7.3.13). Denote by μ_x the number of steps which the chain X'_k starting from $x \in G \backslash S_h$ spends in the domain $G \backslash S_h$ before X'_k gets into S_h. In connection with the chain X'_k we consider the boundary value problem (cf. (7.1.31)–(7.1.32)):

$$PV - V = -f(x), \quad x \in G\backslash S_h, \tag{7.3.14}$$

$$V(x) = 0, \quad x \in S_h, \tag{7.3.15}$$

where P is the one-step transition operator: $PV(x) = EV(X'_1), X'_0 = x.$
 It is known [450] that the solution of this problem is the function:

$$V(x) = E \sum_{k=0}^{\mu_x - 1} f(X'_k). \tag{7.3.16}$$

If we find the solution $V(x)$ of (7.3.14)–(7.3.15) with a function $f(x)$ which everywhere in $G\backslash S_h$ satisfies the condition

$$f(x) \geq I_{G\backslash S_h}(x), \tag{7.3.17}$$

then, due to (7.3.16),

$$E\mu_x \leq V(x). \tag{7.3.18}$$

We take $V(x)$ of the form [194]:

$$V(x) = \begin{cases} A^2 - |x + B|^{2n}, & x \in G\backslash S_h, \\ 0, & x \in S_h, \end{cases} \tag{7.3.19}$$

where B is a d-dimensional vector such that

$$\min_{x \in \bar{G}} |x + B| \geq C > 0, \tag{7.3.20}$$

n is a sufficiently large natural number (how to choose it is shown below) and $A^2 = \max_{x \in \bar{G}} |x + B|^{2n}$. The function $V(x)$ satisfies the boundary condition (7.3.15).
 Let $x \in G\backslash S_h$, then $X'_1 = x + hb(x) + h^{1/2}\sigma(x)\,\xi$. It is not difficult to obtain
(recall that $a = \sigma\sigma^\mathsf{T}$):

$$PV(x) - V(x) = -hn|x + B|^{2n-4} \tag{7.3.21}$$

$$\times \left[2|x + B|^2(x + B, b(x)) + |x + B|^2 \sum_{i=1}^{d} a^{ii} \right.$$

$$\left. + 2(n-1) \sum_{i,j=1}^{d} a^{ij}(x)(x^i + B^i)(x^j + B^j) \right] + O(h^2).$$

The relations (7.3.21) and (7.3.3) imply

$$PV - V \le -hn|x + B|^{2n-2} \left[2(x + B, b(x)) + (2n - 2 + d) \lambda_1^2 \right] + O(h^2).$$

We select n so that for all $x \in \bar{G}$:

$$2(x + B, b(x)) + (2n - 2 + d) \lambda_1^2 \ge C > 0$$

that is always possible. Then, for a sufficiently small h we obtain $PV - V \le -\gamma h$, where $\gamma > 0$ is independent of h and x.

Obviously, the function $v(x) = V(x)/(\gamma h)$ is the solution of the problem (7.3.14)–(7.3.15) with $f(x) \ge 1$. Therefore, (see (7.3.17)–(7.3.19)) $E\mu_x \le A^2/(\gamma h)$. From here and the inequalities (7.3.12), it is not difficult to obtain the estimate (7.3.13). Lemma 7.3.2 is proved.   □

We extend the definition of the constructed chain for all k by the rule: if $k > \varkappa$, then $(X_k, Y_k, Z_k) = (X_\varkappa, Y_\varkappa, Z_\varkappa)$. Introduce

$$d_k = u(X_k) Y_k + Z_k - u(X_k') Y_k - Z_k,$$
$$d_k' = u(X_{k+1}') Y_{k+1} + Z_{k+1} - u(X_k) Y_k - Z_k, \quad k = 0, 1, \dots.$$

The lemma on one-step errors is proved analogously to Lemma 7.2.3.

Lemma 7.3.3 *The following inequalities hold:*

$$|E(d_k \mid X_k', Y_k, Z_k)| \le Ch Y_k I_{S_h}(X_k') \chi_{\varkappa \ge k}, \tag{7.3.22}$$

$$|E(d_k' \mid X_k, Y_k, Z_k)| \le Ch^2 Y_k \chi_{\varkappa > k}. \tag{7.3.23}$$

Theorem 7.3.4 *Algorithm 7.3.1 has weak order of accuracy $O(h)$:*

$$|E(\varphi(X_\varkappa) Y_\varkappa + Z_\varkappa) - u(x_0)| \le Ch, \tag{7.3.24}$$

where C is independent of x_0, h.

Proof We have

$$
\begin{aligned}
R &:= E(\varphi(X_\varkappa) Y_\varkappa + Z_\varkappa) - u(x_0) = E(u(X_\varkappa) Y_\varkappa + Z_\varkappa) - u(x_0) \\
&= E(u(X_0) Y_0 + Z_0) - u(x_0) \\
&\quad + E \sum_{k=0}^{\varkappa-1} (u(X_{k+1}) Y_{k+1} + Z_{k+1} - u(X_k) Y_k - Z_k) \\
&= E \sum_{k=0}^{\infty} (d_k + d_k') = \sum_{k=0}^{\infty} EE(d_k' \mid X_k, Y_k, Z_k) + \sum_{k=0}^{\infty} EE(d_k \mid X_k', Y_k, Z_k).
\end{aligned}
$$

It is obvious that $Y_k > 0$ and (since $c(x) \leq 0$) $Y_k \leq 1, k = 0, 1, \ldots$. Using Lemmas 7.3.2 and 7.3.3, we obtain

$$|R| \leq Ch^2 \sum_{k=0}^{\infty} E(Y_k \chi_{\varkappa > k}) + Ch \sum_{k=0}^{\infty} E(Y_k I_{S_h}(X_k) \chi_{\varkappa \geq k})$$

$$\leq Ch^2 E\varkappa + ChE\nu_{x_0} \leq Ch,$$

where C is independent of x_0, h. Theorem 7.3.4 is proved. $\qquad\square$

Remark 7.3.5 It is possible to construct algorithms of random walks analogous to Algorithms 7.2.6 and 7.2.9 for the elliptic problem (7.3.1)–(7.3.2).

7.3.2 Other Methods for Elliptic Problems

In this subsection we construct two methods for the elliptic problem (7.3.1)–(7.3.2) which are similar to the methods for the parabolic problem from Sect. 7.1. Here we use the probabilistic representation (7.3.4)–(7.3.7) with

$$\mu(x) = \sigma^{-1}(x)b(x) , \tag{7.3.25}$$

and we restrict ourselves to the case $F(x) = 0$. Then, the system (7.3.5)–(7.3.7) has the form

$$dX = \sigma(X)dw(t), \quad X(0) = x, \tag{7.3.26}$$
$$dY = c(X)Ydt + (\sigma^{-1}(X)b(X))^{\mathsf{T}}Ydw(t), \quad Y(0) = y, \tag{7.3.27}$$
$$dZ = g(X)Ydt, \quad Z(0) = z. \tag{7.3.28}$$

Denote by $U_r \subset \mathbf{R}^d$ a sphere of radius r with center at the origin. Let $x \in G$. Consider the one-step approximation of the solution to the system (7.3.26)–(7.3.28):

$$X_1 = x + \sigma(x)w(\vartheta), \tag{7.3.29}$$

$$Y_1 = y + yc(x)\frac{r^2}{d} + y\mu^{\mathsf{T}}(x)w(\vartheta), \tag{7.3.30}$$

$$Z_1 = z + yg(x)\frac{r^2}{d} , \tag{7.3.31}$$

where $w(\vartheta)$ has the uniform distribution on the sphere ∂U_r and r is such that the ellipsoid $(\sigma^{-1}(x)(X - x), \sigma^{-1}(x)(X - x)) = r^2$ belongs to $\bar{G}$.

It is not difficult to prove the following lemma (see details in [281]).

Lemma 7.3.6 *The one-step order of the approximation (7.3.29)–(7.3.31) with respect to r is equal to 4:*

$$|E[u(X_1)\, Y_1 + Z_1] - [u(x)\, y + z]| \leq C y r^4,$$

where C depends on derivatives of $u(x)$ up to fourth order.

Denote by Γ_δ the interior of a δ-neighborhood of the boundary ∂G belonging to $\bar{G}$. On the basis of the one-step approximation (7.3.29)–(7.3.31) we construct the numerical algorithm.

Algorithm 7.3.7 *Let $X_0 = x \in \bar{G}$, $Y_0 = 1$, $Z_0 = 0$, $k = 0$ and r be a small positive number. Let ξ be a point uniformly distributed on a unit sphere with center at the origin and $\xi_1, \xi_2, \ldots$ be independent random points distributed as ξ. If $X_k \in G \backslash \Gamma_{\lambda_{dr}}$ then we set $r_{k+1} = r$, and if $X_k \in \Gamma_{\lambda_{dr}} \backslash \Gamma_{r^2}$ then we find a number r_{k+1} such that the ellipsoid $(\sigma^{-1}(X_k)(X - X_k), \sigma^{-1}(X_k)(X - X_k)) = r_{k+1}^2$ touches the boundary ∂G. In both cases we evaluate X_{k+1}, Y_{k+1}, Z_{k+1} according to the system*

$$X_{k+1} = X_k + \sigma(X_k) r_{k+1} \xi_{k+1}, \tag{7.3.32}$$

$$Y_{k+1} = Y_k + Y_k c(X_k) \frac{r_{k+1}^2}{d} + Y_k \mu^{\mathsf{T}}(X_k) r_{k+1} \xi_{k+1}, \tag{7.3.33}$$

$$Z_{k+1} = Z_k + Y_k g(X_k) \frac{r_{k+1}^2}{d}, \tag{7.3.34}$$

where $\mu(x)$ is from (7.3.25).

If $X_{k+1} \in \Gamma_{r^2}$ then the algorithm is stopped and we take $(\bar{X}_\varkappa, Y_\varkappa, Z_\varkappa)$ with $\varkappa = k + 1$, $\bar{X}_\varkappa = X_\varkappa^\pi$ as the final state of the Markov chain (recall that $x^\pi \in \partial G$ is the projection of the point x on the boundary of the domain G). If $X_{k+1} \in G \backslash \Gamma_{r^2}$ then we put $k := k + 1$ and continue the algorithm.

The above algorithm needs a sampling method for points uniformly distributed on a unit sphere, for which various algorithms are available e.g. in [253, 332, 436].

It is possible to prove (see [281] and also Sect. 7.4.3) that the average number of steps for Algorithm 7.3.7 is equal to $O(1/r^2)$ and if $c(x) \leq -c_0 < 0$ then this algorithm has the second order of convergence with respect to r.

Theorem 7.3.8 *Let $c(x) \leq -c_0 < 0$. Then Algorithm 7.3.7 has the second order of convergence with respect to r, i.e., for all sufficiently small r*

$$|E(\varphi(\bar{X}_\varkappa) Y_\varkappa - Z_\varkappa) - u(x)| \leq K r^2.$$

Now let us propose another algorithm.

Algorithm 7.3.9 *Let $X_0 = x \in G$, $Y_0 = 1$, $Z_0 = 0$, $k = 0$ and r be a small positive number. Introduce the random vector η which coordinates η^i, $i = 1, \ldots, d$, are* *mutually independent random variables taking values $\pm 1/\sqrt{d}$ with probability $1/2$. Let $\eta_1, \eta_2, \ldots$ be independent random points distributed as η. If $X_k \in G\backslash\Gamma_{\lambda_{dr}}$ then we set $r_{k+1} = r$, and if $X_k \in \Gamma_{\lambda_{dr}}\backslash\partial G$ then we find a minimal number r_{k+1} such that one of the points from the set $\{X : X = X_k + \sigma(X_k)r_{k+1}\eta_{k+1}\}$ belongs to ∂G. In both cases we evaluate $X_{k+1}, Y_{k+1}, Z_{k+1}$ according to the system*

$$X_{k+1} = X_k + \sigma(X_k)r_{k+1}\eta_{k+1} , \tag{7.3.35}$$

$$Y_{k+1} = Y_k + Y_k c(X_k)\frac{r_{k+1}^2}{d} + Y_k \mu^{\mathsf{T}}(X_k)r_{k+1}\eta_{k+1} , \tag{7.3.36}$$

$$Z_{k+1} = Z_k + Y_k g(X_k)\frac{r_{k+1}^2}{d} , \tag{7.3.37}$$

where $\mu(x)$ is from (7.3.25).

In the second case the point X_{k+1} with probability $1/2^d$ falls on ∂G. Let $\varkappa = \varkappa_x$ be the first number at which $X_\varkappa \in \partial G$. The random walk is stopped at this random step $\varkappa$ and $(X_\varkappa, Y_\varkappa, Z_\varkappa)$ is taken as the final state of the Markov chain.

Note that the random walk X_k obtained in Algorithm 7.3.9 gets a finite number of values at each step (it is equal to 2^d) in contrast to the random walk in Algorithm 7.3.7 and it does not require any neighborhood Γ_δ of the boundary ∂G.

It is not difficult to prove that the average number of steps for Algorithm 7.3.9 is equal to $O(1/r^2)$ and if $c(x) \leq -c_0 < 0$ then this algorithm has the second order of convergence with respect to r [281].

Theorem 7.3.10 *Let $c(x) \leq -c_0 < 0$. Then Algorithm 7.3.9 has the second order of convergence with respect to r, i.e., for all sufficiently small r*

$$|E(\varphi(X_\varkappa)Y_\varkappa - Z_\varkappa) - u(x)| \leq Kr^2 .$$

7.3.3 Numerical Tests

Consider the Dirichlet problem for the elliptic equation (cf. (7.2.23)–(7.2.24)):

$$\frac{1}{2}(1.21 - x_2^2 - x_3^2)\frac{\partial^2 u}{\partial x_1^2} + \frac{1}{2}\frac{\partial^2 u}{\partial x_2^2} + \frac{1}{2}\frac{\partial^2 u}{\partial x_3^2} \tag{7.3.38}$$
$$+ 6x_1^2(1.21 - x_2^2 - x_3^2) + 6x_2^2 = 0, \ x \in U_1 ,$$
$$u\,|_{\partial U_1} = 1.21 - x_1^4 - x_2^4. \tag{7.3.39}$$

The solution of this problem is

$$u(x) = 1.21 - x_1^4 - x_2^4.$$

Table 7.3.1 Elliptic problem. The results of approximate solution of the problem (7.3.38)–(7.3.39) according to Algorithm 7.3.1 for $F \equiv 0$. The exact solution $u(0, 0, 0) = 1.21$, $u(0.5, 0, 0) = 1.1475$.

h	M	$\bar{u}(0, 0, 0)$	$\approx E\varkappa$	$\bar{u}(0.5, 0, 0)$	$\approx E\varkappa$
0.04	10^3	1.0337 ± 0.0180	7	0.9361 ± 0.0239	5.1
0.01	4×10^4	1.1504 ± 0.0035	32	1.0810 ± 0.0042	23.6
0.0016	10^6	1.1999 ± 0.0007	207	1.1359 ± 0.0009	155
0.0001	4×10^6	1.2093 ± 0.0004	3325	1.1467 ± 0.0004	2477

Table 7.3.2 Elliptic problem. The results of approximate solution of the problem (7.3.38)–(7.3.39) according to Algorithm 7.3.7. The exact solution $u(0, 0, 0) = 1.21$, $u(0.5, 0, 0) = 1.1475$.

r	M	$\bar{u}(0, 0, 0)$	$\approx E\varkappa$	$\bar{u}(0.5, 0, 0)$	$\approx E\varkappa$
0.4	10^4	1.0346 ± 0.0062	5.8	0.9382 ± 0.0075	4.0
0.2	10^4	1.1578 ± 0.0071	23.8	1.0912 ± 0.0082	17.8
0.1	10^4	1.1910 ± 0.0073	99.0	1.1289 ± 0.0084	74.0
0.08	10^4	1.1983 ± 0.0074	153.5	1.1354 ± 0.0085	113.6

The results of simulation of $u(0, 0, 0)$ and $u(0.5, 0, 0)$ by Algorithms 7.3.1 and 7.3.7 are given in Tables 7.3.1 and 7.3.2, respectively. The values in Tables 7.3.1 (Table 7.3.2) are approximations of $\bar{u} = E(\varphi(X_\varkappa) Y_\varkappa + Z_\varkappa)$ ($\bar{u} = E(\varphi(\bar{X}_\varkappa) Y_\varkappa + Z_\varkappa)$) and evaluated by a formula analogous to (7.2.25). The Monte Carlo error for values in the tables is not greater than errors of numerical integration. Analyzing the results presented in Table 7.3.1, we can conclude that the error of numerical integration by Algorithm 7.3.1 is proportional to Ch and the mean number of steps $E\varkappa$ is proportional to K/h. We see from Table 7.3.2 that the error of Algorithm 7.3.7 is proportional to Cr^2 and the mean number of steps $E\varkappa$ is proportional to K/r^2. To avoid a confusion, we recall that h is step in time and r is step in space and $h \sim r^2$. Comparing the results of Tables 7.3.1 and 7.3.2, we obtain that the constant C at h in the error of numerical integration is larger for Algorithm 7.3.1 while the constant K at $1/h$ in the estimate of the mean number of steps $E\varkappa$ is larger for Algorithm 7.3.7.

We can reduce the Monte Carlo error in simulations by Algorithm 7.3.1. As it was mentioned in Sect. 7.1, it is possible to select a function F so that the variance $\bar{D}$ of the random variable $\bar{\eta} = \varphi(X_\varkappa) Y_\varkappa + Z_\varkappa$ related to the discrete system will decrease. Such a choice of F allows us to reduce computational costs. For instance, if we take

$$F^{\mathsf{T}} = \left(4x_1^3 \sqrt{1.21 - x_2^2 - x_3^2}, \; 4x_2^3, \; 0 \right),$$

then it is not difficult to show that the variance D of the random variable $\eta = \varphi(X_x(\tau)) Y_{x,1}(\tau) + Z_{x,1,0}(\tau)$ related to the system of SDEs is equal to zero (see Theorem 2.3.3). Since the accuracy order of the method is $O(h)$, the variance $\bar{D}$ satisfies the inequality $\bar{D} \leq Ch$, where C is independent of x, h. Therefore, the

Monte Carlo error is bounded from above by $C\sqrt{h/M}$. In numerical experiments, making use of the function F selected as above leads to a significant decrease of computational costs in comparison with simulations when $F \equiv 0$.

7.4 Specific Random Walks for Elliptic Equations and Boundary Layer

In this section, we construct some effective methods for the Dirichlet problem for the elliptic equation (cf. (7.3.1)–(7.3.2)):

$$\frac{a^2}{2}\Delta u + \sum_{i=1}^{d} b^i(x)\frac{\partial u}{\partial x^i} + c(x)u + g(x) = 0, \quad x \in G, \quad u\,|_{\partial G} = \varphi(x), \quad (7.4.1)$$

We assume that

(i) G is an open bounded set with twice continuously differentiable boundary ∂G;
(ii) the coefficients $a^{ij}(x)$, $b^i(x)$, $c(x)$, $g(x)$ belong to the class $C^2(\bar{G})$, $c(x) \le 0$, $\varphi \in C^4(\partial G)$.

These conditions ensure existence of the unique solution $u(x)$ of the problem (7.4.1) belonging to the class $C^4(\bar{G})$ [330].

The solution of the problem (7.4.1) has the probabilistic representation (cf. (7.3.4), (7.3.26)–(7.3.28)):

$$u(x) = E(\varphi(X_x(\tau))Y_{x,1}(\tau) + Z_{x,1,0}(\tau)), \quad (7.4.2)$$

where $X_x(t)$, $Y_{x,y}(t)$, $Z_{x,y,z}(t)$, $t \ge 0$, is the solution of the system

$$dX = a\,dw(t), \quad X(0) = x, \quad (7.4.3)$$

$$dY = c(X)Y\,dt + \frac{1}{a}b^{\mathsf{T}}(X)Y\,dw(t), \quad Y(0) = y, \quad (7.4.4)$$

$$dZ = g(X)Y\,dt, \quad Z(0) = z. \quad (7.4.5)$$

In (7.4.3)–(7.4.5), $b(x) = (b^1(x), \ldots, b^d(x))^{\mathsf{T}}$, $w(t) = (w^1(t), \ldots, w^d(t))^{\mathsf{T}}$ is a standard Wiener process which is defined on a probabilistic space $(\Omega, \mathcal{F}, P)$ and measurable with respect to the flow $\mathcal{F}_t$, $t \ge 0$, and τ is the first passage time of the trajectory $X_x(t)$ to the boundary ∂G. For definiteness, we always set $y > 0$. The simplicity of the equation (7.4.3) allows us to simulate its solutions exactly.

Let Γ_δ (Γ_{ar}) be the interior of a δ-neighborhood (of an ar-neighborhood) of the boundary ∂G belonging to $\bar{G}$. Usually, r is taken sufficiently small and δ is taken smaller than ar, $\delta = O(r^q)$, $q > 1$.

Consider the following random walk over small spheres which starts at $x \in G\backslash\Gamma_\delta$. We set $X_0 = x$. If $X_0 \in G\backslash\Gamma_{ar}$ then the next point X_1 is found as a uniformly distributed point on the sphere of radius ar with center at X_0. If $X_0 \in \Gamma_{ar} \setminus \Gamma_\delta$ then X_1 has the uniform distribution on the tangent sphere of radius $\rho(X_0, \partial G)$. Clearly, X_1 can be interpreted in both cases as

$$X_1 = X_0 + aw(\vartheta_1),$$

where ϑ_1 is the first passage time of the process $X_0 + aw(t)$ to the corresponding sphere. We can repeat this by starting from X_1 and so on. Let $\varkappa = \varkappa_x$ be the first number at which $X_\varkappa \in \Gamma_\delta$ and set $\vartheta_\kappa = 0$ for $k > \varkappa$ and $X_k = X_\varkappa$ for $k \geq \varkappa$. As a result, we obtain the random walk

$$X_0 = x$$
$$X_1 = X_0 + aw(\vartheta_1)$$
$$\cdots\cdots\cdots\cdots$$
$$X_k = X_{k-1} + a(w(\vartheta_1 + \cdots + \vartheta_k) - w(\vartheta_1 + \cdots + \vartheta_{k-1})), \ \ k = 1, \ldots, \varkappa,$$
$$X_k = X_\varkappa, \ k \geq \varkappa,$$

which stops at a random step $\varkappa$. It is a Markov chain. We emphasize that, fortunately, we do not need to solve a quite difficult problem of simulating the first passage times ϑ_k to find X_k (see the corresponding discussion in the introduction to Chap. 6).

Let $\mathcal{B}_k = \sigma(X_0, X_1, \ldots, X_k), k = 1, 2, \ldots$, be the sequence of σ-algebras generated by the random walk $X_0, X_1, \ldots, X_k, \ldots$. Presuppose that a method approximating the solution of (7.4.3)–(7.4.5) is proposed and the sequences $Y_0, Y_1, \ldots, Y_k, \ldots$, $Z_0, Z_1, \ldots, Z_k, \ldots$ which approximate $Y_{x,y}(\vartheta_1 + \cdots + \vartheta_k)$, $Z_{x,y,z}(\vartheta_1 + \cdots + \vartheta_k)$, $k = 1, 2, \ldots$, respectively, are constructed so that Y_k, Z_k are $\mathcal{B}_k$-measurable and they are stopped at the random step $\varkappa$. Let $\bar{X}_\varkappa$ be the point on the boundary ∂G closest to $X_\varkappa$. Put $\bar{Y}_\varkappa = Y_\varkappa$, $\bar{Z}_\varkappa = Z_\varkappa$. Introduce the function

$$v(x, y, z) = E(\varphi(X_x(\tau))Y_{x,y}(\tau) + Z_{x,y,z}(\tau)). \tag{7.4.6}$$

Clearly,

$$v(x, y, z) = u(x)y + z. \tag{7.4.7}$$

We are interested in the difference

$$\mathcal{R} = Ev(\bar{X}_\varkappa, \bar{Y}_\varkappa, \bar{Z}_\varkappa) - Ev(X_x(\tau), Y_{x,y}(\tau), Z_{x,y,z}(\tau))$$
$$= Ev(\bar{X}_\varkappa, \bar{Y}_\varkappa, \bar{Z}_\varkappa) - v(x, y, z)$$

since $Ev(\bar{X}_\varkappa, \bar{Y}_\varkappa, \bar{Z}_\varkappa) = E(\varphi(\bar{X}_\varkappa)\bar{Y}_\varkappa + \bar{Z}_\varkappa) = E(\varphi(\bar{X}_\varkappa)Y_\varkappa + Z_\varkappa)$ is taken as an approximation of $v(x, y, z) = u(x)y + z$. We have

$$v(\bar{X}_\varkappa, \bar{Y}_\varkappa, \bar{Z}_\varkappa) - v(x, y, z) = (v(\bar{X}_\varkappa, \bar{Y}_\varkappa, \bar{Z}_\varkappa) - v(X_\varkappa, Y_\varkappa, Z_\varkappa))$$

$$+(v(X_\varkappa, Y_\varkappa, Z_\varkappa) - v(X_{\varkappa-1}, Y_{\varkappa-1}, Z_{\varkappa-1})) + \cdots + (v(X_1, Y_1, Z_1) - v(x, y, z))$$

$$= (u(\bar{X}_\varkappa) - u(X_\varkappa))Y_\varkappa + \sum_{k=1}^{\infty}(v(X_k, Y_k, Z_k) - v(X_{k-1}, Y_{k-1}, Z_{k-1}))\chi_{\varkappa \geq k}$$

and, consequently,

$$\mathcal{R} = E(u(\bar{X}_\varkappa) - u(X_\varkappa))Y_\varkappa \qquad (7.4.8)$$

$$+ \sum_{k=1}^{\infty} E(v(X_k, Y_k, Z_k) - v(X_{k-1}, Y_{k-1}, Z_{k-1}))\chi_{\varkappa \geq k}.$$

An estimate of $\mathcal{R}$ depends on a bound of the first term and on a one-step approximation which gives bounds for the summands in right-hand side of (7.4.8). Our aim is to find one-step approximations which do not require simulation of ϑ_k and errors of which can be bounded without using any derivatives or at least without using high-order derivatives of the solution $u(x)$ of the original problem (7.4.1). The latter is very important for problems with small parameter at higher derivatives because of presence of a boundary layer, and the higher derivatives the larger their values. Such approximations are based on simulation of some conditional expectations like

$$\xi^i = E\left(\int_0^\vartheta w^i(s)ds/w(\vartheta)\right), \qquad \xi^{ij} = E\left(\int_0^\vartheta w^i(s)dw^j(s)/w(\vartheta)\right).$$

Section 7.4.1 is devoted to auxiliary lemmas and to simulation of ξ^i and ξ^{ij}. Various one-step approximations are constructed in Sect. 7.4.2.

The estimate of the first term in (7.4.8) depends on δ. The average number of steps $E\varkappa$ also depends on δ. A choice of δ is related to accuracy of a one-step approximation. As usual, $\delta = O(r^k)$ if the order of one-step approximation is equal to $O(r^{k+2})$. The convergence theorems are given in Sects. 7.4.3 and 7.4.4 which start with theorems on the average number of steps $E\varkappa$ and with other results relevant for evaluation of the sum in (7.4.8) (see Sect. 7.4.3).

In the case of a small parameter at second derivatives (this case is treated in Sect. 7.5), the system (7.4.3)–(7.4.5) becomes a system with small noise and we construct some specific methods for its approximate integration. Another way rests on the fact that in almost the entire domain G except a narrow boundary layer the solution of the Dirichlet problem can be found with high accuracy and simply by analytical tools (this part of the solution is known as an external expansion). Having this in mind, we propose a method of random walk in the narrow layer to find the remaining part of the solution (known as an interior expansion). The effectiveness of this analytic-numerical method is achieved because of small average number of steps for the random paths in the very narrow domain.

7.4.1   Conditional Expectation of Ito Integrals Associated with Wiener Process in the Ball

Here we will exploit both a probabilistic representation and an explicit form of the solution of the Dirichlet problem for the Poisson equation in the ball $U_r = \{x = (x^1, \ldots, x^d): |x|^2 = (x^1)^2 + \cdots + (x^d)^2 \le r^2\}$:

$$\frac{1}{2}\Delta u + g(x) = 0, \quad |x| < r, \tag{7.4.9}$$

$$u\,|_{|x|=r} = \varphi(x). \tag{7.4.10}$$

In (7.4.9)–(7.4.10), $g(x) \in C^1(|x| \le r)$, $\varphi(x) \in C(|x| = r)$.

The probabilistic representation of the solution of the problem (7.4.9)–(7.4.10) has the form

$$u(x) = E\varphi(x + w(\vartheta_x)) + E \int_0^{\vartheta_x} g(x + w(s))ds, \tag{7.4.11}$$

where $w(t) = (w^1(t), \ldots, w^d(t))^{\mathsf{T}}$ is a d-dimensional standard Wiener process and ϑ_x is the first passage time of the process $x + w(t)$ to the sphere ∂U_r.

The explicit formula for the solution has the following form [330]:

$$u(x) = \int_{|\xi|=r} P_r(x, \xi)\varphi(\xi)\,dS_\xi + \int_{|\xi|<r} G_r(x, \xi)g(\xi)\,d\xi, \tag{7.4.12}$$

where P_r is the Poisson kernel:

$$P_r(x, \xi) = \frac{r^2 - |x|^2}{\sigma_d\,r\,|x - \xi|^d} \tag{7.4.13}$$

and G_r is the Green function which for $d = 2$ is equal to

$$G_r(x, \xi) = \frac{1}{2\pi} \ln \frac{|x|\,|(r/|x|)^2 x - \xi|}{r|x - \xi|}, \quad d = 2, \tag{7.4.14}$$

and for $d > 2$ is equal to

$$G_r(x, \xi) = \frac{1}{(d - 2)\sigma_d} \left(\frac{1}{|x - \xi|^{d-2}} - \frac{(r/|x|)^{d-2}}{|(r/|x|)^2 x - \xi|^{d-2}} \right), \quad d > 2. \tag{7.4.15}$$

In (7.4.13) and (7.4.15), σ_d is the area of the unit sphere in $\mathbf{R}^d$: $\sigma_d = 2\pi^{d/2}/\Gamma(d/2)$. Recall that $\sigma_d\,r^{d-1}$ is the area of the sphere ∂U_r and $\sigma_d\,r^d/d$ is the volume of the ball U_r.

Proceeding to simulation of the conditional expectation $E(\int_0^\vartheta w^i(s)ds/w(\vartheta))$ where $\vartheta = \vartheta_0$ is the first passage time of the Wiener process $w(t)$ to the sphere ∂U_r, we assume that

$$E\left(\int_0^\vartheta w^i(s)ds/w(\vartheta)\right) = \alpha w^i(\vartheta), \quad i = 1, \ldots, d. \tag{7.4.16}$$

If (7.4.16) is true then the constant α can be found from the condition

$$E\left(\int_0^\vartheta w^i(s)ds - \alpha w^i(\vartheta)\right)^2 \longrightarrow_\alpha \min,$$

i.e.,

$$\alpha = \frac{Ew^i(\vartheta)\int_0^\vartheta w^i(s)ds}{E[w^i(\vartheta)]^2}. \tag{7.4.17}$$

Lemma 7.4.1 *For every $i = 1, \ldots, d$ the following formulas hold:*

$$E(w^i(\vartheta))^2 = \frac{r^2}{d}, \tag{7.4.18}$$

$$Ew^i(\vartheta)\int_0^\vartheta w^i(s)ds = E\int_0^\vartheta (w^i(s))^2 ds = \frac{r^4}{2d(d+2)}, \tag{7.4.19}$$

and, consequently, α from (7.4.17) is equal to

$$\alpha = \frac{r^2}{2(d+2)}. \tag{7.4.20}$$

Proof The relation (7.4.18) evidently follows from the identity $(w^1(\vartheta))^2 + \cdots + (w^d(\vartheta))^2 = r^2$. Further, Ito's formula implies

$$dw^i(t)\int_0^t w^i(s)ds = \int_0^t w^i(s)ds \times dw^i(t) + (w^i(t))^2 dt$$

and, therefore,

$$Ew^i(\vartheta)\int_0^\vartheta w^i(s)ds = E\int_0^\vartheta (w^i(s))^2 ds.$$

It is not difficult to verify that the function $u = r^4 - |x|^4$ is a solution to the problem

$$\frac{1}{2}\Delta u + 2(d+2)|x|^2 = 0, \qquad u\,|_{|x|=r} = 0.$$

Therefore (see (7.4.11))

$$u(0) = r^4 = 2(d+2)\,E \int_0^\vartheta \sum_{k=1}^d (w^k(s))^2 ds = 2d(d+2)\,E \int_0^\vartheta (w^i(s))^2 ds \quad (7.4.21)$$

that gives (7.4.19). $\square$

Thus, we have shown that the assertion (7.4.16) is true. The following theorems are proved in [281].

Theorem 7.4.2 *For every $i = 1, \ldots, d$ the equality*

$$E\left(\int_0^\vartheta w^i(s)ds / w(\vartheta)\right) = \frac{r^2}{2(d+2)} w^i(\vartheta) \qquad (7.4.22)$$

holds.

Theorem 7.4.3 *For every $i, j = 1, \ldots, d$, the following formulas hold:*

$$E\left[\int_0^\vartheta w^i(s)dw^i(s)/w(\vartheta)\right] = \frac{1}{2}[w^i(\vartheta)]^2 - \frac{r^2}{2d}, \qquad (7.4.23)$$

$$E\left[\int_0^\vartheta w^i(s)dw^j(s)/w(\vartheta)\right] = \frac{1}{2}w^i(\vartheta)w^j(\vartheta), \; i \neq j. \qquad (7.4.24)$$

Introduce the functions

$$h_m(x) = E\vartheta_x^m, \quad m = 1, 2, \ldots,$$

where $x \in U_r$ and ϑ_x is the first passage time of the process $x + w(t)$ to the sphere ∂U_r.

As it follows from one of Dynkin's theorems (see [86, Theorem 13.17]), the function $h_m(x)$ is the only solution of the Dirichlet problem

$$\frac{1}{2}\Delta h_1 + 1 = 0, \quad h_1 \mid_{\partial U_r} = 0, \tag{7.4.25}$$

$$\frac{1}{2}\Delta h_m + m h_{m-1}(x) = 0, \quad h_m \mid_{\partial U_r} = 0, \quad m = 2, 3, \ldots.$$

The solution of this problem is obviously a function of the variable $\chi = (x, x)^{1/2} = |x|, \ 0 \le \chi \le r$. We denote this function as $q_m(\chi)$. We easily obtain the following boundary value problem for $d > 1$ (we recall that d is a dimension of the Wiener process $w(t)$):

$$\frac{1}{2}q_1'' + \frac{d-1}{2\chi}q_1' + 1 = 0, \quad q_1(0) < \infty, \quad q_1(r) = 0, \tag{7.4.26}$$

$$\frac{1}{2}q_m'' + \frac{d-1}{2\chi}q_m' + m q_{m-1}(\chi) = 0, \quad q_m(0) < \infty, \quad q_m(r) = 0.$$

We note that if $d = 1$ then (7.4.25) can be rewritten in the form

$$\frac{1}{2}h_m'' + m h_{m-1}(x) = 0, \quad h_m(-r) = h_m(r) = 0.$$

The equations (7.4.26) are solvable by quadratures. Their solution has the form

$$q_m(\chi) = \alpha_0 \chi^{2m} + \alpha_1 \chi^{2(m-1)}r^2 + \alpha_2 \chi^{2(m-2)}r^4 + \cdots + \alpha_m r^{2m}.$$

Then, we can sequentially obtain:

$$h_1(x) = \frac{r^2 - |x|^2}{d},$$

$$h_2(x) = \frac{|x|^4}{d(d+2)} - \frac{2r^2|x|^2}{d^2} + \frac{(d+4)r^4}{d^2(d+2)},$$

and so on. In particular

$$E\vartheta = \frac{r^2}{d}, \quad E\vartheta^2 = \frac{d+4}{d^2(d+2)}r^4, \quad D\vartheta = \frac{2}{d^2(d+2)}r^4. \tag{7.4.27}$$

But such formulas become complicated with growth of m. For example,

$$E\vartheta^3 = \frac{d^2 + 12d + 48}{d^3(d+2)(d+4)}r^6.$$

Therefore, it is useful to obtain some simple bounds for $h_m(x)$ [281].

Lemma 7.4.4 *The functions $h_m(x)$ are bounded as*

$$\frac{1}{d^m}(r^2 - |x|^2)^m \le h_m(x) \le \frac{m!}{d^m}r^{2m-2}(r^2 - |x|^2), \quad m = 1, 2, \ldots. \qquad (7.4.28)$$

Consequently,

$$\frac{1}{d^m}r^{2m} \le E\vartheta^m \le \frac{m!}{d^m}r^{2m} \qquad (7.4.29)$$

and for $\lambda < d/r^2$

$$E\exp(\lambda\vartheta) \le \frac{d}{d - \lambda r^2}. \qquad (7.4.30)$$

7.4.2   Specific One-Step Approximations for Elliptic Equations

In this subsection we construct some one-step approximations of the solution of the system (7.4.3)–(7.4.5).

For definiteness, let $X_0 = x \in G \backslash \Gamma_{ar}$. Then, the next point X_1 has the uniform distribution on the sphere of radius ar with center at X_0:

$$X_1 = x + aw(\vartheta), \qquad (7.4.31)$$

where ϑ is the first passage time of the solution $X_x(t)$ of the equation (7.4.3) to the sphere $\partial U_r(x)$ of radius r with center at x.

Consider the solution $X_x(t)$, $Y_{x,y}(t)$, $Z_{x,y,z}(t)$ of the system (7.4.3)–(7.4.5) at the time ϑ: $X_x(\vartheta)$, $Y_{x,y}(\vartheta)$, $Z_{x,y,z}(\vartheta)$. As we have just seen, $X_1 = X_x(\vartheta)$ can be simulated exactly. Our aim is to construct an approximation Y_1, Z_1 of $Y_{x,y}(\vartheta)$, $Z_{x,y,z}(\vartheta)$ so that the difference

$$\begin{aligned}
d &= E(v(X_1, Y_1, Z_1) - v(X_x(\vartheta), Y_{x,y}(\vartheta), Z_{x,y,z}(\vartheta))) \qquad (7.4.32) \\
&= E(u(X_1)Y_1 + Z_1 - u(X_x(\vartheta))Y_{x,y}(\vartheta) - Z_{x,y,z}(\vartheta)) \\
&= Eu(X_x(\vartheta))(Y_1 - Y_{x,y}(\vartheta)) + E(Z_1 - Z_{x,y,z}(\vartheta))
\end{aligned}$$

is small.

Repeatedly applying Ito's formula like the Wagner-Platen expansion (see Sect. 1.2.2), we obtain

$$Y_{x,y}(\vartheta) = y + \frac{1}{a} y \sum_{i=1}^{d} b^i(x) w^i(\vartheta) + c(x) y \vartheta \tag{7.4.33}$$

$$+ \frac{1}{a^2} y \sum_{i=1}^{d} \sum_{j=1}^{d} b^i(x) b^j(x) \int_{0}^{\vartheta} w^j(t) \, dw^i(t)$$

$$+ y \sum_{i=1}^{d} \sum_{j=1}^{d} \frac{\partial b^i}{\partial x^j}(x) \int_{0}^{\vartheta} w^j(t) \, dw^i(t) + \rho_{11} + \rho_{12} + \rho_{13},$$

where

$$\rho_{11} = a \sum_{i=1}^{d} \int_{0}^{\vartheta} \int_{0}^{t} \frac{\partial c}{\partial x^i} (X_x(s)) \, Y_{x,y}(s) \, dw^i(s) \, dt \tag{7.4.34}$$

$$+ \frac{1}{a} \sum_{i=1}^{d} \int_{0}^{\vartheta} \int_{0}^{t} c(X_x(s)) b^i(X_x(s)) \, Y_{x,y}(s) \, dw^i(s) \, dt$$

$$+ \frac{1}{a} \sum_{i=1}^{d} \int_{0}^{\vartheta} \int_{0}^{t} c(X_x(s)) b^i(X_x(s)) \, Y_{x,y}(s) \, ds \, dw^i(t)$$

$$+ \sum_{i=1}^{d} \sum_{j=1}^{d} \int_{0}^{\vartheta} \int_{0}^{t} \left(\frac{a}{2} \frac{\partial^2 b^i}{(\partial x^j)^2}(X_x(s)) + \frac{1}{a} \frac{\partial b^i}{\partial x^j}(X_x(s)) \, b^j(X_x(s)) \right)$$

$$\times Y_{x,y}(s) \, ds \, dw^i(t),$$

$$\rho_{12} = \int_{0}^{\vartheta} \int_{0}^{t} \left(c^2(X_x(s)) + \frac{a^2}{2} \sum_{i=1}^{d} \frac{\partial^2 c}{(\partial x^i)^2}(X_x(s)) \right) Y_{x,y}(s) \, ds \, dt \tag{7.4.35}$$

$$+ \sum_{i=1}^{d} \int_{0}^{\vartheta} \int_{0}^{t} \frac{\partial c}{\partial x^i}(X_x(s)) \, b^i(X_x(s)) \, Y_{x,y}(s) \, ds \, dt,$$

and ρ_{13} contains a sum of integrals like

$$I_{i_1,i_2,i_3} = \int_{0}^{\vartheta} \int_{0}^{t} \int_{0}^{s} f^{i_1 i_2 i_3}(X_x(s_1)) Y_{x,y}(s_1) \, dw^{i_1}(s_1) \, dw^{i_2}(s) \, dw^{i_3}(t) \tag{7.4.36}$$

$$i_1 = 0, \ldots, n, \quad i_2 \neq 0, \quad i_3 \neq 0,$$

where $f^{i_1 i_2 i_3}$ is a finite sum of products and each product does not have more than three factors of the form b^i, $\partial b^i/\partial x^j$, $\partial^2 b^i/\partial x^j \partial x^k$, $\partial^3 b^i/\partial x^j \partial x^k \partial x^l$, and c. We underline that $\rho_{13} = 0$ if $b = 0$.

We have for Z:

$$Z_{x,y,z}(\vartheta) = z + g(x)\, y\, \vartheta + \rho_{21} + \rho_{22}, \tag{7.4.37}$$

where

$$\rho_{21} = a \sum_{i=1}^{d} \int_0^\vartheta \int_0^t \frac{\partial g}{\partial x^i}(X_x(s)) Y_{x,y}(s)\, dw^i(s)\, dt \tag{7.4.38}$$

$$+ \frac{1}{a} \sum_{i=1}^{d} \int_0^\vartheta \int_0^t g(X_x(s))\, b^i(X_x(s))\, Y_{x,y}(s)\, dw^i(s)\, dt$$

and

$$\rho_{22} = \sum_{i=1}^{d} \int_0^\vartheta \int_0^t \frac{\partial g}{\partial x^i}(X_x(s))\, b^i(X_x(s))\, Y_{x,y}(s)\, ds\, dt \tag{7.4.39}$$

$$+ \int_0^\vartheta \int_0^t \left(g(X_x(s))c(X_x(s)) + \frac{a^2}{2} \sum_{i=1}^{d} \frac{\partial^2 g}{(\partial x^i)^2}(X_x(s)) \right) Y_{x,y}(s)\, ds\, dt.$$

Let us put

$$Y_1 = y + \frac{1}{a} y \sum_{i=1}^{d} b^i(x) w^i(\vartheta) + c(x) y \frac{r^2}{d} \tag{7.4.40}$$

$$+ \frac{1}{2a^2} y \sum_{i=1}^{d} \sum_{j=1}^{d} b^i(x) b^j(x) w^i(\vartheta) w^j(\vartheta)$$

$$- \frac{1}{2a^2} y \frac{r^2}{d} \sum_{i=1}^{d} [b^i(x)]^2 + \frac{1}{2} y \sum_{i=1}^{d} \sum_{j=1}^{d} \frac{\partial b^i}{\partial x^j}(x)\, w^i(\vartheta)\, w^j(\vartheta)$$

$$- \frac{1}{2} y \frac{r^2}{d} \sum_{i=1}^{d} \frac{\partial b^i}{\partial x^i}(x),$$

$$Z_1 = z + g(x)\, y\, \frac{r^2}{d}. \tag{7.4.41}$$

We note that $Y_1 > 0$ for a sufficiently small r since we supposed that $y > 0$.

We have

$$d = E\left[u(X_x(\vartheta))E(Y_1 - Y_{x,y}(\vartheta) \mid w(\vartheta))\right] + E(Z_1 - Z_{x,y,z}(\vartheta)). \qquad (7.4.42)$$

Then, due to the relation $E(\vartheta \mid w(\vartheta)) = E\vartheta = r^2/d$, Theorem 7.4.3, and the formulas (7.4.33) and (7.4.37), we have

$$d = -E[u(X_x(\vartheta))E(\rho_{11} + \rho_{12} + \rho_{13} \mid w(\vartheta))] - E(\rho_{21} + \rho_{22}) \quad (7.4.43)$$
$$= -E[u(X_x(\vartheta))(\rho_{11} + \rho_{12} + \rho_{13})] - E(\rho_{21} + \rho_{22}).$$

Let

$$M_0(x) = \max_{\xi \in \bar{U}(x)} |u(\xi)|, \quad M_1(x) = \max_{\xi \in \bar{U}(x),\, 1 \le i \le d} |\frac{\partial u}{\partial x^i}(\xi)|.$$

By a thorough analysis of the one-step error (7.4.43), the following theorem is proved in [281].

Theorem 7.4.5 *The one-step error $d = d(x, y, r)$ of the approximation (7.4.31), (7.4.40)–(7.4.41) is of the form*

$$|d| \le (K_0 M_0(x) + K_1 M_1(x) + K_2)\, yr^4, \qquad (7.4.44)$$

where K_0, K_1, K_2 are constants depending on a, b, c, and g only. That is, the degree of smallness of this approximation with respect to r is equal to 4.

Let us consider now the simpler approximation

$$Y_1 = y + \frac{1}{a} y \sum_{i=1}^{d} b^i(x) w^i(\vartheta) + c(x)\, y\, \frac{r^2}{d}, \qquad (7.4.45)$$

$$Z_1 = z + g(x)\, y\, \frac{r^2}{d}. \qquad (7.4.46)$$

Introduce

$$M_2(x) := \max_{\xi \in \bar{U}(x),\, 1 \le i,j \le d} \left| \frac{\partial^2 u}{\partial x^i \partial x^j}(\xi) \right|.$$

We have the following result [281].

Theorem 7.4.6 *The one-step error $d = d(x, y, r)$ of the approximation (7.4.31), (7.4.45)–(7.4.46) has the form*

$$|d| \le (K_0 M_0(x) + K_1 M_1(x) + K_2 M_2(x) + K_3)\, y\, r^4. \qquad (7.4.47)$$

Remark 7.4.7 The degree of smallness of both approximations (7.4.40)–(7.4.41) and (7.4.45)–(7.4.46) is equal to 4. But the estimate (7.4.44) does not depend on second derivatives of the function u.

Now consider the Dirichlet problem for the nonhomogeneous Helmholtz equation

$$\frac{1}{2}a^2 \Delta u + c(x)u + g(x) = 0, \quad x \in G, \tag{7.4.48}$$

$$u \mid_{\partial G} = \varphi(x), \tag{7.4.49}$$

i.e., when $b^i(x) = 0$, $i = 1, \ldots, d$, in (7.4.1).

In this case we construct two one-step approximation given in the next theorems (see their proofs in [281]).

Theorem 7.4.8 *Consider the one-step approximation*

$$X_1 = x + aw(\vartheta), \tag{7.4.50}$$

$$Y_1 = y + c(x) y \frac{r^2}{d} + a \sum_{i=1}^{d} \frac{\partial c}{\partial x^i}(x) y \frac{r^2}{2(d+2)} w^i(\vartheta) \tag{7.4.51}$$

$$+ \frac{1}{2}c_1(x) y \frac{4+d}{d^2(2+d)} r^4,$$

$$Z_1 = z + g(x) y \frac{r^2}{d} + \frac{1}{2}g_1(x) y \frac{4+d}{d^2(2+d)} r^4, \tag{7.4.52}$$

where

$$c_1(x) = c^2(x) + \frac{a^2}{2} \sum_{i=1}^{d} \frac{\partial^2 c}{(\partial x^i)^2}(x),$$

$$g_1(x) = g(x) c(x) + \frac{a^2}{2} \sum_{i=1}^{d} \frac{\partial^2 g}{(\partial x^i)^2}(x).$$

The one-step error $d = d(x, y, r)$ of this approximation has the form

$$|d| \leq (K_0 M_0(x) + a^2 K_1 M_1(x) + a^4 K_2 M_2(x) + K_3) y r^6, \tag{7.4.53}$$

where K_0, K_1, K_2, K_3 depend on c and g only.

Theorem 7.4.9 *The one-step error $d = d(x, y, r)$ of the approximation*

$$X_1 = x + aw(\vartheta), \quad Y_1 = y + c(x) y\frac{r^2}{d}, \quad Z_1 = z + g(x) y\frac{r^2}{d} \tag{7.4.54}$$

has the form

$$|d| \leq [K_0 M_0(x) + a^2 K_1 M_1(x) + K_2] y r^4. \tag{7.4.55}$$

7.4.3 The Average Number of Steps

In this subsection, we consider the question about average characteristics of the number of steps $\varkappa$ of the homogeneous Markov chain X_k (see the beginning of the current section) before it gets into a neighborhood of the boundary ∂G. Then, in the next subsection, we construct a number of algorithms for the Dirichlet problem which are based on the one-step approximations proposed in Sect. 7.4.2. Finally, we prove some convergence theorems using the results of Sects. 7.4.2 and 7.4.3.

In connection with the homogeneous Markov chain X_k from the introduction to Sect. 7.4, we introduce the one-step transition function (cf. Sect. 6.1.2):

$$P(x, B) = P(X_1 \in B \mid X_0 = x),$$

where B is a Borel set belonging to $\bar{G}$. If $x \in G \backslash \Gamma_{ar}$ then $P(x, B)$ is concentrated on the sphere $\partial U_{ar}(x)$ of radius ar with center at x. If $x \in \Gamma_{ar} \backslash \Gamma_\delta$ then $P(x, B)$ is concentrated on the sphere $\partial U_\rho(x)$ of radius $\rho(x, \partial G)$ with center at x. And if $x \in \Gamma_\delta$ then $P(x, B)$ is concentrated at the point x. Define an operator P acting on functions $v(x)$, $x \in \bar{G}$, by the formula

$$Pv(x) = \int_{\bar{G}} P(x, dy)v(y) = Ev(X_1), \quad X_0 = x .$$

Consider the boundary value problem in $\bar{G}$ (cf. (6.1.13)–(6.1.14) and (7.3.14)–(7.3.15)):

$$Pv(x) - v(x) = -f(x), \quad x \in G \backslash \Gamma_\delta , \tag{7.4.56}$$

$$v(x) = 0, \quad x \in \Gamma_\delta , \tag{7.4.57}$$

which is connected with the chain X_k.

In (7.4.56), $f(x)$ is a continuous function defined on the compact set $G \backslash \Gamma_\delta$: $f \in C(G \backslash \Gamma_\delta)$. It is not difficult to prove that there exists a unique solution of the problem (7.4.56)–(7.4.57) which is a continuous function on $G \backslash \Gamma_\delta$. This solution is known to be the following function [450] (cf. (6.1.15)):

$$v(x) = E \sum_{k=0}^{\varkappa_x - 1} f(X_k), \quad X_0 = x , \tag{7.4.58}$$

where $\varkappa_x$ relates to the chain starting at x. If $v(x)$ is the solution of the boundary value problem (7.4.56)–(7.4.57) with a function $f(x)$ such that $f(x) \geq 1$ for $x \in G \backslash \Gamma_\delta$, then, thanks to (7.4.58), we have

$$E\varkappa_x \leq v(x). \tag{7.4.59}$$

Consider the function (cf. Remark 6.1.5):

$$V_1(x) = \begin{cases} A^2 + (\alpha, x) - x^2 \, , \, x \in G \backslash \Gamma_\delta \, , \\ 0 \, , \qquad\qquad\qquad x \in \Gamma_\delta \, , \end{cases}$$

where the constant A^2 and the vector α are such that the inequality $A^2 + (\alpha, x) - x^2 \geq 0$ is valid for all $x \in \bar{G}$. The next lemma is proved by arguments similar to those used in the proof of Lemma 6.1.4 and in Remark 6.1.5 (see further details in [281]).

Lemma 7.4.10 *The inequalities*

$$P V_1(x) - V_1(x) \leq -a^2 r^2 \, , \quad x \in G \backslash \Gamma_{\alpha r} \, , \tag{7.4.60}$$

$$P V_1(x) - V_1(x) \leq 0 \, , \quad x \in \Gamma_{\alpha r} \backslash \Gamma_\delta \, , \tag{7.4.61}$$

hold.

We also introduce the function

$$V_2(x) = \begin{cases} \ln(\alpha r / \delta) + 1 \, , & x \in G \backslash \Gamma_{\alpha r} \, , \\ \ln(\rho(x) / \delta) + 1 \, , & x \in \Gamma_{\alpha r} \backslash \Gamma_\delta \, , \\ 0 \, , & x \in \Gamma_\delta \, , \end{cases}$$

where $\rho(x) = \rho(x, \partial G)$.

Lemma 7.4.11 *(see [281]). For a sufficiently small $r > 0$ the following inequalities hold:*

$$P V_2(x) - V_2(x) \leq 0 \, , \qquad x \in G \backslash \Gamma_{\alpha r} \, , \tag{7.4.62}$$

$$P V_2(x) - V_2(x) \leq -C_d \, , \quad x \in \Gamma_{\alpha r} \backslash \Gamma_\delta \, , \tag{7.4.63}$$

where C_d does not depend on x.

If the domain G is convex, the assumption on smallness of r can be omitted.

Theorem 7.4.12 *There exist constants B and C such that for any x and sufficiently small $r > 0$:*

$$E \varkappa_x \leq \frac{B}{a^2 r^2} + C \ln \frac{\alpha r}{\delta} \, . \tag{7.4.64}$$

If $\delta = O(r^p)$, $p > 1$, then

$$E \varkappa_x \leq \frac{B + 1}{a^2 r^2} \, . \tag{7.4.65}$$

If G is convex and $r \geq 1/2$ where l is a diameter of G, then the Markov chain becomes the random walk over touching spheres and the average number of steps is estimated as

$$E \varkappa_x \leq C \ln \frac{l}{2\delta} \ . \tag{7.4.66}$$

Proof The inequalities (7.4.64)–(7.4.66) follow from Lemma 7.4.10, Lemma 7.4.11, and (7.4.59) if we take $v(x)$ of the form:

$$v(x) = \frac{V_1(x)}{a^2 r^2} + \frac{V_2(x)}{C_d} \ . \tag{7.4.67}$$

$\square$

Remark 7.4.13 We emphasize that the value of p does not play any essential role in the upper bound of the average number of steps $E\varkappa$.

Lemma 7.4.14 *Let* $q(x) > 0$, $q \in C(G \backslash \Gamma_\delta)$, $f(x) \geq 0$, $f \in C(G \backslash \Gamma_\delta)$, $f(x) = 0$ *for* $x \in \Gamma_\delta$. *Let* $z(x)$ *be a solution to the boundary value problem*

$$q(x) P z(x) - z(x) = -f(x) \ , \quad x \in G \backslash \Gamma_\delta \ , \tag{7.4.68}$$

$$z(x) = 0 \ , \quad x \in \Gamma_\delta \ . \tag{7.4.69}$$

Then for $x \in G \backslash \Gamma_\delta$

$$z(x) = f(x) + E \sum_{k=1}^{\varkappa_x - 1} f(X_k) \Pi_{i=0}^{k-1} q(X_i) \ . \tag{7.4.70}$$

Proof We have for $x \in G \backslash \Gamma_\delta$:

$$
\begin{aligned}
z(x) &= f(x) + q(x) P z(x) = f(x) + q(x) E z(X_1) \\
&= f(x) + q(x) E(f(X_1) + q(X_1) P z(X_1)) \\
&= f(x) + q(x) E(\chi_{\varkappa_x > 1} f(X_1)) + q(x) E(\chi_{\varkappa_x > 1} q(X_1) E(z(X_2)/X_1)) \\
&= f(x) + q(x) E(\chi_{\varkappa_x > 1} f(X_1)) + q(x) E(\chi_{\varkappa_x > 2} q(X_1) z(X_2)) \\
&= f(x) + q(x) E(\chi_{\varkappa_x > 1} f(X_1)) + q(x) E(\chi_{\varkappa_x > 2} q(X_1) f(X_2)) \\
&\quad + q(x) E(\chi_{\varkappa_x > 3} q(X_1) q(X_2) z(X_3)) \\
&= \cdots = f(x) + q(x) E(\chi_{\varkappa_x > 1} f(X_1)) \\
&\quad + \cdots + q(x) E(\chi_{\varkappa_x > N} q(X_1) \cdots q(X_{N-1}) f(X_N)) \\
&\quad + q(x) E(\chi_{\varkappa_x > N+1} q(X_1) \cdots q(X_N) z(X_{N+1})) \ .
\end{aligned}
$$

Now we tend N to infinity and obtain (7.4.70). $\square$

Corollary 7.4.15 *Let the conditions of Lemma 7.4.14 be satisfied. If* $q = const > 1$ *and* $f(x) = 1$ *for* $x \in G \backslash \Gamma_\delta$ *then*

$$z(x) = E(1 + q + q^2 + \cdots + q^{\varkappa_x - 1}) = \frac{1}{q-1}(E q^{\varkappa_x} - 1) \ .$$

If $q = const > 1$ and $f(x) \geq c$ for $x \in G\backslash\Gamma_\delta$ then

$$Eq^{\varkappa_x} < \infty, \quad E(1 + q + \cdots + q^{\varkappa_x - 1}) \leq \frac{1}{c}z(x) .$$

Lemma 7.4.16 *Let $\delta = O(r^p)$, $p > 1$. Then there exist constants $\beta > 0$ and $K > 0$ such that for all sufficiently small r*

$$E \sum_{k=0}^{\varkappa_x - 1} (1 + \beta r^2)^k = O\left(\frac{1}{r^2}\right) , \qquad (7.4.71)$$

$$E(1 + \beta r^2)^{\varkappa_x} < K , \qquad (7.4.72)$$

$$P(\varkappa_x \geq k) \leq K(1 - \beta r^2)^k . \qquad (7.4.73)$$

Proof For the function $v(x)$ from (7.4.67) we have

$$(1 + \beta r^2)Pv - (1 + \beta r^2)v \leq -(1 + \beta r^2), \quad x \in G\backslash\Gamma_\delta ,$$

or

$$(1 + \beta r^2)Pv - v \leq \beta r^2 v - (1 + \beta r^2), \quad x \in G\backslash\Gamma_\delta ,$$

$$v = 0, \quad x \in \Gamma_\delta .$$

Thus, the function $v(x)$ is a solution to the problem (7.4.68)–(7.4.69) with $q(x) = 1 + \beta r^2$ and with $f(x)$ which satisfies the inequality

$$f(x) \geq 1 + \beta r^2 - \beta r^2 v = 1 + \beta r^2 - \frac{\beta}{a^2} V_1(x) - \frac{\beta r^2}{C_d} V_2(x) .$$

Clearly, $f(x) \geq 1/2$ for sufficiently small β and r. Using Corollary 7.4.15, we obtain

$$E \sum_{k=0}^{\varkappa_x - 1} (1 + \beta r^2)^k \leq 2v(x)$$

and, consequently, (7.4.71) is proved.

The relation (7.4.71) implies (7.4.72). The relation (7.4.73) is obtained from (7.4.72) by the Chebyshev inequality.   □

7.4.4 Numerical Algorithms and Convergence Theorems

Here we construct a number of algorithms for the Dirichlet problem (7.4.1), which are based on the one-step approximations proposed in Sect. 7.4.2, and we prove their convergence.

We assume that the domain G, the coefficients $b^i(x)$, $c(x)$, $g(x)$, and the function $\varphi(x)$ in (7.4.1) satisfy the conditions (i)–(ii) (see them at the beginning of the current section). We recall that Γ_δ is the interior of a δ-neighborhood of the boundary ∂G belonging to $\bar{G}$. Let $U_1 \in \mathbf{R}^d$ be an open unit ball with center at the origin and with the boundary ∂U_1. Let ξ be a point uniformly distributed on the sphere ∂U_1 and ξ_1, $\xi_2, \ldots$ be independent random points distributed as ξ (see e.g. [253, 332, 436] for their modeling).

Using the one-step approximation (7.4.45)–(7.4.46), we construct the following algorithm.

Algorithm 7.4.17 *If $X_k \in G \setminus \Gamma_{ar}$, we set $r_{k+1} = r$. If $X_k \in \Gamma_{ar} \setminus \Gamma_{r^2}$, we set $r_{k+1} = \rho(X_k, \partial G)/a$. And in both cases*

$$X_{k+1} = X_k + a r_{k+1} \xi_{k+1}, \quad X_0 = x, \tag{7.4.74}$$

$$Y_{k+1} = Y_k [1 + \frac{r_{k+1}}{a} \sum_{i=1}^{d} b^i(X_k)\xi_{k+1}^i + c(X_k)\frac{r_{k+1}^2}{d}], \quad Y_0 = 1, \tag{7.4.75}$$

$$Z_{k+1} = Z_k + Y_k \, g(X_k)\frac{r_{k+1}^2}{d}, \quad Z_0 = 0. \tag{7.4.76}$$

Let $\varkappa = \varkappa_x$ be the first number at which $X_\varkappa \in \Gamma_{r^2}$. Then we set $X_k = X_\varkappa$ for $k \geq \varkappa$, i.e., our algorithm is stopped at the random step $\varkappa$. Having obtained $X_\varkappa$, we find the point $\bar{X}_\varkappa \in \partial G$ which is the closest to $X_\varkappa$. We take $(\bar{X}_\varkappa, Y_\varkappa, Z_\varkappa)$ as the final state of the Markov chain.

Using the obtained $\bar{X}_\varkappa, Y_\varkappa, Z_\varkappa$, we evaluate

$$v(\bar{X}_\varkappa, Y_\varkappa, Z_\varkappa) = u(\bar{X}_\varkappa)Y_\varkappa + Z_\varkappa = \varphi(\bar{X}_\varkappa)Y_\varkappa + Z_\varkappa.$$

The solution of the problem (7.4.1) is approximately equal to

$$u(x) \approx E\left(\varphi(\bar{X}_\varkappa)Y_\varkappa + Z_\varkappa\right) \approx \frac{1}{M} \sum_{m=1}^{M} \left(\varphi(\bar{X}_\varkappa^{(m)})Y_\varkappa^{(m)} + Z_\varkappa^{(m)}\right), \tag{7.4.77}$$

where $\bar{X}_\varkappa^{(m)}, Y_\varkappa^{(m)}, Z_\varkappa^{(m)}$, $m = 1, \ldots, M$, are independent realizations of the algorithm (7.4.74)–(7.4.76). The first approximate equality in (7.4.77) involves an error due to replacing $X_x(\tau), Y_{x,1}(\tau), Z_{x,1,0}(\tau)$ by $\bar{X}_\varkappa, Y_\varkappa, Z_\varkappa$; in the second approximate equality the error comes from the Monte Carlo method. The first error is estimated by $O(r^2)$ (see Theorem 7.4.20 below) and the second one is $O(1/\sqrt{M})$.

The algorithm based on the one-step approximation (7.4.40)–(7.4.41) takes the form.

Algorithm 7.4.18 *Define r_{k+1} as in Algorithm 7.4.17. Then*

$$X_{k+1} = X_k + ar_{k+1}\xi_{k+1}, \quad X_0 = x, \tag{7.4.78}$$

$$Y_{k+1} = Y_k[1 + \frac{r_{k+1}}{a} \sum_{i=1}^{d} b^i(X_k)\xi_{k+1}^i + c(X_k)\frac{r_{k+1}^2}{d} \tag{7.4.79}$$

$$+ \gamma(X_k, r_{k+1}, \xi_{k+1})], \quad Y_0 = 1,$$

$$Z_{k+1} = Z_k + Y_k \, g(X_k)\frac{r_{k+1}^2}{d}, \quad Z_0 = 0, \tag{7.4.80}$$

where

$$\gamma(x, r, \xi) = \frac{r^2}{2a^2} \sum_{i=1}^{d} \sum_{j=1}^{d} b^i(x)b^j(x)\xi^i\xi^j - \frac{r^2}{2a^2 d} \sum_{i=1}^{d} [b^i(x)]^2$$

$$+ \frac{r^2}{2} \sum_{i=1}^{d} \sum_{j=1}^{d} \frac{\partial b^i}{\partial x^j}(x)\xi^i\xi^j - \frac{r^2}{2d} \sum_{i=1}^{d} \frac{\partial b^i}{\partial x^i}(x).$$

The algorithm is stopped at a random step $\varkappa$ when $X_\varkappa \in \Gamma_{r^2}$. Having obtained $X_\varkappa$, we find the point $\bar{X}_\varkappa \in \partial G$ which is the closest to $X_\varkappa$. We take $(\bar{X}_\varkappa, Y_\varkappa, Z_\varkappa)$ as the final state of the Markov chain.

Now consider the Dirichlet problem for the nonhomogeneous Helmholtz equation (7.4.48)–(7.4.49). In this case we can also suggest two algorithms. One of them is based on the one-step approximation (7.4.54). Here we write down the other one which is based on the one-step approximation (7.4.51)–(7.4.52).

Algorithm 7.4.19 *We set $r_{k+1} = r$ if $X_k \in G\backslash\Gamma_{ar}$ and $r_{k+1} = \rho(X_k, \partial G)/a$ if $X_k \in \Gamma_{ar}\backslash\Gamma_{r^4}$. Then*

$$X_{k+1} = X_k + ar_{k+1}\xi_{k+1}, \quad X_0 = x, \tag{7.4.81}$$

$$Y_{k+1} = Y_k [1 + c(X_k)\frac{r_{k+1}^2}{d} + \frac{a \, r_{k+1}^3}{2(d+2)} \sum_{i=1}^{d} \frac{\partial c}{\partial x^i}(X_k) \, \xi_{k+1}^i \tag{7.4.82}$$

$$+ \frac{4+d}{2d^2(2+d)} c_1(X_k) \, r_{k+1}^4], \quad Y_0 = 1,$$

$$Z_{k+1} = Z_k + Y_k[g(X_k)\frac{r_{k+1}^2}{d} + \frac{4+d}{2d^2(2+d)} \, g_1(X_k) \, r_{k+1}^4], \quad Z_0 = 0. \tag{7.4.83}$$

The algorithm is stopped at a random step $\varkappa$ when $X_\varkappa \in \Gamma_{r^4}$. Having obtained $X_\varkappa$, we find the point $\bar{X}_\varkappa \in \partial G$ which is the closest to $X_\varkappa$. We take $(\bar{X}_\varkappa, Y_\varkappa, Z_\varkappa)$ as the final state of the Markov chain.

We note that by Theorem 7.4.12 the average number of steps for all the algorithms presented here is $O(1/r^2)$.

Proceeding to convergence theorems, let us use the relation (7.4.8):

$$|E(\varphi(\bar{X}_\varkappa)Y_\varkappa - Z_\varkappa) - u(x)| = |\mathcal{R}| \le |E(u(\bar{X}_\varkappa) - u(X_\varkappa))Y_\varkappa| + \sum_{k=1}^\infty |\tilde{d}_k| ,$$

where

$$\tilde{d}_k = E\chi_{\varkappa > k-1}(v(X_k, Y_k, Z_k) - v(X_{k-1}, Y_{k-1}, Z_{k-1})) .$$

For brevity, we write $\varkappa$ instead of $\varkappa_x$ everywhere.

Clearly,

$$u(X_{k-1})Y_{k-1} - Z_{k-1} = v(X_{k-1}, Y_{k-1}, Z_{k-1})$$
$$= E(v(X_{X_{k-1}}(\vartheta_k), Y_{X_{k-1}, Y_{k-1}}(\vartheta_k), Z_{X_{k-1}, Y_{k-1}, Z_{k-1}}(\vartheta_k))/\mathcal{B}_{k-1}) .$$

Therefore

$$\tilde{d}_k = E\chi_{\varkappa > k-1}d_k ,$$

where

$$d_k = E[v(X_k, Y_k, Z_k) \tag{7.4.84}$$
$$-v(X_{X_{k-1}}(\vartheta_k), Y_{X_{k-1}, Y_{k-1}}(\vartheta_k), Z_{X_{k-1}, Y_{k-1}, Z_{k-1}}(\vartheta_k))/\mathcal{B}_{k-1}]$$

is a one-step error at the point $(X_{k-1}, Y_{k-1}, Z_{k-1})$. Thus

$$|\mathcal{R}| \le |E(u(\bar{X}_\varkappa) - u(X_\varkappa))Y_\varkappa| + \sum_{k=1}^\infty |E\chi_{\varkappa > k-1} d_k| . \tag{7.4.85}$$

The following convergence theorems were proved in [281].

Theorem 7.4.20 *Let $c(x) \le -c_0 < 0$. Then Algorithms 7.4.17 and 7.4.18 have the second order of convergence with respect to r, i.e., for all sufficiently small r*

$$|E(\varphi(\bar{X}_\varkappa)Y_\varkappa - Z_\varkappa) - u(x)| \le Kr^2 . \tag{7.4.86}$$

In addition, the constant K for Algorithm 7.4.17 depends on first and second derivatives of the required solution $u(x)$ while this constant for Algorithm 7.4.18 depends on first derivatives only.

Theorem 7.4.21 *Let $c(x) \le 0$. Then Algorithm 7.4.19 has the fourth order of convergence with respect to r:*

$$|E(\varphi(\bar{X}_{\varkappa})Y_{\varkappa} - Z_{\varkappa}) - u(x)| \le Kr^4 . \tag{7.4.87}$$

The constant K depends on first and second derivatives of $u(x)$.

Remark 7.4.22 The simpler algorithm based on the one-step approximation (7.4.54) has the second order of convergence with respect to r and the constant K depends on first derivatives of $u(x)$ only.

Remark 7.4.23 In this section we have proposed a number of algorithms for (7.4.1), i.e., for (7.3.1) with $a^{ij}(x) = a^2 \delta_{ij}$, where δ_{ij} is the Kronecker delta. All the results obtained here can be carried over to the case of elliptic equation (7.3.1) with constant coefficients a^{ij}.

7.5   Methods for Elliptic Equations with Small Parameter at Higher Derivatives

Proceeding to numerical investigation of a boundary layer, let us consider the model problem for the nonhomogeneous Helmholtz equation :

$$\frac{1}{2}\varepsilon^2 \Delta u + c(x)u = g(x), \quad x \in U_R , \tag{7.5.1}$$

$$u \mid_{\partial U_R} = 0 , \tag{7.5.2}$$

where $\varepsilon \ll 1$, $U_R \subset \mathbf{R}^d$ is an open ball of radius R with center at the origin, $c(x)$ and $g(x)$ belong to $C^\infty(\bar{U}_R)$, and $c(x) \le -c_0 < 0$, $x \in \bar{U}_R$.

    A solution $u(x, \varepsilon)$ of this problem varies slowly everywhere in U_R except a small neighborhood of ∂U_R, which is called boundary layer and which decreases in size with decrease of ε. The solution $u(x, \varepsilon)$ varies sharply in the boundary layer. It is well known (see [178] and references therein) that the width of the boundary layer for the problem (7.5.1)–(7.5.2) is proportional to ε, i.e., the boundary layer has the form $\Gamma_{l\varepsilon}$ (l is a positive number). Moreover, it is known that

$$|u(x, \varepsilon)| \le K, \quad |\frac{\partial u}{\partial x^i}(x, \varepsilon)| \le K, \quad |\frac{\partial^2 u}{\partial x^i \partial x^j}(x, \varepsilon)| \le K, \quad x \in U_R \backslash \Gamma_{l\varepsilon}, \tag{7.5.3}$$

$$|u(x, \varepsilon)| \le K, \quad |\frac{\partial u}{\partial x^i}(x, \varepsilon)| \le \frac{K}{\varepsilon}, \quad |\frac{\partial^2 u}{\partial x^i \partial x^j}(x, \varepsilon)| \le \frac{K}{\varepsilon^2}, \quad x \in \Gamma_{l\varepsilon}.$$

An analytical approach to the problem (7.5.1)–(7.5.2) consists in construction of an external asymptotic expansion $V(x, \varepsilon)$ and of an interior asymptotic expansion $W(x, \varepsilon)$ [178]. They give the solution in $U_R \backslash \Gamma_{l\varepsilon}$ and $\Gamma_{l\varepsilon}$, respectively. The external expansion can be written as

$$V(x, \varepsilon) = \sum_{k=0}^{\infty} \varepsilon^{2k} v_k(x),$$

where

$$v_0(x) = \frac{g(x)}{c(x)}, \quad v_k(x) = -\frac{\Delta v_{k-1}(x)}{c(x)}, \quad k \geq 1.$$

The function $V(x, \varepsilon)$ is an asymptotic solution of (7.5.1)–(7.5.2) in $U_R \backslash \Gamma_{l\varepsilon}$, i.e., the function

$$V_m(x, \varepsilon) = \sum_{k=0}^{m} \varepsilon^{2k} v_k(x) \tag{7.5.4}$$

differs from the solution in $U_R \backslash \Gamma_{l\varepsilon}$ by $O(\varepsilon^{2m+2})$.

The interior expansion $W(x, \varepsilon)$ is needed to compensate a discrepancy in the boundary conditions. It turned out that outside the boundary layer $W(x, \varepsilon) = O(\varepsilon^N)$, $\varepsilon \to 0$, for any N. The sum $V + W$ is an asymptotic solution of the problem (7.5.1)–(7.5.2). The interior expansion is constructed in a more complicated way than the external one, and it is not given here.

It should be mentioned that the problem (7.5.1)–(7.5.2) is one of the simplest in the theory of boundary layer. If, for instance, the condition $c(x) \leq -c_0 < 0$, $x \in \bar{U}_R$, is violated so that the function $c(x)$ may take zero values then analytical investigation of the corresponding problem becomes exceedingly intricate. Therefore, numerical solution of problems with small parameter at higher derivatives is challenging and also of applicable interest. Here one can use general numerical methods (for example, the methods from Sects. 7.3 to 7.4) without taking into account the smallness of the parameter at higher derivatives. Principal difficulties lie in the fact that derivatives of the solution in the boundary layer are large and the average number of steps evaluated in Theorem 7.4.12 is as big as $O(1/\varepsilon^2 r^2)$. Let us analyze these and some other difficulties in the case of the model problem (7.5.1)–(7.5.2).

As before, we consider a random walk over spheres with radius εr in $U_R \backslash \Gamma_{\varepsilon r}$ (we have ε instead of a now) and over spheres tangent to ∂U_R in $\Gamma_{\varepsilon r} \backslash \Gamma_\delta$, where δ is sufficiently small (in any case $\delta < \varepsilon r / 2$).

Now it is convenient to present the error $\mathcal{R}$ (see (7.4.85)) in the following form

$$|\mathcal{R}| \leq |E(u(\bar{X}_{\varkappa}) - u(X_{\varkappa}))Y_{\varkappa}| + \sum_{k=1}^{\infty} |E\chi_{\varkappa > k-1} d_k| \qquad (7.5.5)$$

$$\leq |E(u(\bar{X}_{\varkappa}) - u(X_{\varkappa}))Y_{\varkappa}| + \sum_{k=1}^{\infty} |E\chi_{\Gamma_{l\varepsilon} \backslash \Gamma_{\delta}}(X_{k-1}) d_k|$$

$$+ \sum_{k=1}^{\infty} |E\chi_{U_R \backslash \Gamma_{l\varepsilon}}(X_{k-1}) d_k|$$

because

$$\chi_{\varkappa > k-1} = \chi_{\Gamma_{l\varepsilon} \backslash \Gamma_{\delta}}(X_{k-1}) + \chi_{U_R \backslash \Gamma_{l\varepsilon}}(X_{k-1}).$$

Let the one-step error d_k be bounded by $\delta_0(r, \varepsilon)Y_{k-1}$ in the part $\Gamma_{l\varepsilon} \backslash \Gamma_{\delta}$ of the boundary layer $\Gamma_{l\varepsilon}$ and by $\delta_1(r, \varepsilon)Y_{k-1}$ outside the boundary layer, i.e., in $U_R \backslash \Gamma_{l\varepsilon}$. We note that the method (7.4.74)–(7.4.76) under $b(x) = 0$ and the method (7.4.81)–(7.4.83) have $Y_k \leq 1$ for sufficiently small r if $c(x) \leq -c_0 < 0$. It follows from (7.5.5) that for these methods

$$|\mathcal{R}| \leq |E(u(\bar{X}_{\varkappa}) - u(X_{\varkappa}))Y_{\varkappa}| + \delta_0(r, \varepsilon)E\varkappa_0 + \delta_1(r, \varepsilon)E\varkappa_1, \qquad (7.5.6)$$

where $\varkappa_0$ and $\varkappa_1$ are random numbers of steps inside and outside the boundary layer, respectively. Clearly, $\varkappa_0$ and $\varkappa_1$ depend on x. Due to Theorem 7.4.12, we have $E\varkappa_1 \leq K/\varepsilon^2 r^2$. Fortunately, as it is shown in the lemma below, $E\varkappa_0 \leq K/r^2$ (see its proof in [281]).

Lemma 7.5.1 *There exists a constant $K > 0$ such that for any $x \in U_R \backslash \Gamma_{\delta}$ and sufficiently small ε and r*

$$E\varkappa_0 \leq \frac{K}{r^2}. \qquad (7.5.7)$$

We return to the inequality (7.5.6). The first term in the right-hand side of (7.5.6) is bounded by $K\delta/\varepsilon$ according to (7.5.3) (we note in passing that for the problem (7.5.1)–(7.5.2) we do not need to find $\bar{X}_{\varkappa}$ as $u(\bar{X}_{\varkappa}) = 0$). If we choose $\delta = O(r^p)$, then the first term can be made sufficiently small. At the same time, due to Theorem 7.4.12, the average number of steps depends on p insignificantly and, as before, it is estimated as $O(1/\varepsilon^2 r^2)$. The factor $E\varkappa_0 = O(1/r^2)$ in the second term (Lemma 7.5.1) is comparatively small, and the other factor $\delta_0(r, \varepsilon)$ depends on behavior of the solution in the boundary layer and it may take large values. But the errors of the methods from the previous section do not contain too higher order derivatives of the solution. Therefore, the second term can also be small. The third term in (7.5.6) has the very large factor $E\varkappa_1 = O(1/\varepsilon^2 r^2)$ and, consequently, this term can be decreased by means of $\delta_1(r, \varepsilon)$ only. Thus, the principal problem is to construct a sufficiently accurate and effective one-step approximation in the larger domain $U_R \backslash \Gamma_{l\varepsilon}$. Let us take into account that in the case of the problem (7.5.1)–(7.5.2) the system (7.4.3)–(7.4.5) is a system with small noise:

$$dX = \varepsilon dw(t), \tag{7.5.8}$$

$$dY = c(X)Y dt, \tag{7.5.9}$$

$$dZ = g(X)Y dt. \tag{7.5.10}$$

In Chap. 4, we constructed specific methods based on time disretization for systems with small noise. The errors of these methods have the form $O(h^p + \varepsilon^k h^q)$, $q < p$, where h is a step with respect to time. The time-step order of such a method is equal to q which is comparatively low and, due to this fact, one may reach a certain efficiency. Moreover, thanks to a large p and the factor ε^k at h^q, the method error becomes sufficiently small, and the method reaches high exactness. These ideas can be carried over to approximation based on a space discretization considered in this section. We shall construct an efficient one-step approximation in the main domain $U_R \backslash \Gamma_{l\varepsilon}$ with an error of the form $O(r^{2p} + \varepsilon^k r^{2q})$. We recall that the solution $u(x, \varepsilon)$ varies slowly in $U_R \backslash \Gamma_{l\varepsilon}$.

First, we analyze the method based on the one-step approximation (7.4.54). According to (7.5.3), $M_0(x)$ and $M_1(x)$, $x \in \Gamma_{l\varepsilon}$, in (7.4.55) are bounded by K and K/ε, respectively. Hence $\delta_0(r, \varepsilon) \leq Kr^4$ (of course, we have to take ε instead of a in (7.4.55)) and, due to Lemma 7.5.1, the second term in (7.5.6) has the acceptable bound $O(r^2)$. Clearly, the third term has the following bound:

$$\delta_1(r, \varepsilon) E \varkappa_1 \leq Kr^4 \frac{K}{\varepsilon^2 r^2} \leq \frac{Kr^2}{\varepsilon^2},$$

and, to obtain an acceptable accuracy, we have to take a very small r. Moreover, this circumstance leads to an increase of the average number of steps.

We analogously get for the method (7.4.81)–(7.4.83):

$$\delta_0(r, \varepsilon) E \varkappa_0 \leq Kr^4, \quad \delta_1(r, \varepsilon) \leq Kr^6, \quad \delta_1(r, \varepsilon) E \varkappa_1 \leq \frac{Kr^4}{\varepsilon^2}.$$

This method can be simplified without essential loss of accuracy. To this end, consider the following algorithm.

Algorithm 7.5.2 We set $r_{k+1} = r$ if $X_k \in U_R \backslash \Gamma_{\varepsilon r}$ and $r_{k+1} = (R - |X_k|)/\varepsilon$ if $X_k \in \Gamma_{\varepsilon r} \backslash \Gamma_\delta$ for $\delta < \varepsilon r$. Then

$$X_{k+1} = X_k + \varepsilon r_{k+1}\xi_{k+1}, \quad X_0 = x, \tag{7.5.11}$$

$$Y_{k+1} = Y_k \left[1 + c(X_k)\frac{r_{k+1}^2}{d} + \frac{\varepsilon r_{k+1}^3}{2(d+2)}\sum_{i=1}^{d}\frac{\partial c}{\partial x^i}(X_k)\xi_{k+1}^i \right. \tag{7.5.12}$$

$$\left. + \frac{4+d}{2d^2(2+d)}c^2(X_k)r_{k+1}^4\right], \quad Y_0 = 1,$$

$$Z_{k+1} = Z_k + Y_k \left[g(X_k) \frac{r_{k+1}^2}{d} + \frac{4+d}{2d^2(2+d)} c(X_k) g(X_k) r_{k+1}^4 \right], \quad Z_0 = 0. \quad (7.5.13)$$

Let $\varkappa = \varkappa_x$ be the first number at which $X_\varkappa \in \Gamma_\delta$. Then we set $X_k = X_\varkappa$ for $k \geq \varkappa$, i.e., our algorithm is stopped at the random step $\varkappa$. Having obtained $X_\varkappa$, we find the point $\bar{X}_\varkappa \in \partial G$ which is the nearest to $X_\varkappa$. We take $(\bar{X}_\varkappa, Y_\varkappa, Z_\varkappa)$ as the final state of the Markov chain.

Note that a choice of δ in this algorithm is specified below depending on the problem considered.

The method (7.5.11)–(7.5.13) does not require evaluation of the second derivatives $\partial^2 c/(\partial x^i)^2$ and $\partial^2 g/(\partial x^i)^2$ at each step in contrast to the method (7.4.81)–(7.4.83).

One can prove that

$$|d| \leq K \left(M_0(x) + \varepsilon^2 M_1(x) \right) y \, r^6 + K \, \varepsilon^4 M_2(x) \, y \, r^4 + K \left(\varepsilon^2 r^4 + r^6 \right) y.$$

Therefore (see (7.5.3)),

$$|\delta_i(r, \varepsilon)| \leq K(\varepsilon^2 r^4 + r^6), \quad i = 0, 1. \quad (7.5.14)$$

The error (7.5.14) is only of the fourth order with respect to r (due to this fact the method (7.5.11)–(7.5.13) is quite simple) but at the same time it is sufficiently small due to the factor ε^2. Using (7.5.14) with regard to $\delta = r^3$, it is not difficult to obtain the following result (we note that now the first term in (7.5.6) is $O(r^3/\varepsilon) \leq Kr^2 + Kr^4/\varepsilon^2$).

Theorem 7.5.3 *Let $\delta = r^3$. The error of Algorithm 7.5.2 is estimated as*

$$|\mathcal{R}| \leq Kr^2 + K \frac{r^4}{\varepsilon^2} \quad (7.5.15)$$

and the average number of steps for this method is equal to $O(1/\varepsilon^2 r^2)$.

We emphasize that the large average number of steps leads to extraordinary computational costs. At the same time, we can find the solution of the problem (7.5.1)–(7.5.2) in $U_R \backslash \Gamma_{l\varepsilon}$ with high accuracy using the truncated external expansion (7.5.4). We use this fact and construct below an analytic-numerical method.

We set

$$u(x, \varepsilon) \approx V_m(x, \varepsilon), \quad x \in U_R \backslash \Gamma_{l\varepsilon},$$

and, instead of (7.5.1)–(7.5.2), we introduce the problem for the Helmholtz equation with small parameter:

$$\frac{1}{2} \varepsilon^2 \Delta u + c(x) u = g(x), \qquad R - l\varepsilon < |x| < R, \quad (7.5.16)$$

$$u \mid_{|x|=R-l\varepsilon} = V_m(x, \varepsilon), \quad u \mid_{|x|=R} = 0. \quad (7.5.17)$$

Consider the random walk defined by $r < \max(\varepsilon, l\varepsilon)$ and $\delta \ll r$ in the layer $R - l\varepsilon \le |x| \le R$: if $R - l\varepsilon \le |X_k| < R - l\varepsilon + \delta$ or $R - \delta < |X_k| \le R$, then $X_{k+1} = X_k$; if $R - l\varepsilon + \delta \le |X_k| < R - l\varepsilon + \varepsilon r$ or $R - \varepsilon r < |X_k| \le R - \delta$, then r_{k+1} is equal to $(|X_k| - (R - l\varepsilon))/\varepsilon$ or $(R - |X_k|)/\varepsilon$, respectively; if $R - l\varepsilon + \varepsilon r \le |X_k| \le R - \varepsilon r$ then $r_{k+1} = r$. In the second and third cases we put

$$X_{k+1} = X_k + \varepsilon r_{k+1} \xi_{k+1}. \tag{7.5.18}$$

It is not difficult to prove the following theorem.

Theorem 7.5.4 Let $\delta = r^5$. Then the error of Algorithm 7.5.2 for the problem (7.5.16)–(7.5.17) is estimated as

$$|\mathcal{R}| \le K(\varepsilon^2 r^2 + r^4) + K \frac{r^5}{\varepsilon} \tag{7.5.19}$$

and the average number of steps is equal to $O(1/r^2)$.

It is clear that the error of the analytic-numerical method for the original problem (7.5.1)–(7.5.2) is larger than (7.5.19) by $O(\varepsilon^{2m+2})$. The proposed analytic-numerical method is very effective: it is more accurate (compare the errors (7.5.19) and (7.5.15)) and it has the smaller average number of steps.

Remark 7.5.5 Undoubtedly, many results obtained here for the model problem (7.5.1)–(7.5.2) can be used for more general problems. In particular, they can be carried over to the problem (7.4.1) with $a = \varepsilon$, $b(x) = 0$, $c(x) \le -c_0 < 0$ without any essential change.

7.6 Methods for the Neumann Problem for Parabolic Equations

In this section we construct Markov chains such that the expectation of a certain functional of chain paths is close to the solution of the Neumann (Robin) problem for parabolic equations. These Markov chains weakly approximate the solution of the system of SDEs which are (stochastic) characteristics for the Neumann problem.

Let G be a bounded domain in $\mathbf{R}^d$ and $Q = [T_0, T_1) \times G$ be a cylinder in $\mathbf{R}^{d+1}$. Consider the parabolic equation:

$$\frac{\partial u}{\partial t} + \frac{1}{2} \sum_{i,j=1}^{d} a^{ij}(t, x) \frac{\partial^2 u}{\partial x^i \partial x^j} + \sum_{i=1}^{d} b^i(t, x) \frac{\partial u}{\partial x^i} + c(t, x)u \tag{7.6.1}$$

$$+ g(t, x) = 0, \quad (t, x) \in Q,$$

with the initial condition

$$u(T_1, x) = f(x) \qquad (7.6.2)$$

and the Neumann boundary condition

$$\frac{\partial u}{\partial \nu} + \varphi(t, x)u = \psi(t, x), \quad t \in [T_0, T_1], \quad x \in \partial G , \qquad (7.6.3)$$

where ν is the direction of the inner normal to the surface ∂G at a point $x \in \partial G$.

We assume that the coefficients $a^{ij} = a^{ji}$ satisfy the strict ellipticity condition in $\bar{Q}$ and that the surface ∂G and the coefficients of the problem (7.6.1)–(7.6.3) are sufficiently smooth. In what follows, we require existence of a sufficiently smooth classical solution of the problem (7.6.1)–(7.6.3) in $\bar{Q}$ [222, 241, 267].

The solution of the problem (7.6.1)–(7.6.3) has the following probabilistic representation [111, 123, 177]:

$$u(t, x) = E \left[f(X_{t,x}(T_1))Y_{t,x,1}(T_1) + Z_{t,x,1,0}(T_1) \right] , \qquad (7.6.4)$$

where $X_{t,x}(s)$, $Y_{t,x,y}(s)$, $Z_{t,x,y,z}(s)$, $s \geq t$, is the solution of the Cauchy problem for the system of SDEs:

$$dX = b(s, X) \, ds + \sigma(s, X) \, dw(s) + \nu(X) \, I_{\partial G}(X) \, dL(s), \quad X(t) = x, \qquad (7.6.5)$$
$$dY = c(s, X) \, Y \, ds + \varphi(s, X) \, I_{\partial G}(X) Y \, dL(s), \quad Y(t) = y, \qquad (7.6.6)$$
$$dZ = g(s, X) \, Y \, ds - \psi(s, X) \, I_{\partial G}(X) Y \, dL(s), \quad Z(t) = z. \qquad (7.6.7)$$

In (7.6.5)–(7.6.7), $(t, x) \in Q$, $w(s) = (w^1(s), \ldots, w^d(s))^\mathsf{T}$ is a standard Wiener process, $b(s, x)$ is a d-dimensional column-vector composed of the coefficients $b^i(s, x)$, the $d \times d$ matrix $\sigma(s, x)$ is obtained from the formula $\sigma(s, x)\sigma^\mathsf{T}(s, x) = a(s, x)$, $a(s, x) = \{a^{ij}(s, x)\}$, $i, j = 1, \ldots, d$, $L(s)$ is the local time of the process X on the boundary ∂G, i.e., it is a scalar increasing process which grows only when $X(s) \in \partial G$, Y and Z are scalars, and $I_A(x)$ is the indicator function of a set A.

7.6.1 Methods Based on Changing Local Coordinates near the Boundary

Our aim is to construct a Markov chain (t_k, X_k, Y_k, Z_k) which approximates in the weak sense the solution of the system (7.6.5)–(7.6.7). Due to the reflection, such a Markov chain is terminated on the upper base of the cylinder $\bar{Q}$. We shall construct this chain analogously to the chains from Sect. 7.1. We require that the conditional expectation of the difference

$$d = u(t_{k+1}, X_{k+1})Y_{k+1} + Z_{k+1} - u(t_k, X_k)Y_k - Z_k \qquad (7.6.8)$$

is small at each step provided X_k, Y_k, Z_k are known (for definiteness, we assume that all t_k are deterministic).

Let $r > 0$ be a sufficiently small number. If the distance $\rho(X_k, \partial G)$ between the point X_k and the boundary ∂G is greater than r, then given the point (t_k, X_k) we construct the point (t_{k+1}, X_{k+1}) using one of the methods for solving the Dirichlet problem from Sect. 7.1 which one-step order of accuracy is $O(r^4)$.

Now let X_k be a point near the boundary (a boundary point), i.e., $\rho(X_k, \partial G) \leq r$. We suppose that $t_k \leq T_1 - r^2$. Otherwise, we terminate the chain at the point $(t_{k+1}, X_{k+1}, Y_{k+1}, Z_{k+1}) = (T_1, X_k, Y_k, Z_k)$. In this case the error on the last step is $O(r^2)$. For convenience, we denote by (t_0, x_0, y_0, z_0), $t_0 \leq T_1 - r^2$, $\rho(x_0, \partial G) \leq r$, the point (t_k, X_k, Y_k, Z_k). Our nearest aim is to construct a point $t_1 = t_0 + r^2 = t_0 + h$, X_1, Y_1, Z_1 such that the expectation Ed of the difference (7.6.8) is of order $O(r^3)$.

7.6.1.1 One-Step Approximation for Boundary Points

Let x_0^π be the projection of x_0 on ∂G and $r_0 := \rho(x_0, \partial G) = \rho(x_0, x_0^\pi)$, $r_0 \leq r$. In a neighborhood of the point x_0 we introduce a new orthogonal coordinate system such that its origin coincides with x_0 and one of the coordinate vectors (for definiteness, the first one) coincides with $\nu(x_0)$. We denote by $\chi = (\chi^1, \ldots, \chi^d)^\mathsf{T}$ the coordinates of the vector $x - x_0$ in the new coordinate system. This transformation of variables has the form

$$\chi = Q(x - x_0), \quad x = x_0 + Q^\mathsf{T}\chi, \tag{7.6.9}$$

where $Q = \{q_{ij}\}$ is an orthogonal matrix.

In the new variables the point x_0 has zero coordinates, the point x_0^π has the coordinates $(-r_0, 0, \ldots, 0)$, and the equation (7.6.1) is written as

$$\frac{\partial u}{\partial t} + \frac{1}{2}\sum_{i,j=1}^d \alpha^{ij}(t,\chi)\frac{\partial^2 u}{\partial \chi^i \partial \chi^j} + \sum_{i=1}^d \beta^i(t,\chi)\frac{\partial u}{\partial \chi^i} + c(t,\chi)u \tag{7.6.10}$$

$$+ g(t,\chi) = 0, \quad (t,x) \in Q,$$

and the boundary condition (7.6.3) takes the form

$$\frac{\partial u}{\partial \nu} + \varphi(t,\chi)u = \psi(t,\chi), \quad t \in [T_0, T_1], \quad x \in \partial G. \tag{7.6.11}$$

In (7.6.10)–(7.6.11) we preserve the notation for the functions u, c, g, φ, ψ which depend on t, χ according to (7.6.9). Further, the coefficients $\alpha^{ij} = \sum_{k,m=1}^d q_{ik}a^{km}q_{jm}$ form the matrix $QAQ^\mathsf{T} = \{\alpha^{ij}\}$, and the coefficients $\beta^i = \sum_{k=1}^d q_{ik}b^k$ form the vector Qb.

Let us write two relations for the function u which are important for further analysis. For brevity, we will write $(-r_0, 0)$ instead of $(-r_0, 0, \ldots, 0)$. We have

$$u(t_0, -r_0, 0) = u(t_0, 0, 0) + \frac{\partial u}{\partial \chi^1}(t_0, 0, 0) \times (-r_0) + O(r^2),$$

$$\frac{\partial u}{\partial \chi^1}(t_0, -r_0, 0) = \frac{\partial u}{\partial \chi^1}(t_0, 0, 0) + \frac{\partial^2 u}{(\partial \chi^1)^2}(t_0, 0, 0) \times (-r_0) + O(r^2).$$

The condition (7.6.11) at the point $(t_0, -r_0, 0)$ can be written in the form

$$\frac{\partial u}{\partial \chi^1}(t_0, -r_0, 0) = -\varphi(t_0, -r_0, 0)\, u(t_0, -r_0, 0) + \psi(t_0, -r_0, 0).$$

The above three relations imply

$$\left(\frac{\partial u}{\partial \chi^1}\right)_0 = -\varphi_0\,(1 + \varphi_0 r_0)u_0 + r_0\left(\frac{\partial^2 u}{(\partial \chi^1)^2}\right)_0 \tag{7.6.12}$$

$$+(1 + \varphi_0 r_0)\psi_0 + O(r^2),$$

where u_0, φ_0, and ψ_0 mean $u(t_0, 0, 0)$, $\varphi(t_0, -r_0, 0)$, and $\psi(t_0, -r_0, 0)$, respectively. The symbol $(\cdot)_0$ means that the corresponding function is evaluated at the point $(t_0, 0, 0)$.

Further, differentiating (7.6.11) with respect to $\chi^2, \ldots, \chi^d$ at the point $(t_0, -r_0, 0)$, we obtain

$$\frac{\partial^2 u}{\partial \chi^1 \partial \chi^j}(t_0, -r_0, 0) + \frac{\partial \varphi}{\partial \chi^j}(t_0, -r_0, 0)\, u(t_0, -r_0, 0)$$

$$+\varphi(t_0, -r_0, 0)\frac{\partial u}{\partial \chi^j}(t_0, -r_0, 0) = \frac{\partial \psi}{\partial \chi^j}(t_0, -r_0, 0),$$

whence it follows that

$$\varphi_0\left(\frac{\partial u}{\partial \chi^j}\right)_0 = -\varphi'_{j0}\, u_0 - \left(\frac{\partial^2 u}{\partial \chi^1 \partial \chi^j}\right)_0 + \psi'_{j0} + O(r), \tag{7.6.13}$$

where φ'_{j0} and ψ'_{j0} are the values of the corresponding derivatives at the point $(t_0, -r_0, 0)$.

Let a point (t_0, x_0, y_0, z_0) be such that $\rho(x_0, \partial G) \le r$, $t_0 \le T_1 - r^2$. Now we form $(t_1, X_1, Y_1, Z_1) = (t_0 + r^2, x_0 + \Delta X, y_0 + \Delta Y, z_0 + \Delta Z)$ so that the point (t_1, X_1) belongs to $\bar{Q}$, the random variable ΔX is of order $O(r)$, and the equality

$$E\left(u(t_1, X_1)\, Y_1 + Z_1 - u(t_0, x_0)\, y_0 - z_0\right) = O(r^3) \tag{7.6.14}$$

holds. This relation is written in the new coordinates as

$$E\left(u(t_1, \chi)\, Y_1 + Z_1 - u(t_0, x_0)\, y_0 - z_0\right) = O(r^3), \tag{7.6.15}$$

where $\chi = Q\,\Delta X$.

Theorem 7.6.1 *Let*

$$t_1 = t_0 + r^2, \quad \chi^1 = \chi_0^1 + \chi_1^1, \quad \chi_0^1 = \beta_0^1 r^2, \tag{7.6.16}$$

$$\chi_1^1 = \left[\alpha_0^{11} r^2 + r_0^2\right]^{1/2} - r_0 - \beta_0^1 r^2, \tag{7.6.17}$$

$$\chi^j = \chi_0^j + \varphi_0 \chi_1^1 + \chi_2^j, \quad \chi_0^j = \beta_0^j r^2, \quad \chi_1^j = -\alpha_0^{1j} r^2, \tag{7.6.18}$$

$$\chi_2^j = \sum_{l=2}^d \lambda^{jl} \nu_l r, \quad j = 2, \dots, d,$$

where ν_l are independent random variables taking the values ± 1 with probability $1/2$, the numbers λ^{jl}, $j, l = 2, \dots, d$, satisfy the equalities

$$\sum_{k=2}^d \lambda^{ik} \lambda^{jk} = \alpha^{ij}, \quad i, j = 2, \dots, d, \tag{7.6.19}$$

(i.e., they form the $(d-1) \times (d-1)$-matrix λ satisfying the equality $\lambda \lambda^\mathsf{T} = \{\alpha^{i,j}\}_{i,j=2\div d}$ which results in $E\chi_2^i \chi_2^j = \alpha^{ij} r^2$), and

$$\Delta Y = \left(c_0 r^2 + \varphi_0 (1 + \varphi_0 r_0)\chi_1^1 + \sum_{j=2}^d \varphi_{j0}' \chi_1^j + \varphi_0^2 \left(\chi_1^1\right)^2\right) y_0, \tag{7.6.20}$$

$$\Delta Y = \left(g_0 r^2 - \psi_0 (1 + \varphi_0 r_0)\chi_1^1 - \sum_{j=2}^d \psi_{j0}' \chi_1^j - \varphi_0 \psi_0 \left(\chi_1^1\right)^2\right) y_0. \tag{7.6.21}$$

Then the relation (7.6.15) (the relation (7.6.14) in the old coordinates) holds.

Proof We have

$$d = u(t_1, \chi) Y_1 + Z_1 - u_0 y_0 - z_0 \tag{7.6.22}$$

$$= \left[u_0 + \left(\frac{\partial u}{\partial t}\right)_0 r^2 + \left(\frac{\partial u}{\partial \chi^1}\right)_0 \chi^1 + \sum_{j=2}^d \left(\frac{\partial u}{\partial \chi^j}\right)_0 \chi^j + \frac{1}{2} \left(\frac{\partial^2 u}{(\partial \chi^1)^2}\right)_0 (\chi^1)^2 \right.$$

$$\left. + \sum_{j=2}^d \left(\frac{\partial^2 u}{\partial \chi^1 \partial \chi^j}\right)_0 \chi^1 \chi^j + \frac{1}{2} \sum_{i,j=2}^d \left(\frac{\partial^2 u}{\partial \chi^i \partial \chi^j}\right)_0 \chi^i \chi^j + O(r^3)\right] (y_0 + \Delta Y)$$

$$+ \Delta Z - u_0 y_0.$$

Using (7.6.12), (7.6.13), (7.6.17), (7.6.18), we get

$$\left(\frac{\partial u}{\partial \chi^1}\right)_0 \chi^1 = \left(\frac{\partial u}{\partial \chi^1}\right)_0 \chi_0^1$$

$$+ \left[-\varphi_0\,(1+\varphi_0 r_0)u_0 + r_0 \left(\frac{\partial^2 u}{(\partial \chi^1)^2}\right)_0 + (1+\varphi_0 r_0)\psi_0\right]\chi_1^1 + O(r^3)\,,$$

$$\sum_{j=2}^{d}\left(\frac{\partial u}{\partial \chi^j}\right)_0 \chi^j = \sum_{j=2}^{d}\left(\frac{\partial u}{\partial \chi^j}\right)_0 (\chi_0^j + \chi_2^j) + \varphi_0 \sum_{j=2}^{d}\left(\frac{\partial u}{\partial \chi^j}\right)_0 \chi_1^j$$

$$= \sum_{j=2}^{d}\left(\frac{\partial u}{\partial \chi^j}\right)_0 (\chi_0^j + \chi_2^j)$$

$$+ \sum_{j=2}^{d}\left(-\varphi_{j0}'\,u_0 - \left(\frac{\partial^2 u}{\partial \chi^1 \partial \chi^j}\right)_0 + \psi_{j0}'\right)\chi_1^j + O(r^3)\,,$$

$$\sum_{j=2}^{d}\left(\frac{\partial^2 u}{\partial \chi^1 \partial \chi^j}\right)_0 \chi^1 \chi^j = \sum_{j=2}^{d}\left(\frac{\partial^2 u}{\partial \chi^1 \partial \chi^j}\right)_0 \chi^1 \chi_2^j + O(r^3)\,,$$

$$\frac{1}{2}\sum_{i,j=2}^{d}\left(\frac{\partial^2 u}{\partial \chi^i \partial \chi^j}\right)_0 \chi^i \chi^j = \frac{1}{2}\sum_{i,j=2}^{d}\left(\frac{\partial^2 u}{\partial \chi^i \partial \chi^j}\right)_0 \chi_2^i \chi_2^j + O(r^3)\,.$$

Substituting these expressions in (7.6.22) and making some elementary calculations (note that $2r_0\chi_1^1 + (\chi^1)^2 = \alpha_0^{11}r^2 + O(r^3)$), we obtain

$$d = \left[\left(\frac{\partial u}{\partial t}\right)_0 r^2 + \left(\frac{\partial u}{\partial \chi^1}\right)_0 \beta_0^1 r^2 + + \frac{1}{2}\left(\frac{\partial^2 u}{(\partial \chi^1)^2}\right)_0 \alpha_0^{11}\,r^2\right.$$

$$+ (-\varphi_0\,(1+\varphi_0 r_0)u_0 + \psi_0(1+\varphi_0 r_0))\,\chi_1^1 + \sum_{j=2}^{d}\left(\frac{\partial u}{\partial \chi^j}\right)_0 \beta_0^j\,r^2$$

$$+ \sum_{j=2}^{d}\left(\frac{\partial u}{\partial \chi^j}\right)_0 \chi_2^j + \sum_{j=2}^{d}\left(\frac{\partial^2 u}{\partial \chi^1 \partial \chi^j}\right)_0 \alpha_0^{1j}r^2 + \sum_{j=2}^{d}\alpha_0^{1j}\,(\varphi_{j0}'\,u_0 - \psi_{j0}')\,r^2$$

$$\left.+ \sum_{j=2}^{d}\left(\frac{\partial^2 u}{\partial \chi^1 \partial \chi^j}\right)_0 \chi^1 \chi_2^j + \frac{1}{2}\sum_{i,j=2}^{d}\left(\frac{\partial^2 u}{\partial \chi^i \partial \chi^j}\right)_0 \chi_2^i \chi_2^j + O(r^3)\right]y_0$$

$$+ u_0\Delta Y + (-\varphi_{j0}'\,u_0 + \psi_{j0}')\chi_1^1\Delta Y + \sum_{j=2}^{d}\left(\frac{\partial u}{\partial \chi^j}\right)_0 \chi_2^j\Delta Y + \Delta Z + O(r^3)\,.$$

Further, taking into account that $E\chi_2^j = 0$, $j = 2, \ldots, d$, we get

$$
\begin{aligned}
Ed = &\left[\left(\frac{\partial u}{\partial t} \right)_0 r^2 + \sum_{i=1}^d \beta_0^i \left(\frac{\partial u}{\partial \chi^i} \right)_0 r^2 + \frac{1}{2} \sum_{i,j=1}^d \alpha_0^{ij} \left(\frac{\partial^2 u}{\partial \chi^i \partial \chi^j} \right)_0 r^2 \right. \\
& + \left(-\varphi_0 \left(1 + \varphi_0 r_0 \right) \chi_1^1 + \sum_{j=2}^d \alpha_0^{1j} \varphi_{j0}' \, r^2 \right) u_0 + \psi_0 (1 + \varphi_0 r_0) \chi_1^1 \\
& \left. - \sum_{j=2}^d \alpha_0^{1j} \psi_{j0}' \, r^2 \right] y_0 \\
& + u_0 \left[c_0 r^2 + \varphi_0 \left(1 + \varphi_0 r_0 \right) \chi_1^1 - \sum_{j=2}^d \alpha_0^{1j} \varphi_{j0}' \, r^2 + \varphi_0^2 \left(\chi_1^1 \right)^2 \right] y_0 \\
& - \varphi_0^2 u_0 \left(\chi_1^1 \right)^2 y_0 + \psi_0 \left(\chi_1^1 \right)^2 \varphi_0 + g_0 r^2 y_0 - \psi_0 (1 + \varphi_0 r_0) \chi_1^1 \\
& + \sum_{j=2}^d \alpha_0^{1j} \psi_{j0}' \, r^2 y_0 - \varphi_0 \psi_0 \left(\chi_1^1 \right)^2 y_0 + O(r^3) \\
= &\left[\left(\frac{\partial u}{\partial t} \right)_0 + \sum_{i=1}^d \beta_0^i \left(\frac{\partial u}{\partial \chi^i} \right)_0 + \frac{1}{2} \sum_{i,j=1}^d \alpha_0^{ij} \left(\frac{\partial^2 u}{\partial \chi^i \partial \chi^j} \right)_0 + c_0 u_0 + g_0 \right] \\
& \times r^2 y_0 + O(r^3) = O(r^3) .
\end{aligned}
$$

Theorem 7.6.1 is proved. $\qquad\square$

7.6.1.2 Convergence Theorems

Theorem 7.6.2 *Assume that the one-step transition for nonboundary points is made by a method with one-step error $O(r^4)$ whose time step h_k satisfies the inequality $h_k \geq C_0 r^2$ and whose increment ΔX satisfies the inequality $|\Delta X| \leq C_1 r$, where C_0 and C_1 are some positive constants (for instance, one can exploit methods from Sect. 7.1 here). Let the one-step transition for boundary points (t_0, x_0), $t_0 < T_1 - r^2$, be made in accordance with Theorem 7.6.1. Let the last step $\varkappa$, when $T_1 - r^2 \leq t_{\varkappa-1} < T_1$, result in the transition to the point $(t_\varkappa, X_\varkappa, Y_\varkappa, Z_\varkappa) = (T_1, X_{\varkappa-1}, Y_{\varkappa-1}, Z_{\varkappa-1})$. Then such a method has weak order of accuracy $O(r^2)$, i.e.,*

$$
u(t, x) = E\left[f(X_\varkappa) Y_\varkappa + Z_\varkappa \right] + O(r^2), \quad t_0 = t, \ \ X_0 = x. \tag{7.6.23}
$$

The average number of steps is equal to $E\varkappa = O(1/r^2)$.

Proof The proof of this theorem is analogous to the one of Theorems 7.1.12 and 7.1.13. The following arguments are used here. It is obvious that the number of steps for nonboundary points does not exceed $(T_1 - T_0)/Cr^2 = O(1/r^2)$. Since the error at each such step is $O(r^4)$, the total error over these steps is $O(r^2)$. It turns out (see Lemma 7.6.3 below) that the average number of steps for boundary points is estimated by $O(1/r)$. Since the one-step error for boundary points is $O(r^3)$, total error over these steps is $O(r^2)$ as well. Finally, the error on the last step does not exceed $O(r^2)$. Thus, the total error of the method is $O(r^2)$. Theorem 7.6.2 is proved.   □

An implementation of the algorithm described in Theorem 7.6.2 can be found e.g. in [28]. Now we turn our attention to the formulation and proof of Lemma 7.6.3 mentioned above. Along with the surface ∂G, we consider the surface $\partial_l G$ which is concentric to ∂G and belongs to G, and the distance l between ∂G and $\partial_l G$ is fixed and small so that any ball of radius l with a center on $\partial_l G$ entirely belongs to G. The layer between $\partial_l G$ and ∂G is denoted as G_l. Consider a Markov chain (t_k, X_k) generated by one or another method specified in Theorem 7.6.2 which starts from the point (t, x), i.e., $t_0 = t$ and $X_0 = x$.

Let (t_k, X_k), $k = 0, \ldots, \varkappa$, be a certain path of this chain. Consider all the time moments $t_{k_1}, \ldots, t_{k_q}$ which the path spends in the cylindric layer $\bar{Q}_r = [T_0, T_1 - r^2] \times \bar{G}_r$. The random number q depends on t and x: $q = q(t, x)$ (of course, q also depends on r).

Lemma 7.6.3 *The average time $v(t, x) = Eq(t, x)$ which the chain (t_k, X_k) spends in the cylindric layer $\bar{Q}_r = [T_0, T_1 - r^2] \times \bar{G}_r$ uniformly in $(t, x) \in \bar{Q}$ satisfies the estimate*

$$v(t, x) \leq \frac{C}{r} . \tag{7.6.24}$$

Proof We use here the same technique as, e.g., in the proof of Theorem 7.1.13. In connection with the Markov chain (t_k, X_k), we consider the boundary value problem

$$PV - V = -g(t, x), \quad (t, x) \in [T_0, T_1 - r^2] \times \bar{G}, \tag{7.6.25}$$
$$V(t, x) = 0, \quad (t, x) \in (T_1 - r^2, T_1] \times \bar{G},$$

where P is the one-step transition operator $Pv(t, x) = Ev(t_1, X_1)$, $t_0 = t$, $X_0 = x$. As is known (see, e.g., Sect. 7.1.2), if $V(x)$ is the solution of the boundary value problem (7.6.25) with a function $g(t, x)$ such that

$$g(t, x) \geq I_{\bar{Q}_r}(t, x), \qquad (t, x) \in [T_0, T_1 - r^2] \times \bar{G}, \tag{7.6.26}$$

then

$$v(t, x) \leq V(t, x). \tag{7.6.27}$$

Introduce the functions

$$w(x) = \begin{cases} \rho^2(x, \partial_l G), & x \in \bar{G}_l, \\ 0, & x \in \bar{G} \backslash \bar{G}_l, \end{cases}$$

and

$$W(t, x) = \begin{cases} 0, & (t, x) \in (T_1 - r^2, T_1] \times \bar{G}, \\ K(T_1 - t) + w(x), & (t, x) \in [T_0, T_1 - r^2] \times \bar{G}. \end{cases} \quad (7.6.28)$$

This function satisfies the boundary condition in (7.6.25), $K \geq 0$ is a constant whose value is specified below. Now we evaluate $PW - W$ (i.e., the function $-g(t, x)$) for various points (t, x).

Let a point (t, x) be such that $T_0 \leq t \leq T_1 - r^2$ and x belongs to the layer between the surfaces $\partial_l G$ and $\partial_r G$, $r \ll l$ and, besides, $\rho(x, \partial_l G) \geq C_1 r$. But virtue of this choice of (t, x), the point $x + \Delta X \in \bar{G}$ (see the conditions of Theorem 7.6.2). Therefore, $w(x) = \rho^2(x, \partial_l G)$, $w(x + \Delta X) = \rho^2(x + \Delta X, \partial_l G)$. Let p_x be the projection of x on the surface $\partial_l G$. We have $\rho(x + \Delta X, \partial_l G) = \rho(x, \partial_l G) + (\Delta X, (x - p_x)/|x - p_x|) + O(r^2)$, where $(\cdot, \cdot)$ is the scalar product. Hence

$$E\left[w(x + \Delta X) - w(x)\right]$$

$$= E\left[\left(\rho(x, \partial_l G) + \left(\Delta X, \frac{x - p_x}{|x - p_x|}\right)\right)^2 - \rho^2(x, \partial_l G)\right] + O(r^2)$$

$$= 2\rho(x, \partial_l G)E\left(\Delta X, \frac{x - p_x}{|x - p_x|}\right) + E\left(\Delta X, \frac{x - p_x}{|x - p_x|}\right)^2 + O(r^2) = O(r^2)$$

since $|E\Delta X| = O(r^2)$ and $E|\Delta X|^2 = O(r^2)$ for the methods exploited in Theorem 7.6.2. Therefore

$$PW - W = -K\Delta t + O(r^2). \quad (7.6.29)$$

Now let $\rho(x, \partial_l G) < C_1 r$. In this case both $w(x) = O(r^2)$ and $w(x + \Delta X) = O(r^2)$ and, hence, (7.6.29) holds again.

We obviously have

$$PW - W = -K\Delta t$$

for points x which are inside $\partial_l G$ and for which the distance $\rho(x, \partial_l G) > C_1 r$.

Since $O(r^2)$ in (7.6.29) satisfies the inequality $|O(r^2)| \leq Cr^2$ uniformly in (t, x), where $C > 0$ is a constant, and since $\Delta t \geq C_0 r^2$, we can find a constant K such that the right-hand side of (7.6.29) is non-positive. Thus, the function $f(t, x) = -(PW - W)$ is nonnegative.

Finally, we consider a point (t, x) such that $T_0 \leq t \leq T_1 - r^2$, $x \in \bar{G}_r$. In connection with the notation used in Theorem 7.6.1 it is convenient to use (t_0, x_0) instead of (t, x) here. We have

$$\rho(x_0, \partial_l G) = l - r_0, \quad \rho(x_0 + \Delta X, \partial_l G) = l - r_0 - \chi^l + O(r^2).$$

Further,

$$E\left[w(x_0 + \Delta X) - w(x_0)\right] = \left(l - r_0 - \chi^1\right)^2 - (l - r_0)^2 + O(r^2)$$
$$= -2l\chi^1 + O(r^2) \, .$$

It is not difficult to see that χ^1 has a uniform lower bound proportional to r. Hence, we obtain for such points (t, x):

$$PW(t, x) - W(t, x) = -f(t, x), \quad f(t, x) \geq qr.$$

As a result, we have constructed the function W which is a solution of the problem (7.6.25) with the nonnegative everywhere function $f(t, x)$ satisfying the inequality $f \geq qr$ in $\bar{Q}_r$. Now we consider the function $V(t, x) = W(t, x)/qr$. It is evident that $V(t, x)$ is the solution of the problem (7.6.25) with the function $g(t, x) = f(t, x)/qr$ which satisfies the inequality (7.6.26). Therefore, (7.6.27) holds and $v(t, x)$ satisfies the estimate (7.6.24). $\qquad\square$

The proof of Theorem 7.6.2 inspires the following ideas. Since the time which the chain spends in $\bar{G}_r$ is $O(1/r)$, it is sufficient for convergence of a method to require that the one-step error in $\bar{G}_r$ is $O(r^2)$ instead of $O(r^3)$. However, in this case the accuracy order of the method is $O(r)$ only, but at this expense we can construct a simpler method.

Theorem 7.6.4 *Let*

$$t_1 = t_0 + r^2, \tag{7.6.30}$$

$$\chi^1 = qr \, , \tag{7.6.31}$$

where q is a positive number,

$$\chi^j = 0, \quad j = 2, \ldots, d \, , \tag{7.6.32}$$

$$\Delta Y = \varphi_0 \, \chi^1 \, y_0 = \varphi_0 \, qr \, y_0 \, , \tag{7.6.33}$$

$$\Delta Z = -\psi_0 \, \chi^1 \, y_0 = -\psi_0 \, qr \, y_0 \, . \tag{7.6.34}$$

Then the relations (7.6.14)–(7.6.15) hold with the replacement of $O(r^3)$ by $O(r^2)$, and the method from Theorem 7.6.2 with (7.6.16)–(7.6.21) replaced by (7.6.30)– (7.6.34) has the order of accuracy $O(r)$, i.e., the relation (7.6.23) holds with the replacement of $O(r^2)$ by $O(r)$.

Proof We have

$$d = u(t_1, \chi)Y_1 + Z_1 - u_0 y_0 - z_0$$
$$= \left[u_0 + \left(\frac{\partial u}{\partial \chi^1}\right)_0 \chi^1 + O(r^2)\right] (y_0 + \varphi_0 \, \chi^1 \, y_0) - \psi_0 \, \chi^1 \, y_0 - u_0 \, y_0$$
$$= u_0 \, \varphi_0 \, \chi^1 \, y_0 + \left(\frac{\partial u}{\partial \chi^1}\right)_0 \chi^1 \, y_0 - \psi_0 \, \chi^1 \, y_0 + O(r^2).$$

At the same time, it follows from (7.6.12) that

$$\left(\frac{\partial u}{\partial \chi^1}\right)_0 + \varphi_0\, u_0 - \psi_0 = O(r)\,,$$

whence $d = O(r^2)$.

The last assertion of this theorem is proved by reference to Lemma 7.6.3 as we did in Theorem 7.6.2. Theorem 7.6.4 is proved. $\qquad\Box$

7.6.2 Simple Random Walk with Symmetric Reflection

In this section we consider a weak approximation of (7.6.5)–(7.6.7) to solve the Neumann (Robin) problem (7.6.1)–(7.6.3), which does not require any orthogonal transformation of diffusion matrix or change of local coordinates, thus it is easy to implement. This method is based on the idea of symmetrized reflection on the boundary and making use of the weak Euler scheme with bounded random variables. The latter makes sure that discretized sample paths cannot move beyond the boundary outside the domain by more that $O(\sqrt{h})$, where h is the time step. This method was proposed in [233]. We note that the method from [35] is also based on symmetrized reflection but uses the Euler scheme with Gaussian random variables to model Wiener increments.

Let $(t_0, x) \in Q$. Introduce the uniform discretization of the time interval $[t_0, T]$ so that $t_0 < \cdots < t_N = T, h := (T_1 - T_0)/N$ and $t_{k+1} = t_k + h$.

We consider a Markov chain $(t_k, X_k)_{k \geq 0}$ with $X_0 = x$ approximating the solution $X_{t_0,x}(t)$ of the reflected SDEs (7.6.5). The chain has an auxiliary (intermediate) step X'_{k+1} every time the chain moves from the time layer t_k to t_{k+1}, for which the weak Euler scheme is used:

$$X'_{k+1} = X_k + hb_k + h^{1/2}\sigma_k\xi_{k+1}, \tag{7.6.35}$$

where $b_k = b(t_k, X_k)$, $\sigma_k = \sigma(t_k, X_k)$ and $\xi_{k+1} = (\xi^1_{k+1}, \ldots, \xi^d_{k+1})^\top$, ξ^i_{k+1}, $i = 1, \ldots, d, k = 0, \ldots, N - 1$, are mutually independent random variables taking values ± 1 with probability $1/2$.

Taking this auxiliary step X'_{k+1} while moving from X_k to X_{k+1} gives us an opportunity to check whether the realized value of X'_{k+1} is inside the domain G or not. If $X'_{k+1} \in \bar{G}$ then on the same time layer we assign values to X_{k+1} as

$$X_{k+1} = X'_{k+1}.$$

However, if the realized value of X'_{k+1} goes outside of $\bar{G}$ then we need an additional construction so that $X_{k+1} \in G$. First, we find the projection of X'_{k+1} onto ∂G which we denote as X_{k+1}^{π} and we calculate $r_{k+1} = \text{dist}(X'_{k+1}, X_{k+1}^{\pi})$ which is the shortest distance between X'_{k+1} and X_{k+1}^{π}. Note that $\text{dist}(X_k, X'_{k+1}) = \mathcal{O}(h^{1/2})$, therefore under smoothness assumptions made at the beginning of this section (see more precise assumptions in [233]) and for sufficiently small $h > 0$, the projection X_{k+1}^{π} of X'_{k+1} on ∂G is unique. Moreover, the projection X_{k+1}^{π} and the shortest distance r_{k+1} satisfy the equation $X_{k+1}^{\pi} = X'_{k+1} + r_{k+1}\nu(X_{k+1}^{\pi})$, where $\nu(X_{k+1}^{\pi})$ is the inward normal vector to the boundary ∂G at X_{k+1}^{π}. Thereafter, we add $r_{k+1}\nu(X_{k+1}^{\pi})$ to X_{k+1}^{π} to arrive at a point which we take as X_{k+1}. This transition from intermediate step X'_{k+1} to X_{k+1} makes sure that $X_{k+1} \in G$. We also highlight that X'_{k+1} and X_{k+1} are symmetrical around X_{k+1}^{π} along the direction $\nu(X_{k+1}^{\pi})$. Therefore, combining the above steps of calculating X_{k+1}^{π} from X'_{k+1} and then X_{k+1} from X_{k+1}^{π}, we have

$$X_{k+1} = X'_{k+1} + 2r_{k+1}\nu(X_{k+1}^{\pi}). \tag{7.6.36}$$

The equations (7.6.6)–(7.6.7) are approximated as follows. If the intermediate step X'_{k+1} belongs to $\bar{G}$ then we use the Euler scheme:

$$Y_{k+1} = Y_k + hc(t_k, X_k)Y_k, \tag{7.6.37}$$

$$Z_{k+1} = Z_k + hg(t_k, X_k)Y_k. \tag{7.6.38}$$

If $X'_{k+1} \notin \bar{G}$ then

$$\begin{aligned} Y_{k+1} &= Y_k + hc(t_k, X_k)Y_k + 2r_{k+1}\varphi(t_{k+1}, X_{k+1}^{\pi})Y_k \\ &\quad + 2r_{k+1}^2\varphi^2(t_{k+1}, X_{k+1}^{\pi})Y_k, \end{aligned} \tag{7.6.39}$$

$$\begin{aligned} Z_{k+1} &= Z_k + hg(t_k, X_k)Y_k - 2r_{k+1}\psi(t_{k+1}, X_{k+1}^{\pi})Y_k \\ &\quad - 2r_{k+1}^2\psi(t_{k+1}, X_{k+1}^{\pi})\varphi(t_{k+1}, X_{k+1}^{\pi})Y_k. \end{aligned} \tag{7.6.40}$$

The corresponding algorithm is formally written below.

Algorithm 7.6.5 *Algorithm to approximate (7.6.5)–(7.6.7).*

STEP 1. Set $X_0 = x$, $Y_0 = 1$, $Z_0 = 0$, $X'_0 = x$, $k = 0$.
STEP 2. Simulate ξ_{k+1} and find X'_{k+1} using (7.6.35).
STEP 3. If $X'_{k+1} \in \bar{G}$ then $X_{k+1} = X'_{k+1}$ and calculate Y_{k+1} and Z_{k+1} according to (7.6.37) and (7.6.38), respectively, **else** find X_{k+1}, Y_{k+1} and Z_{k+1} according to (7.6.36), (7.6.39) and (7.6.40), respectively.
STEP 4. $k + 1 = N$ then **stop**, **else** put $k := k + 1$ and **return** to STEP 2.

Following the same ideas as in the previous subsections, the finite-time convergence theorem was proved in [233].

Theorem 7.6.6 *The weak order of accuracy of Algorithm 7.6.5 is $O(h)$, i.e., for sufficiently small $h > 0$*

$$| E(f(X_N)Y_N + Z_N) - u(t_0, X_0) | \leq Ch, \qquad (7.6.41)$$

where $u(t, x)$ is solution of (7.6.1)–(7.6.3) and C is a positive constant independent of h.

We remark that in [233] the following aspects were also considered (i) computing ergodic limits in the domain G and on the boundary ∂G using the weak approximation as in Algorithm 7.6.5 of the reflected SDEs (7.6.1), (ii) sampling from distributions with compact support using reflected stochastic gradient systems (cf. Sect. 5.11), and (iii) solving linear elliptic equations with Robin boundary conditions.

Chapter 8
Probabilistic Approach to Numerical Solution of the Cauchy Problem for Nonlinear Parabolic Equations

Nonlinear partial differential equations (PDEs) of parabolic type are of great interest in both theoretical and applied aspects. To mention just a few, the Burgers equation, Fisher-Kolmogorov-Petrovskii-Piskunov (FKPP) and Ginzburg-Landau equations, equations with nonlinear diffusion and with blow-up solutions, reaction-diffusion systems are examples of parabolic nonlinear PDEs. They are suggested as mathematical models of problems in many fields such as fluid dynamics, filtration, combustion, biochemistry, dynamics of populations, etc. Nonlinear PDEs are usually not susceptible of analytic solution and mostly investigated by means of numerical methods. They are usually treated using a variety of deterministic approaches (see, e.g., [68, 144, 222, 396, 408, 437] and references therein), while probabilistic approaches to nonlinear PDEs has received only a limited attention (see [111, 220, 434] and references therein and also Chap. 10 on FBSDEs here).

An original probabilistic approach to constructing layer methods for solving nonlinear PDEs of parabolic type was proposed in [284]. It is based on the well-known probabilistic representations of solutions of linear parabolic equations and on the ideas of weak sense numerical integration of SDEs (see Chap. 2). In spite of the probabilistic nature these methods are nevertheless deterministic.

We start this chapter (Sect. 8.1) with the probabilistic approach to the Cauchy problem for linear parabolic equations. The approach cannot be carried over to nonlinear problems directly. However, its local version can be generalized to the Cauchy problem for semilinear parabolic equations. This is done in Sect. 8.2, where some layer methods are constructed in the nonlinear case. A practical realization of layer methods requires space discretization and an interpolation. We propose a number of numerical algorithms and prove their convergence. The principal ideas are demonstrated in the one-dimensional case. The multi-dimensional case is shortly discussed in Sect. 8.3. In addition, we show how the results obtained can be extended to reaction-diffusion systems. Some numerical tests are presented in Sect. 8.4.

The second part of this chapter (Sects. 8.5-8.7) deals with application of the probabilistic approach to the Cauchy problem for semilinear parabolic equations with small parameter [305]. Nonlinear parabolic equations with small parameter arise in

© Springer Nature Switzerland AG 2021

G. N. Milstein and M. V. Tretyakov, *Stochastic Numerics for Mathematical Physics*, Scientific Computation, https://doi.org/10.1007/978-3-030-82040-4_8

a variety of applications (see, e.g., [68, 107, 178, 192, 384] and references therein). For instance, they are used in gas dynamics, when one has to take into account small viscosity and small heat conductivity. Some problems of combustion are described by PDEs with small parameter. They also arise as the result of introducing artificial viscosity in systems of first-order hyperbolic equations that is one of the popular approaches to numerical solution of inviscid problems of gas dynamics [333, 378, 454]. The probabilistic representations of the solution to the Cauchy problem for semilinear parabolic equations with small parameter are connected with systems of SDEs with small noise. To construct effective layer methods for this problem, we exploit special weak approximations for SDEs with small noise proposed in Chap. 4.

The probabilistic approach considered in this chapter takes into account a coefficient dependence on the space variables and a relationship between diffusion and advection in an intrinsic manner. In particular, the layer methods allow us to avoid difficulties stemming from essentially variable coefficients and strong advection.

Other probabilistic numerical methods for nonlinear PDEs are available, e.g., in [76, 220, 247, 320, 434] and also in Chap. 10 of this book. Layer methods for boundary value problems for parabolic PDEs are considered in the next chapter. They were also used for solving Navier-Stokes equations (see [325, 327] and references therein). Layer methods for SPDEs are presented in Chap. 11.

8.1   Probabilistic Approach to Linear Parabolic Equations

Consider the Cauchy problem for linear parabolic equation

$$\frac{\partial u}{\partial t} + \frac{1}{2}\sum_{i,j=1}^{d} a^{ij}(t,x)\frac{\partial^2 u}{\partial x^i \partial x^j} + \sum_{i=1}^{d} b^i(t,x)\frac{\partial u}{\partial x^i} + c(t,x)u + g(t,x) = 0,$$

$$(8.1.1)$$

$$t_0 \le t < T, \quad x \in \mathbf{R}^d,$$

with the initial condition

$$u(T,x) = \varphi(x). \qquad (8.1.2)$$

The matrix $a(t,x) = \{a^{ij}(t,x)\}$ is assumed to be symmetric and positive semidefinite.

Let $\sigma(t,x)$ be a matrix obtained from the equation

$$a(t,x) = \sigma(t,x)\sigma^\mathsf{T}(t,x).$$

This equation is solvable with respect to σ (for instance, by a lower triangular matrix) at least for a positively definite a.

The solution to the problem (8.1.1)–(8.1.2) has various probabilistic representations (cf. Sect. 2.3.2):

$$u(t, x) = E[\varphi(X_{t,x}(T)) Y_{t,x,1}(T) + Z_{t,x,1,0}(T)], \quad t \le T, \ x \in \mathbf{R}^d, \quad (8.1.3)$$

where $X_{t,x}(s)$, $Y_{t,x,y}(s)$, $Z_{t,x,y,z}(s)$, $s \ge t$, is the solution of the Cauchy problem for the system of SDEs

$$dX = b(s, X)ds - \sigma(s, X)\mu(s, X)ds + \sigma(s, X)dw(s), \quad X(t) = x, \quad (8.1.4)$$
$$dY = c(s, X)Yds + \mu^{\mathsf{T}}(s, X)Ydw(s), \quad Y(t) = y, \quad (8.1.5)$$
$$dZ = g(s, X)Yds + F^{\mathsf{T}}(s, X)Ydw(s), \quad Z(t) = z. \quad (8.1.6)$$

Here $w(s) = (w^1(s), \ldots, w^d(s))^{\mathsf{T}}$ is a d-dimensional standard Wiener process, $b(s, x)$ is the column-vectors of dimension d composed from the coefficients $b^i(s, x)$, $\mu(s, x)$ and $F(s, x)$ are arbitrary column-vectors of dimension d, Y and Z are scalars. The usual representation (see [86]) can be seen in (8.1.3)–(8.1.6) if $\mu = 0$, $F = 0$. The case $F = 0$ rests on Girsanov's theorem. For $F \ne 0$, the representation (8.1.3) is evidently true as well. We note that the representations (8.1.3)–(8.1.6) are the well-known Feynman–Kac formula.

In what follows it is assumed that all the coefficients of (8.1.1) and (8.1.4)–(8.1.6) and the solution of the problem (8.1.1)–(8.1.2) (which is assumed to exist and to be unique) are sufficiently smooth and satisfy some conditions of growth under large $|x|$ so that these conditions are sufficient for applying the theory of weak methods (see Chap. 2).

Let us consider a time discretization, for definiteness the equidistant one:

$$T = t_N > t_{N-1} > \cdots > t_0 = t, \quad \frac{T - t_0}{N} = h.$$

Recall that a weak approximation of the system (8.1.4)–(8.1.6) consists in construction of the system of stochastic difference equations

$$X_0 = x, \ X_{m+1} = X_m + A(t_m, X_m, h; \xi_m), \quad (8.1.7)$$
$$Y_0 = 1, \ Y_{m+1} = Y_m + \alpha(t_m, X_m, h; \xi_m)Y_m, \quad (8.1.8)$$
$$Z_0 = 0, \ Z_{m+1} = Z_m + \beta(t_m, X_m, h; \xi_m)Y_m, \quad (8.1.9)$$
$$m = 0, 1, \ldots, N - 1,$$

where X_m is a d-dimensional vector, Y_m and Z_m are scalars, ξ_m is a random vector of a certain dimension, A is a d-dimensional vector function, α and β are scalar functions, ξ_m is independent of $X_0, \ldots, X_m$ and $\xi_0, \ldots, \xi_{m-1}$.

Let the system (8.1.7)–(8.1.9) be a weak scheme of order p for the system (8.1.4)–(8.1.6), i.e., (see Chap. 2)

$$\bar{u}(t_0, x) = \bar{u}(t, x) := E(\varphi(X_N)Y_N + Z_N) = u(t, x) + R_N, \quad (8.1.10)$$

where
$$|R_N| \le K(1 + |x|^{\varkappa})h^p,$$

and $K > 0$, $\varkappa \ge 0$ are some constants.

Standard numerical methods for PDEs, including the finite difference ones (see, e.g., [370, 375, 395, 417, 455]), can successfully be applied to solve the problem (8.1.1)–(8.1.2) provided the dimension d of the space variable x is comparatively small ($d \le 3$) while for larger dimensions these numerical procedures become unrealistic due to a huge volume of computations. Fortunately, in many cases functionals only or even individual values of a solution have to be found. A probabilistic approach for such problems has an essential advantage as long as the problem under consideration can be reduced to solving the corresponding system of ordinary stochastic differential equations.

Let us recall (see details in Chap. 2) that the probabilistic representation (8.1.3)–(8.1.6) and its approximation (8.1.10), (8.1.7)–(8.1.9) give an example of the approach which allows to find the individual values $u(t, x)$ of the solution of problem (8.1.1)–(8.1.2) even in high-dimensional ($d > 3$) cases. The value $\bar{u}(t, x)$ is evaluated by applying the Monte-Carlo technique:

$$\bar{u}(t, x) \approx \frac{1}{M} \sum_{m=1}^{M} (\varphi(X_N^{(m)}) Y_N^{(m)} + Z_N^{(m)}),$$

where $(X_N^{(m)}, Y_N^{(m)}, Z_N^{(m)})$, $m = 1, \ldots, M$, are independent realizations of the process defined by (8.1.7)–(8.1.9).

But the probabilistic approach is useful not only in this respect. Here we exploit it to construct some *layer methods*. To demonstrate this, let us consider the Cauchy problem

$$X_k = x, \quad X_{l+1} = X_l + A(t_l, X_l, h; \xi_l), \tag{8.1.11}$$
$$Y_k = y, \quad Y_{l+1} = Y_l + \alpha(t_l, X_l, h; \xi_l) Y_l, \tag{8.1.12}$$
$$Z_k = z, \quad Z_{l+1} = Z_l + \beta(t_l, X_l, h; \xi_l) Y_l, \tag{8.1.13}$$
$$l = k, k+1, \ldots, N-1; \quad 0 \le k \le N-1,$$

which is related to the system (8.1.7)–(8.1.9). Denote by $\bar{X}_{t_k, x}(t_l)$, $\bar{Y}_{t_k, x, y}(t_l)$, $\bar{Z}_{t_k, x, y, z}(t_l)$, $t_l \ge t_k$, the solution of this problem.

Introduce the function (recall that $T = t_N$):

$$\bar{u}(t_k, x, y, z) = E(\varphi(\bar{X}_{t_k, x}(T)) \bar{Y}_{t_k, x, y}(T) + \bar{Z}_{t_k, x, y, z}(T)).$$

Clearly, the function $\bar{u}(t_k, x, y, z)$ can be written as

$$\bar{u}(t_k, x, y, z) = \bar{u}(t_k, x)y + z,$$

where

$$\bar{u}(t_k, x) = E(\varphi(\bar{X}_{t_k,x}(T))\bar{Y}_{t_k,x,1}(T) + \bar{Z}_{t_k,x,1,0}(T)).$$

Let $t = t_0 \le t_k < t_l \le T$. Since

$$\bar{X}_{t_k,x}(T) = \bar{X}_{t_l, \bar{X}_{t_k,x}(t_l)}(T)$$

$$\bar{Y}_{t_k,x,1}(T) = \bar{Y}_{t_l, \bar{X}_{t_k,x}(t_l), \bar{Y}_{t_k,x,1}(t_l)}(T)$$

$$\bar{Z}_{t_k,x,1,0}(T) = \bar{Z}_{t_l, \bar{X}_{t_k,x}(t_l), \bar{Y}_{t_k,x,1}(t_l), \bar{Z}_{t_k,x,1,0}(t_l)}(T),$$

we have

$$\bar{u}(t_k, x) = EE[\varphi(\bar{X}_{t_l, \bar{X}_{t_k,x}(t_l)}(T))\bar{Y}_{t_l, \bar{X}_{t_k,x}(t_l), \bar{Y}_{t_k,x,1}(t_l)}(T) \tag{8.1.14}$$

$$+\bar{Z}_{t_l, \bar{X}_{t_k,x}(t_l), \bar{Y}_{t_k,x,1}(t_l), \bar{Z}_{t_k,x,1,0}(t_l)}(T) / \bar{X}_{t_k,x}(t_l), \bar{Y}_{t_k,x,1}(t_l), \bar{Z}_{t_k,x,1,0}(t_l)]$$

$$= E(\bar{u}(t_l, \bar{X}_{t_k,x}(t_l))\bar{Y}_{t_k,x,1}(t_l) + \bar{Z}_{t_k,x,1,0}(t_l)), \quad \bar{u}(t_N, x) = \varphi(x).$$

Using (8.1.14) sequentially with $l = k + 1$:

$$\bar{u}(t_k, x) = E(\bar{u}(t_{k+1}, \bar{X}_{t_k,x}(t_{k+1}))\bar{Y}_{t_k,x,1}(t_{k+1}) + \bar{Z}_{t_k,x,1,0}(t_{k+1})), \tag{8.1.15}$$

$$k = N - 1, \ldots, 0,$$

one can recurrently find the approximate solution $\bar{u}(t_{N-1}, x), \bar{u}(t_{N-2}, x), \ldots, \bar{u}(t_0, x)$ of the problem (8.1.1)–(8.1.2) starting from

$$\bar{u}(t_N, x) = \varphi(x). \tag{8.1.16}$$

This method is deterministic if we are able to calculate the expectations explicitly (see, e.g., the formulas (8.1.20) or (8.1.23) below). To realize (8.1.15) numerically, it is sufficient to calculate the functions $\bar{u}(t_k, x)$ at some nodes x_i applying some kind of interpolation at every layer.

Further, it is more convenient to clarify further ideas on simple examples. To this end, let us consider the following one-dimensional ($d = 1$) problem for the heat equation:

$$\frac{\partial u}{\partial t} + \frac{\sigma^2}{2} \frac{\partial^2 u}{\partial x^2} = 0, \quad t < 0, \quad x \in \mathbf{R}, \quad u(0, x) = \varphi(x). \tag{8.1.17}$$

Since $c = 0$, $g = 0$, we omit the equations for Y and Z. We have

$$dX = \sigma dw(s), \quad X(t_0) = x, \quad t_0 < 0. \tag{8.1.18}$$

Example 8.1.1 Consider the weak Euler scheme

$$X_{k+1} = X_k + \sigma\sqrt{h}\xi_k, \quad X_0 = x, \tag{8.1.19}$$

where $P(\xi_k = \pm 1) = 1/2$.

If we set $l = k + 1$ in (8.1.14), we obtain

$$\bar{u}(t_k, x) = E\bar{u}(t_{k+1}, \bar{X}_{t_k,x}(t_{k+1})) \tag{8.1.20}$$

$$= \frac{1}{2}\bar{u}(t_{k+1}, x - \sigma\sqrt{h}) + \frac{1}{2}\bar{u}(t_{k+1}, x + \sigma\sqrt{h}),$$

$$\bar{u}(t_N, x) = \varphi(x).$$

Here $t_N = 0$, $h = -t_0/N$, $t_k = -h(N - k) = t_{k+1} - h$, $k = N - 1, \ldots, 0$. The relation (8.1.20) is a linear difference equation. The equation (8.1.19) can be considered as a characteristic one for (8.1.20), and the formula

$$\bar{u}(t_k, x) = E\varphi(\bar{X}_{t_k,x}(t_N)) \tag{8.1.21}$$

gives the probabilistic representation of the solution to the equation (8.1.20). It is well known (see Chap. 2) that this solution differs from the solution of the problem (8.1.17) by a quantity of order $O(h)$.

It is not difficult to see that evaluation of the values $\bar{u}(t_k, x_i)$ by layers due to the formula (8.1.20) coincides with the simplest explicit finite difference scheme for solving (8.1.17) if we set $h_t = h$, $h_x = \sigma\sqrt{h}$ and consider the equidistant space discretization : $x_i = x_0 + i\sigma\sqrt{h}$, $i = 0, \pm 1, \pm 2, \ldots$, x_0 is a point belonging to **R**.

If we need the solution of (8.1.17) for all points (t_k, x_i), we can use (8.1.20) to find $\bar{u}(t_k, x_i)$ layerwise. But if we need the solution at a separate point (t_k, x), the formula (8.1.21) is more convenient. Of course, in the last case the Monte Carlo error arises in addition.

Example 8.1.2 Now consider a more general scheme than (8.1.19):

$$X_{k+1} = X_k + \alpha\sqrt{h}\eta_k, \quad X_0 = x, \tag{8.1.22}$$

where the constant $\alpha \geq \sigma$, $P(\eta = \pm 1) = \sigma^2/2\alpha^2$, $P(\eta = 0) = 1 - \sigma^2/\alpha^2$. Instead of (8.1.20), we get

$$\bar{u}(t_k, x) = E\bar{u}(t_{k+1}, \bar{X}_{t_k,x}(t_{k+1})) = \left(1 - \frac{\sigma^2}{\alpha^2}\right)\bar{u}(t_{k+1}, x) \tag{8.1.23}$$

$$+ \frac{\sigma^2}{2\alpha^2}\bar{u}(t_{k+1}, x - \alpha\sqrt{h}) + \frac{\sigma^2}{2\alpha^2}\bar{u}(t_{k+1}, x + \alpha\sqrt{h}),$$

$$\bar{u}(t_N, x) = \varphi(x).$$

Again due to the theory of weak methods for SDEs, the formula (8.1.21) with $\bar{X}$ from (8.1.22) approximates the solution of the problem (8.1.17) with accuracy $O(h)$. The formula (8.1.21) can be realized either by the Monte Carlo method or layerwise according to (8.1.23). The layer realization (8.1.23) is deterministic and coincides (if we choose the corresponding space grid) with the following finite difference scheme

$$\frac{\bar{u}(t_k, x_i) - \bar{u}(t_{k+1}, x_i)}{h_t} = \frac{\sigma^2}{2} \frac{\bar{u}(t_{k+1}, x_{i+1}) - 2\bar{u}(t_{k+1}, x_i) + \bar{u}(t_{k+1}, x_{i-1})}{h_x^2},$$

$$h_t = h, \quad h_x = \alpha\sqrt{h}.$$

$$(8.1.24)$$

Due to the Lax-Richtmyer equivalence theorem [375, 395], the method (8.1.24) (or, what is the same, the method (8.1.23)) converges with the rate $O(h)$ if $\alpha \geq \sigma$. If $\alpha < \sigma$, the numerical approximation (8.1.24) is not stable from the point of view of the theory of finite difference methods, and the method (8.1.24) diverges. We underline that there is no probabilistic scheme of the form (8.1.11)–(8.1.13), (8.1.15) corresponding to (8.1.24) with $\alpha < \sigma$. Convergence theorems for weak methods (in comparison with the theory of finite difference methods) do not contain any conditions on stability of their approximations. Methods of the probabilistic nature like (8.1.11)–(8.1.13), (8.1.15) are intrinsic to a problem considered (especially when its coefficients are nonconstant) because the suitable choice of h_x is achieved automatically.

Let us note that the methods (8.1.20) and (8.1.23) do not need any interpolation because the layer $\bar{u}(t_k, x_i)$ makes use of the previous layer $\bar{u}(t_{k+1}, x)$ at the nodes x_j only. But such a property of layer methods under consideration is rather exception than a rule. In conclusion, let us give two other examples.

Example 8.1.3 Consider the scheme (8.1.22) with $\alpha = \sigma\sqrt{3}$:

$$X_{k+1} = X_k + \sigma\sqrt{3h}\eta_k, \quad X_0 = x,$$

$$(8.1.25)$$

where $P(\eta = \pm 1) = 1/6$, $P(\eta = 0) = 2/3$. Since

$$E\eta = E\eta^3 = 0, \quad E(\sqrt{3}\eta)^2 = 1, \quad E(\sqrt{3}\eta)^4 = 3,$$

this scheme has the second order of accuracy (cf. Sect. 2.2).

We obtain the following finite difference method from (8.1.25):

$$\bar{u}(t_k, x_i) = \frac{1}{6}(\bar{u}(t_{k+1}, x_{i+1}) + \bar{u}(t_{k+1}, x_{i-1})) + \frac{2}{3}\bar{u}(t_{k+1}, x_i),$$

$$(8.1.26)$$

where $x_{i+1} - x_i = \sigma\sqrt{3h}$.

Since the scheme (8.1.25) is of the second order, the method (8.1.26) is also of order 2, i.e., $|u(t_k, x_i) - \bar{u}(t_k, x_i)| = O(h^2)$.

Example 8.1.4 Consider another scheme

$$X_{k+1} = X_k + \sigma\sqrt{h}\zeta_k, \quad X_0 = x, \tag{8.1.27}$$

where $P(\zeta = 0) = p, \ P(\zeta = \pm\alpha) = q, \ P(\zeta = \pm\beta) = r$.

If, for example, $\alpha = 1, \ \beta = \sqrt{6}, \ p = 1/3, \ q = 3/10, \ r = 1/30$, then

$$E\zeta = E\zeta^3 = E\zeta^5 = 0, \ E\zeta^2 = 1, \ E\zeta^4 = 3, \ E\zeta^6 = 15,$$

and the scheme is of order 3 (cf. Sect. 3.1.3). The corresponding method

$$\bar{u}(t_k, x) = E\bar{u}(t_{k+1}, \bar{X}_{t_k,x}(t_{k+1})) = E\bar{u}(t_{k+1}, x + \sigma\sqrt{h}\zeta_k) \tag{8.1.28}$$

$$= \frac{1}{30}\bar{u}(t_{k+1}, x - \sigma\sqrt{6h}) + \frac{3}{10}\bar{u}(t_{k+1}, x - \sigma\sqrt{h}) + \frac{1}{3}\bar{u}(t_{k+1}, x)$$

$$+ \frac{3}{10}\bar{u}(t_{k+1}, x + \sigma\sqrt{h}) + \frac{1}{30}\bar{u}(t_{k+1}, x + \sigma\sqrt{6h})$$

is of order 3 too. But an interpolation is necessary for numerical realization of (8.1.28) on some grid of nodes x_i because of incommensurability of $\sigma\sqrt{h}$ and $(\sigma\sqrt{6h} - \sigma\sqrt{h})$.

Let us note in passing that, for example, the scheme

$$X_{k+1} = X_k + \sigma\sqrt{h}\nu_k, \quad X_0 = x,$$

where $P(\nu = 0) = 7/18, \ P(\nu = \pm 1) = 1/4, \ P(\nu = \pm 2) = 1/20, \ P(\nu = \pm 3) = 1/180$, also induces a method of order 3. Evidently, this method has the form

$$\bar{u}(t_k, x_i) = \frac{1}{180}\bar{u}(t_{k+1}, x_{i-3}) + \frac{1}{20}\bar{u}(t_{k+1}, x_{i-2}) + \frac{1}{4}\bar{u}(t_{k+1}, x_{i-1}) \tag{8.1.29}$$

$$+ \frac{7}{18}\bar{u}(t_{k+1}, x_i) + \frac{1}{4}\bar{u}(t_{k+1}, x_{i+1}) + \frac{1}{20}\bar{u}(t_{k+1}, x_{i+2})$$

$$+ \frac{1}{180}\bar{u}(t_{k+1}, x_{i+3}),$$

where $x_{i+1} - x_i = \sigma\sqrt{h}$.

Remark 8.1.5 Consider the Cauchy problem for an autonomous linear parabolic equation in its usual form (with positive direction of time θ):

$$\frac{\partial v}{\partial \theta} = \frac{1}{2}\sum_{i,j=1}^{d} a^{ij}(x)\frac{\partial^2 v}{\partial x^i \partial x^j} + \sum_{i=1}^{d} b^i(x)\frac{\partial v}{\partial x^i} + c(x)v + g(x), \ \theta > 0, \ x \in \mathbf{R}^d,$$

$$\tag{8.1.30}$$

$$v(0, x) = \varphi(x). \tag{8.1.31}$$

Changing the variables $t = \theta$, $u(t, x) = v(-t, x)$, we get the Cauchy problem of the form (8.1.1)–(8.1.2) for the function $u(t, x)$ where $t < 0$, $x \in \mathbf{R}^d$, $T = 0$, $u(T, x) = \varphi(x)$. In the considered case the system (8.1.4)–(8.1.6) is autonomous as well (we suppose the function $\mu(s, x)$ in (8.1.4)–(8.1.6) to be independent of s). Therefore (see (8.1.3))

$$v(0, x) = u(-0, x) = E(\varphi(X_{-\theta,x}(0))Y_{-\theta,x,1}(0) + Z_{-\theta,x,1,0}(0))$$
$$= E(\varphi(X_{0,x}(\theta))Y_{0,x,1}(\theta) + Z_{0,x,1,0}(\theta)) , \ \theta > 0, \ x \in \mathbf{R}^d,$$

i.e., we can consider the positive direction of time for both the parabolic equation and its characteristic system of SDEs. According to this fact, we can write the following more convenient procedure in place of (8.1.15), (8.1.16):

$$\bar{v}(0, x) = \varphi(x), \tag{8.1.32}$$

$$\bar{v}(\theta_{k+1}, x) = E(\bar{v}(\theta_k, \bar{X}_{0,x}(h))\bar{Y}_{0,x,1}(h) + \bar{Z}_{0,x,1,0}(h)) , \ k = 0, \dots, N - 1,$$

where $0 = \theta_0 < \theta_1 < \cdots < \theta_N = \theta$; $h = \theta/N$ (of course, we consider A, α, and β in the scheme (8.1.11)–(8.1.13) to be independent of t_m). At the same time, we preferred to follow the general style of our exposition in Examples 8.1.1-8.1.4.

8.2 Layer Methods for Semilinear Parabolic Equations

For simplicity, we restrict ourselves to the one-dimensional case $d = 1$ in this section.

8.2.1 The Construction of Layer Methods

Consider the Cauchy problem

$$\frac{\partial u}{\partial t} + \frac{1}{2}\sigma^2(t, x, u)\frac{\partial^2 u}{\partial x^2} + b(t, x, u)\frac{\partial u}{\partial x} + g(t, x, u) = 0, \ t_0 \le t < T , \ x \in \mathbf{R},$$
$$\tag{8.2.1}$$

$$u(T, x) = \varphi(x). \tag{8.2.2}$$

Let $u = u(t, x)$ be the solution of the problem (8.2.1)–(8.2.2), which is assumed to exist, to be unique, to be sufficiently smooth, and to satisfy some conditions of boundedness. One can find many theoretical results on this topic in [144, 222, 396, 408, 437] (see also references therein). If we substitute $u = u(t, x)$ in the coefficients σ^2, b, g, we obtain a linear parabolic equation. We assume that all the requirements mentioned in the previous section in connection with the equation

(8.1.1) are fulfilled for the obtained linear equation as well. Let us note that in comparison with (8.1.1) this linear equation does not contain a term linear in u. It is so due to the general form of g in (8.2.1). Sometimes it may be preferable to represent $g(t, x, u)$ as $g(t, x, u) = c(t, x)u + g_0(t, x, u)$ (for instance, in the case of small $g_0(t, x, u)$) and to substitute $u = u(t, x)$ in the function g_0 only. Clearly, in this case we obtain another linear equation and another probabilistic representation. For definiteness, we shall consider the case without term linear in u, i.e., we take $c(s, x) \equiv 0$, and we also put $\mu(s, x) \equiv 0$, $F(s, x) \equiv 0$ in the system (8.1.4)–(8.1.6). We have (see (8.1.3) with $Y \equiv 1$)

$$u(t, x) = E(\varphi(X_{t,x}(T)) + \int_t^T g(s, X_{t,x}(s), u(s, X_{t,x}(s)))ds) , \quad t \le T, \quad x \in \mathbf{R},$$

(8.2.3)

where $X_{t,x}(s)$ is the solution of the Cauchy problem for the equation

$$dX = b(s, X, u(s, X))ds + \sigma(s, X, u(s, X))dw(s), \quad X(t) = x .$$

Consider the equidistant time discretization

$$T = t_N > t_{N-1} > \cdots > t_0 = t , \quad \frac{T - t_0}{N} = h .$$

Due to (8.2.3), we have

$$u(t_k, x) = E(u(t_{k+1}, X_{t_k,x}(t_{k+1}))$$ (8.2.4)

$$+ \int_{t_k}^{t_{k+1}} g(s, X_{t_k,x}(s), u(s, X_{t_k,x}(s)))ds)$$

$$= E(u(t_{k+1}, X_{t_k,x}(t_{k+1})) + Z_{t_k,x,0}(t_{k+1})) ,$$

where X, Z satisfy the following system

$$dX = b(s, X, u(s, X))ds + \sigma(s, X, u(s, X))dw(s), \quad X(t_k) = x ,$$ (8.2.5)

$$dZ = g(s, X, u(s, X))ds, \quad Z(t_k) = 0 .$$ (8.2.6)

Applying the explicit weak Euler scheme with the simplest simulation of noise to the system (8.2.5)–(8.2.6), we get

$$X_{t_k,x}(t_{k+1}) \simeq \bar{X}_{t_k,x}(t_{k+1}) = x + b(t_k, x, u(t_k, x))h + \sigma(t_k, x, u(t_k, x))\sqrt{h}\xi_k ,$$

(8.2.7)

$$Z_{t_k,x,0}(t_{k+1}) \simeq \bar{Z}_{t_k,x,0}(t_{k+1}) = g(t_k, x, u(t_k, x))h ,$$ (8.2.8)

where $\xi_{N-1}, \xi_{N-2}, \ldots, \xi_0$ are i.i.d. random variables which are distributed by the law $P(\xi = \pm 1) = 1/2$.

Using (8.2.4), we obtain

$$u(t_k, x) \simeq E(u(t_{k+1}, \bar{X}_{t_k,x}(t_{k+1})) + \bar{Z}_{t_k,x,0}(t_{k+1})) \qquad (8.2.9)$$

$$= \frac{1}{2}u(t_{k+1}, x + b(t_k, x, u(t_k, x))h + \sigma(t_k, x, u(t_k, x))\sqrt{h})$$

$$+ \frac{1}{2}u(t_{k+1}, x + b(t_k, x, u(t_k, x))h - \sigma(t_k, x, u(t_k, x))\sqrt{h})$$

$$+ g(t_k, x, u(t_k, x))h.$$

Following (8.2.9), one can write for the approximations $\bar{u}(t_k, x)$:

$$\bar{u}(t_N, x) = \varphi(x), \qquad (8.2.10)$$

$$\bar{u}(t_k, x) = \frac{1}{2}\bar{u}(t_{k+1}, x + b(t_k, x, \bar{u}(t_k, x))h + \sigma(t_k, x, \bar{u}(t_k, x))\sqrt{h})$$

$$+ \frac{1}{2}\bar{u}(t_{k+1}, x + b(t_k, x, \bar{u}(t_k, x))h - \sigma(t_k, x, \bar{u}(t_k, x))\sqrt{h})$$

$$+ g(t_k, x, \bar{u}(t_k, x))h,$$

$$k = N - 1, \ldots, 1, 0.$$

The method (8.2.10) is an implicit layer method for solution of the Cauchy problem (8.2.1)–(8.2.2). This method is deterministic though the probabilistic approach is used for its derivation. Recall that it rests on the explicit Euler scheme.

Now let us use the following implicit scheme instead of (8.2.7)–(8.2.8):

$$\bar{X}_{t_k,x}(t_{k+1}) := \bar{X}_{k+1} = x + b(t_{k+1}, \bar{X}_{k+1}, u(t_{k+1}, \bar{X}_{k+1}))h \qquad (8.2.11)$$

$$+ \sigma(t_{k+1}, \bar{X}_k, u(t_{k+1}, \bar{X}_k))\sqrt{h}\xi_k,$$

$$\bar{Z}_{t_k,x,0}(t_{k+1}) := \bar{Z}_{k+1} = g(t_{k+1}, \bar{X}_{k+1}, u(t_{k+1}, \bar{X}_{k+1}))h, \qquad (8.2.12)$$

where $\xi_{N-1}, \xi_{N-2}, \ldots, \xi_0$ are the same as in (8.2.7).

Let $\bar{X}_{k+1} = \bar{X}_{k+1}(\xi_k)$ be the solution of (8.2.11) (recall that the function $u(t_{k+1}, x)$ is assumed to be known). The variable ξ_k takes two different values. Denote by $\bar{X}^1_{k+1}$, $\bar{X}^2_{k+1}$ the corresponding values of $\bar{X}_{k+1}$. Introduce the analogous notation for the two values of $\bar{Z}_{k+1}$. As a result, we obtain the method

$$\bar{u}(t_N, x) = \varphi(x), \qquad (8.2.13)$$

$$\bar{u}(t_k, x) = \frac{1}{2}(\bar{u}(t_{k+1}, \bar{X}^1_{k+1}) + \bar{Z}^1_{k+1}) + \frac{1}{2}(\bar{u}(t_{k+1}, \bar{X}^2_{k+1}) + \bar{Z}^2_{k+1}).$$

It is deterministic just as the method (8.2.10).

The formula (8.2.13) is explicit but to find $\bar{X}_{k+1}$ we have to use the implicit scheme (8.2.11). Therefore, both the method (8.2.10) and the method (8.2.13) are implicit.

To find $\bar{u}(t_k, x)$ from (8.2.10), one can apply the method of simple iteration. If we take $\bar{u}(t_{k+1}, x)$ as a null iteration, we get the following first iteration (we denote this iteration as $\bar{u}(t_k, x)$ again):

$$\bar{u}(t_N, x) = \varphi(x), \tag{8.2.14}$$

$$\bar{u}(t_k, x) = \frac{1}{2}\bar{u}(t_{k+1}, x + b(t_k, x, \bar{u}(t_{k+1}, x))h + \sigma(t_k, x, \bar{u}(t_{k+1}, x))\sqrt{h})$$

$$+ \frac{1}{2}\bar{u}(t_{k+1}, x + b(t_k, x, \bar{u}(t_{k+1}, x))h - \sigma(t_k, x, \bar{u}(t_{k+1}, x))\sqrt{h})$$

$$+ g(t_k, x, \bar{u}(t_{k+1}, x))h,$$

$$k = N - 1, \ldots, 1, 0.$$

The formula (8.2.14) gives an explicit method for recurrent layerwise solution of the problem (8.2.1)–(8.2.2). We note that if we apply another approximate method to solve (8.2.10) (for example, taking the second iteration), we obtain another explicit method which can possess better properties than (8.2.14) (just as in numerical integration of ordinary differential equations).

Analogously, applying the method of simple iteration to (8.2.11) with x as a null iteration and substituting the obtained first iteration in (8.2.12) and (8.2.13), we obtain the following explicit method which slightly differs from (8.2.14):

$$\bar{u}(t_N, x) = \varphi(x), \tag{8.2.15}$$

$$\bar{u}(t_k, x) = \frac{1}{2}\bar{u}(t_{k+1}, \ x + b(t_{k+1}, x, \bar{u}(t_{k+1}, x))h + \sigma(t_{k+1}, x, \bar{u}(t_{k+1}, x))\sqrt{h})$$

$$+ \frac{1}{2}\bar{u}(t_{k+1}, \ x + b(t_{k+1}, x, \bar{u}(t_{k+1}, x))h - \sigma(t_{k+1}, x, \bar{u}(t_{k+1}, x))\sqrt{h})$$

$$+ g(t_{k+1}, x, \bar{u}(t_{k+1}, x))h,$$

$$k = N - 1, \ldots, 1, 0.$$

Now we make use of a higher-order method of numerical integration of SDEs in the case of Eq. (8.2.1) with constant σ. Let us apply the second order (in the weak sense) Runge–Kutta scheme (see Sect. 2.2.2) to the system (8.2.5)–(8.2.6) with constant σ. We obtain (instead of (8.2.7)–(8.2.8)):

$$X_{t_k,x}(t_{k+1}) \simeq \bar{X}_{t_k,x}(t_{k+1}) = x + \sigma\sqrt{h}\xi_k + \frac{1}{2}b(t_k, x, u(t_k, x))h \tag{8.2.16}$$

$$+ \frac{1}{2}b(t_{k+1}, x + b(t_k, x, u(t_k, x))h$$

$$+ \sigma\sqrt{h}\xi_k, u(t_{k+1}, x + b(t_k, x, u(t_k, x))h + \sigma\sqrt{h}\xi_k))h,$$

$$Z_{t_k,x,0}(t_{k+1}) \simeq \bar{Z}_{t_k,x,0}(t_{k+1}) \tag{8.2.17}$$

$$= \frac{1}{2}g(t_k, x, u(t_k, x))h + \frac{1}{2}g(t_{k+1}, x + b(t_k, x, u(t_k, x))h$$

$$+ \sigma\sqrt{h}\xi_k, u(t_{k+1}, x + b(t_k, x, u(t_k, x))h + \sigma\sqrt{h}\xi_k))h,$$

where $\xi_{N-1}, \xi_{N-2}, \ldots, \xi_0$ are i.i.d. random variables distributed by the law $P(\xi = 0) = 2/3$, $P(\xi = \pm\sqrt{3}) = 1/6$.

Then, we obtain the following implicit layer method instead of (8.2.10):

$$\bar{u}(t_N, x) = \varphi(x), \tag{8.2.18}$$

$$\bar{u}(t_k, x) = \frac{2}{3}\bar{u}\left(t_{k+1}, x + \frac{1}{2}\bar{b}h + \frac{1}{2}b(t_{k+1}, x + \bar{b}h, \bar{u}(t_{k+1}, x + \bar{b}h))h\right)$$

$$+ \frac{1}{6}\bar{u}(t_{k+1}, x + \sigma\sqrt{3h} + \frac{1}{2}\bar{b}h$$

$$+ \frac{1}{2}b(t_{k+1}, x + \sigma\sqrt{3h} + \bar{b}h, \bar{u}(t_{k+1}, x + \sigma\sqrt{3h} + \bar{b}h))h)$$

$$+ \frac{1}{6}\bar{u}(t_{k+1}, x - \sigma\sqrt{3h} + \frac{1}{2}\bar{b}h$$

$$+ \frac{1}{2}b(t_{k+1}, x - \sigma\sqrt{3h} + \bar{b}h, \bar{u}(t_{k+1}, x - \sigma\sqrt{3h} + \bar{b}h))h)$$

$$+ \frac{1}{2}g(t_k, x, \bar{u}(t_k, x))h + \frac{1}{3}g(t_{k+1}, x + \bar{b}h, \bar{u}(t_{k+1}, x + \bar{b}h))h$$

$$+ \frac{1}{12}g(t_{k+1}, x + \sigma\sqrt{3h} + \bar{b}h, \bar{u}(t_{k+1}, x + \sigma\sqrt{3h} + \bar{b}h))h$$

$$+ \frac{1}{12}g(t_{k+1}, x - \sigma\sqrt{3h} + \bar{b}h, \bar{u}(t_{k+1}, x - \sigma\sqrt{3h} + \bar{b}h))h,$$

where $\bar{b} = b(t_k, x, \bar{u}(t_k, x))$.

This method has the one-step error of order 3. If we take $\bar{u}(t_{k+1}, x)$ as a null iteration, we obtain the first iteration differing from the solution of (8.2.18) by a quantity of order $O(h^2)$, and only starting from the second iteration we attain the needed order of accuracy. So, the implicit method (8.2.18) becomes the explicit one of the same order after two simple iterations.

Clearly, resting on the ideas leading to the methods obtained, one can construct a lot of new methods using some other probabilistic representations or other methods of numerical integration of SDEs.

8.2.2 Convergence Theorem for a Layer Method

We continue to treat the problem (8.2.1)–(8.2.2).

We assume (recall that for simplicity in writing the case $d = 1$ is considered):

(i) The coefficients $b(t, x, u)$, $\sigma(t, x, u)$, $g(t, x, u)$ are uniformly bounded:

$$|b| \leq K, \ |\sigma| \leq K, \ |g| \leq K, \ t_0 \leq t \leq T, \ x \in \mathbf{R}, \ u_\circ < u < u^\circ, \quad (8.2.19)$$

where $-\infty \leq u_\circ < u^\circ \leq \infty$ are some constants.

(ii) The coefficients $b(t, x, u)$, $\sigma(t, x, u)$, $g(t, x, u)$ uniformly satisfy the Lipschitz condition with respect to x and u :

$$|b(t, x_2, u_2) - b(t, x_1, u_1)| + |\sigma(t, x_2, u_2) - \sigma(t, x_1, u_1)| \quad (8.2.20)$$
$$+ \ |g(t, x_2, u_2) - g(t, x_1, u_1)|$$
$$\leq K(|x_2 - x_1| + |u_2 - u_1|), \ t_0 \leq t \leq T, \ x_1, x_2 \in \mathbf{R}, \ u_\circ < u_1, u_2 < u^\circ.$$

(iii) There exists a unique bounded solution $u(t, x)$ of the problem (8.2.1)–(8.2.2) such that

$$u_\circ < u_* \leq u(t, x) \leq u^* < u^\circ, \ t_0 \leq t \leq T, \ x \in \mathbf{R}, \quad (8.2.21)$$

and there exist the uniformly bounded derivatives:

$$\left| \frac{\partial^m u}{\partial t^i \partial x^l} \right| \leq K, \ i = 0, \ l = 1, 2, 3, 4; \ i = 1, \ l = 0, 1, 2; \ i = 2, \ l = 0;$$
$$(8.2.22)$$

$$t_0 \leq t \leq T, \ x \in \mathbf{R}.$$

First of all let us evaluate the one-step error of the method (method (8.2.14)):

$$\bar{u}(t_N, x) = \varphi(x), \quad (8.2.23)$$

$$\bar{u}(t_k, x) = \frac{1}{2}\bar{u}(t_{k+1}, x + b(t_k, x, \bar{u}(t_{k+1}, x))h + \sigma(t_k, x, \bar{u}(t_{k+1}, x))\sqrt{h})$$

$$+ \frac{1}{2}\bar{u}(t_{k+1}, x + b(t_k, x, \bar{u}(t_{k+1}, x))h - \sigma(t_k, x, \bar{u}(t_{k+1}, x))\sqrt{h})$$

$$+ \ g(t_k, x, \bar{u}(t_{k+1}, x))h \,,$$

$$k = N - 1, \ldots, 1, 0.$$

This error on the k-th layer (on the $(N - k)$-th step) is evidently equal to $v(t_k, x) - u(t_k, x)$, where

$$v(t_k, x) = \frac{1}{2}u(t_{k+1}, x + b(t_k, x, u(t_{k+1}, x))h + \sigma(t_k, x, u(t_{k+1}, x))\sqrt{h})$$

$$\text{(8.2.24)}$$

$$+ \frac{1}{2}u(t_{k+1}, x + b(t_k, x, u(t_{k+1}, x))h - \sigma(t_k, x, u(t_k, x))\sqrt{h})$$

$$+ g(t_k, x, u(t_{k+1}, x))h.$$

Lemma 8.2.1 *Under the assumptions (i)–(iii) the one-step error of the method (8.2.23) has the second order of smallness with respect to h :*

$$|v(t_k, x) - u(t_k, x)| \le Ch^2, \qquad \text{(8.2.25)}$$

where $C > 0$ does not depend on x, h, k.

Proof Expanding the functions $u(t_k + h, x + bh \pm \sigma\sqrt{h})$ at (t_k, x) in powers of h and $bh \pm \sigma\sqrt{h}$ and using the assumptions on boundedness (8.2.19) and (8.2.22), we get

$$v(t_k, x) = u(t_k, x) + \frac{\partial u}{\partial t}(t_k, x)h + \frac{\partial u}{\partial x}(t_k, x)bh \qquad \text{(8.2.26)}$$

$$+ \frac{1}{2}\frac{\partial^2 u}{\partial x^2}(t_k, x)\sigma^2 h + gh + O(h^2).$$

In (8.2.26) the coefficients b, σ^2, g have t_k, x, $u(t_{k+1}, x)$ as their arguments, and

$$|O(h^2)| \le Ch^2, \qquad \text{(8.2.27)}$$

where C does not depend on x, h, k. Now applying the Lipschitz condition (8.2.20) with respect to the variable u, it is not difficult to obtain

$$v(t_k, x) = u(t_k, x) + \frac{\partial u}{\partial t}(t_k, x)h + \frac{\partial u}{\partial x}(t_k, x)b(t_k, x, u(t_k, x))h \qquad \text{(8.2.28)}$$

$$+ \frac{1}{2}\frac{\partial^2 u}{\partial x^2}(t_k, x)\sigma^2(t_k, x, u(t_k, x))h + g(t_k, x, u(t_k, x))h + O(h^2),$$

where $O(h^2)$ satisfies the relation (8.2.27) again.

Because $u(t, x)$ is a solution of Eq. (8.2.1), the inequality (8.2.25) follows from (8.2.28). □

Theorem 8.2.2 *Under the assumptions (i)–(iii) the method (8.2.23) has the first order of convergence, i.e.,*

$$|\bar{u}(t_k, x) - u(t_k, x)| \le Kh, \qquad \text{(8.2.29)}$$

where $K > 0$ does not depend on x, h, k.

Proof Denote the error of the method (8.2.23) on the k-th layer as $R(t_k, x) := \bar{u}(t_k, x) - u(t_k, x)$. Then, we have

$$\bar{u}(t_k, x) = u(t_k, x) + R(t_k, x), \quad \bar{u}(t_{k+1}, x) = u(t_{k+1}, x) + R(t_{k+1}, x). \quad (8.2.30)$$

By (8.2.23) and (8.2.30), we get

$$u(t_k, x) + R(t_k, x) \qquad\qquad\qquad\qquad\qquad\qquad\qquad (8.2.31)$$

$$= \bar{u}(t_k, x) = \frac{1}{2}\bar{u}(t_{k+1}, x + \bar{b}h + \bar{\sigma}\sqrt{h}) + \frac{1}{2}\bar{u}(t_{k+1}, x + \bar{b}h - \bar{\sigma}\sqrt{h}) + \bar{g}h$$

$$= \frac{1}{2}u(t_{k+1}, x + \bar{b}h + \bar{\sigma}\sqrt{h}) + \frac{1}{2}u(t_{k+1}, x + \bar{b}h - \bar{\sigma}\sqrt{h}) + \bar{g}h$$

$$+ \frac{1}{2}\varepsilon(t_{k+1}, x + \bar{b}h + \bar{\sigma}\sqrt{h}) + \frac{1}{2}\varepsilon(t_{k+1}, x + \bar{b}h - \bar{\sigma}\sqrt{h}),$$

where $\bar{b}$, $\bar{\sigma}$, $\bar{g}$ are the coefficients $b(t, x, u)$, $\sigma(t, x, u)$, $g(t, x, u)$ evaluated at $t = t_k$, $x = x$, $u = \bar{u}(t_{k+1}, x) = u(t_{k+1}, x) + R(t_{k+1}, x)$. For example, $\bar{b} = b(t_k, x, u(t_{k+1}, x) + R(t_{k+1}, x))$.

Here we have to assume for a while that the value $u(t_{k+1}, x) + R(t_{k+1}, x)$ remains in the interval $(u_\circ, u^\circ)$ (see the conditions (8.2.19) and (8.2.20)). Clearly, $R(t_N, x) = 0$, and below we prove recurrently that $R(t_k, x)$ is sufficiently small under a sufficiently small h. Thereupon, thanks to (8.2.21), the above assumption will be justified.

We have

$$\bar{b} = b(t_k, x, u(t_{k+1}, x) + \varepsilon(t_{k+1}, x)) = b(t_k, x, u(t_{k+1}, x)) + \Delta b = b + \Delta b,$$

where $b := b(t_k, x, u(t_{k+1}, x))$ and Δb satisfies the inequality (thanks to (8.2.20))

$$|\Delta b| \leq K|\varepsilon(t_{k+1}, x)|. \qquad\qquad\qquad\qquad\qquad (8.2.32)$$

Analogously,

$$\bar{\sigma} = \sigma + \Delta\sigma, \; |\Delta\sigma| \leq K|\varepsilon(t_{k+1}, x)|, \; \bar{g} = g + \Delta g, \; |\Delta g| \leq K|\varepsilon(t_{k+1}, x)|.$$
$$(8.2.33)$$

It is not difficult to see that (8.2.32) and (8.2.33) imply the equalities

$$u(t_{k+1}, x + \bar{b}h \pm \bar{\sigma}\sqrt{h}) = u(t_{k+1}, x + bh \pm \sigma\sqrt{h}) \qquad\qquad (8.2.34)$$

$$+ \frac{\partial u}{\partial x}(t_{k+1}, x + bh)\,(\Delta bh \pm \Delta\sigma\sqrt{h}) + \Delta_\pm h,$$

where $\Delta_\pm$ satisfy the inequality of the type (8.2.32). Substituting this in (8.2.31), we obtain

$$u(t_k, x) + R(t_k, x) \tag{8.2.35}$$

$$= \frac{1}{2}u(t_{k+1}, x + bh + \sigma\sqrt{h}) + \frac{1}{2}u(t_{k+1}, x + bh - \sigma\sqrt{h}) + gh$$

$$+ \frac{1}{2}R(t_{k+1}, x + \bar{b}h + \bar{\sigma}\sqrt{h}) + \frac{1}{2}R(t_{k+1}, x + \bar{b}h - \bar{\sigma}\sqrt{h}) + r_k$$

$$= v(t_k, x) + \frac{1}{2}R(t_{k+1}, x + \bar{b}h + \bar{\sigma}\sqrt{h}) + \frac{1}{2}R(t_{k+1}, x + \bar{b}h - \bar{\sigma}\sqrt{h}) + r_k,$$

where

$$|r_k| \leq K|\varepsilon(t_{k+1}, x)|h. \tag{8.2.36}$$

Finally, using Lemma 8.2.1, we get

$$R(t_k, x) = \frac{1}{2}R(t_{k+1}, x + \bar{b}h + \bar{\sigma}\sqrt{h}) + \frac{1}{2}R(t_{k+1}, x + \bar{b}h - \bar{\sigma}\sqrt{h}) + r_k + O(h^2). \tag{8.2.37}$$

Now introduce

$$R_k := \sup_{-\infty < x < \infty} |R(t_k, x)|. \tag{8.2.38}$$

It follows from (8.2.36) and (8.2.37) (in addition recall that $R(t_N, x) = 0$):

$$R_N = 0, \ R_k \leq R_{k+1} + K R_{k+1}h + Ch^2, \ k = N-1, \ldots, 1, 0, \tag{8.2.39}$$

which implies

$$R_k \leq \frac{C}{K}\left(e^{K(T-t_0)} - 1\right)h, \ k = N, \ldots, 0.$$

Theorem 8.2.2 is proved. □

Remark 8.2.3 The result (8.2.29) for the method (8.2.23) can be justified under some other conditions as well. For instance, it is possible to allow a linear growth of the coefficients b, σ, g under $|x| \to \infty$ instead of the condition (i) if we assume in addition that the derivatives of the solution $u(t, x)$ from (8.2.22) are not only bounded but some of them go to zero under $|x| \to \infty$ (namely, if the expressions $\left|\frac{\partial^m u}{\partial t^i \partial x^l}\right|(1 + |x|^l)$, $i = 0, l = 1, 2, 3, 4$; $i = 1, l = 1, 2$, are uniformly bounded). Besides, we emphasize that the conditions of Theorem 8.2.2 are not necessary and the method (8.2.23) can be applied much broader than it is determined by (i)–(iii). At the same time, the conditions (i)–(iii) are acceptable in many situations.

8.2.3 Numerical Algorithms

A recursive procedure can be applied to implement the method (8.2.23). But for large $T - t_0$ and small h such a procedure is computationally too expensive. To

avoid recursive calculations and to construct a numerical algorithm, the method
(8.2.23) (just as other layer methods) needs a discretization in the variable x. Consider
the equidistant space discretization: $x_j = x_0 + j\alpha h$, $j = 0, \pm 1, \pm 2, \ldots$, $x_0 \in \mathbf{R}$,
$\alpha > 0$ is a number, i.e., h_x is taken to be equal to $\alpha h = \alpha h_t$. Using, for example, the
linear interpolation, we construct the following algorithm.

Algorithm 8.2.4 *The algorithm is defined by the following formulas*

$$\bar{u}(t_N, x) = \varphi(x), \qquad (8.2.40)$$

$$\bar{u}(t_k, x_j) = \frac{1}{2}\bar{u}(t_{k+1}, x_j + b(t_{k+1}, x_j, \bar{u}(t_{k+1}, x_j))h$$
$$+ \sigma(t_{k+1}, x_j, \bar{u}(t_{k+1}, x_j))\sqrt{h})$$
$$+\frac{1}{2}\bar{u}(t_{k+1}, x_j + b(t_{k+1}, x_j, \bar{u}(t_{k+1}, x_j))h - \sigma(t_{k+1}, x_j, \bar{u}(t_{k+1}, x_j))\sqrt{h})$$
$$+g(t_k, x_j, \bar{u}(t_{k+1}, x_j))h, \ \ x_j = x_0 + j\alpha h, \ \ j = 0, \pm 1, \pm 2, \ldots,$$

$$\bar{u}(t_k, x) = \frac{x_{j+1} - x}{\alpha h}\bar{u}(t_k, x_j) + \frac{x - x_j}{\alpha h}\bar{u}(t_k, x_{j+1}), \ \ x_j \le x \le x_{j+1}, \qquad (8.2.41)$$
$$k = N - 1, \ldots, 1, 0.$$

Theorem 8.2.5 *Under the assumptions* (i)–(iii) *Algorithm 8.2.4 has the first order
of convergence, i.e., the approximation* $\bar{u}(t_k, x)$ *from the formula* $(8.2.41)$ *satisfies
the relation*

$$|\bar{u}(t_k, x) - u(t_k, x)| \le Kh, \qquad (8.2.42)$$

where K does not depend on x, h, k.

Proof Let us introduce the error of Algorithm 8.2.4 on the k-th layer

$$R(t_k, x) := \bar{u}(t_k, x) - u(t_k, x)$$

and R_k in accordance with (8.2.38):

$$R_k := \sup_{-\infty < x < \infty} |R(t_k, x)|.$$

Of course, these new $R(t_k, x)$ and R_k differ from the old ones. Just as earlier, we are
able to obtain for the nodes x_j (cf. (8.2.37)):

$$R(t_k, x_j) = \frac{1}{2}R(t_{k+1}, x_j + \bar{b}h + \bar{\sigma}\sqrt{h}) + \frac{1}{2}R(t_{k+1}, x_j + \bar{b}h - \bar{\sigma}\sqrt{h})$$
$$+ r_k + O(h^2).$$

Hence

$$|R(t_k, x_j)| \leq R_{k+1} + K R_{k+1} h + C h^2 . \tag{8.2.43}$$

We have

$$u(t_k, x) = \frac{x_{j+1} - x}{\alpha h} u(t_k, x_j) + \frac{x - x_j}{\alpha h} u(t_k, x_{j+1}) + O(h^2), \quad x_j \leq x \leq x_{j+1}, \tag{8.2.44}$$

where the interpolation error $O(h^2)$ satisfies the inequality of the form (8.2.27).

We get from (8.2.44) and (8.2.41):

$$R(t_k, x) = \frac{x_{j+1} - x}{\alpha h} R(t_k, x_j) + \frac{x - x_j}{\alpha h} R(t_k, x_{j+1}) + O(h^2), \quad x_j \leq x \leq x_{j+1},$$

whence due to (8.2.43) for all x :

$$|R(t_k, x)| \leq R_{k+1} + K R_{k+1} h + C h^2, \tag{8.2.45}$$

of course, with another constant C.

The inequality (8.2.45) implies (8.2.39). Theorem 8.2.5 is proved. □

Remark 8.2.6 Along with the linear interpolation (8.2.41) it is natural to use the spline approximation of the form

$$\bar{u}(t_k, x) = \sum_{i=-\infty}^{\infty} \bar{u}(t_k, x_i) B \left(\frac{x - x_i}{\alpha h} \right), \quad x_i = x_0 + i\alpha h, \ x \in \mathbf{R}, \tag{8.2.46}$$

$$k = N - 1, \dots, 1, 0,$$

where $B(x)$ is the standard cubic B-spline

$$B(x) = \begin{cases} \frac{2}{3} - x^2 + \frac{1}{2}|x|^3, & |x| \leq 1, \\ \frac{1}{6}(2 - |x|)^3, & 1 \leq |x| \leq 2, \\ 0, & |x| \geq 2. \end{cases}$$

The spline (8.2.46) is twice continuously differentiable, and because $B(x)$ is locally supported, the series (8.2.46) does not have more than four nonzero terms for any $x \in \mathbf{R}$.

It is known (see, e.g., [32]) that the spline $\Lambda(x) = \sum_{i=-\infty}^{\infty} f(x_i) B \left(\frac{x - x_i}{\alpha h} \right)$ possesses good approximating and smoothing properties. In particular, if there exists a third derivative of $f(x)$ and it is bounded then there exist constants C_1 and C_2 such that

$$|f(x) - \Lambda(x)| \leq C_1 h^2, \ |f'(x) - \Lambda'(x)| \leq C_2 h, \ x \in \mathbf{R} .$$

And since the sequence $B_i(x) = B\left(\dfrac{x - x_i}{\alpha h}\right)$ provides a nonnegative partition of unity:

$$\sum_{i=-\infty}^{\infty} B_i(x) = 1, \quad B_i(x) \geq 0, \quad \text{all } i,$$

the proof of Theorem 8.2.5 can be carried over for the case of the approximation (8.2.46).

Remark 8.2.7 To reduce the amount of nodes x_j, it is natural to take advantage of cubic interpolation with step $h_x = \beta\sqrt{h}$ instead of linear interpolation with step $h_x = \alpha h$. Then we obtain the following algorithm:

$$\bar{u}(t_N, x) = \varphi(x), \tag{8.2.47}$$

$$\bar{u}(t_k, x_j) = \frac{1}{2}\bar{u}(t_{k+1}, x_j + b(t_{k+1}, x_j, \bar{u}(t_{k+1}, x_j))h$$

$$+ \sigma(t_{k+1}, x_j, \bar{u}(t_{k+1}, x_j))\sqrt{h})$$

$$+ \frac{1}{2}\bar{u}(t_{k+1}, \; x_j + b(t_{k+1}, x_j, \bar{u}(t_{k+1}, x_j))h - \sigma(t_{k+1}, x_j, \bar{u}(t_{k+1}, x_j))\sqrt{h})$$

$$+ g(t_k, x_j, \bar{u}(t_{k+1}, x_j))h, \; x_j = x_0 + j\beta\sqrt{h}, \; j = 0, \pm 1, \pm 2, \dots,$$

$$\bar{u}(t_k, x) = \sum_{i=0}^{3} \Phi_{j,i}(x)\,\bar{u}(t_k, x_{j+i}), \; x_{j+1} \leq x \leq x_{j+2}, \tag{8.2.48}$$

$$\Phi_{j,i}(x) = \prod_{k=0, k\neq i}^{3} \frac{x - x_{j+k}}{x_{j+i} - x_{j+k}}, \; k = N - 1, \dots, 1, 0.$$

It can be shown as earlier that the inequality (8.2.43) holds for the algorithm (8.2.47)–(8.2.48). But in place of (8.2.44) we get

$$u(t_k, x) = \sum_{i=0}^{3} \Phi_{j,i}(x)\,u(t_k, x_{j+i}) + O(h_x^4), \; x_{j+1} \leq x \leq x_{j+2},$$

and, consequently,

$$R(t_k, x) = \sum_{i=0}^{3} \Phi_{j,i}(x)\,R(t_k, x_{j+i}) + O(h^2), \; x_{j+1} \leq x \leq x_{j+2}. \tag{8.2.49}$$

Though $\sum_{i=0}^{3} \Phi_{j,i}(x) = 1$ for any x, the sum of the absolute values $\sum_{i=0}^{3} |\Phi_{j,i}(x)|$ can take values greater than one. And instead of the inequality (8.2.45), we can obtain the following one only:

$$|R(t_k, x)| \leq AR_{k+1} + KR_{k+1}h + Ch^2,$$

where the constant A is unfortunately greater than one. Therefore, our proof of Theorem 8.2.5 cannot be carried over for the case of cubic interpolation. At the same time, a number of numerical experiments showed a good quality of the procedure (8.2.47)–(8.2.48) (see, e.g., Sect. 8.7).

Remark 8.2.8 Consider the Cauchy problem for an autonomous semilinear parabolic equation with positive direction of time t:

$$\frac{\partial u}{\partial t} = \frac{1}{2}\sigma^2(x, u)\frac{\partial^2 u}{\partial x^2} + b(x, u)\frac{\partial u}{\partial x} + g(x, u), \ t > 0, \ x \in \mathbf{R}, \tag{8.2.50}$$

$$u(0, x) = \varphi(x). \tag{8.2.51}$$

If we substitute a solution $u(t, x)$ of the problem (8.2.50)–(8.2.51) in the coefficients σ, b, g, Eq. (8.2.50) becomes nonautonomous and that is why the reasoning of Remark 8.1.5 cannot be carried over for the problem (8.2.50)–(8.2.51). Nevertheless, the following procedure with positive direction of time can be obtained from (8.2.40)–(8.2.41):

$$\bar{u}(0, x) = \varphi(x), \tag{8.2.52}$$

$$\bar{u}(t_{k+1}, x_j) = \frac{1}{2}\bar{u}(t_k, x_j + b(x_j, \bar{u}(t_k, x_j))h + \sigma(x_j, \bar{u}(t_k, x_j))\sqrt{h})$$

$$+ \frac{1}{2}\bar{u}(t_k, x_j + b(x_j, \bar{u}(t_k, x_j))h - \sigma(x_j, \bar{u}(t_k, x_j))\sqrt{h})$$

$$+ g(x_j, \bar{u}(t_k, x_j))h,$$

$$x_j = x_0 + j\alpha h, \ j = 0, \pm 1, \pm 2, \ldots, \ t_k = kh, \ h = t/N,$$

$$\bar{u}(t_k, x) = \frac{x_{j+1} - x}{\alpha h}\bar{u}(t_k, x_j) + \frac{x - x_j}{\alpha h}\bar{u}(t_k, x_{j+1}), \ x_j \leq x \leq x_{j+1}, \tag{8.2.53}$$

$$k = 0, 1, \ldots, N - 1.$$

This procedure is used in numerical examples in Sect. 8.4.

8.3 Multi-dimensional Case

8.3.1 Multi-dimensional Parabolic Equation

Consider the Cauchy problem for multi-dimensional semilinear parabolic equation

$$\frac{\partial u}{\partial t} + \frac{1}{2} \sum_{i,j=1}^{d} a^{ij}(t,x,u) \frac{\partial^2 u}{\partial x^i \partial x^j} + \sum_{i=1}^{d} b^i(t,x,u) \frac{\partial u}{\partial x^i} + g(t,x,u) = 0, \quad (8.3.1)$$

$$t_0 \leq t < T, \ x \in \mathbf{R}^d,$$

$$u(T,x) = \varphi(x). \quad (8.3.2)$$

As in Sect. 8.2, we can write the same relations (8.2.3)–(8.2.8) here but with the distinction that x, X, and b are d-dimensional vectors, σ is a $d \times d$-dimensional matrix such that $\sigma\sigma^{\mathsf{T}} = a = \{a^{ij}\}$, and $\xi_{N-1}, \xi_{N-2}, \ldots, \xi_0$ in (8.2.7) are i.i.d. d-dimensional vectors with i.i.d. components ξ_k^i, $i = 1, \ldots, d$, and each component ξ^i is distributed by the law $P(\xi = \pm 1) = 1/2$.

Using (8.2.4), we obtain (here we restrict ourselves to the two-dimensional case for simplicity in writing):

$$u(t_k, x) = u(t_k, x^1, x^2) \quad (8.3.3)$$

$$\simeq Eu(t_{k+1}, \bar{X}^1_{t_k,x}(t_{k+1}), \bar{X}^2_{t_k,x}(t_{k+1})) + E\bar{Z}_{t_k,x,0}(t_{k+1})$$

$$= \frac{1}{4} u(t_{k+1}, x^1 + b^1 h + \sigma^{11}\sqrt{h} + \sigma^{12}\sqrt{h}, x^2 + b^2 h + \sigma^{21}\sqrt{h} + \sigma^{22}\sqrt{h})$$

$$+ \frac{1}{4} u(t_{k+1}, x^1 + b^1 h + \sigma^{11}\sqrt{h} - \sigma^{12}\sqrt{h}, x^2 + b^2 h + \sigma^{21}\sqrt{h} - \sigma^{22}\sqrt{h})$$

$$+ \frac{1}{4} u(t_{k+1}, x^1 + b^1 h - \sigma^{11}\sqrt{h} + \sigma^{12}\sqrt{h}, x^2 + b^2 h - \sigma^{21}\sqrt{h} + \sigma^{22}\sqrt{h})$$

$$+ \frac{1}{4} u(t_{k+1}, x^1 + b^1 h - \sigma^{11}\sqrt{h} - \sigma^{12}\sqrt{h}, x^2 + b^2 h - \sigma^{21}\sqrt{h} - \sigma^{22}\sqrt{h}) + gh,$$

where $b^i = b^i(t_k, x^1, x^2, u(t_k, x^1, x^2))$, $\sigma^{ij} = \sigma^{ij}(t_k, x^1, x^2, u(t_k, x^1, x^2))$, $i, j = 1, 2$, $g = g(t_k, x^1, x^2, u(t_k, x^1, x^2))$.

Then a method analogous to (8.2.23) has the form:

$$\bar{u}(t_N, x^1, x^2) = \varphi(x^1, x^2), \quad (8.3.4)$$

$$\bar{u}(t_k, x^1, x^2)$$

$$= \frac{1}{4}\bar{u}(t_{k+1}, x^1 + \bar{b}^1 h + \bar{\sigma}^{11}\sqrt{h} + \bar{\sigma}^{12}\sqrt{h}, x^2 + \bar{b}^2 h + \bar{\sigma}^{21}\sqrt{h} + \bar{\sigma}^{22}\sqrt{h})$$

$$+ \frac{1}{4}\bar{u}(t_{k+1}, x^1 + \bar{b}^1 h + \bar{\sigma}^{11}\sqrt{h} - \bar{\sigma}^{12}\sqrt{h}, x^2 + \bar{b}^2 h + \bar{\sigma}^{21}\sqrt{h} - \bar{\sigma}^{22}\sqrt{h})$$

$$+ \frac{1}{4}\bar{u}(t_{k+1}, x^1 + \bar{b}^1 h - \bar{\sigma}^{11}\sqrt{h} + \bar{\sigma}^{12}\sqrt{h}, x^2 + \bar{b}^2 h - \bar{\sigma}^{21}\sqrt{h} + \bar{\sigma}^{22}\sqrt{h})$$

$$+ \frac{1}{4}\bar{u}(t_{k+1}, x^1 + \bar{b}^1 h - \bar{\sigma}^{11}\sqrt{h} - \bar{\sigma}^{12}\sqrt{h}, x^2 + \bar{b}^2 h - \bar{\sigma}^{21}\sqrt{h} - \bar{\sigma}^{22}\sqrt{h}) + \bar{g}h ,$$

where $\bar{b}^i = b^i(t_k, x^1, x^2, \bar{u}(t_{k+1}, x^1, x^2))$, $\bar{\sigma}^{ij} = \sigma^{ij}(t_k, x^1, x^2, \bar{u}(t_{k+1}, x^1, x^2))$, $i, j = 1, 2$, $\bar{g} = g(t_k, x^1, x^2, \bar{u}(t_{k+1}, x^1, x^2))$, $k = N - 1, \ldots, 1, 0$.

Consider the equidistant space discretization: $x_j^1 = x_0^1 + j\alpha^1 h$, $x_l^2 = x_0^2 + l\alpha^2 h$, $j, l = 0, \pm 1, \pm 2, \ldots$, (x_0^1, x_0^2) is a point belonging to $\mathbf{R}^2$ and $\alpha^1 > 0$, $\alpha^2 > 0$ are some numbers, i.e., h_{x^1}, h_{x^2} are taken to be equal to $\alpha^1 h$, $\alpha^2 h$, respectively. Using the sequential linear interpolation, we construct the following algorithm based on the method (8.3.4):

$$\bar{u}(t_N, x^1, x^2) = \varphi(x^1, x^2), \tag{8.3.5}$$

$$\bar{u}(t_k, x_j^1, x_l^2)$$

$$= \frac{1}{4}\bar{u}(t_{k+1}, x_j^1 + \bar{b}^1 h + \bar{\sigma}^{11}\sqrt{h} + \bar{\sigma}^{12}\sqrt{h}, x_l^2 + \bar{b}^2 h + \bar{\sigma}^{21}\sqrt{h} + \bar{\sigma}^{22}\sqrt{h})$$

$$+ \frac{1}{4}\bar{u}(t_{k+1}, x_j^1 + \bar{b}^1 h + \bar{\sigma}^{11}\sqrt{h} - \bar{\sigma}^{12}\sqrt{h}, x_l^2 + \bar{b}^2 h + \bar{\sigma}^{21}\sqrt{h} - \bar{\sigma}^{22}\sqrt{h})$$

$$+ \frac{1}{4}\bar{u}(t_{k+1}, x_j^1 + \bar{b}^1 h - \bar{\sigma}^{11}\sqrt{h} + \bar{\sigma}^{12}\sqrt{h}, x_l^2 + \bar{b}^2 h - \bar{\sigma}^{21}\sqrt{h} + \bar{\sigma}^{22}\sqrt{h})$$

$$+ \frac{1}{4}\bar{u}(t_{k+1}, x_j^1 + \bar{b}^1 h - \bar{\sigma}^{11}\sqrt{h} - \bar{\sigma}^{12}\sqrt{h}, x_l^2 + \bar{b}^2 h - \bar{\sigma}^{21}\sqrt{h} - \bar{\sigma}^{22}\sqrt{h}) + \bar{g}h ,$$

where all the coefficients $\bar{b}$ and $\bar{\sigma}$ are evaluated at $t_k, x_j^1, x_l^2, \bar{u}(t_{k+1}, x_j^1, x_l^2)$,

$$\bar{u}(t_k, x^1, x^2) = \frac{x_{j+1}^1 - x^1}{\alpha^1 h}\frac{x_{l+1}^2 - x^2}{\alpha^2 h}\bar{u}(t_k, x_j^1, x_l^2) \tag{8.3.6}$$

$$+ \frac{x_{j+1}^1 - x^1}{\alpha^1 h}\frac{x^2 - x_l^2}{\alpha^2 h}\bar{u}(t_k, x_j^1, x_{l+1}^2)$$

$$+ \frac{x^1 - x_j^1}{\alpha^1 h}\frac{x_{l+1}^2 - x^2}{\alpha^2 h}\bar{u}(t_k, x_{j+1}^1, x_l^2)$$

$$+ \frac{x^1 - x_j^1}{\alpha^1 h}\frac{x^2 - x_l^2}{\alpha^2 h}\bar{u}(t_k, x_{j+1}^1, x_{l+1}^2) ,$$

$$x_j^1 \le x^1 \le x_{j+1}^1, \quad x_l^2 \le x^2 \le x_{l+1}^2, \quad (x^1, x^2) \ne (x_i^1, x_m^2),$$

$$i, m = 0, \pm 1, \pm 2, \ldots, \quad k = N - 1, \ldots, 1, 0 .$$

Remark 8.3.1 The sequential linear interpolation in (8.3.6) is not linear with respect to both variables x^1 and x^2. The following triangular interpolation of $u(t, x^1, x^2)$

$$\bar{u}(t_k, x^1, x^2) = \left(1 - \frac{x^1 - x_j^1}{\alpha^1 h} - \frac{x^2 - x_l^2}{\alpha^2 h}\right) \bar{u}(t_k, x_j^1, x_l^2) \tag{8.3.7}$$

$$+ \frac{x^2 - x_l^2}{\alpha^2 h} \bar{u}(t_k, x_j^1, x_{l+1}^2) + \frac{x^1 - x_j^1}{\alpha^1 h} \bar{u}(t_k, x_{j+1}^1, x_l^2)$$

is linear and it has an error of $O(h^2)$ just as the interpolation (8.3.6). This interpolation is not suitable for the all points (x^1, x^2) from the rectangle $\Pi_{j,l} = \{(x^1, x^2) : x_j^1 \leq x^1 \leq x_{j+1}^1, \; x_l^2 \leq x^2 \leq x_{l+1}^2\}$ by the same reason as it was mentioned in Remark 8.2.7. But for the points from the triangle with the corners (x_j^1, x_l^2), (x_j^1, x_{l+1}^2), (x_{j+1}^1, x_l^2), the interpolation (8.3.7) is suitable because

$$\frac{x^1 - x_j^1}{\alpha^1 h} + \frac{x^2 - x_l^2}{\alpha^2 h} \leq 1 \tag{8.3.8}$$

for these points.

For the other points of the rectangle $\Pi_{j,l}$, we have

$$\frac{x_{j+1}^1 - x^1}{\alpha^1 h} + \frac{x_{l+1}^2 - x^2}{\alpha^2 h} < 1, \tag{8.3.9}$$

and we can use the formula

$$\bar{u}(t_k, x^1, x^2) = \left(1 - \frac{x_{j+1}^1 - x^1}{\alpha^1 h} - \frac{x_{l+1}^2 - x^2}{\alpha^2 h}\right) \bar{u}(t_k, x_{j+1}^1, x_{l+1}^2) \tag{8.3.10}$$

$$+ \frac{x_{j+1}^1 - x^1}{\alpha^1 h} \bar{u}(t_k, x_j^1, x_{l+1}^2) + \frac{x_{l+1}^2 - x^2}{\alpha^2 h} \bar{u}(t_k, x_{j+1}^1, x_l^2).$$

Thus, the formulas (8.3.7) and (8.3.10) for $(x^1, x^2) \in \Pi_{j,l}$ satisfy (8.3.8) and (8.3.9), respectively, and they give the other suitable rule of interpolation.

Convergence theorems for the method (8.3.4) and for the algorithm (8.3.5) with the interpolation (8.3.6) or (8.3.7)–(8.3.10) are analogous to Theorems 8.2.2 and 8.2.5.

8.3.2 Probabilistic Approach to Reaction-Diffusion Systems

Reaction-diffusion systems have received a great deal of attention, motivated by both their widespread occurrence in models from physics, chemistry, and biology, and by the richness of the structure of their solution sets. Reaction-diffusion systems can model a number of interesting phenomena like, e.g., threshold behavior, multiple steady states and hysteresis, spatial patterns, moving fronts or pulses and oscillations (see, e.g. [144, 408]).

The methods constructed in previous sections can also be applied to the Cauchy problem for systems of reaction-diffusion equations of the form (for simplicity we write them for the one-dimensional x):

$$\frac{\partial u_q}{\partial t} + L_q u_q + g_q(t, x, u) = 0, \quad t_0 \leq t < T, \ x \in \mathbf{R}, \ q = 1, \ldots, n, \quad (8.3.11)$$

$$u_q(T, x) = \varphi_q(x), \quad (8.3.12)$$

where

$$u := (u_1, \ldots, u_n),$$

$$L_q := \frac{1}{2} \sigma_q^2(t, x, u) \frac{\partial^2}{\partial x^2} + b_q(t, x, u) \frac{\partial}{\partial x}.$$

It is not difficult to derive the method which is analogous to (8.2.23):

$$\bar{u}_q(t_N, x) = \varphi_q(x), \quad (8.3.13)$$

$$\bar{u}_q(t_k, x) = \frac{1}{2} \bar{u}_q(t_{k+1}, x + b_q(t_k, x, \bar{u}(t_{k+1}, x))h + \sigma_q(t_k, x, \bar{u}(t_{k+1}, x))\sqrt{h})$$
$$+ \frac{1}{2} \bar{u}_q(t_{k+1}, x + b_q(t_k, x, \bar{u}(t_{k+1}, x))h - \sigma_q(t_k, x, \bar{u}(t_{k+1}, x))\sqrt{h})$$
$$+ g_q(t_k, x, \bar{u}(t_{k+1}, x))h,$$

$$k = N - 1, \ldots, 1, 0,$$

and then the corresponding algorithm (see (8.2.40)–(8.2.41)).

The system (8.3.11) is such that the linear system of parabolic equations obtained after substituting $u = u(t, x)$ in the coefficients σ_q, b_q, g_q splits and, therefore, every parabolic equation can be solved separately. In connection with this fact, one can consider n separate simple systems of the type (8.2.5)–(8.2.6). Such a recipe is impossible for reaction-diffusion systems containing equations with derivatives of different functions among $u_1, \ldots, u_n$. Consider, for example, the system

$$\frac{\partial u_q}{\partial t} + \frac{1}{2} \sigma^2(t, x, u) \frac{\partial^2 u_q}{\partial x^2} + \sum_{j=1}^{n} \sigma(t, x, u) b_{jq}(t, x, u) \frac{\partial u_j}{\partial x} + g_q(t, x, u) = 0$$

$$(8.3.14)$$

with the conditions (8.3.12) (we pay attention that σ in (8.3.14) does not depend on q). In this case one can use the following probabilistic representation [271]:

$$u_q(t_k, x) = E \sum_{l=1}^{n} u_l(t_{k+1}, X_{t_k, x}(t_{k+1})) Y_{t_k, x, q}^l(t_{k+1}) + E Z_{t_k, x, q, 0}(t_{k+1}), \quad (8.3.15)$$

where $X_{t_k,x}(s)$, $Y^l_{t_k,x,q}(s)$, $Z_{t_k,x,q,0}(s)$ is the solution of the Cauchy problem for the system of SDEs

$$dX = \sigma(s, X, u(s, X))dw(s), \quad X(t_k) = x, \tag{8.3.16}$$

$$dY^j = \sum_{l=1}^{n} b_{jl}(s, X, u(s, X))Y^l dw(s), \quad Y^j(t_k) = \delta_{jq} = \begin{cases} 0, & j \neq q, \\ 1, & j = q, \end{cases}$$

$$dZ = \sum_{l=1}^{n} g_l(s, X, u(s, X))Y^l ds, \quad Z(t_k) = 0.$$

Now it is not difficult to derive the method which is analogous to (8.2.23):

$$\bar{u}_q(t_N, x) = \varphi_q(x), \tag{8.3.17}$$

$$\bar{u}_q(t_k, x) = \frac{1}{2} \sum_{l=1}^{n} \bar{u}_l(t_{k+1}, x + \sigma(t_k, x, \bar{u}(t_{k+1}, x))\sqrt{h})$$

$$\times (\delta_{lq} + b_{lq}(t_k, x, \bar{u}(t_{k+1}, x))\sqrt{h})$$

$$+ \frac{1}{2} \sum_{l=1}^{n} \bar{u}_l(t_{k+1}, x - \sigma(t_k, x, \bar{u}(t_{k+1}, x))\sqrt{h})$$

$$\times (\delta_{lq} - b_{lq}(t_k, x, \bar{u}(t_{k+1}, x))\sqrt{h}) + g_q(t_k, x, \bar{u}(t_{k+1}, x))h,$$

$$k = N - 1, \ldots, 1, 0,$$

and then the corresponding algorithm.

Convergence Theorems 8.2.2 and 8.2.5 can be carried over to these method and algorithm.

8.4 Numerical Examples

Example 8.4.1 Consider the quasilinear equation with power law nonlinearities (see, e.g., [396] and also Sect. 10.5.5 in this book):

$$\frac{\partial u}{\partial t} = \frac{\partial}{\partial x}(u^\alpha \frac{\partial u}{\partial x}) + u^{\alpha+1}, \quad t > 0, \ x \in \mathbf{R}, \tag{8.4.1}$$

where $\alpha > 0$ is a constant.

Equation (8.4.1) has blow-up solutions. In particular, it has the following blow-up automodelling solution (see Fig. 8.4.1 as well):

Fig. 8.4.1 The solution of (8.4.1)–(8.4.2) with $\alpha = 2$, $T_0 = 1$

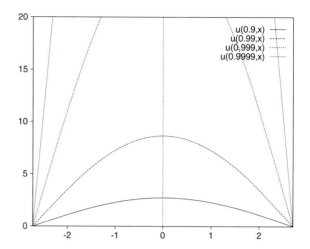

$$u(t, x) = \begin{cases} (T_0 - t)^{-1/\alpha} \left(\frac{2(\alpha+1)}{\alpha(\alpha+2)} \cos^2 \frac{\pi x}{L} \right)^{1/\alpha}, & |x| < \frac{L}{2}, \\ 0, & |x| \geq \frac{L}{2}, \ 0 < t < T_0, \end{cases} \tag{8.4.2}$$

where

$$L = \frac{2\pi}{\alpha}(\alpha + 1)^{1/2}.$$

The temperature $u(t, x)$ grows infinitely when $t \to T_0$. At the same time, the heat is localized in the interval $(-L/2, \ L/2)$. The function

$$v = \frac{1}{\alpha + 1} u^{\alpha+1}$$

satisfies the equation

$$\frac{\partial v}{\partial t} = \frac{1}{2} 2(\alpha + 1)^{\alpha/(\alpha+1)} v^{\alpha/(\alpha+1)} \frac{\partial^2 v}{\partial x^2} + (\alpha + 1)^{(2\alpha+1)/(\alpha+1)} v^{(2\alpha+1)/(\alpha+1)} \tag{8.4.3}$$

which is of the form (8.2.50).

We use the algorithm (8.2.52)–(8.2.53) to find the solution of (8.4.3) with $\alpha = 2$ and with the initial condition

$$v(0, x) = \begin{cases} \frac{\sqrt{3}}{8} \cos^3 \frac{\pi x}{L}, & |x| < \frac{L}{2}, \\ 0, & |x| \geq \frac{L}{2}. \end{cases} \tag{8.4.4}$$

In this case $T_0 = 1$.

Table 8.4.1 The absolute errors of the algorithm (8.2.52)–(8.2.53) for the Cauchy problem (8.4.3)–(8.4.4) at $t = 0.5$

	$h = 10^{-1}$	$h = 10^{-2}$	$h = 10^{-3}$	$h = 10^{-4}$
$err_{\bar{v}}$	$0.9931 \cdot 10^{-1}$	$1.422 \cdot 10^{-2}$	$1.489 \cdot 10^{-3}$	$1.500 \cdot 10^{-4}$
$err_{\bar{u}}$	$0.9572 \cdot 10^{-1}$	$1.643 \cdot 10^{-2}$	$4.432 \cdot 10^{-3}$	$13.77 \cdot 10^{-4}$
$err_{\bar{u}}[-2, 2]$	$0.7015 \cdot 10^{-1}$	$0.9552 \cdot 10^{-2}$	$1.215 \cdot 10^{-3}$	$1.003 \cdot 10^{-4}$

A dependence of the following errors on h

$$err_{\bar{v}} = \max_{x_i} |\bar{v}(t, x_i) - v(t, x_i)|,$$

$$err_{\bar{u}} = \max_{x_i} |\bar{u}(t, x_i) - u(t, x_i)|, \quad \bar{u}(t, x_i) = (3\bar{v}(t, x_i))^{1/3},$$

$$err_{\bar{u}}[-2, 2] = \max_{|x_i| \le 2} |\bar{u}(t, x_i) - u(t, x_i)|$$

is given in Table 8.4.1 for $t = 0.5$. We take $h_t = h_x = h$ here.

Rather large values of $err_{\bar{u}}$ are due to the fact that the values $v(t, x_i)$ for x_i close to the ends of the interval $(-L/2, L/2)$ are very small and, consequently, for such x_i

$$|\bar{u}(t, x_i) - u(t, x_i)| = |(3\bar{v}(t, x_i))^{1/3} - (3v(t, x_i))^{1/3}|$$
$$\simeq 3^{1/3} |\bar{v}(t, x_i) - v(t, x_i)|^{1/3},$$

i.e., $err_{\bar{u}} = O(h^{1/3})$. But the difference $\bar{u}(t, x_i) - u(t, x_i)$ on a subinterval $(-a, a)$, $a < L/2$, behaves as $O(h)$ (see the row $err_{\bar{u}}[-2, 2]$ in Table 8.4.1).

For times t which are close to the explosion time T_0, the errors $err_{\bar{v}}$ become fairly large (we pay attention that v in our example is proportional to u^3). However, if we are interested in finding the explosion time, it is natural to consider another characteristic when $t \to T_0$. Table 8.4.2 presents the values

Table 8.4.2 The relative errors $\delta(t, h)$ of the algorithm (8.2.52)–(8.2.53) for the Cauchy problem (8.4.3)–(8.4.4) and the explosion time

	$h = 10^{-1}$	$h = 10^{-2}$	$h = 10^{-3}$	$h = 10^{-4}$
$t = 0.9$	$3.644 \cdot 10^{-1}$	$1.024 \cdot 10^{-1}$	$1.313 \cdot 10^{-2}$	$1.353 \cdot 10^{-3}$
$t = 0.99$	–	$5.298 \cdot 10^{-1}$	$1.815 \cdot 10^{-1}$	$2.585 \cdot 10^{-2}$
$t = 0.999$	–	–	$6.167 \cdot 10^{-1}$	$2.436 \cdot 10^{-1}$
$t = 0.9999$	–	–	–	$6.704 \cdot 10^{-1}$
t^*	1.5	1.07	1.008	1.0001

$$\delta(t, h) = \frac{err_{\bar{u}}}{u(t, 0)}$$

and the time t^* at which the values of $\bar{u}$ become larger than 10^4, i.e., this time evaluates the explosion time.

8.5 Probabilistic Approach to Semilinear Parabolic Equations with Small Parameter

In this section we apply the probabilistic approach described in previous sections to the Cauchy problem for semilinear parabolic equations with small parameter of the form

$$\frac{\partial u}{\partial t} + \frac{\varepsilon^2}{2} \sum_{i,j=1}^d a^{ij}(t, x, u) \frac{\partial^2 u}{\partial x^i \partial x^j} + \sum_{i=1}^d (b^i(t, x, u) + \varepsilon^2 c^i(t, x, u)) \frac{\partial u}{\partial x^i} \quad (8.5.1)$$

$$+ g(t, x, u) = 0, \quad t \in [t_0, T), \ x \in \mathbf{R}^d,$$

$$u(T, x) = \varphi(x). \quad (8.5.2)$$

The probabilistic representations of the solution to this problem are connected with systems of SDEs with small noise. Let the Cauchy problem (8.5.1)–(8.5.2) have the unique solution $u = u(t, x)$ which is sufficiently smooth and satisfies some conditions of boundedness (see the corresponding theoretical results, e.g., in [119, 222, 408, 437]). If we substitute $u = u(t, x)$ in the coefficients of (8.5.1), we obtain a linear parabolic equation with small parameter. The solution of this linear equation has the following probabilistic representation

$$u(t, x) = E(\varphi(X_{t,x}(T)) + Z_{t,x,0}(T)), \quad t \le T, \ x \in \mathbf{R}^d, \quad (8.5.3)$$

where $X_{t,x}(s)$, $Z_{t,x,z}(s)$, $s \ge t$, is the solution of the Cauchy problem for the system of SDEs with small noise:

$$dX = (b(s, X, u(s, X)) + \varepsilon^2 c(s, X, u(s, X))) ds \quad (8.5.4)$$
$$+ \varepsilon \sigma(s, X, u(s, X)) dw, \quad X(t) = x,$$

$$dZ = g(s, X, u(s, X)) ds, \quad Z(t) = z. \quad (8.5.5)$$

Here $w(s) = (w^1(s), \ldots, w^d(s))^\mathsf{T}$ is a d-dimensional standard Wiener process, $b(s, x, u)$ and $c(s, x, u)$ are d-dimensional column-vectors composed from the coefficients $b^i(s, x, u)$ and $c^i(s, x, u)$ of (8.5.1), $\sigma(s, x, u)$ is a $d \times d$-matrix obtained from the equation $a(s, x, u) = \sigma(s, x, u)\sigma^\mathsf{T}(s, x, u)$, where $a = \{a^{ij}\}$; the equation is solvable with respect to σ (for instance, by a lower triangular matrix) at least in the case of a positively definite a.

Just for systems like (8.5.4)–(8.5.5), special weak approximations are proposed in Chap. 4. Applying these special approximations, we get new layer methods intended to solve the Cauchy problem (8.5.1)–(8.5.2).

To simplify the notation, we write $u(t, x)$ and $X_{t,x}(s)$ instead of $u(t, x; \varepsilon)$ and $X_{t,x}^{\varepsilon}(s)$ here and in what follows.

In Sects. 8.5.1–8.5.3, specific implicit and explicit layer methods for the problem (8.5.1)–(8.5.2) are proposed. Some theorems on their rates of convergence both in the regular and in the singular cases are given. For simplicity, we deal with the one-dimensional case of the problem (8.5.1)–(8.5.2). Extensions to the multi-dimensional case and systems of reaction-diffusion equations can be done as in Sect. 8.3 (see details in [305]). In Sect. 8.6, we propose both two-layer and three-layer methods in a particular case of the problem (8.5.1)–(8.5.2). Some results of numerical tests on the Burgers equation with small viscosity and on the generalized FKPP-equation with small parameter are given in Sect. 8.7.

For simplicity, we consider the Cauchy problem (8.5.1)–(8.5.2) for $d = 1$:

$$\frac{\partial u}{\partial t} + \frac{\varepsilon^2}{2}\sigma^2(t, x, u)\frac{\partial^2 u}{\partial x^2} + (b(t, x, u) + \varepsilon^2 c(t, x, u))\frac{\partial u}{\partial x} + g(t, x, u) = 0, \quad (8.5.6)$$

$$t \in [t_0, T), \ x \in \mathbf{R},$$

$$u(T, x) = \varphi(x). \tag{8.5.7}$$

The probabilistic representation of the solution $u(t, x)$ to this problem has the form (8.5.3)–(8.5.5) with $d = 1$.

8.5.1 Implicit Layer Method and Its Convergence

Applying the Runge–Kutta scheme (4.5.10) to the system (8.5.4)–(8.5.5), we get

$$X_{t_k,x}(t_{k+1}) \simeq \bar{X}_{t_k,x}(t_{k+1}) = x + \frac{1}{2}hb_k \tag{8.5.8}$$

$$+ \frac{1}{2}hb(t_{k+1}, x + hb_k, u(t_{k+1}, x + hb_k)) + \varepsilon^2 hc_k + \varepsilon h^{1/2}\sigma_k\xi_k,$$

$$Z_{t_k,x,z}(t_{k+1}) \simeq \bar{Z}_{t_k,x,z}(t_{k+1}) = z + \frac{1}{2}hg_k$$

$$+ \frac{1}{2}hg(t_{k+1}, x + hb_k, u(t_{k+1}, x + hb_k)),$$

where b_k, c_k, σ_k, and g_k are the coefficients b, c, σ, and g evaluated at the point $(t_k, x, u(t_k, x))$.

Using a probabilistic representation of the form (8.2.4), we obtain

$$u(t_k, x)$$
$$\simeq E(u(t_{k+1}, \bar{X}_{t_k,x}(t_{k+1})) + \bar{Z}_{t_k,x,0}(t_{k+1}))$$
$$= \frac{1}{2}u(t_{k+1}, x + h[b_k + b(t_{k+1}, x + hb_k, u(t_{k+1}, x + hb_k))]/2$$
$$+ \varepsilon^2 hc_k + \varepsilon h^{1/2}\sigma_k)$$
$$+ \frac{1}{2}u(t_{k+1}, x + h[b_k + b(t_{k+1}, x + hb_k, u(t_{k+1}, x + hb_k))]/2$$
$$+ \varepsilon^2 hc_k - \varepsilon h^{1/2}\sigma_k)$$
$$+ \frac{1}{2}hg_k + \frac{1}{2}hg(t_{k+1}, x + hb_k, u(t_{k+1}, x + hb_k)).$$

We can approximate $u(t_k, x)$ by $v(t_k, x)$ found from

$$v(t_k, x) \qquad (8.5.9)$$
$$= \frac{1}{2}u(t_{k+1}, x + h[\tilde{b}_k + b(t_{k+1}, x + h\tilde{b}_k, u(t_{k+1}, x + h\tilde{b}_k))]/2$$
$$+ \varepsilon^2 h\tilde{c}_k + \varepsilon h^{1/2}\tilde{\sigma}_k)$$
$$+ \frac{1}{2}u(t_{k+1}, x + h[\tilde{b}_k + b(t_{k+1}, x + h\tilde{b}_k, u(t_{k+1}, x + h\tilde{b}_k))]/2$$
$$+ \varepsilon^2 h\tilde{c}_k - \varepsilon h^{1/2}\tilde{\sigma}_k)$$
$$+ \frac{1}{2}h\tilde{g}_k + \frac{1}{2}hg(t_{k+1}, x + h\tilde{b}_k, u(t_{k+1}, x + h\tilde{b}_k)),$$

where $\tilde{b}_k$, $\tilde{c}_k$, $\tilde{\sigma}_k$, and $\tilde{g}_k$ are the coefficients b, c, σ, and g evaluated at the point $(t_k, x, v(t_k, x))$.

The corresponding implicit layer method has the form

$$\bar{u}(t_N, x) = \varphi(x), \qquad (8.5.10)$$

$$\bar{u}(t_k, x)$$
$$= \frac{1}{2}\bar{u}(t_{k+1}, x + h[\bar{b}_k + b(t_{k+1}, x + h\bar{b}_k, \bar{u}(t_{k+1}, x + h\bar{b}_k))]/2$$
$$+ \varepsilon^2 h\bar{c}_k + \varepsilon h^{1/2}\bar{\sigma}_k)$$
$$+ \frac{1}{2}\bar{u}(t_{k+1}, x + h[\bar{b}_k + b(t_{k+1}, x + h\bar{b}_k, \bar{u}(t_{k+1}, x + h\bar{b}_k))]/2$$
$$+ \varepsilon^2 h\bar{c}_k - \varepsilon h^{1/2}\bar{\sigma}_k)$$
$$+ \frac{1}{2}h\bar{g}_k + \frac{1}{2}hg(t_{k+1}, x + h\bar{b}_k, \bar{u}(t_{k+1}, x + h\bar{b}_k)),$$

$$k = N - 1, \ldots, 0,$$

where $\bar{b}_k$, $\bar{c}_k$, $\bar{\sigma}_k$, and $\bar{g}_k$ are the coefficients b, c, σ, and g evaluated at the point $(t_k, x, \bar{u}(t_k, x))$.

Let us assume that

(i) The coefficients $b(t, x, u)$ and $g(t, x, u)$ and their first and second derivatives are continuous and uniformly bounded, the coefficients $c(t, x, u)$ and $\sigma(t, x, u)$ and their first derivatives are continuous and uniformly bounded:

$$\left| \frac{\partial^{i+j+l} b}{\partial t^i \partial x^j \partial u^l} \right| \le K, \quad \left| \frac{\partial^{i+j+l} g}{\partial t^i \partial x^j \partial u^l} \right| \le K, \quad 0 \le i+j+l \le 2, \tag{8.5.11}$$

$$\left| \frac{\partial^{i+j+l} c}{\partial t^i \partial x^j \partial u^l} \right| \le K, \quad \left| \frac{\partial^{i+j+l} \sigma}{\partial t^i \partial x^j \partial u^l} \right| \le K, \quad 0 \le i+j+l \le 1,$$

$$t_0 \le t \le T, \; x \in \mathbf{R}, \; u_\circ < u < u^\circ,$$

where $-\infty \le u_\circ$, $u^\circ \le \infty$ are some constants.

(ii) There exists a unique bounded solution $u(t, x)$ of the problem (8.5.6)–(8.5.7) such that

$$u_\circ < u_* \le u(t, x) \le u^* < u^\circ, \tag{8.5.12}$$

where u_*, u^* are some constants, and there exist the uniformly bounded derivatives:

$$\left| \frac{\partial^{i+j} u}{\partial t^i \partial x^j} \right| \le K, \; i = 0, \; j = 1, 2, 3, 4; \; i = 1, \; j = 0, 1, 2; \tag{8.5.13}$$

$$i = 2, \; j = 0, 1; \; i = 3, \; j = 0; \quad t_0 \le t \le T, \; x \in \mathbf{R}, \; 0 < \varepsilon \le \varepsilon^*.$$

For the singular case, when the condition (8.5.13) is not fulfilled, see [305] and [314, Sect. 7.5.3].

Lemma 8.5.1 (see [305]) *Under the assumptions (i)–(ii), the one-step error of the implicit layer method (8.5.10) is estimated by $O(h^3 + \varepsilon^2 h^2)$, i.e.,*

$$|v(t_k, x) - u(t_k, x)| \le C \, (h^3 + \varepsilon^2 h^2),$$

where $v(t_k, x)$ is found from (8.5.9), $C > 0$ does not depend on h, k, x, ε.

Theorem 8.5.2 *Under the assumptions (i)–(ii), the global error of the implicit layer method (8.5.10) is estimated by $O(h^2 + \varepsilon^2 h)$:*

$$|u(t_k, x) - \bar{u}(t_k, x)| \le K \, (h^2 + \varepsilon^2 h),$$

where the constant $K > 0$ does not depend on h, k, x, ε.

A proof of this theorem is analogous to the proof of Theorem 8.2.2 (see details in [305]).

Remark 8.5.3 For linear parabolic equations, i.e., when the coefficients of (8.5.6) do not depend on u, the method (8.5.10) becomes the explicit one with the global error $O(h^2 + \varepsilon^2 h)$ and can be applied to solving linear parabolic equations with small parameter. Note also that if the dimension d of the linear problem is high ($d > 3$ in practice) and it is enough to find the solution at a few points only, the Monte Carlo technique is preferable.

8.5.2 Explicit Layer Methods

To implement the implicit method (8.5.10), one can use the method of simple iteration. If we take $u(t_{k+1}, x)$ as a null iteration, in the case of $b(t, x, u) \neq b(t, x)$ or $g(t, x, u) \neq g(t, x)$, the first iteration provides the one-step error $O(h^2)$ only. One can show that applying the second iteration, we get $O(h^3 + \varepsilon^2 h^2)$ as the one-step error. However it is possible to reach the same one-step accuracy by some modification of the first iteration that reduces the number of recalculations. The explicit layer method obtained this way has the form (we use the same notation $\bar{u}(t_k, x)$ again):

$$\bar{u}(t_N, x) = \varphi(x), \qquad\qquad (8.5.14)$$

$$\hat{b}_k = b(t_k, x, \bar{u}(t_{k+1}, x)), \quad \hat{c}_k = c(t_k, x, \bar{u}(t_{k+1}, x)), \quad \hat{\sigma}_k = \sigma(t_k, x, \bar{u}(t_{k+1}, x)),$$

$$\bar{u}^{(1)}(t_k, x) = \bar{u}(t_{k+1}, x + h\hat{b}_k) + hg(t_k, x, \bar{u}(t_{k+1}, x)),$$

$$\begin{aligned}
&\bar{u}(t_k, x)\\
&= \frac{1}{2}\bar{u}(t_{k+1}, x + h[b(t_k, x, \bar{u}^{(1)}(t_k, x)) + b(t_{k+1}, x + h\hat{b}_k, \bar{u}(t_{k+1}, x + h\hat{b}_k))]/2\\
&\quad + \varepsilon^2 h\hat{c}_k + \varepsilon h^{1/2}\hat{\sigma}_k)\\
&\quad + \frac{1}{2}\bar{u}(t_{k+1}, x + h[b(t_k, x, \bar{u}^{(1)}(t_k, x)) + b(t_{k+1}, x + h\hat{b}_k, \bar{u}(t_{k+1}, x + h\hat{b}_k))]/2\\
&\quad + \varepsilon^2 h\hat{c}_k - \varepsilon h^{1/2}\hat{\sigma}_k)\\
&\quad + \frac{1}{2}hg(t_k, x, \bar{u}^{(1)}(t_k, x)) + \frac{1}{2}hg(t_{k+1}, x + h\hat{b}_k, \bar{u}(t_{k+1}, x + h\hat{b}_k)),
\end{aligned}$$

$$k = N - 1, \ldots, 0.$$

The following theorem can be proved by the arguments like those used for Theorem 8.2.2.

Theorem 8.5.4 *Under the assumptions (i)–(ii), the global error of the explicit layer method (8.5.14) is estimated by $O(h^2 + \varepsilon^2 h)$:*

$$|u(t_k, x) - \bar{u}(t_k, x)| \leq K\,(h^2 + \varepsilon^2 h),$$

where the constant $K > 0$ does not depend on h, k, x, ε.

Remark 8.5.5 Naturally, we can take other weak approximations (more accurate than the ones we have used above) of SDEs with small noise from Chap. 4 to construct the corresponding high-order (with respect to h and ε) methods for the problem (8.5.6)–(8.5.7). In Sect. 8.6 we give high-order methods in some particular cases of the equation (8.5.6).

8.5.3   Numerical Algorithms Based on Interpolation

Introduce an equidistant space discretization: $\{x_j = x_0 + jh_x, \ j = 0, \pm 1, \pm 2, \ldots\}$, $x_0 \in \mathbf{R}$, h_x is a sufficiently small positive number. Since it does not lead to any misunderstanding, we use the old notation $\bar{u}$, $\bar{u}^{(1)}$, etc. for new values here.

Theorem 8.5.6 *Under the assumptions (i)–(ii), the numerical algorithm based on the explicit method (8.5.14) and on the linear interpolation:*

$$\bar{u}(t_N, x) = \varphi(x), \qquad\qquad (8.5.15)$$

$$\hat{b}_{k,j} = b(t_k, x_j, \bar{u}(t_{k+1}, x_j)), \ \ \hat{c}_{k,j} = c(t_k, x_j, \bar{u}(t_{k+1}, x_j)),$$
$$\hat{\sigma}_{k,j} = \sigma(t_k, x_j, \bar{u}(t_{k+1}, x_j)),$$

$$\bar{u}^{(1)}(t_k, x_j) = \bar{u}(t_{k+1}, x_j + h\hat{b}_{k,j}) + hg(t_k, x_j, \bar{u}(t_{k+1}, x_j)),$$

$$\bar{u}(t_k, x_j)$$
$$= \frac{1}{2}\bar{u}(t_{k+1}, x_j + h[b(t_k, x_j, \bar{u}^{(1)}(t_k, x_j))$$
$$+ b(t_{k+1}, x_j + h\hat{b}_{k,j}, \bar{u}(t_{k+1}, x_j + h\hat{b}_{k,j}))]/2 + \varepsilon^2 h\hat{c}_{k,j} + \varepsilon h^{1/2}\hat{\sigma}_{k,j})$$
$$+ \frac{1}{2}\bar{u}(t_{k+1}, x_j + h[b(t_k, x_j, \bar{u}^{(1)}(t_k, x_j))$$
$$+ b(t_{k+1}, x_j + h\hat{b}_{k,j}, \bar{u}(t_{k+1}, x_j + h\hat{b}_{k,j}))]/2 + \varepsilon^2 h\hat{c}_{k,j} - \varepsilon h^{1/2}\hat{\sigma}_{k,j})$$
$$+ \frac{1}{2}hg(t_k, x_j, \bar{u}^{(1)}(t_k, x_j))$$
$$+ \frac{1}{2}hg(t_{k+1}, x_j + h\hat{b}_{k,j}, \bar{u}(t_{k+1}, x_j + h\hat{b}_{k,j})),$$

$$\bar{u}(t_k, x) = \frac{x_{j+1} - x}{h_x}\bar{u}(t_k, x_j) + \frac{x - x_j}{h_x}\bar{u}(t_k, x_{j+1}), \ \ x_j < x < x_{j+1},$$

$$j = 0, \pm 1, \pm 2, \ldots, \ \ k = N - 1, \ldots, 0,$$

has global error of order $O(h^2 + \varepsilon^2 h)$ if the value of h_x is selected as $h_x = \alpha \min(h^{3/2}, \varepsilon h)$, where α is a positive constant.

This theorem can be proved by the arguments like those used for Theorem 8.2.5 (see details in [305]). Remark 8.2.6 is applicable here.

8.6 High-Order Methods for Semilinear Equation with Small Constant Diffusion and Zero Advection

Here we restrict ourselves to the case of $d = 1$ for simplicity again. Consider the Cauchy problem for the semilinear heat equation

$$\frac{\partial u}{\partial t} + \frac{\varepsilon^2}{2} \frac{\partial^2 u}{\partial x^2} + g(t, x, u) = 0, \ t \in [t_0, T), \quad x \in \mathbf{R}, \tag{8.6.1}$$

$$u(T, x) = \varphi(x). \tag{8.6.2}$$

We assume that $g(t, x, u)$ is a uniformly bounded and sufficiently smooth function and conditions like (ii) from Sect. 8.5.1 are fulfilled for the solution $u(t, x)$ to (8.6.1)–(8.6.2). Note that to construct high-order methods we need uniform boundedness of the derivatives of $u(t, x)$ with higher orders than in the assumption (8.5.13). To realize the methods of this section, we can avoid any interpolation choosing a special space discretization. The methods of this section are tested by simulation of the generalized FKPP-equation with a small parameter (see Sect. 8.7.2).

The probabilistic representation of the solution to (8.6.1)–(8.6.2) has the form (see (8.5.3)–(8.5.5)):

$$u(t, x) = E(\varphi(X_{t,x}(T)) + Z_{t,x,0}(T)), \tag{8.6.3}$$

where $X_{t,x}(s)$, $Z_{t,x,z}(s)$, $s \geq t$, satisfies the system

$$dX = \varepsilon dw(s), \quad X(t) = x, \tag{8.6.4}$$
$$dZ = g(s, X, u(s, X))ds, \quad Z(t) = z.$$

Note that the system (8.6.4) is a system of differential equations with small *additive* noise.

8.6.1 Two-Layer Methods

Let us write down the layer methods (8.2.23) and (8.5.14) and the second-order method (8.2.18) in the case of the problem (8.6.1)–(8.6.2).

The explicit layer method (8.2.23) *with error* $O(h)$ *has the form*

$$\bar{u}(t_N, x_j) = \varphi(x_j), \tag{8.6.5}$$

$$\bar{u}(t_k, x_j) = \frac{1}{2}\bar{u}(t_{k+1}, x_j + \varepsilon h^{1/2}) + \frac{1}{2}\bar{u}(t_{k+1}, x_j - \varepsilon h^{1/2})$$
$$+ hg(t_{k+1}, x_j, \bar{u}(t_{k+1}, x_j)),$$

$$x_j = x_0 + j\varepsilon h^{1/2}, \quad j = 0, \pm 1, \pm 2, \ldots, \quad k = N - 1, \ldots, 0.$$

Note that it coincides with the well-known finite-difference scheme under the special relation of time and space steps ($h_x = \varepsilon h_t^{1/2}$) in the scheme.

In the case of the problem (8.6.1)–(8.6.2) *the explicit layer method* (8.5.14) *with error* $O(h^2 + \varepsilon^2 h)$ takes the form

$$\bar{u}(t_N, x_j) = \varphi(x_j), \tag{8.6.6}$$

$$\bar{u}^{(1)}(t_k, x_j) = \bar{u}(t_{k+1}, x_j) + hg(t_k, x_j, \bar{u}(t_{k+1}, x_j)),$$

$$\bar{u}(t_k, x_j) = \frac{1}{2}\bar{u}(t_{k+1}, x_j + \varepsilon h^{1/2}) + \frac{1}{2}\bar{u}(t_{k+1}, x_j - \varepsilon h^{1/2})$$
$$+ \frac{1}{2}h[g(t_k, x_j, \bar{u}^{(1)}(t_k, x_j)) + g(t_{k+1}, x_j, \bar{u}(t_{k+1}, x_j))],$$

$$x_j = x_0 + j\varepsilon h^{1/2}, \quad j = 0, \pm 1, \pm 2, \ldots, \quad k = N - 1, \ldots, 0.$$

In the case of the problem (8.6.1)–(8.6.2) *the implicit layer method* (8.2.18) *with error* $O(h^2)$ has the form

$$\bar{u}(t_N, x_j) = \varphi(x_j), \tag{8.6.7}$$

$$\bar{u}(t_k, x_j) = \frac{1}{6}\bar{u}(t_{k+1}, x_j + \sqrt{3}\varepsilon h^{1/2}) + \frac{2}{3}\bar{u}(t_{k+1}, x_j) + \frac{1}{6}\bar{u}(t_{k+1}, x_j - \sqrt{3}\varepsilon h^{1/2})$$
$$+ \frac{h}{2}g(t_k, x_j, \bar{u}(t_k, x_j)) + \frac{h}{3}g(t_{k+1}, x_j, \bar{u}(t_{k+1}, x_j))$$
$$+ \frac{h}{12}g(t_{k+1}, x_j + \sqrt{3}\varepsilon h^{1/2}, \bar{u}(t_{k+1}, x_j + \sqrt{3}\varepsilon h^{1/2}))$$
$$+ \frac{h}{12}g(t_{k+1}, x_j - \sqrt{3}\varepsilon h^{1/2}, \bar{u}(t_{k+1}, x_j - \sqrt{3}\varepsilon h^{1/2})),$$

$$x_j = x_0 + j\sqrt{3}\varepsilon h^{1/2}, \quad j = 0, \pm 1, \pm 2, \ldots, \quad k = N - 1, N - 2, \ldots, 0.$$

To solve the algebraic equations obtained at each step of the method (8.6.7), one can use the Newton method or the method of simple iteration.

Remark 8.6.1 In the singular case the natural bounds for derivatives of the solution to (8.6.1)–(8.6.2) have the form (see e.g. [119, 178, 408]):

$$\left|\frac{\partial^{i+j} u}{\partial t^i \partial x^j}\right| \le \frac{K}{\varepsilon^j}, \ t \in [t_0, T], \ x \in \mathbf{R}, \ 0 < \varepsilon \le \varepsilon^*. \tag{8.6.8}$$

These bounds are obtained using the following change of variables: $t = t'$, $x = \varepsilon x'$.

One can prove under (8.6.8) that the errors of both methods (8.6.5) and (8.6.6) are estimated as

$$|u(t_k, x) - \bar{u}(t_k, x)| \le Kh,$$

where the constant K does not depend on x, k, h, ε. Nevertheless, the method (8.6.6) gives better results than (8.6.5) in our experiments. One can explain this by the fact that the constant K in (8.6.6) is much smaller than the K in (8.6.5). Under (8.6.8) the error of the method (8.6.7) remains $O(h^2)$.

Note that in Sect. 8.7.2 we present results of testing these methods (instead of (8.6.5) we use a modification of it) on an equation in which the term g depends on ε and the derivatives of the solution have other bounds than (8.6.8) (see details in Sect. 8.7.2).

8.6.2 Three-Layer Methods

Here we obtain two three-layer methods. We estimate their one-step errors but do not prove their convergence that requires stability analysis of multi-layer methods. We test these methods in our experiments, and they give fairly good results.

To calculate $\bar{u}(t_{k+1}, x)$ by a three-layer method, two previous layers are used. So, to start simulations we should know $\bar{u}(t_N, x)$ and $\bar{u}(t_{N-1}, x)$. To simulate $\bar{u}(t_{N-1}, x)$ one can use, e.g., the two-layer method (8.6.7) with a sufficiently small step. Below we consider this layer to be known and denote $\psi(x) := \bar{u}(t_{N-1}, x)$.

Apply the special Runge–Kutta scheme (4.5.18) to approximate (8.6.4):

$$X_{t_k,x}(t_{k+1}) \simeq \bar{X}_{t_k,x}(t_{k+1}) = x + \varepsilon h^{1/2}\xi_k, \tag{8.6.9}$$

$$
\begin{aligned}
Z_{t_k,x,z}(t_{k+1}) \simeq \bar{Z}_{t_k,x,z}(t_{k+1}) = z + \frac{h}{6}(&g(t_k, x, u(t_k, x)) \\
+ &2g(t_{k+1/2}, x, u(t_{k+1/2}, x)) \\
+ &2g(t_{k+1/2}, x + \varepsilon h^{1/2}\xi_k, u(t_{k+1/2}, x + \varepsilon h^{1/2}\xi_k)) \\
+ &g(t_{k+1}, x + \varepsilon h^{1/2}\xi_k, u(t_{k+1}, x + \varepsilon h^{1/2}\xi_k))),
\end{aligned}
$$

where ξ_k are i.i.d. variables with the law $P(\xi = 0) = 2/3$, $P(\xi = \pm\sqrt{3}) = 1/6$.

The implicit method with one-step error $O(h^5 + \varepsilon^2 h^3)$ has the form (to get the method we use the scheme (8.6.9) with the time step $2h$) :

$$\bar{u}(t_N, x_j) = \varphi(x_j), \ \bar{u}(t_{N-1}, x_j) = \psi(x_j), \tag{8.6.10}$$

$$\bar{u}(t_k, x_j) = \frac{1}{6}\bar{u}(t_{k+2}, x_j + \sqrt{6}\varepsilon h^{1/2}) + \frac{2}{3}\bar{u}(t_{k+2}, x_j) + \frac{1}{6}\bar{u}(t_{k+2}, x_j - \sqrt{6}\varepsilon h^{1/2})$$

$$+ \frac{h}{3}g(t_k, x_j, \bar{u}(t_k, x_j)) + \frac{10h}{9}g(t_{k+1}, x_j, \bar{u}(t_{k+1}, x_j))$$

$$+ \frac{h}{9}g(t_{k+1}, x_j + \sqrt{6}\varepsilon h^{1/2}, \bar{u}(t_{k+1}, x_j + \sqrt{6}\varepsilon h^{1/2}))$$

$$+ \frac{h}{9}g(t_{k+1}, x_j - \sqrt{6}\varepsilon h^{1/2}, \bar{u}(t_{k+1}, x_j - \sqrt{6}\varepsilon h^{1/2}))$$

$$+ \frac{h}{18}g(t_{k+2}, x_j + \sqrt{6}\varepsilon h^{1/2}, \bar{u}(t_{k+2}, x_j + \sqrt{6}\varepsilon h^{1/2}))$$

$$+ \frac{2h}{9}g(t_{k+2}, x_j, \bar{u}(t_{k+2}, x_j))$$

$$+ \frac{h}{18}g(t_{k+2}, x_j - \sqrt{6}\varepsilon h^{1/2}, \bar{u}(t_{k+2}, x_j - \sqrt{6}\varepsilon h^{1/2})),$$

$$x_j = x_0 + j\sqrt{6}\varepsilon h^{1/2}, \ \ j = 0, \pm 1, \pm 2, \ldots, \ \ k = N - 2, N - 3, \ldots, 0.$$

Let us look at the stability properties of this method in the simple case when u and g in (8.6.1) do not depend on x, i.e., apply the method (8.6.10) to the ordinary differential equation

$$\frac{du}{dt} + g(t, u) = 0, \ \ t \le T, \ u(T) = \varphi. \tag{8.6.11}$$

Recall (see, e.g., [155]) that a linear n-step method for (8.6.11)

$$\alpha_n u_k + \alpha_{n-1}u_{k+1} + \cdots + \alpha_0 u_{k+n} = h\left(\beta_n g_k + \cdots + \beta_0 g_{k+n}\right),$$

$$g_i = g(t_i, u_i), \ \ \alpha_n \ne 0, \ |\alpha_0| + |\beta_0| > 0,$$

is zero-stable (D-stable) if the generating polynomial

$$\alpha_n\lambda^n + \alpha_{n-1}\lambda^{n-1} + \cdots + \alpha_0 = 0 \tag{8.6.12}$$

satisfies the root condition: the roots of (8.6.12) lie on or within the unit circle, and the roots on the unit circle are simple.

In the case of (8.6.11) the method (8.6.10) coincides with the Milne two-step method which is of the order $O(h^4)$ and is zero-stable. Its generating polynomial has two roots: 1 and -1. As is known [155], the root -1 can be dangerous for some differential equations. The method (8.6.10) has unstable behavior in our numerical tests on the generalized FKPP-equation with a small parameter (8.7.8)–(8.7.10) (Sect. 8.7.2). One can see that the method (8.6.10) does not preserve the property $u \le 1$ of the problem (8.7.8)–(8.7.10) that leads to an unstable behavior of the approximate solutions. We modify the method (8.6.10) in the experiments: if $\bar{u}(t_k, x_j) > 1$, we put $\bar{u}(t_k, x_j) = 1$. Since locally, in a single step, the arising difference $0 < \bar{u}(t_k, x_j) - 1$ is not greater than the one-step error of this method, this modification does not change the one-step accuracy order of the method. The modified method turned out to work fairly well when applied to the generalized FKPP-equation. However, the modification is based on the knowledge of the properties of the solution, and it may be difficult to find such a modification for another problem. Fortunately, we are able to approximate the system (8.6.4) by another weak scheme and obtain a method for (8.6.1)–(8.6.2) with better stability properties in the sense considered above but with the one-step error of lower order (see the method (8.6.13) below). Let us note that it is possible to reach both the same one-step accuracy $O(h^5 + \varepsilon^2 h^3)$ and the better stability properties by a four-layer method.

Approximate the system (8.6.4) by the special scheme with one-step order $O(h^4 + \varepsilon^2 h^3)$:

$$X_{t_k,x}(t_{k+1}) \simeq \bar{X}_{t_k,x}(t_{k+1}) = x + \varepsilon h^{1/2}\xi_k,$$

$$\begin{aligned}
Z_{t_k,x,z}(t_{k+1}) \simeq \bar{Z}_{t_k,x,z}(t_{k+1}) &= z + \frac{h}{12}[5g(t_k, x, u(t_k, x)) \\
&\quad + g(t_{k+1}, x, u(t_{k+1}, x))] \\
&\quad + 7g(t_{k+1}, x + \varepsilon h^{1/2}\xi_k, u(t_{k+1}, x + \varepsilon h^{1/2}\xi_k)) \\
&\quad - g(t_{k+2}, x + \varepsilon h^{1/2}\xi_k, u(t_{k+2}, x + \varepsilon h^{1/2}\xi_k))),
\end{aligned}$$

where ξ_k are i.i.d. variables with the law $P(\xi = 0) = 2/3$, $P(\xi = \pm\sqrt{3}) = 1/6$.
The three-layer implicit method with one-step error $O(h^4 + \varepsilon^2 h^3)$ has the form

$$\bar{u}(t_N, x_j) = \varphi(x_j), \quad \bar{u}(t_{N-1}, x_j) = \psi(x_j), \tag{8.6.13}$$

$$\begin{aligned}
\bar{u}(t_k, x_j) &= \frac{1}{6}\bar{u}(t_{k+1}, x_j + \sqrt{3}\varepsilon h^{1/2}) + \frac{2}{3}\bar{u}(t_{k+1}, x_j) + \frac{1}{6}\bar{u}(t_{k+1}, x_j - \sqrt{3}\varepsilon h^{1/2}) \\
&\quad + \frac{5h}{12}g(t_k, x_j, \bar{u}(t_k, x_j)) + \frac{17h}{36}g(t_{k+1}, x_j, \bar{u}(t_{k+1}, x_j)) \\
&\quad + \frac{7h}{72}g(t_{k+1}, x_j + \sqrt{3}\varepsilon h^{1/2}, \bar{u}(t_{k+1}, x_j + \sqrt{3}\varepsilon h^{1/2})) \\
&\quad + \frac{7h}{72}g(t_{k+1}, x_j - \sqrt{3}\varepsilon h^{1/2}, \bar{u}(t_{k+1}, x_j - \sqrt{3}\varepsilon h^{1/2})) \\
&\quad - \frac{h}{72}g(t_{k+2}, x_j + \sqrt{3}\varepsilon h^{1/2}, \bar{u}(t_{k+2}, x_j + \sqrt{3}\varepsilon h^{1/2})) \\
&\quad - \frac{h}{18}g(t_{k+2}, x_j, \bar{u}(t_{k+2}, x_j)) \\
&\quad - \frac{h}{72}g(t_{k+2}, x_j - \sqrt{3}\varepsilon h^{1/2}, \bar{u}(t_{k+2}, x_j - \sqrt{3}\varepsilon h^{1/2})),
\end{aligned}$$

$$x_j = x_0 + j\sqrt{3}\varepsilon h^{1/2}, \quad j = 0, \pm1, \pm2, \ldots, \quad k = N - 2, N - 3, \ldots, 0.$$

For (8.6.11) this method coincides with one of the implicit two-step Adams methods of order $O(h^3)$, and the roots of its generating polynomial are 1 and 0. One can expect that in the case of the problem (8.6.1)–(8.6.2) the method (8.6.13) also possesses better stability properties than (8.6.10). In our numerical tests on the generalized FKPP-equation with small parameter (Sect. 8.7.2) the method (8.6.13) has stable behavior.

To solve the algebraic equations, obtained at each step of the methods (8.6.10) and (8.6.13), one can use the Newton method or the method of simple iteration.

Remark 8.6.2 The methods of this section can be extended for problems of a higher dimension or for systems of reaction-diffusion equations. Additionally using other weak approximations to SDEs with small additive noise, new layer methods can be constructed. For instance, three- and four-layer methods with the one-step error $O(h^5 + \varepsilon^2 h^2)$ can be obtained. It is also not difficult to get an implicit four-layer method with the one-step error $O(h^5 + \varepsilon^2 h^3)$ for (8.6.1)–(8.6.2) possessing good stability properties in the above sense, or an explicit four-layer method with the one-step error $O(h^4 + \varepsilon^2 h^3)$, and so on.

8.7   Numerical Tests

We noted in Remark 8.2.7 that the algorithms rested on *cubic interpolation* give quite good results. We use the advantage of the cubic interpolation in our numerical tests on the Burgers equation. Let us recall that a sufficiently smooth function $f(x)$, $x \in \mathbf{R}$, can be interpolated by cubic interpolation as

$$f(x) \simeq \bar{f}(x) = \sum_{i=0}^{3} \Phi_{j,i}(x)\, f(x_{j+i}), \quad x_{j+1} < x < x_{j+2}, \tag{8.7.1}$$

$$\Phi_{j,i}(x) = \prod_{m=0, m \neq i}^{3} \frac{x - x_{j+m}}{x_{j+i} - x_{j+m}},$$

where $x_j = x_0 + j \times h_x$, $x_0 \in \mathbf{R}$, $j = 0, \pm 1, \pm 2, \ldots$, h_x is a positive number.
The error of the cubic interpolation (8.7.1) is estimated by

$$|\bar{f}(x) - f(x)| \leq \frac{3}{128} \max_{x_j < x < x_{j+3}} \left| \frac{\partial^4 u}{\partial x^4} \right| \times h_x^4, \quad x_{j+1} < x < x_{j+2}.$$

Recall (see Theorem 8.5.6) that the algorithm (8.5.15), based on the layer method (8.5.14) and on linear interpolation, has the error estimated by $O(h^2 + \varepsilon^2 h)$ provided $h_x = \min(h^{3/2}, \varepsilon h)$. One can expect that under the assumptions (i)–(ii) from Sect. 8.5.1 the algorithm based on the layer method (8.5.14) and on the cubic interpolation (8.7.1) can achieve the same accuracy $O(h^2 + \varepsilon^2 h)$ with h_x taken equal to $\min(h^{3/4}, \sqrt{\varepsilon h})$ only. Our numerical tests on the Burgers equation support this supposition.

8.7.1 The Burgers Equation with Small Viscosity

The one-dimensional Burgers equation with small viscosity has the form

$$\frac{\partial u}{\partial t} = \frac{\varepsilon^2}{2}\frac{\partial^2 u}{\partial x^2} - u\frac{\partial u}{\partial x}, \quad t > 0, \ x \in \mathbf{R}, \tag{8.7.2}$$

$$u(0, x) = \varphi(x). \tag{8.7.3}$$

Let $v(t, x)$ satisfy the heat equation

$$\frac{\partial v}{\partial t} = \frac{\varepsilon^2}{2}\frac{\partial^2 v}{\partial x^2}, \quad t > 0, \ x \in \mathbf{R},$$

$$v(0, x) = \exp\left(-\frac{1}{\varepsilon^2}\int_0^x \varphi(\xi)d\xi\right).$$

Then the Cole-Hopf transformation [63, 170]

$$u(t, x) = -\frac{\varepsilon^2}{v}\frac{\partial v}{\partial x}$$

satisfies (8.7.2)–(8.7.3).

The explicit solution of the problem (8.7.2)–(8.7.3) takes the form

$$u(t, x) = \frac{\int_{-\infty}^{\infty} K(t, x, y)\varphi(y)\exp\left(-\frac{1}{\varepsilon^2}\int_0^y \varphi(\xi)d\xi\right)dy}{\int_{-\infty}^{\infty} K(t, x, y)\exp\left(-\frac{1}{\varepsilon^2}\int_0^y \varphi(\xi)d\xi\right)dy},$$

$$K(t, x, y) = \frac{1}{\sqrt{2\pi\varepsilon^2 t}}\exp\left(-\frac{(x-y)^2}{2\varepsilon^2 t}\right).$$

Let us take the initial condition $\varphi(x)$ of the form

$$\varphi(x) = \begin{cases} c, & x < l_0, \\ \lambda(x), & l_0 \le x \le l_0 + l, \\ d, & x > l_0 + l, \end{cases} \tag{8.7.4}$$

where c, d, l_0, l are some numbers, $c > d$, $l \geq 0$; $\lambda(x)$ is a bounded measurable function, and $d \leq \lambda(x) \leq c$.

Recall some theoretical facts concerning the problem (8.7.2)–(8.7.3), (8.7.4) (see details, e.g., in [178, 408]). The solution $u(t, x)$ to (8.7.2)–(8.7.3), (8.7.4) is uniformly bounded:

$$d \leq u(t, x) \leq c, \ x \in \mathbf{R}, \ 0 \leq t, \ 0 \leq \varepsilon \leq \varepsilon^*.$$

Let the initial condition $\varphi(x)$ be a sufficiently smooth function. Introduce the time moment T such that the solution of the hyperbolic problem obtained from (8.7.2)–(8.7.3), (8.7.4) for $\varepsilon = 0$ is smooth at $t < T$ and discontinuous at $t \geq T$. The solution $u(t, x)$ to (8.7.2)–(8.7.3), (8.7.4) is regular for $t \leq t_* < T$:

$$\left| \frac{\partial^{i+j} u}{\partial t^i \partial x^j}(t, x) \right| \leq K, \ x \in \mathbf{R}, \ 0 \leq t \leq t_*, \ 0 \leq \varepsilon \leq \varepsilon^*.$$

If $t \geq T$ then the solution is singular in an interval $(x_*(t), x^*(t))$ with width $|x^*(t) - x_*(t)| \sim \varepsilon^2$:

$$\left| \frac{\partial^{i+j} u}{\partial t^i \partial x^j}(t, x) \right| \leq \frac{K}{\varepsilon^{2(i+j)}}, \ x \in (x_*(t), x^*(t)), \ t \geq T, \ 0 < \varepsilon \leq \varepsilon^*,$$

$$\left| \frac{\partial^{i+j} u}{\partial t^i \partial x^j}(t, x) \right| \leq K, \ x \notin (x_*(t), x^*(t)), \ t \geq T, \ 0 < \varepsilon \leq \varepsilon^*.$$

In our experiments we take $\lambda(x)$ equal to

$$\lambda(x) = a - b \sin \frac{\pi x}{\mu}, \quad \mu > 0, \quad b > 0, \tag{8.7.5}$$

$$c = a + b, \quad d = a - b, \quad l = \mu, \quad l_0 = -\frac{\mu}{2}.$$

For this $\lambda(x)$ the moment T can easily be found: $T = \mu/(\pi b)$.

We compare the behavior of two algorithms. The first one is based on the layer method (8.5.14) with the cubic interpolation (8.7.1). In the case of the problem (8.7.2)–(8.7.3) it has the form

$$\bar{u}(0, x) = \varphi(x), \tag{8.7.6}$$

$$\bar{u}(t_{k+1}, x_j) = \frac{1}{2} \bar{u}(t_k, x_j - h\bar{u}(t_k, x_j - h\bar{u}(t_k, x_j)) + \varepsilon h^{1/2})$$
$$+ \frac{1}{2} \bar{u}(t_k, x_j - h\bar{u}(t_k, x_j - h\bar{u}(t_k, x_j)) - \varepsilon h^{1/2}),$$

$$\bar{u}(t_k, x) = \sum_{i=0}^{3} \Phi_{j,i}(x)\, \bar{u}(t_k, x_{j+i}), \quad x_{j+1} < x < x_{j+2},$$

$$\Phi_{j,i}(x) = \prod_{m=0, m\neq i}^{3} \frac{x - x_{j+m}}{x_{j+i} - x_{j+m}},$$

$$j = 0, \pm 1, \pm 2, \dots, \quad k = 0, \dots, N-1,$$

where $x_j = x_0 + j \times h_x$.

The second algorithm is based on the layer method (8.2.23) and on the cubic interpolation (8.7.1):

$$\bar{u}(0, x) = \varphi(x), \tag{8.7.7}$$

$$\bar{u}(t_{k+1}, x_j) = \frac{1}{2}\bar{u}(t_k, x_j - h\bar{u}(t_k, x_j) + \varepsilon h^{1/2})$$
$$+ \frac{1}{2}\bar{u}(t_k, x_j - h\bar{u}(t_k, x_j) - \varepsilon h^{1/2}),$$

$$\bar{u}(t_k, x) = \sum_{i=0}^{3} \Phi_{j,i}(x)\, \bar{u}(t_k, x_{j+i}), \quad x_{j+1} < x < x_{j+2},$$

$$\Phi_{j,i}(x) = \prod_{m=0, m\neq i}^{3} \frac{x - x_{j+m}}{x_{j+i} - x_{j+m}},$$

$$j = 0, \pm 1, \pm 2, \dots, \quad k = 0, \dots, N-1.$$

Table 8.7.1 gives the results of simulation of the problem (8.7.2)–(8.7.3) with $\varphi(x)$ from (8.7.4), (8.7.5) in the case of the regular solution. In this case the assumptions (i)–(ii) from Sect. 8.5.1 are fulfilled, and the algorithm (8.7.6) has the error estimated by $O(h^2 + \varepsilon^2 h)$ and the algorithm (8.7.7) has the error estimated by $O(h)$. The value of h_x is taken equal to $h^{3/4}$. We present the errors of the approximate solutions $\bar{u}$ in the discrete Chebyshev norm and in the l^1-norm:

$$err^c(t) = \max_{x_i} |\bar{u}(t, x_i) - u(t, x_i)|,$$

$$err^l(t) = \sum_{i} |\bar{u}(t, x_i) - u(t, x_i)| \times h_x.$$

One can infer from Table 8.7.1 that the proposed special algorithm (8.7.6) with error $O(h^2 + \varepsilon^2 h)$ requires less computational effort than the algorithm (8.7.7) with error $O(h)$ and that the experimental data confirm the orders of accuracy of the algorithms given by the theoretical results.

Table 8.7.1 The Burgers equation (regular solution). Dependence of the errors $err^c(t_*)$ and $err^l(t_*)$ on h and ε for the algorithms (8.7.6) and (8.7.7) when $a = b = 0.5$, $\mu = 8$, and $t_* = 4$ ($T \approx 5.09$)

		Algorithm (8.7.6)		Algorithm (8.7.7)	
ε	h	$err^c(t_*)$	$err^l(t_*)$	$err^c(t_*)$	$err^l(t_*)$
	0.3	$0.1351 \cdot 10^{-1}$	$0.1531 \cdot 10^{-1}$	0.1130	0.1397
0.3	0.1	$0.2146 \cdot 10^{-2}$	$0.3347 \cdot 10^{-2}$	$0.3978 \cdot 10^{-1}$	$0.4628 \cdot 10^{-1}$
	0.01	$0.2295 \cdot 10^{-3}$	$0.3874 \cdot 10^{-3}$	$0.4221 \cdot 10^{-2}$	$0.4799 \cdot 10^{-2}$
	0.001	$0.2265 \cdot 10^{-4}$	$0.3947 \cdot 10^{-4}$	$0.4244 \cdot 10^{-3}$	$0.4814 \cdot 10^{-3}$
	0.3	$0.2325 \cdot 10^{-1}$	$0.2051 \cdot 10^{-1}$	0.1539	0.1519
	0.1	$0.4255 \cdot 10^{-2}$	$0.2287 \cdot 10^{-2}$	$0.6084 \cdot 10^{-1}$	$0.5007 \cdot 10^{-1}$
0.1	0.03	$0.3489 \cdot 10^{-3}$	$0.2396 \cdot 10^{-3}$	$0.2029 \cdot 10^{-1}$	$0.1553 \cdot 10^{-1}$
	0.01	$0.4444 \cdot 10^{-4}$	$0.5442 \cdot 10^{-4}$	$0.6751 \cdot 10^{-2}$	$0.5169 \cdot 10^{-2}$
	0.001	$0.5529 \cdot 10^{-5}$	$0.6374 \cdot 10^{-5}$	$0.6806 \cdot 10^{-3}$	$0.5189 \cdot 10^{-3}$

To find the solution $u(t, x)$ to the problem (8.7.2)–(8.7.3), (8.7.4) for $t > T$, when the solution is singular, we realize the following numerical procedure: we simulate the problem by the algorithms (8.7.6) and (8.7.7) with a sufficiently big time step h and with $h_x = h^{3/4}$ up to the time moment $t_* < T$; then we change the time step h to a smaller one h_*, take $h_x = h_*$, and continue the simulations.

Table 8.7.2 gives the results of simulation of the problem (8.7.2)–(8.7.3) with $\varphi(x)$ from (8.7.4), (8.7.5) when $t > T$. One can see that in the singular case the behavior of the algorithm (8.7.6) is also better than the behavior of (8.7.7).

In connection with this example, see also numerical experiments in [284] and [36].

8.7.2 The Generalized FKPP-Equation with a Small Parameter

Consider the problem

$$\frac{\partial u}{\partial t} = \frac{\varepsilon^2}{2} \frac{\partial^2 u}{\partial x^2} + g(x, u; \varepsilon), \quad t > 0, \ x \in \mathbf{R}, \tag{8.7.8}$$

$$u(0, x) = \chi_-(x) = \begin{cases} 1, & x < 0 \\ 1/2, & x = 0 \\ 0, & x > 0, \end{cases} \tag{8.7.9}$$

Table 8.7.2 The Burgers equation (singular solution). The errors $err^c(t)$ and $err^l(t)$ for $t = 8$ ($T \approx 5.09$). Other parameter values are the same as in Table 8.7.1. The time steps h and h_* are used when $t \leq t_*$ and $t > t_*$, respectively

			Algorithm (8.7.6)		Algorithm (8.7.7)	
ε	h	h_*	$err^c(t)$	$err^l(t)$	$err^c(t)$	$err^l(t)$
	0.1	0.01	$0.6322 \cdot 10^{-2}$	$0.2713 \cdot 10^{-2}$	0.1693	$0.6555 \cdot 10^{-1}$
0.3	0.01	0.001	$0.4036 \cdot 10^{-3}$	$0.2482 \cdot 10^{-3}$	$0.1771 \cdot 10^{-1}$	$0.6782 \cdot 10^{-2}$
	0.001	0.0001	$0.5977 \cdot 10^{-4}$	$0.3760 \cdot 10^{-4}$	$0.1776 \cdot 10^{-2}$	$0.6931 \cdot 10^{-3}$
	0.1	0.001	$0.5553 \cdot 10^{-1}$	$0.2351 \cdot 10^{-2}$	> 0.5	$0.6594 \cdot 10^{-1}$
	0.03	0.001	$0.1219 \cdot 10^{-1}$	$0.4699 \cdot 10^{-3}$	> 0.5	$0.3189 \cdot 10^{-1}$
0.1	0.03	0.0001	$0.3955 \cdot 10^{-2}$	$0.1718 \cdot 10^{-3}$	0.4029	$0.1716 \cdot 10^{-1}$
	0.01	0.0001	$0.7047 \cdot 10^{-3}$	$0.3007 \cdot 10^{-4}$	0.1687	$0.6828 \cdot 10^{-2}$
	0.001	0.0001	$0.4139 \cdot 10^{-3}$	$0.2312 \cdot 10^{-4}$	$0.5461 \cdot 10^{-1}$	$0.2185 \cdot 10^{-2}$

and take

$$g(x, u; \varepsilon) = \frac{1}{\varepsilon^2} c(x) u(1 - u), \qquad (8.7.10)$$

$$c(x) = c + \frac{a}{\pi} \arctan \alpha(x - b).$$

Here $\varepsilon > 0$ is a small parameter, $\alpha > 0$ is a big number, c, a, and b are positive constants, and $a/2 < c < 3a/2$.

The problem (8.7.8)–(8.7.10) is a generalization of the FKPP-equation. The theoretical results for this problem obtained in [111] give the following. For $t < T_0 \approx \frac{b\sqrt{2a}}{c + 0.5a}$, the wave propagates to the right of the domain $G_0 = \{x < 0\}$ with the velocity $\sqrt{2c - a}$, "taking no notice" of the fact that after $x = b$ the coefficient $c(x)$ takes a larger value $c + a/2$. But at the time T_0, a new "source" arises at the point $x = b$, away from which the front starts propagating in both directions: to the left with the velocity close to $\sqrt{2c - a}$ and to the right with the velocity close to $\sqrt{2c + a}$. Fig. 8.7.1, obtained in our numerical experiments, demonstrates this phenomenon.

In the experiments we observe that for our values of the parameters ($\varepsilon = 0.1$, see the caption to Fig. 8.7.1)

$$\min_{-\infty < x < b} \bar{u}(5.75, x) \approx 10^{-71}$$

while there is already the new front at $x = b$ (see Fig. 8.7.1c). So, the "channel" through which the new "source" is initialized is very narrow. This fact has to be

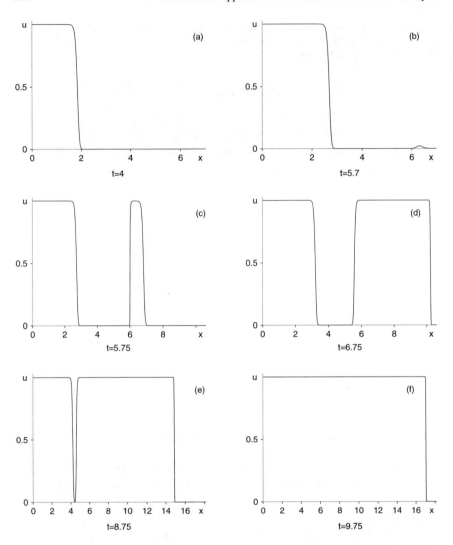

Fig. 8.7.1 The FKPP-equation. Evolution of the solution $u(t, x)$ to the problem (8.7.8)–(8.7.10) for $\varepsilon = 0.1, c = 1.125, a = 2, b = 6, \alpha = 150$ simulated by the method (8.6.13) with $h = 0.0001$

taken into account for realizing numerical procedures on a computer. For instance, when $\varepsilon = 0.04$

$$\min_{-\infty < x < b} \bar{u}(5.75, x) \approx 10^{-439},$$

which is less than the smallest positive number $(\sim 10^{-308})$ realizable by many compilers. To observe the phenomenon in this case, one has to compose a special numerical procedure or use a special compiler.

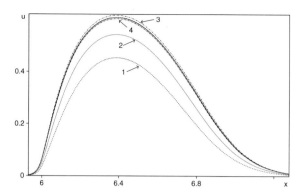

Fig. 8.7.2 The FKPP-equation (new source appearance). Comparison of the methods. The solid curve is simulated by (8.6.7) and (8.6.13) with $h = 0.0001$, and it visually coincides with the exact solution. The curves 1, 2 are simulated by (8.7.14) and (8.6.6) with $h = 0.0001$; the curves 3, 4 by (8.6.7) and (8.6.13) with $h = 0.001$. Here $\varepsilon = 0.2$, $t = 5.8$, and the other parameter values are the same as in Figure 8.7.1

The following simple calculations (which, of course, do not pretend to be an exhaustive explanation) clarify such a high sensitivity. We see from Figure 8.7.2 that for time $t \in [4.8, 5.8]$ and $\varepsilon = 0.2$ the arising new source is such that the second derivative $\partial^2 u / \partial x^2$ for, e.g., $x \geq 4$ is not too large and therefore the equation (8.7.8) can be approximately replaced by

$$\frac{du}{dt} = \frac{1}{\varepsilon^2} c(x) u (1 - u), \quad \varepsilon = 0.2, \quad x \geq 4. \tag{8.7.11}$$

Then

$$u(t, x) = \frac{u(t_0, x) e^{c(x)(t-t_0)/\varepsilon^2}}{1 + u(t_0, x)(e^{c(x)(t-t_0)/\varepsilon^2} - 1)}. \tag{8.7.12}$$

Let $x = 6.4$, $t_0 = 4.8$, $t = 5.8$. In this case $c(x) \approx 2.125$ and $u(5.8, 6.4) \approx 0.6$ (see Fig. 8.7.2). Then by (8.7.12) we get

$$u(4.8, 6.4) \approx 1.3 \times 10^{-23}. \tag{8.7.13}$$

At the same time, if such a small source arises at a point $x < b$ where $c(x) \approx 1.125$, we obtain the following very small value $u(5.8, x)|_{x<b} \approx 2.1 \times 10^{-11}$, i.e., u grows at $x = 6.4$ much faster than at $x < b$.

An additional confirmation of the high sensitiveness of the considered model is given, for instance, by the following experiment. If we put $u(0, x) = \mu$ for $x > 0$ with a small positive μ, e.g. 10^{-15}, in the initial condition (8.7.9) and take the other parameters as in Fig. 8.7.1, the new "source" arises at a moment $t \ll 1$.

Here we compare five numerical methods: the methods (8.6.6), (8.6.7), (8.6.10), (8.6.13) given in Sect. 8.6 (of course, we take their versions adapted to problems

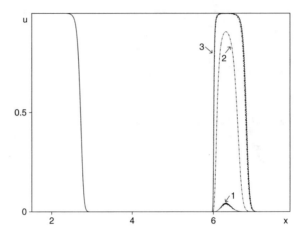

Fig. 8.7.3 The FKPP-equation. Comparison of the methods. The curve 3 is simulated by (8.6.7) and (8.6.13) with $h = 0.0005$ and $h = 0.0001$, and it visually coincides with the exact solution. The curves 1 and 2 are simulated by (8.7.14) and (8.6.6) with $h = 0.0001$. Here $t = 5.75$, and the other parameter values are the same as in Fig. 8.7.1

with positive direction of time, cf. Remark 8.2.8) and the first-order method written below. The results of the numerical tests are given on Figs. 8.7.2 and 8.7.3.

The first-order method:

$$\bar{u}(0, x_j) = \chi_-(x_j), \tag{8.7.14}$$

$$\bar{u}(t_{k+1}, x_j) = \frac{1}{2}\bar{u}(t_k, x_j + \varepsilon h^{1/2}) + \frac{1}{2}\bar{u}(t_k, x_j - \varepsilon h^{1/2})$$
$$+ \frac{h}{2}(g(x_{j-1}, \bar{u}(t_k, x_{j-1})) + g(x_{j+1}, \bar{u}(t_k, x_{j+1}))),$$

$$x_j = x_0 + j\varepsilon h^{1/2}, \ \ j = 0, \pm 1, \dots, \ \ k = 0, \dots, N - 1.$$

It can be checked that for a sufficiently small h this method preserves the monotonicity property of the solution. The first-order method (8.6.5), which has $hg(x_j, \bar{u}(t_k, x_j))$ instead of $h(g(x_{j-1}, \bar{u}(t_k, x_{j-1})) + g(x_{j+1}, \bar{u}(t_k, x_{j+1})))/2$, does not preserve the monotonicity property and has unstable behavior for the problem considered.

The algebraic equations arising in the implementation of the methods (8.6.7), (8.6.10), and (8.6.13) at each step are quadratic and are solved exactly. The results of testing of the three-layer method (8.6.10) were discussed in Sect. 8.6.2.

Chapter 9
Numerical Solution of the Nonlinear Dirichlet and Neumann Problems Based on the Probabilistic Approach

In this chapter we apply the probabilistic approach of Chap. 8 to nonlinear problems with Dirichlet and Neumann boundary conditions [309, 310]. We recall that the probabilistic approach is based on making use of the well-known probabilistic representations of solutions to linear parabolic equations and the ideas of weak sense numerical integration of SDEs. Despite their probabilistic nature these methods are nevertheless deterministic. The probabilistic approach takes into account the coefficients' dependence on the space variables and a relationship between diffusion and advection in an intrinsic manner. In particular, the layer methods allow us to avoid difficulties stemming from rapidly varying coefficients and strong advection.

Sections 9.1–9.2 are devoted to the nonlinear Dirichlet problems while Sects. 9.3–9.4 deal with the nonlinear Neumann problems. Combining methods from Sects. 9.1 and 9.3, we can solve mixed boundary value problems, i.e., when we have the Dirichlet condition on a part of the boundary ∂G and the Neumann condition on the rest of ∂G. A layer algorithm was used in [446] to simulate FKPP-equation with random advection and Dirichlet boundary conditions. Layer methods for Navier–Stokes equations with Dirichlet boundary conditions were proposed in [325].

9.1 Layer Methods for the Dirichlet Problem for Semilinear Parabolic Equations

Let G be a bounded domain in $\mathbf{R}^d$, $Q = [t_0, T) \times G$ be a cylinder in $\mathbf{R}^{d+1}$, $\Gamma = \bar{Q} \setminus Q$. The set Γ is a part of the boundary of the cylinder Q consisting of the upper base and the lateral surface. Consider the Dirichlet problem for the semilinear parabolic equation

© Springer Nature Switzerland AG 2021
G. N. Milstein and M. V. Tretyakov, *Stochastic Numerics for Mathematical Physics*,
Scientific Computation, https://doi.org/10.1007/978-3-030-82040-4_9

$$\frac{\partial u}{\partial t} + \frac{1}{2} \sum_{i,j=1}^{d} a^{ij}(t, x, u)\frac{\partial^2 u}{\partial x^i \partial x^j} + \sum_{i=1}^{d} b^i(t, x, u)\frac{\partial u}{\partial x^i} \tag{9.1.1}$$

$$+ g(t, x, u) = 0, \quad (t, x) \in Q,$$

$$u(t, x)|_{\Gamma} = \varphi(t, x). \tag{9.1.2}$$

The form of Eq. (9.1.1) is relevant to a probabilistic approach, i.e., the equation is considered under $t < T$, and the "initial" conditions are prescribed at $t = T$. Using the well-known probabilistic representation of the solution to (9.1.1)–(9.1.2) (see [86, 111]), we get

$$u(t, x) = E(\varphi(\tau, X_{t,x}(\tau)) + Z_{t,x,0}(\tau)). \tag{9.1.3}$$

In (9.1.3), $X_{t,x}(s)$, $Z_{t,x,z}(s)$, $(t, x) \in Q$, $s \geq t$, is the solution of the Cauchy problem for the Ito system of SDEs

$$dX = b(s, X, u(s, X)) ds + \sigma(s, X, u(s, X)) dw(s), \quad X(t) = x, \tag{9.1.4}$$

$$dZ = g(s, X, u(s, X)) ds, \quad Z(t) = z,$$

where $w(s) = (w^1(s), \ldots, w^d(s))^{\mathsf{T}}$ is a standard Wiener process, $b(s, x, u) = (b^1(s, x, u), \ldots, b^d(s, x, u))^{\mathsf{T}}$ is a column vector, the matrix $\sigma = \sigma(s, x, u)$ is obtained from the equation

$$\sigma\sigma^{\mathsf{T}} = a, \quad \sigma = \{\sigma^{ij}(s, x, u)\}, \quad a = \{a^{ij}(s, x, u)\},$$

and $\tau = \tau_{t,x}$ is the first exit time of the trajectory $(s, X_{t,x}(s))$ from the domain Q.

If Eq. (9.1.1) is linear, the system (9.1.4) does not contain the unknown function $u(s, x)$ and therefore one can use weak approximation schemes for solving (9.1.4) with the Monte Carlo realization of representation (9.1.3). A number of constructive schemes are presented in Chap. 7. The procedures of Chap. 7 together with the Monte Carlo approach allow us to find a value $u(t, x)$ at a single point even when the domain G is high dimensional.

Of course, the nonlinear case is much more complicated. But we are aimed to construct layer methods and due to this fact it becomes possible to use a one-step (local) version of the representation (9.1.3) (see formula (9.1.7) below). Introduce a time discretization, for definiteness the equidistant one:

$$T = t_N > t_{N-1} > \cdots > t_0, \quad h := \frac{T - t_0}{N}.$$

The methods proposed in this chapter give an approximation $\bar{u}(t_k, x)$ of the solution $u(t_k, x)$, $k = N, \ldots, 0$, $x \in \bar{G}$, i.e., step by step everywhere in the domain $\bar{G}$. It is feasible if the dimension of the domain G is small ($d \leq 3$). To construct the layer methods, we exploit the ideas of weak sense numerical integration of SDEs in bounded domain and obtain some approximate relations on the basis of (9.1.7), (9.1.4). The relations allow us to express $\bar{u}(t_k, x)$, $k = N - 1, \ldots, 0$, recurrently in terms of $\bar{u}(t_{k+1}, x)$. Despite their probabilistic nature these methods are deterministic as in the previous chapter.

In Sects. 9.1.1–9.1.3, we derive layer methods relying on the numerical integration of SDEs and state theorems on their convergence, which proofs using both deterministic and probabilistic type arguments can be found in [309]. To realize layer methods in practice, we need a discretization in the variable x with an interpolation at every step to turn a method into an algorithm. Such numerical algorithms are constructed in Sect. 9.1.4. The ideas can be demonstrated at $d = 1$, and we restrict ourselves to this case in the current section. Layer methods in the case of $d \geq 2$ can be found in [309]. Numerical tests are presented in Sect. 9.2 where, in particular, we compare the proposed layer methods with the well-known finite-difference schemes.

9.1.1 Construction of a Layer Method of First Order

The boundary value problem (9.1.1)–(9.1.2) in the one-dimensional case has the form

$$\frac{\partial u}{\partial t} + \frac{1}{2}\sigma^2(t, x, u)\frac{\partial^2 u}{\partial x^2} + b(t, x, u)\frac{\partial u}{\partial x} + g(t, x, u) = 0, \quad (t, x) \in Q, \quad (9.1.5)$$

$$u(t, x)|_\Gamma = \varphi(t, x). \quad (9.1.6)$$

In this case Q is the partly open rectangle: $Q = [t_0, T) \times (\alpha, \beta)$, and Γ consists of the upper base $\{T\} \times [\alpha, \beta]$ and two vertical intervals: $[t_0, T) \times \{\alpha\}$ and $[t_0, T) \times \{\beta\}$. We assume that $\sigma(t, x, u) \geq \sigma_* > 0$ for $(t, x) \in \bar{Q}$, $-\infty < u < \infty$.

Let $u = u(t, x)$ be the solution of the problem (9.1.5)–(9.1.6), which is assumed to exist, to be unique, and to be sufficiently smooth. One can find many theoretical results on this topic in [144, 222, 396, 408, 437] (see also references therein).

Analogously to (9.1.3), we have

$$u(t_k, x) = E\left(u(\vartheta_{t_k,x}, X_{t_k,x}(\vartheta_{t_k,x})) + Z_{t_k,x,0}(\vartheta_{t_k,x})\right), \quad (9.1.7)$$

where $\vartheta_{t_k,x} = \vartheta_{t_k,x}(t_{k+1}) := \tau_{t_k,x} \wedge t_{k+1}$, and X, Z satisfy the system (9.1.4).

Let us suppose for a while that it is possible to extend the coefficients of Eq. (9.1.5) so that the new equation has a solution $u(t, x)$ on $[t_0, T) \times \mathbf{R}$ which is an extension of the solution to the boundary value problem (9.1.5)–(9.1.6). Then, instead of (9.1.7), we obtain (we assume that the layer $u(t_{k+1}, x)$ is known)

$$u(t_k, x) = E[u(t_{k+1}, X_{t_k,x}(t_{k+1})) + Z_{t_k,x,0}(t_{k+1})]. \quad (9.1.8)$$

9.1.1.1   Approximation for Internal Points

Applying the explicit weak Euler scheme with the simplest simulation of noise to the system (9.1.4), we get

$$\bar{X}_{t_k,x}(t_{k+1}) = x + h\, b(t_k, x, u(t_k, x)) + \sqrt{h}\, \sigma(t_k, x, u(t_k, x))\, \xi\,, \qquad (9.1.9)$$

$$\bar{Z}_{t_k,x,0}(t_{k+1}) = hg(t_k, x, u(t_k, x))\,, \qquad (9.1.10)$$

where the ξ is distributed by the law: $P(\xi = \pm 1) = 1/2$.

Using (9.1.8), we get to within $O(h^2)$:

$$
\begin{aligned}
u(t_k, x) &\simeq E[u(t_{k+1}, \bar{X}_{t_k,x}(t_{k+1})) + \bar{Z}_{t_k,x,0}(t_{k+1})] \qquad (9.1.11)\\
&= \frac{1}{2}u(t_{k+1}, x + hb(t_k, x, u(t_k, x)) - \sqrt{h}\sigma(t_k, x, u(t_k, x)))\\
&\quad + \frac{1}{2}u(t_{k+1}, x + hb(t_k, x, u(t_k, x)) + \sqrt{h}\sigma(t_k, x, u(t_k, x)))\\
&\quad + hg(t_k, x, u(t_k, x))\,.
\end{aligned}
$$

Now we can obtain an implicit relation for an approximation of $u(t_k, x)$. Applying the method of simple iteration to the implicit relation and taking $u(t_{k+1}, x)$ as a null iteration, we get the following explicit one-step approximation $v(t_k, x)$ of $u(t_k, x)$:

$$v(t_k, x) = \frac{1}{2}u(t_{k+1}, x + hb_k - \sqrt{h}\sigma_k) + \frac{1}{2}u(t_{k+1}, x + hb_k + \sqrt{h}\sigma_k) + hg_k\,,$$
$$(9.1.12)$$

where b_k, σ_k, g_k are the coefficients $b(t, x, u)$, $\sigma(t, x, u)$, $g(t, x, u)$ evaluated at the point $(t_k, x, u(t_{k+1}, x))$.

But in reality we know the layer $u(t_{k+1}, x)$ for $\alpha \le x \le \beta$ only. At the same time the argument $x + hb_k - \sqrt{h}\sigma_k$ for x close to α is less than α and the argument $x + hb_k + \sqrt{h}\sigma_k$ for x close to β is more than β. Thus, we need to extend the layer $u(t_{k+1}, x)$ in a constructive manner.

9.1.1.2   Approximation for Points near the Boundary

Using the explicit weak Euler scheme for the initial point (t, α) with $t_k \le t \le t_{k+1}$, we put (cf. (9.1.9)–(9.1.10)):

$$\bar{X}_{t,\alpha}(t_{k+1}) = x + b(t, \alpha, u(t, \alpha)) \cdot (t_{k+1} - t) + \sigma(t, \alpha, u(t, \alpha))\sqrt{t_{k+1} - t}\,\xi\,,$$

$$\bar{Z}_{t,\alpha,0}(t_{k+1}) = g(t, \alpha, u(t, \alpha)) \cdot (t_{k+1} - t)\,. \qquad (9.1.13)$$

Analogously, we define $\bar{X}_{t,\beta}(t_{k+1})$, $\bar{Z}_{t,\beta,0}(t_{k+1})$.

We have (see (9.1.11) and (9.1.13)) for $t_k \le t \le t_{k+1}$:

$$u(t, \alpha) \simeq E[u(t_{k+1}, \bar{X}_{t,\alpha}(t_{k+1})) + \bar{Z}_{t,\alpha,0}(t_{k+1})] \qquad (9.1.14)$$

$$= \frac{1}{2} u(t_{k+1}, \alpha + b(t, \alpha, u(t, \alpha)) \cdot (t_{k+1} - t) - \sigma(t, \alpha, u(t, \alpha))\sqrt{t_{k+1} - t})$$

$$+ \frac{1}{2} u(t_{k+1}, \alpha + b(t, \alpha, u(t, \alpha)) \cdot (t_{k+1} - t) + \sigma(t, \alpha, u(t, \alpha))\sqrt{t_{k+1} - t})$$

$$+ g(t, \alpha, u(t, \alpha)) \cdot (t_{k+1} - t) .$$

If we replace (recall that $u(t, \alpha) = \varphi(t, \alpha)$ due to (9.1.6)) the argument $(t, \alpha, u(t, \alpha)) = (t, \alpha, \varphi(t, \alpha))$ by $(t_k, \alpha, \varphi(t_{k+1}, \alpha))$, the right-hand side of (9.1.14) is changed by a quantity of order $O(h^2)$. Since the approximation in (9.1.14) is also of order $O(h^2)$, we get

$$\frac{1}{2} u(t_{k+1}, \alpha + b(t_k, \alpha, \varphi(t_{k+1}, \alpha)) \cdot (t_{k+1} - t) \qquad (9.1.15)$$

$$-\sigma(t_k, \alpha, \varphi(t_{k+1}, \alpha))\sqrt{t_{k+1} - t})$$

$$= \varphi(t, \alpha) - \frac{1}{2} u(t_{k+1}, \alpha + b(t_k, \alpha, \varphi(t_{k+1}, \alpha)) \times (t_{k+1} - t)$$

$$+\sigma(t_k, \alpha, \varphi(t_{k+1}, \alpha))\sqrt{t_{k+1} - t})$$

$$-g(t_k, \alpha, \varphi(t_{k+1}, \alpha)) \cdot (t_{k+1} - t) + O(h^2) .$$

Introduce

$$\alpha_0 := \alpha + hb(t_k, \alpha, \varphi(t_{k+1}, \alpha)) - \sqrt{h}\sigma(t_k, \alpha, \varphi(t_{k+1}, \alpha)) .$$

Clearly $\alpha_0 < \alpha$ and

$$\alpha_0 \le \alpha + b(t_k, \alpha, \varphi(t_{k+1}, \alpha)) \cdot (t_{k+1} - t) - \sigma(t_k, \alpha, \varphi(t_{k+1}, \alpha))\sqrt{t_{k+1} - t} \le \alpha$$

for $t_k \le t \le t_{k+1}$ under a sufficiently small $h > 0$.
Analogously

$$\frac{1}{2} u(t_{k+1}, \beta + b(t_k, \beta, \varphi(t_{k+1}, \beta)) \cdot (t_{k+1} - t) \qquad (9.1.16)$$

$$+\sigma(t_k, \beta, \varphi(t_{k+1}, \beta))\sqrt{t_{k+1} - t})$$

$$= \varphi(t, \beta) - \frac{1}{2} u(t_{k+1}, \beta + b(t_k, \beta, \varphi(t_{k+1}, \beta)) \times (t_{k+1} - t)$$

$$-\sigma(t_k, \beta, \varphi(t_{k+1}, \beta))\sqrt{t_{k+1} - t})$$

$$-g(t_k, \beta, \varphi(t_{k+1}, \beta)) \cdot (t_{k+1} - t) + O(h^2),$$

$$\beta_0 := \beta + hb(t_k, \beta, \varphi(t_{k+1}, \beta)) + \sqrt{h}\sigma(t_k, \beta, \varphi(t_{k+1}, \beta)) .$$

The relations (9.1.15)–(9.1.16) give the desired extension of the function $u(t_{k+1}, x)$ on the interval $[\alpha_0, \beta_0]$.

Let us return to the formula (9.1.12). The arguments $x + hb_k - \sqrt{h}\sigma_k$ and $x + hb_k + \sqrt{h}\sigma_k$ are monotone increasing functions in $x \in [\alpha, \beta]$ for a sufficiently small $h > 0$. Their values belong to $[\alpha_0, \beta_0]$, and $x + hb_k + \sqrt{h}\sigma_k$ is always (for $x \in [\alpha, \beta]$) more than α while $x + hb_k - \sqrt{h}\sigma_k$ is less than β. Let $x + hb_k - \sqrt{h}\sigma_k < \alpha$ (clearly it is possible for x close to α). Due to the above, there exists a unique root $\gamma_k(x)$, $0 < \gamma_k(x) \le 1$, of the quadratic equation

$$\alpha + \gamma_k h\, b(t_k, \alpha, \varphi(t_{k+1}, \alpha)) - \sqrt{\gamma_k h}\sigma(t_k, \alpha, \varphi(t_{k+1}, \alpha)) \qquad (9.1.17)$$
$$= x + hb_k - \sqrt{h}\sigma_k.$$

Analogously, if $x + hb_k + \sqrt{h}\sigma_k > \beta$, then there exists a unique root $\delta_k(x)$, $0 < \delta_k(x) \le 1$, of the quadratic equation

$$\beta + \delta_k h\, b(t_k, \beta, \varphi(t_{k+1}, \beta)) + \sqrt{\delta_k h}\sigma(t_k, \beta, \varphi(t_{k+1}, \beta)) \qquad (9.1.18)$$
$$= x + hb_k + \sqrt{h}\sigma_k.$$

If, for instance, $x + hb_k - \sqrt{h}\sigma_k < \alpha$, then one can replace the value $u(t_{k+1}, x + hb_k - \sqrt{h}\sigma_k)/2$ in (9.1.12) by the value due to (9.1.17) and (9.1.15):

$$\frac{1}{2}u(t_{k+1}, x + hb_k - \sqrt{h}\sigma_k)$$
$$= \frac{1}{2}u(t_{k+1}, \alpha + \gamma_k hb(t_k, \alpha, \varphi(t_{k+1}, \alpha)) - \sqrt{\gamma_k h}\sigma(t_k, \alpha, \varphi(t_{k+1}, \alpha)))$$
$$\approx \varphi(t_{k+1-\gamma_k}, \alpha)$$
$$- \frac{1}{2}\gamma_k h\, u(t_{k+1}, \alpha + b(t_k, \alpha, \varphi(t_{k+1}, \alpha)) + \sqrt{\gamma_k h}\sigma(t_k, \alpha, \varphi(t_{k+1}, \alpha)))$$
$$- \gamma_k h\, g(t_k, \alpha, \varphi(t_{k+1}, \alpha)),$$

where $t_{k+1-\gamma_k} = t_k + h(1 - \gamma_k)$.

9.1.1.3 The Layer Method

As a result, we obtain the following one-step approximation $v(t_k, x)$ for $u(t_k, x)$:

$$v(t_k, x) = \frac{1}{2}u(t_{k+1}, x + hb_k - \sqrt{h}\sigma_k) + \frac{1}{2}u(t_{k+1}, x + hb_k + \sqrt{h}\sigma_k) \qquad (9.1.19)$$
$$+ hg_k, \quad \text{if } x + hb_k \pm \sqrt{h}\sigma_k \in [\alpha, \beta];$$

$$v(t_k, x) = \varphi(t_{k+1-\gamma_k}, \alpha) - \gamma_k h g(t_k, \alpha, \varphi(t_{k+1}, \alpha))$$

$$-\frac{1}{2} u(t_{k+1}, \alpha + \gamma_k h b(t_k, \alpha, \varphi(t_{k+1}, \alpha)) + \sqrt{\gamma_k h} \sigma(t_k, \alpha, \varphi(t_{k+1}, \alpha)))$$

$$+\frac{1}{2} u(t_{k+1}, x + h b_k + \sqrt{h} \sigma_k) + h g_k, \quad \text{if } x + h b_k - \sqrt{h} \sigma_k < \alpha;$$

$$v(t_k, x) = \frac{1}{2} u(t_{k+1}, x + h b_k - \sqrt{h} \sigma_k) + \varphi(t_{k+1-\delta_k}, \beta)$$

$$- \delta_k h\, g(t_k, \beta, \varphi(t_{k+1}, \beta))$$

$$-\frac{1}{2} u(t_{k+1}, \beta + \delta_k h\, b(t_k, \beta, \varphi(t_{k+1}, \beta)) - \sqrt{\delta_k h} \sigma(t_k, \beta, \varphi(t_{k+1}, \beta)))$$

$$+ h g_k, \quad \text{if } x + h b_k + \sqrt{h} \sigma_k > \beta, \ k = N-1, \dots, 1, 0,$$

where b_k, σ_k, g_k are the coefficients $b(t, x, u)$, $\sigma(t, x, u)$, $g(t, x, u)$ evaluated at the point $(t_k, x, u(t_{k+1}, x))$ and γ_k, δ_k are the corresponding roots of Eqs. (9.1.17) and (9.1.18).

Thus *the layer method* acquires the form

$$\bar{u}(t_N, x) = \varphi(t_N, x), \quad x \in [\alpha, \beta], \tag{9.1.20}$$

$$\bar{u}(t_k, x) = \frac{1}{2} \bar{u}(t_{k+1}, x + h\bar{b}_k - \sqrt{h}\bar{\sigma}_k) + \frac{1}{2} \bar{u}(t_{k+1}, x + h\bar{b}_k + \sqrt{h}\bar{\sigma}_k) + h\bar{g}_k,$$

$$\text{if } x + h\bar{b}_k \pm \sqrt{h}\bar{\sigma}_k \in [\alpha, \beta];$$

$$\bar{u}(t_k, x) = \varphi(t_{k+1-\bar{\gamma}_k}, \alpha) - \bar{\gamma}_k h g(t_k, \alpha, \varphi(t_{k+1}, \alpha))$$

$$-\frac{1}{2} \bar{u}(t_{k+1}, \alpha + \bar{\gamma}_k h b(t_k, \alpha, \varphi(t_{k+1}, \alpha)) + \sqrt{\bar{\gamma}_k h} \sigma(t_k, \alpha, \varphi(t_{k+1}, \alpha)))$$

$$+\frac{1}{2} \bar{u}(t_{k+1}, x + h\bar{b}_k + \sqrt{h}\bar{\sigma}_k) + h\bar{g}_k, \quad \text{if } x + h\bar{b}_k - \sqrt{h}\bar{\sigma}_k < \alpha;$$

$$\bar{u}(t_k, x) = \frac{1}{2} \bar{u}(t_{k+1}, x + h\bar{b}_k - \sqrt{h}\bar{\sigma}_k) + \varphi(t_{k+1-\bar{\delta}_k}, \beta)$$

$$-\bar{\delta}_k h g(t_k, \beta, \varphi(t_{k+1}, \beta))$$

$$-\frac{1}{2} \bar{u}(t_{k+1}, \beta + \bar{\delta}_k h b(t_k, \beta, \varphi(t_{k+1}, \beta)) - \sqrt{\bar{\delta}_k h} \sigma(t_k, \beta, \varphi(t_{k+1}, \beta)))$$

$$+ h\bar{g}_k, \quad \text{if } x + h\bar{b}_k + \sqrt{h}\bar{\sigma}_k > \beta;$$

$$k = N-1, \dots, 1, 0,$$

where $\bar{b}_k$, $\bar{\sigma}_k$, $\bar{g}_k$ are the coefficients $b(t, x, u)$, $\sigma(t, x, u)$, $g(t, x, u)$ evaluated at the point $(t_k, x, \bar{u}(t_{k+1}, x))$ and $\bar{\gamma}_k$, $\bar{\delta}_k$ are the corresponding roots of Eqs. (9.1.17) and (9.1.18) with the right-hand sides $x + h\bar{b}_k - \sqrt{h}\bar{\sigma}_k$ and $x + h\bar{b}_k + \sqrt{h}\bar{\sigma}_k$.

The method (9.1.20) is an explicit layer method for solving the Dirichlet problem (9.1.5)–(9.1.6). This method is deterministic, even though it is constructed by a probabilistic approach. The method is of the first order of convergence with respect to h (see Theorem 9.1.3 below).

Remark 9.1.1 Let us briefly discuss some differences between the layer methods obtained here and the well-known finite-difference methods. Finite-difference methods also allow us to express an approximate solution on the layer $t = t_k$ recurrently in terms of the solution on the layer $t = t_{k+1}$. For their construction, both the time step Δt and the space step Δx are used. Moreover, the nodes of the layer $t = t_{k+1}$ used to evaluate $\bar{u}(t_k, x_j)$ are apriori prescribed. In our methods we use the time step h only, and the points from the layer $t = t_{k+1}$ to evaluate $\bar{u}(t_k, x)$ arise automatically. A location of these points depends on the coefficients of the problem considered and on the weak scheme chosen. As a result, the probabilistic approach takes into account a coefficient dependence on the space variables and a relationship between diffusion and advection in an intrinsic manner. In particular, the layer methods allow us to avoid difficulties stemming from rapidly varying coefficients and strong advection. We should also note that the probabilistic approach gives a natural way to derive a lot of various new methods.

9.1.2   Convergence Theorem

We make the following assumptions.

(i) There exists a unique solution $u(t, x)$ of problem (9.1.5)–(9.1.6) such that

$$u_\circ < u_* \le u(t, x) \le u^* < u^\circ, \ t_0 \le t \le T, \ x \in [\alpha, \beta], \qquad (9.1.21)$$

where $u_\circ$, u_*, u^*, u° are some constants, and there exist uniformly bounded derivatives:

$$\left| \frac{\partial^{i+j} u}{\partial t^i \partial x^j} \right| \le K, \ i = 0, \ j = 1, 2, 3, 4; \ i = 1, \ j = 0, 1, 2; \ i = 2, \ j = 0;$$

$$(9.1.22)$$

$$t_0 \le t \le T, \ x \in [\alpha, \beta].$$

(ii) The coefficients $b(t, x, u)$, $\sigma(t, x, u)$, $g(t, x, u)$ and their first and second derivatives in x and u are uniformly bounded:

$$\left| \frac{\partial^{i+j} b}{\partial x^i \partial u^j} \right| \le K, \ \left| \frac{\partial^{i+j} \sigma}{\partial x^i \partial u^j} \right| \le K, \ \left| \frac{\partial^{i+j} g}{\partial x^i \partial u^j} \right| \le K, \ 0 \le i + j \le 2, \qquad (9.1.23)$$

$$t_0 \le t \le T, \ x \in [\alpha, \beta], \ u_\circ < u < u^\circ.$$

The lemma on the one-step error $\rho(t_k, x)$ of method (9.1.20) and the global convergence theorem are proved analogously to Lemma 8.2.1 and Theorem 8.2.2, respectively (see the proofs in [309]).

Lemma 9.1.2 *Under the assumptions* (i) *and* (ii), *the one-step error* $\rho(t_k, x)$ *of the method* (9.1.20) *has the second order of smallness with respect to h, i.e.,*

$$|\rho(t_k, x)| = |v(t_k, x) - u(t_k, x)| \le Ch^2,$$

where $v(t_k, x)$ *is defined by* (9.1.19) *and* $C > 0$ *does not depend on h, k, x.*

Theorem 9.1.3 *Under the assumptions* (i) *and* (ii), *the method* (9.1.20) *has the first order of convergence with respect to h, i.e.,*

$$|\bar{u}(t_k, x) - u(t_k, x)| \le Kh,$$

where $K > 0$ *does not depend on h, k, x.*

9.1.3 A Layer Method with a Simpler Approximation near the Boundary

Without exploiting the idea used above of involving the points outside the interval $[\alpha, \beta]$ while constructing a layer method, it is possible to get a layer method that is simpler but with a larger one-step error near the boundary than (9.1.20) (see Lemma 9.1.4 below). Let us note that in spite of the greater one-step boundary error the global error of this method will be $O(h)$ again (see Theorem 9.1.5). Here we approximate the solution $u(t_k, x)$, when the point x is close to α (or β), using values of the solution at a point $(t_{k+\lambda_k}, \alpha)$ with some $\lambda_k \in (0, 1)$ (or at a point $(t_{k+\mu_k}, \beta)$ with some $\mu_k \in (0, 1)$) and at the point $(t_{k+1}, x + hb_k + \sqrt{h}\sigma_k)$ (or $(t_{k+1}, x + hb_k - \sqrt{h}\sigma_k)$) with some (positive) weights. These two weights may be interpreted as probabilities of reaching and not reaching of α (or β). The method obtained on this way has the form

$$\bar{u}(t_N, x) = \varphi(t_N, x), \quad x \in [\alpha, \beta], \tag{9.1.24}$$

$$\bar{u}(t_k, x) = \frac{1}{2}\bar{u}(t_{k+1}, x + h\bar{b}_k - \sqrt{h}\bar{\sigma}_k) + \frac{1}{2}\bar{u}(t_{k+1}, x + h\bar{b}_k + \sqrt{h}\bar{\sigma}_k) + h\bar{g}_k,$$
$$\text{if } x + h\bar{b}_k \pm \sqrt{h}\bar{\sigma}_k \in [\alpha, \beta];$$

$$\bar{u}(t_k, x) = \frac{1}{1 + \sqrt{\bar{\lambda}_k}}\varphi(t_{k+\bar{\lambda}_k}, \alpha) + \frac{\sqrt{\bar{\lambda}_k}}{1 + \sqrt{\bar{\lambda}_k}}\bar{u}(t_{k+1}, x + h\bar{b}_k + \sqrt{h}\bar{\sigma}_k)$$
$$+ \sqrt{\bar{\lambda}_k}h\bar{g}_k, \text{ if } x + h\bar{b}_k - \sqrt{h}\bar{\sigma}_k < \alpha;$$

$$\bar{u}(t_k, x) = \frac{1}{1 + \sqrt{\bar{\mu}_k}} \varphi(t_{k+\bar{\mu}_k}, \beta) + \frac{\sqrt{\bar{\mu}_k}}{1 + \sqrt{\bar{\mu}_k}} \bar{u}(t_{k+1}, x + h\bar{b}_k - \sqrt{h}\bar{\sigma}_k)$$

$$+ \sqrt{\bar{\mu}_k} h \bar{g}_k, \quad \text{if } x + h\bar{b}_k + \sqrt{h}\bar{\sigma}_k > \beta;$$

$$k = N - 1, \dots, 1, 0,$$

where $\bar{b}_k$, $\bar{\sigma}_k$, $\bar{g}_k$ are the coefficients $b(t, x, u)$, $\sigma(t, x, u)$, $g(t, x, u)$ evaluated at the point $(t_k, x, \bar{u}(t_{k+1}, x))$ and $0 < \bar{\lambda}_k, \bar{\mu}_k < 1$ are roots of the quadratic equations (it is not difficult to verify that the roots exist and are unique)

$$\alpha = x + \bar{\lambda}_k h \bar{b}_k - \sqrt{\bar{\lambda}_k h}\, \bar{\sigma}_k, \quad \beta = x + \bar{\mu}_k h \bar{b}_k + \sqrt{\bar{\mu}_k h}\, \bar{\sigma}_k.$$

This method involves one value of the function $\varphi(t, x)$ and one value of the approximate solution $\bar{u}(t_{k+1}, y)$ on the previous layer in contrast to the method (9.1.20) which requires evaluating one value of the function $\varphi(t, x)$ and two values of the approximate solution $\bar{u}(t_{k+1}, y)$ on the previous layer. The method (9.1.24) is based on Algorithm 7.2.9 for linear Dirichlet problems.

Lemma 9.1.4 *Under the assumptions* (i) *and* (ii), *the one-step error* $\rho(t_k, x)$ *of the method* (9.1.24) *is estimated as*

$$|\rho(t_k, x)| \leq Ch^2 \quad \text{if } x + hb_k \pm \sqrt{h}\sigma_k \in [\alpha, \beta];$$

$$|\rho(t_k, x)| \leq Ch^{3/2} \quad \text{if } x + hb_k - \sqrt{h}\sigma_k < \alpha \text{ or } x + hb_k + \sqrt{h}\sigma_k > \beta.$$

The proof is analogous to that of Lemma 8.2.1 (see also the proof of Lemma 9.1.2 in [309]). The following convergence theorem for the method (9.1.24) takes place, which probabilistic proof is available in [309].

Theorem 9.1.5 *Under the assumptions* (i) *and* (ii), *the method* (9.1.24) *has the global error estimated as*

$$|\bar{u}(t_k, x) - u(t_k, x)| \leq Kh, \tag{9.1.25}$$

where $K > 0$ *does not depend on* h, k, x.

Remark 9.1.6 It follows from the proof of Theorem 9.1.5 that to construct a first-order method we can use an approximation of $u(t_k, x)$ which one-step error near the boundary (i.e., when $(t_k, x) \in \partial\Gamma$) is estimated as $O(h)$ only (cf. Lemma 9.1.4). For instance, we can approximate the solution $u(t_k, x)$, when x is close to α, by values of the solution at the point (t_{k+1}, α) and at a point $(t_{k+1}, \hat{x}_k) \in \bar{Q}\backslash\partial\Gamma$ (e.g., one can take $\hat{x}_k = \alpha + h^{1/2} \max \sigma + h \max |b|$, where the maxima are taken over $(t, x) \in \bar{Q}, u \in [u_\circ, u^\circ]$) with the weights $p = \dfrac{\hat{x}_k - x}{\hat{x}_k - \alpha}$ and $q = 1 - p$ respectively. Analogously, we can approximate $u(t_k, x)$ when x is close to β. Making use of this approximation for $(t_k, x) \in \partial \Gamma$ and the Euler approximation for $(t_k, x) \in \bar{Q} \backslash \partial \Gamma$, we will get the new layer method with the first order of convergence (see also Sect. 7.2.1, where such a construction is used in Algorithm 7.2.1 for solving linear Dirichlet problems by the Monte Carlo approach). This layer method is slightly easier than (9.1.24) and its generalization to the multi-dimensional case is also easier. But we prefer to present the method (9.1.24) (see its generalization to the two-dimensional case in [309]) because methods with a more accurate approximation near the boundary give more accurate results even if they have the same order of convergence as methods less accurate near the boundary (see, e.g., numerical tests in Sect. 9.2).

9.1.4 Numerical Algorithms and Their Convergence

To become a numerical algorithm, the method (9.1.20) (just as other layer methods) needs a discretization in the variable x. Consider an equidistant space discretization with a space step h_x (recall that the notation for time step is h): $x_j = \alpha + jh_x$, $j = 0, 1, 2, \ldots, M$, $h_x = (\beta - \alpha)/M$. Using, for example, linear interpolation, we construct the following algorithm (we denote it as $\bar{u}(t_k, x)$ again, since this does not cause any confusion).

Algorithm 9.1.7 The algorithm is defined by the following formulas

$$\bar{u}(t_N, x) = \varphi(t_N, x), \quad x \in [\alpha, \beta], \tag{9.1.26}$$

$$\bar{u}(t_k, x_j) = \frac{1}{2}\bar{u}(t_{k+1}, x_j + h\bar{b}_{k,j} - \sqrt{h}\bar{\sigma}_{k,j}) + \frac{1}{2}\bar{u}(t_{k+1}, x_j + h\bar{b}_{k,j} + \sqrt{h}\bar{\sigma}_{k,j})$$
$$+ h\bar{g}_{k,j}, \quad \text{if } x_j + h\bar{b}_{k,j} \pm \sqrt{h}\bar{\sigma}_{k,j} \in [\alpha, \beta];$$

$$\bar{u}(t_k, x_j) = \varphi(t_{k+1-\bar{\gamma}_{k,j}}, \alpha) - \bar{\gamma}_{k,j} hg(t_k, \alpha, \varphi(t_{k+1}, \alpha))$$
$$- \frac{1}{2}\bar{u}(t_{k+1}, \alpha + \bar{\gamma}_{k,j} hb(t_k, \alpha, \varphi(t_{k+1}, \alpha)) + \sqrt{\bar{\gamma}_{k,j} h}\sigma(t_k, \alpha, \varphi(t_{k+1}, \alpha)))$$
$$+ \frac{1}{2}\bar{u}(t_{k+1}, x_j + h\bar{b}_{k,j} + \sqrt{h}\bar{\sigma}_{k,j}) + h\bar{g}_{k,j}, \quad \text{if } x_j + h\bar{b}_{k,j} - \sqrt{h}\bar{\sigma}_{k,j} < \alpha;$$

$$\bar{u}(t_k, x_j) = \frac{1}{2}\bar{u}(t_{k+1}, x_j + h\bar{b}_{k,j} - \sqrt{h}\bar{\sigma}_{k,j}) + \varphi(t_{k+1-\bar{\delta}_{k,j}}, \beta)$$
$$- \bar{\delta}_{k,j} hg(t_k, \beta, \varphi(t_{k+1}, \beta))$$
$$- \frac{1}{2}\bar{u}(t_{k+1}, \beta + \bar{\delta}_{k,j} hb(t_k, \beta, \varphi(t_{k+1}, \beta)) - \sqrt{\bar{\delta}_{k,j} h}\sigma(t_k, \beta, \varphi(t_{k+1}, \beta)))$$
$$+ h\bar{g}_{k,j}, \quad \text{if } x_j + h\bar{b}_{k,j} + \sqrt{h}\bar{\sigma}_{k,j} > \beta;$$

$$j = 1, 2, \ldots, M - 1,$$

$$\bar{u}(t_k, x) = \frac{x_{j+1} - x}{h_x} \bar{u}(t_k, x_j) + \frac{x - x_j}{h_x} \bar{u}(t_k, x_{j+1}), \quad x_j < x < x_{j+1}, \quad (9.1.27)$$

$$j = 0, 1, 2, \ldots, M - 1, \ k = N - 1, \ldots, 1, 0,$$

where $\bar{b}_{k,j}$, $\bar{\sigma}_{k,j}$, $\bar{g}_{k,j}$ are the coefficients $b(t, x, u)$, $\sigma(t, x, u)$, $g(t, x, u)$ calculated at the point $(t_k, x_j, \bar{u}(t_{k+1}, x_j))$ and $0 < \bar{\gamma}_{k,j}, \bar{\delta}_{k,j} \leq 1$ are roots of Eqs. (9.1.17) and (9.1.18) with the right-hand sides $x_j + h\bar{b}_{k,j} - \sqrt{h}\bar{\sigma}_{k,j}$ and $x_j + h\bar{b}_{k,j} + \sqrt{h}\bar{\sigma}_{k,j}$, respectively.

Theorem 9.1.8 *If the value of h_x is taken equal to $\varkappa h$, $\varkappa$ is a positive constant, then under the assumptions (i) and (ii) Algorithm 9.1.7 has the first order of convergence, i.e., for the approximation $\bar{u}(t_k, x)$ from (9.1.26)–(9.1.27) the following inequality holds*

$$|\bar{u}(t_k, x) - u(t_k, x)| \leq Kh, \quad (9.1.28)$$

where $K > 0$ does not depend on x, h, k.

The proof of Theorem 9.1.8 differs only little from the proof of Theorem 8.2.5 and is therefore omitted.

Remark 9.1.9 As it was mentioned in Remark 8.2.7, it is natural to consider cubic interpolation instead of the linear one for constructing numerical algorithms. The use of cubic interpolation allows us to take the space step $h_x = \varkappa \sqrt{h}$ (in contrast to $h_x = \varkappa h$ for linear interpolation) and, thus, to reduce the volume of computations. Moreover, if we use cubic interpolation, we can avoid special formulas near the boundary choosing some appropriate $\varkappa$ (indeed, we can take, e.g., $\varkappa = 2 \max\limits_{t \in [t_0, T], x \in \overline{G}, u \in [u^\circ, u^\circ]} \sigma(t, x, u)$, then for a sufficiently small h the points $x_j + h\bar{b}_{k,j} \pm \sqrt{h}\bar{\sigma}_{k,j}$, $j = 1, 2, \ldots, M - 1$, always belong to $[\alpha, \beta]$). We have not succeeded in proving a convergence theorem in the case of cubic interpolation. In Sect. 9.2 we test an algorithm based on cubic interpolation. The tests give fairly good results. See also theoretical explanations and numerical tests in Chap. 8.

Remark 9.1.10 Clearly, the algorithms can be considered with variable time steps and space steps. An algorithm with variable space steps is used in our numerical tests (see Sect. 9.2.1).

9.2   Numerical Tests of Layer Methods for the Dirichlet Problems

In the previous section we have dealt with semilinear parabolic equations with negative direction of time t: the equations are considered under $t < T$ and the "initial" conditions are given at $t = T$. This form of equations is suitable for the probabilistic approach which we use to construct numerical methods. Of course, the proposed methods are adaptable to semilinear parabolic equations with positive direction of time, and this adaptation is particularly easy in the autonomous case (cf. Remark 8.2.8). In our numerical tests we use algorithms with positive direction of time (see, e.g., (9.2.15)–(9.2.16) below).

9.2.1 The Burgers Equation

Consider the Dirichlet problem for the one-dimensional Burgers equation:

$$\frac{\partial u}{\partial t} = \frac{\varepsilon^2}{2} \frac{\partial^2 u}{\partial x^2} - u \frac{\partial u}{\partial x}, \quad t > 0, \ x \in (-1, 1), \tag{9.2.1}$$

$$u(0, x) = -A \sin \pi x, \quad x \in [-1, 1], \tag{9.2.2}$$

$$u(t, \pm 1) = 0, \ t > 0. \tag{9.2.3}$$

This problem was used for testing various numerical methods in, e.g., [5, 21, 106] (see also references therein). By means of the Cole–Hopf transformation (see Sect. 8.7.1), one can find the explicit solution of the problem (9.2.1)–(9.2.3) in the form

$$u(t, x) = -A \frac{\displaystyle\int_{-\infty}^{\infty} \sin \pi(x - y) \exp\left(-\frac{A}{\pi \varepsilon^2} \cos \pi(x - y) - \frac{y^2}{2\varepsilon^2 t}\right) dy}{\displaystyle\int_{-\infty}^{\infty} \exp\left(-\frac{A}{\pi \varepsilon^2} \cos \pi(x - y) - \frac{y^2}{2\varepsilon^2 t}\right) dy} \tag{9.2.4}$$

or

$$u(t, x) = \frac{\pi \varepsilon^2}{2} \frac{\displaystyle\sum_{n=1}^{\infty} n a_n \exp\left(-\frac{1}{8}\varepsilon^2 \pi^2 n^2 t\right) \sin \frac{1}{2}\pi n(x + 1)}{\frac{1}{2}a_0 + \displaystyle\sum_{n=1}^{\infty} a_n \exp\left(-\frac{1}{8}\varepsilon^2 \pi^2 n^2 t\right) \cos \frac{1}{2}\pi n(x + 1)} \tag{9.2.5}$$

with

$$a_n = \int_{-1}^{1} \exp\left(-\frac{A}{\pi \varepsilon^2} \cos \pi x\right) \cos \frac{1}{2}\pi n(x + 1) \, dx.$$

We simulate the problem (9.2.1)–(9.2.3) on relatively small time intervals $[0, T]$, where the formula (9.2.4) is more convenient. For a small ε, there is a thin internal

Fig. 9.2.1 A typical solution $u(t, x)$ of the problem (9.2.1)–(9.2.3) for $\varepsilon = 0.1$, $A = 2$ and various time moments

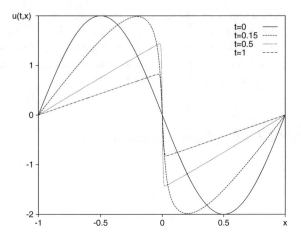

layer with width $\sim \varepsilon^2$, where the solution to (9.2.1)–(9.2.3) has singular behavior (see, e.g., [178] and references therein). Derivatives of the solution go to infinity as $\varepsilon \to 0$. A typical behavior of the solution is demonstrated in Fig. 9.2.1.

Here we test three algorithms. The first one is algorithm (9.1.26)–(9.1.27). The second one is the algorithm based on layer method (9.1.24) and linear interpolation. In these two algorithms (they both use linear interpolation), we take the space step h_x being equal to the time step h. The third algorithm is based on cubic interpolation (see Remark 9.1.9). In the case of the problem (9.2.1)–(9.2.3) it has the form

$$\bar{u}(0, x) = -A \sin \pi x, \ x \in [-1, 1], \qquad (9.2.6)$$

$$\bar{u}(t_{k+1}, x_0) = \bar{u}(t_{k+1}, -1) = 0,$$

$$\bar{u}(t_{k+1}, x_M) = \bar{u}(t_{k+1}, 1) = 0,$$

$$\bar{u}(t_{k+1}, x_j) = \frac{1}{2}\bar{u}(t_k, x_j - h\bar{u}(t_k, x_j) - \varepsilon h^{1/2}) + \frac{1}{2}\bar{u}(t_k, x_j - h\bar{u}(t_k, x_j) + \varepsilon h^{1/2}),$$

$$j = 1, \ldots, M - 1,$$

$$\bar{u}(t_k, x) = \sum_{i=0}^{3} \Phi_{j,i}(x)\bar{u}(t_k, x_{j+i}), \ x_j < x < x_{j+3},$$

$$\Phi_{j,i}(x) = \prod_{m=0, m \neq i}^{3} \frac{x - x_{j+m}}{x_{j+i} - x_{j+m}},$$

$$k = 0, \ldots, N - 1.$$

Table 9.2.1 The Burgers equation. Dependence of the errors $err^c(t, h)$ and $err^l(t, h)$ in h for algorithms (9.1.26)–(9.1.27) and (9.2.6) when $t = 0.5$, $\varepsilon = 0.1$, and $A = 2$

h	Algorithm (9.1.26)–(9.1.27)		Algorithm (9.2.6)	
	$err^c(t, h)$	$err^l(t, h)$	$err^c(t, h)$	$err^l(t, h)$
0.01	$1.239 \cdot 10^{-1}$	$3.035 \cdot 10^{-2}$	$1.854 \cdot 10^{-1}$	$3.081 \cdot 10^{-2}$
0.0016	$4.574 \cdot 10^{-2}$	$5.311 \cdot 10^{-3}$	$5.855 \cdot 10^{-2}$	$5.481 \cdot 10^{-3}$
0.0001	$2.673 \cdot 10^{-3}$	$3.288 \cdot 10^{-4}$	$3.737 \cdot 10^{-3}$	$3.466 \cdot 10^{-4}$
0.000016	$4.261 \cdot 10^{-4}$	$5.259 \cdot 10^{-5}$	$5.919 \cdot 10^{-4}$	$5.527 \cdot 10^{-5}$

Here we use a nonequidistant discretization of the interval $[-1, 1]$. We take $h_x := x_{j+1} - x_j = \varepsilon\sqrt{h}$ in $[-0.1, 0.1]$ and $h_x = \sqrt{h}$ outside $[-0.1, 0.1]$. Such a choice of h_x is dictated by the fact that if h is comparatively large (e.g., $h = \varepsilon^2$), the equidistant discretization with $h_x = \sqrt{h}$ has not more than one node in the thin internal layer. For a sufficiently small h, it is possible to use the equidistant discretization for cubic interpolation as well.

Since $\varepsilon \ll 1$, the points $x_j - h\bar{u}(t_k, x_j) \pm \varepsilon h^{1/2}$, $j = 1, \ldots, M - 1$, belong to the interval $(-1, 1)$. Thus, we avoid using special formulas near the boundary in (9.2.6) (see Remark 9.1.9 as well).

Table 9.2.1 gives numerical results obtained by using the algorithms (9.1.26)–(9.1.27) and (9.2.6). The algorithm based on the layer method (9.1.24) and linear interpolation gives results practically identical to the ones for (9.1.26)–(9.1.27). We present the errors of the approximate solutions $\bar{u}$ in the discrete Chebyshev norm and in the l^1-norm:

$$err^c(t, h) = \max_{x_i} |\bar{u}(t, x_i) - u(t, x_i)|,$$

$$err^l(t, h) = \sum_i h_x |\bar{u}(t, x_i) - u(t, x_i)|.$$

The algorithms based on linear interpolation require both larger volume of computations per time layer and larger amount of memory than the algorithm (9.2.6) based on cubic interpolation. For instance, in the considered case the algorithm (9.1.26)–(9.1.27) with $h = 0.0001$ needs 2×10^4 evaluations of $\bar{u}(t_k, x)$ per layer $t = t_k$ and to store an array of 2×10^4 elements, and the algorithm (9.2.6) with the same step $h = 0.0001$ requires only 380 evaluations of $\bar{u}(t_k, x)$ per layer and an array of 380 elements (see also Remark 9.1.9).

9.2.2 Comparison Analysis

In this section we give some comparison analysis of the layer methods proposed in this chapter and the well-known finite-difference schemes (see also Remark 9.1.1). We use (9.2.1)–(9.2.3) as a test problem again. Here we compare the algorithm (9.2.6) with two explicit finite-difference schemes (9.2.7) and (9.2.8) of order $O(\Delta t, \Delta x^2)$, where Δt is a time step and Δx is a space step. These finite-difference schemes are used for simulation of the Burgers equation in [5, 107].

The method of differences forward in time and central differences in space (the FTCS scheme) applied to the divergent form of the Burgers equation is written as

$$\bar{u}(0, x) = -A \sin \pi x, \quad x \in [-1, 1], \tag{9.2.7}$$

$$\bar{u}(t_{k+1}, x_0) = \bar{u}(t_{k+1}, -1) = 0, \quad \bar{u}(t_{k+1}, x_M) = \bar{u}(t_{k+1}, 1) = 0,$$

$$\bar{u}(t_{k+1}, x_j) = \bar{u}(t_k, x_j) - \frac{\Delta t}{4\Delta x}(\bar{u}^2(t_k, x_{j+1}) - \bar{u}^2(t_k, x_{j-1}))$$

$$+ \frac{\varepsilon^2}{2} \frac{\Delta t}{\Delta x^2}(\bar{u}(t_k, x_{j+1}) - 2\bar{u}(t_k, x_j) + \bar{u}(t_k, x_{j-1})),$$

$$j = 1, \ldots, M - 1, \quad k = 0, \ldots, N - 1,$$

where the step of a time discretization $\Delta t := T/N$ and $t_k = k \, \Delta t$ and the step of space discretization $\Delta x := 2/M$ and $x_j = -1 + j \, \Delta x$.

In the case of the problem (9.2.1)–(9.2.3) the Brailovskaya scheme has the form (see [5])

$$\bar{u}(0, x) = -A \sin \pi x, \quad x \in [-1, 1], \tag{9.2.8}$$

$$\bar{u}(t_{k+1}, x_0) = \hat{u}(t_{k+1}, x_0) = 0, \quad \bar{u}(t_{k+1}, x_M) = \hat{u}(t_{k+1}, x_M) = 0,$$

$$\hat{u}(t_{k+1}, x_j) = \bar{u}(t_k, x_j) - \frac{\Delta t}{4\Delta x}(\bar{u}^2(t_k, x_{j+1}) - \bar{u}^2(t_k, x_{j-1}))$$

$$+ \frac{\varepsilon^2}{2} \frac{\Delta t}{\Delta x^2}(\bar{u}(t_k, x_{j+1}) - 2\bar{u}(t_k, x_j) + \bar{u}(t_k, x_{j-1})),$$

$$\bar{u}(t_{k+1}, x_j) = \bar{u}(t_k, x_j) - \frac{\Delta t}{4\Delta x}(\hat{u}^2(t_k, x_{j+1}) - \hat{u}^2(t_k, x_{j-1}))$$

$$+ \frac{\varepsilon^2}{2} \frac{\Delta t}{\Delta x^2}(\bar{u}(t_k, x_{j+1}) - 2\bar{u}(t_k, x_j) + \bar{u}(t_k, x_{j-1})),$$

$$j = 1, \ldots, M - 1, \quad k = 0, \ldots, N - 1.$$

The space step Δx in the finite-difference schemes (9.2.7) and (9.2.8) is selected as $\Delta x = \varkappa \times \varepsilon \sqrt{\Delta t}$. The results of Tables 9.2.2 and 9.2.3 correspond to $\varkappa = 4$.

As in Sect. 9.2.1, we realize the algorithm (9.2.6) using a nonequidistant discretization of the interval $[-1, 1]$. For the time step $h = 0.0016$ (Table 9.2.2), we take $h_x := x_{j+1} - x_j = \varepsilon \sqrt{h}$ in $[-0.1, 0.1]$ and $h_x = 2\sqrt{h}$ outside $[-0.1, 0.1]$. And for $h = 0.0001$ (Table 9.2.3) we choose $h_x = \varepsilon \sqrt{h}$ in $[-0.02, 0.02]$ and $h_x = 2\sqrt{h}$ outside $[-0.02, 0.02]$ (see also the explanations in Sect. 9.2.1).

Table 9.2.2 The Burgers equation. The relative errors $\delta^l(t, h)$ (top position) and $\delta^c(t, h)$ (lower position) of algorithm (9.2.6) and finite-difference schemes (9.2.7) and (9.2.8) are given for $t = 0.08$, $\varepsilon = 0.1$, $h = \Delta t = 0.0016$, and various A

A	Algorithm (9.2.6)	Scheme (9.2.7)	Scheme (9.2.8)
5	$7.79 \cdot 10^{-3}$	$2.22 \cdot 10^{-2}$	$2.01 \cdot 10^{-2}$
	$5.28 \cdot 10^{-2}$	$2.05 \cdot 10^{-1}$	$1.66 \cdot 10^{-1}$
		Oscillations	*Oscillations*
10	$1.87 \cdot 10^{-2}$	$\gg 100$	$2.35 \cdot 10^{-2}$
	$9.96 \cdot 10^{-1}$	$\gg 100$	$6.94 \cdot 10^{-2}$
			Oscillations
15	$2.70 \cdot 10^{-2}$	*Overflow*	$3.64 \cdot 10^{-1}$
	$9.84 \cdot 10^{-1}$		$3.42 \cdot 10^{0}$
			Big oscillations

Table 9.2.3 The Burgers equation. The relative errors $\delta^l(t, h)$ (top position) and $\delta^c(t, h)$ (lower position) of algorithm (9.2.6) and finite-difference schemes (9.2.7) and (9.2.8) are given for $h = \Delta t = 0.0001$, the other parameters are as in Table 9.2.2

A	Algorithm (9.2.6)	Scheme (9.2.7)	Scheme (9.2.8)
5	$1.26 \cdot 10^{-3}$	$8.88 \cdot 10^{-4}$	$7.81 \cdot 10^{-4}$
	$4.68 \cdot 10^{-2}$	$1.55 \cdot 10^{-2}$	$1.10 \cdot 10^{-2}$
10	$1.24 \cdot 10^{-3}$	$3.56 \cdot 10^{-3}$	$3.58 \cdot 10^{-3}$
	$9.25 \cdot 10^{-2}$	$1.54 \cdot 10^{-1}$	$1.54 \cdot 10^{-1}$
		Oscillations	*Oscillations*
15	$1.91 \cdot 10^{-3}$	$5.07 \cdot 10^{-3}$	$5.11 \cdot 10^{-3}$
	$1.99 \cdot 10^{-1}$	$1.81 \cdot 10^{-1}$	$1.84 \cdot 10^{-1}$
		Oscillations	*Oscillations*

Tables 9.2.2 and 9.2.3 present the relative errors $\delta^c(t, h)$ and $\delta^l(t, h)$. The error $\delta^c(t, h)$ is equal to

$$\delta^c(t, h) = \frac{\max_{x_i} |\bar{u}(t, x_i) - u(t, x_i)|}{\max_{x_i} |u(t, x_i)|}$$

for all three methods. The error $\delta^l(t, h)$ is equal to

$$\delta^l(t, h) = \frac{1}{\max_{x_i} |u(t, x_i)|} \sum_i h_x |\bar{u}(t, x_i) - u(t, x_i)|$$

for the algorithm (9.2.6) while for the schemes (9.2.7) and (9.2.8) it is given by

$$\delta^l(t, h) = \frac{1}{\max_{x_i} |u(t, x_i)|} \sum_i \Delta x |\bar{u}(t, x_i) - u(t, x_i)|.$$

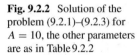

Fig. 9.2.2 Solution of the problem (9.2.1)–(9.2.3) for $A = 10$, the other parameters are as in Table 9.2.2

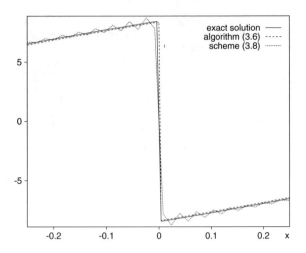

The comment *"oscillations"* means that the numerical solution has oscillations in a neighborhood of $x = 0$. An illustration of such oscillations is given in Fig. 9.2.2. The comment *"overflow"* indicates that overflow error occurs during simulation.

Let us observe that if we take $\varkappa = 2$ in order to improve accuracy of the results obtained by the Brailovskaya scheme (9.2.8), e.g., for $\Delta t = 0.0016$, $A = 10$ (see Table 9.2.2 and Fig. 9.2.2), the numerical solution becomes more unstable and overflow error occurs. If we take $\varkappa = 8$ in this case, the errors and amplitude of oscillations become greater than for $\varkappa = 4$.

We can conclude from the results presented in Tables 9.2.2 and 9.2.3 and in Fig. 9.2.2 that the algorithm (9.2.6) based on the layer method demonstrates a more stable behavior than the finite-difference schemes when the parameter A is sufficiently large. In the considered problem (9.2.1)–(9.2.3) large values of A lead, in particular, to large advection in a neighborhood of $x = 0$. Our experiments confirm that the layer methods allow us to avoid difficulties stemming from strong advection (see Remark 9.1.1). It should also be mentioned that the algorithms based on layer methods require more CPU time than finite-difference schemes. For example, in the case of parameters as in Table 9.2.3 to solve (9.2.1)–(9.2.3) by the algorithm (9.2.6) we need ≈ 2 s. while the schemes (9.2.7) and (9.2.8) require ≈ 0.3 and ≈ 0.4 s. correspondingly. But the algorithm (9.2.6) gives us quite appropriate results with the greater step $h = 0.0016$ (see Table 9.2.2 and Fig. 9.2.2) and in this case it requires ≈ 0.06 s. Simulations were made on PC with Intel Pentium 233 MHz processor using Borland C compiler.

9.2.3 Quasilinear Equation with Power Law Nonlinearities

Consider the Dirichlet problem for quasilinear parabolic equation with power law nonlinearities [395, 396]:

$$\frac{\partial u}{\partial t} = \frac{1}{2}\frac{\partial}{\partial x}\left(u^q\frac{\partial u}{\partial x}\right), \quad t \in (0,1), \quad x > 0, \quad q > 0, \tag{9.2.9}$$

with the initial condition

$$u(0,x) = (1 - x/L)^{2/q}, \quad x \in [0,L], \tag{9.2.10}$$

$$u(0,x) = 0, \quad x > L,$$

and the boundary regime

$$u(t,0) = (1-t)^{-1/q}, \quad t \in [0,1), \tag{9.2.11}$$

where $L = \sqrt{(q+2)/q}$.

The exact solution of this problem has the form [395, 396]:

$$u(t,x) = \left(\frac{1 - x/L}{\sqrt{1 - t}}\right)^{2/q} \quad \text{for } x \in [0,L]$$

and

$$u(t,x) = 0 \ \text{ for } x > L.$$

The temperature $u(t,x)$ grows infinitely as $t \to 1$. At the same time the heat remains localized in the interval $[0,L)$. Figure 9.2.3 presents a typical behavior of the solution to (9.2.9)–(9.2.11).

Equation (9.2.9) is not of the form (9.1.5). The function

$$v = u^{q+1}$$

satisfies the problem

$$\frac{\partial v}{\partial t} = \frac{1}{2}v^{q/(q+1)}\frac{\partial^2 v}{\partial x^2}, \quad t \in (0,1), \quad x > 0, \tag{9.2.12}$$

$$v(0,x) = (1 - x/L)^{2(q+1)/q}, \quad x \in [0,L], \tag{9.2.13}$$

$$v(0,x) = 0, \quad x > L,$$

$$v(t,0) = (1-t)^{-(q+1)/q}, \quad t \in [0,1). \tag{9.2.14}$$

Equation (9.2.12) has the form of (9.1.5) but with positive direction of time.

Fig. 9.2.3 A typical solution $u(t, x)$ of problem (9.2.9)–(9.2.11) for $q = 1.5$ and various time moments

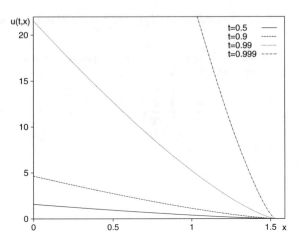

We simulate the solution of (9.2.12)–(9.2.14) by two algorithms: the algorithm (9.1.26)–(9.1.27) and the algorithm based on the layer method (9.1.24) and linear interpolation. The last one in the case of the problem (9.2.12)–(9.2.14) has the form

$$\bar{v}(0, x) = \begin{cases} (1 - x/L)^{2(q+1)/q}, & x \in [0, L], \\ 0, & x \in (L, \infty), \end{cases} \tag{9.2.15}$$

$$\bar{v}(t_{k+1}, x_j) = \frac{1}{2}\bar{v}(t_k, x_j - \sqrt{h}\,(\bar{v}(t_k, x_j))^{q/2(q+1)})$$

$$+ \frac{1}{2}\bar{v}(t_k, x_j + \sqrt{h}\,(\bar{v}(t_k, x_j))^{q/2(q+1)}),$$

$$\text{if } x_j - \sqrt{h}(\bar{v}(t_k, x_j))^{q/2(q+1)} \geq 0;$$

$$\bar{v}(t_{k+1}, x_j) = \frac{1}{1 + \sqrt{\bar{\lambda}_k}}(1 - t_{k+1-\bar{\lambda}_k})^{-(q+1)/q}$$

$$+ \frac{\sqrt{\bar{\lambda}_k}}{1 + \sqrt{\bar{\lambda}_k}}\bar{v}(t_k, x_j + \sqrt{h}(\bar{v}(t_k, x_j))^{q/2(q+1)}),$$

$$\bar{\lambda}_k = \left(\frac{x_j}{(\bar{v}(t_k, x_j))^{q/2(q+1)} \cdot \sqrt{h}}\right)^2, \quad \text{if } x_j - (\bar{v}(t_k, x_j))^{q/2(q+1)} \cdot \sqrt{h} < 0;$$

$$\bar{v}(t_{k+1}, x) = \frac{x_{j+1} - x}{h_x}\bar{v}(t_{k+1}, x_j) + \frac{x - x_j}{h_x}\bar{v}(t_{k+1}, x_{j+1}), \quad x_j \leq x \leq x_{j+1},$$

$$\tag{9.2.16}$$

$$j = 0, 1, 2, \ldots, \quad k = 1, \ldots, N,$$

where $x_j = j\,h_x$, $t_k = k\,h$.

Table 9.2.4 Quasilinear equation with power law nonlinearities. Dependence of errors $err_{\bar{v}}(t, h)$ (top position) and $err_{\bar{u}}(t, h)$ (lower position) in h and t for the algorithm (9.2.15)–(9.2.16) under $q = 1.5$

	$h = 10^{-1}$	$h = 10^{-2}$	$h = 10^{-3}$	$h = 10^{-4}$
$t = 0.5$	$0.8664 \cdot 10^{-1}$	$0.8786 \cdot 10^{-2}$	$0.9705 \cdot 10^{-3}$	$1.018 \cdot 10^{-4}$
	$0.3542 \cdot 10^{-1}$	$0.7693 \cdot 10^{-2}$	$1.685 \cdot 10^{-3}$	$3.622 \cdot 10^{-4}$
$t = 0.9$	>5	$8.094 \cdot 10^{-1}$	$8.265 \cdot 10^{-2}$	$8.817 \cdot 10^{-3}$
	$5.910 \cdot 10^{-1}$	$8.109 \cdot 10^{-2}$	$8.656 \cdot 10^{-3}$	$8.918 \cdot 10^{-4}$

Table 9.2.5 Quasilinear equation with power law nonlinearities. Dependence of the relative error $\delta(t, h)$ in h and t for the algorithm (9.2.15)–(9.2.16) under $q = 1.5$

	$h = 10^{-1}$	$h = 10^{-2}$	$h = 10^{-3}$	$h = 10^{-4}$
$t = 0.9$	$1.273 \cdot 10^{-1}$	$1.747 \cdot 10^{-2}$	$1.865 \cdot 10^{-3}$	$1.921 \cdot 10^{-4}$
$t = 0.99$	$-$	$1.392 \cdot 10^{-1}$	$1.789 \cdot 10^{-2}$	$1.913 \cdot 10^{-3}$
$t = 0.999$	$-$	$-$	$1.398 \cdot 10^{-1}$	$1.801 \cdot 10^{-2}$
$t = 0.9999$	$-$	$-$	$-$	$1.400 \cdot 10^{-1}$

In our tests we take $h_x = h$. Tables 9.2.4 and 9.2.5 give numerical results obtained by using the algorithm (9.2.15)–(9.2.16). The algorithm (9.1.26)–(9.1.27) gives similar results and they are omitted here. Table 9.2.4 presents the errors

$$err_{\bar{v}}(t, h) := \max_j |\bar{v}(t, x_j) - v(t, x_j)|,$$

$$err_{\bar{u}}(t, h) := \max_j |\bar{u}(t, x_j) - u(t, x_j)|, \quad \bar{u}(t, x_j) = (\bar{v}(t, x_j))^{1/(q+1)}.$$

For times t which are close to the explosion time $t = 1$, the functions $u(t, x)$ and $v(t, x)$ take large values and the absolute errors become fairly large. In Table 9.2.5 we present the relative error

$$\delta(t, h) := \frac{err_{\bar{u}}(t, h)}{u(t, 0)}$$

at times close to the explosion.

In the experiments, the tested algorithms converge as $O(h)$ which is in complete agreement with our theoretical results.

9.3 Layer Methods for the Neumann Problem for Semilinear Parabolic Equations

Let G be a bounded domain in $\mathbf{R}^d$, $Q = [t_0, T) \times G$ be a cylinder in $\mathbf{R}^{d+1}$, $\Gamma = \bar{Q} \setminus Q$. The set Γ is a part of the boundary of the cylinder Q consisting of the upper base and the lateral surface. Consider the Neumann problem for the semilinear parabolic equation

$$\frac{\partial u}{\partial t} + \frac{1}{2} \sum_{i,j=1}^{d} a^{ij}(t, x, u) \frac{\partial^2 u}{\partial x^i \partial x^j} + \sum_{i=1}^{d} b^i(t, x, u) \frac{\partial u}{\partial x^i} + g(t, x, u) = 0, \quad (9.3.1)$$

$$(t, x) \in Q,$$

with the initial condition

$$u(T, x) = \varphi(x) \tag{9.3.2}$$

and the boundary condition

$$\frac{\partial u}{\partial \nu} = \psi(t, x, u), \ \ t \in [t_0, T], \ \ x \in \partial G, \tag{9.3.3}$$

where ν is the direction of the internal normal to the boundary ∂G at the point $x \in \partial G$.

Using the well known probabilistic representation of the solution to (9.3.1)–(9.3.3) (see [111, 123]), we get

$$u(t, x) = E(\varphi(X_{t,x}(T)) + Z_{t,x,0}(T)), \tag{9.3.4}$$

where $X_{t,x}(s)$, $Z_{t,x,z}(s)$, $t_0 \leq t < T$, $s \geq t$, $x \in \bar{G}$, is a solution of the Cauchy problem to the Ito system of SDEs

$$dX = b(s, X, u(s, X))ds + \sigma(s, X, u(s, X))dw(s) \tag{9.3.5}$$
$$+\nu(X)I_{\partial G}(X)dL(s), \ \ X(t) = x,$$

$$dZ = g(s, X, u(s, X))ds - \psi(s, X, u(s, X))I_{\partial G}(X)dL(s), \ \ Z(t) = z.$$

Here $w(s) = (w^1(s), \dots, w^d(s))^{\mathsf{T}}$ is a standard Wiener process, $b(s, x, u) = (b^1(s, x, u), \dots, b^d(s, x, u))^{\mathsf{T}}$ is a column vector, the matrix $\sigma = \sigma(s, x, u)$ is obtained from the equation

$$\sigma \sigma^{\mathsf{T}} = a, \ \sigma = \{\sigma^{ij}(s, x, u)\}, \ a = \{a^{ij}(s, x, u)\}, \ i, j = 1, \dots, d,$$

$L(s)$ is a local time of the process X on ∂G, and $I_A(x)$ is the indicator of a set A.

Layer methods proposed in this section for the problem (9.3.1)–(9.3.3) are based on weak schemes for integration of SDEs with reflection from Chap. 7.

In Sect. 9.3.1, two layer methods for the nonlinear Neumann problem are constructed. Some convergence theorems are given in Sect. 9.3.2. To realize a layer method in practice, a discretization in the variable x with interpolation at every step is needed to turn the method into an algorithm. Such numerical algorithms are given in Sect. 9.3.3. We restrict ourselves to the one-dimensional ($d = 1$) case in this section. Layer methods in the case of $d \geq 2$ are available in [310]. Numerical tests are presented in Sect. 9.4. Their results are in complete agreement with theoretical ones. In particular, it was demonstrated in numerical tests that layer methods may be preferable to finite-difference ones in the case of strong advection.

9.3.1 Construction of Layer Methods

The Neumann boundary value problem in the one-dimensional case has the form

$$\frac{\partial u}{\partial t} + \frac{1}{2}\sigma^2(t, x, u)\frac{\partial^2 u}{\partial x^2} + b(t, x, u)\frac{\partial u}{\partial x} + g(t, x, u) = 0, \tag{9.3.6}$$

$$t_0 \leq t < T, \ \alpha < x < \beta,$$

$$u(T, x) = \varphi(x), \ \alpha \leq x \leq \beta; \tag{9.3.7}$$

$$\frac{\partial u}{\partial x}(t, \alpha) = \psi_1(t, u(t, \alpha)), \quad \frac{\partial u}{\partial x}(t, \beta) = \psi_2(t, u(t, \beta)), \ t_0 \leq t \leq T. \tag{9.3.8}$$

In this case Q is a partly open rectangle: $Q = [t_0, T) \times (\alpha, \beta)$, and Γ consists of the upper base $\{T\} \times [\alpha, \beta]$ and two vertical intervals: $[t_0, T) \times \{\alpha\}$ and $[t_0, T) \times \{\beta\}$. We assume that $\sigma(t, x, u) \geq \sigma_* > 0$ for $(t, x) \in \bar{Q}$, $-\infty < u < \infty$.

Let $u = u(t, x)$ be a solution of the problem (9.3.6)–(9.3.8) which is assumed to exist, to be unique, and to be sufficiently smooth. Theoretical results on this topic are available in [114, 222, 241, 437] (see also references therein).

Analogously to (9.3.4), we have the local representation

$$u(t_k, x) = E(u(t_{k+1}, X_{t_k, x}(t_{k+1})) + Z_{t_k, x, 0}(t_{k+1})), \tag{9.3.9}$$

where $X_{t,x}(s)$, $Z_{t,x,z}(s)$, $t_0 \leq t < T$, $s \geq t$, $x \in [\alpha, \beta]$, satisfy (9.3.5).

9.3.1.1 Method of Order $O(h)$

Applying a slightly modified weak scheme with one-step boundary order $O(h^{3/2})$ from Sect. 7.6 to the system (9.3.5), it is not difficult to obtain

$$X_{t_k,x}(t_{k+1}) \simeq \bar{X}_{t_k,x}(t_{k+1}) = x + h\tilde{b}_k + h^{1/2}\tilde{\sigma}_k\xi_k, \qquad (9.3.10)$$

$$Z_{t_k,x,z}(t_{k+1}) \simeq \bar{Z}_{t_k,x,z}(t_{k+1}) = z + h\tilde{g}_k, \quad \text{if } x + h\tilde{b}_k \pm h^{1/2}\tilde{\sigma}_k \in [\alpha, \beta];$$

$$\bar{X}_{t_k,x}(t_{k+1}) = x + (\alpha - x) + \sqrt{h\tilde{\sigma}_k^2 + (\alpha - x)^2},$$

$$\bar{Z}_{t_k,x,z}(t_{k+1}) = z + h\tilde{g}_k - \psi_1(t_k, u(t_k, \alpha)) \cdot (\alpha - x - h\tilde{b}_k + \sqrt{h\tilde{\sigma}_k^2 + (\alpha - x)^2}),$$

$$\text{if } x + h\tilde{b}_k - h^{1/2}\tilde{\sigma}_k < \alpha;$$

$$\bar{X}_{t_k,x}(t_{k+1}) = x + (\beta - x) - \sqrt{h\tilde{\sigma}_k^2 + (\beta - x)^2},$$

$$\bar{Z}_{t_k,x,z}(t_{k+1}) = z + h\tilde{g}_k - \psi_2(t_k, u(t_k, \beta)) \cdot (\beta - x - h\tilde{b}_k - \sqrt{h\tilde{\sigma}_k^2 + (\beta - x)^2}),$$

$$\text{if } x + h\tilde{b}_k + h^{1/2}\tilde{\sigma}_k > \beta.$$

Here $\tilde{b}_k$, $\tilde{\sigma}_k$, $\tilde{g}_k$ are the coefficients $b(t, x, u)$, $\sigma(t, x, u)$, $g(t, x, u)$ evaluated at the point $(t_k, x, u(t_k, x))$ and $\xi_{N-1}, \xi_{N-2}, \dots, \xi_0$ are i.i.d. random variables with the law $P(\xi = \pm 1) = 1/2$.

One can see that using the approximation (9.3.10) and the representation (9.3.9), we get an implicit one-step approximation for $u(t_k, x)$. Applying the method of simple iteration to this implicit approximation with $u(t_{k+1}, x)$ as a null iteration, we come to the explicit one-step approximation $v(t_k, x)$ of $u(t_k, x)$:

$$v(t_k, x) = \frac{1}{2}u(t_{k+1}, x + hb_k - h^{1/2}\sigma_k) + \frac{1}{2}u(t_{k+1}, x + hb_k + h^{1/2}\sigma_k) \quad (9.3.11)$$

$$+ hg_k, \quad \text{if } x + hb_k \pm h^{1/2}\sigma_k \in [\alpha, \beta];$$

$$v(t_k, x) = u(t_{k+1}, \alpha + \sqrt{h\sigma_k^2 + (\alpha - x)^2})$$

$$- \psi_1(t_{k+1}, u(t_{k+1}, \alpha)) \cdot (\alpha - x - hb_k + \sqrt{h\sigma_k^2 + (\alpha - x)^2}) + hg_k,$$

$$\text{if } x + hb_k - h^{1/2}\sigma_k < \alpha;$$

$$v(t_k, x) = u(t_{k+1}, \beta - \sqrt{h\sigma_k^2 + (\beta - x)^2})$$

$$- \psi_2(t_{k+1}, u(t_{k+1}, \beta)) \cdot (\beta - x - hb_k - \sqrt{h\sigma_k^2 + (\beta - x)^2}) + hg_k,$$

$$\text{if } x + hb_k + h^{1/2}\sigma_k > \beta;$$

$$k = N - 1, \dots, 1, 0,$$

where b_k, σ_k, g_k are the coefficients b, σ, g evaluated at the point $(t_k, x, u(t_{k+1}, x))$. Let us observe that within the limits of the considered accuracy it is possible to take

t_{k+1} instead of t_k. That is why, one can take, for instance, $\psi_1(t_{k+1}, u(t_{k+1}, \alpha))$ instead of $\psi_1(t_k, u(t_{k+1}, \alpha))$ in (9.3.11).

The corresponding explicit layer method for solving the Neumann problem (9.3.6)–(9.3.8) has the form

$$\bar{u}(t_N, x) = \varphi(t_N, x), \quad x \in [\alpha, \beta], \tag{9.3.12}$$

$$\bar{u}(t_k, x) = \frac{1}{2}\bar{u}(t_{k+1}, x + h\bar{b}_k - h^{1/2}\bar{\sigma}_k) + \frac{1}{2}\bar{u}(t_{k+1}, x + h\bar{b}_k + h^{1/2}\bar{\sigma}_k)$$
$$+ h\bar{g}_k, \quad \text{if } x + h\bar{b}_k \pm h^{1/2}\bar{\sigma}_k \in [\alpha, \beta];$$

$$\bar{u}(t_k, x) = \bar{u}(t_{k+1}, \alpha + \sqrt{h\bar{\sigma}_k^2 + (\alpha - x)^2})$$
$$- \psi_1(t_{k+1}, \bar{u}(t_{k+1}, \alpha)) \cdot (\alpha - x - h\bar{b}_k + \sqrt{h\bar{\sigma}_k^2 + (\alpha - x)^2}) + h\bar{g}_k,$$
$$\text{if } x + h\bar{b}_k - h^{1/2}\bar{\sigma}_k < \alpha;$$

$$\bar{u}(t_k, x) = \bar{u}(t_{k+1}, \beta - \sqrt{h\bar{\sigma}_k^2 + (\beta - x)^2})$$
$$- \psi_2(t_{k+1}, \bar{u}(t_{k+1}, \beta)) \cdot (\beta - x - h\bar{b}_k - \sqrt{h\bar{\sigma}_k^2 + (\beta - x)^2}) + h\bar{g}_k,$$
$$\text{if } x + h\bar{b}_k + h^{1/2}\bar{\sigma}_k > \beta;$$

$$k = N - 1, \ldots, 1, 0,$$

where $\bar{b}_k = \bar{b}_k(x) = b(t_k, x, \bar{u}(t_{k+1}, x))$, $\bar{\sigma}_k = \bar{\sigma}_k(x) = \sigma(t_k, x, \bar{u}(t_{k+1}, x))$, $\bar{g}_k = \bar{g}_k(x) = g(t_k, x, \bar{u}(t_{k+1}, x))$.

This layer method has the one-step error near the boundary estimated by $O(h^{3/2})$ and for internal points estimated by $O(h^2)$ (see Lemma 9.3.1 in the next subsection). We prove that its order of convergence is $O(h)$ when the boundary condition does not depend on the solution (see Theorem 9.3.2). Apparently, this is so in the general case as well (see Remark 9.3.3).

Another method with the same one-step error is given in Sect. 9.3.4.

9.3.1.2   Method of Order $O(\sqrt{h})$

Applying the weak scheme with one-step boundary order $O(h)$ from Sect. 7.6 to the system (9.3.5), it is not difficult to obtain

$$X_{t_k,x}(t_{k+1}) \simeq \bar{X}_{t_k,x}(t_{k+1}) = x + h\tilde{b}_k + h^{1/2}\tilde{\sigma}_k\xi_k, \tag{9.3.13}$$
$$Z_{t_k,x,z}(t_{k+1}) \simeq \bar{Z}_{t_k,x,z}(t_{k+1}) = z + h\tilde{g}_k, \quad \text{if } x + h\tilde{b}_k \pm h^{1/2}\tilde{\sigma}_k \in [\alpha, \beta];$$

$$\bar{X}_{t_k,x}(t_{k+1}) = x + qh^{1/2}, \quad \bar{Z}_{t_k,x,z}(t_{k+1}) = z - \psi_1(t_k, u(t_k, \alpha))qh^{1/2},$$
$$\text{if } x + h\tilde{b}_k - h^{1/2}\tilde{\sigma}_k < \alpha;$$

$$\bar{X}_{t_k,x}(t_{k+1}) = x - qh^{1/2}, \quad \bar{Z}_{t_k,x,z}(t_{k+1}) = z + \psi_2(t_k, u(t_k, \beta))qh^{1/2},$$
$$\text{if } x + h\tilde{b}_k + h^{1/2}\tilde{\sigma}_k > \beta.$$

Here $\tilde{b}_k$, $\tilde{\sigma}_k$, $\tilde{g}_k$ are the coefficients $b(t, x, u)$, $\sigma(t, x, u)$, $g(t, x, u)$ evaluated at the point $(t_k, x, u(t_k, x))$, $\xi_{N-1}, \xi_{N-2}, \ldots, \xi_0$ are i.i.d. random variables with the law $P(\xi = \pm 1) = 1/2$, and q is a positive number (see Remark 9.3.6 below, where a discussion on choosing q is given).

As before, we obtain the following explicit one-step approximation $v(t_k, x)$ of $u(t_k, x)$:

$$v(t_k, x) = \frac{1}{2}u(t_{k+1}, x + hb_k - h^{1/2}\sigma_k) + \frac{1}{2}u(t_{k+1}, x + hb_k + h^{1/2}\sigma_k) + hg_k,$$
$$\tag{9.3.14}$$

$$\text{if } x + hb_k \pm h^{1/2}\sigma_k \in [\alpha, \beta];$$

$$v(t_k, x) = u(t_{k+1}, x + qh^{1/2}) - \psi_1(t_{k+1}, u(t_{k+1}, \alpha))qh^{1/2},$$
$$\text{if } x + hb_k - h^{1/2}\sigma_k < \alpha;$$

$$v(t_k, x) = u(t_{k+1}, x - qh^{1/2}) + \psi_2(t_{k+1}, u(t_{k+1}, \beta))qh^{1/2},$$
$$\text{if } x + hb_k + h^{1/2}\sigma_k > \beta;$$

$$k = N - 1, \ldots, 1, 0.$$

The corresponding explicit layer method for solving the Neumann problem (9.3.6)–(9.3.8) has the form

$$\bar{u}(t_N, x) = \varphi(t_N, x), \quad x \in [\alpha, \beta], \tag{9.3.15}$$

$$\bar{u}(t_k, x) = \frac{1}{2}\bar{u}(t_{k+1}, x + h\bar{b}_k - h^{1/2}\tilde{\sigma}_k) + \frac{1}{2}\bar{u}(t_{k+1}, x + h\bar{b}_k + h^{1/2}\tilde{\sigma}_k)$$
$$+ h\bar{g}_k, \quad \text{if } x + h\bar{b}_k \pm h^{1/2}\tilde{\sigma}_k \in [\alpha, \beta];$$

$$\bar{u}(t_k, x) = \bar{u}(t_{k+1}, x + qh^{1/2}) - \psi_1(t_{k+1}, \bar{u}(t_{k+1}, \alpha))qh^{1/2},$$
$$\text{if } x + h\bar{b}_k - h^{1/2}\tilde{\sigma}_k < \alpha;$$

$$\bar{u}(t_k, x) = \bar{u}(t_{k+1}, x - qh^{1/2}) + \psi_2(t_{k+1}, \bar{u}(t_{k+1}, \beta))qh^{1/2},$$
$$\text{if } x + h\bar{b}_k + h^{1/2}\tilde{\sigma}_k > \beta;$$

$$k = N - 1, \ldots, 1, 0,$$

where $\bar{b}_k = \bar{b}_k(x) = b(t_k, x, \bar{u}(t_{k+1}, x))$, $\bar{\sigma}_k = \bar{\sigma}_k(x) = \sigma(t_k, x, \bar{u}(t_{k+1}, x))$, $\bar{g}_k = \bar{g}_k(x) = g(t_k, x, \bar{u}(t_{k+1}, x))$.

This layer method is simpler but less accurate than (9.3.12). Its one-step error near the boundary is $O(h)$ and for internal points is $O(h^2)$ (see Lemma 9.3.4 in the next subsection). We prove that its order of convergence is $O(h^{1/2})$ when the boundary condition does not depend on the solution (see Theorem 9.3.5). Apparently, this is so in the general case as well.

A method of the same convergence order is proposed for the linear Neumann problem in [64]. This method is extended to the nonlinear problem in Sect. 9.3.4.

9.3.2 Convergence Theorems

We make the following assumptions.
(i) There exists a unique solution $u(t, x)$ of the problem (9.3.6)–(9.3.8) such that

$$- \infty \leq u_\circ < u_* \leq u(t, x) \leq u^* < u^\circ \leq \infty, \ t_0 \leq t \leq T, \ x \in [\alpha, \beta], \quad (9.3.16)$$

where $u_\circ$, u_*, u^*, u° are some constants, and there exist the uniformly bounded derivatives:

$$\left| \frac{\partial^{i+j} u}{\partial t^i \partial x^j} \right| \leq K, \ i = 0, \ j = 1, 2, 3, 4; \ i = 1, \ j = 0, 1, 2; \ i = 2, \ j = 0;$$

$$(9.3.17)$$

$$t_0 \leq t \leq T, \ x \in [\alpha, \beta].$$

(ii) The coefficients $b(t, x, u)$, $\sigma(t, x, u)$, $g(t, x, u)$ are uniformly bounded and uniformly satisfy the Lipschitz condition with respect to x and u:

$$|b| \leq K, \ |\sigma| \leq K, \ |g| \leq K, \quad (9.3.18)$$

$$|b(t, x_2, u_2) - b(t, x_1, u_1)| + |\sigma(t, x_2, u_2) - \sigma(t, x_1, u_1)|$$
$$+ |g(t, x_2, u_2) - g(t, x_1, u_1)| \leq K \left(|x_2 - x_1| + |u_2 - u_1| \right),$$

$$t_0 \leq t \leq T, \ x \in [\alpha, \beta], \ u_\circ < u < u^\circ.$$

Lemma 9.3.1 *Under the assumptions* (i) *and* (ii), *the one-step error* $\rho(t_k, x)$ *of the method* (9.3.12) *is estimated as*

$$|\rho(t_k, x)| = |v(t_k, x) - u(t_k, x)| \leq Ch^2, \ x + hb_k \pm h^{1/2}\sigma_k \in [\alpha, \beta]; \quad (9.3.19)$$

$$|\rho(t_k, x)| = |v(t_k, x) - u(t_k, x)| \leq Ch^{3/2}, \tag{9.3.20}$$

$$x + hb_k - h^{1/2}\sigma_k < \alpha \ or \ x + hb_k + h^{1/2}\sigma_k > \beta \,,$$

where $v(t_k, x)$ is the corresponding one-step approximation and $C > 0$ does not depend on h, k, x.

Proof If both points $x + hb_k \pm \sqrt{h}\sigma_k$ belong to $[\alpha, \beta]$, (9.3.19) follows directly from Lemma 8.2.1.

Let us consider the case when the point $x + hb_k - h^{1/2}\sigma_k < \alpha$. Due to (9.3.11), we get

$$v(t_k, x) = u(t_{k+1}, x + \Delta X^\alpha) - \psi_1(t_{k+1}, u(t_{k+1}, \alpha)) \cdot (\Delta X^\alpha - hb_k) + hg_k, \tag{9.3.21}$$

where

$$\Delta X^\alpha := \alpha - x + \sqrt{h\sigma_k^2 + (\alpha - x)^2} \,.$$

It is clear that

$$|\alpha - x| \leq Ch^{1/2}, \quad |\Delta X^\alpha| \leq Ch^{1/2}. \tag{9.3.22}$$

Taking into account that $\psi_1(t_{k+1}, u(t_{k+1}, \alpha)) = u'_x(t_{k+1}, \alpha)$ (see (9.3.8)), then expanding the functions $u(t_{k+1}, x + \Delta X^\alpha)$ and $u'_x(t_{k+1}, x + (\alpha - x))$ at the point (t_k, x), and using the assumptions (i), (ii), and the inequalities (9.3.22), we get

$$v(t_k, x) = u + \frac{\partial u}{\partial t}h + \frac{\partial u}{\partial x}\Delta X^\alpha + \frac{1}{2}\frac{\partial^2 u}{\partial x^2}(\Delta X^\alpha)^2 \tag{9.3.23}$$

$$- \frac{\partial u}{\partial x}(\Delta X_\alpha - hb_k) - \frac{\partial^2 u}{\partial x^2}(\alpha - x)\Delta X^\alpha + g_k h + O(h^{3/2})$$

$$= u + h\left(\frac{\partial u}{\partial t} + b_k\frac{\partial u}{\partial x} + g_k\right) + \frac{1}{2}\frac{\partial^2 u}{\partial x^2}\Delta X^\alpha(\Delta X^\alpha - 2(\alpha - x))$$

$$+ O(h^{3/2}),$$

where the function u and its derivatives are evaluated at the point (t_k, x). The expression $\Delta X^\alpha(\Delta X^\alpha - 2(\alpha - x))$ is equal to $h\sigma_k^2$.

Due to the assumptions (i) and (ii), we obtain

$$b_k = b(t_k, x, u(t_{k+1}, x)) = \tilde{b}_k + O(h),$$

$$\sigma_k^2 = \tilde{\sigma}_k^2 + O(h), \quad g_k = \tilde{g}_k + O(h),$$

where $\tilde{b}_k$, $\tilde{\sigma}_k$, $\tilde{g}_k$ are evaluated at the point $(t_k, x, u(t_k, x))$.

Then we get from (9.3.23) that

$$v(t_k, x) = u + h\left(\frac{\partial u}{\partial t} + b\frac{\partial u}{\partial x} + \frac{\sigma^2}{2}\frac{\partial^2 u}{\partial x^2} + g\right) + O(h^{3/2}). \tag{9.3.24}$$

Since $u(t, x)$ is the solution of problem (9.3.6)–(9.3.8), the relation (9.3.24) implies

$$v(t_k, x) = u(t_k, x) + O(h^{3/2}).$$

The case $x + hb_k + h^{1/2}\sigma_k > \beta$ can be considered analogously. □

Theorem 9.3.2 *Let the Neumann problem for Eq. (9.3.6) with condition (9.3.7) have the following boundary conditions*

$$\frac{\partial u}{\partial x}(t, \alpha) = \psi_1(t), \quad \frac{\partial u}{\partial x}(t, \beta) = \psi_2(t), \quad t_0 \le t \le T. \tag{9.3.25}$$

Under the assumptions (i) and (ii), the method (9.3.12) has the first order of convergence with respect to h, i.e.,

$$|\bar{u}(t_k, x) - u(t_k, x)| \le Kh,$$

where $K > 0$ does not depend on h, k, x.

To prove this theorem, we exploit ideas of proving convergence theorems for probabilistic methods from Chap. 7 (see details in [310]).

Remark 9.3.3 Most likely, the conclusion of Theorem 9.3.2 is true under the boundary conditions (9.3.8). We have not succeeded in proving such a general theorem but we can prove it in the case of the linear (Robin) boundary conditions

$$\frac{\partial u}{\partial x}(t, \alpha) = \varphi_1(t)u(t, \alpha) + \psi_1(t), \tag{9.3.26}$$

$$\frac{\partial u}{\partial x}(t, \beta) = \varphi_2(t)u(t, \beta) + \psi_2(t), \quad t_0 \le t \le T.$$

Moreover, numerical experiments confirm the above conjecture (see Sect. 9.4.1).

It turns out that the method (9.3.15) in the case (9.3.25) (and in the case (9.3.26) as well) is convergent with order $O(h^{1/2})$. As above, this fact is apparently true for the case of general boundary conditions.

Let us formulate the corresponding results (see their proofs in [310]).

Lemma 9.3.4 *Under the assumptions (i) and (ii), the one-step error $\rho(t_k, x)$ of the method (9.3.15) is estimated as*

$$|\rho(t_k, x)| = |v(t_k, x) - u(t_k, x)| \le Ch^2, \quad x + hb_k \pm h^{1/2}\sigma_k \in [\alpha, \beta];$$

$$|\rho(t_k, x)| = |v(t_k, x) - u(t_k, x)| \le Ch,$$
$$x + hb_k - h^{1/2}\sigma_k < \alpha \text{ or } x + hb_k + h^{1/2}\sigma_k > \beta,$$

where $v(t_k, x)$ is defined by (9.3.14) and $C > 0$ does not depend on h, k, x.

Theorem 9.3.5 *Under the assumptions (i) and (ii), the method (9.3.15) for the Neumann problem (9.3.6)–(9.3.7), (9.3.25) is of order $O(h^{1/2})$, i.e.,*

$$|\bar{u}(t_k, x) - u(t_k, x)| \le K h^{1/2}, \tag{9.3.27}$$

where $K > 0$ does not depend on h, k, x.

Remark 9.3.6 The layer method (9.3.15) has the parameter q, which, in principle, can be any positive number. Naturally, the value of q affects the method's accuracy: K in (9.3.27) depends on q. By an extended analysis of the one-step boundary error and of the mean number of steps of the corresponding Markov chain in the boundary layer $\partial \Gamma$, we get

$$K \le C_1 \left(\frac{1}{q} \max_{(t,x) \in \bar{Q}} \left| \frac{\partial u}{\partial t} \right| + \frac{q}{2} \max_{(t,x) \in \bar{Q}} \left| \frac{\partial^2 u}{\partial x^2} \right| \right) + C_2,$$

where $C_i > 0$, $i = 1, 2$, do not depend on h, k, x, and q.

Evidently, both large and small values of q are not appropriate. If we know estimates of derivatives of the solution to the considered problem, it is not difficult to find an appropriate q. But generally the choice of q requires a special consideration.

Let $b(t, x, u) \equiv 0$ and $g(t, x, u) \equiv 0$. In this case the one-step boundary error $\rho(t_k, x)$ of the method (9.3.15) near α is evaluated as

$$\rho(t_k, x) = \frac{1}{2} \frac{\partial^2 u}{\partial x^2}(t_k, x) \cdot \left(q^2 h + 2(x - \alpha) q h^{1/2} - h \sigma_k^2 \right) + O(h^{3/2}),$$

$$x - h^{1/2} \sigma_k < \alpha,$$

and analogously near β. Taking $q h^{1/2} = \alpha - x + \sqrt{h \sigma_k^2 + (\alpha - x)^2}$, we obtain $\rho(t_k, x) = O(h^{3/2})$. Substitution of this q (depending on k and x) in (9.3.15) gives us a method with convergence order $O(h)$, which coincides with the method (9.3.12). Such an analysis also suggests that it is preferable to take $q \approx \sigma$.

9.3.3 Numerical Algorithms

Consider the equidistant space discretization with space step h_x (recall that the notation for time step is h): $x_j = \alpha + j h_x$, $j = 0, 1, 2, \ldots, M$, $h_x = (\beta - \alpha)/M$.

Using linear interpolation, we construct the following algorithm on the basis of the method (9.3.12) (we denote it as $\bar{u}(t_k, x)$ again, since this should not cause any confusion).

Algorithm 9.3.7 The algorithm is defined by the following formulas

$$\bar{u}(t_N, x) = \varphi(t_N, x), \ x \in [\alpha, \beta], \tag{9.3.28}$$

$$\bar{u}(t_k, x_j) = \frac{1}{2}\bar{u}(t_{k+1}, x_j + h\bar{b}_{k,j} - h^{1/2}\bar{\sigma}_{k,j}) + \frac{1}{2}\bar{u}(t_{k+1}, x_j + h\bar{b}_{k,j} + h^{1/2}\bar{\sigma}_{k,j})$$
$$+ h\bar{g}_{k,j}, \ \text{if } x_j + h\bar{b}_{k,j} \pm h^{1/2}\bar{\sigma}_{k,j} \in [\alpha, \beta];$$

$$\bar{u}(t_k, x_j) = \bar{u}\left(t_{k+1}, \alpha + \sqrt{h\bar{\sigma}_{k,j}^2 + (\alpha - x_j)^2}\right)$$
$$- \psi_1(t_{k+1}, \bar{u}(t_{k+1}, \alpha)) \cdot \left(\alpha - x_j - h\bar{b}_{k,j} + \sqrt{h\bar{\sigma}_{k,j}^2 + (\alpha - x_j)^2}\right) + h\bar{g}_{k,j},$$
$$\text{if } x_j + h\bar{b}_{k,j} - h^{1/2}\bar{\sigma}_{k,j} < \alpha;$$

$$\bar{u}(t_k, x_j) = \bar{u}\left(t_{k+1}, \beta - \sqrt{h\bar{\sigma}_{k,j}^2 + (\beta - x_j)^2}\right)$$
$$- \psi_2(t_{k+1}, \bar{u}(t_{k+1}, \beta)) \cdot \left(\beta - x_j - h\bar{b}_{k,j} - \sqrt{h\bar{\sigma}_{k,j}^2 + (\beta - x_j)^2}\right) + h\bar{g}_{k,j},$$
$$\text{if } x_j + h\bar{b}_{k,j} + h^{1/2}\bar{\sigma}_{k,j} > \beta; \ j = 1, 2, \dots, M - 1,$$

$$\bar{u}(t_k, x) = \frac{x_{j+1} - x}{h_x}\bar{u}(t_k, x_j) + \frac{x - x_j}{h_x}\bar{u}(t_k, x_{j+1}), \ x_j < x < x_{j+1}, \tag{9.3.29}$$

$$j = 0, 1, 2, \dots, M - 1, \ k = N - 1, \dots, 1, 0,$$

where $\bar{b}_{k,j}$, $\bar{\sigma}_{k,j}$, $\bar{g}_{k,j}$ are the coefficients b, σ, g evaluated at the point $(t_k, x_j, \bar{u}(t_{k+1}, x_j))$.

Using probabilistic arguments, we proved the convergence theorem (see details in [310]).

Theorem 9.3.8 *Consider the problem (9.3.6)–(9.3.7), (9.3.25). If h_x is taken equal to $\varkappa h$, $\varkappa$ is a positive constant, then under the assumptions (i) and (ii) Algorithm 9.3.7 has the first order of convergence, i.e., for the approximation $\bar{u}(t_k, x)$ from the formulas (9.3.28)–(9.3.29) the following inequality holds*

$$|\bar{u}(t_k, x) - u(t_k, x)| \leq Kh,$$

where $K > 0$ does not depend on x, h, k.

Remarks 9.1.9 and 9.1.10 are valid here.

On the basis of linear interpolation and the layer method (9.3.15), we get the following algorithm.

Algorithm 9.3.9 The algorithm is defined by the following formulas

$$\bar{u}(t_N, x) = \varphi(t_N, x), \quad x \in [\alpha, \beta], \tag{9.3.30}$$

$$\bar{u}(t_k, x_j) = \frac{1}{2}\bar{u}(t_{k+1}, x_j + h\bar{b}_{k,j} - h^{1/2}\bar{\sigma}_{k,j}) + \frac{1}{2}\bar{u}(t_{k+1}, x_j + h\bar{b}_{k,j} + h^{1/2}\bar{\sigma}_{k,j})$$
$$+ h\bar{g}_{k,j}, \quad \text{if } x_j + h\bar{b}_{k,j} \pm h^{1/2}\bar{\sigma}_{k,j} \in [\alpha, \beta];$$

$$\bar{u}(t_k, x_j) = \bar{u}(t_{k+1}, x_j + q\sqrt{h}) - \psi_1(t_{k+1}, \bar{u}(t_{k+1}, \alpha)) \cdot qh^{1/2},$$
$$\text{if } x_j + h\bar{b}_{k,j} - h^{1/2}\bar{\sigma}_{k,j} < \alpha;$$

$$\bar{u}(t_k, x_j) = \bar{u}(t_{k+1}, x_j - q\sqrt{h}) + \psi_2(t_{k+1}, \bar{u}(t_{k+1}, \beta)) \cdot qh^{1/2},$$
$$\text{if } x_j + h\bar{b}_{k,j} + h^{1/2}\bar{\sigma}_{k,j} > \beta; \quad j = 1, 2, \ldots, M-1,$$

$$\bar{u}(t_k, x) = \frac{x_{j+1} - x}{h_x}\bar{u}(t_k, x_j) + \frac{x - x_j}{h_x}\bar{u}(t_k, x_{j+1}), \quad x_j < x < x_{j+1}, \tag{9.3.31}$$

$$j = 0, 1, 2, \ldots, M-1, \quad k = N-1, \ldots, 1, 0,$$

where $\bar{b}_{k,j}$, $\bar{\sigma}_{k,j}$, $\bar{g}_{k,j}$ are the coefficients b, σ, g evaluated at the point $(t_k, x_j, \bar{u}(t_{k+1}, x_j))$.

Theorem 9.3.10 *Consider the problem (9.3.6)–(9.3.7), (9.3.25). If h_x is taken equal to $\varkappa h^{3/4}$, $\varkappa$ is a positive constant, then under the assumptions (i) and (ii) Algorithm 9.3.9 has order of convergence $O(\sqrt{h})$, i.e., for the approximation $\bar{u}(t_k, x)$ from the formulas (9.3.30)–(9.3.31) the following inequality holds*

$$|\bar{u}(t_k, x) - u(t_k, x)| \leq K\sqrt{h},$$

where $K > 0$ does not depend on x, h, k.

9.3.4   Some Other Layer Methods

In this subsection two additional methods in the case of $d = 1$ are given.

Using the concept of fictitious nodes, we obtain the following method (see details in the preprint [307]):

$$\bar{u}(t_N, x) = \varphi(x), \quad x \in [\alpha, \beta], \tag{9.3.32}$$

$$\bar{u}(t_k, x) = \frac{1}{2}\bar{u}(t_{k+1}, x + h\bar{b}_k - h^{1/2}\bar{\sigma}_k) + \frac{1}{2}\bar{u}(t_{k+1}, x + h\bar{b}_k + h^{1/2}\bar{\sigma}_k)$$
$$+ h\bar{g}_k, \quad \text{if } x + h\bar{b}_k \pm h^{1/2}\bar{\sigma}_k \in [\alpha, \beta],$$

$$\bar{u}(t_k, x) = \frac{1}{2}\bar{u}(t_{k+1}, 2\alpha - x - h\bar{b}_k + h^{1/2}\bar{\sigma}_k)$$

$$-\psi_1(t_{k+1}, \bar{u}(t_{k+1}, \alpha)) \cdot (\alpha - x - h\bar{b}_k + h^{1/2}\bar{\sigma}_k)$$

$$+\frac{1}{2}\bar{u}(t_{k+1}, x + h\bar{b}_k + h^{1/2}\bar{\sigma}_k) + h\bar{g}_k, \text{ if } x + h\bar{b}_k - h^{1/2}\bar{\sigma}_k < \alpha,$$

$$\bar{u}(t_k, x) = \frac{1}{2}\bar{u}(t_{k+1}, x + h\bar{b}_k - h^{1/2}\bar{\sigma}_k) + \frac{1}{2}\bar{u}(t_{k+1}, 2\beta - x - h\bar{b}_k - h^{1/2}\bar{\sigma}_k)$$

$$+\psi_2(t_{k+1}, \bar{u}(t_{k+1}, \beta)) \cdot (x + h\bar{b}_k + h^{1/2}\bar{\sigma}_k - \beta) + h\bar{g}_k,$$

$$\text{if } x + h\bar{b}_k + h^{1/2}\bar{\sigma}_k > \beta;$$

$$k = N - 1, \dots, 1, 0,$$

where $\bar{b}_k$, $\bar{\sigma}_k$, $\bar{g}_k$ are the coefficients b, σ, g evaluated at the point $(t_k, x, \bar{u}(t_{k+1}, x))$.

The method (9.3.32) is an explicit layer method for solving the Neumann problem (9.3.6)–(9.3.8). We prove that its one-step error near the boundary is $O(h^{3/2})$ and for internal points is $O(h^2)$. This method has order of convergence $O(h)$. For the Monte Carlo algorithm which uses fictitious nodes, see Algorithm 7.6.5 of Sect. 7.6.2 and the corresponding discussion in [233].

The method (9.3.32) is more complicated than the method (9.3.12). At the same time, it demonstrates more accurate results than (9.3.12) in our numerical tests (see Sect. 9.4.1).

The method (9.3.15) is an extension of the method of order $O(h^{1/2})$ from Sect. 7.6 to the nonlinear case. In [64] another method of order $O(h^{1/2})$ for linear Neumann problems is proposed. Its extension to the nonlinear Neumann problem (9.3.6)–(9.3.8) has the form

$$\bar{u}(t_N, x) = \varphi(t_N, x), \ x \in [\alpha, \beta], \tag{9.3.33}$$

$$\bar{u}(t_k, x) = \frac{1}{2}\bar{u}(t_{k+1}, x + h\bar{b}_k - h^{1/2}\bar{\sigma}_k) + \frac{1}{2}\bar{u}(t_{k+1}; x + h\bar{b}_k + h^{1/2}\bar{\sigma}_k)$$

$$+h\bar{g}_k, \quad \text{if } x + h\bar{b}_k \pm h^{1/2}\bar{\sigma}_k \in [\alpha, \beta];$$

$$\bar{u}(t_k, x) = \frac{1}{2}\bar{u}(t_{k+1}, \alpha) + \frac{1}{2}\bar{u}(t_{k+1}, x + h\bar{b}_k + h^{1/2}\bar{\sigma}_k)$$

$$-\frac{1}{2}\psi_1(t_{k+1}, \bar{u}(t_{k+1}, \alpha))(\alpha - x - h\bar{b}_k + h^{1/2}\bar{\sigma}_k) + h\bar{g}_k,$$

$$\text{if } x + h\bar{b}_k - h^{1/2}\bar{\sigma}_k < \alpha;$$

$$\bar{u}(t_k, x) = \frac{1}{2}\bar{u}(t_{k+1}, x + h\bar{b}_k - h^{1/2}\bar{\sigma}_k) + \frac{1}{2}\bar{u}(t_{k+1}, \beta)$$

$$-\frac{1}{2}\psi_2(t_{k+1}, \bar{u}(t_{k+1}, \beta))(\beta - x - h\bar{b}_k - h^{1/2}\bar{\sigma}_k) + h\bar{g}_k,$$

$$\text{if } x + h\bar{b}_k + h^{1/2}\bar{\sigma}_k > \beta;$$

$$k = N - 1, \ldots, 1, 0,$$

where $\bar{b}_k = \bar{b}_k(x) = b(t_k, x, \bar{u}(t_{k+1}, x))$, $\bar{\sigma}_k = \bar{\sigma}_k(x) = \sigma(t_k, x, \bar{u}(t_{k+1}, x))$, $\bar{g}_k = \bar{g}_k(x) = g(t_k, x, \bar{u}(t_{k+1}, x))$.

We proved that the one-step error of this method near the boundary is $O(h)$ and for internal points is $O(h^2)$. Apparently, this layer method has order of convergence $O(h^{1/2})$. It is more complicated near the boundary than (9.3.15). At the same time, the method (9.3.33) demonstrates more accurate results than (9.3.15) in our numerical tests (see Sect. 9.4.1).

Algorithms based on linear interpolation and on the layer methods of this section can be written as in Sect. 9.3.3.

9.4   Numerical Tests for the Neumann Problem

9.4.1   Comparison of Various Layer Methods

Consider the Neumann problem for the one-dimensional Burgers equation:

$$\frac{\partial u}{\partial t} = \frac{\sigma^2}{2}\frac{\partial^2 u}{\partial x^2} - u\frac{\partial u}{\partial x}, \quad t > 0, \ x \in (-4, 4), \tag{9.4.1}$$

$$u(0, x) = -\frac{\sigma^2 \sinh x}{\cosh x + A}, \quad x \in [-4, 4], \tag{9.4.2}$$

$$\frac{\partial u}{\partial x}(t, \pm 4) = -\sigma^2 \frac{1 + A \exp\left(-\sigma^2 t/2\right)\cosh 4}{[\cosh 4 + A \exp\left(-\sigma^2 t/2\right)]^2}, \quad t \geq 0. \tag{9.4.3}$$

Here A is a positive constant.

The exact solution to this problem has the form (see [27])

$$u(t, x) = -\frac{\sigma^2 \sinh x}{\cosh x + A \exp\left(-\sigma^2 t/2\right)}.$$

In the tests we use cubic interpolation (cf. Remark 9.1.9)

Table 9.4.1 Dependence of the errors $err^c(t)$ (bottom position) and $err^l(t)$ (top position) in h when $t = 2$, $\sigma = 1.5$, and $A = 2$

h	Algorithm (9.3.28)–(9.3.29)	Algorithm (9.3.28), (9.4.4)	Algorithm (9.3.32), (9.4.4)	Algorithm (9.3.30), (9.4.4)	Algorithm (9.3.33), (9.4.4)
0.16	$5.216 \cdot 10^{-1}$	$7.434 \cdot 10^{-1}$	$5.967 \cdot 10^{-2}$	>1	$7.333 \cdot 10^{-1}$
	$8.509 \cdot 10^{-2}$	$1.177 \cdot 10^{-1}$	$1.380 \cdot 10^{-2}$	$3.328 \cdot 10^{-1}$	$1.098 \cdot 10^{-1}$
0.01	$3.170 \cdot 10^{-2}$	$1.888 \cdot 10^{-2}$	$3.867 \cdot 10^{-3}$	$3.722 \cdot 10^{-1}$	$1.346 \cdot 10^{-1}$
	$5.748 \cdot 10^{-3}$	$3.737 \cdot 10^{-3}$	$1.224 \cdot 10^{-3}$	$6.161 \cdot 10^{-2}$	$2.192 \cdot 10^{-2}$
0.0016	$4.479 \cdot 10^{-3}$	$3.835 \cdot 10^{-3}$	$7.124 \cdot 10^{-4}$	$1.653 \cdot 10^{-1}$	$4.909 \cdot 10^{-2}$
	$8.149 \cdot 10^{-4}$	$7.444 \cdot 10^{-4}$	$2.127 \cdot 10^{-4}$	$2.750 \cdot 10^{-2}$	$8.172 \cdot 10^{-3}$
0.0001	$2.387 \cdot 10^{-4}$	$2.711 \cdot 10^{-4}$	$4.639 \cdot 10^{-5}$	$4.378 \cdot 10^{-2}$	$1.168 \cdot 10^{-2}$
	$4.479 \cdot 10^{-5}$	$5.213 \cdot 10^{-5}$	$1.357 \cdot 10^{-5}$	$7.307 \cdot 10^{-3}$	$1.968 \cdot 10^{-3}$

$$\bar{u}(t_k, x) = \sum_{i=0}^{3} \Phi_{j,i}(x)\, \bar{u}(t_k, x_{j+i}), \quad x_j < x < x_{j+3}, \tag{9.4.4}$$

$$\Phi_{j,i}(x) = \prod_{m=0, m \neq i}^{3} \frac{x - x_{j+m}}{x_{j+i} - x_{j+m}}.$$

Here we test the following five algorithms: (i) the algorithm (9.3.28)–(9.3.29) (Algorithm 9.3.7), (ii) the algorithm based on layer method (9.3.12) and cubic interpolation (9.4.4) (algorithm (9.3.28), (9.4.4)), (iii) the algorithm based on layer method (9.3.32) and cubic interpolation (9.4.4), (iv) the algorithm based on layer method (9.3.15) and cubic interpolation (9.4.4) (algorithm (9.3.30), (9.4.4)), and (v) the algorithm based on layer method (9.3.33) and cubic interpolation (9.4.4). We take the space step $h_x = h$ for linear interpolation and $h_x = \sqrt{h}$ for cubic interpolation. The parameter q of algorithm (9.3.30), (9.4.4) is taken equal to 1.

Table 9.4.1 and Fig. 9.4.1 give numerical results obtained by these algorithms. In the table the errors of the approximate solutions $\bar{u}$ are presented in the discrete Chebyshev norm (bottom position) and in the l^1-norm (top position):

$$err^c(t) = \max_{x_i} |\bar{u}(t, x_i) - u(t, x_i)|, \tag{9.4.5}$$

$$err^l(t) = \sum_{i} h_x |\bar{u}(t, x_i) - u(t, x_i)|.$$

In the experiments the algorithm (9.3.30), (9.4.4) and the algorithm (9.3.33), (9.4.4) converge as $O(h^{1/2})$, the other algorithms converge as $O(h)$. We note that the algorithm (9.3.33), (9.4.4) gives more accurate results than the algorithm (9.3.30), (9.4.4), and the algorithm (9.3.32), (9.4.4) is more accurate than the algorithms

Fig. 9.4.1 Solution of the problem (9.4.1)–(9.4.3). Here $h = 0.16$, the other parameters are as in Table 9.4.1

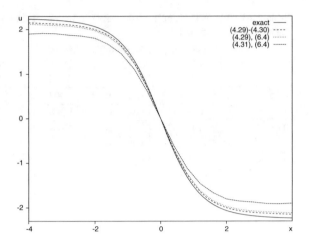

(9.3.28). The algorithms (9.3.28)–(9.3.29) and (9.3.28), (9.4.4) demonstrate almost the same accuracy. But the algorithm (9.3.28)–(9.3.29) (as well as other algorithms based on linear interpolation) requires both larger volume of computations per time layer and larger amount of memory than the algorithm (9.3.28), (9.4.4) based on cubic interpolation (see also Remark 9.1.9 and numerical tests in Sect. 9.2).

Further, the boundary condition (9.4.3) can be rewritten in the form

$$\frac{\partial u}{\partial x}(t, \pm 4) = u(t, \pm 4)\left(\frac{1}{\sigma^2}u(t, \pm 4) - 1\right) - \frac{\sigma^2 \exp(\pm 4)}{\cosh 4 + A \exp\left(-\sigma^2 t/2\right)}, \quad t \geq 0.$$
(9.4.6)

In order to provide an experimental verification of the conjecture from Remark 9.3.3 we apply the algorithm (9.3.28), (9.4.4)–(9.4.1) with the initial condition (9.4.2) and the nonlinear boundary condition (9.4.6). Taking the same values of parameters as in Table 9.4.1, we obtain, in particular, that for $h = 0.01$ the error $err^c(2) = 1.103 \times 10^{-3}$ and for $h = 0.0001$ the error $err^c(2) = 1.138 \times 10^{-5}$ that experimentally confirms the conjecture.

In Appendix A.5 of [314] the algorithm (9.3.28), (9.4.4) is implemented for (9.4.1)–(9.4.3).

9.4.2   A Comparison Analysis of Layer Methods and Finite-Difference Schemes

Here the test problem is the Burgers equation

$$\frac{\partial u}{\partial t} = \frac{\sigma^2}{2}\frac{\partial^2 u}{\partial x^2} - u\frac{\partial u}{\partial x}, \quad t > 0, \ x \in (-2, 8),$$
(9.4.7)

with the following initial and boundary conditions

$$u(0, x) = \varphi(x) := \begin{cases} a, & x \in [-2, 0), \\ (a+b)/2, & x = 0, \\ b, & x \in (0, 8], \end{cases} \tag{9.4.8}$$

$$\frac{\partial u}{\partial x}(t, x) = \psi(t, x), \; t > 0, \; x \in \{-2, 8\}, \tag{9.4.9}$$

where

$$\psi(t, x) = -\frac{(a-b)^2 J_1(t, x) J_2(t, x)}{\sigma^2 \times (J_1(t, x) + J_2(t, x))^2}$$
$$+ \sqrt{\frac{2}{\pi \sigma^2 t}} \exp\left(\frac{-x^2}{2\sigma^2 t}\right) \frac{b-a}{J_1(t, x) + J_2(t, x)},$$

and

$$J_1(t, x) = \exp\left(\frac{a(at - 2x)}{2\sigma^2}\right) \operatorname{erfc}\left(\frac{x - at}{\sqrt{2\sigma^2 t}}\right),$$

$$J_2(t, x) = \exp\left(\frac{b(bt - 2x)}{2\sigma^2}\right) \operatorname{erfc}\left(\frac{bt - x}{\sqrt{2\sigma^2 t}}\right).$$

The exact solution of this problem is

$$u(t, x) = \frac{a J_1(t, x) + b J_2(t, x)}{J_1(t, x) + J_2(t, x)}.$$

We compare the algorithm (9.3.28), (9.4.4) with the method of differences forward in time and central differences in space applied to the divergent form of the Burgers equation. This finite-difference scheme in application to the problem (9.4.7)–(9.4.9) is written as

$$\bar{u}(0, x) = \varphi(x), \; x \in [-2, 8], \tag{9.4.10}$$

$$\bar{u}(t_{k+1}, x_{-1}) = \bar{u}(t_{k+1}, x_1) - 2\Delta x \times \psi(t_{k+1}, x_0),$$

$$\bar{u}(t_{k+1}, x_{M+1}) = \bar{u}(t_{k+1}, x_{M-1}) + 2\Delta x \times \psi(t_{k+1}, x_M),$$

$$\bar{u}(t_{k+1}, x_j) = \bar{u}(t_k, x_j) - \frac{\Delta t}{4\Delta x}(\bar{u}^2(t_k, x_{j+1}) - \bar{u}^2(t_k, x_{j-1}))$$
$$+ \frac{\sigma^2}{2} \frac{\Delta t}{\Delta x^2}(\bar{u}(t_k, x_{j+1}) - 2\bar{u}(t_k, x_j) + \bar{u}(t_k, x_{j-1})),$$

$$j = 0, \ldots, M, \; k = 0, \ldots, N - 1,$$

Table 9.4.2 Comparison analysis. Dependence of the errors $err^c(t)$ (bottom position) and $err^l(t)$ (top position) in h when $t = 0.6$, $\sigma = 0.4$, $a = 11$, and $b = 9$

h	0.01	0.0016	0.0004	0.0001
Algorithm	4.859×10^{-1}	1.031×10^{-1}	2.659×10^{-2}	6.792×10^{-3}
(9.3.28), (9.4.4)	1.208	3.425×10^{-1}	8.980×10^{-2}	2.308×10^{-2}
Scheme (9.4.10)	Overflow	2.531	5.057×10^{-2}	1.234×10^{-2}
		6.375	1.261×10^{-1}	2.766×10^{-2}
		Oscillations		

Fig. 9.4.2 Solution of the problem (9.4.7)–(9.4.9). Solid line is the exact solution, dotted line is the algorithm (9.3.28), (9.4.4), and dashed line is the scheme (9.4.10). Here $h = 0.0016$, the other parameters are as in Table 9.4.2

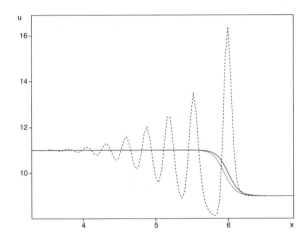

where the step of time discretization $\Delta t := T/N$ and $t_k = k \times \Delta t$ and the step of space discretization $\Delta x := 10/M$ and $x_j = -2 + j \times \Delta x$.

The explicit scheme (9.4.10) is of order $O(\Delta t, \Delta x^2)$. It is used for simulation of the Burgers equation in [5, 107].

In the experiments we take the space step $h_x = \sqrt{h}$ for the algorithm (9.3.28), (9.4.4) and the space step $\Delta x = \sqrt{\Delta t}$ for the finite-difference scheme (9.4.10).

Table 9.4.2 presents the errors in the discrete Chebyshev norm (bottom position in the table) and in the l^1-norm (top position) (see (9.4.5)). The comment "overflow" indicates that overflow error occurs during simulation. The comment "oscillations" means that the numerical solution has oscillations (see Fig. 9.4.2). We see that the algorithm (9.3.28), (9.4.4) demonstrates a more stable behavior than the finite-difference scheme (9.4.10). In the test problem (9.4.7)–(9.4.9) large values of a, b lead, in particular, to large advection in a neighborhood of the front. These experiments confirm that the layer methods allow us to avoid difficulties stemming from strong advection (see also comparison analysis in Sect. 9.2.2). We note that the algorithms based on layer methods require more CPU time than finite-difference schemes when one uses the same step h. For example, in the case of parameters as in Table 9.4.2 to solve (9.4.7)–(9.4.9) by the algorithm (9.3.28), (9.4.4) with $h = 0.0004$ we needed ≈ 5 s. while the scheme (9.4.10) required ≈ 2.8 s. But the algorithm (9.3.28), (9.4.4) gives us quite appropriate results with the greater step $h = 0.0016$ (see Table 9.4.2 and Fig. 9.4.2) and in this case it required ≈ 0.7 s. Simulations were made on PC with Intel Pentium 233 MHz processor using Borland C compiler.

Chapter 10
Solving FBSDEs Using Layer Methods

Forward-backward stochastic differential equations (FBSDEs) have numerous applications in stochastic control theory and mathematical finance (see, e.g. [66, 83, 90, 244, 350, 358, 442, 465] and references therein). Consider a FBSDE of the form

$$X = a(t, X, Y)dt + \sigma(t, X, Y)dw(t), \quad X(t_0) = x, \tag{10.0.1}$$

$$dY = -g(t, X, Y)dt - f^{\mathsf{T}}(t, X, Y)Zdt + Z^{\mathsf{T}}dw(t), \quad Y(T) = \varphi(X(T)). \tag{10.0.2}$$

Here $X = X(t)$ and $a = a(t, x, y)$ are d-dimensional vectors; $\sigma = \sigma(t, x, y)$ is a $d \times n$-matrix; $Y = Y(t)$, $g = g(t, x, y)$, and $\varphi = \varphi(x)$ are scalars; $Z = Z(t)$ and $f = f(t, x, y)$ are n-dimensional vectors; $w(t)$ is an n-dimensional $\{\mathcal{F}_t\}_{t\geq0}$-adapted Wiener process, where $(\Omega, \mathcal{F}, \mathcal{F}_t, P)$, $t_0 \leq t \leq T$, is a filtered probability space. It is known (see, e.g. [7, 243, 244, 347, 350, 465] and references therein) that there exists a unique $\{\mathcal{F}_t\}_{t\geq0}$-adapted solution $(X(t), Y(t), Z(t))$ of the system (10.0.1)–(10.0.2) under some appropriate smoothness and boundedness conditions on its coefficients.

Due to the four-step scheme from [243] (we recall it in Sect. 10.1.1), the solution of (10.0.1)–(10.0.2) is connected with the Cauchy problem for the semilinear PDE:

$$\frac{\partial u}{\partial t} + \sum_{i=1}^{d} a^i(t, x, u)\frac{\partial u}{\partial x^i} + \frac{1}{2}\sum_{i,j=1}^{d} a^{ij}(t, x, u)\frac{\partial^2 u}{\partial x^i \partial x^j} \tag{10.0.3}$$

$$= -g(t, x, u) - \sum_{k=1}^{n} f^k(t, x, u)\sum_{i=1}^{d}\sigma^{ik}(t, x, u)\frac{\partial u}{\partial x^i}, \quad t < T, \ x \in \mathbf{R}^d,$$

$$u(T, x) = \varphi(x), \tag{10.0.4}$$

where

© Springer Nature Switzerland AG 2021
G. N. Milstein and M. V. Tretyakov, *Stochastic Numerics for Mathematical Physics*,
Scientific Computation, https://doi.org/10.1007/978-3-030-82040-4_10

$$a^{ij} := \sum_{k=1}^{d} \sigma^{ik} \sigma^{jk}.$$

In turn the corresponding solution of the semilinear PDE has a probabilistic representation using the FBSDE (10.0.1)–(10.0.2), which is a generalization of the Feynman–Kac formula (see, e.g. [244, 347, 348, 350, 352, 357, 465]). We note that the PDE problem (10.0.3)–(10.0.4) coincides with 8.3.1)–(8.3.2) if we choose

$$b^i (t, x, y) = a^i (t, x, u) + \sum_{k=1}^{n} f^k (t, x, u) \sigma^{ik} (t, x, u), \ i = 1, \ldots, d.$$

Over the last two decades there have been many works on numerical approximation of FBSDEs, here we cite only a few of them [17, 18, 25, 55, 81, 133, 135, 247, 317, 320] (see also references therein). In this chapter we first (Sect. 10.1) consider an effective numerical algorithm for solving (10.0.1)–(10.0.2) in the mean-square sense. To this end, we exploit the four-step scheme due to [243] for numerical solution of (10.0.1)–(10.0.2) together with layer methods from Chap. 8. Along with the algorithms for FBSDEs on a fixed finite time interval, we also construct algorithms for FBSDEs with random terminal time (see e.g. [347, 350, 465] and references therein) in Sects. 10.2 and 10.3. In these sections we have to consider the Dirichlet boundary value problem for semilinear parabolic PDE instead of (10.0.3)–(10.0.4). For solving the Dirichlet problem, we use the probabilistic approach again and layer methods from Chap. 9. To simulate solutions of FBSDEs with random terminal time, we use the approximations of SDEs in space-time bounded domains from Chap. 6 (see also [304]). We also consider the case of unbounded random terminal time, which is connected with the Dirichlet boundary value problem for semilinear elliptic PDE. For clarity of exposition, we state and prove our results for the one-dimensional case ($d = 1, n = 1$) although it is not difficult to generalize them to any dimension. In this respect the bottle-neck (the same as for layer methods of Chaps. 8 and 9) of the presented algorithms is linear interpolation which is used to restore the function $\bar{u}(t_k, x), x \in \mathbf{R}^d$, by its values $\bar{u}(t_k, x_j)$ at a set of nodes x_j. However, layer methods can be turned into numerical algorithms by exploiting other approximations of functions. Recent developments in the theory of multidimensional approximation and deep learning [79, 136, 462] give approximations which work well in relatively high dimension. Making use of such approximations of functions would allow to realize layer methods for high-dimensional nonlinear PDEs. In Sect. 10.4 we extend the probabilistic layer methods of Chap. 8 to quasilinear PDEs (recall that in Chaps. 8 and 9 we considered semi-linear PDEs) and construct numerical methods for more general FBSDEs than (10.0.1)– (10.0.2). The results are supported by numerical experiments which are presented in Sect. 10.5. This chapter is mainly based on [317, 320].

10.1 Discretization of FBSDEs and Related Semilinear Parabolic PDEs

In this section we consider a numerical algorithm to solve FBSDEs (10.0.1)–(10.0.2). In Sect. 10.1.1 we recall the four-step scheme due to [243]. In Sect. 10.1.2 we obtain some results concerning accuracy of approximating derivatives $\partial u/\partial x^i$ by finite differences. In Sect. 10.1.3 we prove the mean-square convergence of the Euler method for FBSDEs when using approximations of u by a layer method from Chap. 8 and of $\partial u/\partial x$ of Sect. 10.1.3.

10.1.1 Four-Step Scheme for Solving FBSDEs

In this subsection we recall the four-step scheme for solving FBSDEs (10.0.1)–(10.0.2) [243]. Assume that the solution $u(t, x)$ of the Cauchy problem (10.0.3)–(10.0.4) is known. Consider the following SDE

$$dX = a(t, X, u(t, X))dt + \sigma(t, X, u(t, X))dw(t), \quad X(t_0) = x. \qquad (10.1.1)$$

Let $X(t) = X_{t_0, x}(t)$ be a solution of the Cauchy problem (10.1.1). Introduce

$$Y(t) = u(t, X_{t_0, x}(t)), \qquad (10.1.2)$$

$$Z^j(t) = \sum_{i=1}^{d} \sigma^{ij}(t, X_{t_0, x}(t), Y(t)) \frac{\partial u}{\partial x^i}(t, X_{t_0, x}(t)), \quad j = 1, \ldots, n.$$

It turns out that $(X(t), Y(t), Z(t))$ defined by (10.1.1)–(10.1.2) is the solution of the FBSDE (10.0.1)–(10.0.2). Indeed, it is $\{\mathcal{F}_t\}_{t \geq 0}$-adapted and (10.0.1) is evidently satisfied. To verify (10.0.2), it suffices to apply Ito's formula to $Y(t) = u(t, X_{t_0, x}(t))$.

10.1.2 Approximation of the Derivative $\partial u / \partial x$ by Finite Differences

For simplicity in writing, we restrict ourselves to a one-dimensional version of the problem ($d = 1$, $n = 1$). Introducing

$$b(t, x, y) := a(t, x, y) + f(t, x, y)\sigma(t, x, y),$$

we get

$$\frac{\partial u}{\partial t} + b(t, x, u)\frac{\partial u}{\partial x} + \frac{1}{2}\sigma^2(t, x, u)\frac{\partial^2 u}{\partial x^2} + g(t, x, u) = 0, \; t < T, \; x \in \mathbf{R},$$
(10.1.3)

$$u(T, x) = \varphi(x).$$
(10.1.4)

For convenience, we shall assume throughout the paper that the standing assumptions of [81] are fulfilled. We prefer to state them here in a less specific way since different numerical methods will require different, e.g., level of smoothness of the coefficients.

Assumption 10.1.1 It is assumed that the coefficients b, σ, g and the function φ are sufficiently smooth and all these functions together with their derivatives up to some order are bounded on $[t_0, T] \times \mathbf{R} \times \mathbf{R}$. Besides, it is supposed that σ is bounded away from zero: $\sigma \geq \sigma_0$, where σ_0 is a positive constant.

Assumption 10.1.1 ensures the existence of a unique bounded solution $u(t, x)$ with bounded derivatives up to some order. We note that these assumptions are more than necessary (e.g., another type of assumptions are given on pages 570 and 573 in Sect. 8.2.2) and the methods considered in this chapter can be used under broader conditions.

Consider the solution $u(t, x)$ of the Cauchy problem for semilinear parabolic Eqs. (10.1.3)–(10.1.4) and its approximations $\tilde{u}(t_k, x)$ by the layer method (8.2.23) (it is convenient in this chapter to denote the layer method as $\tilde{u}(t_k, x)$ instead of $\bar{u}(t_k, x)$ in (8.2.23), which should not lead to any confusion) and $\bar{u}(t_k, x)$ by Algorithm 8.2.4.

Proposition 10.1.2 *The following formula holds:*

$$\frac{\partial u}{\partial x}(t_k, x) = \frac{\bar{u}(t_k, x + \gamma\sqrt{h}) - \bar{u}(t_k, x - \gamma\sqrt{h})}{2\gamma\sqrt{h}} + O(h^{1/2}), \qquad (10.1.5)$$

where γ is a positive number and h is the time step which is the same as in Algorithm 8.2.4.
The analogous formula is valid if the function $\bar{u}$ is substituted by $\tilde{u}$.

Proof Since the solution u has third bounded derivatives with respect to x (we assume that the functions from Assumption 10.1.1 have continuous bounded first derivative with respect to t and second derivatives with respect to x and u including the mixed ones), we have

$$\frac{\partial u}{\partial x}(t_k, x) = \frac{u(t_k, x + \gamma\sqrt{h}) - u(t_k, x - \gamma\sqrt{h})}{2\gamma\sqrt{h}} + O(h).$$

Now (10.1.5) immediately follows from the inequality (8.2.42) from Theorem 8.2.5.   □

Remark 10.1.3 Analogous to (10.1.5), we also get

$$\frac{\partial u}{\partial x}(t_k, x) = \frac{\bar{u}(t_k, x + \gamma h^{1/3}) - \bar{u}(t_k, x - \gamma h^{1/3})}{2\gamma h^{1/3}} + O(h^{2/3}). \tag{10.1.6}$$

In fact, it is possible to prove a more accurate result than (10.1.5) or (10.1.6) for the layer approximation $\tilde{u}$ from (8.2.23).

Theorem 10.1.4 *The following formula holds:*

$$\frac{\partial u}{\partial x}(t_k, x) = \frac{\tilde{u}(t_k, x + \gamma\sqrt{h}) - \tilde{u}(t_k, x - \gamma\sqrt{h})}{2\gamma\sqrt{h}} + O(h), \tag{10.1.7}$$

where γ is a positive number and h is the time step which is the same as in (8.2.23).

Proof Clearly, the pair of functions $u(t, x)$ and $v(t, x) := \dfrac{\partial u}{\partial x}(t, x)$ satisfy the Cauchy problem for two parabolic equations consisting of (10.1.3)–(10.1.4) and of

$$\begin{aligned}
&\frac{\partial v}{\partial t} + b(t, x, u)\frac{\partial v}{\partial x} + \frac{1}{2}\sigma^2(t, x, u)\frac{\partial^2 v}{\partial x^2} + \left(\sigma\frac{\partial\sigma}{\partial x} + \sigma\frac{\partial\sigma}{\partial u}v\right)\frac{\partial v}{\partial x} \\
&+ \left(\frac{\partial b}{\partial x} + \frac{\partial b}{\partial u}v\right)v + \frac{\partial g}{\partial x} + \frac{\partial g}{\partial u}v = 0, \ t < T, \ x \in \mathbf{R},
\end{aligned} \tag{10.1.8}$$

$$v(T, x) = \varphi'(x). \tag{10.1.9}$$

To solve the problem (10.1.3)–(10.1.4), (10.1.8)–(10.1.9), we use a layer method based on a local probabilistic representation. To this aim, introduce the system of SDEs with respect to X and scalars P, Q, R :

$$dX = b(s, X, u(s, X))ds + \sigma(s, X, u(s, X))dw(s), \quad X(t_k) = x, \tag{10.1.10}$$
$$dP = g(s, X, u(s, X))ds, \quad P(t_k) = 0,$$
$$dQ = \left(\frac{\partial b}{\partial x} + \frac{\partial b}{\partial u}v(s, X)\right)Qds + \left(\frac{\partial\sigma}{\partial x} + \frac{\partial\sigma}{\partial u}v(s, X)\right)Qdw(s), \quad Q(t_k) = 1,$$
$$dR = \left(\frac{\partial g}{\partial x} + \frac{\partial g}{\partial u}v(s, X)\right)Qds, \quad R(t_k) = 0,$$

where $\partial b/\partial x$, $\partial b/\partial u$ and the other derivatives are known functions of $s, X, u(s, X)$. One can verify that the following local probabilistic representation holds (cf. (8.2.4), (8.2.5)–(8.2.6)):

$$u(t_k, x) = E\left[u(t_{k+1}, X_{t_k,x}(t_{k+1})) + P_{t_k,x,0}(t_{k+1})\right] \tag{10.1.11}$$
$$v(t_k, x) = E\left[v(t_{k+1}, X_{t_k,x}(t_{k+1}))Q_{t_k,x,1}(t_{k+1}) + R_{t_k,x,1,0}(t_{k+1})\right].$$

The corresponding layer method $\tilde{u}(t_k, x)$, $\tilde{v}(t_k, x)$ is given by the formulas (8.2.23) for $\tilde{u}$ while $\tilde{v}$ is found from:

$$\tilde{v}(t_N, x) = \varphi'(x), \qquad (10.1.12)$$

$$\tilde{v}(t_k, x) = \frac{1}{2}\tilde{v}(t_{k+1}, x + h\tilde{b}_k - \sqrt{h}\tilde{\sigma}_k)$$

$$\times \left[1 + h\left(\frac{\partial \tilde{b}_k}{\partial x} + \frac{\partial \tilde{b}_k}{\partial u}\tilde{v}(t_{k+1}, x) \right) - h^{1/2}\left(\frac{\partial \tilde{\sigma}_k}{\partial x} + \frac{\partial \tilde{\sigma}_k}{\partial u}\tilde{v}(t_{k+1}, x) \right) \right]$$

$$+ \frac{1}{2}\tilde{v}(t_{k+1}, x + h\tilde{b}_k + \sqrt{h}\tilde{\sigma}_k)$$

$$\times \left[1 + h\left(\frac{\partial \tilde{b}_k}{\partial x} + \frac{\partial \tilde{b}_k}{\partial u}\tilde{v}(t_{k+1}, x) \right) + h^{1/2}\left(\frac{\partial \tilde{\sigma}_k}{\partial x} + \frac{\partial \tilde{\sigma}_k}{\partial u}\tilde{v}(t_{k+1}, x) \right) \right]$$

$$+ h\left(\frac{\partial \tilde{g}_k}{\partial x} + \frac{\partial \tilde{g}_k}{\partial u}\tilde{v}(t_{k+1}, x) \right), \quad k = N - 1, \dots, 1, 0,$$

where $\tilde{b}_k := b(t_k, x, \tilde{u}(t_{k+1}, x))$, $\tilde{\sigma}_k := \sigma(t_k, x, \tilde{u}(t_{k+1}, x))$, and the notation $\partial \tilde{b}_k / \partial x$ means $\dfrac{\partial \tilde{b}_k}{\partial x} := \dfrac{\partial b}{\partial x}(t_k, x, \tilde{u}(t_{k+1}, x))$ and so on. The layer method (8.2.23), (10.1.12) for the system (10.1.3)–(10.1.4), (10.1.8)–(10.1.9) is of order one. The order of convergence can be proved (see (see Theorem 8.2.2) if we assume that the functions from Assumption 10.1.1 have continuous bounded second mixed derivatives with respect to t, x and t, u and third derivatives with respect to x and u including the mixed ones. This also implies that $\tilde{u}(t_k, x)$ has, in particular, continuous bounded third derivative with respect to x.

Further, it is straightforward to verify

$$\frac{\partial \tilde{u}}{\partial x}(t_k, x) = \tilde{v}(t_k, x).$$

Thus we get

$$\frac{\partial u}{\partial x}(t_k, x) = v(t_k, x) = \tilde{v}(t_k, x) + O(h) = \frac{\partial \tilde{u}}{\partial x}(t_k, x) + O(h). \qquad (10.1.13)$$

Since $\tilde{u}(t_k, x)$ has bounded third derivative with respect to x, we obtain

$$\frac{\partial \tilde{u}}{\partial x}(t_k, x) = \frac{\tilde{u}(t_k, x + \gamma\sqrt{h}) - \tilde{u}(t_k, x - \gamma\sqrt{h})}{2\gamma\sqrt{h}} + O(h). \qquad (10.1.14)$$

The formulas (10.1.13) and (10.1.14) imply (10.1.7).   □

In the multi-dimensional case ((10.0.3)–(10.0.4) with $d > 1$) we use the approximation

$$\frac{\partial u}{\partial x^i}(t_k, x) \approx \frac{\bar{u}(t_k, x^1, \dots, x^i + \gamma\sqrt{h}, \dots, x^d) - \bar{u}(t_k, x^1, \dots, x^i - \gamma\sqrt{h}, \dots, x^d)}{2\gamma\sqrt{h}},$$

$$(10.1.15)$$

$$i = 1, \ldots, d,$$

where $\bar{u}$ is found by a multi-dimensional algorithm analogous to Algorithm 8.2.4 (see Sect. 8.3.1). Proposition 10.1.2 is valid for (10.1.15) as well.

10.1.3 Numerical Integration of FBSDEs

Let $\bar{u}(t_k, x)$ be defined by Algorithm 8.2.4 and introduce the notation

$$\frac{\Delta \bar{u}}{\Delta x}(t_k, x) := \frac{\bar{u}(t_k, x + \gamma\sqrt{h}) - \bar{u}(t_k, x - \gamma\sqrt{h})}{2\gamma\sqrt{h}} \quad \text{for some} \gamma > 0, \qquad (10.1.16)$$

and also $\Delta_k w := w(t_k + h) - w(t_k)$. In practice it is advisable to choose the parameter γ close to the diffusion σ at the point (t_k, x).

Consider two numerical schemes for the FBSDE (10.0.1)–(10.0.2) with $d = n = 1$:

the Euler scheme

$$X_0 = x,$$
$$X_{k+1} = X_k + a(t_k, X_k, \bar{u}(t_k, X_k))h + \sigma(t_k, X_k, \bar{u}(t_k, X_k))\Delta_k w, \qquad (10.1.17)$$
$$k = 0, \ldots, N - 1;$$

and *the first-order scheme*

$$X_0 = x, \qquad\qquad\qquad\qquad\qquad\qquad\qquad (10.1.18)$$
$$X_{k+1} = X_k + a(t_k, X_k, \bar{u}(t_k, X_k))h + \sigma(t_k, X_k, \bar{u}(t_k, X_k))\Delta_k w$$
$$+ \frac{1}{2}\sigma(t_k, X_k, \bar{u}(t_k, X_k))\left(\frac{\partial\sigma}{\partial x}(t_k, X_k, \bar{u}(t_k, X_k))\right.$$
$$\left. + \frac{\partial\sigma}{\partial u}(t_k, X_k, \bar{u}(t_k, X_k))\frac{\Delta\bar{u}}{\Delta x}(t_k, X_k)\right) \cdot \left(\Delta_k^2 w - h\right),$$
$$k = 0, \ldots, N - 1;$$

the components Y and Z of the solution to (10.0.1)–(10.0.2) are approximated as

$$Y_k = \bar{u}(t_k, X_k), \quad Z_k = \sigma(t_k, X_k, Y_k)\frac{\Delta\bar{u}}{\Delta x}(t_k, X_k), \qquad (10.1.19)$$

where X_k is either from (10.1.17) or (10.1.18).

Let us note that the first-order scheme (10.1.18) becomes the Euler scheme in the case of additive noise in (10.0.1).

Theorem 10.1.5 (i) *The Euler scheme (10.1.17), (10.1.19) has the mean-square order of convergence 1/2, i.e.,*

$$E\left[(X(t_k) - X_k)^2 + (Y(t_k) - Y_k)^2 + (Z(t_k) - Z_k)^2\right]^{1/2} \le K\,(1 + x^2)^{1/2}h^{1/2},$$

$$(10.1.20)$$

where K does not depend on x, k, and h.
(ii) *The scheme (10.1.18), (10.1.19) has the first mean-square order of convergence for X and Y, i.e.,*

$$E\left[(X(t_k) - X_k)^2 + (Y(t_k) - Y_k)^2\right]^{1/2} \le K\,(1 + x^2)^{1/2}h;$$

$$(10.1.21)$$

and if

$$\left|\frac{\Delta\bar{u}}{\Delta x}(t_k, x) - \frac{\partial u}{\partial x}(t_k, x)\right| \le Ch,$$

$$(10.1.22)$$

then

$$E\left[(Z(t_k) - Z_k)^2\right]^{1/2} \le K\,(1 + x^2)^{1/2}h,$$

$$(10.1.23)$$

where K does not depend on x, k, and h.

***Proof* (i)** Let us prove that the Euler scheme satisfies

$$E\left[(X(t_k) - X_k)^2\right]^{1/2} \le K\,(1 + x^2)^{1/2}h^{1/2}.$$

$$(10.1.24)$$

Assume for a while that the solution $u(t, x)$ to (10.0.3)–(10.0.4) is known exactly. Then the coefficients in (10.0.1) are known functions of t and x and we can apply the standard mean-square Euler scheme:

$$\hat{X}_{k+1} = \hat{X}_k + a(t_k, \hat{X}_k, u(t_k, \hat{X}_k))h + \sigma(t_k, \hat{X}_k, u(t_k, \hat{X}_k))\Delta_k w,$$
$$k = 0, \ldots, N - 1,$$

$$(10.1.25)$$

which is of mean-square order $1/2$, i.e., $\hat{X}_k$ from (10.1.25) satisfies a relation like (10.1.24).

Now we compare X_k and $\hat{X}_k$. To this end, we exploit the fundamental convergence theorem (Theorem 1.1.1 of Chap. 1). It tells that if a one-step approximation $\bar{X}_{t,x}(t + h)$ of the solution $X_{t,x}(t + h)$ satisfies the conditions

$$|E(X_{t,x}(t + h) - \bar{X}_{t,x}(t + h))| \le K(1 + |x|^2)^{1/2}h^{p_1},$$

$$(10.1.26)$$

$$\left[E|X_{t,x}(t + h) - \bar{X}_{t,x}(t + h)|^2\right]^{1/2} \le K(1 + |x|^2)^{1/2}h^{p_2}$$

$$(10.1.27)$$

with $p_2 > 1/2$ and $p_1 \ge p_2 + 1/2$, then the corresponding mean-square method X_k has order of convergence $p_2 - 1/2$, i.e.,

$$E\left[(X(t_k) - X_k)^2\right]^{1/2} \le K\,(1 + x^2)^{1/2}h^{p_2 - 1/2}.$$

Introduce the one-step approximations corresponding to X_k and $\hat{X}_k$:

$$X(t + h) \approx \bar{X} = x + a(t, x, \bar{u}(t, x))h + \sigma(t, x, \bar{u}(t, x))\Delta w \qquad (10.1.28)$$

and

$$X(t + h) \approx \hat{X} = x + a(t, x, u(t, x))h + \sigma(t, x, u(t, x))\Delta w. \qquad (10.1.29)$$

It is known (see Sect. 1.1.5) that $\hat{X}$ from (10.1.29) satisfies (10.1.26) with $p_1 = 2$ and (10.1.27) with $p_2 = 1$. Due to Assumption 10.1.1 and the relation (8.2.42), we get

$$E(\hat{X} - \bar{X}) = a(t, x, u(t, x))h - a(t, x, \bar{u}(t, x))h = O(h^2)$$

and

$$E(\hat{X} - \bar{X})^2 = [a(t, x, u(t, x)) - a(t, x, \bar{u}(t, x))]^2 h^2$$
$$+ [\sigma(t, x, u(t, x)) - \sigma(t, x, \bar{u}(t, x))]^2 h = O(h^3),$$

whence it follows that $\bar{X}$ from (10.1.28) also satisfies (10.1.26) with $p_1 = 2$ and (10.1.27) with $p_2 = 1$. Then, applying the fundamental convergence theorem, we prove (10.1.24). Now the rest of (10.1.20) follows from (10.1.2), (10.1.24), (8.2.42), and (10.1.5).

(ii) To prove (10.1.21), we just repeat all the arguments as above. The only difference is that this time we compare the one-step approximation corresponding to (10.1.18) with the one corresponding to the standard first-order mean-square scheme (see Chap. 1) applied to (10.0.1) assuming again that $u(t, x)$ is known exactly. We note that the estimate (10.1.5) is enough to obtain (10.1.21) but it would imply the mean-square order $1/2$ for Z instead of (10.1.23). At the same time, it is clear that by (10.1.2), (10.1.21) together with (10.1.22) we obtain (10.1.23). □

Remark 10.1.6 Theorem 10.1.5 is also valid for other Euler-type methods (e.g., for the implicit Euler method) as well as for other first-order mean-square methods. Further, in the case of additive noise this theorem can be extended to constructive mean-square methods of order $3/2$ (see Sect. 1.6) using second-order methods from Sect. 8.6 for the semilinear parabolic problem (10.1.3)–(10.1.4).

Remark 10.1.7 Let us consider the weak Euler scheme

$$X_0 = x, \qquad (10.1.30)$$
$$X_{k+1} = X_k + a(t_k, X_k, \bar{u}(t_k, X_k))h + \sigma(t_k, X_k, \bar{u}(t_k, X_k))\xi_k,$$
$$k = 0, \ldots, N - 1,$$

where ξ_k, $k = 0, \ldots, N - 1$, are i.i.d. random variables with the distribution $P(\xi = \pm 1) = 1/2$ and $\bar{u}(t_k, x)$ is defined by Algorithm 8.2.4. It is possible to prove that for a sufficiently smooth function $F(x, y)$ satisfying some boundedness conditions the Euler method (10.1.30) and (10.1.19) is of weak order one, i.e.,

$$EF(X(t_k), Y(t_k)) - EF(X_k, Y_k) = O(h).$$

The proof is based on the main theorem on convergence of weak approximations, see Theorem 2.2.1 from Sect. 2.2.1.

Remark 10.1.8 It is not difficult to generalize both the numerical algorithm considered and Theorem 10.1.5 to the case $d > 1$ using the multi-dimensional version of the Euler method (10.1.17) and (10.1.19), a multi-dimensional algorithm analogous to Algorithm 8.2.4 (see Sect. 8.3.1) to solve (10.0.3)–(10.0.4), and (10.1.15) to approximate the derivatives.

10.2   FBSDEs with Random Terminal Time

In this section we deal with FBSDEs with random terminal time and their links to the Dirichlet problems for semi-linear parabolic (Sect. 10.2.1) and elliptic (Sect. 10.2.2) PDEs. We also consider approximations of solutions to these PDEs and of their spatial derivatives.

10.2.1   The Parabolic Case

Let G be a bounded domain in $\mathbf{R}^d$, $Q = [t_0, T) \times G$ be a cylinder in $\mathbf{R}^{d+1}$, $\Gamma = \overline{Q} \setminus Q$. The set Γ is a part of the boundary of the cylinder Q consisting of the upper base and the lateral surface. Let $\varphi(t, x)$ be a function defined on Γ.

Consider the FBSDE with random terminal time (see e.g. [347, 350, 465]):

$$dX = a(t, X, Y)dt + \sigma(t, X, Y)dw(t), \quad X(t_0) = x \in G, \tag{10.2.1}$$

$$dY = -g(t, X, Y)dt - f^\mathsf{T}(t, X, Y)Zdt + Z^\mathsf{T}dw(t), \quad Y(\tau) = \varphi(\tau, X(\tau)), \tag{10.2.2}$$

where $\tau = \tau_{t_0,x}$ is the first exit time of the trajectory $(t, X_{t_0,x}(t))$ from the domain Q, i.e., the point $(\tau, X(\tau))$ belongs to Γ. A solution to (10.2.1)–(10.2.2) is defined as an $\{\mathcal{F}_t\}_{t \geq 0}$-adapted vector $(X(t), Y(t), Z(t))$ together with the Markov moment τ, which satisfy (10.2.1)–(10.2.2). This solution is connected with the Dirichlet boundary value problem for the semilinear parabolic equation

$$\frac{\partial u}{\partial t} + \sum_{i=1}^{d} a^i(t, x, u)\frac{\partial u}{\partial x^i} + \frac{1}{2}\sum_{i,j=1}^{d} a^{ij}(t, x, u)\frac{\partial^2 u}{\partial x^i \partial x^j} \tag{10.2.3}$$

$$= -g(t, x, u) - \sum_{k=1}^{n} f^k(t, x, u)\sum_{i=1}^{d} \sigma^{ik}(t, x, u)\frac{\partial u}{\partial x^i}, \quad (t, x) \in Q,$$

$$u(t, x)|_\Gamma = \varphi(t, x). \tag{10.2.4}$$

Let $u(t, x)$ be the solution of (10.2.3)–(10.2.4), which is supposed to exist, to be unique, and to be sufficiently smooth. One can find many theoretical results on this topic in [222] (see also references therein and in Chap. 9). To be definite, we assume here the conditions like Assumption 10.1.1 together with sufficient smoothness of the boundary ∂G and of the function φ to be fulfilled.

Consider the following SDE in Q :

$$dX = a(t, X, u(t, X))dt + \sigma(t, X, u(t, X))dw(t), \quad X(t_0) = x, \tag{10.2.5}$$

with random terminal time τ which is defined as the first exit time of the trajectory $(t, X_{t_0,x}(t))$ of (10.2.5) from the domain Q. Introduce

$$Y(t) = u(t, X_{t_0,x}(t)), \quad t_0 \le t \le \tau, \tag{10.2.6}$$

$$Z^j(t) = \sum_{i=1}^{d} \sigma^{ij}(t, X_{t_0,x}(t), Y(t)) \frac{\partial u}{\partial x^i}(t, X_{t_0,x}(t)), \quad j = 1, \dots, n, \quad t_0 \le t \le \tau.$$

Clearly, the four-tuple $(X_{t_0,x}(t), Y(t), Z(t), \tau)$ is a solution of (10.2.1)–(10.2.2).

In sequel we restrict ourselves to the one-dimensional version of (10.2.1)–(10.2.2) $(d = 1, n = 1)$. Introducing

$$b(t, x, y) := a(t, x, y) + f(t, x, y)\sigma(t, x, y),$$

we get

$$\frac{\partial u}{\partial t} + b(t, x, u)\frac{\partial u}{\partial x} + \frac{1}{2}\sigma^2(t, x, u)\frac{\partial^2 u}{\partial x^2} + g(t, x, u) = 0, \quad (t, x) \in Q, \tag{10.2.7}$$

$$u(t, x)|_\Gamma = \varphi(t, x). \tag{10.2.8}$$

In this case Q is the partly open rectangle: $Q = [t_0, T) \times (\alpha, \beta)$, and Γ consists of the upper base $\{T\} \times [\alpha, \beta]$ and two vertical intervals: $[t_0, T) \times \{\alpha\}$ and $[t_0, T) \times \{\beta\}$.

In Chap. 9 a number of algorithms for solving the problem (10.2.7)–(10.2.8) is presented. Here we will use Algorithm 9.1.7 from Sect. 9.1.4. To construct Z due to (10.2.6), we need an approximation of $\partial u/\partial x$. To this end, we propose to use the formulas (cf. (10.1.5)):

$$\begin{aligned}\frac{\partial u}{\partial x}(t_k, x) &= \frac{\bar{u}(t_k, x + \gamma\sqrt{h}) - \bar{u}(t_k, x - \gamma\sqrt{h})}{2\gamma\sqrt{h}} + O(h^{1/2}) \\ &:= \frac{\Delta\bar{u}}{\Delta x}(t_k, x) + O(h^{1/2}) \\ &\text{if } x \pm \gamma\sqrt{h} \in [\alpha, \beta];\end{aligned} \tag{10.2.9}$$

$$\frac{\partial u}{\partial x}(t_k, x) = \frac{4\bar{u}(t_k, x + \gamma\sqrt{h}) - 3\bar{u}(t_k, x) - \bar{u}(t_k, x + 2\gamma\sqrt{h})}{2\gamma\sqrt{h}} + O(h^{1/2})$$
$$:= \frac{\Delta\bar{u}}{\Delta x}(t_k, x) + O(h^{1/2})$$
$$\text{if } x - \gamma\sqrt{h} < \alpha;$$

(10.2.10)

and

$$\frac{\partial u}{\partial x}(t_k, x) = \frac{\bar{u}(t_k, x - 2\gamma\sqrt{h}) - 4\bar{u}(t_k, x - \gamma\sqrt{h}) + 3\bar{u}(t_k, x)}{2\gamma\sqrt{h}} + O(h^{1/2})$$
$$:= \frac{\Delta\bar{u}}{\Delta x}(t_k, x) + O(h^{1/2})$$
$$\text{if } x + \gamma\sqrt{h} > \beta.$$

(10.2.11)

Most probably, the accuracy in (10.2.9)–(10.2.11) is $O(h)$ rather than $O(h^{1/2})$.

Let us note that if we apply the method of differentiation to the boundary value problem (10.2.7)–(10.2.8), we obtain Eq. (10.1.8) for $v = \partial u/\partial x$ in Q and the Neumann boundary condition. Namely, this boundary condition is of the form:

$$\text{on the upper base of } Q : \ v(T, x) = \frac{\partial\varphi}{\partial x}(T, x)$$

and, for example, on the vertical interval $[t_0, T] \times \{\alpha\}$:

$$b(t, \alpha, \varphi(t, \alpha))v(t, \alpha) + \frac{1}{2}\sigma^2(t, \alpha, \varphi(t, \alpha))\frac{\partial v}{\partial x}(t, \alpha)$$
$$= -\frac{\partial\varphi}{\partial t}(t, \alpha) - g(t, \alpha, \varphi(t, \alpha)).$$

Thus, in the case of FBSDEs with random terminal time the approach of [81] leads to a complicated system of boundary value problems.

10.2.2 The Elliptic Case

The random terminal time in FBSDE (10.2.1)–(10.2.2) is bounded by the time T. Now we consider FBSDEs with unbounded random terminal time. Let G be a bounded domain in $\mathbf{R}^d$ and $Q = [0, \infty) \times G$ be a cylinder in $\mathbf{R}^{d+1}$.

Consider the FBSDE with random terminal time (see e.g. [347, 350, 465]):

$$dX = a(X, Y)dt + \sigma(X, Y)dw(t), \quad X(0) = x \in G, \tag{10.2.12}$$
$$dY = -g(X, Y)dt - f^{\mathsf{T}}(X, Y)Zdt + Z^{\mathsf{T}}dw(t), \quad Y(\tau) = \varphi(X(\tau)), \tag{10.2.13}$$

where $\tau = \tau_x$ is the first exit time of the trajectory $X_x(t)$ from the domain G, i.e., the point $X(\tau)$ belongs to the boundary ∂G of G. A solution to (10.2.12)–(10.2.13) is defined as an $\{\mathcal{F}_t\}_{t\geq 0}$-adapted vector $(X(t), Y(t), Z(t))$ together with the Markov moment τ, which satisfy (10.2.12)–(10.2.13). This solution is connected with the Dirichlet boundary value problem for the semilinear elliptic equation

$$\sum_{i=1}^{d} a^i(x, u)\frac{\partial u}{\partial x^i} + \frac{1}{2}\sum_{i,j=1}^{d} a^{ij}(x, u)\frac{\partial^2 u}{\partial x^i \partial x^j} \qquad (10.2.14)$$

$$= -g(x, u) - \sum_{k=1}^{n} f^k(x, u)\sum_{i=1}^{d} \sigma^{ik}(x, u)\frac{\partial u}{\partial x^i}, \quad x \in G,$$

$$u(x)|_{\partial G} = \varphi(x). \qquad (10.2.15)$$

Let $u(x)$ be the solution of (10.2.14)–(10.2.15), which is assumed to exist, to be unique, and to be sufficiently smooth. Consider the following SDE in G :

$$dX = a(X, u(X))dt + \sigma(X, u(X))dw(t), \quad X(0) = x, \qquad (10.2.16)$$

with random terminal time τ which is defined as the first exit time of the trajectory $X_x(t)$ of (10.2.16) from the domain G. Introduce

$$Y(t) = u(X_x(t)), \quad 0 \le t \le \tau, \qquad (10.2.17)$$

$$Z^j(t) = \sum_{i=1}^{d} \sigma^{ij}(X_x(t), Y(t))\frac{\partial u}{\partial x^i}(X_x(t)), \quad j = 1, \ldots, n, \; 0 \le t \le \tau.$$

Clearly, the four-tuple $(X_x(t), Y(t), Z(t), \tau)$ is a solution of (10.2.12)–(10.2.13). Note that in the one-dimensional case the problem (10.2.14)–(10.2.15) is the boundary value problem just for second-order ordinary differential equation which numerical solution does not cause any problem. To solve (10.2.14)–(10.2.15) for $d > 1$, one can use finite-difference methods or apply a multi-dimensional layer method analogous to Algorithm 9.1.7 (see [309]) using ideas of the relaxation method.

10.3 Numerical Integration of FBSDEs with Random Terminal Time

Trajectories $(t, X(t))$ of the SDE (10.2.1) belong to the space-time bounded domain $\bar{Q}$ and a corresponding approximation (ϑ_k, X_k) should possess the same property. It is obvious that, e.g., the standard Euler scheme (10.1.17) does not satisfy this requirement and specific methods are needed. Such approximations were considered in Chap. 6. Here we will use algorithms for space-time diffusions in space-time bounded domains from Sects. 6.3–6.5. In the one-dimensional case they take a simpler form which is presented below. We note that here the notation partly differs from that used in Chap. 6.

Consider the one-dimensional SDE

$$dX = \chi_{\tau_{t,x}>s}b(s, X)ds + \chi_{\tau_{t,x}>s}\sigma(s, X)dw(s), \quad X(t) = X_{t,x}(t) = x, \qquad (10.3.1)$$

in a space-time bounded domain $Q = [t_0, T) \times (\alpha, \beta)$; the Markov moment $\tau_{t,x}$ is the first-passage time of the process $(s, X_{t,x}(s))$, $s \geq t$, to $\Gamma = \bar{Q} \backslash Q$.

Let $I_r := [-r, r]$, $r > 0$, $\Pi := [0, l) \times I_1$ for some $l > 0$, and $\Pi_h := [0, lh) \times I_{\sqrt{h}}$. Take a point $(s, y) \in Q$ and introduce another interval $I(s, y; h) := [x + hb(s, y) - \sigma(s, y)\sqrt{h}, x + hb(s, y) + \sigma(s, y)\sqrt{h}]$ and also the space-time rectangle $\Pi(s, y; h) = [s, s + lh) \times I(s, y; h)$. Let Γ_δ be an intersection of a δ-neighborhood of the set Γ with the domain Q. Below we take δ equal to $\lambda h^{(1-\varepsilon)/2}$ with $0 < \varepsilon \leq 1$ and $\lambda = 2 \max_{(s,y) \in \bar{Q}} |\sigma(s, y)|$. Now we construct a random walk over small space-time rectangles.

Algorithm 10.3.1 (*Random walk over small space-time rectangles*) *Choose a time step $h > 0$ and numbers $0 < \varepsilon \leq 1$ and $L > 0$.*

Step 0. $X_0 = x$, $\vartheta_0 = t$, $(t, x) \in Q$, $k = 0$.

Step 1.   *If $(\vartheta_k, X_k) \in \Gamma_{\lambda h^{(1-\varepsilon)/2}}$ or $k \geq L/h$, then Stop and*
  (i) put $\nu = k$, $(\vartheta_\nu, X_\nu) = (\vartheta_k, X_k)$;
  (ii) if $\vartheta_\nu \geq T - \lambda h^{(1-\varepsilon)/2}$ then $\bar{\tau}_{t,x} = T$ and $\xi_{t,x} = X_\nu \in (\alpha, \beta)$, otherwise $\bar{\tau}_{t,x} = \vartheta_\nu$ and $\xi_{t,x}$ is the end of the interval $[\alpha, \beta]$ nearest to X_ν.

Step 2.   *Put $k := k + 1$. Simulate the first exit point $(\theta_k, w(\vartheta_{k-1} + \theta_k) - w(\vartheta_{k-1}))$ of the process $(s - \vartheta_{k-1}, w(s) - w(\vartheta_{k-1}))$, $s > \vartheta_{k-1}$, from the rectangle Π_h. Put*

$$\vartheta_k = \vartheta_{k-1} + \theta_k, \tag{10.3.2}$$
$$X_k = X_{k-1} + b(\vartheta_{k-1}, X_{k-1})\theta_k + \sigma(\vartheta_{k-1}, X_{k-1})(w(\vartheta_k) - w(\vartheta_{k-1})). \tag{10.3.3}$$

Go to Step 1.

The sequence (ϑ_k, X_k) obtained by Algorithm 10.3.1 is a Markov chain stopping at the Markov moment ν in the neighborhood $\Gamma_{\lambda h^{(1-\varepsilon)/2}}$ of the boundary Γ. At each step $(\vartheta_k, X_k) \in \partial \Pi(\vartheta_{k-1}, X_{k-1}; h)$ and $\overline{\Pi}(\vartheta_{k-1}, X_{k-1}; h) \subset Q$, i.e., the chain belongs to the space-time bounded domain Q with probability one. The simulated points (ϑ_k, X_k) are close in the mean-square sense to $(\vartheta_k, X(\vartheta_k))$ and the point $(\bar{\tau}_{t,x}, \xi_{t,x})$ is close to $(\tau_{t,x}, X(\tau_{t,x}))$. It is proved in Chap. 6 (see Theorems 6.4.11, 6.4.14 and 6.4.15) that

$$\left(E\,|X(\vartheta_k) - X_k|^2\right)^{1/2} \leq K\,(\sqrt{h} + e^{-c_h L}), \quad k = 1, \ldots, \nu, \tag{10.3.4}$$
$$E|\tau_{t,x} - \bar{\tau}_{t,x}| \leq K\,(h^{(1-\varepsilon)/2} + e^{-c_h L}),$$
$$\left(E\,|X_{t,x}(\tau_{t,x}) - \xi_{t,x}|^2\right)^{1/2} \leq K\,(h^{(1-\varepsilon)/4} + e^{-c_h L/2}),$$

where the constant K is independent of h, k, t, x and c_h tends to a positive constant independent of L as $h \to 0$. We note that the accuracy of the algorithm depends on the choice of h, ε, and L. Clearly, we reach higher accuracy by decreasing h or/and ε and increasing L.

Algorithm 10.3.1 in its turn requires an algorithm of simulating the first exit point $(\theta, w(\theta))$ of the process $(s, w(s))$, $s > 0$, from the rectangle Π_h which was considered in Theorem 6.3.9 from Sect. 6.3.4 and which one-dimensional version is given in Algorithm 10.3.1, where $\mathcal{P}(t)$ and $\mathcal{Q}(\mu; t)$ are as in Theorem 6.3.9 and are available in Sect. 6.3.3.

Algorithm 10.3.2 (*simulating exit point of* $(t, W(t))$ *from space-time rectangle* Π) *Let* ι, ν, *and* γ *be independent random variables. Let* ι *be simulated by the law* $P(\iota = -1) = \mathcal{P}(l)$, $P(\iota = 1) = 1 - \mathcal{P}(l)$, ν *be simulated by the law* $P(\nu = \pm 1) = \dfrac{1}{2}$, *and* γ *be uniformly distributed on* $[0, 1]$.

Then a random point (τ, ξ), *distributed as the exit point* $(\tau, W(\tau))$, *is simulated as follows. If the simulated value of* ι *is equal to* -1, *then the point* (τ, ξ) *belongs to the lateral sides of* Π *and*

$$\tau = \mathcal{P}^{-1}(\gamma \mathcal{P}(l)), \ \xi = \nu,$$

otherwise, when $\iota = 1$, *the point* (τ, ξ) *belongs to the upper base of* Π *and*

$$\tau = l, \ \xi = \mathcal{Q}^{-1}(\gamma; l).$$

Also, by Corollary 6.3.10, we have the following statement.

Corollary 10.3.3 *Let* θ *be the first-passage time of the process* $(s, w(s))$, $s > 0$, *to the boundary* $\partial \Pi_h$. *Then the point*

$$(\theta, w) = (h\tau, \sqrt{h}\, \xi),$$

where (τ, ξ) *is simulated by Algorithm 10.3.2, has the same distribution as* $(\theta, w(\theta))$.

Algorithm 10.3.1 together with Algorithm 10.3.2 and Corollary 10.3.3 gives the constructive procedure for modelling the Markov chain (ϑ_k, X_k) which approximates trajectories $(t, X(t))$ of the SDE (10.3.1) in the space-time bounded domain Q.

Remark 10.3.4 In the one-dimensional case one can construct a random walk which terminates on the boundary Γ rather than in a boundary layer. Indeed, fix a sufficiently small $h > 0$ and define the function $\rho(t, x; h)$, $(t, x) \in Q$, in the following way. If $\Pi(t, x; h) \in Q$, set $\rho \equiv \rho(t, x; h) = h$. Otherwise, find $\rho(t, x; h) < h$ such that $\Pi(t, x; \rho)$ touches the boundary Γ, i.e., either $t + \rho = T$ or one of the ends of the interval $I(t, x; \rho)$ coincides with α or β. At each iteration of the algorithm we find $h_k = \rho(\vartheta_{k-1}, X_{k-1}; h)$ and simulate the first exit point $(\theta_k, w(\vartheta_{k-1} + \theta_k) - w(\vartheta_{k-1}))$ of the process $(s - \vartheta_{k-1}, w(s) - w(\vartheta_{k-1}))$, $s > \vartheta_{k-1}$, from the rectangle Π_{h_k}. Then we evaluate (ϑ_k, X_k) due to (10.3.2)–(10.3.3). We stop the algorithm when $(\vartheta_k, X_k) \in \Gamma$ and put $\nu = k$, $(\vartheta_\nu, X_\nu) = (\vartheta_k, X_k)$, $\bar{\tau}_{t,x} = \vartheta_\nu$, $\xi_{t,x} = X_\nu$. In comparison with Algorithm 10.3.1, the algorithm of this remark allows us to simulate a one-dimensional space-time Brownian motion $(s, w(s))$ exactly. We pay attention that this algorithm cannot be generalized even to the two-dimensional case while Algorithm 10.3.1 is available for any dimension (see Theorem 6.3.9 in Sect. 6.3.4).

Now we are in position to propose a numerical algorithm for solving the FBSDE with random terminal time (10.2.1)–(10.2.2). Let $\bar{u}(t_k, x)$ be defined by Algorithm 9.1.7 and $\dfrac{\Delta \bar{u}}{\Delta x}(t_k, x)$ by (10.2.9)–(10.2.11). Further, we define $\bar{u}(t, x)$ by linear interpolation:

$$\bar{u}(t, x) = \frac{t_k - t}{h} \bar{u}(t_{k-1}, x) + \frac{t - t_{k-1}}{h} \bar{u}(t_k, x), \quad t_{k-1} \le x \le t_k, \tag{10.3.5}$$

and analogously we define $\dfrac{\Delta \bar{u}}{\Delta x}(t, x)$. It is clear (see Theorem 9.1.3 and (10.2.9)–(10.2.11)) that

$$|u(t, x) - \bar{u}(t, x)| \le Kh \tag{10.3.6}$$

and

$$\left| \frac{\partial u}{\partial x}(t, x) - \frac{\Delta \bar{u}}{\Delta x}(t, x) \right| \le K\sqrt{h}. \tag{10.3.7}$$

We approximate $X(t)$ from (10.2.1)–(10.2.2) by Algorithm 10.3.1 in which (10.3.3) is substituted by

$$X_k = X_{k-1} + b(\vartheta_{k-1}, X_{k-1}, \bar{u}(\vartheta_{k-1}, X_{k-1}))\theta_k \tag{10.3.8}$$
$$+ \sigma(\vartheta_{k-1}, X_{k-1}, \bar{u}(\vartheta_{k-1}, X_{k-1}))(w(\vartheta_k) - w(\vartheta_{k-1})).$$

The algorithm also gives us the approximation $(\bar{\tau}_{t,x}, \xi_{t,x})$ for the first exit point $(\tau_{t,x}, X_{t,x}(\tau_{t,x}))$ of the trajectory $(s, X_{t,x}(s))$ from Q. Further, we compute the components Y and Z as

$$Y_k = \bar{u}(\vartheta_k, X_k), \quad Z_k = \sigma(\vartheta_k, X_k, Y_k) \frac{\Delta \bar{u}}{\Delta x}(t_k, X_k), \quad k = 1, \ldots, \nu, \tag{10.3.9}$$

$$\bar{Y}_\nu = \bar{u}(\bar{\tau}_{t,x}, \xi_{t,x}), \quad \bar{Z}_\nu = \sigma(\bar{\tau}_{t,x}, \xi_{t,x}, \bar{Y}_\nu) \frac{\Delta \bar{u}}{\Delta x}(\bar{\tau}_{t,x}, \xi_{t,x}).$$

It is possible to prove (cf. (10.3.4) and (10.3.6)–(10.3.7)) that

$$E\left[(X(\vartheta_k) - X_k)^2 + (Y(\vartheta_k) - Y_k)^2 + (Z(\vartheta_k) - Z_k)^2 \right]^{1/2}$$
$$\le K\left(\sqrt{h} + e^{-c_h L}\right),$$
$$k = 1, \ldots, \nu,$$
$$E\left[(X_{t,x}(\tau_{t,x}) - \xi_{t,x})^2 + (Y(\tau_{t,x}) - \bar{Y}_\nu)^2 + (Z(\tau_{t,x}) - \bar{Z}_\nu)^2 \right]^{1/2} \tag{10.3.10}$$
$$\le K\left(h^{(1-\varepsilon)/4} + e^{-c_h L/2}\right),$$
$$E|\tau_{t,x} - \bar{\tau}_{t,x}| \le K\left(h^{(1-\varepsilon)/2} + e^{-c_h L}\right).$$

Using Algorithm 10.3.1, we can also simulate the FBSDE with unbounded terminal time (10.2.12)–(10.2.13) analogously to the approximation of the FBSDE (10.2.1)–(10.2.2) considered in this section.

10.4 A Probabilistic Approach to Solving Quasilinear Parabolic Equations

Consider the more general FBSDEs than (10.0.1)–(10.0.2):

$$dX = a(t, X, Y, Z)dt + \sigma(t, X, Y)dw(t), \quad X(t_0) = x, \qquad (10.4.1)$$
$$dY = -e(t, X, Y, Z)dt + Z^\mathsf{T}dw(t), \quad Y(T) = \varphi(X(T)), \qquad (10.4.2)$$

where $w(t)$ is an n-dimensional $\{\mathcal{F}_t\}_{t\geq 0}$-adapted Wiener process; $X = X(t)$ and $a = a(t, x, y, z)$ are d-dimensional vectors; $\sigma = \sigma(t, x, y)$ is a $d \times n$-matrix; $Y = Y(t)$, $e = e(t, x, y, z)$, and $\varphi = \varphi(x)$ are scalars; $Z = Z(t)$ is an n-dimensional vector. It is known (see, e.g. [7, 75, 243, 347, 350, 465] and references therein) that there exists a unique $\{\mathcal{F}_t\}_{t\geq 0}$-adapted solution $(X(t), Y(t), Z(t))$ of the system (10.4.1)–(10.4.2) under some appropriate smoothness and boundedness conditions on its coefficients.

Due to the four-step scheme from [243], the solution of (10.4.1)–(10.4.2) is connected with the Cauchy problem for the quasilinear partial differential equation (PDE):

$$\frac{\partial u}{\partial t} + \sum_{i=1}^{d} a^i (t, x, u, \nabla u\sigma(t, x, u)) \frac{\partial u}{\partial x^i} + \frac{1}{2} \sum_{i,j=1}^{d} a^{ij}(t, x, u)\frac{\partial^2 u}{\partial x^i \partial x^j} \qquad (10.4.3)$$

$$= -e(t, x, u, \nabla u\sigma(t, x, u)), \quad t_0 \leq t < T, \ x \in \mathbf{R}^d, \qquad (10.4.4)$$
$$u(T, x) = \varphi(x),$$

where $a^{ij} := \sum_{k=1}^{d} \sigma^{ik}\sigma^{jk}$ and $\nabla u\sigma$ is the n-dimensional row-vector with components $\sum_{j=1}^{d}\sigma^{ji}\dfrac{\partial u}{\partial x^j}$. In turn the corresponding solution of the quasilinear PDE (10.4.3)–(10.4.4) has a probabilistic representation using the FBSDE (10.4.1)–(10.4.2), which is a generalization of the Feynman–Kac formula (see, e.g. [244, 347, 348, 350, 352, 357, 465]).

For clarity of the exposition, let consider (10.4.1)–(10.4.2) in the case $d = n = 1$:

$$dX = a(t, X, Y, Z)dt + \sigma(t, X, Y)dw(t), \quad X(t_0) = x, \tag{10.4.5}$$
$$dY = -g(t, X, Y, Z)dt - f(t, X, Y, Z)Zdt - c(t, X, Y, Z)Ydt + Zdw(t), \tag{10.4.6}$$

$$Y(T) = \varphi(X(T)).$$

It coincides with (10.4.1)–(10.4.2) having

$$e(t, x, y, z) = g(t, x, y, z) + f(t, x, y, z)z + c(t, x, y, z)y.$$

This form is preferred here due to the assumptions made below on the coefficients. The accompanying quasilinear PDE has the form (cf. (10.4.3)):

$$\frac{\partial u}{\partial t} + \frac{1}{2}\sigma^2(t, x, u)\frac{\partial^2 u}{\partial x^2} + b\left(t, x, u, \sigma(t, x, u)\frac{\partial u}{\partial x}\right)\frac{\partial u}{\partial x} \tag{10.4.7}$$

$$+c\left(t, x, u, \sigma(t, x, u)\frac{\partial u}{\partial x}\right)u = -g\left(t, x, u, \sigma(t, x, u)\frac{\partial u}{\partial x}\right), \quad t < T, \ x \in \mathbf{R},$$

$$u(T, x) = \varphi(x), \tag{10.4.8}$$

where
$$b := a + \sigma f.$$

The solution to this problem is assumed to exist, be unique, be sufficiently smooth, and satisfy some conditions on boundedness. For convenience, we shall assume throughout this section that the assumptions of [81] are fulfilled. We prefer to state them here in a less specific way.

Assumption 10.4.1 It is assumed that the coefficients a, σ, g, f, c and the function φ are sufficiently smooth and all these functions together with their derivatives up to some order are uniformly bounded with respect to their arguments. Besides, it is supposed that σ is bounded away from zero: $\sigma \geq \sigma_0$, where σ_0 is a positive constant.

Assumption 10.4.1 ensure the existence of a unique bounded solution $u(t, x)$ with bounded derivatives up to some order (see the corresponding theoretical results in [144, 222, 396, 437] and references therein). We emphasize that although we prove convergence results under Assumption 10.4.1, these assumptions are not necessary and the methods of this section can be used under broader conditions. For instance, this is demonstrated by our numerical tests in Sect. 10.5.5, where we simulate a quasilinear equation with power law nonlinearities.

Consider the system of SDEs

$$dX = \sigma(s, X, u(s, X))dw(s), \quad X(t_k) = x,$$
$$dQ = c\left(s, X, u(s, X), \sigma(s, X, u(s, X))\tfrac{\partial u}{\partial x}(s, X)\right)Qds$$
$$+\frac{1}{\sigma(s,X,u(s,X))}b\left(s, X, u(s, X), \sigma(s, X, u(s, X))\tfrac{\partial u}{\partial x}(s, X)\right)Qdw(s), \qquad (10.4.9)$$
$$Q(t_k) = 1,$$
$$dR = g\left(s, X, u(s, X), \sigma(s, X, u(s, X))\tfrac{\partial u}{\partial x}(s, X)\right)Qds, \quad R(t_k) = 0,$$

and introduce the function

$$U = u(s, X_{t_k,x}(s))Q_{t_k,x,1}(s) + R_{t_k,x,1,0}(s),$$

where $u(t, x)$ is a solution of (10.4.7)–(10.4.8). Due to the Ito formula, we get

$$d\left(u(s, X)Q + R\right) = \frac{\partial u}{\partial s}Qds + \frac{\partial u}{\partial x}\sigma Qdw + \frac{\sigma^2}{2}\frac{\partial^2 u}{\partial x^2}Qds + ucQds \quad (10.4.10)$$
$$+ b\frac{\partial u}{\partial x}Qds + gQds + \frac{u}{\sigma}bQdw,$$

where $X := X_{t_k,x}(s)$, $u := u(s, X_{t_k,x}(s))$, $\sigma := \sigma(s, X_{t_k,x}(s), u(s, X_{t_k,x}(s)))$, and b, c, g have $\left(s, X, u, \sigma(s, X, u)\tfrac{\partial u}{\partial x}\right)$ as their arguments. Since u satisfies (10.4.7), we obtain from (10.4.10):

$$d\left(u(s, X)Q + R\right) = \frac{\partial u}{\partial x}\sigma Qdw + \frac{u}{\sigma}bQdw,$$

whence

$$u(t_{k+1}, X_{t_k,x}(t_{k+1}))Q_{t_k,x,1}(t_{k+1}) + R_{t_k,x,1,0}(t_{k+1}) - u(t_k, x)$$
$$= \int_{t_k}^{t_{k+1}}\left(\frac{\partial u}{\partial x}\sigma Q + \frac{u}{\sigma}bQ\right)dw.$$

Consequently, we obtain the local probabilistic representation of the solution to (10.4.7)–(10.4.8):

$$u(t_k, x) = E\left[u(t_{k+1}, X_{t_k,x}(t_{k+1}))Q_{t_k,x,1}(t_{k+1}) + R_{t_k,x,1,0}(t_{k+1})\right], \qquad (10.4.11)$$

where X, Q, and R satisfy (10.4.9).

Consider the layer method given by the formulas:

$$\tilde{u}(t_N, x) = \varphi(x), \quad \tilde{u}(t_k, x) = \frac{1}{2}\tilde{u}\left(t_{k+1}, x - \sqrt{h}\tilde{\sigma}_k\right) \times \left[1 + h\tilde{c}_k - \sqrt{h}\frac{1}{\tilde{\sigma}_k}\tilde{b}_k\right]$$

$$\qquad\qquad\qquad\qquad\qquad\qquad\qquad\qquad\qquad\qquad\qquad\qquad (10.4.12)$$

$$+ \frac{1}{2}\tilde{u}\left(t_{k+1}, x + \sqrt{h}\tilde{\sigma}_k\right) \times \left[1 + h\tilde{c}_k + \sqrt{h}\frac{1}{\tilde{\sigma}_k}\tilde{b}_k\right] + h\tilde{g}_k, \quad k = N - 1, \ldots, 1, 0,$$

where

$$\tilde{\sigma}_k = \sigma(t_k, x, \tilde{u}(t_{k+1}, x)), \tag{10.4.13}$$

$$\tilde{b}_k = b\left(t_k, x, \tilde{u}(t_{k+1}, x), \tilde{\sigma}_k \frac{\Delta \tilde{u}}{\Delta x}(t_{k+1}, x)\right),$$

$$\frac{\Delta \tilde{u}}{\Delta x}(t_{k+1}, x) := \frac{\tilde{u}(t_{k+1}, x + \sqrt{h}\tilde{\sigma}_k) - \tilde{u}(t_{k+1}, x - \sqrt{h}\tilde{\sigma}_k)}{2\sqrt{h}\tilde{\sigma}_k},$$

and the notation $\tilde{c}_k$ and $\tilde{g}_k$ are analogous to $\tilde{b}_k$. The following theorem is proved in [320].

Theorem 10.4.2 *The method (10.4.12) has the first order of convergence, i.e.,*

$$|\tilde{u}(t_k, x) - u(t_k, x)| \le Kh, \tag{10.4.14}$$

where K does not depend on x, h, k.

Remark 10.4.3 The following layer method can also be proposed for (10.4.7)–(10.4.8):

$$\tilde{u}(t_N, x) = \varphi(x), \quad \tilde{u}(t_k, x) = \frac{1}{2}\tilde{u}(t_{k+1}, x + h\tilde{b}_k - \sqrt{h}\tilde{\sigma}_k) \times \left[1 + h\tilde{c}_k\right] \tag{10.4.15}$$

$$+ \frac{1}{2}\tilde{u}(t_{k+1}, x + h\tilde{b}_k + \sqrt{h}\tilde{\sigma}_k) \times \left[1 + h\tilde{c}_k\right] + h\tilde{g}_k, \quad k = N - 1, \dots, 1, 0,$$

where

$$\tilde{\sigma}_k = \sigma(t_k, x, \tilde{u}(t_{k+1}, x)), \tag{10.4.16}$$

$$\tilde{b}_k = b\left(t_k, x, \tilde{u}(t_{k+1}, x), \tilde{\sigma}_k \frac{\Delta \tilde{u}}{\Delta x}(t_{k+1}, x)\right),$$

$$\frac{\Delta \tilde{u}}{\Delta x}(t_{k+1}, x) := \frac{\tilde{u}(t_{k+1}, x + \gamma\sqrt{h}) - \tilde{u}(t_{k+1}, x - \gamma\sqrt{h})}{2\gamma\sqrt{h}},$$

the notation $\tilde{c}_k$ and $\tilde{g}_k$ are analogous to $\tilde{b}_k$, and γ is a positive number. Numerical tests (see Sect. 10.5) suggest that the method (10.4.15)–(10.4.16) is of first order as well.

10.4.1   Numerical Algorithms

To become a numerical algorithm, the method (10.4.12) needs a discretization in the variable x. Consider the equidistant space discretization:

$$x_j = x_0 + j\alpha h, \quad j = 0, \pm 1, \pm 2, \dots,$$

where x_0 is a point on $\mathbf{R}$ and α is a positive number, i.e., the step h_x of space discretization is equal to αh, where $h = h_t$ is the step of time discretization. Using, for instance, linear interpolation, we construct the following algorithm on the basis of the layer method (10.4.12):

$$\bar{u}(t_N, x) = \varphi(x), \quad \bar{u}(t_k, x_j) = \frac{1}{2}\bar{u}(t_{k+1}, x_j - \sqrt{h}\bar{\sigma}_{k,j}) \tag{10.4.17}$$

$$\times \left[1 + h\bar{c}_{k,j} - \sqrt{h}\frac{1}{\bar{\sigma}_{k,j}}\bar{b}_{k,j}\right] + \frac{1}{2}\bar{u}(t_{k+1}, x_j + \sqrt{h}\bar{\sigma}_{k,j})$$

$$\times \left[1 + h\bar{c}_{k,j} + \sqrt{h}\frac{1}{\bar{\sigma}_{k,j}}\bar{b}_{k,j}\right] + h\bar{g}_{k,j}, \ j = 0, \pm 1, \pm 2, \ldots,$$

where

$$\begin{aligned}
\bar{\sigma}_{k,j} &= \sigma(t_k, x_j, \bar{u}(t_{k+1}, x_j)), \tag{10.4.18}\\
\bar{b}_{k,j} &= b\left(t_k, x_j, \bar{u}(t_{k+1}, x_j), \bar{\sigma}_{k,j}\frac{\Delta\bar{u}}{\Delta x}(t_{k+1}, x_j)\right),\\
\frac{\Delta\bar{u}}{\Delta x}(t_{k+1}, x_j) &:= \frac{\bar{u}(t_{k+1}, x_j + \sqrt{h}\bar{\sigma}_{k,j}) - \bar{u}(t_{k+1}, x_j - \sqrt{h}\bar{\sigma}_{k,j})}{2\sqrt{h}\bar{\sigma}_{k,j}},
\end{aligned}$$

the notation $\bar{c}_{k,j}$ and $\bar{g}_{k,j}$ are analogous to $\bar{b}_{k,j}$, and

$$\bar{u}(t_k, x) = \frac{x_{j+1} - x}{\alpha h}\bar{u}(t_k, x_j) + \frac{x - x_j}{\alpha h}\bar{u}(t_k, x_{j+1}), \ x_j \le x \le x_{j+1}, \tag{10.4.19}$$

$$k = N - 1, \ldots, 1, 0.$$

The following theorem is proved in [320].

Theorem 10.4.4 *Algorithm (10.4.17)–(10.4.19) has the first order of convergence, i.e.,*

$$|\bar{u}(t_k, x) - u(t_k, x)| \le Kh, \tag{10.4.20}$$

where K does not depend on x, h, k.

Along with linear interpolation, a spline approximation can also be considered and the cubic interpolation with step $h_x = \alpha\sqrt{h}$ can be used to reduce the amount of nodes x_j (see Remark 8.2.6 on page 575). Clearly, both method and algorithm can be considered with variable time and space steps. Algorithms based on the layer method (10.4.15)–(10.4.16) can be written as well.

Corollary 10.4.5 *Define*

$$\bar{u}(t, x) = \frac{t_{k+1} - t}{h}\bar{u}(t_k, x) + \frac{t - t_k}{h}\bar{u}(t_{k+1}, x), \ t_k \le t \le t_{k+1}, \ x \in \mathbf{R}.$$

Then

$$|\bar{u}(t, x) - u(t, x)| \le Kh. \tag{10.4.21}$$

10.4.2 Solution of the FBSDE (10.4.5)–(10.4.6)

This subsection extends the result from Sect. 10.1.1 to the quasi-linear case.

Assume that the solution $u(t, x)$ of the Cauchy problem (10.4.7)– (10.4.8) is known. Consider the equation

$$dX = a(t, X, u(t, X), \sigma(t, X, u(t, X))\tfrac{\partial u}{\partial x}(t, X))dt + \sigma(t, X, u(t, X))dw(t),$$
$$X(t_0) = x. \tag{10.4.22}$$

Let $X(t) = X_{t_0, x}(t)$ be a solution of (10.4.22). Introduce

$$Y(t) = u(t, X_{t_0, x}(t)), \quad Z(t) = \sigma(t, X, u(t, X))\frac{\partial u}{\partial x}(t, X). \tag{10.4.23}$$

It turns out that $(X(t), Y(t), Z(t))$ defined by (10.4.22) and (10.4.23) is the solution of the FBSDE (10.4.5)–(10.4.6). Indeed, it is $\{\mathcal{F}_t\}_{t \ge 0}$-adapted and (10.4.5) is evidently satisfied. To verify (10.4.6), it suffices to apply Ito's formula to $Y(t) = u(t, X_{t_0, x}(t))$ and take into account that u is the solution of the problem (10.4.7)–(10.4.8).

10.4.3 Numerical Integration of FBSDEs

Let $\bar{u}(t_k, x)$ be defined by the algorithm (10.4.17)–(10.4.19). Introduce the notation

$$\frac{\Delta\bar{u}}{\Delta x}(t, x) := \frac{\bar{u}(t, x + \gamma\sqrt{h}) - \bar{u}(t, x - \gamma\sqrt{h})}{2\gamma\sqrt{h}} \quad \text{for some } \gamma > 0,$$

and also $\Delta_k w := w(t_k + h) - w(t_k)$.

For FBSDE (10.4.5)–(10.4.6) consider the Euler scheme:

the component X is approximated as

$$X_0 = x, \tag{10.4.24}$$
$$X_{k+1} = X_k + a\left(t_k, X_k, \bar{u}(t_k, X_k), \bar{\sigma}_k\frac{\Delta\bar{u}}{\Delta x}(t_k, x)\right)h + \bar{\sigma}_k \Delta_k w;$$

the components Y and Z are approximated as

$$Y_k = \bar{u}(t_k, X_k), \quad Z_k = \bar{\sigma}_k\frac{\Delta\bar{u}}{\Delta x}(t_k, X_k), \quad k = 0, \ldots, N - 1, \tag{10.4.25}$$

where $\bar{\sigma}_k := \sigma(t_k, X_k, \bar{u}(t_k, X_k))$.

The following theorem is proved in [320].

Theorem 10.4.6 *The Euler scheme (10.4.24), (10.4.25) has the mean-square order of convergence* $1/2$, *i.e.,*

$$\left(E\left[(X(t_k) - X_k)^2 + (Y(t_k) - Y_k)^2 + (Z(t_k) - Z_k)^2 \right] \right)^{1/2}$$
$$\leq K\, (1 + x^2)^{1/2} h^{1/2},$$

where K does not depend on x, k, *and* h.

10.5   Numerical Experiments

In this section we test the proposed algorithms on model problems. In Sects. 10.5.1 and 10.5.2, we describe and present results of numerical experiments for the case of semilinear PDEs. In Sects. 10.5.3–10.5.5, we describe and present results of numerical experiments for the case of quasilinear PDEs. The test problems of Sects. 10.5.1 and 10.5.4 are such that the solution of both PDEs and the related FBSDEs are known explicitly. In Sect. 10.5.5 we simulate a quasilinear equation with power law nonlinearities.

10.5.1   Description of the Test Problems: Semilinear Case

Consider the FBSDE

$$dX = \frac{X\left(1 + X^2\right)}{\left(2 + X^2\right)^3} dt + \frac{1 + X^2}{2 + X^2} \sqrt{\frac{1 + 2Y^2}{1 + Y^2 + \exp\left(-\frac{2X^2}{t+1}\right)}} \, dw(t),$$

$$X(0) = x, \tag{10.5.1}$$

$$dY = -g(t, X, Y)dt - f(t, X, Y)Zdt + Z\,dw(t),$$
$$Y(T) = \exp\left(-\frac{X^2(T)}{T+1}\right), \tag{10.5.2}$$

where

$$g(t, x, u) = \frac{1}{t+1} \exp\left(-\frac{x^2}{t+1}\right) \tag{10.5.3}$$

$$\times \left[\frac{4x^2\left(1+x^2\right)}{\left(2+x^2\right)^3} + \left(\frac{1+x^2}{2+x^2}\right)^2 \left(1 - \frac{2x^2}{t+1}\right) - \frac{x^2}{t+1}\right],$$

$$f(t, x, u) = \frac{x}{\left(2+x^2\right)^2} \sqrt{\frac{1 + u^2 + \exp\left(-\frac{2x^2}{t+1}\right)}{1 + 2u^2}}.$$

Note that Assumption 10.1.1 is satisfied.

The corresponding Cauchy problem (see (10.1.3)–(10.1.4)) has the form

$$\frac{\partial u}{\partial t} + \frac{1}{2}\left(\frac{1+x^2}{2+x^2}\right)^2 \frac{1 + 2u^2}{1 + u^2 + \exp\left(-\frac{2x^2}{t+1}\right)} \frac{\partial^2 u}{\partial x^2} + \frac{2x\left(1+x^2\right)}{\left(2+x^2\right)^3} \frac{\partial u}{\partial x}$$

$$= \frac{1}{t+1} \exp\left(-\frac{x^2}{t+1}\right)\left[\frac{x^2}{t+1} - \frac{4x^2\left(1+x^2\right)}{\left(2+x^2\right)^3} - \left(\frac{1+x^2}{2+x^2}\right)^2 \left(1 - \frac{2x^2}{t+1}\right)\right],$$

$$t < T, \ x \in \mathbf{R}, \tag{10.5.4}$$

$$u(T, x) = \exp\left(-\frac{x^2}{T+1}\right). \tag{10.5.5}$$

We use the problem (10.5.1)–(10.5.2) to test the numerical algorithms proposed in Sect. 10.1.3. To this end, we need to know the exact solution of this problem. First, it can easily be verified that the solution of the problem (10.5.4)–(10.5.5) is the function

$$u(t, x) = \exp\left(-\frac{x^2}{t+1}\right). \tag{10.5.6}$$

Now we find the solution of (10.5.1)–(10.5.2). Substituting

$$Y(t) = u(t, X(t)) = \exp\left(-\frac{X(t)^2}{t+1}\right)$$

in (10.5.1), we get

$$dX = \frac{X\left(1+X^2\right)}{\left(2+X^2\right)^3} dt + \frac{1+X^2}{2+X^2} dw(t), \quad X(0) = x, \tag{10.5.7}$$

which solution can be expressed by the formula

$$X(t) = \Lambda(x + \arctan x + w(t)), \tag{10.5.8}$$

where the function $\Lambda(z)$ is defined by the equation

$$\Lambda + \arctan \Lambda = z. \tag{10.5.9}$$

Indeed, $X(0) = \Lambda(x + \arctan x) = x$. Further, due to the Ito formula, we have

$$dX = \Lambda'(x + \arctan x + w(t)) \, dw + \frac{1}{2}\Lambda''(x + \arctan x + w(t)) \, dt$$

and by (10.5.9) we get

$$\Lambda' = \frac{1 + \Lambda^2}{2 + \Lambda^2}, \quad \Lambda'' = \frac{2\Lambda(1 + \Lambda^2)}{(2 + \Lambda^2)^3},$$

whence it follows that (10.5.8) satisfies (10.5.7).

Thus, the solution of (10.5.1)–(10.5.2) is

$$X(t) = \Lambda(x + \arctan x + w(t)), \quad Y(t) = \exp\left(-\frac{X(t)^2}{t+1}\right), \tag{10.5.10}$$

$$Z(t) = -\frac{2X(t)\left(1 + X^2(t)\right)}{(t+1)\left(2 + X^2(t)\right)} \exp\left(-\frac{X(t)^2}{t+1}\right),$$

where $\Lambda(z)$ is defined by (10.5.9).

Now consider the test problem for numerical algorithms for FBSDEs with random terminal time (cf. (10.5.1)–(10.5.2)):

$$dX = \frac{X\left(1 + X^2\right)}{\left(2 + X^2\right)^3} dt + \frac{1 + X^2}{2 + X^2}\sqrt{\frac{1 + 2Y^2}{1 + Y^2 + \exp\left(-\frac{2X^2}{t+1}\right)}} \, dw(t),$$

$$X(t_0) = x \in (0, \beta), \tag{10.5.11}$$

$$dY = -g(t, X, Y)dt - f(t, X, Y)Zdt + Z \, dw(t), \tag{10.5.12}$$

$$Y(\tau_x) = \exp\left(-\frac{X^2(\tau_{t_0,x})}{\tau_{t_0,x} + 1}\right),$$

where $g(t, x, y)$ and $f(t, x, y)$ are from (10.5.3) and $\tau_{t_0,x}$ is the first exit time of the trajectory $(t, X_{t_0,x}(t))$, $t > t_0 > -1$, from the space-time rectangle $[t_0, T) \times (0, \beta)$, i.e., either $\tau_{t_0,x} = T$ or $X(\tau_{t_0,x})$ is equal to 0 or β. The corresponding Dirichlet problem (see (10.2.3)–(10.2.4) and also (10.5.4)–(10.5.5)) has the form

$$\frac{\partial u}{\partial t} + \frac{1}{2}\left(\frac{1+x^2}{2+x^2}\right)^2 \frac{1+2u^2}{1+u^2+\exp\left(-\dfrac{2x^2}{t+1}\right)} \frac{\partial^2 u}{\partial x^2} + \frac{2x\left(1+x^2\right)}{\left(2+x^2\right)^3} \frac{\partial u}{\partial x}$$

$$= \frac{1}{t+1}\exp\left(-\frac{x^2}{t+1}\right)\left[\frac{x^2}{t+1} - \frac{4x^2\left(1+x^2\right)}{\left(2+x^2\right)^3} - \left(\frac{1+x^2}{2+x^2}\right)^2\left(1-\frac{2x^2}{t+1}\right)\right],$$

$$t < T, \ x \in (0, \beta),$$

$$(10.5.13)$$

$$u(t, 0) = 1, \quad u(t, \beta) = \exp\left(-\frac{\beta^2}{T+1}\right), \tag{10.5.14}$$

$$u(T, x) = \exp\left(-\frac{x^2}{T+1}\right). \tag{10.5.15}$$

Obviously, the solution of this problem is given by (10.5.6) again. The exact solution of (10.5.11)–(10.5.12) can be simulated using formulas (10.5.10).

10.5.2   Numerical Results

We simulate (10.5.1)–(10.5.2) using the Euler scheme (10.1.17) and (10.1.19), where the solution u of (10.5.4)–(10.5.5) is approximated by Algorithm 8.2.4. Of course, practical realization of such algorithms always requires a truncation of the infinite space domain using the knowledge of behavior of solutions at infinity. In this example we restrict simulation to the space interval $[-20, 20]$. To check that this truncation does not affect accuracy, we performed control simulation for the interval $[-30, 30]$.

Figure 10.5.1 presents comparison of the exact sample trajectories $X(t)$, $Y(t)$, $Z(t)$ found due to (10.5.10) and the approximate trajectories obtained by the Euler scheme (10.1.17) and (10.1.19). Table 10.5.1 gives errors in simulation of the test problem (10.5.1)– (10.5.2) by the Euler scheme (10.1.17) and (10.1.19). The "$\pm$" reflects the Monte Carlo error only, it does not reflect the error of the method. More precisely, the averages presented in the table are computed in the following way:

$$E\left(X(T) - X_N\right)^2 \doteq \frac{1}{M}\sum_{m=1}^{M}\left(X^{(m)}(T) - X_N^{(m)}\right)^2 \pm 2\sqrt{\frac{\bar{D}_M}{M}},$$

where

$$\bar{D}_M = \frac{1}{M}\sum_{m=1}^{M}\left(X^{(m)}(T) - X_N^{(m)}\right)^4 - \left[\frac{1}{M}\sum_{m=1}^{M}\left(X^{(m)}(T) - X_N^{(m)}\right)^2\right]^2,$$

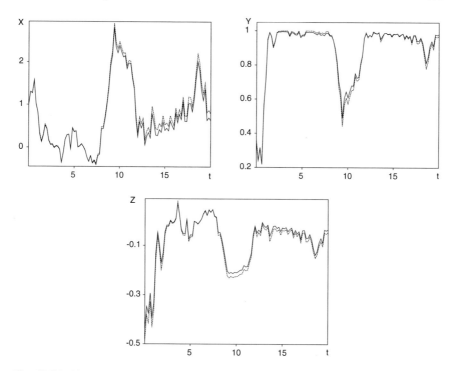

Fig. 10.5.1 Simulation of the FBSDE (10.5.1)–(10.5.2) using the layer method from Algorithm 8.8.2.4 and the Euler scheme (10.1.17) and (10.1.19) (*solid lines*) with $h = 0.2$, $\alpha = 1$, and $x = 1$. The corresponding exact trajectory (*dashed lines*) is found due to (10.5.10). The upper left figure gives the sample trajectories for $X(t)$, the upper right figure – for $Y(t)$, and the lower figure – for $Z(t)$

Table 10.5.1 Errors in simulation of the FBSDE (10.5.1)–(10.5.2) by the Euler scheme (10.1.17) and (10.1.19) with $\alpha = 1$ and various time steps h. The corresponding exact solution is found due to (10.5.10). Here $T = 20$ and $x = 1$. The expectations are computed by the Monte Carlo technique simulating $M = 1000$ independent realizations of $X(T)$ and X_N. The "$\pm$" reflects the Monte Carlo error only, it does not reflect the error of the method

h	$\max\limits_{k,j}\left| u(t_k, x_j) - \bar{u}(t_k, x_j) \right|$	$\left[E\left(X(T) - X_N\right)^2 \right]^{1/2}$	$\left[E\left(Y(T) - Y_N\right)^2 \right]^{1/2}$	$\left[E\left(Z(T) - Z_N\right)^2 \right]^{1/2}$
0.5	0.15×10^0	0.249 ± 0.014	0.0330 ± 0.0021	0.0127 ± 0.0008
0.2	0.58×10^{-1}	0.162 ± 0.010	0.0220 ± 0.0014	0.0080 ± 0.0005
0.05	0.14×10^{-1}	0.080 ± 0.005	0.0109 ± 0.0007	0.0041 ± 0.0003
0.02	0.53×10^{-2}	0.051 ± 0.003	0.0069 ± 0.0004	0.0024 ± 0.0002
0.005	0.12×10^{-2}	0.025 ± 0.002	0.0034 ± 0.0002	0.0012 ± 0.0001

$X^{(m)}(T)$ and $X_N^{(m)}$ are independent realizations of $X(T)$ and X_N, respectively. The numerical results are in a good agreement with the theoretical ones proved for the Euler method: convergence of X_N, Y_N, Z_N is of mean-square order $1/2$. We also see that $\bar{u}$ has the first order convergence.

The algorithms from Sect. 10.3 were tested on the FBSDE with random terminal time (10.5.11)–(10.5.12). The tests supported the obtained theoretical results.

10.5.3   Description of the Test Problem: Quasi-Linear Case

We select a problem which solution can be simulated exactly and then we use it to test the numerical algorithms proposed in Sect. 10.4. Consider a FBSDE of the form (cf. (10.5.1)–(10.5.2)):

$$dX = a(t, X, Z)dt + \sigma(X)\,dw(t), \quad X(0) = x, \tag{10.5.16}$$

$$dY = -g(t, X)dt + Z\,dw(t), \quad Y(T) = \varphi(X(T)), \tag{10.5.17}$$

where

$$a(t, x, z) = \frac{x\left(1+x^2\right)}{\left(2+x^2\right)^3} \cdot \frac{1+2z^2}{1+z^2+\dfrac{4x^2}{(t+1)^2}\left(\dfrac{1+x^2}{2+x^2}\right)^2 \exp\left(-\dfrac{2x^2}{t+1}\right)},$$

$$\sigma(x) = \frac{1+x^2}{2+x^2},$$

$$g(t, x) = \frac{1}{t+1} \exp\left(-\frac{x^2}{t+1}\right)$$
$$\times \left[\frac{2x^2\left(1+x^2\right)}{\left(2+x^2\right)^3} + \left(\frac{1+x^2}{2+x^2}\right)^2 \left(1 - \frac{2x^2}{t+1}\right) - \frac{x^2}{t+1}\right],$$

$$\varphi(x) = \exp\left(-\frac{x^2}{T+1}\right).$$

Note that Assumption 10.4.1 is satisfied.

The corresponding Cauchy problem (see (10.4.7)–(10.4.8)) has the form

$$\frac{\partial u}{\partial t} + \frac{1}{2}\left(\frac{1+x^2}{2+x^2}\right)^2 \frac{\partial^2 u}{\partial x^2}$$

$$+\frac{x\left(1+x^2\right)}{\left(2+x^2\right)^3} \frac{1+2\left(\frac{1+x^2}{2+x^2}\right)^2\left(\frac{\partial u}{\partial x}\right)^2}{1+\left(\frac{1+x^2}{2+x^2}\right)^2\left[\left(\frac{\partial u}{\partial x}\right)^2 + \frac{4x^2}{(t+1)^2}\exp\left(-\frac{2x^2}{t+1}\right)\right]}\frac{\partial u}{\partial x}$$

$$= \frac{1}{t+1}\exp\left(-\frac{x^2}{t+1}\right)\left[\frac{x^2}{t+1} - \frac{2x^2\left(1+x^2\right)}{\left(2+x^2\right)^3} - \left(\frac{1+x^2}{2+x^2}\right)^2\left(1-\frac{2x^2}{t+1}\right)\right],$$

$$t < T, \ x \in \mathbf{R},$$

(10.5.18)

$$u(T, x) = \exp\left(-\frac{x^2}{T+1}\right).$$ (10.5.19)

It can easily be verified that the solution of the problem (10.5.18)– (10.5.19) is the function

$$u(t, x) = \exp\left(-\frac{x^2}{t+1}\right).$$ (10.5.20)

Now we find the solution of (10.5.16)–(10.5.17). Substituting

$$Z(t) = \sigma(X(t)) \cdot \frac{\partial u}{\partial x}(t, X(t)) = \frac{-2X(t)}{t+1}\frac{1+X^2(t)}{2+X^2(t)}\exp\left(-\frac{X^2(t)}{t+1}\right)$$ (10.5.21)

in (10.5.16), we get

$$dX = \frac{X\left(1+X^2\right)}{\left(2+X^2\right)^3}dt + \frac{1+X^2}{2+X^2}dw(t), \ \ X(0) = x,$$ (10.5.22)

which solution can be expressed by the formula

$$X(t) = \Lambda(x + \arctan x + w(t)),$$ (10.5.23)

where the function $\Lambda(z)$ is defined by the equation (see Sect. 10.5.1):

$$\Lambda + \arctan \Lambda = z.$$ (10.5.24)

We also have (cf. (10.5.20)):

$$Y(t) = \exp\left(-\frac{X^2(t)}{t+1}\right).$$ (10.5.25)

10.5.4 Numerical Results

We test the algorithm based on the layer method (10.4.15)–(10.4.16) and on linear interpolation which applied to (10.5.18)–(10.5.19) takes the form

$$\bar{u}(t_N, x) = \varphi(x), \qquad (10.5.26)$$

$$\bar{u}(t_k, x_j) = \frac{1}{2}\bar{u}(t_{k+1}, x_j + h\bar{a}_{k,j} - \sqrt{h}\bar{\sigma}_{k,j}) + \frac{1}{2}\bar{u}(t_{k+1}, x_j + h\bar{a}_{k,j} + \sqrt{h}\bar{\sigma}_{k,j})$$

$$+ h\bar{g}_{k,j}, \quad j = 0, \pm 1, \pm 2, \ldots,$$

$$\bar{u}(t_k, x) = \frac{x_{j+1} - x}{\alpha h}\bar{u}(t_k, x_j) + \frac{x - x_j}{\alpha h}\bar{u}(t_k, x_{j+1}), \quad x_j \le x \le x_{j+1}, \qquad (10.5.27)$$

$$k = N - 1, \ldots, 1, 0,$$

where

$$\bar{\sigma}_{k,j} = \sigma(t_k, x_j), \quad \bar{a}_{k,j} = a\left(t_k, x_j, \bar{u}(t_{k+1}, x_j), \bar{\sigma}_{k,j}\frac{\Delta\bar{u}}{\Delta x}(t_{k+1}, x_j)\right),$$

$$\bar{g}_{k,j} = g(t_k, x_j), \qquad (10.5.28)$$

$$\frac{\Delta\bar{u}}{\Delta x}(t_{k+1}, x_j) := \frac{\bar{u}(t_{k+1}, x_j + \gamma\sqrt{h}) - \bar{u}(t_{k+1}, x_j - \gamma\sqrt{h})}{2\gamma\sqrt{h}}.$$

We recall that we do not have a proof of convergence for this algorithm. In our numerical experiments it demonstrates the first-order convergence. The numerical results obtained by (10.5.26)–(10.5.28) are similar to those obtained by the algorithm (10.4.17)–(10.4.19) which order is theoretically proved in Theorem 10.4.4.

To realize the proposed numerical algorithms (e.g., (10.5.26)–(10.5.28)) in practice, we truncate the infinite space domain using knowledge of solutions' behavior at infinity. In practice we check that the truncation does not significantly affect accuracy of the results by running control simulation with a different truncation and comparing the results. Further, since we use the PDE solver to simulate coupled FBSDEs, we need to ensure that the truncated domain is wide enough for accurate simulation of the forward equation. In the case of the test problem (10.5.18)–(10.5.19) the solution $u(t, x)$ tends to zero exponentially fast (see (10.5.20)). Then it does not significantly affect the solution if we modify the problem and put $u(t, x)$ equal to zero for x outside a sufficiently large interval. In our experiments we restrict simulation to the space interval $[-20, 20]$. To check that this truncation does not affect accuracy, we performed control simulation for the interval $[-30, 30]$. The averages in Table 10.5.2 are computed by the Monte Carlo technique (see Sect. 10.5.2).

We also simulate the FBSDE (10.5.16)–(10.5.17) using the Euler scheme (10.4.24)–(10.4.25), where the solution u of (10.5.18)–(10.5.19) is approximated by the algorithm (10.4.17)–(10.4.19). The corresponding results are very similar to those from Table 10.5.2 obtained by the algorithm (10.5.26)–(10.5.28). The numer-

Table 10.5.2 Errors in simulation of the FBSDE (10.5.16)–(10.5.17) by the Euler scheme (10.4.24)–(10.4.25) with $\bar{u}(t, x)$ found by the algorithm (10.5.26)–(10.5.28). The corresponding exact solution is found due to (10.5.6), (10.5.23), (10.5.25), (10.5.21). Here $T = 20$, $x = 1$, $\alpha = 1$, and $\gamma = 1$. The expectations are computed by the Monte Carlo technique simulating $M = 1000$ independent realizations of $X(T)$ and X_N. The "$\pm$" reflects the Monte Carlo error only, it does not reflect the error of the method

h	$\max_{k,j} |u(t_k, x_j) - \bar{u}(t_k, x_j)|$	$\left[E\left(X(T) - X_N\right)^2\right]^{1/2}$	$\left[E\left(Y(T) - Y_N\right)^2\right]^{1/2}$	$\left[E\left(Z(T) - Z_N\right)^2\right]^{1/2}$
0.5	0.19×10^0	0.247 ± 0.014	0.0328 ± 0.0021	0.0125 ± 0.0008
0.2	0.74×10^{-1}	0.161 ± 0.010	0.0219 ± 0.0014	0.0079 ± 0.0005
0.05	0.19×10^{-1}	0.079 ± 0.005	0.0108 ± 0.0007	0.0041 ± 0.0003
0.02	0.69×10^{-2}	0.051 ± 0.003	0.0068 ± 0.0004	0.0024 ± 0.0002
0.005	0.16×10^{-2}	0.025 ± 0.002	0.0034 ± 0.0002	0.0012 ± 0.0001

ical results are in a good agreement with the theoretical ones, i.e., the observed convergence of X_N, Y_N, Z_N is of mean-square order 1/2 and of $\bar{u}$ is of order one. We note that this is not only in agreement with Theorems 10.4.4 and 10.4.6 and Remark 10.4.3.

10.5.5 Quasilinear Equation with Power Law Nonlinearities

Consider the quasilinear PDE with power law nonlinearities (see, e.g., Sects. 8.4 and 9.2.3 and [396]):

$$\frac{\partial v}{\partial s} = \frac{\partial}{\partial x}\left(v^\alpha \frac{\partial v}{\partial x}\right) + v^{\alpha+1}, \quad s > 0, \ x \in \mathbf{R}, \tag{10.5.29}$$

where $\alpha > 0$ is a constant. As is well known, this equation has blow-up solutions. In particular, it has the following blow-up automodelling solution for time $0 \le s < T_0$:

$$v(s, x) = \begin{cases} (T_0 - s)^{-1/\alpha}\left(\frac{2(\alpha+1)}{\alpha(\alpha+2)} \cos^2 \frac{\pi x}{L}\right)^{1/\alpha}, & |x| < \frac{L}{2}, \\ 0, & |x| \ge \frac{L}{2}, \end{cases} \tag{10.5.30}$$

where

$$L = \frac{2\pi}{\alpha}(\alpha + 1)^{1/2}.$$

The temperature $v(s, x)$ grows infinitely when $s \to T_0$ while heat is localized in the interval $(-L/2, \ L/2)$.

The function $V = v^{\alpha+1}/(\alpha + 1)$ satisfies the corresponding semilinear PDE and in Chap. 8 this fact was exploited to test layer methods for semilinear PDEs (see

Fig. 10.5.2 Simulation of the quasilinear PDE with power law nonlinearities (10.5.31)–(10.5.32) by the algorithm based on the layer method (10.4.15)–(10.4.16). Here $\alpha = \gamma = 1$

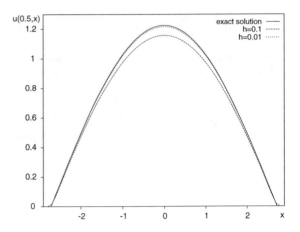

Sect. 8.4). The layer methods proposed in this chapter (Sect. 10.4) allow to simulate the quasilinear PDE (10.5.29) directly.

For definiteness, let us take $\alpha = 2$ and $T_0 = 1$. Consider the Cauchy problem with negative direction of time:

$$\frac{\partial u}{\partial t} + u^2 \frac{\partial^2 u}{\partial x^2} + 2u \left(\frac{\partial u}{\partial x} \right)^2 = -u^3, \ 0 < t < 1, \ x \in \mathbf{R}, \tag{10.5.31}$$

$$u(1, x) = \varphi(x) = \begin{cases} \frac{\sqrt{3}}{2} \cos \frac{\pi x}{L}, & |x| < \frac{L}{2}, \\ 0, & |x| \geq \frac{L}{2}. \end{cases} \tag{10.5.32}$$

As the related FBSDE, one can take

$$dX = 2YZdt + \sqrt{2}Ydw(t), \quad X(t_0) = x, \tag{10.5.33}$$

$$dY = -Y^3dt + Zdw(t), \quad Y(T) = \varphi(X(T)), \tag{10.5.34}$$

i.e., we put $a(t, x, y, z) = 2yz$, $\sigma(t, x, y) = \sqrt{2}y$, $g = 0$, $f = 0$, $c(t, x, y, z) = y^2$ in (10.4.5)–(10.4.6).

We simulate the problem (10.5.31)–(10.5.32) by the algorithm based on the layer method (10.4.15)–(10.4.16) and on linear interpolation. As in the case of the test problem from Sect. 6.2, we observe first order of convergence of this algorithm. Some numerical results are presented on Fig. 10.5.2. We note that the problem

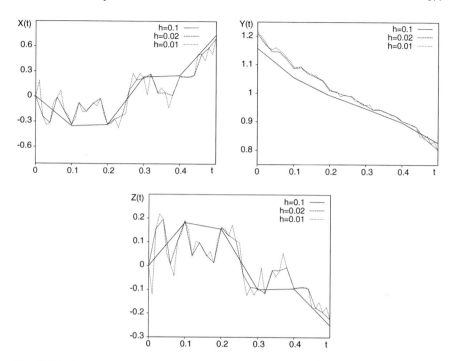

Fig. 10.5.3 Simulation of the FBSDE (10.5.33)–(10.5.34) using the layer method (10.4.15)–(10.4.16) and the Euler scheme (10.1.17)–(10.1.19). Here $\alpha = \gamma = 1$ and $x = 0$. The upper left figure gives the sample trajectories for $X(t)$, the upper right figure – for $Y(t)$, and the lower figure – for $Z(t)$

(10.5.31)–(10.5.32) does not satisfy the assumptions on boundedness of the coefficients (Assumption 10.4.1) used in our theoretical considerations but the algorithm gives quite accurate results. Figure 10.5.3 presents sample trajectories $X(t)$, $Y(t)$, $Z(t)$ obtained by the Euler scheme (10.4.24)–(10.4.25). Simulations are done along the same Wiener trajectory.

Chapter 11
Solving Parabolic SPDEs by Averaging Over Characteristics

The interest in linear stochastic PDEs (SPDEs) of parabolic type is mainly due to their well-known relation to the nonlinear filtering problem (see Sect. 11.7 and [219, 238, 345, 363, 390, 391]) although they have other applications as well (see, e.g., [391] and references therein). In this chapter we exploit the method of characteristics (the averaging over the characteristic formula – the generalization of the Feynman-Kac formula) and numerical integration of (ordinary) SDEs together with the Monte Carlo technique to construct numerical methods for linear SPDEs of parabolic type. The Monte Carlo methods are efficient for solving high-dimensional SPDEs when functionals or individual values of the solution have to be found.

The probabilistic approach based on the method of characteristics is also exploited here to construct layer methods for SPDEs (this idea in the case of deterministic PDEs was presented in Chap. 8). The layer methods are competitive with finite difference schemes (see [147] and references therein), they can be used when one needs to find the SPDE solution everywhere in the space-time domain $[t_0, T] \times \mathbf{R}^d$. It was shown in the deterministic case (see Chaps. 8–10) that layer methods have some advantages in comparison with conventional PDE solvers. We expect that they possess remarkable properties in the SPDE case as well. We construct layer methods both for linear and semilinear SPDEs. Semilinear SPDEs are used for modelling in physics, biology and chemistry (see [67, 209] and references therein). For other numerical approaches to linear and semilinear SPDEs, see, among many works e.g., [47, 141, 149, 150, 211, 212, 326, 328, 329, 467–470] and references therein.

This chapter is mainly based on [321, 323]. In Sect. 11.1 we recall probabilistic representations (the method of characteristics) for SPDEs of parabolic type from [215, 219, 345, 390]. In Sect. 11.2 we propose a number of approximation methods for the SPDEs and study their mean-square and almost sure (a.s.) convergence. The methods are based on approximating the characteristic SDEs, for which we exploit both the mean-square and weak-sense numerical integration. Section 11.3 deals with variance reduction methods that are important for any Monte Carlo procedures. We propose layer methods for linear SPDEs in Sect. 11.4 and for semilinear SPDEs in

© Springer Nature Switzerland AG 2021

G. N. Milstein and M. V. Tretyakov, *Stochastic Numerics for Mathematical Physics*, Scientific Computation, https://doi.org/10.1007/978-3-030-82040-4_11

Sect. 11.5. Some results of numerical experiments are presented in Sect. 11.6. In Sect. 11.7 the nonlinear filtering problem is introduced and linked to linear SPDEs.

11.1   Conditional Probabilistic Representations of Solutions to Linear SPDEs

Let $(\Omega, \mathcal{F}, P)$ be a probability space, $\mathcal{F}_t$, $T_0 \leq t \leq T$, be a nondecreasing family of σ-subalgebras of $\mathcal{F}$, $(w(t), \mathcal{F}_t) = ((w_1(t), \ldots, w_q(t))^\mathsf{T}, \mathcal{F}_t)$ be a q-dimensional standard Wiener process. Consider the Cauchy problem for the backward SPDE

$$-dv = [\mathcal{L}v + f(t,x)]\,dt + \sum_{r=1}^{q} [\mathcal{M}_r v + \gamma_r(t,x)] * dw_r(t), \quad (t,x) \in [T_0, T) \times \mathbf{R}^d,$$

$$\text{(11.1.1)}$$

$$v(T, x) = \varphi(x), \quad x \in \mathbf{R}^d, \tag{11.1.2}$$

where

$$\mathcal{L}v(t,x) := \frac{1}{2} \sum_{i,j=1}^{d} a^{ij}(t,x) \frac{\partial^2}{\partial x^i \partial x^j} v(t,x) + \sum_{i=1}^{d} b^i(t,x) \frac{\partial}{\partial x^i} v(t,x) \quad \text{(11.1.3)}$$
$$+ c(t,x)v(t,x),$$

$$\mathcal{M}_r v(t,x) := \sum_{i=1}^{d} \alpha_r^i(t,x) \frac{\partial}{\partial x^i} v(t,x) + \beta_r(t,x)v(t,x), \quad r = 1, \ldots, q.$$

The notation "$*dw_r$" in (11.1.1) means backward Ito integral. We recall [390] that to define this integral one introduces the "backward" Wiener processes

$$\tilde{w}_r(t) := w_r(T) - w_r(T - (t - T_0)), \quad r = 1, \ldots, q, \quad T_0 \leq t \leq T, \tag{11.1.4}$$

and a decreasing family of σ-subalgebras $\mathcal{F}_T^t$, $T_0 \leq t \leq T$, induced by the increments $w_r(T) - w_r(t')$, $r = 1, \ldots, q$, $t' \geq t$. A σ-algebra induced by $\tilde{w}_r(t')$, $t' \leq t$, coincides with $\mathcal{F}_T^{T-(t-T_0)}$. Then the backward Ito integral is defined as the Ito integral with respect to $\tilde{w}(s)$:

$$\int_t^{t'} \psi(t'') * dw_r(t'') := \int_{T-(t'-T_0)}^{T-(t-T_0)} \psi(T - (t'' - T_0))d\tilde{w}_r(t''), \quad T_0 \leq t \leq t' \leq T,$$

where $\psi(T - (t - T_0))$, $t \leq T$, is an $\mathcal{F}_T^{T-(t-T_0)}$-adapted square-integrable function. The process $v(t,x)$ is $\mathcal{F}_T^t$-adapted, it depends on $w(s) - w(t)$, $t \leq s \leq T$. We pay attention that the more precise notation for the solution of (11.1.1)–(11.1.2) is $v(t, x; \omega)$, $\omega \in \Omega$, but we use, as a rule, the shorter one $v(t, x)$. In what follows we will need the three assumptions.

Assumption 11.1.1 The coefficients in (11.1.1) are sufficiently smooth and their derivatives up to some order are bounded (in particular, it follows from here that the coefficients are globally Lipschitz). Besides, c, β_r, f, and γ_r are bounded.

Assumption 11.1.2 The function $\varphi(x)$ is sufficiently smooth and that $\varphi(x)$ and its derivatives up to some order belong to the class of functions satisfying an inequality of the form

$$|\varphi(x)| \leq K(1 + |x|^{\varkappa}), \quad x \in \mathbf{R}^d, \tag{11.1.5}$$

where K and $\varkappa$ are positive constants.

Assumption 11.1.3 The matrix $a = \{a^{ij}\}$ is symmetric and the matrix $a - \alpha\alpha^\mathsf{T}$ is nonnegative definite (the coercivity condition).

Assumptions 11.1.1–11.1.3 ensure the existence of a unique classical solution $v(t, x)$ of (11.1.1)–(11.1.2) which has derivatives in x^i, $i = 1, \ldots, d$, up to a sufficiently high order satisfying inequality of the form (11.1.5) a.s. with a positive random variable $K = K(\omega)$, which moments of a sufficiently high order are bounded (see [214, 390]). They are sufficient for all the statements in Sects. 11.2 and 11.3 (some additional assumptions are needed in Sect. 11.4). At the same time, they are not necessary and the methods presented in this chapter can be used under broader conditions.

Let a $d \times p$-matrix $\sigma(t, x)$ be obtained from the equation

$$\sigma(t, x)\sigma^\mathsf{T}(t, x) = a(t, x) - \alpha(t, x)\alpha^\mathsf{T}(t, x).$$

The solution of the problem (11.1.1)–(11.1.2) has the following conditional probabilistic representation (the conditional Feynman-Kac formula) [215, 219, 345, 390]:

$$v(t, x) = E^w \left[\varphi(X_{t,x}(T))Y_{t,x,1}(T) + Z_{t,x,1,0}(T)\right], \quad T_0 \leq t \leq T, \tag{11.1.6}$$

where $X_{t,x}(s)$, $Y_{t,x,y}(s)$, $Z_{t,x,y,z}(s)$, $t \leq s \leq T$, is the solution of the SDEs

$$dX = \left[b(s, X) - \sum_{r=1}^{q} \alpha_r(s, X)\beta_r(s, X)\right] ds + \sum_{r=1}^{p} \sigma_r(s, X)dW_r(s)$$
$$+ \sum_{r=1}^{q} \alpha_r(s, X)dw_r(s),$$
$$X(t) = x, \tag{11.1.7}$$
$$dY = c(s, X)Y ds + \sum_{r=1}^{q} \beta_r(s, X)Y dw_r(s), \quad Y(t) = y,$$
$$dZ = f(s, X)Y ds + \sum_{r=1}^{q} \gamma_r(s, X)Y dw_r(s), \quad Z(t) = z,$$

and $W(s) = (W_1(s), \ldots, W_p(s))^\mathsf{T}$ is a p-dimensional standard Wiener process independent of $w(s)$ and the expectation E^w in (11.1.6) is taken over the realizations of $W(s)$, $t \le s \le T$, for a fixed $w(s)$, $t \le s \le T$; in other words, $E^w(\cdot)$ means the conditional expectation $E(\cdot|w(s) - w(t), t \le s \le T)$.

Remark 11.1.4 Introduce the new time variable $s := T - (t - T_0)$ and introduce the functions $u(s, x) := v(T + T_0 - s, x)$, $\tilde{a}^{ij}(s, x) := a^{ij}(T + T_0 - s, x)$, $\tilde{b}^i(s, x) := b^i(T + T_0 - s, x)$, and analogously $\tilde{c}(s, x)$, $\tilde{f}(s, x)$, $\tilde{\alpha}_r^i(s, x)$, $\tilde{\beta}_r(s, x)$, $\tilde{\gamma}_r(s, x)$. Then one can show that $u(s, x)$ is the solution of the Cauchy problem for the forward SPDE (see [390, p. 173]):

$$du = \left[\frac{1}{2} \sum_{i,j=1}^{d} \tilde{a}^{ij}(s, x) \frac{\partial^2}{\partial x^i \partial x^j} u + \sum_{i=1}^{d} \tilde{b}^i(s, x) \frac{\partial}{\partial x^i} u + \tilde{f}(s, x) \right] ds \qquad (11.1.8)$$

$$+ \sum_{r=1}^{q} \left[\sum_{i=1}^{d} \tilde{\alpha}_r^i(s, x) \frac{\partial}{\partial x^i} u + \tilde{\beta}_r(s, x) u + \tilde{\gamma}_r(s, x) \right] d\tilde{w}_r(s), \quad (s, x) \in (T_0, T] \times \mathbf{R}^d,$$

$$u(T_0, x) = \varphi(x), \quad x \in \mathbf{R}^d, \qquad (11.1.9)$$

where $\tilde{w}_r(s)$ are defined in (11.1.4). The process $u(s, x)$ is $\mathcal{F}_T^{T+T_0-t}$-adapted, it depends on $\tilde{w}_r(s')$, $r = 1, \ldots, q$, $T_0 \le s' \le s$. Analogously, for a given forward SPDE one can write down the corresponding backward SPDE. Thus, the methods for backward SPDEs considered in this chapter can be used for solving forward SPDEs as well.

Remark 11.1.5 Consider an infinite dimensional Wiener process $B(t)$ taking values in some Hilbert space H and with covariance operator Q (which is assumed to be a nuclear operator). Let e_r be unit eigenvectors of Q with nonzero eigenvalues $\lambda_r = (Qe_r, e_r)$ then (see, e.g. [67, 214, 390]) $w_r(t) := (e_r, B(t))/\sqrt{\lambda_r}$ are independent standard Wiener processes and the infinite dimensional Wiener process is represented as

$$B(t) = \sum_{r=1}^{\infty} \sqrt{\lambda_r} w_r(t) e_r. \qquad (11.1.10)$$

Further, for any H-valued process $\psi(s)$ for which the integral $\int_0^t \psi(s) dB(s)$ is defined, one has [214, 390]

$$\int_0^t \psi(s) dB(s) = \sum_{r=1}^{\infty} \int_0^t \psi_r(s) dw_r(t) \qquad (11.1.11)$$

with $\psi_r(s) = (\psi(s), e_r)\sqrt{\lambda_r}$. The Wiener process $B(t)$ and the integral $\int_0^t \psi(s) dB(s)$ can be approximated by truncating the expansions in (11.1.10) and (11.1.11) (see, e.g. [67]). Then it is possible to view the SPDEs (11.1.1) and (11.1.8) as approximations of

SPDEs driven by the infinite dimensional Wiener process. Consequently, the methods of this chapter for (11.1.1) can, in principle, be used for approximating SPDEs with infinite dimensional Wiener process but we do not consider this aspect any further here.

11.2 Numerical Methods Based on the Conditional Feynman-Kac Formula (11.1.6)–(11.1.7)

Our purpose is to simulate approximately the random variable $v(t, x, \omega)$ under any fixed (t, x) using the probabilistic representation (11.1.6)–(11.1.7). In this section we construct mean-square approximations $\bar{v}(t, x, \omega)$ (see Sect. 11.2.1) and $\tilde{v}(t, x, \omega)$ (see Sect. 11.2.2) for $v(t, x, \omega)$, i.e., we construct $\bar{v}(t, x, \omega)$ (or $\tilde{v}(t, x, \omega)$) to be close to $v(t, x, \omega)$ in the mean-square sense. For realization of probabilistic representation (11.1.6), one can use numerical integration of SDEs (11.1.7) with respect to the Wiener process $W(s)$ both in the mean-square sense (Sect. 11.2.1; the approximation $\bar{v}(t, x)$) and in the weak sense (Sect. 11.2.2; the approximation $\tilde{v}(t, x)$). We prove a.s. convergence of the proposed methods.

11.2.1 Mean-Square Simulation of the Representation (11.1.6)–(11.1.7)

Consider the one-step approximation for (11.1.7):

$$\bar{X}_{t,x}(t + h) = x + h\left[b(t, x) - \sum_{r=1}^{q} \alpha_r(t, x)\beta_r(t, x) \right] + \sum_{r=1}^{q} \alpha_r(t, x)\Delta_t w_r$$
$$+ \sum_{r=1}^{p} \sigma_r(t, x)\Delta_t W_r,$$

$$\bar{Y}_{t,x,y}(t + h) = y + hc(t, x)y + \sum_{r=1}^{q} \beta_r(t, x)y\Delta_t w_r,$$

$$\bar{Z}_{t,x,y,z}(t + h) = z + hf(t, x)y + \sum_{r=1}^{q} \gamma_r(t, x)y\Delta_t w_r,$$

$$(11.2.1)$$

where $\Delta_t w_r := w_r(t + h) - w_r(t)$, $\Delta_t W_r := W_r(t + h) - W_r(t)$.

The approximation (11.2.1) generates the mean-square Euler method in the usual way. We introduce a partition of the time interval $[t, T]$, for simplicity the equidistant one: $t = t_0 < \cdots < t_N = T$, with step size $h = (T - t)/N$. An approximation of $(X_{t,x}(t_k), Y_{t,x,1}(t_k), Z_{t,x,1,0}(t_k))$ is denoted by $(\bar{X}_k, \bar{Y}_k, \bar{Z}_k)$.

The Euler scheme takes the form

$$\bar{X}_0 = x, \quad \bar{X}_{k+1} = \bar{X}_k + h \left[b(t_k, \bar{X}_k) - \sum_{r=1}^{q} \alpha_r(t_k, \bar{X}_k)\beta_r(t_k, \bar{X}_k) \right] \qquad (11.2.2)$$

$$+ \sum_{r=1}^{q} \alpha_r(t_k, \bar{X}_k)\Delta_k w_r + \sum_{r=1}^{p} \sigma_r(t_k, \bar{X}_k)\Delta_k W_r \,,$$

$$\bar{Y}_0 = 1, \quad \bar{Y}_{k+1} = \bar{Y}_k + hc(t_k, \bar{X}_k)\bar{Y}_k + \sum_{r=1}^{q} \beta_r(t_k, \bar{X}_k)\bar{Y}_k \Delta_k w_r \,,$$

$$\bar{Z}_0 = 0, \quad \bar{Z}_{k+1} = \bar{Z}_k + hf(t_k, \bar{X}_k)\bar{Y}_k + \sum_{r=1}^{q} \gamma_r(t_k, \bar{X}_k)\bar{Y}_k \Delta_k w_r,$$

$$k = 0, \dots, N-1,$$

where $\Delta_k w_r := w_r(t_{k+1}) - w_r(t_k)$, $\Delta_k W_r := W_r(t_{k+1}) - W_r(t_k)$.

Let

$$\bar{v}(t, x) := E^w \left[\varphi(\bar{X}_N)\bar{Y}_N + \bar{Z}_N \right], \qquad (11.2.3)$$

where $\bar{X}_N, \bar{Y}_N, \bar{Z}_N$ are from (11.2.2).

Below we use the same letter $C = C(x)$ or $C = C(x, \omega)$ for various expressions of the form $K(1 + |x|^{\varkappa})$ (see (11.1.5)) with K being a positive constant or $K = K(\omega)$ being a positive random variable, respectively.

Theorem 11.2.1 *Let Assumptions 11.1.1–11.1.3 hold. The method (11.2.2)–(11.2.3) satisfies the inequality for $p \geq 1$:*

$$\left(E \left| \bar{v}(t, x) - v(t, x) \right|^{2p} \right)^{1/2p} \leq C(x)h^{1/2}, \qquad (11.2.4)$$

where C does not depend on the discretization step h, i.e., in particular, (11.2.2)–(11.2.3) is of mean-square order $1/2$.

For almost every trajectory $w(\cdot)$ and any $\varepsilon > 0$ there exists $C(x, \omega) > 0$ such that

$$|\bar{v}(t, x) - v(t, x)| \leq C(x, \omega)h^{1/2-\varepsilon}, \qquad (11.2.5)$$

where C does not depend on the discretization step h, i.e., the method (11.2.2)–(11.2.3) converges with order $1/2 - \varepsilon$ a.s.

Proof Introduce $\mathbf{x} = (x, y, z)$, $\mathbf{X}(s) := (X(s), Y(s), Z(s))^{\mathsf{T}}$, $\bar{\mathbf{X}}_k := (\bar{X}_k, \bar{Y}_k, \bar{Z}_k)^{\mathsf{T}}$, and $\psi(\mathbf{x}) := \varphi(x)y + z$.

It is known that $\mathbf{X}(s)$ (see [123]) and $\bar{\mathbf{X}}_k$ (see Lemma 1.1.5) have bounded moments of any order and also that (see Sect. 1.1.6 and also [145, 448]) for $p \geq 1$

$$\left(E \left| \bar{\mathbf{X}}_k - \mathbf{X}(t_k) \right|^{2p} \right)^{1/2p} \leq C(x) h^{1/2}, \tag{11.2.6}$$

where C does not depend on h.

By the conditional version of Jensen's inequality, we get

$$
\begin{aligned}
E \left| \bar{v}(t, x) - v(t, x) \right|^{2p} &= E \left| E^w \psi(\bar{\mathbf{X}}_N) - E^w \psi(\mathbf{X}(t_N)) \right|^{2p} \\
&\leq E \left[E^w \left| \psi(\bar{\mathbf{X}}_N) - \psi(\mathbf{X}(t_N)) \right| \right]^{2p} \\
&\leq E \left[E^w \left| \psi(\bar{\mathbf{X}}_N) - \psi(\mathbf{X}(t_N)) \right|^{2p} \right] \\
&= E \left| \psi(\bar{\mathbf{X}}_N) - \psi(\mathbf{X}(t_N)) \right|^{2p}.
\end{aligned}
$$

Using the smoothness of ψ and the assumption that derivatives of ψ satisfy an inequality of the form (11.1.5), we obtain

$$\left| \psi(\bar{\mathbf{X}}_N) - \psi(\mathbf{X}(t_N)) \right| \leq K \left(1 + |\mathbf{X}(t_N)|^{\varkappa} + |\bar{\mathbf{X}}_N|^{\varkappa} \right) \left| \bar{\mathbf{X}}_N - \mathbf{X}(t_N) \right|,$$

where K is a positive constant. Then due to boundedness of the moments, the Cauchy-Bunyakovskii inequality, and (11.2.6), we get

$$
\begin{aligned}
E \left| \bar{v}(t, x) - v(t, x) \right|^{2p} &\leq K E (1 + |\mathbf{X}(t_N)|^{\varkappa} + |\bar{\mathbf{X}}_N|^{\varkappa})^{2p} \left| \bar{\mathbf{X}}_N - \mathbf{X}(t_N) \right|^{2p} \\
&\leq K \sqrt{E(1 + |\mathbf{X}(t_N)|^{\varkappa} + |\bar{\mathbf{X}}_N|^{\varkappa})^{4p}} \sqrt{E \left| \bar{\mathbf{X}}_N - \mathbf{X}(t_N) \right|^{4p}} \leq C(x) h^p,
\end{aligned}
$$

whence (11.2.4) follows.

Now denote $R := \left| \bar{v}(t, x) - v(t, x) \right|$. The Markov inequality together with (11.2.4) implies

$$P(R > h^{\gamma}) \leq \frac{E R^{2p}}{h^{2p\gamma}} \leq C(x) h^{p(1-2\gamma)}.$$

Then for any $\gamma = 1/2 - \varepsilon$ there is a sufficiently large $p \geq 1$ such that (recall that $h = T/N$)

$$\sum_{N=1}^{\infty} P\left(R > \frac{T^{\gamma}}{N^{\gamma}} \right) \leq C(x) T^{p(1-2\gamma)} \sum_{N=1}^{\infty} \frac{1}{N^{p(1-2\gamma)}} < \infty.$$

Hence, due to the Borel-Cantelli lemma, the random variable $\varsigma := \sup_{h>0} h^{-\gamma} R$ is a.s. finite which implies (11.2.5). $\qquad\square$

11.2.2   Weak Simulation of the Representation (11.1.6)–(11.1.7)

Now consider another Euler-type scheme for (11.1.7):

$$
\tilde{X}_0 = x, \quad \tilde{X}_{k+1} = \tilde{X}_k + h \left[b(t_k, \tilde{X}_k) - \sum_{r=1}^{q} \alpha_r(t_k, \tilde{X}_k)\beta_r(t_k, \tilde{X}_k) \right] \tag{11.2.7}
$$

$$
+ \sum_{r=1}^{q} \alpha_r(t_k, \tilde{X}_k)\Delta_k w_r + h^{1/2} \sum_{r=1}^{p} \sigma_r(t_k, \tilde{X}_k)\xi_{rk} \,,
$$

$$
\tilde{Y}_0 = 1, \quad \tilde{Y}_{k+1} = \tilde{Y}_k + hc(t_k, \tilde{X}_k)\tilde{Y}_k + \sum_{r=1}^{q} \beta_r(t_k, \tilde{X}_k)\tilde{Y}_k \Delta_k w_r \,,
$$

$$
\tilde{Z}_0 = 0, \quad \tilde{Z}_{k+1} = \tilde{Z}_k + hf(t_k, \tilde{X}_k)\tilde{Y}_k + \sum_{r=1}^{q} \gamma_r(t_k, \tilde{X}_k)\tilde{Y}_k \Delta_k w_r,
$$

$$
k = 0, \dots, N-1,
$$

where $\Delta_k w_r := w_r(t_{k+1}) - w_r(t_k)$ and ξ_{rk} are i.i.d. random variables with the moments

$$
E\xi = 0, \quad E\xi^2 = 1, \quad E\xi^3 = 0, \quad E\xi^m < \infty, \tag{11.2.8}
$$

for a sufficiently large integer $m \geq 4$. For instance, one can take ξ with the law

$$
P(\xi = \pm 1) = 1/2. \tag{11.2.9}
$$

We note that here a part of the SDE system (11.1.7) (which involves the Wiener process $W(t)$) is simulated weakly while the other part (which involves the Wiener process $w(t)$) is simulated in the mean-square sense.

Let

$$
\tilde{v}(t, x) := E^w \left[\varphi(\tilde{X}_N)\tilde{Y}_N + \tilde{Z}_N \right], \tag{11.2.10}
$$

where $\tilde{X}_N$, $\tilde{Y}_N$, $\tilde{Z}_N$ are from (11.2.7).

The same theorem as Theorem 11.2.1 is valid for $\tilde{v}(t, x)$ as well, however, with another proof (see it in [323]).

Theorem 11.2.2 *Let Assumptions 11.1.1–11.1.3 hold. The method (11.2.7), (11.2.10) satisfies the inequality for $p \geq 1$:*

$$
\left(E \left| \tilde{v}(t, x) - v(t, x) \right|^{2p} \right)^{1/2p} \leq C(x) h^{1/2}, \tag{11.2.11}
$$

where C does not depend on the discretization step h, i.e., in particular, (11.2.7), (11.2.10) is of mean-square order $1/2$.

For almost every trajectory $w(\cdot)$ and any $\varepsilon > 0$ there exists $C(x, \omega) > 0$ such that

$$|\tilde{v}(t, x) - v(t, x)| \leq C(x, \omega)h^{1/2-\varepsilon}, \tag{11.2.12}$$

where C does not depend on the discretization step h, i.e., the method (11.2.7), (11.2.10) converges with order $1/2 - \varepsilon$ a.s.

Remark 11.2.3 In some particular cases of the SPDE (11.1.1)–(11.1.2) the order of a.s. convergence of the Euler schemes (11.2.2), (11.2.3) and (11.2.7), (11.2.10) is higher. For instance, it is not difficult to modify the proofs of Theorems 11.2.1 and 11.2.2 to get that the Euler schemes have a.s. order of convergence $1 - \varepsilon$ if a and α_r are constant and $\beta_r = 0$ and $\gamma_r = 0$ (note that in this case (11.1.7) is a system with additive noise for which the standard Euler scheme for SDEs is of mean-square order 1, see Sect. 1.1.5). Further, in [323] it is proved that the considered schemes are also of a.s. order $1 - \varepsilon$ when $\alpha_r = 0$, $\gamma_r = 0$, $c = 0$, $f = 0$, $\beta_r \neq 0$.

Remark 11.2.4 Using the weak-sense numerical integration of SDEs in bounded domains (see Chap. 7), the proposed approach can be carried over to the boundary value problems for SPDEs, see e.g. the case of the Dirichlet problem in [413].

11.3 Other Probabilistic Representations, Monte Carlo Error, and Variance Reduction

Using the Monte Carlo technique, we approximate the solution of the backward SPDE (11.1.1)–(11.1.2) as (see (11.1.6)):

$$v(t, x) := v(t, x; \omega) = E^w \left[\varphi(X_{t,x}(T))Y_{t,x,1}(T) + Z_{t,x,1,0}(T) \right] \tag{11.3.1}$$
$$\approx \bar{v}(t, x) := E^w \left[\varphi(X_N)Y_N + Z_N \right]$$
$$\approx \hat{v}(t, x) := \frac{1}{M} \sum_{m=1}^{M} [\varphi(_m X_N)_m Y_N + {_m}Z_N],$$

where the first approximate equality involves an error due to replacing X, Y, Z by X_N, Y_N, Z_N (the error of numerical integration of (11.1.7) by (11.2.7) or (11.2.2)) and the error in the second approximate equality comes from the Monte Carlo technique; $_m X_N$, $_m Y_N$, $_m Z_N$, $m = 1, \ldots, M$, are independent realizations of X_N, Y_N, Z_N.

The error of numerical integration was analyzed in the previous section. Now let us consider the Monte Carlo error. The error of the Monte Carlo method in (11.3.1) is evaluated by

$$\bar{\rho} = c \frac{\left[Var^w \left\{ \varphi(\bar{X}_{t,x}(T))\bar{Y}_{t,x,1}(T) + \bar{Z}_{t,x,1,0}(T) \right\} \right]^{1/2}}{M^{1/2}},$$

where, e.g., the values $c = 1, 2, 3$ correspond to the probabilities 0.68, 0.95, 0.997, respectively, that $E^w [\varphi(X_N)Y_N + Z_N]$ falls in the corresponding credible interval. Since $Var^w \{\varphi(\bar{X}_{t,x}(T))\bar{Y}_{t,x,1}(T) + \bar{Z}_{t,x,1,0}(T)\}$ is close to the variance

$$V := Var^w \left[\varphi(X_{t,x}(T))Y_{t,x,1}(T) + Z_{t,x,1,0}(T)\right],$$

we can assume that the error of the Monte Carlo method is estimated by

$$\rho = c \, \frac{V^{1/2}}{M^{1/2}}.$$

If V is large then we have to simulate a very large number of trajectories to achieve a satisfactory accuracy. Fortunately, there exist other probabilistic representations for $v(t, x)$ which allow us to reduce the variance.

The solution of the problem (11.1.1)–(11.1.2) also has the following probabilistic representations:

$$v(t, x) = E^w \left[\varphi(X_{t,x}(T))Y_{t,x,1}(T) + Z_{t,x,1,0}(T)\right], \quad t_0 \leq t \leq T, \qquad (11.3.2)$$

where $X_{t,x}(s)$, $Y_{t,x,y}(s)$, $Z_{t,x,y,z}(s)$, $t \leq s \leq T$, is the solution of the SDEs

$$dX = \left[b(s, X) - \sum_{r=1}^{q} \alpha_r(s, X)\beta_r(s, X) - \sum_{r=1}^{p} \sigma_r(s, X)\mu_r(s, X)\right]ds \qquad (11.3.3)$$

$$+ \sum_{r=1}^{p} \sigma_r(s, X)dW_r(s) + \sum_{r=1}^{q} \alpha_r(s, X)dw_r(s), \quad X(t) = x,$$

$$dY = c(s, X)Y ds + \sum_{r=1}^{p} \mu_r(s, X)Y dW_r(s) + \sum_{r=1}^{q} \beta_r(s, X)Y dw_r(s), \quad Y(t) = y,$$

$$dZ = f(s, X)Y ds + \sum_{r=1}^{p} \lambda_r(s, X)Y dW_r(s) + \sum_{r=1}^{q} \gamma_r(s, X)Y dw_r(s), \quad Z(t) = z,$$

$\mu = (\mu_1, \ldots, \mu_p)^{\mathsf{T}}$ and $\lambda = (\lambda_1, \ldots, \lambda_p)^{\mathsf{T}}$ are arbitrary p-dimensional vector functions satisfying some regularity assumptions. When $\mu = 0$ and $\lambda = 0$, we have the usual representation (11.1.6)–(11.1.7). The representation for $\mu = 0$, $\lambda \neq 0$ follows from the equality

$$E^w \int_t^T \lambda^{\mathsf{T}}(s, X(s))Y(s)dW(s) = 0.$$

For $\mu \neq 0$, it can be proved by arguments similar to those in [219, pp. 308–309] making use of Theorem 4.4.5 [219, p. 152] for the forward flow. We should note that X, Y, Z in (11.3.2)–(11.3.3) differ from X, Y, Z in (11.1.6)–(11.1.7), however, this does not lead to any ambiguity. While $v(t, x)$ does not depend on the choice of μ and λ, the variance

$$V = Var^w \left[\varphi(X_{t,x}(T))Y_{t,x,1}(T) + Z_{t,x,1,0}(T) \right]$$

does. Then one may hope to find μ and λ such that the variance V is relatively low and thus the Monte Carlo error can be reduced. Theorem 11.3.1 is helpful in this respect. Its use for variance reduction gives the method of importance sampling if $\lambda = 0$ and the method of control variates if $\mu = 0$. The form of Theorem 11.3.1 is very similar to the corresponding result in the case of deterministic PDEs (see Sect. 2.3). However, its proof is much more complicated in the case of SPDEs than for PDEs, it can be found in [323].

Theorem 11.3.1 *Let Assumptions 11.1.1–11.1.3 hold. Let μ_r and λ_r be such that for any $x \in \mathbf{R}^d$ there exists a solution to the system (11.3.3) on the interval $[t, T]$. Then*

$$Var^w \left[\varphi(X_{t,x}(T))Y_{t,x,1}(T) + Z_{t,x,1,0}(T) \right] \tag{11.3.4}$$

$$= E^w \int_t^T Y_{t,x,1}^2 \sum_{r=1}^p \left[\sum_{i=1}^d \sigma_r^i \frac{\partial v}{\partial x^i} + \mu_r v + \lambda_r \right]^2 d\theta$$

provided that the expectation in (11.3.4) exists. In (11.3.4) all the functions in the integrand have $(\theta, X_{t,x}(\theta))$ as their argument.

Theorem 11.3.1 implies that if μ_r and λ_r are such that

$$\sum_{i=1}^d \sigma_r^i \frac{\partial v}{\partial x^i} + \mu_r v + \lambda_r = 0, \ r = 1, \ldots, p, \tag{11.3.5}$$

then the right-hand side of (11.3.4) is zero and, consequently, the variance is zero. We recall that $v(s, x)$ depends on $w(\theta) - w(s)$, $s \leq \theta \leq T$. However, we need to require that $\mu_r(s, x)$ does not depend on $w(\theta) - w(s)$, $s \leq \theta \leq T$, otherwise the coefficients in (11.3.3) depend on $w(\theta) - w(s)$, $s \leq \theta \leq T$, and we are facing the difficulty with interpreting (11.3.3) except the case when α_r is independent of x and $\beta_r = \gamma_r = 0$. In the mentioned case all the integrals appearing in the integral form of (11.3.3) can be given the usual meaning in the sense of Ito calculus. At the same time, dependence of the function $\lambda_r(s, x)$ on $w(\theta) - w(s)$, $s \leq \theta \leq T$, does not cause any trouble in interpreting the third equation in (11.3.3), and the identity (11.3.5) can, in principle, be achieved. We note that Theorem 11.3.1 and its proof are valid for $\lambda_r(s, x)$ depending on $w(\theta) - w(s)$, $s \leq \theta \leq T$. Further, when (11.3.3) is discretely simulated, one can simulate the process $w(s)$ in advance and fix it. Then it is possible to get approximate solutions of (11.3.3) for μ_r depending on a current position and on the fixed $w(\theta) - w(s)$, $s \leq \theta \leq T$. Thus we obtain a heuristic interpretation of

(11.3.3) for μ_r depending on future of w and, apparently, the result of Theorem 11.3.1 can be used for such μ_r as well.

Of course, μ_r and λ_r satisfying (11.3.5) cannot be constructed without knowing $v(s, x)$, $t \le s \le T$, $x \in \mathbf{R}^d$. Nevertheless, Theorem 11.3.1 claims a general possibility of variance reduction by a proper choice of the functions μ_r and λ_r. Theorem 11.3.1 can be used, for example, if we know a function $\tilde{v}(s, x)$ being close to $v(s, x)$ (see a practical approach to constructing such approximate functions in the case of deterministic PDEs in Sect. 2.4). Then we take any μ_r and λ_r satisfying

$$\sum_{i=1}^{d} \sigma_r^i \frac{\partial \tilde{v}}{\partial x^i} + \mu_r \tilde{v} + \lambda_r = 0 \qquad (11.3.6)$$

and we expect that the variance $Var^w \left[\varphi(X_{t,x}(T)) Y_{t,x,1}(T) + Z_{t,x,1,0}(T) \right]$ is although not zero but small.

11.4   Layer Methods for Linear SPDEs

In the previous sections we use the averaging over the characteristic formula to propose Monte Carlo methods for linear SPDEs. In this section we exploit probabilistic representations to construct some layer methods for linear SPDEs (see them in the case of deterministic PDEs in Chap. 8). Layer methods for semilinear SPDEs are considered in the next section. Layer methods are competitive with finite difference schemes [147] and can be used for relatively low dimensional SPDEs when one needs to find the solution $v(t, x)$ everywhere in $[t_0, T] \times \mathbf{R}^d$. The bottleneck for using the layer methods in the case of higher dimensional SPDEs is interpolation which should be used to restore the function $\bar{v}(t_k, x)$, $x \in \mathbf{R}^d$, by its values $\bar{v}(t_k, x_j)$ at a set of nodes x_j. The linear interpolation, which is successfully applied for low dimensional problems, is not effective for the higher dimensional ones. However, layer methods can be turned into numerical algorithms by exploiting other approximations of functions. Developments in the theory of multidimensional approximation (see, e.g. [136, 462] and references therein) give approximations which can work well in relatively high dimension.

The method (11.2.7), (11.2.10) can be turned into a layer method. Indeed, analogously to the probabilistic representation (11.1.6)–(11.1.7) one can write down the local probabilistic representation of the solution to (11.1.1)–(11.1.2):

$$v(t_k, x) = E^w \left[v(t_{k+1}, X_{t_k,x}(t_{k+1})) Y_{t_k,x,1}(t_{k+1}) + Z_{t_k,x,1,0}(t_{k+1}) \right], \qquad (11.4.1)$$

where $X_{t,x}(s)$, $Y_{t,x,y}(s)$, $Z_{t,x,y,z}(s)$, $t \le s \le T$, is the solution of the SDEs (11.1.7). Replacing $X_{t_k,x}(t_{k+1})$, $Y_{t_k,x,1}(t_{k+1})$, $Z_{t_k,x,1,0}(t_{k+1})$, by the one-step Euler approximation $\bar{X}_{t_k,x}(t_{k+1})$, $\bar{Y}_{t_k,x,1}(t_{k+1})$, $\bar{Z}_{t_k,x,1,0}(t_{k+1})$ as in (11.2.7), (11.2.9), we obtain the one-step approximation $\check{v}(t_k, x)$ for $v(t_k, x)$:

$$\check{v}(t_k, x) = E^w \left[v(t_{k+1}, \bar{X}_{t_k,x}(t_{k+1})) \bar{Y}_{t_k,x,1}(t_{k+1}) + \bar{Z}_{t_k,x,1,0}(t_{k+1}) \right]. \qquad (11.4.2)$$

Now for simplicity in writing let us consider the case $d = 1$ and $q = 1$. Thanks to the use of the discrete random variables (11.2.9), the expectation in (11.4.2) can be evaluated exactly:

$$\check{v}(t_k, x) = \{ \tfrac{1}{2} v(t_{k+1}, x + h [b(t_k, x) - \alpha(t_k, x)\beta(t_k, x)]$$
$$+ h^{1/2}\sigma(t_k, x) + \alpha(t_k, x)\Delta_k w)$$
$$+ \tfrac{1}{2} v \left(t_{k+1}, x + h [b(t_k, x) - \alpha(t_k, x)\beta(t_k, x)] - h^{1/2}\sigma(t_k, x) + \alpha(t_k, x)\Delta_k w \right) \}$$
$$\times [1 + hc(t_k, x) + \beta(t_k, x)\Delta_k w] + hf(t_k, x) + \gamma(t_k, x)\Delta_k w. \qquad (11.4.3)$$

Based on the one-step approximation (11.4.3), we obtain the layer method for (11.1.1)–(11.1.2) with $d = 1$, $q = 1$:

$$\tilde{v}(t_N, x) = \varphi(x), \qquad (11.4.4)$$

$$\tilde{v}(t_k, x) = \{ \tfrac{1}{2} \tilde{v}(t_{k+1}, x + h [b(t_k, x) - \alpha(t_k, x)\beta(t_k, x)]$$
$$+ h^{1/2}\sigma(t_k, x) + \alpha(t_k, x)\Delta_k w)$$
$$+ \tfrac{1}{2} \tilde{v} \left(t_{k+1}, x + h [b(t_k, x) - \alpha(t_k, x)\beta(t_k, x)] - h^{1/2}\sigma(t_k, x) + \alpha(t_k, x)\Delta_k w \right)$$
$$\times [1 + hc(t_k, x) + \beta(t_k, x)\Delta_k w] + hf(t_k, x) + \gamma(t_k, x)\Delta_k w, \ k = N - 1, \ldots, 0. \qquad (11.4.5)$$

It is not difficult to see that $\tilde{v}(t_k, x)$ from (11.4.4) coincides with the $\tilde{v}(t_k, x)$ due to (11.2.10). Then according to Theorem 11.2.2, the estimates (11.2.11) and (11.2.12) are valid for $\tilde{v}(t_k, x)$ from (11.4.4).

To realize (11.4.4) numerically, it is sufficient to calculate the functions $\tilde{v}(t_k, x)$ at some nodes x_j applying an interpolation at every layer. So, to become a numerical algorithm, the layer method (11.4.4) needs a discretization in the variable x. Consider the equidistant space discretization:

$$x_j = x_0 + jh_x, \ j = 0, \pm 1, \pm 2, \ldots, \qquad (11.4.6)$$

where x_0 is a point on $\mathbf{R}$ and $h_x > 0$ is a step of the space discretization. Using, for instance, linear interpolation, we construct the following algorithm on the basis of the layer method (11.4.4):

$$\bar{v}(t_N, x) = \varphi(x), \tag{11.4.7}$$

$$\bar{v}(t_k, x_j) = \tfrac{1}{2}\{\bar{v}(t_{k+1}, x_j + h\left[b(t_k, x_j) - \alpha(t_k, x_j)\beta(t_k, x_j)\right]$$
$$+ h^{1/2}\sigma(t_k, x_j) + \alpha(t_k, x_j)\Delta_k w)$$
$$+ \bar{v}(t_{k+1}, x_j + h\left[b(t_k, x_j) - \alpha(t_k, x_j)\beta(t_k, x_j)\right]$$
$$- h^{1/2}\sigma(t_k, x_j) + \alpha(t_k, x_j)\Delta_k w)\}$$
$$\times \left[1 + hc(t_k, x_j) + \beta(t_k, x_j)\Delta_k w\right] + hf(t_k, x_j) + \gamma(t_k, x_j)\Delta_k w,$$
$$j = 0, \pm 1, \pm 2, \dots,$$

$$\bar{v}(t_k, x) = \frac{x_{j+1} - x}{h_x}\bar{v}(t_k, x_j) + \frac{x - x_j}{h_x}\bar{v}(t_k, x_{j+1}), \quad x_j \le x \le x_{j+1}, \tag{11.4.8}$$

$$k = N - 1, \dots, 1, 0.$$

We use the same notation $\bar{v}$ for the two different functions: one is defined by the layer algorithm (11.4.7)–(11.4.8) and the other – in Sect. 11.2.1 for (11.2.3) but this should not cause any confusion since we do not use the approximation (11.2.3) in the current section.

If we need the solution of (11.1.1)–(11.1.2) for all points (t_k, x_i), we can use (11.4.7)–(11.4.8) to find $\bar{v}(t_k, x_j)$ layerwise. But if we need the solution at a particular point (t_k, x), the formula (11.2.10) is more convenient.

Remark 11.4.1 Thanks to the probabilistic approach, we do not need any stability criteria for the layer methods in comparison with finite-difference schemes, on which the Lax-Richtmyer equivalence theorem [375, 395] imposes a requirement on relation between the time step h and the space step h_x. E.g., for $b = 0$, $\alpha = 0$, $c = 0$, $\beta = 0$, and σ is a constant, the method (11.4.4) (we note that it does not need an interpolation in this case) coincides with the stable finite-difference scheme (see [147] and references therein)

$$\frac{\bar{v}(t_k, x_j) - \bar{v}(t_{k+1}, x_j)}{h} = \frac{\sigma^2}{2}\frac{\bar{v}(t_{k+1}, x_{j+1}) - 2\bar{v}(t_{k+1}, x_j) + \bar{v}(t_{k+1}, x_{j-1})}{h_x^2}$$
$$+ f(t_k, x_j) + \gamma(t_k, x_j)\frac{\Delta_k w}{h}$$

with $h_x = \sigma\sqrt{h}$. In layer methods the suitable choice of points at which $\bar{v}$ has to be evaluated is achieved automatically taking into account coefficient dependence on the space variables and a relationship between various terms (driving noise, diffusion, and advection) in an intrinsic manner.

The next convergence theorem was proved in [323].

Theorem 11.4.2 *Let Assumptions 11.1.1 and 11.1.3 hold and in addition also assume that the coefficients a, b, α_r are uniformly bounded. Let the function φ have* *bounded derivatives up to a sufficiently high order. Then the algorithm (11.4.7)–(11.4.8) with $h_x = \varkappa h^{3/4}$, $\varkappa > 0$, satisfies the inequality for $p \geq 1$:*

$$\left(E \, |\bar{v}(t_k, x) - v(t_k, x)|^{2p} \right)^{1/2p} \leq K h^{1/2}, \qquad (11.4.9)$$

where $K > 0$ does not depend on x, h, k, i.e., in particular, (11.4.7)–(11.4.8) is of mean-square order $1/2$.

And for almost every trajectory $w(\cdot)$ and any $\varepsilon > 0$, the algorithm (11.4.7)–(11.4.8) with $h_x = \varkappa h^{3/4}$ converges with order $1/2 - \varepsilon$, i.e.,

$$|v(t_k, x) - \bar{v}(t_k, x)| \leq C(\omega) h^{1/2-\varepsilon} \quad a.s., \qquad (11.4.10)$$

where C does not depend on x, h, k.

We note that for some particular SPDEs (see Remark 11.2.3) the algorithm (11.4.7)–(11.4.8) with $h_x = \varkappa h$ converges with the weak order $1 - \varepsilon$.

Remark 11.4.3 The generalization to the multidimensional SPDE, $d > 1$, $q \geq 1$, is straightforward (see it in the case of deterministic PDEs in Chap. 8). We also note that other types of interpolation can be exploited here to construct layer algorithms as in Chap. 8.

11.5 Layer Methods for Semilinear SPDEs

In this section, we generalize layer methods for linear SPDEs considered in Sect. 11.4 to the case of semilinear SPDEs (see layer methods for semilinear deterministic PDEs in Chap. 8 and for quasilinear deterministic PDEs in Sect. 10.4). As in the previous section, we restrict ourselves to the one-dimensional ($d = 1$) and one noise ($q = 1$) case for simplicity in writing only. There is no difficulty to generalize the methods of this section to the multi-dimensional multi-noise case (see how this is done in the case of deterministic PDEs in Chap. 8).

We consider the Cauchy problem for the backward semilinear SPDE

$$-dv = [Lv + f(t, x, v)] \, dt + \left[\alpha(t, x, v) \frac{\partial}{\partial x} v + \gamma(t, x, v) \right] * dw(t), \qquad (11.5.1)$$

$$(t, x) \in [T_0, T) \times \mathbf{R},$$

$$v(T, x) = \varphi(x), \quad x \in \mathbf{R}, \qquad (11.5.2)$$

where

$$Lv(t, x) := \frac{1}{2} a(t, x, v(t, x)) \frac{\partial^2}{\partial x^2} v(t, x) + b(t, x, v(t, x)) \frac{\partial}{\partial x} v(t, x), \qquad (11.5.3)$$

$$f(t, x, v) := f_0(t, x)v + f_1(t, x, v), \quad \gamma(t, x, v) := \gamma_0(t, x)v + \gamma_1(t, x, v).$$

Assumption 11.5.1 The coefficients $a(t, x, v)$, $b(t, x, v)$, $f_0(t, x)$, $f_1(t, x, v)$, $\alpha(t, x, v)$, $\gamma_0(t, x)$, and $\gamma_1(t, x, v)$ and the function $\varphi(x)$ are bounded and have bounded derivatives up to some order and that the coercivity condition is satisfied, i.e., $\sigma^2(t, x, v) := a(t, x, v) - \alpha^2(t, x, v) \geq 0$ for all t, x, v. Also, the problem (11.5.1)–(11.5.2) has the classical solution $v(t, x)$, which has derivatives in x up to a sufficiently high order; and the solution and its spatial derivatives have bounded moments up to some order.

We note that under Assumption 11.5.1 the existence of the classical solution is proved in the case $a(t, x, v) = a(t, x)$, $b(t, x, v) = b(t, x)$, $\alpha(t, x, v) = \alpha(t, x)$ in [214] and in the case $a(t, x, v) = a(t, x)$ in [345, 349].

Formally generalizing the one-step approximation (11.4.3), we get the one-step approximation for the semilinear problem (11.5.1)–(11.5.2):

$$\check{v}(t_k, x) = \frac{1}{2} v(t_{k+1}, x + hb(t_k, x, v(t_{k+1}, x)) + h^{1/2}\sigma(t_k, x, v(t_{k+1}, x)) \quad (11.5.4)$$

$$+\alpha(t_k, x, v(t_{k+1}, x))\Delta_k w)$$
$$+\frac{1}{2} v(t_{k+1}, x + hb(t_k, x, v(t_{k+1}, x)) - h^{1/2}\sigma(t_k, x, v(t_{k+1}, x))$$
$$+\alpha(t_k, x, v(t_{k+1}, x))\Delta_k w)$$
$$+hf(t_k, x, v(t_{k+1}, x)) + \gamma(t_k, x, v(t_{k+1}, x))\Delta_k w.$$

In the rest of this section we use the same letter K for various deterministic constants and $C = C(\omega)$ for various positive random variables. The next one-step error lemma was proved in [323].

Lemma 11.5.2 *Let Assumption 11.5.1 hold. The error $\rho = \rho(t_k) = \check{v}(t_k, x) - v(t_k, x)$ of the one-step approximation (11.5.4) is estimated as*

$$|E(\rho|\mathcal{F}_T^{t_{k+1}})| \leq C(\omega)h^2, \ |E(\Delta_k w\rho|\mathcal{F}_T^{t_{k+1}})| \leq C(\omega)h^2, \quad (11.5.5)$$

$$\left(E\,[\rho]^{2p}\right)^{1/2p} \leq Kh, \ p \geq 1, \quad (11.5.6)$$

where $C(\omega)$ and K do not depend on h, k, x and $EC^2 < \infty$. In the case of additive noise, $\alpha = 0$ and $\gamma(t, x, v) = \gamma(t, x)$ (i.e., $\gamma_0(t, x) = 0$, $\gamma_1(t, x, v) = \gamma(t, x)$), the error ρ satisfies (11.5.5) and

$$\left(E\,[\rho]^{2p}\right)^{1/2p} \leq Kh^{3/2}, \ p \geq 1. \quad (11.5.7)$$

Based on the one-step approximation (11.5.4), we obtain the layer method:

$$\bar{v}(t_N, x) = \varphi(x), \quad (11.5.8)$$

$$\bar{v}(t_k, x) = \tfrac{1}{2}\bar{v}(t_{k+1}, x + hb(t_k, x, \bar{v}(t_{k+1}, x)) + h^{1/2}\sigma(t_k, x, \bar{v}(t_{k+1}, x))$$
$$+\alpha(t_k, x, \bar{v}(t_{k+1}, x))\Delta_k w)$$
$$+\tfrac{1}{2}\bar{v}(t_{k+1}, x + hb(t_k, x, \bar{v}(t_{k+1}, x)) - h^{1/2}\sigma(t_k, x, \bar{v}(t_{k+1}, x))$$
$$+\alpha(t_k, x, \bar{v}(t_{k+1}, x))\Delta_k w)$$
$$+hf(t_k, x, \bar{v}(t_{k+1}, x)) + \gamma(t_k, x, \bar{v}(t_{k+1}, x))\Delta_k w, \quad k = N-1, \dots, 1, 0.$$

Let us prove the following technical lemma which will be exploited in proving convergence Theorem 11.5.4.

Lemma 11.5.3 *Let Assumption 11.5.1 hold. Then*

$$R(t_k, x) := \bar{v}(t_k, x) - v(t_k, x)$$
$$= \tfrac{1}{2}R(t_{k+1}, x + \bar{b}h + \bar{\sigma}h^{1/2} + \bar{\alpha}\Delta_k w) + \tfrac{1}{2}R(t_{k+1}, x + \bar{b}h - \bar{\sigma}h^{1/2} + \bar{\alpha}\Delta_k w)$$
$$+\Delta f h + \Delta\gamma\Delta_k w + \frac{\partial v}{\partial x}(t_{k+1}, x)(\Delta bh + \Delta\alpha\Delta_k w) + r + \rho,$$

$$(11.5.9)$$

where ρ is the one-step error as in Lemma 11.5.2 and

$$|r| \le C |R(t_{k+1}, x)|(h + \Delta_k^2 w).$$

Proof From (11.5.8) and due to the equality $\bar{v}(t_{k+1}, \cdot) = v(t_{k+1}, \cdot) + R(t_{k+1}, \cdot)$, we get

$$\bar{v}(t_k, x) = \frac{1}{2}v(t_{k+1}, x + \bar{b}h + \bar{\sigma}h^{1/2} + \bar{\alpha}\Delta_k w) \qquad (11.5.10)$$
$$+ \frac{1}{2}v(t_{k+1}, x + \bar{b}h - \bar{\sigma}h^{1/2} + \bar{\alpha}\Delta_k w)$$
$$+\frac{1}{2}R(t_{k+1}, x + \bar{b}h + \bar{\sigma}h^{1/2} + \bar{\alpha}\Delta_k w) + \frac{1}{2}R(t_{k+1}, x + \bar{b}h - \bar{\sigma}h^{1/2} + \bar{\alpha}\Delta_k w)$$
$$+ \bar{f}h + \bar{\gamma}\Delta_k w, \quad k = N-1, \dots, 1, 0,$$

where $\bar{b}, \bar{\sigma}, \bar{\alpha}, \bar{f}, \bar{\gamma}$ are the coefficients $b(t, x, v), \dots, \gamma(t, x, v)$ calculated at $t = t_k, \ x = x, \ v = \bar{v}(t_{k+1}, x) = v(t_{k+1}, x) + R(t_{k+1}, x)$.
    We have

$$\bar{b} = b(t_k, x, v(t_{k+1}, x) + R(t_{k+1}, x)) = b(t_k, x, v(t_{k+1}, x)) + \Delta b := b + \Delta b,$$
$$(11.5.11)$$

where Δb satisfies the inequality

$$|\Delta b| \le K |R(t_{k+1}, x)| \qquad (11.5.12)$$

with a deterministic constant K. Due to uniform boundedness of b, we also get

$$|\Delta b|^2 \le |\Delta b| \times K |R(t_{k+1}, x)| \le K |R(t_{k+1}, x)|. \qquad (11.5.13)$$

Analogously,

$$\bar{\sigma} = \sigma + \Delta\sigma, \ | \Delta\sigma | \leq K \ | \ R(t_{k+1}, x) \ |, \ | \Delta\sigma |^2 \leq K \ | \ R(t_{k+1}, x) \ |, \quad (11.5.14)$$
$$\bar{\alpha} = \alpha + \Delta\alpha, \ | \Delta\alpha | \leq K \ | \ R(t_{k+1}, x) \ |, \ | \Delta\alpha |^2 \leq K \ | \ R(t_{k+1}, x) \ |.$$

For Δf and $\Delta\gamma$, we have

$$| \Delta f | \leq K \ | \ R(t_{k+1}, x) \ |, \ | \Delta\gamma | \leq K \ | \ R(t_{k+1}, x) \ |. \quad (11.5.15)$$

It is not difficult to see that (11.5.12)–(11.5.14) imply the equalities

$$v(t_{k+1}, x + \bar{b}h \pm \bar{\sigma}h^{1/2} + \bar{\alpha}\Delta_k w) = v(t_{k+1}, x + bh \pm \sigma h^{1/2} + \alpha\Delta_k w)$$
$$+ \tfrac{\partial v}{\partial x}(t_{k+1}, x + bh \pm \sigma h^{1/2} + \alpha\Delta_k w)(\Delta bh \pm \Delta\sigma h^{1/2} + \Delta\alpha\Delta_k w) + r,$$
$$(11.5.16)$$

where

$$| r | \leq C \ | \ \Delta bh \pm \Delta\sigma h^{1/2} + \Delta\alpha\Delta_k w \ |^2 \leq C \ | \ R(t_{k+1}, x) \ | \ (h + \Delta_k^2 w), \quad (11.5.17)$$

C is a random variable with uniformly bounded second moment EC^2.
Further,

$$\frac{\partial v}{\partial x}(t_{k+1}, x + bh \pm \sigma h^{1/2} + \alpha\Delta_k w) = \frac{\partial v}{\partial x}(t_{k+1}, x) + Ch^{1/2} + C\Delta_k w,$$
$$(11.5.18)$$

$$EC^2 < \infty.$$

Substituting (11.5.18) in (11.5.16) and then the new (11.5.16) in (11.5.10) and taking into account (11.5.12)–(11.5.14) with (11.5.17), we obtain

$$\bar{v}(t_k, x) = \frac{1}{2}v(t_{k+1}, x + bh + \sigma h^{1/2} + \alpha\Delta_k w) \quad (11.5.19)$$

$$+ \tfrac{1}{2}v(t_{k+1}, x + bh - \sigma h^{1/2} + \alpha\Delta_k w)$$
$$+ fh + \gamma\Delta_k w + \Delta fh + \Delta\gamma\Delta_k w + \tfrac{\partial v}{\partial x}(t_{k+1}, x)(\Delta bh + \Delta\alpha\Delta_k w)$$
$$+ \tfrac{1}{2}R(t_{k+1}, x + \bar{b}h + \bar{\sigma}h^{1/2} + \bar{\alpha}\Delta_k w)$$
$$+ \tfrac{1}{2}R(t_{k+1}, x + \bar{b}h - \bar{\sigma}h^{1/2} + \bar{\alpha}\Delta_k w) + r$$
$$= \check{v}(t_k, x) + \tfrac{1}{2}R(t_{k+1}, x + \bar{b}h + \bar{\sigma}h^{1/2} + \bar{\alpha}\Delta_k w)$$
$$+ \tfrac{1}{2}R(t_{k+1}, x + \bar{b}h - \bar{\sigma}h^{1/2} + \bar{\alpha}\Delta_k w)$$
$$+ \Delta fh + \Delta\gamma\Delta_k w + \tfrac{\partial v}{\partial x}(t_{k+1}, x)(\Delta bh + \Delta\alpha\Delta_k w) + r,$$

where the new r also satisfies

$$| \, r \, | \leq C \mid R(t_{k+1}, x) \mid (h + \Delta_k^2 w).$$

At last, using Lemma 11.5.2, we obtain

$$R(t_k, x) = \bar{v}(t_k, x) - v(t_k, x) = \check{v}(t_k, x) - v(t_k, x) \tag{11.5.20}$$

$$+ \tfrac{1}{2} R(t_{k+1}, x + \bar{b}h + \bar{\sigma}h^{1/2} + \bar{\alpha}\Delta_k w) + \tfrac{1}{2} R(t_{k+1}, x + \bar{b}h - \bar{\sigma}h^{1/2} + \bar{\alpha}\Delta_k w)$$

$$+ \Delta f h + \Delta\gamma\Delta_k w + \tfrac{\partial v}{\partial x}(t_{k+1}, x)(\Delta bh + \Delta\alpha\Delta_k w) + r$$

$$= \tfrac{1}{2} R(t_{k+1}, x + \bar{b}h + \bar{\sigma}h^{1/2} + \bar{\alpha}\Delta_k w) + \tfrac{1}{2} R(t_{k+1}, x + \bar{b}h - \bar{\sigma}h^{1/2} + \bar{\alpha}\Delta_k w)$$

$$+ \Delta f h + \Delta\gamma\Delta_k w + \tfrac{\partial v}{\partial x}(t_{k+1}, x)(\Delta bh + \Delta\alpha\Delta_k w) + r + \rho,$$

whence (11.5.9) follows.   □

Due to Lemmas 11.5.2 and 11.5.3, one can expect that the method (11.5.8) converges with the mean-square order $1/2$. Here we restrict ourselves to proving the following theorem.

Theorem 11.5.4 *Let Assumptions 11.5.1 hold. Let $\alpha = 0$ and b and σ be independent of v. Then*

$$(E(\bar{v}(t_k, x) - v(t_k, x))^2)^{1/2} \leq Kh^{1/2}, \tag{11.5.21}$$

where K does not depend on h, k, x. If, in addition, $\gamma(t, x, v) = \gamma(t, x)$ (i.e., in the additive noise case) then

$$(E(\bar{v}(t_k, x) - v(t_k, x))^2)^{1/2} \leq Kh. \tag{11.5.22}$$

Proof In the case $\alpha = 0$ and b and σ independent of v the remainder r in (11.5.20) is equal to zero (see (11.5.17)). Then we get

$$R(t_k, x) = \frac{1}{2} R(t_{k+1}, x + bh + \sigma h^{1/2}) + \frac{1}{2} R(t_{k+1}, x + bh - \sigma h^{1/2}) \tag{11.5.23}$$
$$+ \Delta f h + \Delta\gamma\Delta_k w + \rho,$$

where ρ satisfies (11.5.6) in the case of general γ and (11.5.7) in the case $\gamma(t, x, v) = \gamma(t, x)$. Let us square (11.5.23) and then take conditional expectation with respect to σ-algebra $\mathcal{F}_T^{t_{k+1}}$. We get

$$E(R^2(t_k, x) \mid \mathcal{F}_T^{t_{k+1}}) = \frac{1}{4} R^2(t_{k+1}, x + bh + \sigma h^{1/2}) \tag{11.5.24}$$

$$+\tfrac{1}{4}R^2(t_{k+1}, x + bh - \sigma h^{1/2})$$

$$+\tfrac{1}{2}R(t_{k+1}, x + bh + \sigma h^{1/2})R(t_{k+1}, x + bh - \sigma h^{1/2}) + h^2(\Delta f)^2 + h(\Delta\gamma)^2$$

$$+E(\rho^2 \mid \mathcal{F}_T^{t_{k+1}})$$

$$+(R(t_{k+1}, x + bh + \sigma h^{1/2}) + R(t_{k+1}, x + bh - \sigma h^{1/2}))(\Delta fh + E(\rho \mid \mathcal{F}_T^{t_{k+1}}))$$

$$+2\Delta fh E(\rho \mid \mathcal{F}_T^{t_{k+1}}) + 2\Delta\gamma E(\Delta_k w\rho \mid \mathcal{F}_T^{t_{k+1}}).$$

Using Lemma 11.5.2 and the inequalities (11.5.15), one can estimate the terms in (11.5.24). For instance,

$$\mid \Delta\gamma E(\Delta_k w\rho \mid \mathcal{F}_T^{t_{k+1}}) \mid \le K \mid R(t_{k+1}, x) \mid C(\omega)h^2$$

$$\le \frac{1}{2}K^2 R^2(t_{k+1}, x)h + \frac{1}{2}C^2(\omega)h^3.$$

Introduce the notation

$$m(t_k) := \sup_x E R^2(t_k, x).$$

After taking expectations in (11.5.24), we easily obtain

$$E R^2(t_k, x) \le (1 + Kh)m(t_{k+1}) + E\rho^2(t_{k+1}) + Kh^3,$$

where the constant K does not depend on h, k, x. Therefore

$$m(t_k) \le (1 + Kh)m(t_{k+1}) + E\rho^2(t_{k+1}) + Kh^3,$$

whence (11.5.21) follows because of $E\rho^2 \le Kh^2$ in the multiplicative noise case (i.e., when γ depends on v) and (11.5.22) follows because of $E\rho^2 \le Kh^3$ in the additive noise case.   □

Introduce the equidistant space discretization as in (11.4.6). Using linear interpolation, we construct the following algorithm on the basis of the layer method (11.5.8):

$$\bar{v}(t_N, x) = \varphi(x), \tag{11.5.25}$$

$$\bar{v}(t_k, x_j) = \tfrac{1}{2}\bar{v}(t_{k+1}, x_j + hb(t_k, x_j, \bar{v}(t_{k+1}, x_j))$$

$$+h^{1/2}\sigma(t_k, x_j, \bar{v}(t_{k+1}, x_j)) + \alpha(t_k, x_j, \bar{v}(t_{k+1}, x_j))\Delta_k w)$$

$$+\tfrac{1}{2}\bar{v}(t_{k+1}, x_j + hb(t_k, x_j, \bar{v}(t_{k+1}, x_j))$$

$$-h^{1/2}\sigma(t_k, x_j, \bar{v}(t_{k+1}, x_j)) + \alpha(t_k, x_j, \bar{v}(t_{k+1}, x_j))\Delta_k w)$$

$$+hf(t_k, x_j, \bar{v}(t_{k+1}, x_j)) + \gamma(t_k, x_j, \bar{v}(t_{k+1}, x_j))\Delta_k w, \quad j = 0, \pm 1, \pm 2, \dots,$$

$$\bar{v}(t_k, x) = \frac{x_{j+1} - x}{h_x}\bar{v}(t_k, x_j) + \frac{x - x_j}{h_x}\bar{v}(t_k, x_{j+1}), \quad x_j \le x \le x_{j+1}, \tag{11.5.26}$$

$$k = N - 1, \ldots, 1, 0.$$

The following convergence theorem is proved for this algorithm using ideas from the above proof of Theorem 11.5.4 and also from the proof of an analogous theorem for an algorithm in the deterministic case (see Chap. 8).

Theorem 11.5.5 *Let Assumptions 11.5.1 hold. Let $\alpha = 0$ and b and σ be independent of v. Then the error of the one-step approximation (11.5.25)–(11.5.26) with $h_x = \varkappa h^{3/4}$, $\varkappa > 0$, is estimated as*

$$(E(\bar{v}(t_k, x) - v(t_k, x))^2)^{1/2} \le K h^{1/2}, \tag{11.5.27}$$

where K does not depend on h, k, x. If, in addition, $\gamma(t, x, v) = \gamma(t, x)$ (i.e., in the additive noise case) and $h_x = \varkappa h$, $\varkappa > 0$, then

$$(E(\bar{v}(t_k, x) - v(t_k, x))^2)^{1/2} \le K h. \tag{11.5.28}$$

We note that Remark 11.4.3 is valid here.

11.6 Numerical Experiments

11.6.1 Model Problem: Ornstein-Uhlenbeck Equation

We consider the problem

$$-dv = \left[\frac{a^2}{2}\frac{\partial^2 v}{\partial x^2} + bx\frac{\partial v}{\partial x}\right]dt + \alpha\frac{\partial v}{\partial x} * dw(t), \quad (t, x) \in [T_0, T) \times \mathbf{R}, \tag{11.6.1}$$

$$v(T, x) = \varphi(x), \quad x \in \mathbf{R}, \tag{11.6.2}$$

where $w(t)$ is a standard scalar Wiener process, a, b, α are constants and

$$\sigma^2 = a^2 - \alpha^2 \ge 0. \tag{11.6.3}$$

The solution of the problem is given by (see (11.1.6)–(11.1.7)):

$$v(t, x) = E^w \varphi(X_{t,x}(T)), \tag{11.6.4}$$

$$dX = bX ds + \sigma dW(s) + \alpha dw(s). \tag{11.6.5}$$

We have

$$X_{t,x}(T) = e^{b(T-t)}x + \int_t^T e^{b(T-\vartheta)}[\sigma dW(\vartheta) + \alpha dw(\vartheta)]. \tag{11.6.6}$$

Clearly, the conditional distribution of $X_{t,x}(T)$ under $w(s)$, $t \leq s \leq T$, is Gaussian with the parameters

$$m(t, x) := E\left(X_{t,x}(T)/w(s) - w(t), t \leq s \leq T\right) \tag{11.6.7}$$

$$= e^{b(T-t)}x + \int_t^T e^{b(T-\vartheta)}\alpha dw(\vartheta)$$

$$= e^{b(T-t)}x + \alpha\left(w(T) - w(t)\right) + \alpha b \int_t^T e^{b(T-\vartheta)}\left(w(\vartheta) - w(t)\right) d\vartheta,$$

$$\delta^2(t) := Var\left(X_{t,x}(T)/w(s) - w(t), t \leq s \leq T\right) \tag{11.6.8}$$

$$= \frac{\sigma^2}{2b}\left(e^{2b(T-t)} - 1\right).$$

From here, we get the following explicit solution of the Ornstein-Uhlenbeck equation:

$$v(t, x) = E^w \varphi(X_{t,x}(T)) \tag{11.6.9}$$

$$= \frac{1}{\sqrt{2\pi}\delta(t)} \int_{-\infty}^{\infty} \varphi(\xi) \exp\left(-\frac{(\xi - m(t, x))^2}{2\delta^2(t)}\right) d\xi.$$

We have

$$\frac{\partial v}{\partial x}(t, x) = \frac{e^{b(T-t)}}{\sqrt{2\pi}\delta^3(t)} \int_{-\infty}^{\infty} \varphi(\xi)[\xi - m(t, x)] \exp\left(-\frac{(\xi - m(t, x))^2}{2\delta^2(t)}\right) d\xi. \tag{11.6.10}$$

Now consider the problem which is a perturbation of (11.6.1)–(11.6.2):

$$-du = \left[\frac{a^2}{2}\frac{\partial^2 u}{\partial x^2} + (bx + \varepsilon b_1(x))\frac{\partial u}{\partial x}\right] dt + \alpha\frac{\partial u}{\partial x} * dw(t), \quad (t, x) \in [T_0, T) \times \mathbf{R}, \tag{11.6.11}$$

$$u(T, x) = \varphi(x), \quad x \in \mathbf{R}. \tag{11.6.12}$$

Aiming to reduce variance in the Monte Carlo procedure, let us use the representation (11.3.2)–(11.3.3) with $\mu = 0$ (hence $Y \equiv 1$) and with

$$\lambda(s, x) = -\sigma\frac{\partial v}{\partial x}(s, x), \tag{11.6.13}$$

where $\partial v/\partial x$ is from (11.6.10). We have

$$u(t, x) = E^w \left[\varphi(X_{t,x}(T)) + Z_{t,x,1,0}(T)\right], \tag{11.6.14}$$

where X, Z satisfy the system

$$dX = [bX + \varepsilon b_1(X)]\,ds + \sigma dW(s) + \alpha dw(s), \qquad (11.6.15)$$

$$dZ = -\sigma \frac{\partial v}{\partial x}(s, X)dW(s).$$

To realize (11.6.14), we use both the mean-square and weak Euler procedures for solving (11.6.15) with respect to the Wiener process $W(t)$. Due to Theorem 11.3.1, the variance is expected to be small if ε is relatively small. Here we exploit the control variates method (we put $\mu = 0$) for variance reduction. Since in the SPDEs (11.6.1)–(11.6.2) and (11.6.11)–(11.6.12) the coefficient α is constant and $\beta = \gamma = 0$, the method of important sampling can also be used without any theoretical difficulties (see the corresponding comment at the end of Sect. 11.3).

11.6.2 Numerical Results

For definiteness, we simulate (11.6.1)–(11.6.2) with

$$\varphi(x) = x^2. \qquad (11.6.16)$$

In the tests we fix a trajectory $w(t)$, $0 \le t \le T$, which is obtained with a small time step equal to 0.0001. To evaluate the exact solution given by (11.6.9), we simulate the integral in (11.6.7) by the trapezoidal rule with the step 0.0001. In Table 11.6.1 we present the results of simulating the solution of (11.6.1)–(11.6.2) , (11.6.16) by the weak Euler-type scheme (11.2.7), (11.2.9). One can observe convergence with order one that is in good agreement with our theoretical results (see Remark 11.2.3 and note that in this example α and σ are constant). In the table the "$\pm$" reflects the Monte Carlo error only, it gives the confidence interval for the corresponding value with probability 0.95. Similar results are obtained by the mean-square Euler scheme (11.2.2). For instance, for $h = 0.1$ (the other parameters are the same as in Table 11.6.1) we get $\hat{v}(0, 0) = 0.7550 \pm 0.0019$.

To demonstrate the variance reduction technique from Sect. 11.3, we first simulate (11.6.1)–(11.6.2), (11.6.16) using a probabilistic representation of the form (11.6.14), (11.6.15) with $\varepsilon = 0$. In particular, we obtain that for $M = 10^4$, $h = 0.1$ (the other parameters are the same as in Table 11.6.1) $\hat{v}(0, 0) = 0.7562 \pm 0.0012$ and for $M = 100$, $h = 0.01$ the approximate value $\hat{v}(0, 0) = 0.7383 \pm 0.0010$. Recalling that the results in Table 11.6.1 are obtained with $M = 10^6$ Monte Carlo runs, one can see that we reach the same level of the Monte Carlo error with significantly less number of Monte Carlo runs. We note that although we use the optimal $\lambda(s, x)$ here, the variance (and, consequently the Monte Carlo error) is not zero, which is due to the error of numerical integration of the equation for Z in (11.6.15).

Table 11.6.1 *Ornstein-Uhlenbeck equation.* Evaluation of $v(t, x)$ from (11.6.1) to (11.6.2), (11.6.16) with various time steps h. Here $a = 1$, $b = -1$, $\alpha = 0.5$, and $T = 10$. The expectations are computed by the Monte Carlo technique simulating $M = 10^6$ independent realizations. The "$\pm$" reflects the Monte Carlo error only, it does not reflect the error of the method. All simulations are done along the same sample path $w(t)$. The corresponding reference value is 0.73726, which is found due to (11.6.9)

h	$\hat{v}(0, 0)$
0.2	0.7735 ± 0.0019
0.1	0.7557 ± 0.0018
0.05	0.7456 ± 0.0018
0.02	0.7412 ± 0.0018
0.01	0.7391 ± 0.0018

Table 11.6.2 *Perturbed Ornstein-Uhlenbeck equation.* Evaluation of $u(t, x)$ from (11.6.11) to (11.6.12), (11.6.16), (11.6.17) at $(0, 0)$ with various time steps h. Here $a = 1$, $b = -1$, $\alpha = 0.5$, $\varepsilon = 0.1$, and $T = 10$. The expectations are computed by the Monte Carlo technique simulating $M = 10^4$ independent realizations. The corresponding reference value is 0.6006 ± 0.0004, which is found by simulation with $h = 0.001$ and $M = 10^6$

h	Without variance reduction	With variance reduction
0.2	0.614 ± 0.014	0.6014 ± 0.0042
0.1	0.611 ± 0.014	0.6037 ± 0.0040
0.02	0.604 ± 0.014	0.6001 ± 0.0041

Now we evaluate the solution of the perturbed Ornstein-Uhlenbeck equation (11.6.11)–(11.6.12). We take $\varphi(x)$ from (11.6.16) and we choose a small $\varepsilon > 0$ and

$$b_1(x) = -x^3. \tag{11.6.17}$$

We simulate (11.6.14), (11.6.15), (11.6.16), (11.6.17) both without employing the variance reduction technique (i.e., we put $Z(T) = 0$ in (11.6.14)) and with variance reduction (i.e., using $\lambda(s, x)$ from (11.6.13) to (11.6.10)) by the Euler-type scheme (11.2.7), (11.2.9). The results of the experiments are presented in Table 11.6.2. When the variance reduction technique is used, the Monte Carlo error is 3.5 times less than in the standard simulation without variance reduction. In other words, to reach the same accuracy, we can run 12 times less trajectories in the case with variance reduction than without one, which is a significant gain of computational efficiency.

11.7 Nonlinear Filtering

In this section, we briefly discuss one of the main applications of SPDEs – nonlinear
filtering [16, 184, 219, 238, 390, 416, 445]. Let $(\Omega, \mathcal{F}, P)$ be a complete probability
space, $\mathcal{F}_t$, $0 \le t \le T$, be a filtration satisfying the usual hypotheses, $(w(t), \mathcal{F}_t)$
and $(v(t), \mathcal{F}_t)$ be d_1-dimensional and r-dimensional independent standard Wiener
processes, respectively. We consider the classical filtering scheme

$$dX = \alpha(X)ds + \sigma(X)dw(s), \quad X(0) = x, \tag{11.7.1}$$
$$dy = \beta(X)ds + dv(s), \quad y(0) = 0, \tag{11.7.2}$$

where $X(t) \in \mathbf{R}^d$ is the unobservable signal process, $y(t) \in \mathbf{R}^r$ is the observation process, $\alpha(x)$ and $\beta(x)$ are d-dimensional and r-dimensional vector functions,
respectively; $\sigma(x)$ is a $d \times d_1$ dimensional matrix function. We assume that the functions α, β, and σ are bounded and have bounded derivatives up to some order. The
vector $X(0) = x$ in (11.7.1) can be random, it is independent of both w and v and
its density $\varphi(\cdot)$ is supposed to be known.

Let $f(x)$ be a function on $\mathbf{R}^d$ with the same properties as those of α, β, σ. The
filtering problem consists in constructing the estimate $\hat{f}(t) = \hat{f}(X(t))$ based on the
observation $y(s)$, $0 \le s \le t$, which is the best in the mean-square sense:

$$\hat{f}(t) = \hat{f}(t, y(\cdot)) := E[f(X(t)) \mid y(s), \ 0 \le s \le t]. \tag{11.7.3}$$

There are two PDE formulations to solve the filtering problem. The first one
is based on backward pathwise filtering equations [59, 74, 220, 346, 390]. We
emphasize that the pathwise filtering equations are ordinary (not stochastic) partial
differential equations. Due to this fact, nonlinear filtering problems are reduced to
comparatively standard problems for linear equations of parabolic type and probabilistic representations of their solutions can be obtained. To numerically solve these
PDE problems, the ideas of the weak-sense numerical integration of SDEs can be
used together with the Monte Carlo technique [321]. In particular, it is proved in
[321] that the Euler method converges with weak order $1 - \varepsilon$ for almost every observation $y(s)$, $0 \le s \le t$. The proof is nontrivial since here a part of the SDE system is
approximated in the weak sense while the other part (which involves the observation)
is simulated pathwisely.

The second formulation is related to parabolic SPDEs of nonlinear filtering [16,
219, 220, 346, 390]: either linear SPDE (known as the Zakai equation) or nonlinear
SPDE (usually called the Kushner-Stratonovich equation or the Fujisaki-Kallianpur-
Kunita equation). The linear filtering SPDE can be solved by the numerical methods
as discussed earlier in this chapter (see also [16, 321, 323, 324, 363] and references
therein). In this section we limit ourselves to an expository material on the backward
pathwise filtering equations and linear backward SPDEs of nonlinear filtering.

11.7.1   Kallianpur-Striebel Formula

We define the stochastic process $\eta(t)$ as

$$\eta^{-1}(t) := \exp\left\{ -\int_0^t \beta^{\mathsf{T}}(X(s))dv(s) - \frac{1}{2}\int_0^t |\beta(X(s))|^2\, ds \right\} \qquad (11.7.4)$$

$$= \exp\left\{ -\int_0^t \beta^{\mathsf{T}}(X(s))dy(s) + \frac{1}{2}\int_0^t |\beta(X(s))|^2\, ds \right\}.$$

According to our assumptions, we have

$$E\eta^{-1}(t) = 1,\ 0 \le t \le T.$$

Using $\eta^{-1}(T)$, we introduce the new probability measure $\tilde{P}$ on $(\Omega, \mathcal{F})$

$$\tilde{P}(\Gamma) = \int_\Gamma \eta^{-1}(T) dP(\omega).$$

The measures P and $\tilde{P}$ are mutually absolutely continuous and the Radon-Nikodym derivative $dP/d\tilde{P}$ is equal to $\eta(T)$:

$$\frac{dP}{d\tilde{P}}(X(\cdot), y(\cdot)) = \eta(T) = \exp\left\{ \int_0^T \beta^{\mathsf{T}}(X(s))dy(s) - \frac{1}{2}\int_0^T |\beta(X(s))|^2\, ds \right\}.$$
$$(11.7.5)$$

One can show by the Girsanov theorem that there exists a standard Wiener process $(w(t), \tilde{v}(t))$ on $(\Omega, \mathcal{F}, \mathcal{F}_t, \tilde{P})$ such that its part $w(t)$ coincides with that in (11.7.1) and that the process $(X(s), y(s))$ satisfies the system of Ito equations

$$dX = \alpha(X)ds + \sigma(X)dw(s),\ X(0) = x, \qquad (11.7.6)$$
$$dy = d\tilde{v}(s),\ y(0) = 0. \qquad (11.7.7)$$

Thus, the processes $X(s)$ and $y(s)$ are independent on $(\Omega, \mathcal{F}, \mathcal{F}_s, \tilde{P})$ and $y(s)$ is a Wiener process.

The following formula holds for a function $F(x, y)$:

$$EF(X(t), y(t)) = \tilde{E}\,(F(X(t), y(t))\eta(T)) = \tilde{E}(\tilde{E}\,(F(X(t), y(t))\eta(T)\mid \mathcal{F}_t))$$
$$= \tilde{E}(F(X(t), y(t))\tilde{E}(\eta(T)\mid \mathcal{F}_t)) = \tilde{E}\,(F(X(t), y(t))\eta(t))\,,$$
$$(11.7.8)$$

where $\tilde{E}$ is the expectation with respect to $\tilde{P}$.

The formula (11.7.8) can be rewritten:

$$EF(X(t), y(t)) \mid_{(11.7.1)-(11.7.2)}$$

$$= E \left(F(X(t), y(t)) \exp \left\{ \int_0^t \beta^\mathsf{T}(X(s))dy(s) \right. \right.$$

$$\left. \left. -\frac{1}{2} \int_0^t |\beta(X(s))|^2 ds \right\} \right) \mid_{(11.7.6)-(11.7.7)}$$

(11.7.9)

or

$$EF(X(t), y(t)) \mid_{(11.7.1)-(11.7.2)} = E \left(F(X(t), y(t))\eta(t) \right) \mid_{(11.7.11)}, \qquad (11.7.10)$$

where the system (11.7.11) has the form

$$dX = \alpha(X)ds + \sigma(X)dw(s), \quad X(0) = x, \qquad (11.7.11)$$
$$dy = d\tilde{v}(s), \quad y(0) = 0,$$
$$d\eta = \beta^\mathsf{T}(X)\eta d\tilde{v}(s), \quad \eta(0) = 1.$$

The representations (11.7.9) and (11.7.10) are constructive: unlike (11.7.8) they admit direct application of the Monte Carlo technique using methods of weak approximation of SDEs (see e.g. [321]).

Let us recall a general version of Bayes' formula (see, e.g. [238, Ch. 7, Sec. 9], [390, Ch. 6, Sec. 1.1]).

Lemma 11.7.1 *Suppose two mutually absolutely continuous measures Q and $\tilde{Q}$ are given on $(\Omega, \mathcal{F})$. If ξ is a Q-integrable random variable and $\mathcal{G}$ is some σ-subalgebra of $\mathcal{F}$, then almost surely*

$$E[\xi \mid \mathcal{G}] = \frac{\tilde{E}[\xi dQ/d\tilde{Q} \mid \mathcal{G}]}{\tilde{E}[dQ/d\tilde{Q} \mid \mathcal{G}]}. \qquad (11.7.12)$$

Clearly, the formula (11.7.8) is a particular case of (11.7.12). Another particular case is the well-known Kallianpur-Striebel formula for the mean (11.7.3):

$$E[f(X(t)) \mid y(s), \ 0 \le s \le t] = \frac{\tilde{E}[f(X(t))\eta(t) \mid y(s), \ 0 \le s \le t]}{\tilde{E}[\eta(t) \mid y(s), \ 0 \le s \le t]}, \qquad (11.7.13)$$

where

$$\eta(t) = \exp\left\{\int_0^t \beta^{\mathsf{T}}(X_{0,x}(s))dy(s) - \frac{1}{2}\int_0^t |\beta(X_{0,x}(s))|^2 ds\right\} \qquad (11.7.14)$$

and $X(s) = X_{0,x}(s)$ is the solution of (11.7.6).

The numerator and denominator in the Kallianpur-Striebel formula (11.7.13) can be written in the form:

$$\tilde{E}[f(X(t))\eta(t) \mid y(s),\ 0 \le s \le t]$$

$$= E\left[f(X_{0,x}(t))\exp\left\{\int_0^t \beta^{\mathsf{T}}(X_{0,x}(s))dy(s) - \frac{1}{2}\int_0^t |\beta(X_{0,x}(s))|^2 ds\right\}\right]\Bigg|_{(11.7.6)},$$

$$(11.7.15)$$

$$\tilde{E}[\eta(t) \mid y(s),\ 0 \le s \le t]$$

$$= E\left[\exp\left\{\int_0^t \beta^{\mathsf{T}}(X_{0,x}(s))dy(s) - \frac{1}{2}\int_0^t |\beta(X_{0,x}(s))|^2 ds\right\}\right]\Bigg|_{(11.7.6)}.$$

$$(11.7.16)$$

The sign $|_{(11.7.6)}$ means that $X_{0,x}(s)$ is the solution of (11.7.6) under a fixed trajectory y, i.e., the averaging here is carried out with respect to X only. These formulas are true due to independence of $X(\cdot)$ and $y(\cdot)$ on $(\Omega, \mathcal{F}, \tilde{P})$. In what follows we will also use the following notation for the conditional expectation:

$$\tilde{E}^y F(X(\cdot), y(\cdot)) := \tilde{E}\left[F(X(\cdot), y(\cdot)) | y(s'),\ 0 \le s' \le s\right],$$

where $F(X(\cdot), y(\cdot))$ is a functional of $X(s'), y(s'), 0 \le s' \le s$.

Remark 11.7.2 Due to the Ito formula, we have

$$\int_0^t \beta^{\mathsf{T}}(X_{0,x}(s))dy(s) = \beta^{\mathsf{T}}(X_{0,x}(t))y(t) - \int_0^t \sum_{i=1}^d \frac{\partial \beta^{\mathsf{T}}}{\partial x^i}\alpha_i y(s)ds$$

$$- \frac{1}{2}\int_0^t \sum_{i,l=1}^d \sum_{k=1}^{d_1} \sigma_{ik}\sigma_{lk}\frac{\partial^2 \beta^{\mathsf{T}}}{\partial x^i \partial x^l}y(s)ds - \int_0^t \sum_{i=1}^d \frac{\partial \beta^{\mathsf{T}}}{\partial x^i}y(s)\sum_{j=1}^{d_1}\sigma_{ij}dw_j(s)$$

which allows us to consider $y(\cdot)$ in (11.7.15) and (11.7.16) to be fixed.

11.7.2 Backward Pathwise Filtering Equations and Probabilistic Representation of Their Solutions

Define the $d \times d$ dimensional matrix $a = \{a_{ij}\}$:

$$a(x) = \sigma(x)\sigma^{\mathsf{T}}(x), \quad a_{ij}(x) = \sum_{k=1}^{d_1} \sigma_{ik}(x)\sigma_{jk}(x). \tag{11.7.17}$$

Due to the Ito formula, the process $\eta(t)$ can be transformed:

$$\eta(t) = \exp\{\beta^{\mathsf{T}}(X_{0,x}(t))y(t)\}$$

$$\times \exp\left\{-\int_0^t \sum_{i=1}^d \frac{\partial \beta^{\mathsf{T}}}{\partial x^i}\alpha_i y(s)ds - \frac{1}{2}\int_0^t \sum_{i,l=1}^d a_{il}\frac{\partial^2 \beta^{\mathsf{T}}}{\partial x^i \partial x^l}y(s)ds - \frac{1}{2}\int_0^t |\beta|^2 ds\right\}$$

$$\times \exp\left\{-\int_0^t \sum_{i=1}^d \frac{\partial \beta^{\mathsf{T}}}{\partial x^i}y(s)\sum_{j=1}^{d_1}\sigma_{ij}dw_j(s))\right\} := \exp\{\beta^{\mathsf{T}}(X_{0,x}(t))y(t)\} \cdot \zeta(t),$$

$$\tag{11.7.18}$$

where all the functions α, β, $\partial\beta/\partial x^i$, $\partial^2\beta/\partial x^i\partial x^l$, σ in the integrands have $X_{0,x}(s)$ as their arguments.

Introduce the system of SDEs for X and the scalar ζ :

$$dX = \alpha(X)ds' + \sigma(X)dw(s'), \quad X(s) = x, \quad s' \geq s, \tag{11.7.19}$$

$$d\zeta = -\zeta\left(\sum_{i=1}^d \frac{\partial \beta^{\mathsf{T}}}{\partial x^i}\alpha_i y(s') + \frac{1}{2}\sum_{i,l=1}^d a_{il}\frac{\partial^2 \beta^{\mathsf{T}}}{\partial x^i \partial x^l}y(s')\right)ds' \tag{11.7.20}$$

$$+\frac{1}{2}\zeta\left(|\sigma^{\mathsf{T}}\mathrm{grad}(\beta^{\mathsf{T}}y(s'))|^2 - |\beta|^2\right)ds' - \zeta(\sigma^{\mathsf{T}}\mathrm{grad}(\beta^{\mathsf{T}}y(s')))^{\mathsf{T}}dw(s'),$$

$$\zeta(s) = z.$$

In (11.7.20) we use the short notation

$$\frac{\partial^2 \beta^{\mathsf{T}}}{\partial x^i \partial x^l}y(s) = \sum_{k=1}^r \frac{\partial^2 \beta_k}{\partial x^i \partial x^l}y^k(s), \quad \sigma^{\mathsf{T}}\mathrm{grad}(\beta^{\mathsf{T}}y) = \sum_{i=1}^d \sigma_{ij}\sum_{k=1}^r \frac{\partial \beta_k}{\partial x^i}y^k(s).$$

Let us fix the obtained observation $y(s)$, $0 \leq s \leq t$. Consider the function

$$u_g(s, x) = E[g(X_{s,x}(t))\exp\{\beta^{\mathsf{T}}(X_{s,x}(t))y(t)\}\zeta_{s,x,1}(t)]\,|_{(11.7.19)-(11.7.20)}, \tag{11.7.21}$$

where g is a scalar function on $\mathbf{R}^d$, $x \in \mathbf{R}^d$ is deterministic now, and $X_{s,x}(s')$, $\zeta_{s,x,1}(s')$ is the solution of the system (11.7.19)–(11.7.20) starting from $(x, 1)$ at the instant s. Clearly, for a deterministic $X(0) = x$ the values $u_f(0, x)$ and $u_1(0, x)$ (here the function $\mathbf{1} = \mathbf{1}(x)$ is identically equal to 1) coincide with the numerator and denominator of (11.7.13), respectively. Hence

$$\hat{f}(t, y(\cdot)) = E[f(X(t)) \mid y(s),\ 0 \le s \le t]$$
$$= \frac{E[f(X_{0,x}(t))\exp\{\beta^{\mathsf{T}}(X_{0,x}(t))y(t)\}\zeta_{0,x,1}(t)]\mid_{(11.7.19)-(11.7.20)}}{E[\exp\{\beta^{\mathsf{T}}(X_{0,x}(t))y(t)\}\zeta_{0,x,1}(t)]\mid_{(11.7.19)-(11.7.20)}} = \frac{u_f(0, x)}{u_1(0, x)},$$
$$(11.7.22)$$

where x is deterministic.

If $X(0) = \xi$ is random with density $\varphi(\cdot)$, then the numerator is equal to the integral $\int u_f(0, x)\varphi(x)dx$ and the denominator is equal to the integral $\int u_1(0, x)\varphi(x)dx$, i.e.,

$$\hat{f}(t, y(\cdot)) = E[f(X(t)) \mid y(s),\ 0 \le s \le t]$$
$$= \frac{E[f(X_{0,\xi}(t))\exp\{\beta^{\mathsf{T}}(X_{0,\xi}(t))y(t)\}\zeta_{0,\xi,1}(t)]\mid_{(11.7.19)-(11.7.20)}}{E[\exp\{\beta^{\mathsf{T}}(X_{0,\xi}(t))y(t)\}\zeta_{0,\xi,1}(t)]\mid_{(11.7.19)-(11.7.20)}} \qquad (11.7.23)$$
$$= \frac{\int u_f(0, x)\varphi(x)dx}{\int u_1(0, x)\varphi(x)dx}.$$

It is well known that the function $u_g(s, x)$ is the solution to the Cauchy problem for the linear parabolic equation

$$\frac{\partial u}{\partial s} + \frac{1}{2}\sum_{i,j=1}^{d} a_{ij}(x)\frac{\partial^2 u}{\partial x^i \partial x^j} + \sum_{i=1}^{d} b_i(s, x)\frac{\partial u}{\partial x^i} + c(s, x)u = 0,\ s < t, \quad (11.7.24)$$

with the condition on the right-hand side of the time interval $[0, t]$:

$$u(t, x) = g(x)\exp\{\beta^{\mathsf{T}}(x)y(t)\} := G(t, x). \qquad (11.7.25)$$

In (11.7.24) we have

$$b(s, x) = \{b_i(s, x)\} = \alpha(x) - a(x)\mathrm{grad}(\beta^{\mathsf{T}}(x)y(s)), \qquad (11.7.26)$$

$$c(s, x) = -\alpha^{\mathsf{T}}(x)\mathrm{grad}(\beta^{\mathsf{T}}(x)y(s)) - \frac{1}{2}\sum_{i,l=1}^{d} a_{il}(x)\frac{\partial^2 \beta^{\mathsf{T}}(x)}{\partial x^i \partial x^l}y(s) \qquad (11.7.27)$$
$$- \frac{1}{2}\mid \beta(x)\mid^2 + \frac{1}{2}\mid \sigma^{\mathsf{T}}(x)\mathrm{grad}(\beta^{\mathsf{T}}(x)y(s))\mid^2 .$$

The equation (11.7.24) is known as the backward pathwise filtering equation. The probabilistic representation (the Feynman-Kac formula) of its solution satisfying the condition (11.7.25) is given by (11.7.21), (11.7.19)–(11.7.20).

Thus, finding the estimate $\hat{f}(t)$ amounts to evaluating means of the form

$$u_{g,\varphi} := \int u_g(0, x)\varphi(x)dx = \tilde{E}^y[G(t, X_{0,\xi}(t))\zeta_{0,\xi,1}(t)]. \tag{11.7.28}$$

11.7.3 Backward SPDE for Nonlinear Filtering

We fix a time moment t and introduce the function

$$v_g(s, x) = \tilde{E}^y\left[g(X_{s,x}(t))\eta_{s,x,1}(t)\right], \tag{11.7.29}$$

where g is a scalar function on $\mathbf{R}^d$, $x \in \mathbf{R}^d$ is deterministic, and $X_{s,x}(s')$, $\eta_{s,x,1}(s')$ is the solution of the system

$$dX = \alpha(X)ds' + \sigma(X)dw(s'), \quad X(s) = x, \tag{11.7.30}$$
$$d\eta = \beta^{\mathsf{T}}(X)\eta dy(s'), \quad \eta(s) = 1.$$

Here $w(s')$ and $y(s')$ are independent standard Wiener processes on $(\Omega, \mathcal{F}, \mathcal{F}_s, \tilde{P})$.

The function $v_g(s, x)$ is the solution of the Cauchy problem for the backward linear SPDE (see, e.g. [220, 346] and [390, Ch. 6, Sec. 3]):

$$-dv = \left[\frac{1}{2}\sum_{i,j=1}^d a_{ij}(x)\frac{\partial^2 v}{\partial x^i \partial x^j} + \sum_{i=1}^d \alpha_i(x)\frac{\partial v}{\partial x^i}\right]ds + \beta^{\mathsf{T}}(x)v * dy, \quad s < t,$$

$$\tag{11.7.31}$$
$$v(t, x) = g(x). \tag{11.7.32}$$

The notation "$*dy$" means backward Ito integral as in Sect. 11.1.

The solution $v(s, x)$ to the SPDE (11.7.31)–(11.7.32) relates to the solution $u(s, x)$ of the backward pathwise PDE (11.7.24)–(11.7.25) [59, 238, 346, 390]:

$$v(s, x) = \exp(-\beta^{\mathsf{T}}(x)y(s))u(s, x), \tag{11.7.33}$$

and, in particular, $v(0, x) = u(0, x)$. Indeed, by the same arguments as those used in (11.7.18), we get

$$\eta_{s,x,1}(t) = \exp\left(\beta^{\mathsf{T}}(X_{s,x}(t))y(t) - \beta^{\mathsf{T}}(x)y(s)\right)\zeta_{s,x,1}(t).$$

We pay attention that in (11.7.18) $\eta(t) = \eta_{0,x,1}(t)$ and $\zeta = \zeta_{0,x,1}(t))$. Comparing (11.7.21) and (11.7.29), we arrive at (11.7.33). We note that the relation (11.7.33) can also be verified by direct substitution of $v(s, x)$ from (11.7.33) in (11.7.31)–(11.7.32).

As it was already pointed out before, the equation (11.7.24) has the form of a usual (not stochastic) PDE whose coefficients depend on $y(s)$. Clearly, (11.7.24)–(11.7.25) is well-defined for any continuous function $y(s)$. Then the solution of (11.7.24)–(11.7.25) can be interpreted in the pathwise sense, i.e., we can consider the PDE problem (11.7.24)–(11.7.25) with the fixed observation sample path $y(s)$, $0 \le s \le t$. In its turn, the pathwise interpretation of the PDE (11.7.24)–(11.7.25) together with the relation (11.7.33) allows us to interpret the solution to the SPDE (11.7.31)–(11.7.32) in the pathwise sense as well [59, 74]. We also note that the relation (11.7.33) implies some properties of $v(s, x)$. Under the above assumptions, the function $v(s, x)$ is smooth in x and the function and its derivatives with respect to x are continuous in s.

If $X(0) = \xi$ is a random variable with density $\varphi(\cdot)$, then we can write (cf. (11.7.13), (11.7.15)–(11.7.16), and (11.7.29)):

$$\hat{f}(t, y(\cdot)) = E[f(X(t)) \mid y(s), \ 0 \le s \le t] = \frac{v_{f,\varphi}}{v_{1,\varphi}}, \qquad (11.7.34)$$

where

$$v_{g,\varphi} := \int v_g(0, x)\varphi(x)dx = \tilde{E}^y[g(X_{0,\xi}(t))\eta_{0,\xi,1}(t)].$$

References

1. Abdulle, A., Cirilli, S.: S-ROCK: Chebyshev methods for stiff stochastic differential equations. SIAM J. Sci. Comput. **30**, 997–1014 (2008)
2. Abdulle, A., Li, T.: S-ROCK methods for stiff Ito SDEs. Commun. Math. Sci. **6**, 845–868 (2008)
3. Abdulle, A., Vilmart, G., Zygalakis, K.: Weak second order explicit stabilized methods for stiff stochastic differential equations. SIAM J. Sci. Comput. **35**, A1792–A1814 (2013)
4. Abdulle, A., Vilmart, G., Zygalakis, K.C.: High order numerical approximation of the invariant measure of ergodic SDEs. SIAM J. Numer. Anal. **52**(4), 1600–1622 (2014)
5. Anderson, D.A., Tannehill, J.C., Pletcher, R.H.: Computational Fluid Mechanics and Heat Transfer. CRC Press, New York (1984)
6. Anderson, S.L.: Random number generators on vector supercomputers and other advanced architectures. SIAM Rev. **32**, 221–251 (1990)
7. Antonelli, F.: Backward-forward stochastic differential equations. Ann. Appl. Probab. **3**, 777–793 (1993)
8. Antonov, A., Konikov, M., Spector, M.: Modern SABR Analytics Formulas and Insights for Quants. Former Physicists and Mathematicians. Springer, Berlin (2019)
9. Applebaum, D.: Lévy Processes and Stochastic Calculus. Cambridge University Press, Cambridge (2009)
10. Arnold, L.: Random Dynamical Systems. Springer, Berlin (1998)
11. Arnold, L., Crauel, H., Eckmann, J.P. (eds.): Lyapunov Exponents. Lecture Notes in Math, vol. 1486. Springer, Berlin (1991)
12. Arnold, L., Wihstutz, V. (eds.): Lyapunov Exponents. Lecture Notes in Math, vol. 1186. Springer, Berlin (1986)
13. Arnold, V.I.: Mathematical Methods of Classical Mechanics. Springer, Berlin (1989)
14. Auslender, E.I., Milstein, G.N.: Asymptotic expansions of the Lyapunov index for linear stochastic systems with small noises. Prikl. Matem. Mekhan. **46**, 358–365 (1982)
15. Bailey, N.Y., Hibberd, S., Tretyakov, M.V., Power, H.: Effect of random forcing on fluid-lubricated bearing. IMA J. Appl. Math. **84**(3), 632–649 (2019)
16. Bain, A., Crisan, D.: Fundamentals of Stochastic Filtering. Springer, Berlin (2009)
17. Bally, V.: Approximation scheme for solutions of BSDE. In: El Karoui, N., Mazliak, L. (eds.) Backward Stochastic Differential Equations, pp. 177–191. Pitman, London (1997)
18. Bally, V., Pages, G., Printems, J.: A stochastic quantization method for nonlinear problems. Monte Carlo Methods Appl. **7**, 21–34 (2001)
19. Bally, V., Talay, D.: The law of the Euler scheme for stochastic differential equations I. Convergence rate of the distribution function. Probab. Theory Relat. Fields **104**, 43–60 (1996)
20. Banas, L.: Numerical methods for the Landau-Lifshitz-Gilbert equation. In: Li, Z., Vulkov, L., Waśniewski, J. (eds.) Numerical Analysis and Its Applications. Lecture Notes in Computer Science, pp. 158–165. Springer, Berlin (2005)

© Springer Nature Switzerland AG 2021

G. N. Milstein and M. V. Tretyakov, *Stochastic Numerics for Mathematical Physics*,
Scientific Computation, https://doi.org/10.1007/978-3-030-82040-4

21. Basdevant, C., Deville, M., Haldenwang, P., Lacroix, J.M., Onazzani, J., Peyret, R., Orlandi, P., Patera, A.T.: Spectral and finite difference solutions of the Burgers equations. Comput. Fluids **14**, 23–41 (1986)
22. Baxendale, P.: Moment stability and large deviations for linear stochastic differential equations. In: N. Ikeda (ed.) Proc. Taniguchi Symp. on Probab. Meth. in Math. Physics. Katata and Kyoto, 1985, pp. 31–54. Kinokuniya, Tokyo (1987)
23. Belomestny, D., Milstein, G.N.: Monte Carlo evaluation of American options using consumption processes. Int. J. Theor. Appl. Finance **9**(4), 455–481 (2006)
24. Belomestny, D., Milstein, G.N., Schoenmakers, J.G.M.: Sensitivities for Bermudan options by regression methods. Dec. Econ. Finance **33**, 117–138 (2010)
25. Bender, C., Denk, R.: A forward scheme for backward SDEs. Stoch. Proc. Appl. **117**, 1793–1812 (2007)
26. Bensoussan, A., Glowinski, R., Răşcanu, A.: Approximation of some stochastic differential equations by the splitting up method. Appl. Math. Optim. **25**, 81–106 (1992)
27. Benton, E.R., Platzman, G.: A table of solutions of the one-dimensional Burgers equation. Quart. Appl. Math. **30**, 195–212 (1972)
28. Bernal, F.: An implementation of Milstein's method for general bounded diffusions. J. Sci. Comput. **79**, 867–890 (2019)
29. Beskos, A., Roberts, G.O.: Exact simulation of diffusions. Ann. Appl. Probab. **15**, 2422–2444 (2005)
30. Bismut, J.M.: Mécanique aléatoire. Lecture Notes in Math, vol. 866. Springer, Berlin (1981)
31. Bjork, T.: Arbitrage Theory In Continuous Time. Oxford University Press, Oxford (2009)
32. de Boor, C.: A Practical Guide to Splines. Springer, Berlin (1978)
33. Borodin, A.N., Salminen, P.: Handbook of Brownian Motion-Facts and Formulae. Birkhäuser, Basel (1996)
34. Bossy, M., Fezoui, L., Piperno, S.: Comparison of a stochastic particle method and a finite volume deterministic method applied to Burgers equation. Monte Carlo Methods Appl. **3**, 113–140 (1997)
35. Bossy, M., Gobet, E., Talay, D.: A symmetrized Euler scheme for an efficient approximation of reflected diffusions. J. Appl. Probab. **41**(3), 877–889 (2004)
36. Bossy, M., Talay, D.: A stochastic particle method for the McKean-Vlasov and the Burgers equation. Math. Comput. **66**, 157–192 (1997)
37. Bou-Rabee, N., Owhadi, H.: Long-run accuracy of variational integrators in the stochastic context. SIAM J. Numer. Anal. **48**(1), 278–297 (2010)
38. Bou-Rabee, N., Vanden-Eijnden, E.: A patch that imparts unconditional stability to explicit integrators for Langevin-like equations. J. Comput. Phys. **231**(6), 2565–2580 (2012)
39. Bouleau, N., Lépingle, D.: Numerical Methods for Stochastic Processes. Wiley, New York (1994)
40. Box, G.E.P., Muller, M.E.: A note on the generation of random normal deviates. Ann. Math. Stat. **29**, 610–611 (1958)
41. Boyle, P., Broadie, M., Glasserman, P.: Monte Carlo methods for security pricing. J. Econ. Dyn. Control **21**, 1267–1321 (1997)
42. Bratley, P., Fox, B.L., Schrage, E.L.: A Guide to Simulation. Springer, Berlin (1983)
43. Brent, R.P.: Uniform random number generators for supercomputers. In: Ikeda, N. (ed.) Proceedings Fifth Australian Supercomputer Conference, pp. 95–104. Melbourne (1992)
44. Brigo, D., Mercurio, F.: Interest Rate Models–theory and Practice: With Smile. Inflation and Credit. Springer, Berlin (2006)
45. Brigo, D., Morini, M., Pallavicini, A.: Counterparty Credit Risk. Collateral and Funding. Wiley, New York (2013)
46. Broadie, M., Glasserman, P.: Estimating security price derivatives using simulation. Manag. Sci. **42**, 269–285 (1996)
47. Brzezniak, Z., Carelli, E., Prohl, A.: Finite-element-based discretizations of the incompressible Navier-Stokes equations with multiplicative random forcing. IMA J. Numer. Anal. **33**(3), 771–824 (2013)

48. Buckwar, E., Kelly, C.: Towards a systematic linear stability analysis of numerical methods for systems of stochastic differential equations. SIAM J. Numer. Anal. **48**(1), 298–321 (2010)

49. Buckwar, E., Rößler, A., Winkler, R.: Stochastic Runge-Kutta methods for Ito SODEs with small noise. SIAM J. Sci. Comput. **32**(4), 1789–1808 (2010)

50. Buckwar, E., Winkler, R.: Multistep methods for SDEs and their application to problems with small noise. SIAM J. Numer. Anal. **44**, 779–803 (2006)

51. Buckwar, E., Winkler, R.: Improved linear multi-step methods for stochastic ordinary differential equations. J. Comput. Appl. Math. **205**, 912–922 (2007)

52. Castell, F., Gaines, J.: An efficient approximation method for stochastic differential equations by means of the exponential Lie series. Math. Comput. Simul. **38**(1–3), 13–19 (1995)

53. Ceperley, D.M.: Path integrals in the theory of condensed helium. Rev. Mod. Phys. **67**, 279–355 (1995)

54. Channel, P.J., Scovel, C.: Symplectic integration of Hamiltonian systems. Nonlinearity **3**, 231–259 (1990)

55. Chevance, D.: Numerical methods for backward stochastic differential equations. In: Rogers, L., Talay, D. (eds.) Numerical Methods in Finance, pp. 232–244. Cambridge University Press, Cambridge (1997)

56. Chirikjian, G.S.: Stochastic Models, Information Theory, and Lie Groups, vol. 1. Birkhäuser, Basel (2009)

57. Chitashvili, R.J., Lazrieva, N.L.: Strong solutions of stochastic differential equations with boundary conditions. Stochastics **5**, 225–309 (1981)

58. Chorin, A.: Accurate evaluation of Wiener integrals. Math. Comput. **27**(121), 1–15 (1975)

59. Clark, J.: The design of robust approximation to the stochastic differential equations of nonlinear filtering. In: Skwirzynski, J. (ed.) Communication Systems and Random Process Theory, NATO Advanced Study Institute Series E Applied Sciences, vol. 25, pp. 721–734. Sijthoff and Noordhoff (1978)

60. Clark, J.M.C.: The simulation of pinned diffusions. Proc. 29th IEEE Conf. Decision Control **3**, 1418–1420 (1990)

61. Clark, J.M.C., Cameron, R.T.: The maximum rate of convergence of discrete approximations for stochastic differential equations. In: Lecture Notes in Control and Information Sciences, vol. 25, pp. 162–171. Springer, Berlin (1980)

62. Coddington, P.D.: Random Number Generators for Parallel Computers (1996)

63. Cole, J.: On a quasi-linear parabolic equation occurring in aerodynamics. Q. Appl. Math. **9**(3), 225–236 (1951)

64. Costantini, C., Pacchiarotti, B., Sartoretto, F.: Numerical approximation for functionals of reflecting diffusion processes. SIAM J. Appl. Math. **58**, 73–102 (1998)

65. Cruzeiro, A.B., Malliavin, P., Thalmaier, A.: Geometrization of Monte-Carlo numerical analysis of an elliptic operator: strong approximation. C. R. Acad. Sci. Paris, Ser. I(338), 481–486 (2003)

66. Cvitanić, J., Ma, J.: Hedging options for a large investor and forward-backward SDE's. Ann. Appl. Probab. **6**, 379–398 (1996)

67. Da Prato, G., Zabczyk, J.: Stochastic Equations in Infinite Dimensions. Cambridge University Press (1992)

68. Danilov, V.G., Maslov, V.P., Volosov, K.A.: Mathematical Modelling of Heat and Mass Transfer Processes. Kluwer, Dordrecht (engl. transl. from Russian, 1987) (1995)

69. Davidchack, R.L., Handel, R., Tretyakov, M.V.: Langevin thermostat for rigid body dynamics. J. Chem. Phys. **130**(23), 234101 (2009)

70. Davidchack, R.L., Ouldridge, T.E., Tretyakov, M.V.: New Langevin and gradient thermostats for rigid body dynamics. J. Chem. Phys. **142**(14), 144114 (2015)

71. Davidchack, R.L., Ouldridge, T.E., Tretyakov, M.V.: Geometric integrator for Langevin systems with quaternion-based rotational degrees of freedom and hydrodynamic interactions. J. Chem. Phys. **147**(22), 224103 (2017)

72. Davie, A.M.: Pathwise approximation of stochastic differential equations using coupling (2015). https://www.maths.ed.ac.uk/~sandy/coum.pdf

73. Davie, A.M., Gaines, J.G.: Convergence of numerical schemes for the solution of parabolic stochastic partial differential equations. Math. Comput. **70**, 121–134 (2001)
74. Davis, M.: A pathwise solution of the equations of nonlinear filtering. Theory Probab. Appl. **27**, 160–167 (1982)
75. Delarue, F.: On the existence and uniqueness of solutions to FBSDEs in a nondegenerate case. Stoch. Proc. Appl. **99**, 209–286 (2002)
76. Delarue, F., Menozzi, S.: An interpolated stochastic algorithm for quasi-linear PDEs. Math. Comput. **77**(261), 125–158 (2008)
77. Deligiannidis, G., Maurer, S., Tretyakov, M.V.: Random walk algorithm for the Dirichlet problem for parabolic integro-differential equation. BIT Numer. Math. **61**, 1223–1269 (2021). https://doi.org/10.1007/s10543-021-00863-2
78. Delyon, B., Hu, Y.: Simulation of conditioned diffusion and application to parameter estimation. Stoch. Proc. Appl. **116**, 1660–1675 (2006)
79. DeVore, R.A.: Nonlinear approximation. Acta Numerica **7**, 51–150 (1998)
80. Dick, J., Kuo, F., Sloan, I.: High-dimensional integration: the quasi-Monte Carlo way. Acta Numerica **22**, 133–288 (2013)
81. Douglas, J., Ma, J., Protter, P.: Numerical methods for forward-backward stochastic differential equations. Ann. Appl. Probab. **6**, 940–968 (1996)
82. Duffie, D., Glynn, P.: Efficient Monte Carlo simulation of security prices. Ann. Appl. Probab. **5**, 897–905 (1995)
83. Duffie, D., Ma, J., Yong, J.: Black's consol rate conjecture. Ann. Appl. Probab. **5**, 356–382 (1995)
84. Dumas, W.M., Tretyakov, M.V.: Computing conditional Wiener integrals of functionals of a general form. IMA J. Numer. Anal. **31**, 1217–1251 (2011)
85. Dupuis, P., Ishii, H.: On Lipschitz continuity of the solution mapping to the Skorokhod problem, with applications. Stoch. Stoch. Rep. **35**, 31–62 (1991)
86. Dynkin, E.B.: Markov Processes. Springer, Berlin (1965)
87. Dynkin, E.B., Yushkevich, A.A.: Markov Processes: Theorems and Problems. Plenum, New York (1969)
88. Egorov, A.D., Sobolevsky, P.I., Yanovich, L.: Functional Integrals: Approximative Evaluation and Applications. Kluwer, Dordrecht (1993)
89. El Karoui, N., Mazliak, L. (eds.): Backward Stochastic Differential Equations. Pitman, London (1997)
90. El Karoui, N., Peng, S., Quenez, M.S.: Backward stochastic differential equations in finance. Math. Fin. **7**, 1–71 (1997)
91. Elepov, B.S., Kronberg, A.A., Mikhailov, G.A., Sabelfeld, K.K.: Solution of boundary value problems by the Monte Carlo method. Nauka, Novosibirsk (In Russian) (1980)
92. Elworthy, K.D.: Stochastic Differential Equations on Manifolds. Cambridge University Press, Cambridge (1982)
93. Elworthy, K.D., Li, X.M.: Formulae for the derivatives of heat semigroups. J. Funct. Anal. **125**, 252–286 (1994)
94. Eriksson, K., Johnson, C.: Adaptive finite element methods for parabolic problems IV: Nonlinear problems. SIAM J. Numer. Anal. **32**, 1729–1749 (1995)
95. Ermakov, S.M., Nekrutkin, V.V., Sipin, A.S.: Random Processes for Classical Equations of Mathematical Physics. Kluwer, Dordrecht (1989)
96. Ethier, S.N., Kurtz, T.G.: Markov Processes. Wiley, New York (1986)
97. Fan, J., Gijbels, I.: Local Polynomial Modelling and Its Applications. Chapman & Hall, New York (1996)
98. Fang, W., Giles, M.B.: Adaptive Euler-Maruyama method for SDEs with non-globally Lipschitz drift. Ann. Appl. Probab. **30**, 526–560 (2020)
99. Feller, W.: An Introduction To Probability Theory and Its Applications, vol. 1. Willey, New York (1958)
100. Feng, K., Shang, Z.J.: Volume-preserving algorithms for source-free dynamical systems. Numer. Math. **71**, 451–463 (1995)

101. Feynman, R.P.: Statistical Mechanics. Westview Press (1972)
102. Feynman, R.P., Hibbs, A.R.: Quantum Mechanics and Path Integrals. McGraw-Hill, New York (1965)
103. Feynman, R.P., Leighton, R.B., Sands, M.: The Feynman Lectures on Physics, vol. 1. Addison-Wesley, Boston (1966)
104. Fisher, P., Platen, E.: Applications of the balanced method to stochastic differential equations in filtering. Monte Carlo Methods Appl. **5**, 19–38 (1999)
105. Fleming, W.H., Rishel, R.W.: Deterministic and Stochastic Optimal Control. Springer, Berlin (1975)
106. Fletcher, C.A.J.: Computational Galerkin Methods. Springer, Berlin (1984)
107. Fletcher, C.A.J.: Computational Techniques for Fluid Dynamics. Volumes I. Springer, Berlin, II (1991)
108. Fournié, E., Lasry, J.M., Lebuchoux, J., Lions, P.L., Touzi, N.: Applications of Malliavin calculus to Monte Carlo methods in finance. Fin. Stoch. **3**, 391–412 (1999)
109. Fournié, E., Lasry, J.M., Lebuchoux, J., Lions, P.L., Touzi, N.: Applications of Malliavin calculus to Monte Carlo methods in finance II. Fin. Stoch. **5**, 201–236 (2001)
110. Fox, R.F.: Second-order algorithm for the numerical integration of colored-noise problems. Phys. Rev. A **43**, 2649–2654 (1991)
111. Freidlin, M.I.: Functional Integration and Partial Differential Equations. Princeton University Press, Princeton (1985)
112. Freidlin, M.I.: Markov Processes and Differential Equations: Asymptotic Problems. Birkhäuser, Basel (1996)
113. Freidlin, M.I., Wentzell, A.D.: Random Perturbations of Dynamical Systems. Springer, Berlin (1984)
114. Fridman, A.: Partial Differential Equations of Parabolic Type. Prentice-Hall, Englewood Cliffs, NJ (1964)
115. Gaines, J.G.: Numerical experiments with S(P)DE's. In: Etheridge, A. (ed.) Stochastic Partial Differential Equations, London Mathematical Society Lecture Notes Series, vol. 216, pp. 55–71. Cambridge University Press, Cambridge (1995)
116. Gaines, J.G., Lyons, T.J.: Random generation of stochastic area integrals. SIAM J. Appl. Math. **54**, 1132–1146 (1994)
117. Gaines, J.G., Lyons, T.J.: Variable step size control in the numerical solution of stochastic differential equations. SIAM J. Appl. Math. **57**, 1455–1484 (1997)
118. Gardiner, C.W.: Handbook of Stochastic Methods for Physics. Chemistry and Natural Sciences. Springer, Berlin (1983)
119. Gelfand, I.M.: Some problems in the theory of quasi-linear equations. Uspehi Mat. Nauk **14**, 87–158 (1959)
120. Gelfand, I.M., Yaglom, A.M.: Integration in functional spaces and its application in quantum physics. J. Math. Phys. **1**, 48–69 (1960)
121. Gentle, J.E.: Random Number Generation and Monte Carlo Methods. Springer, Berlin (2003)
122. Gerstner, T., Griebel, M.: Numerical integration using sparse grids. Numer. Algorithms **18**, 209–232 (1998)
123. Gichman, I.I., Skorochod, A.V.: Stochastic Differential Equations. Naukova Dumka, Kiev (1968)
124. Gichman, I.I., Skorochod, A.V.: Stochastic Differential Equations and Their Applications. Naukova Dumka, Kiev (1982)
125. Giles, M.B.: Multilevel Monte Carlo path simulation. Oper. Res. **56**, 607–617 (2008)
126. Giles, M.B.: Multilevel Monte Carlo methods. Acta Numerica **24**, 259–328 (2015)
127. Giles, M.B., Szpruch, L.: Antithetic multilevel Monte Carlo estimation for multi-dimensional SDEs without Levy area simulation. Ann. Appl. Probab. **24**, 1585–1620 (2014)
128. Gladyshev, S.A., Milstein, G.N.: The Runge-Kutta method for calculation of Wiener integrals of functionals of exponential type. Zh. Vychisl. Mat. i Mat. Fiz. **24**, 1136–1149 (1984)
129. Glasserman, P.: Monte Carlo Methods in Financial Engineering. Springer, Berlin (2003)

130. Glasserman, P., Yao, D.: Some guidelines and guarantees for common random numbers. Manag. Sci. **38**, 884–908 (1992)
131. Glynn, P.: In: Optimization of Stochastic Systems Via Simulation, pp. 90–105. Society of Computer Simulation, San Diego (1989)
132. Gobet, E.: Weak approximation of killed diffusion using Euler schemes. Stoch. Proc. Appl. **87**(2), 167–197 (2000)
133. Gobet, E., Lemor, J.P., Warin, X.: A regression-based Monte-Carlo method to solve backward stochastic differential equations. Ann. Appl. Probab. **15**(3), 2002–2172 (2005)
134. Gobet, E., Menozzi, S.: Stopped diffusion processes: boundary corrections and overshoot. Stoch. Proc. Appl. **120**(2), 130–162 (2010)
135. Gobet, E., Turkedjiev, P.: Approximation of backward stochastic differential equations using Malliavin weights and least-squares regression. Bernoulli **22**(1), 530–562 (2016)
136. Goodfellow, I., Bengio, Y., Courville, A.: Deep Learning. MIT Press (2016)
137. Gornostyrev, Y.N., Katsnelson, M.I., Trefilov, A.V., Tretjakov, S.V.: Stochastic approach to simulation of lattice vibrations in strongly anharmonic crystals: Anomalous frequency dependence of the dynamic structure factor. Phys. Rev. B **54**, 3286–3294 (1996)
138. Graham, C., Kurtz, T.G., Méléard, S., Protter, P.E., Pulvirenti, M., Talay, D.: Probabilistic models for nonlinear partial differential equations. In: Talay, D., Tubaro, L. (eds.) Lectures Given at the 1st Session and Summer School held in Montecatini Terme. Lecture Notes in Mathematics, vol. 1627. Springer, Berlin (1996)
139. Graham, C., Talay, D.: Stochastic Simulation and Monte Carlo Methods. Springer, Berlin (2013)
140. Graham, R.: Hopf bifurcation with fluctuating control parameter. Phys. Rev. A **25**, 3234–3258 (1982)
141. Grecksch, W., Kloeden, P.E.: Time-discretised Galerkin approximations of parabolic stochastic PDEs. Bull. Austral. Math. Soc. **54**, 79–85 (1996)
142. Greenside, H.S., Helfand, E.: Numerical integration of stochastic differential equations. Bell Syst. Techn. J. **60**, 1927–1940 (1981)
143. Griebel, M., Holtz, M.: Dimension-wise integration of high-dimensional functions with applications to finance. J. Complex. **26**, 455–489 (2010)
144. Grindrod, P.: The Theory and Applications of Reaction-diffusion Equations: Patterns and Waves. Clarendon Press, Oxford (1996)
145. Gyöngy, I.: A note on Euler's approximations. Potential Anal. **8**, 205–216 (1998)
146. Gyöngy, I.: Lattice approximations for stochastic quasi-linear parabolic partial differential equations driven by space-time white noise II. Potential Anal. **11**, 1–37 (1999)
147. Gyöngy, I.: Approximations of stochastic partial differential equations. In: Stochastic Partial Differential Equations and Applications (Trento, 2002), Lecture Notes in Pure and Appl. Math., vol. 227, pp. 287–307. Dekker (2002)
148. Gyöngy, I., Krylov, N.: Existence of strong solutions for Ito's stochastic equations via approximations. Probab. Theory Relat. Fields **105**, 143–158 (1996)
149. Gyöngy, I., Krylov, N.: On the splitting-up method and stochastic partial differential equations. Ann. Appl. Probab. **31**, 564–591 (2003)
150. Gyöngy, I., Millet, A.: On discretization schemes for stochastic evolution equations. Ann. Appl. Probab. **23**, 99–134 (2005)
151. Gyöngy, I., Nualart, D.: Implicit scheme for stochastic parabolic partial differential equations driven by space-time white noise I. Potential Anal. **7**, 725–757 (1997)
152. Györfi, L., Kohler, M., Krzyżak, A., Walk, H.: A Distribution-Free Theory of Nonparametric Regression. Springer, Berlin (2002)
153. Hagan, P.S., Kumar, D., Lesniewski, A.S., Woodward, D.E.: Managing Smile Risk. The Best of Wilmott, p. 249 (2002)
154. Hairer, E., Lubich, C., Wanner, G.: Geometric Numerical Integration: Structure Presesrving Algorithms for Ordinary Differential Equations. Springer, Berlin (2002)
155. Hairer, E., Nørsett, S.P., Wanner, G.: Solving Ordinary Differential Equations. I. Nonstiff Problems. Springer, Berlin (1993)

156. Hairer, E., Wanner, G.: Solving Ordinary Differential Equations. II. Stiff and Differential-Algebraic Problems. Springer, Berlin (1996)

157. Hall, G., Watt, J.M.: Modern Numerical Methods for Ordinary Differential Equations. Clarendon Press, Oxford (1976)

158. Heard, W.B.: Rigid Body Mechanics. Wiley, New York (2006)

159. Heinrich, S.: Monte Carlo complexity of global solution of integral equations. J. Complex. **14**, 151–175 (1998)

160. Hellekalek, P.: Good random number generators are (not so) easy to find. Math. Comput. Simul. **46**, 485–505 (1998)

161. Heston, S.L.: A closed-form solution for options with stochastic volatility with applications to bond and currency options. Rev. Fin. Stud. **6**, 327–343 (1993)

162. Higham, D., Mao, X., Stuart, A.: Exponential mean-square stability of numerical solutions to stochastic differential equations. LMS J. Comput. Math. **6**, 297–313 (2003)

163. Higham, D.J.: Mean-square and asymptotic stability of the stochastic theta method. SIAM J. Numer. Anal. **38**(3), 753–769 (2000)

164. Higham, D.J., Mao, X., Stuart, A.: Strong convergence of Euler-type methods for nonlinear stochastic differential equations. SIAM J. Numer. Anal. **40**, 1041–1063 (2003)

165. Hille, E., Phillips, R.S.: Functional Analysis and Semigroups. AMS, Providence (1957)

166. Hofmann, N., Mathé, P.: On quasi-Monte Carlo simulation of stochastic differential equations. Math. Comput. **66**, 573–589 (1997)

167. Hofmann, N., Müller-Gronbach, T., Ritter, K.: Optimal approximation of stochastic differential equations by adaptive step-size control. Math. Comput. **69**, 1017–1034 (2000)

168. Hong, J., Wang, X.: Invariant measures for stochastic nonlinear Schrödinger equations. Numerical Approximations and Symplectic Structures, Lecture Notes Math., vol. 2251. Springer, Singapore (2019)

169. Höök, L.J., Johnson, T., Hellsten, T.: Randomized quasi-Monte Carlo simulation of fast-ion thermalization. Comput. Sci. Discovery **5**(1), 014010 (2012)

170. Hopf, E.: The partial differential equation $u_t + uu_x = \mu u_{xx}$. Commun. Pure Appl. Math. **3**(3), 201–230 (1950)

171. Horsthemke, W., Lefever, R.: Noise-Induced Transitions: Theory and Applications in Physics. Chemistry and Biology. Springer, Berlin (1984)

172. Hoshino, S., Ichida, K.: Solution of partial differential equations by a modified random walk. Numer. Math. **18**, 61–72 (1971)

173. Hu, Y.: Semi-implicit Euler-Maruyama scheme for stiff stochastic equations. In: Koerezlioglu, H. (ed.) Stochastic Analysis and Related Topics V: The Silvri Workshop, Progress Probability, vol. 38, pp. 183–202. Birkhauser, Boston (1996)

174. Hu, Y., Watanabe, S.: Donsker delta functions and approximations of heat kernels by the time discretization method. J. Math. Kyoto Univ. **36**, 494–518 (1996)

175. Hutzenthaler, M., Jentzen, A.: Numerical approximations of stochastic differential equations with non-globally Lipschitz continuous coefficients. Memoirs of the American Mathematical Society, vol. 236. AMS, Providence (2015)

176. Hutzenthaler, M., Jentzen, A., Kloeden, P.E.: Strong convergence of an explicit numerical method for SDEs with non-globally Lipschitz continuous coefficients. Ann. Appl. Probab. **22**, 1611–1641 (2012)

177. Ikeda, N., Watanabe, S.: Stochastic Differential Equations and Diffusion Processes. North-Holland, Amsterdam (1981)

178. Il'in, A.M.: Matching of Asymptotic Expansions of Solutions of Boundary Value Problems. Nauka, Moscow (1989)

179. Il'in, A.M., Oleinik, O.A.: Asymptotic behavior of solutions of the Cauchy problem for some quasi-linear equations for large values of time. Mat. Sbornik **51**, 191–216 (1960)

180. Ito, K., McKean, H.P.: Diffusion Processes and Their Sample Paths. Springer, Berlin (1965)

181. Izaguirre, J.A., Catarello, D.P., Wozniak, J.M., Skeel, R.D.: Langevin stabilization of molecular dynamics. J. Chem. Phys. **114**, 2090–2098 (2001)

182. Jacod, J., Protter, P.: Asymptotic error distributions for the Euler method for stochastic differential equations. Ann. Probab. **26**, 267–307 (1998)
183. Kalashnikov, V.V., Rachev, S.T.: Mathematical Methods for Construction of Stochastic Queuing Models. Nauka, Moscow (English translation by Pacific Grove, California, 1990) (1988)
184. Kallianpur, G.: Stochastic Filtering Theory. Springer, Berlin (1980)
185. Kalman, R.E., Falb, P.L., Arbib, M.A.: Topics in Mathematical Control Theory. McGraw-Hill, New York (1969)
186. van Kampen, N.: Relative stability in nonuniform temperature. IBM J. Res. Dev. **32**, 107–111 (1988)
187. Kanagawa, S.: On the Rate of Convergence for Maruyama's Approximate Solutions of Stochastic Differential Equations. Yokohama Math. J. **36**, 79–85 (1988)
188. Karatzas, I., Shreve, S.E.: Brownian Motion and Stochastic Calculus. Springer, Berlin (1988)
189. Karatzas, I., Shreve, S.E.: Methods of Mathematical Finance. Springer, New York (1998)
190. Kebaier, A.: Statistical Romberg extrapolation: a new variance reduction method and applications to option pricing. Ann. Appl. Probab. **14**, 2681–2705 (2005)
191. Kelly, C., Lord, G.J.: Adaptive time-stepping strategies for nonlinear stochastic systems. IMA J. Numer. Anal. **38**, 1523–1549 (2018)
192. Kevorkian, J., Cole, J.D.: Multiple Scale and Singular Perturbation Methods. Springer, Berlin (1996)
193. Khasminskii, R.Z.: In: Trydi VI Vses. Sovechaniya on Th. Prob. and Math. Stat., (ed.) Probabilistic representation of solutions of some differential equations, pp. 177–183. Vilnius (1960)
194. Khasminskii, R.Z.: Stochastic Stability of Differential Equations, 2nd edn. Springer, Berlin (original Russian edition, 1969) (2012)
195. Khasminskii, R.Z., Milstein, G.N.: Moment Lyapunov exponents and stability index. In: Khasminskii, R. (ed.) Stochastic Stability of Differential Equations, pp. 281–322. Springer, Berlin (2012)
196. Khasminskii, R.Z., Nevelson, M.B.: Stochastic Approximation and Recursive Estimation. AMS, Providence (translated from Russian, 1972) (1976)
197. Kinderman, A.J., Monahan, J.F.: Computer generation of random variables using the ratio of uniform deviates. ACM Trans. Math. Soft. **3**, 257–260 (1977)
198. Kitsul, P.I.:(In Russian), : On Continuously-discrete filtering of markov processes of diffusion type. Avtomat. i Telemekh. **31**, 29–37 (1970)
199. Klauder, J.R., Petersen, W.P.: Numerical integration of multiplicative-noise stochastic differential equations. SIAM J. Numer. Anal. **22**, 1153–1166 (1985)
200. Kleinert, H.: Path Integrals in Quantum Mechanics. Statistics and Polymer Physics. World Scientific, Singapore (1995)
201. Kloeden, P.E.: The systematic derivation of higher order numerical methods for stochastic differential equations. Milan J. Math. **70**, 187–207 (2002)
202. Kloeden, P.E., Platen, E.: Numerical Solution of Stochastic Differential Equations. Springer, Berlin (1992)
203. Kloeden, P.E., Platen, E., Schurz, H.: Numerical Solution of SDE Through Computer Experiments. Springer, Berlin (1994)
204. Kloeden, P.E., Platen, E., Wright, W.: The approximation of multiple stochastic integrals. Stoch. Anal. Appl. **10**, 431–441 (1992)
205. Kloeden, P.E., Shott, S.: Linear-implicit strong schemes for Ito-Galerkin approximations of stochastic PDEs. J. Appl. Math. Stoch. Anal. **14**, 47–53 (2001)
206. Knuth, D.E.: The art of computer programming. In: Seminumerical Methods, vol. 2. Addison-Wesley, Reading, Mass (1981)
207. Kohatsu-Higa, A.: High order Ito-Taylor approximations to heat kernels. J. Math. Kyoto Univ. **37**, 129–150 (1997)
208. Kohatsu-Higa, A.: Weak approximations: a Malliavin calculus approach. Math. Comput. **70**, 135–175 (2001)
209. Kotelenez, P.: Stochastic Ordinary and Stochastic Partial Differential Equation: Transition from Microscopic to Macroscopic Equation. Springer (2008)

210. Krivko, M., Tretyakov, M.V.: Application of simplest random walk algorithms for pricing barrier options. In: Gerstner, T., Kloeden, P. (eds.) Recent Developments in Computational Finance, pp. 407–428. World Scentific (2013)

211. Krivko, M., Tretyakov, M.V.: Numerical integration of the Heath-Jarrow-Morton model of interest rates. IMA J. Numer. Anal. **34**(1), 147–196 (2014)

212. Kruse, R.: Strong and Weak Approximation of Semilinear Stochastic Evolution Equations. Springer (2014)

213. Krylov, N.V.: Controllable Processes of Diffusion Type. Nauka, Moscow (1977)

214. Krylov, N.V.: An analytic approach to SPDEs. In: Carmona, R.A., Rozovskii, B. (eds.) Stochastic Partial Differential Equations, Six Perspectives, Mathematical Surveys and Monographs, vol. 64, pp. 185–242. AMS (1999)

215. Krylov, N.V., Rozovskii, B.L.: On the characteristics of degenerate second order parabolic Ito equations. J. Soviet Math. **32**, 336–348 (1986)

216. Kubo, R., Hashitsume, N.: Brownian motion of spins. Prog. Theor. Phys. Suppl. **46**, 210–220 (1970)

217. Kubo, R., Toda, M., Hashitsume, N.: Statistical Physics, vol. II. Springer, Berlin (1985)

218. Kuipers, L., Niederreiter, H.: Uniform Distribution of Sequences. Wiley, NY (1974)

219. Kunita, H.: Stochastic Flows and Stochastic Differential Equations. Cambridge University Press, Cambridge (1990)

220. Kushner, H.: Probability Methods for Approximations in Stochastic Control and For Elliptic Equations. Academic Press, New York (1977)

221. Kuznetsov, D.F.: Stochastic Differential Equations: Theory and Practice of Numerical Solution. University of St.-Peterburg (2017)

222. Ladyzhenskaya, O.A., Solonnikov, V.A., Ural'ceva, N.N.: Linear and Quasilinear Equations of Parabolic Type. AMS, Providence (1988)

223. Lamberton, D., Lapeyre, B.: Introduction to Stochastic Calculus Applied to Finance, 2nd edn. Chapman & Hall/CRC Press, Boca Raton, FL (2007)

224. Landau, L.D., Lifshitz, E.M.: On the theory of the dispersion of magnetic permeability in ferromagnetic bodies. Phys. Zs. Sowjet. **8**, 153–164 (1935)

225. L'Ecuyer, P.: Random number generation. In: Gentle, J.E., Haerdle, W., Mori, Y. (eds.) Handbook of Computational Statistics, pp. 35–71. Springer, Berlin (2012)

226. L'Ecuyer, P.: In: Random Number Generation with Multiple Streams for Sequential and Parallel Computers, pp. 31–44. IEEE Press (2015)

227. L'Ecuyer, P., Perron, G.: On the convergence rates of IPA and FDC derivative estimators. Oper. Res. **42**, 643–656 (1994)

228. Lehmer, D.H.: Mathematical methods in large-scale computing units. Ann. Comput. Lab. Harvard Univ. **26**, 141–146 (1951)

229. Leimkuhler, B., Matthews, C.: Rational construction of stochastic numerical methods for molecular sampling. Appl. Math. Res. Express **2013**(1), 34–56 (2013)

230. Leimkuhler, B., Matthews, C.: Molecular Dynamics with Deterministic and Stochastic Numerical Methods. Springer, Berlin (2015)

231. Leimkuhler, B., Matthews, C., Stoltz, G.: The computation of averages from equilibrium and nonequilibrium Langevin molecular dynamics. IMA J. Numer. Anal. **36**(1), 13–79 (2015)

232. Leimkuhler, B., Matthews, C., Tretyakov, M.V.: On the long-time integration of stochastic gradient systems. Proc. R. Soc. A **470**(2170), 20140120 (2014)

233. Leimkuhler, B., Sharma, A., Tretyakov, M.V.: Simplest random walk for approximating Robin boundary value problems and ergodic limits of reflected diffusions. arXiv: 2006.15670 (under consideration in Ann. Appl. Prob.) (2021)

234. Leliévre, T., Stoltz, G.: Partial differential equations and stochastic methods in molecular dynamics. Acta Numerica **25**, 681–880 (2016)

235. Lepingle, D.: Un schéma d'euler pour équations différentielles stochastiques réfléchies. C.R.A.S. Paris **316**, 601–605 (1993)

236. Lewis, A.L.: Option Valuation Under Stochastic Volatility. Finance Press (2000)

237. Lions, P.L., Sznitman, A.S.: Stochastic differential equations with reflecting boundary conditions. Commun. Pure Appl. Math. **37**(4), 511–537 (1984)
238. Lipster, R.S., Shiryaev, A.N.: Statistics of Random Processes. Nauka, Moscow (1974)
239. Liu, Y.: Numerical Approaches to Stochastic Differential Equations with Boundary Conditions. Purdue University, Thesis (1993)
240. Lobanov, Y.Y.: Functional integrals for nuclear many-particle systems. J. Phys. A **29**, 6653–6669 (1996)
241. Lunardi, A.: Analytic Semigroups and Optimal Regularity in Parabolic Problems. Springer (1995)
242. Ma, J., Protter, P., Martin, J.S., Torres, S.: Numerical methods for backward stochastic differential equations. Ann. Appl. Probab. **12**, 302–316 (2002)
243. Ma, J., Protter, P., Yong, J.: Solving forward-backward stochastic differential equations explicitly - a four step scheme. Probab. Theory Rel. Fields **98**, 339–359 (1994)
244. Ma, J., Yong, J.: Forward-backward stochastic differential equations and their applications. Lecture Notes in Mathematics, vol. 1702. Springer, Berlin (1999)
245. Mackevicius, V.: Second order weak approximations for Stratonovich stochastic differential equations. Liet. Mat. Rink **34**, 226–247 (1994)
246. Mackevicius, V.: Extrapolation of approximations of solutions of stochastic differential equations. In: Probability Theory and Mathematical Statistics (Tokyo, 1995), pp. 276–297. World Sci., River Edge, NJ (1996)
247. Makarov, R.N.: Numerical solution of quasilinear parabolic equations and backward stochastic differential equations. Russian J. Numer. Anal. Math. Model. **18**(5), 397–412 (2003)
248. Malliavin, P., Thalmaier, A.: Stochastic Calculus of Variations in Mathematical Finance. Springer, Berlin (2006)
249. Mannela, R., Palleschi, V.: Fast and precise algorithm for computer simulation of stochastic differential equations. Phys. Rev. Lett. **40**, 3381–3386 (1989)
250. Mao, X.: Stochastic Differential Equations and Applications. Horwood Publishing (2007)
251. Mao, X., Szpruch, L.: Strong convergence rates for backward Euler-Maruyama method for non-linear dissipative-type stochastic differential equations with super-linear diffusion coefficients. Stochastics **85**, 144–171 (2013)
252. Marsaglia, G.: Improving the polar method for generating a pair of random variables. Boeing Sci. Res. Lab. report D1–82-0203 (1962)
253. Marsaglia, G.: Choosing a point from the surface of a sphere. Ann. Math. Stat. **43**(2), 645–646 (1972)
254. Marsaglia, G., Tsang, W.: The Ziggurat method for generating random variables. J. Stat. Soft. **5**, 1–7 (2000)
255. Marsaglia, G., Zaman, A., Marsaglia, J.C.W.: Rapid evaluation of the inverse normal distribution function. Stat. Prob. Lett. **19**, 259–266 (1994)
256. Marsaglia, G.A.: A current view of random number generators. In: Balliard, L. (ed.) Computational Science and Statistics, Sixteenth Symposium on the Interface, pp. 3–10. Elsevier, Amsterdam (1985)
257. Marujama, G.: Continuous Markov processes and stochastic equations. Rend. Mat. Circ. Palermo, Ser. **2**(4), 48–90 (1955)
258. Mascagni, M.: Some methods of parallel pseudorandom number generation. In: Heath, M.T., Ranade, A., Schreiber, R.S. (eds.) Algorithms for parallel processing, pp. 277–288. Springer, Berlin (1999)
259. Mascagni, M., Cuccaro, S.A., Pryor, D.V., Robinson, M.L.: A fast, high-quality, and reproducible lagged-Fibonacci pseudorandom number generator. J. Comput. Phys. **15**, 211–219 (1995)
260. Matsumoto, M., Nishimura, T.: Mersenne twister: a 623-dimensionally equidistributed uniform pseudo-random number generator. ACM Trans. Model. Comput. Simul. **8**, 3–30 (1998)
261. Mattingly, J.C., Stuart, A.M.: Geometric ergodicity of some hypo-elliptic diffusions for particle motions. Markov Proc. Relat. Fields **8**, 199–214 (2002)

262. Mattingly, J.C., Stuart, A.M., Higham, D.J.: Ergodicity for SDEs and approximations: locally Lipschitz vector fields and degenerate noise. Stoch. Proc. Appl. **101**, 185–232 (2002)

263. Mattingly, J.C., Stuart, A.M., Tretyakov, M.V.: Convergence of numerical time-averaging and stationary measures via Poisson equations. SIAM J. Numer. Anal. **48**(2), 552–577 (2010)

264. Menaldi, J.L.: Stochastic variational inequality for reflected diffusion. Indiana Univ. Math. J. **32**, 733–744 (1983)

265. Mentink, J.H., Tretyakov, M.V., Fasolino, A., Katsnelson, M.I., Rasing, T.: Stable and fast semi-implicit integration of the stochastic Landau-Lifshitz equation. J. Phys.: Condens. Matter **22**(17), 176001 (2010)

266. Meyn, S.P., Tweedie, R.L.: Markov Chains and Stochastic Stability. Springer, Berlin (1992)

267. Mikhailov, V.P.: Partial Differential Equations. Nauka, Moscow (1976)

268. Symplectic quaternion scheme for biophysical molecular dynamics: Miller, T., III., Eleftheriou, M., Pattnaik, P., Ndirango, A., Newns, D., Matyna, G. J. Chem. Phys. **116**, 8649–8659 (2002)

269. Milstein, G.N.: Approximate integration of stochastic differential equations. Theor. Probab. Appl. **19**, 583–588 (1974)

270. Milstein, G.N.: A method with second order accuracy for the integration of stochastic differential equations. Theor. Probab. Appl. **23**, 414–419 (1978)

271. Milstein, G.N.: Probabilistic solution of linear systems of elliptic and parabolic equations. Theor. Probab. Appl. **23**, 851–855 (1978)

272. Milstein, G.N.: Weak approximation of solutions of systems of stochastic differential equations. Theor. Probab. Appl. **30**, 706–721 (1985)

273. Milstein, G.N.: A theorem on the order of convergence of mean-square approximations of solutions of systems of stochastic differential equations. Theor. Probab. Appl. **32**, 809–811 (1987)

274. Milstein, G.N.: Numerical integration of stochastic differential equations. Ural State University, Sverdlovsk (Engl. transl. by Kluwer Academic Publishers. Math. Appl. **313**, 1995) (1988)

275. Milstein, G.N.: Solution of the first boundary value problem for equations of parabolic type by means of the integration of stochastic differential equations. Theor. Probab. Appl. **40**, 556–563 (1995)

276. Milstein, G.N.: The solving of boundary value problems by numerical integration of stochastic equations. Math. Comput. Simul. **38**, 77–85 (1995)

277. Milstein, G.N.: Application of the numerical integration of stochastic equations for the solution of boundary value problems with Neumann boundary conditions. Theor. Probab. Appl. **41**, 170–177 (1996)

278. Milstein, G.N.: Evaluation of moment Lyapunov exponents for second order stochastic systems. Random Comput. Dyn. **4**, 301–315 (1996)

279. Milstein, G.N.: The simulation of phase trajectories of a diffusion process in a bounded domain. Stoch. Stoch. Rep. **56**, 103–125 (1996)

280. Milstein, G.N.: Stability index for invariant manifolds of stochastic systems. Random Comput. Dyn. **5**, 177–211 (1997)

281. Milstein, G.N.: Weak approximation of a diffusion process in a bounded domain. Stoch. Stoch. Rep. **62**, 147–200 (1997)

282. Milstein, G.N.: On the mean-square approximation of a diffusion process in a bounded domain. Stoch. Stoch. Rep. **64**, 211–233 (1998)

283. Milstein, G.N.: Orbital stability index for stochastic systems. Stoch. Anal. Appl. **18**, 777–809 (2000)

284. Milstein, G.N.: The probability approach to numerical solution of nonlinear parabolic equations. Numer. Methods Partial Diff. Eq. **18**, 490–522 (2002)

285. Milstein, G.N., Platen, E., Schurz, H.: Balanced implicit methods for stiff stochastic systems. SIAM J. Numer. Anal. **35**, 1010–1019 (1998)

286. Milstein, G.N., P'yanzin, S.A. : Digital modelling of the Kalman-Bucy filter and an optimal filter in quantized arrival of information. Avtomat. i Telemekh. **1**, 59–68 (1985)

287. Milstein, G.N., P'yanzin, S.A. : Regularization and digital modeling of a Kalman-Bucy filter for systems with degenerate noise in the observations. Avtomat. i Telemekh. **11**, 80–92 (1987)

288. Milstein, G.N., Reiß, O., Schoenmakers, J.G.M.: A new Monte Carlo method for American options. Int. J. Theor. Appl. Fin. **7**, 591–614 (2004)

289. Milstein, G.N., Repin, Y.M., Tretyakov, M.V.: Symplectic methods for Hamiltonian systems with additive noise. Preprint No. 640, Weierstraß-Institut für Angewandte Analysis und Stochastik (2001)

290. Milstein, G.N., Repin, Y.M., Tretyakov, M.V.: Symplectic integration of Hamiltonian systems with additive noise. SIAM J. Numer. Anal. **39**, 2066–2088 (2002)

291. Milstein, G.N., Repin, Y.M., Tretyakov, M.V.: Numerical methods for stochastic systems preserving symplectic structure. SIAM J. Numer. Anal. **40**, 1583–1604 (2003)

292. Milstein, G.N., Rybkina, N.F.: An algorithm for random walks over small ellipsoids for solving the general Dirichlet problem. J. Comput. Math. Math. Phys. **33**, 631–647 (1993)

293. Milstein, G.N., Schoenmakers, J.G.M.: Monte Carlo construction of hedging strategies against multi-asset European claims. Stoch. Stoch. Rep. **73**, 125–157 (2002)

294. Milstein, G.N., Schoenmakers, J.G.M.: Uniform approximation of the Cox-Ingersoll-Ross process. Adv. Appl. Probab. **47**, 1132–1156 (2015)

295. Milstein, G.N., Schoenmakers, J.G.M.: Uniform approximation of the Cox-Ingersoll-Ross process via exact simulation at random times. Adv. Appl. Probab. **48**, 1095–1116 (2016)

296. Milstein, G.N., Schoenmakers, J.G.M., Spokoiny, V.: Transition density estimation for stochastic differential equations via forward-reverse representations. Bernoulli **10**, 281–312 (2004)

297. Milstein, G.N., Spokoiny, V.: Construction of mean-self-financing strategies for European options under regime-switching. SIAM J. Finan. Mathem. **5**(1), 532–556 (2014)

298. Milstein, G.N., Tret'yakov, M.V.: Numerical solution of differential equations with colored noise. J. Stat. Phys. **77**, 691–715 (1994)

299. Milstein, G.N., Tretyakov, M.V.: Weak approximation for stochastic differential equations with small noises. Preprint No. 123, Weierstraß-Institut für Angewandte Analysis und Stochastik (1994)

300. Milstein, G.N., Tretyakov, M.V.: Mean-square numerical methods for stochastic differential equations with small noise. SIAM J. Sci. Comput. **18**, 1067–1087 (1997)

301. Milstein, G.N., Tretyakov, M.V.: Numerical methods in the weak sense for stochastic differential equations with small noise. SIAM J. Numer. Anal. **34**, 2142–2167 (1997)

302. Milstein, G.N., Tretyakov, M.V.: Numerical methods for nonlinear parabolic equations with small parameter based on probability approach. Preprint No. 396, Weierstraß-Institut für Angewandte Analysis und Stochastik (1998)

303. Milstein, G.N., Tretyakov, M.V.: Mean velocity of noise-induced transport in the limit of weak periodic forcing. J. Phys. A **32**, 5795–5805 (1999)

304. Milstein, G.N., Tretyakov, M.V.: Simulation of a space-time bounded diffusion. Ann. Appl. Probab. **9**, 732–779 (1999)

305. Milstein, G.N., Tretyakov, M.V.: Numerical algorithms for semilinear parabolic equations with small parameter based on weak approximation of stochastic differential equations. Math. Comput. **60**, 237–267 (2000)

306. Milstein, G.N., Tretyakov, M.V.: Numerical analysis of noise-induced regular oscillations. Physica D **140**, 244–256 (2000)

307. Milstein, G.N., Tretyakov, M.V.: Numerical solution of the Neumann problems for nonlinear parabolic equations by probability approach. Preprint No. 589, Weierstraß-Institut für Angewandte Analysis und Stochastik (2000)

308. Milstein, G.N., Tretyakov, M.V.: Noice-induced unidirectional transport. Stoch. Dyn. **1**, 361–375 (2001)

309. Milstein, G.N., Tretyakov, M.V.: Numerical solution of the Dirichlet problem for nonlinear parabolic equations by a probabilistic approach. IMA J. Numer. Anal **21**, 887–917 (2001)

310. Milstein, G.N., Tretyakov, M.V.: A probabilistic approach to the solution of the Neumann problem for nonlinear parabolic equations. IMA J. Numer. Anal. **22**, 599–622 (2002)

311. Milstein, G.N., Tretyakov, M.V.: The simplest random walks for the Dirichlet problem. Theor. Probab. Appl. **47**, 39–58 (2002)
312. Milstein, G.N., Tretyakov, M.V.: Numerical methods for Langevin type equations based on symplectic integrators. IMA J. Numer. Anal. **23**, 593–626 (2003)
313. Milstein, G.N., Tretyakov, M.V.: Evaluation of conditional Wiener integrals by numerical integration of stochastic differential equations. J. Comput. Phys. **197**, 275–298 (2004)
314. Milstein, G.N., Tretyakov, M.V.: Stochastic Numerics for Mathematical Physics, 1st edn. Springer, Berlin (2004)
315. Milstein, G.N., Tretyakov, M.V.: Numerical analysis of Monte Carlo evaluation of Greeks by finite differences. J. Comput. Fin. **8**, 1–33 (2005)
316. Milstein, G.N., Tretyakov, M.V.: Numerical integration of stochastic differential equations with nonglobally Lipschitz coefficients. SIAM J. Numer. Anal. **43**, 1139–1154 (2005)
317. Milstein, G.N., Tretyakov, M.V.: Numerical algorithms for forward-backward stochastic differential equations. SIAM J. Sci. Comput. **28**(2), 561–582 (2006)
318. Milstein, G.N., Tretyakov, M.V.: Practical variance reduction via regression for simulating diffusions. In: Preprint No. MA 06–019, Department of Mathematics. University of Leicester (2006)
319. Milstein, G.N., Tretyakov, M.V.: Computing ergodic limits for Langevin equations. Physica D **229**, 81–95 (2007)
320. Milstein, G.N., Tretyakov, M.V.: Discretization of forward-backward stochastic differential equations and related quasi-linear parabolic equations. IMA J. Numer Anal. **27**(1), 24–44 (2007)
321. Milstein, G.N., Tretyakov, M.V.: Monte Carlo algorithms for backward equations in nonlinear filtering. Adv. Appl. Probab. **41**, 63–100 (2009)
322. Milstein, G.N., Tretyakov, M.V.: Practical variance reduction via regression for simulating diffusions. SIAM J. Numer. Anal. **47**, 887–910 (2009)
323. Milstein, G.N., Tretyakov, M.V.: Solving parabolic stochastic partial differential equations via averaging over characteristics. Math. Comput. **78**, 2075–2106 (2009)
324. Milstein, G.N., Tretyakov, M.V.: Averaging over characteristics with innovation approach in nonlinear filtering. In: Crisan, D., Rozovskii, B. (eds.) Handbook of Nonlinear Filtering, pp. 892–919. Oxford University Press, Berlin (2011)
325. Milstein, G.N., Tretyakov, M.V.: Solving the Dirichlet problem for Navier-Stokes equations by probabilistic approach. BIT Numer. Math. **52**, 141–153 (2012)
326. Milstein, G.N., Tretyakov, M.V.: Layer methods for Navier-Stokes equations with additive noise. arXiv:1312.5887 (2013)
327. Milstein, G.N., Tretyakov, M.V.: Probabilistic methods for the incompressible Navier-Stokes equations with space periodic conditions. Adv. Appl. Probab. **45**(3), 742–772 (2013)
328. Milstein, G.N., Tretyakov, M.V.: Layer methods for stochastic Navier-Stokes equations using simplest characteristics. J. Comput. Appl. Math. **302**, 1–23 (2016)
329. Milstein, G.N., Tretyakov, M.V.: Mean-square approximation of Navier-Stokes equations with additive noise in vorticity-velocity formulation. Numer. Math. Theor. Meth. Appl. **14**, 1–30 (2021)
330. Miranda, C.: Partial Differential Equations of Elliptic Type. Springer, Berlin (1970)
331. Muller, M.E.: Some continuous Monte Carlo methods for the Dirichlet problem. Ann. Math. Stat. **27**, 569–589 (1956)
332. Muller, M.E.: A note on a method for generating points uniformly on N-dimensional spheres. Commun. ASM **2**, 19–20 (1959)
333. von Neumann, J., Richtmyer, R.: A method for the numerical calculation of hydrodynamic shocks. J. Appl. Phys. **21**, 232–257 (1950)
334. Newton, N.J.: An asymptotically efficient difference formula for solving stochastic differential equations. Stochastics **19**, 175–206 (1986)
335. Newton, N.J.: Variance reduction for simulated diffusions. SIAM J. Appl. Math. **54**, 1780–1805 (1994)

336. Newton, N.J.: Continuous-time Monte Carlo methods and variance reduction. In: Rogers, L., Talay, D. (eds.) Numerical Methods in Finance, pp. 22–42. Cambridge University Press, Cambridge (1997)
337. Niederreiter, H.: Random Number Generation and Quasi-monte Carlo Methods. SIAM, Philadelphia (1992)
338. Niederreiter, H.: Constructions of (t, m, s)-nets and (t, s)-sequences. Finite Fields Appl. **11**, 578–600 (2005)
339. Nikitin, N.N., Razevig, V.D.: Methods of numerical modeling of stochastic differential equations and estimates of their errors. Zh. Vychisl. Mat. i Mat. Fiz. **18**, 106–117 (1978)
340. Novak, E., Ritter, K.: Simple cubature formulas with high polynomial exactness. Constr. Approx. **15**, 499–522 (1999)
341. Nualart, D.: The Malliavin Calculus and Related Topics. Springer, Berlin (2006)
342. Ostrem, K.: Introduction to the Stochastic Theory of Equations. Mir, Moscow (1973)
343. Öttinger, H.C.: Stochastic Processes in Polymeric Fluids: Tools and Examples for Developing Simulation Algorithms. Springer, Berlin (1996)
344. Paley, R.E.A.C., Wiener, N.: Fourier Transforms in the Complex Domain. AMS, New York (1934)
345. Pardoux, E.: Stochastic partial differential equations and filtering of diffusion processes. Stochastics **3**, 127–167 (1979)
346. Pardoux, E.: Nonlinear filtering, prediction and smoothing. In: Hazewinkel, M., Willems, J. (eds.) Stochastic Systems: the Mathematics of Filtering and Identification and Applications, NATO Advanced Study Institute series, D, pp. 529–557. Reidel (1981)
347. Pardoux, E.: Backward stochastic differential equations and viscosity solutions of systems of semilinear parabolic and elliptic PDEs of second order. In: Stochastic Analysis and Related Topics VI, pp. 79–127. Birkhäuser, Basel (1998)
348. Pardoux, E., Peng, S.: Backward stochastic differential equations and quasilinear parabolic partial differential equations. In: Rozovskii, B.L., Sowers, R.B. (eds.) Stochastic partial differential equations and their applications, Lecture Notes in Control and Inform. Sci., vol. 176, pp. 200–217. Springer, Berlin (1992)
349. Pardoux, E., Peng, S.: Backward doubly stochastic differential equations and systems of quasilinear SPDEs. Probab. Theory Relat. Fields **98**, 209–227 (1994)
350. Pardoux, E., Rǎşcanu, A.: Stochastic Differential Equations, Backward SDEs. Partial Differential Equations. Springer, Switzerland (2014)
351. Pardoux, E., Talay, D.: Discretization and simulation of stochastic differential equations. Acta Appl. Math. **3**, 23–47 (1985)
352. Pardoux, E., Tang, S.: Forward backward stochastic differential equations and quasilinear parabolic PDEs. Probab. Theory Rel. Fields **114**, 123–150 (1999)
353. Pardoux, E., Veretennikov, A.Y.: On the Poisson equation and diffusion approximation. I. Ann. Probab. **29**(3), 1061–1085 (2001)
354. Pardoux, E., Veretennikov, A.Y.: On the Poisson equation and diffusion approximation. II. Ann. Probab. **31**(3), 1166–1192 (2003)
355. Pardoux, E., Veretennikov, A.Y.: On the Poisson equation and diffusion approximation. III. Ann. Probab. **33**(3), 1111–1133 (2005)
356. Pavliotis, G.A., Stuart, A.M.: Multiscale Methods. Springer, New York (2008)
357. Peng, S.: Probabilistic interpretation for systems of quasilinear parabolic partial differential equations. Stoch. Stoch. Rep. **37**, 61–74 (1991)
358. Peng, S.: Nonlinear Expectations and Stochastic Calculus Under Uncertainty. Springer, Berlin (2019)
359. Petersen, W.P.: Some experiments on numerical simulations of stochastic differential equations and a new algorithm. J. Comput. Phys. **113**, 75–81 (1994)
360. Petersen, W.P.: A general implicit splitting for stabilizing numerical simulations of Ito stochastic differential equations. SIAM J. Numer. Anal. **35**, 1439–1451 (1998)
361. Pettersson, R.: Approximations for stochastic differential equations with reflecting convex boundaries. Stoch. Proc. Appl. **59**, 295–308 (1995)

362. Pettersson, R.: Penalization schemes for reflecting stochastic differential equations. Bernoulli **3**, 403–414 (1997)
363. Picard, J.: Approximation of nonlinear filtering problems and order of convergence. In: Stochastic systems: the mathematics of filtering and identification and applications, NATO Advanced Study Institute series, D, Lecture Notes in Contr. Inform. ScI. vol. 61, pp. 219–236. Springer (1984)
364. Platen, E.: An approximation method for a class of Ito processes. Lit. Mat. Sb. **21**, 121–133 (1981)
365. Platen, E., Bruti-Liberati, N.: Numerical Solution of Stochastic Differential Equations with Jumps in Finance. Springer, Berlin (2010)
366. Platen, E., Wagner, W.: On a Taylor formula for a class of Ito processes. Probab. Math. Stat. **3**, 37–51 (1982)
367. Pope, S.B.: Simple models of turbulent flows. Phys. Fluids **23**, 011301 (2011)
368. Press, W.H., Teukolsky, S.A., Vetterling, W.T., Flannery, B.P.: Numerical Recipes in C: The Art of Scientific Computing. Cambridge University Press, Cambridge (1992)
369. Protter, P., Talay, D.: The Euler scheme for Levy driven stochastic differential equations. Ann. Probab. **25**, 393–423 (1997)
370. Quarteroni, A., Valli, A.: Numerical Approximation of Partial Differential Equations. Springer, Berlin (1994)
371. Quispel, G.R.W.: Volume-preserving integrators. Phys. Lett. A **206**, 26–30 (1995)
372. Rakitskii, Y.V., Ustinov, S.M., Chernorutskii, I.G.: Numerical Methods for Solving Stiff Systems. Nauka, Moscow (1979)
373. Rao, N.J., Borwankar, J.D., Ramkrishna, D.: Numerical solution of Ito integral equations. SIAM J. Control **12**, 124–139 (1974)
374. Reigada, R., Romero, A.H., Sarmiento, A., Lindenberg, K.: One-dimensional arrays of oscillators: energy localization in thermal equilibrium. J. Chem. Phys. **111**, 1373–1384 (1999)
375. Richtmyer, R.D., Morton, K.W.: Difference Methods for Initial-Value Problems. Interscience, New York (1967)
376. Ripoll, M., Ernst, M.H., Español, P.: Large scale and mesoscopic hydrodynamics for dissipative particle dynamics. J. Chem. Phys. **115**, 7271–7284 (2001)
377. Risken, H.: The Fokker-Planck Equation. Springer, Berlin (1984)
378. Roache, P.J.: Computational Fluid Dynamics. Hermosa, Albuquerque, N.M. (1976)
379. Roberts, G.O., Tweedie, R.L.: Exponential convergence of Langevin distributions and their discrete approximations. Bernoulli **2**, 341–363 (1996)
380. Rodean, H.C.: Stochastic Lagrangian Models of Turbulent Diffusion. American Meteorological Society, Boston (1996)
381. Roepstorff, G.: Path Integral Approach to Quantum Physics. Springer, Berlin (1994)
382. Rogers, L.C.G., Talay, D. (eds.): Numerical Methods in Finance. Cambridge University Press, Cambridge (1997)
383. Rogers, L.C.G., Williams, D.: Diffusions, Markov Processes, and Martingales. Volume 2: Its Calculus. Willey, Chichester (1987)
384. Roos, H.G., Stynes, M., Tobiska, L.: Numerical Methods for Singularity Perturbed Differential Equations: Convection-diffusion and Flow Problems. Springer, Berlin (1996)
385. Ross, S.M.: Simulation. Academic, New York (1997)
386. Rößler, A.: Second order Runge-Kutta methods for Stratonovich stochastic differential equations. BIT Numer. Math. **47**, 657–680 (2007)
387. Rößler, A.: Second order Runge-Kutta methods for Ito stochastic differential equations. SIAM J. Numer. Anal. **47**, 1713–1738 (2009)
388. Rößler, A.: Runge-Kutta methods for strong approximation of stochastic differential equations. SIAM J. Numer. Anal. **48**, 922–952 (2010)
389. Rouah, F.D.: The Heston Model and Its Extensions in *M*atlab and *C*#. Wiley. Finance (2013)
390. Rozovskii, B.L.: Stochastic Evolution Systems. Linear Theory and Application to Nonlinear Filtering. Kluwer Academic Publishers, Dordrecht (1991)
391. Rozovsky, B.L., Lototsky, S.: Stochastic Evolution Systems. Springer, Berlin (2018)

392. Rumelin, W.: Numerical treatment of stochastic differential equations. SIAM J. Numer. Anal. **19**, 604–613 (1982)

393. Ryden, T., Wiktorsson, M.: On the simulation of iterated Ito inegrals. Stoch. Proc. Appl. **91**, 151–168 (2001)

394. Sabelfeld, K.K.: Monte Carlo Methods in Boundary Value Problems. Springer, Berlin (1991)

395. Samarskii, A.A.: Theory of Difference Schemes. Nauka, Moscow (1977)

396. Samarskii, A.A., Galaktionov, V.A., Kurdyumov, S.P., Mikhailov, A.P.: Blow-Up in Quasilinear Parabolic Equations. Walter de Gruyter, Berlin (engl. transl. from Russian, 1987) (1995)

397. Sanz-Serna, J.M.: Symplectic integrators for Hamiltonian problems: an overview. Acta Numerica **1**, 243–286 (1992)

398. Sanz-Serna, J.M., Calvo, M.P.: Numerical Hamiltonian Problems. Chapman & Hall, London (1994)

399. Sawford, B.L.: Turbulence relative dispersion. Annu. Rev. Fluid Mech. **33**, 289–317 (2001)

400. Schenzle, A., Brand, H.: Multiplicative stochastic processes in statistical physics. Phys. Rev. A **20**, 1628–1647 (1979)

401. Schlick, T.: Molecular Modeling and Simulation: An Interdisciplinary Guide. Springer, New York (2010)

402. Seeßelberg, M., Breuer, H.P., Mais, H., Petruccione, F., Honerkamp, J.: Simulation of one-dimensional noisy Hamiltonian systems and their application to particle storage rings. Z. Phys. C **62**, 62–73 (1994)

403. Simon, B.: Functional Integration and Quantum Physics. AMS (2005)

404. Skeel, R.: Integration schemes for molecular dynamics and related applications. In: Ainsworth, M., Levesley, J., Marletta, M. (eds.) Graduate Student's Guide to Numerical Analysis'98, pp. 118–176. Springer, Berlin (1999)

405. Skorokhod, A.: Stochastic equations for diffusion processes in a bounded region. Theor. Probab. Appl. **6**(3), 264–274 (1961)

406. Slominski, L.: On approximation of solutions of multidimensional SDE's with reflecting boundary conditions. Stoch. Proc. Appl. **50**, 197–219 (1994)

407. Slominski, L.: Euler's approximations of solutions of SDEs with reflecting boundary. Stoch. Proc. Appl. **94**, 317–337 (2001)

408. Smoller, J.: Shock Waves and Reaction-diffusion Equations. Springer, Berlin (1983)

409. Smolyak, S.A.: Quadrature and interpolation formulas for tensor products of certain classes of functions. Soviet Math. Dokl. **4**, 240–243 (1963)

410. Snook, I.: The Langevin and Generalised Langevin Approach to the Dynamics of Atomic. Elsevier, Polymeric and Colloidal Systems (2006)

411. Sobol, I.M.: Multidimensional Quadrature Formulas and Haar Functions. Nauka, Moscow (1969)

412. Soize, C.: The Fokker-Planck Equation for Stochastic Dynamical Systems and Its Explicit Steady State Solutions. World Scientific, London (1994)

413. Stanciulescu, V.N., Tretyakov, M.V.: Numerical solution of the Dirichlet problem for linear parabolic SPDEs based on averaging over characteristics. In: Crisan, D. (ed.) Stochastic Analysis 2010, pp. 191–212. Springer, Berlin (2011)

414. Strang, G.: On the construction and comparison of difference schemes. SIAM J. Numer. Anal. **5**, 506–517 (1968)

415. Stratonovich, R.L.: Topics in the Theory of Random Noise. Gordon and Breach, New York (1967)

416. Stratonovich, R.L.: Conditional Markov Processes and Their Applications to Optimal Control Theory. Elsevier (1968)

417. Strikwerda, J.C.: Finite Difference Schemes and Partial Differential Equations. Wadsworth & Brooks, PacificGrove, California (1989)

418. Stuart, A.M., Humphries, A.R.: Dynamical Systems and Numerical Analysis. Cambridge University Press, Cambridge (1996)

419. Suris, Y.B.: The canonicity of mappings generated by Runge-Kutta type methods when integrating the systems $ddotx = -partialupartialx$. U.S.S.R. Comput. Maths. Math. Phys. **29**, 138–144 (1989)

420. Suris, Y.B.: On the canonicity of mappings that can be generalized by methods of Runge-Kutta type for integrating systems $\ddot{x} = -\partial u/\partial x$. Zh. Vychisl. Mat. i Mat. Fiz. **29**, 202–211 (1989)

421. Suris, Y.B.: Hamiltonian methods of Runge-Kutta type and their variational interpretation. Math. Model. **2**, 78–87 (1990)

422. Suris, Y.B.: Preservation of integral invariants in the numerical solution of systems $\ddot{x} = x + f(x)$. Zh. Vychisl. Mat. i Mat. Fiz. **31**, 52–63 (1991)

423. Suris, Y.B.: Partitioned Runge-Kutta methods as phase volume preserving integrators. Phys. Lett. A **220**, 63–69 (1996)

424. Szpruch, L., Zhang, X.: V-integrability, asymptotic stability and comparison property of explicit numerical schemes for non-linear SDEs. Math. Comput. **87**, 755–783 (2018)

425. Takahashi, M., Imada, M.: Monte Carlo calculation of quantum systems. J. Phys. Soc. Jpn. **53**, 963–974 (1984)

426. Talay, D.: How to discretize stochastic differential equations. In: Lecturer Notes in Mathematics, vol. 972, pp. 276–292. Springer, Berlin (1983)

427. Talay, D.: Résolution trajectorillée et analyse numérique des équations différentielles stochastiques. Stochastics **9**, 275–306 (1983)

428. Talay, D.: Efficient numerical schemes for the approximation of expectations of functionals of the solution of an SDE and applications. In: Lecturer Notes in Control and Information Sccience, vol. 61, pp. 294–313. Springer, Berlin (1984)

429. Talay, D.: Second-order discretization schemes for stochastic differential systems for the computation of the invariant law. Stoch. Stoch. Rep. **29**, 13–36 (1990)

430. Talay, D.: Approximation of upper Lyapunov exponents of bilinear stochastic differential equations. SIAM J. Numer. Anal. **28**, 1141–1164 (1991)

431. Talay, D.: Approximation of the invariant probability measure of stochastic Hamiltonian dissipative systems with non-globally Lipschitz coefficients. In: Bouc, R., Soize, C. (eds.) Progress in Stochastic Structural Dynamics, vol. 152, pp. 139–169. Publication du L.M.A.-CNRS (1999)

432. Talay, D.: Stochastic Hamiltonian systems: exponential convergence to the invariant measure, and discretization by the implicit Euler scheme. Markov Proc. Relat. Fields **8**, 163–198 (2002)

433. Talay, D., Tubaro, L.: Expansion of the global error for numerical schemes solving stochastic differential equations. Stoch. Anal. Appl. **8**, 483–509 (1990)

434. Talay, D., Tubaro, L. (eds.): Probabilistic models for nonlinear partial differential equations. Lecture Notes in Mathematics, vol. 1627. Springer, Berlin (1996)

435. Talay, D., Zheng, Z.: Quantiles of the Euler scheme for diffusion processes and financial applications. Math. Fin. **13**, 187–199 (2003)

436. Tashiro, Y.: On methods for generating uniform random points on the surface of a sphere. Ann. Inst. Stat. Math. **29**(1), 295–300 (1977)

437. Taylor, M.E.: Partial Differential Equations III: Nonlinear Equations. Springer, Berlin (1996)

438. Tezuka, S.: Uniform Random Numbers: Theory and Practice. Kluwer, Norwell, Mass (1995)

439. Thomson, D.J.: Criteria for the selection of stochastic models of particle trajectories in turbulent flows. J. Fluid. Mech. **180**, 529–556 (1987)

440. Tocino, A., Vigo-Aguiar, J.: Weak second order conditions for stochastic Runge-Kutta methods. SIAM J. Sci. Comput. **24**, 507–523 (2002)

441. Tolmatz, L.: On the distribution of the square integral of the Brownian bridge. Ann. Appl. Prob. **30**, 253–269 (2002)

442. Touzi, N.: Optimal Stochastic Control, Stochastic Target Problems, and Backward SDE. Springer, New York (2013)

443. Tretyakov, M.V.: Numerical technique for studying stochastic resonance. Phys. Rev. E **57**, 4789–4794 (1998)

444. Tretyakov, M.V.: Introductory Course on Financial Mathematics. Imperial College Press, London (2013)

445. Tretyakov, M.V.: Stochastic filtering. In: Engquist, B. (ed.) Encyclopedia of Applied and Computational Mathematics. Springer, Berlin (2015)

446. Tretyakov, M.V., Fedotov, S.: On FKPP equation with Gaussian shear advection. Phys. D **159**, 191–202 (2001)
447. Tretyakov, M.V., Tretyakov, S.V.: Numerical integration of Hamiltonian systems with external noise. Phys. Lett. A **194**, 371–374 (1994)
448. Tretyakov, M.V., Zhang, Z.: A fundamental mean-square convergence theorem for SDEs with locally Lipschitz coefficients and its applications. SIAM J. Numer. Anal. **51**, 3135–3162 (2013)
449. Tropper, M.M.: Ergodic properties and quasideterministic properties of finite-dimensional stochastic systems. J. Stat. Phys. **17**, 491–509 (1977)
450. Venttsel, A.D.: A Course in the Theory of Random Processes. Nauka, Moscow (1996)
451. Venttsel, A.D., Gladyshev, S.A., Milstein, G.N.: Piecewise constant approximation for Monte Carlo calculation of Wiener integrals. Theor. Probab. Appl. **29**, 715–722 (1984)
452. Veretennikov, A.Y.: On large deviations for approximations of SDEs. Probab. Theory Related Fields **125**, 135–152 (2003)
453. Vilmart, G.: Postprocessed integrators for the high order integration of ergodic SDEs. SIAM J. Sci. Comput. **37**(1), A201–A220 (2015)
454. Vorozhtsov, E.V., Yanenko, N.N.: Methods for the Localization of Singularities in Numerical Solutions of Gas Dynamic Problems. Springer, Berlin (1990 (engl. transl. from Russian, 1985))
455. Vreugdenhil, C.B., Koren, B. (eds.): Numerical Methods for Advection-diffusion Problems, Notes on Numerical Fluid Mechanics, vol. 45. Vieweg, Wiesbaden (1993)
456. Wagner, W.: Monte Carlo evaluation of functionals of solutions of stochastic differential equations. Variance reduction and numerical examples. Stoch. Anal. Appl. **6**, 447–468 (1988)
457. Wagner, W.: Unbiased multi-step estimators for the Monte Carlo evaluation of certain functional integrals. J. Comput. Phys. **79**, 336–352 (1988)
458. Wagner, W.: Unbiased Monte Carlo estimators for functionals of weak solutions of stochastic differential equations. Stoch. Stoch. Rep. **28**, 1–20 (1989)
459. Wagner, W., Platen, E.: Approximation of Ito Integral Equations. Preprint ZIMM. Akad. der Wiss. der DDR, Berlin (1978)
460. Wasilkowski, G.W., Wozniakowski, H.: Explicit cost bounds of algorithms for multivariate tensor product problems. J. Complex. **11**, 1–56 (1995)
461. Watson, E.J.: Primitive polynomials (mod2). Math. Comput. **16**, 368–369 (1962)
462. Wendland, H.: Scattered Data Approximation. Cambridge University Press, Cambridge (2005)
463. Wiktorsson, M.: Joint characteristic function and simultaneous simulation of iterated ito integrals for multiple independent Brownian motions. Ann. Appl. Probab. **11**, 470–487 (2001)
464. Yanenko, N.N.: The Method of Fractional Steps: The Solution of Problems of Mathematical Physics in Several Variables. Springer, Berlin (1971)
465. Yong, J., Zhou, X.Y.: Stochastic Controls. Hamiltonian Systems and HJB Equations. Springer, Berlin (1999)
466. Yoo, H.: Semi-discretzation of stochastic partial differential equations on R^1 by a finite-difference method. Math. Comput. **69**, 653–666 (2000)
467. Zhang, Z., Karniadakis, G.E.: Numerical Methods for Stochastic Partial Differential Equations with White Noise. Springer, Berlin (2017)
468. Zhang, Z., Rozovskii, B.L., Tretyakov, M.V., Karniadakis, G.E.: A multistage Wiener chaos expansion method for stochastic advection-diffusion-reaction equations. SIAM J. Sci. Comput. **34**(2), A914–A936 (2012)
469. Zhang, Z., Tretyakov, M.V., Rozovskii, B.L., Karniadakis, G.E.: A recursive sparse grid collocation method for differential equations with white noise. SIAM J. Sci. Comput. **36**, A1652–A1677 (2014)
470. Zhang, Z., Tretyakov, M.V., Rozovskii, B.L., Karniadakis, G.E.: Wiener chaos versus stochastic collocation methods for linear advection-diffusion-reaction equations with multiplicative white noise. SIAM J. Numer. Anal. **53**(1), 153–183 (2015)

Index

© Springer Nature Switzerland AG 2021
G. N. Milstein and M. V. Tretyakov, *Stochastic Numerics for Mathematical Physics*,
Scientific Computation, https://doi.org/10.1007/978-3-030-82040-4